Clyde Baker

GROUTING, SOIL IMPROVEMENT AND GEOSYNTHETICS

VOLUME 2

Geotechnical Special Publication No. 30

Proceedings of the conference
sponsored by the
Geotechnical Engineering Division
of the American Society of Civil Engineers

In cooperation with the
International Society of Soil Mechanics
and Foundation Engineering — ISSMFE, TC-17

New Orleans, Louisiana
February 25-28, 1992

Edited by Roy H. Borden, Robert D. Holtz and Ilan Juran

ASCE 1852 ®

Published by the
American Society of Civil Engineers
345 East 47th Street
New York, New York 10017-2398

ABSTRACT

The proceedings of the 1992 ASCE Specialty Conference on Grouting, Soil Improvement and Geosynthetics, addresses advances in the state-of-practice, research and technological developments in four rapidly expanding geotechnical areas: 1) Grouting techniques and materials, 2) soil improvement technologies, 3) geosynthetics engineering, and 4) ground modification for geo-environmental applications. The 105 contributed papers and 4 invited keynote papers provide a broad-based technical discussion of design practice, innovative construction technologies and products, and monitored case studies of new engineering applications. Overall, these proceedings represent a significant contribution to the state-of-the-art, which will effectively assist practicing engineers, manufacturers, potential users, and researchers to recognize the limitations of current practice, evaluate the feasibility of innovative technologies, and address the technical challenges that will undoubtedly face the profession in the next century.

Library of Congress Cataloging-in-Publication Data

Grouting, soil improvement, and geosynthetics: proceedings of the conference/sponsored by the Geotechnical Engineering Division of the American Society of Civil Engineers in cooperation with the International Society of Soil Mechanics and Foundation Engineering—ISSMFE, TC-17, New Orleans, Louisiana, February 25-28, 1992; edited by Roy H. Borden, Robert D. Holtz, and Ilan Juran.

p. cm.—(Geotechnical special publication; no. 30)

Includes indexes.

ISBN 0-87262-865-5 (set)

1. Grouting—Congresses. 2. Soil stabilization—Congresses. 3. Geosynthetics—Congresses. I. Borden, Roy H. II. Holtz, R. D. (Robert D.) III. Juran, Ilan. IV. American Society of Civil Engineers. Geotechnical Engineering Division. V. International Society of Soil Mechanics and Foundation Engineering. Technical Committee on Ground Improvement, Reinforcement, and Grouting. VI. Series.

TA755.G76 1992

624.1′51363—dc20 91-47965

CIP

Library of Congress Catalog Card No: 91-47965

ISBN 0-87262-865-5

Manufactured in the United States of America.

CONTENTS

Volume 1

A. Keynote Lectures

B. Jet Grouting

C. Compaction Grouting

D. Grouting Practice

E. Ground Treatment for Tunnels

F. Cement Grouting

G. Remedial Grouting for Dams, Locks and Liquefaction Control

H. Chemical Grouting

Volume 2

I. In-Situ Slope Stabilization

J. Reinforcement of Foundation Soils

K. Densification of Foundation Soils

L. Improvement of Fine Grained Soils

M. Geosynthetics I—Subgrade and Road Layers

N. Geosynthetics II—Embankments and Slopes

O. Geosynthetics III—Walls

P. In-Situ Waste Stabilization

Q. Soil Modification and Environmental Geotechnology

French Research Program CLOUTERRE on Soil Nailing

F. Schlosser[1], P. Unterreiner[2], Associate Members, ASCE, and C. Plumelle[3]

Abstract

In 1986, a four year, four million dollars national research project named CLOUTERRE was initiated by the French Minister of Transport. The main objective was to develop design specifications on soil nailing for temporary and permanent soil nailed walls in excavations. In this article CLOUTERRE design specifications are presented. Four topics are developed: (1) limit state design of soil nailed structures using partial safety factors, (2) estimations of in service deformations of soil nailed walls, (3) design of the facing, and (4) durability and provisions for corrosion. Classical limit equilibrium methods adapted to soil nailed structures and completed with the "Multicriterium Rule" have proved to be efficient in predicting the failure of three full scale experimental soil nailed walls. A new procedure for taking into account safety in the design has been proposed to replace the classical global safety factor. New results have been obtained that allow to design the facing thickness as a function of the spacing between the nails. For permanent structures, recommendations have been developed to dimension the nails as a function of the service life of the structure.

[1] Professor, Ecole Nationale des Ponts et Chaussées, Paris, France. Chairman of TERRASOL, Tour Horizon, 52 Quai de Dion Bouton, 92806 Puteaux, France. Tel.: 33 1 47764314.

[2] Research Engineer, Ecole Nationale des Ponts et Chaussées, CERMES, 93167 Noisy-le-Grand cédex, France. Tel. 33 1 43044098.

[3] Head of Department of Soil Mechanics, Centre Expérimental du Bâtiment et des Travaux Publics, Domaine de Saint Paul, 78470 St-Rémy les Chevreuses, France. Tél.: 33 1 30852496.

Introduction

Soil nailing is a technique which consists of reinforcing in-situ soils with bars or structural members, commonly called nails, which can be either driven or grouted in drilled boreholes. When soil nailing is used in excavations, the nails are placed subhorizontally and withstand primarily tensile forces.

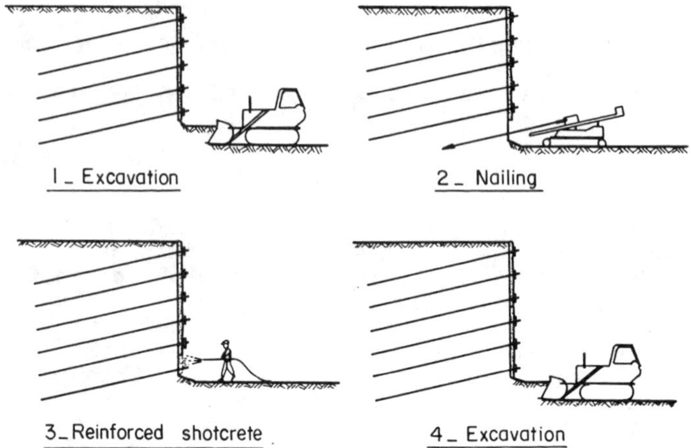

Figure 1 : Construction steps of a soil nailed wall

When compared to classical methods of earth support systems for excavations, soil nailing turns out to be cost effective, which explains its development in the past twenty years. The first soil nailed wall was built in 1972/1973 in Versailles France to retain a cut for a rail road [Rabejac et Toudic, 1974] (Figure 2).

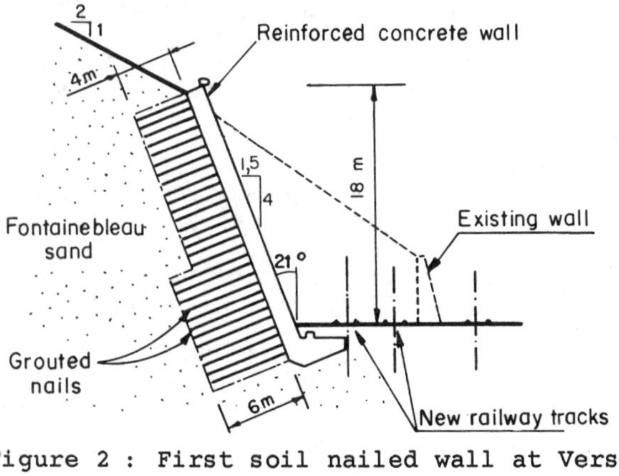

Figure 2 : First soil nailed wall at Versailles

As a result of this first successful project, soil nailing has been used more and more in France as shown in Figure 3.

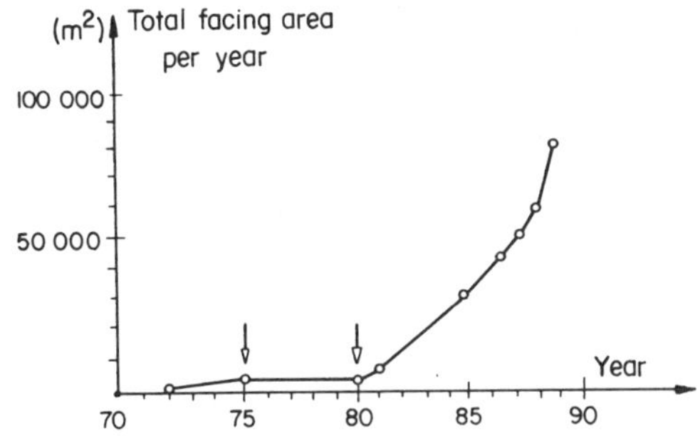

Figure 3 : Amount of soil nailing in France per year

In the mid-eighties, soil nailing applications were developing so quickly in France that the state of practice was ahead of the state of knowledge. There was an urgent need for research, full-scale experimentations and specifications. Therefore, in 1986, a four year, four million dollars national research program CLOUTERRE was initiated by the French Minister of Transport. Twenty-one organizations including private companies and public research laboratories participated directly in the project CLOUTERRE under the scientific direction of F. Schlosser. Based on the results, specifications were developed for soil nailing and published in a final document entitled "Recommendations CLOUTERRE 1991". The objective of this article is to present the specifications developed by the French national project CLOUTERRE for design of both temporary and permanent soil nailed structures for excavations. The following topics will be developed : (1) limit state design of soil nailed structures using partial safety factors, (2) in service deformations of soil nailed walls, (3) design of the facing, and (4) durability and provisions for corrosion.

At Limit State Design with Partial Safety Factors

The main concern when building a soil nailed wall is to ensure it is stable and safe. Three types of global failure have to be studied: internal, mixed and external failures as shown in Figure 4. For the last two failure types, several modes of local failure have to be considered: breakage and pull-out. If the nails can provide some additionnal resistance due to their bending and shear strengths, the soil-nail interaction is more complex and additional failure modes have to be considered: failure of the soil below a nail when the applied pressure exceeds its bearing capacity, failure of the nail by bending and formation of two plastic hinges.

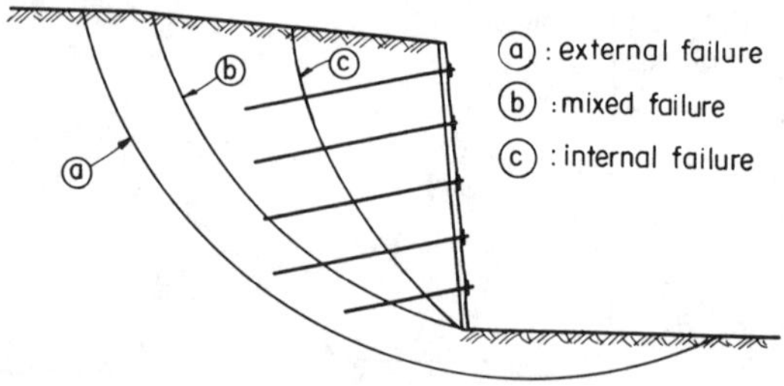

Figure 4 : Different types of failures to be analysed

The design method recommended in France is the "Multi-criterium Method" [Schlosser, 1982, 1983], which is an extension of classical limit equilibrium methods (method of slices) to reinforced soils, allows the bending stiffness and the shear resistance of the nails to be taken into account when necessary. Let's call T_n, the tension/compression force, T_c, the shear force and M, the bending moment of the nails at their intersection point, named O, with a potential failure surface as shown in Figure 5. In the horizontal and vertical equilibrium equations of the mass of soil sliding along the potential failure surface, only T_n and T_c will play a role.

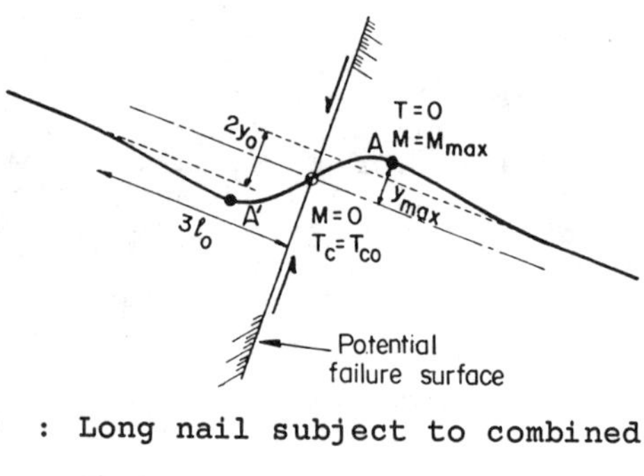

Figure 5 : Long nail subject to combined loading

If the nail can carry only tension or compression, it will fail at 0 either in tension when the maximum tensile force, R_n, is reached or in pull-out when the soil-nail interface resistance, $T_{pull\text{-}out}$, is reached. Therefore in the simplest case, for a given potential failure surface and assuming that the nail facing connection is strong enough, the nail force at 0 at failure will be : $T = \min(R_n, T_{pull\text{-}out})$ with $T_{pull\text{-}out} = \pi\, D\, q_s\, L_a$, D, diameter of driven nails or boreholes of grouted nails, L_a, length of nail behind the failure surface considered, and q_s, limit interface frictionnal resistance. When the nails can carry significant shear forces and bending moments, the calculation of the tension/compression (T_n) and shear force (T_c) in the nails at 0 is more complex.

To understand the behavior of stiff nails, it is helpful to work in the (T_n, T_c) plane, as shown in Figure 6, where T_n, T_c are calculated at the point 0 where the bending moment is zero (0 is a point of inflection). The soil-nail system can fail according to four different failure modes which include the two modes previously described. Each of these failure mode will reduce the admissible domain of the (T_n, T_c) point in the (T_n, T_c) plane.

(1) The nail can fail by simple breakage which reduces the admissible domain to an ellipse of equation: $[T_n/R_n]^2 + [T_c/R_c]^2 \leq 1$ with R_c, shear strength of nail.

(2) The soil-system can fail by pull-out which reduces the admissible domain to a half-plane of equation: $T_n \leq T_{pull-out}$.

(3) The soil can fail below a nail when the pressure its undergoes, exceeds its bearing capacity pressure p_u. Using the classical elastic analysis of a nail or pile subjected to combined loading [Schlosser, 1983], this reduces the admissible domain to a half plane of equation : $T_c \leq l_o \quad p_u D/2$ with $l_o = [4\ E\ I/k_s D]^{1/2}$ (transfer length), E, Young's modulus, I, second moment of area, k_s, modulus of subgrade reaction.

(4) The nail can fail in 0 by shear [T_c is maximum in this point] or by bending moment in A and A', points of maximum bending moments. Those points are symetric and at a distance of $l_s/2$ from the potential failure plane. The distance l_s is equal to $\pi l_o/2$ when the nail first starts to plastify by bending moments in A and A'. If failure is defined when this occurs, the admissible domain is reduced by a parabola as represented in figure 6.

The intersection of the four criteria corresponding to the above four failure modes defines the admissible domain of (T_n, T_c), where this couple is allowed only to be inside. At failure, (T_n, T_c) is on the border of the admissible domain and its location is determined to maximize the work of the nail in the considered potential failure mechanism according to the "multi-criterium" rule.

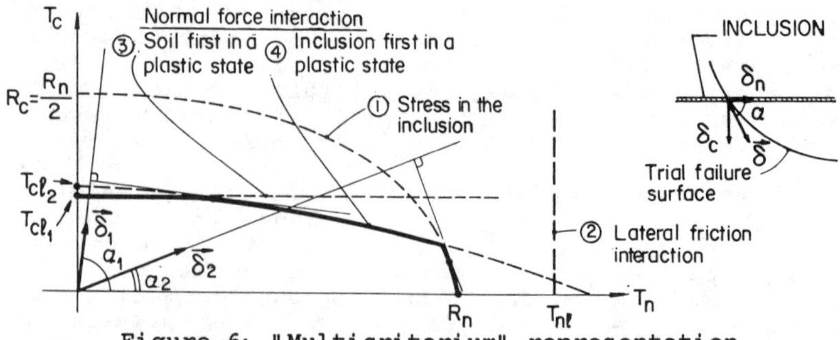

Figure 6: "Multicriterium" representation

For design purposes, bending moments and shear forces are often neglected. This can be acceptable if nails have a low moment of inertia (small cross-sections like in driven nails), are nearly horizontal (inclination to the horizontal lower than 20 degrees) and if there is not any surcharge on top of the wall. For nails carrying or assumed to carry only tension/compression, the admissible domain in the (T_n, T_c) planes reduces to the portion of the T_n axis of equation: $T_n \leq \min(R_n, T_{pull\text{-}out})$. Since the "Multicriterium Method" is a generalization of classical slope stability analyses, all types of failure can be treated continuously from external to mixed and internal. Application of the design method to the full scale experimental walls of CLOUTERRE loaded to failure gave a good agreement. Figure 7 presents the comparison between the predictions and the observations for the first full-scale experimental wall CEBTP #1 loaded to failure by saturation of the soil behind the facing.

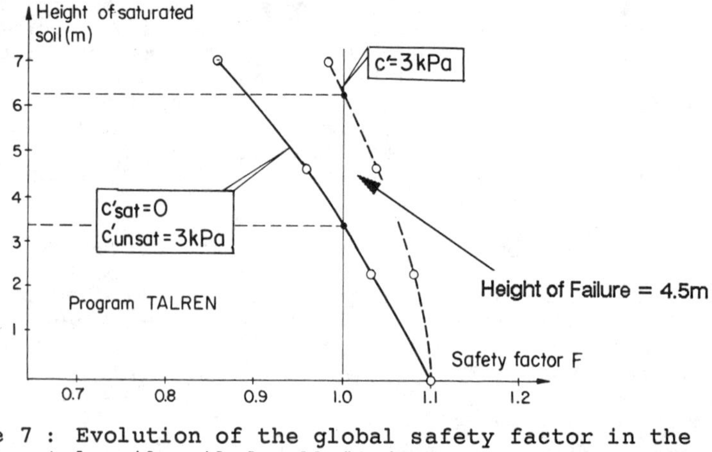

Figure 7 : Evolution of the global safety factor in the experimental soil nailed wall #1 (CEBTP-CLOUTERRE, 1985)

In figure 7, the height of water behind the facing is plotted against the global factor of safety for two different asumptions of the soil cohesion. Before saturation, the soil was partially saturated and had therefore a small apparent cohesion. After saturation, this cohesion changed. In Figure 7, an upper bound of 3 kPa (dashed line) and a lower bound of 0 kPa (solid line) were considered for the cohesion of the saturated soil.

The second concern when designing a structure, after making it stable, is to make it safe by incorporating a factor of safety. Traditionaly, safety is considered globally with the help of a global factor of safety. For simple academic exercice (but not recommended for design purposes), the failure surface can be modelled as a straight line defining a wedge as shown in Figure 8. The equilibrium can be written under the following form: $F_s = T_{max} / T$, with T_{max}, maximum resisting shear forces on the plane, T, projection of external forces on the plane.

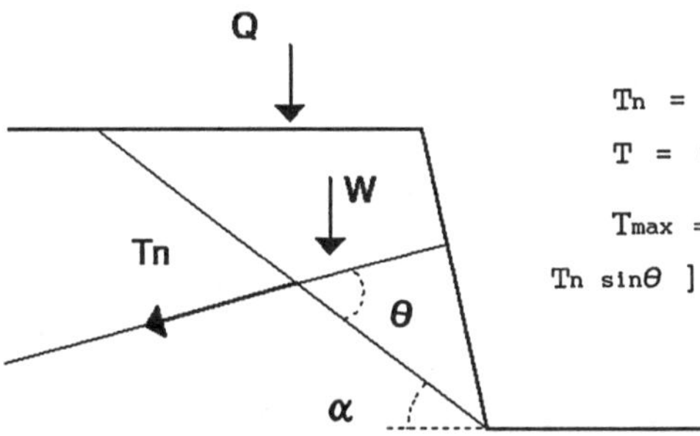

$$T_n = \min (R_n, \pi D L_a q_s)$$

$$T = (W+Q) \sin\alpha - T_n \cos\theta$$

$$T_{max} = [(W+Q) \cos\alpha + T_n \sin\theta] \tan\phi + c H / \sin\alpha$$

$$F_s = T_{max} / T \geq 1.0$$

Figure 8 : Simple analysis of a wedge failure using a global factor of safety

The structure will be stable if F_s is higher than 1.0 and it will be considered to be safe if F_s is higher than 1.3 for a temporary structure and 1.5 for a permanent one. The global factor of safety represents the margin of safety which is taken to account for uncertainties on the materials properties, on the loading conditions and errors inherent to the design method. In the partial safety factors approach, the usual global factor of safety is divided into a series of coefficients. The partial safety factors Γm, applied on the strengths of materials, account for their variabilities and the uncertainties on their measurements. Typically for a soil, the friction angle ϕ' will be better known than the cohesion c'. Therefore, the partial safety factor on ϕ', noted $\Gamma_{m\phi'}$ will have a value of 1.20 lower than the factor on c', noted

$\Gamma_{mc'}$, equal to 1.50. The weighing factors Γ_Q, applied on the external forces Q, account for uncertainties on the loading conditions. On gravity forces G, the partial factor is Γ_G equal to 0.95 or 1.05 to account for variations in the load. A method coefficient Γ_{S3} equal to 1.125 is introduced to account for errors inherent to the method. For the wedge failure mechanism (simplest possible mechanism), the structure will be safe if the following inequality is respected:

$$\Gamma_{S3}\,T \leq T_{max}$$

with

$$T_n = \min\,(R_n/\Gamma_{mRn},\ \pi\,D\,L_a\,q_s/\Gamma_{mqs})$$

$$T = (W\,\Gamma_G + Q\,\Gamma_Q)\sin\alpha - T_n\cos\theta$$

$$T_{max} = [(W\,\Gamma_G + Q\,\Gamma_Q)\cos\alpha + T_n\sin\theta]\tan\phi\,/\,\Gamma_{m\phi} + c'\,(H/\sin\theta)/\Gamma_{mc'}.$$

Ideally, partial safety coefficients and external forces weighing factors should be chosen based on observed dispersion laws of materials parameters and loading conditions while the other factors would be chosen to ensure a given maximum probability of failure [Gässler and Gudehus, 1983]. However, because of the complexity of such statistic-probabilistical analyses and since the current global safety factor method give acceptable design from the point of view of experience, partial factors are chosen to account for uncertainties as well as to give results in agreement with the experience (global safety factor approach) [Table 1].

External Forces	Weighing Factors
Weight G	Γ_W = 0.95 or 1.05*
Surcharge Q	Γ_Q = 0.90 or 1.20*
Soil Strength	Partial Safety Factor
Friction tan	$\Gamma_{m\phi'}$ = 1.20
Cohesion c'	$\Gamma_{mc'}$ = 1.50
Soil-Nail Interface	Partial Safety Factor
Frictional Resistance q_s	Γ_{mqs} = 1.40
Steel Nail	Partial Safety Factor
Tensile Strength R_n	Γ_{mRn} = 1.15
Method Coefficient Γ_{S3} = 1.125	
* or ** depending on the positive or negative effect on stability	

Table 1 : Partial Safety Factors (fundamental combination and ordinary structures)

Comparisons between the usual approach using a global factor of safety of 1.5 and the partial safety factors approach using coefficients in Table 1 indicates that the new approach is conservative for soil with a high cohesion (ϕ' = 20°, c' = 20 kPa) by about 10 % and non conservative for purely frictional soil (ϕ'= 30°, c'= 0 kPa) by about -10 %. However, thanks to the significant efforts of research carried out during CLOUTERRE , the usual approach was considered to be too conservative and the new one was chosen. The most important factor in the design of soil nailed structures is q_s: soil-nail interface frictional resistance. q_s has been observed to be independant of depth for a given soil and nail technology. Based on the 450 pull-out tests data base created in the CLOUTERRE Project, correlations have been developed between q_s and p_l limit pressuremeter pressure to be used for preliminary design. However, for design q_s should be determined by pull-out tests.

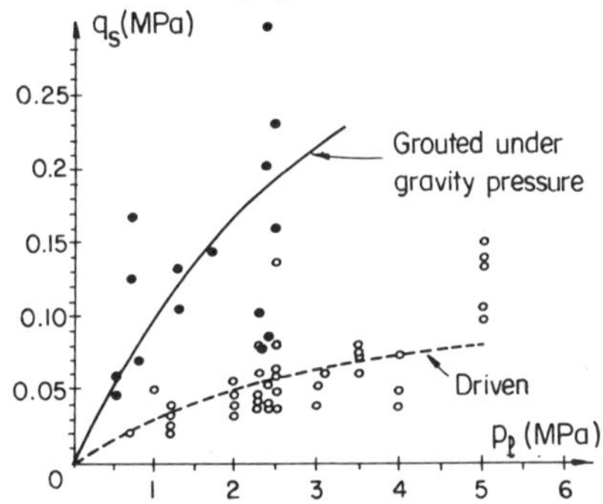

Figure 9 : Example of correlation between q_s, soil nail interface frictional resistance and P_l, limit pressuremeter pressure

In Service Deformations of Soil Nailed Walls

The third major concern, when designing a soil nailed wall in developed areas with existing buildings around, is the estimate of the displacements at the surface of the ground on top of the wall. At the present state of knowledge, best estimates are provided by experience and instrumented walls. At the top of a wall, three displacements can be defined δ_H, δ_V, δ_O (Figure 10a). Typically, δ_H has been observed to be about equal to δ_V. The range of values for δ_H (maximum lateral deflections), defined as a percentage of the wall height H, (Figure 10b), observed in France is 0.1% to 0.3% [Schlosser and Unterreiner, 1991], while in the U.S.A. it is 0.07% to 3.0% [Ferworn and Weatherby, 1991], and in Germany it is 0.25% to 0.3% [Gässler, 1991].

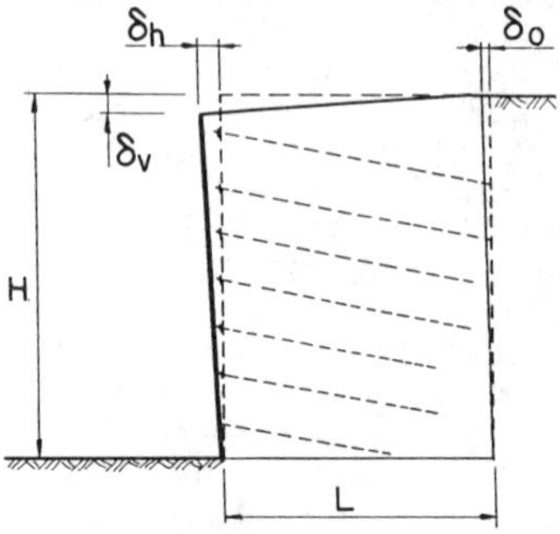

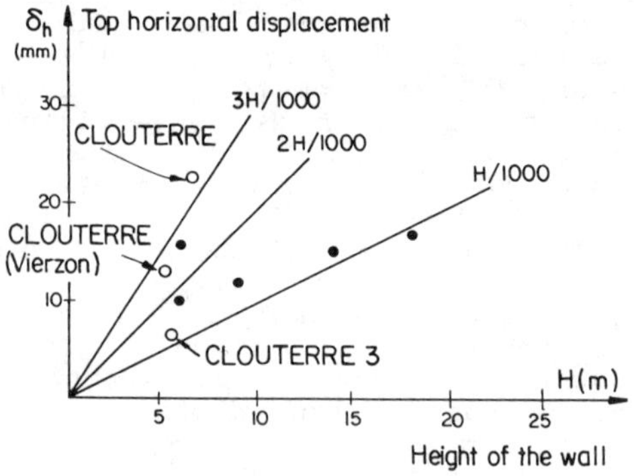

Figure 10a : Definitions of displacements δ_H, δ_V, δ_O.

Fig. 10b : Observed ranges of values for δ_H

On the ground surface at the top of a soil nailed wall, the lateral and vertical displacements which are maximum at the edge of the wall decrease to zero over a length, λ, which is function of the soil type (coefficient κ), the inclination of the wall (η), and the wall height (H) according to the empirical formula $\lambda = \kappa(1 - \tan\eta)$ H. Based on the empirical results, which are summarized in Table 2, one can estimate a priori the amount of differential settlements and extension the foundations of an existing building near a soil nailed wall will have to undergo.

Type of soil	Weathered rocks Stiff soils	Sandy soils	Clayey soils
$\delta_H = \delta_V$	H / 1000	2 H / 1000	3 H /1000
coefficient κ	0.8	1.25	1.5

Table 2 : Summary of data on displacements

Design of Facing

Although the facing of a nailed soil wall does not play a major role in the overall stability, it has an important function concerning the local stability and confinement of the soil between the nails. This role of confinement is highly dependant on the spacing of the nails. On one hand if the spacing were infinite, the facing would have to be a classic thick concrete retaining wall. On the other hand, if the spacing were zero (,i.e., the number of nails were infinite), the facing wouldn't have any role to play. The facing is in equilibrium by balancing the earth pressures and the forces applied. It is usual to consider only a tension T_0 in the nails at the facing and uniform distribution of earth pressure acting on the facing for design purposes. Direct calculation of T_0 is not easy but the ratio T_0/T_{max}, where T_{max} is the maximum tension in the nail in service, is well documented from experiments. CLOUTERRE Recommendations propose to use the following equations :

$T_0/T_{max} = 0.5 + (S-0.5)/5$ for $1m \leq S \leq 3$ m

$T_0/T_{amx}=0.6$ for $S \leq 1$ m

$T_0/T_{max} = 1.0$ for $S \geq 3$ m

with $S = Max(S_V, S_H)$ in .meters, S_V, vertical spacing, S_H, horizontal spacing. Once the ratio T_0/T_{amx} is known, T_{max} has to be estimated. A conservative approach is to use T_{lim}, maximum tension in the nails on the most critical failure surface, since that latter one is an upper bound of T_{max}.

Durability

Durability was of major concern in CLOUTERRE since one of the objectives was to promote soil nailing for permanent retaining structures. Most existing soil nailed structures are built with steel nails. Therefore, corrosion must be accounted for structures with an expected service life of more than 18 months. The most common technique in France is to use nails with thicker cross-sections, which is efficient only if the type of steel used for the nails undergo uniform corrosion and not concentrated corrosion. The sections are calculated so that at the end of the expected service life the remaining non-corroded steel sections are thick enough. In the CLOUTERRE Recommendations, a global index is defined to take into account the type of soil, its resistivity, the soil moisture content and other parameters. Extra-thicknesses of steel are then provided for each type of structures as a function of the global index presented in Table 3.

Class	Service Life ≤ 18 months	Service Life 1,5 to 30 years	Service Life 30 to 100 years
IV	0	2 mm	4 mm
III	0	4 mm	8 mm
II	2 mm	8 mm	plastic barrier
I	compulsory plastic barrier		

Table 3: Extra-Thicknesses Recommended.

Conclusion

With its four years of research, three full scale experimental walls and final specifications, the French national project CLOUTERRE is a major contribution to the present state of knowledge. Classical limit equilibrium methods extended with the "Multicriterium Rule" have proved to be an efficient tool to study the global stability of a soil nailed structure continuously from an external to an internal failure type. Values for the partial safety coefficients have been determined to account for variabilities and uncertainties as well as to match the experience in designing with a global safety factor. A simple method has been proposed to design the facing thickness in function of the nails spacing. Provisions for corrosion (extra-thickness of steel) have been specified in function of the service life, the type of soil and its aggressivity.

References

Ferworn, D. E., Weatherby, D. E., (1991); "A contractor's experience with soil nailing," Proc. of the 70th T.R.B. Annual Meeting, Washington D.C., U.S.A.
Gässler, G., Gudehus, G., (1983); "Soil Nailing - Statistical Design," Proc. 8th ECSMFE, 1983.
Gässler, G. (1991), "German Practice of Soil Nailing," Proc. 70th T.R.B. Meeting, Washington D.C., U.S.A.
Plumelle, C., Schlosser, F., (1991); "Three full-scale experiments of the French project on soil nailing : CLOUTERRE," Proc. 70th T.R.B. Meeting, Washington, U.S.A.
Rabejac, S. and Toudic, P. (1974), "Construction d'un mur de soutènement entre Versailles-Chantiers et Versailles-Matelots," Revue Gén. Chemins de Fer, N° 93.
Recommandations CLOUTERRE 1991, Etat de l'Art et Recommmendations," Presses de l'ENPC.
Schlosser, F., (1982); "Behavior and Design of Soil Nailing," Symp. on Soil and Rock Improvement Techniques including Geotextile, Reinforced Earth, Bangkok.
Schlosser, F., (1983); "Analogie et différences dans le comportement et le calcul des ouvrages de soutènement en Terre Armée et par clouage des sols," ITBTP, No. 418.
Schlosser, F. and Unterreiner, P., (1991); "Soil Nailing in France : Research and Practice," Proc. 70th T.R.B. Meeting, Washington, U.S.A.

Soil Nailing: A Simplified Kinematic Analysis

R. John Byrne[1]

ABSTRACT: A simplified soil-nail interaction analysis is developed for examining the manner in which soil nails develop their reinforcing action under working stress conditions. The predictions of the model are compared with the results from an instrumented 55-foot deep excavation in glacial soils.

INTRODUCTION

Limiting equilibrium methods are convenient design tools for soil nail-reinforced excavations. However, such methods do not directly address the issue of deformation, which is often a primary issue for soil-nailed systems, nor do they address the question of reinforcing loads developed under working stress conditions. A full soil-nail interaction analysis requires consideration of the stress-strain characteristics of the soils, the reinforcing elements (nails), the wall or other face elements to which the reinforcing elements are attached, and the interactions between these components of the overall system. While such analyses are less convenient than limiting equilibrium methods as design tools, they provide an insight into the manner in which soil nails develop their reinforcing action in practice.

This paper discusses a simplified soil-nail interaction model which addresses the parameters controlling the development of nail loads under working stress conditions. The predictions of the model are compared with the results from an instrumented soil nail wall constructed in dense glacial soils.

[1]Principal, Golder Associates Inc., 4104-148th Avenue NE, Redmond, WA 98052.

SIMPLIFIED SOIL-NAIL INTERACTION MODEL

Appendix 1 provides an analytical solution to the simplified one-dimensional problem of an initially unstressed reinforcing element located within a soil column which subsequently undergoes a change in boundary tractions corresponding to the release of a uniform stress (Figure 1). Upon release of the boundary tractions, the soil will tend to expand and induce tensile loads within the reinforcing member. This is analogous to the manner in which the reinforcing load is developed within soil nails (i.e., unstressed nails are installed and loads are mobilized within the nails as the soil mass relaxes and expands sub-horizontally in response to excavation). The released total stress is not generally uniform along the length of the nail, however, as assumed for this simplified analysis. As discussed later, the assumption of a uniform total stress release along the length of the nail is a fair approximation for nails in the upper third of the wall. The problem of release of boundary tractions corresponding to a generalized non-uniform stress release is not amenable to analytical solution but can be handled numerically. The uniform stress change problem is useful, however, in demonstrating the influence of the various material and geometric parameters on the magnitude and distribution of the reinforcing loads developed within the passive reinforcement.

The solution given in Appendix 1 shows that the maximum reinforcing load Tmax, normalized with respect to the equivalent total stress reduction and the average area of reinforced soil per nail, can be expressed in terms of three dimensionless parameters representing the ratios of the soil-nail interface stiffness to the soil stiffness (Ks/Es), the effective nail stiffness to the soil stiffness (E_n/E_s), and the nail length factor to the nail spacing ($(L_-L)/s$) (Figure 2). The dimensionless length factor (L-L)/s will vary between 2 and 4 for most practical designs. The information shown on Figure 2 indicates that the developed nail loads or reinforcing pressure will increase for higher nail stiffnesses, higher nail-soil interface stiffnesses, longer nail lengths and stiffer connections between the nail and the wall facing, smaller nail spacings, and lower soil modulus values.

The soil-nail interface stiffness can be determined from the results of nail pullout or validation testing, which is an essential element of any nailing project. A simplified analytical model for interpreting the results of such field testing to derive an approximate estimate of the interface stiffness, is given in Appendix 2.

The simplified soil nail interaction model described in

Appendix 1 can be extended to examine the response of an embedded reinforcing element when subjected to the

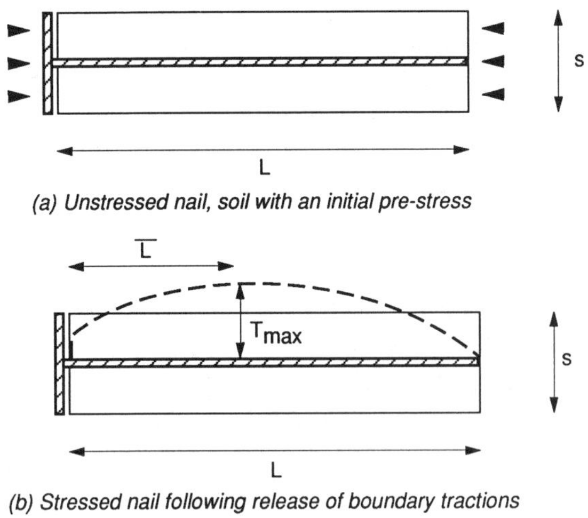

Figure 1. Simplified Soil-Nail Interaction

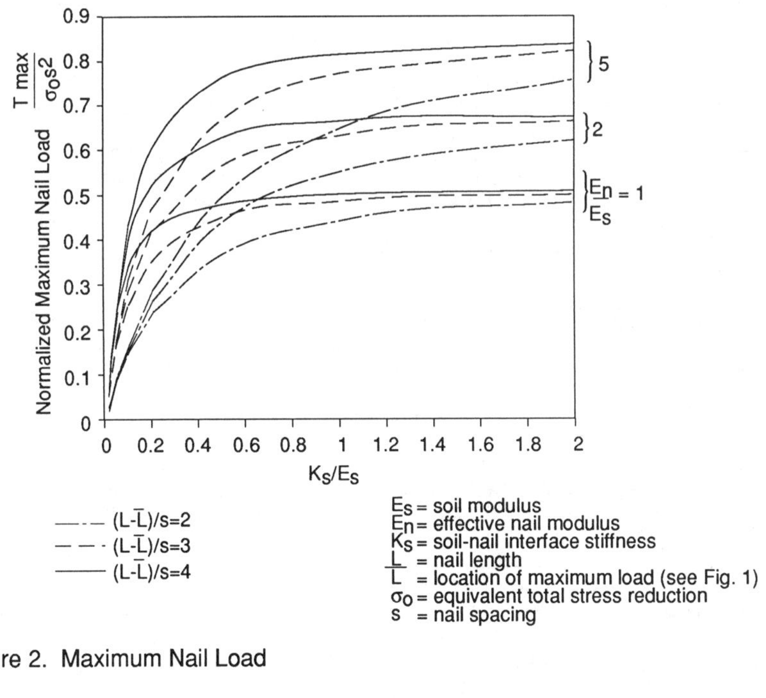

Figure 2. Maximum Nail Load

relaxation of boundary tractions resulting in a non-uniform total stress change along the length of the nail. The non-uniform equivalent total stress reduction can be approximated in a piecewise linear manner, and the resulting equations solved numerically. The derivation of this solution is not presented here because of space limitations. Applications of this more general model will be discussed in connection with the interpretation of the results of an instrumented field project, described below.

INSTRUMENTED SOIL NAILING PROJECT

Thompson and Miller, 1990, discuss results from the monitoring of a 55-foot high vertical cut, reinforced with 35-foot long soil nails (1-1/4 inch-diameter Grade 150 Dywidag bars on six foot vertical and horizontal centers), constructed in downtown Seattle. The soil types consisted of very dense glacial outwash sands and gravels, overlain by typically 8 feet of fill. The nails were grouted into 8-inch diameter holes and were inclined at 15 degrees below horizontal. Instrumentation for this wall included vibrating wire strain gages used to monitor five nails in a vertical section (Figure 3).

A representative set of data is shown on Figure 4, for the monitored nail at level 6 (Figure 3), and indicates the following: nail straining continued at a progressively reducing rate in the post-construction period, such that nail strains approximately doubled in the nine months following excavation; sharp increases in the monitored strain occurred at specific gages, presumably associated with cracking of the grout annulus in the vicinity of the nail gages. Considering that the cross-sectional area of the grout is approximately 50 times that of the steel, it is obvious that the majority of the nail load will initially be carried by the grout. The grout would be expected to creep with time, shedding load to the steel, softening the nail, and probably reducing somewhat the overall load carried by the nail. Hence, the progressive nail straining with time is interpreted to reflect primarily load redistribution within the nail from the grout annulus to the steel bar, rather than transfer of load to the nail by soil creep.

At specific locations, the grout annulus apparently cracked under the sustained tensile load, with an associated localized transfer of the entire nail load to the steel bar. Figure 4 indicates that this cracking apparently occurred at total strain levels within the grout on the order of 150 to 200 microstrain. The locations of the maximum developed strains within the monitored nails are shown on Figure 5. The measured maximum strains, immediately following excavation and some nine months later, are given on Figure 6.

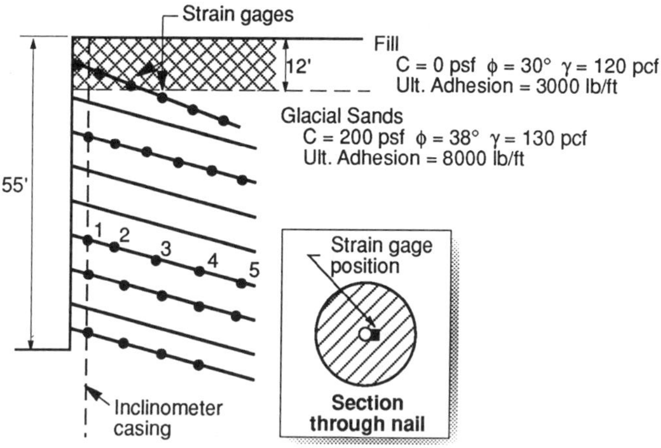

Figure 3. Instrumented Section

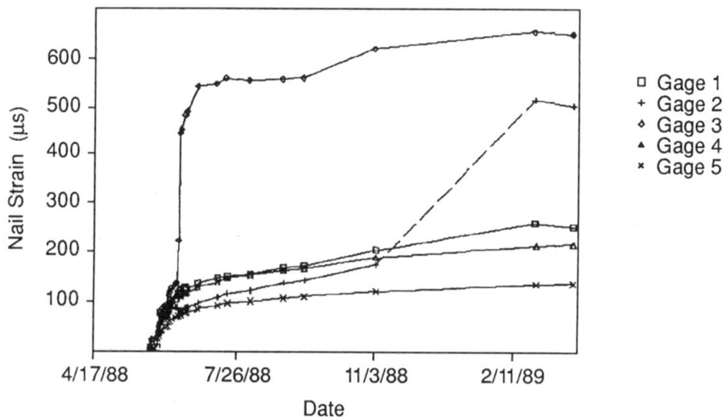

Figure 4. Strain vs. Time - Nail Level 6

SOIL-NAIL INTERACTION ANALYSIS

The simplified models discussed previously were applied to the above project, to determine soil nail loads based on a consideration of the kinematic interactions between the soil and the reinforcing nails. As discussed in Appendix 1, these analyses require information on the initial in situ state of soil stress, the stress changes in the soil in the vicinity of each of the nails following nail installation, the soil modulus, the nail stiffness, and the stiffness of the interface between the soil and the nail. The initial state of soil stress was assumed to be

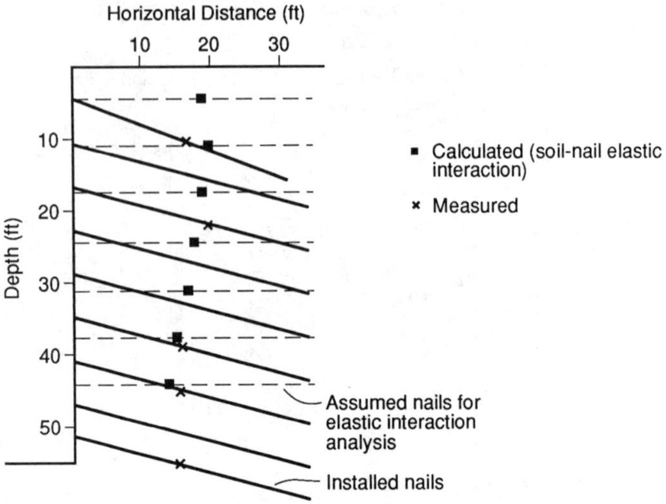

Figure 5. Locations of Maximum Nail Strain

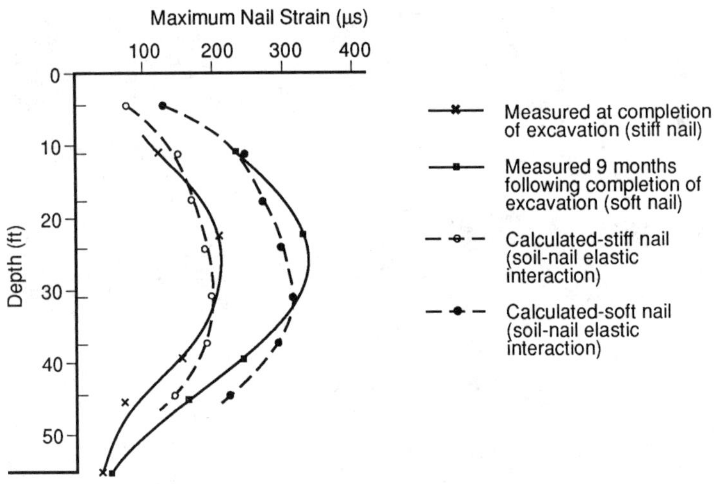

Figure 6. Maximum Nail Strain Distributions

defined by a Ko value of unity, consistent with the glacial history of these Seattle soils. The soil modulus values are indicated on Figure 7a, which shows the soil modulus increasing as a function of the square root of the mean confining pressure. Finite element stress analyses were carried out assuming the above initial stress and soil modulus values, together with a realistic sequence of excavation and nail installation, to determine the equivalent soil stress changes that would be experienced by the nails, following their installation. These normalized stress changes are shown as a function of nail depth and distance from the nail head, on Figure 7b.

The nail stiffness was defined for two time periods - immediately following excavation and approximately nine months later. Immediately following excavation, the stiffness of the composite nail section was defined by assuming a relatively stiff grout modulus ($E_g = 3x10^6$ psi), as indicated on Figure 7c. The long term composite nail stiffness was determined using a grout stiffness of half the initial value, based on the observed magnitudes of the nail creep strains over the nine months following completion of the excavation (Figure 7c).

Soil-nail interface shear stiffness values were determined from examination of field validation and pullout test data, using the simplified model given in Appendix 2. The field data indicated that the interface behaved in a relatively stiff manner and that an interface shear stiffness on the order of 750,000 psf would provide a reasonable representation of the data (Figure 7d).

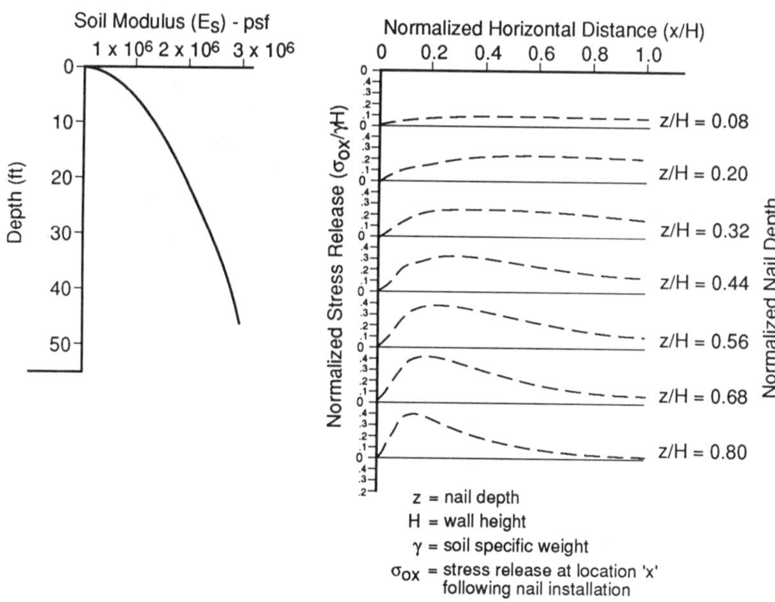

a) Soil Modulus

b) Soil Stress Release Following Nail Installation

Figure 7. Parameter Values for Soil-Nail Interaction Analysis

Stiff Nail

Nail Grout Area A_g	$= 65 in^2$
Nail Grout Modulus E_g	$= 3 \times 10^6$ psi
Nail Steel Area A_{st}	$= 1.23\ in^2$
Nail Steel Modulus E_{st}	$= 29 \times 10^6$ psi
Nail Spacing s	= 6 ft
Effective Nail Modulus E_n (see App 1)	$= \frac{E_g A_g + E_{st} A_{st}}{s^2}$
	$= 6.4 \times 10^6$ psf

Soft Nail

Nail Grout Area Ag	$= 65 in^2$
Nail Grout Modulus E_g	$= 1.5 \times 10^6$ psi
Nail Steel Area A_{st}	$= 1.23\ in^2$
Nail Steel Modulus E_{st}	$= 29 \times 10^6$ psi
Nail Spacing s	= 6 ft
Effective Nail Modulus E_n	$= 3.7 \times 10^6$ psf

c) Effective Nail Modulus

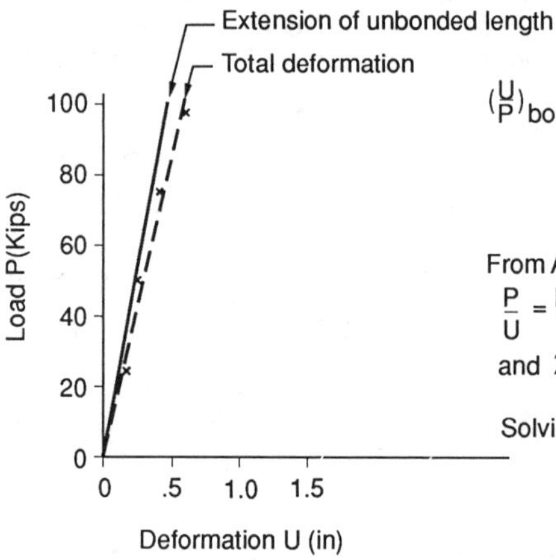

$$\left(\frac{U}{P}\right)_{bonded} = \left(\frac{U}{P}\right)_{total} - \left(\frac{U}{P}\right)_{unbonded}$$

$$= .006 - \frac{L_u}{A_{st}E_{st}}$$

$$= .0012 \text{ in/k}$$

From Appendix 2:

$$\frac{P}{U} = \frac{K_s}{\lambda}\frac{\sinh \lambda L_B}{\cosh \lambda L_B}$$

and $\lambda = \sqrt{\dfrac{K_s}{(E_g A_g + E_{st} A_{st})}}$

Solving iteratively yields $K_s = 750{,}000$ psf

Bonded length L_B	= 18 ft
Unbonded length L_u	= 9 ft
Area of bar A_{st}	$= .785\ in^2$
Steel modulus E_{st}	$= 29 \times 10^3$ Ksi
Area of grout A_g	$= 65\ in^2$
Grout modulus E_g	$= 3 \times 10^3$ Ksi

d) Typical Nail Validation Test

Figure 7. Parameter Values for Soil-Nail Interaction Analysis

Using the above material characterizations (Figure 7) and the simplified model previously described, a soil-nail interaction analysis was performed. Predicted locations of the point of maximum developed strain within the nails are shown on Figure 5, and these show good correlation with the locations inferred from the field measurements. The predicted profiles of maximum strain levels, immediately following construction and nine months later, are shown on Figure 6 and again show relatively good correspondence with the values measured in the field. Figure 6 indicates how the nail strains increase with time

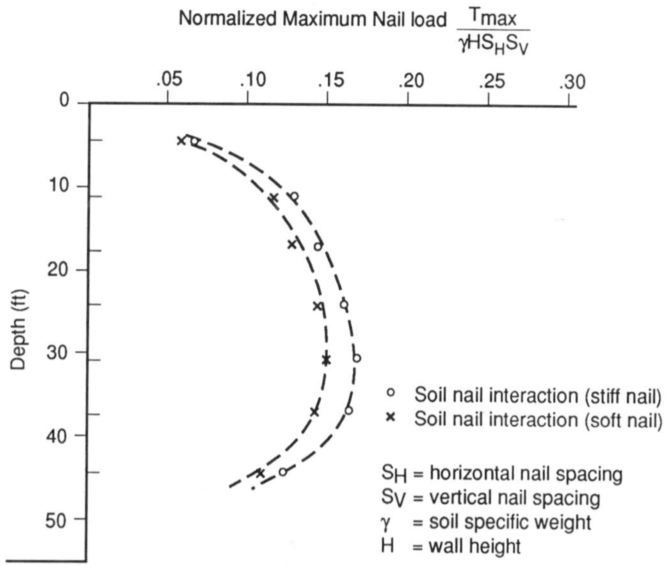

Figure 8. Maximum Nail Load Distributions

as a result of the softening of the nail associated with creep and cracking of the grout annulus. The model also enables determination of the nail loads, and the profiles of predicted maximum nail load are shown, in dimensionless format, on Figure 8. Although the nail strains (Figure 6) are seen to approximately double during the nine-month period following excavation, the change in the reinforcing loads is relatively small and they actually decrease somewhat as a result of the softening of the nail relative to the soil.

CONCLUSIONS

Simplified soil-nail interaction models which explicitly account for factors such as the soil and nail stiffness, and the shear stiffness of the soil-nail interface, appear to be capable of predicting both the magnitudes and distributions of the observed nail strains (or loads) under working conditions, if appropriate material parameters are applied. These models provide a useful insight into the manner in which the reinforcing nails interact with the surrounding soil, within the working stress range.

REFERENCES

Thompson, S., Miller, I. (1990). "Design, Construction and Performance of a Soil Nailed Wall in Seattle, Washington," presented at ASCE Speciality Conference, Cornell University, Ithaca, New York, June 1990.

APPENDIX 1

SIMPLIFIED SOIL-NAIL INTERACTION ANALYSIS

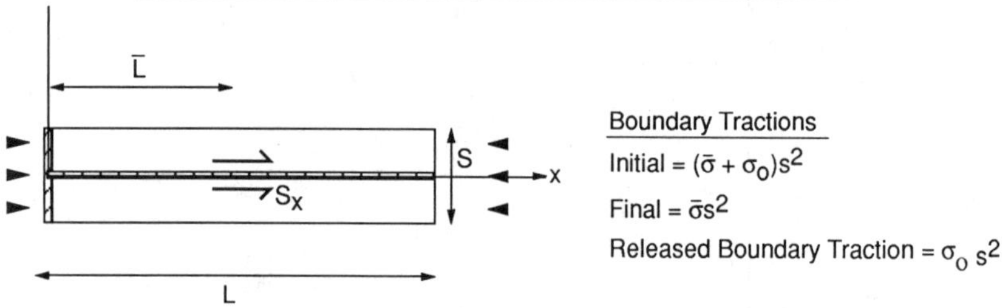

1. Consider a column of soil under uniform initial horizontal stress "$\bar{\sigma}+\sigma_o$" with embedded unstressed nails of length "L" at average spacing "s". The nails are attached to a facing at one end and are free at the other end.

2. The boundary tractions are now reduced from $(\bar{\sigma}+\sigma_o)\ s^2$ to $\bar{\sigma}s^2$ (i.e., $\sigma_o s^2$ = boundary traction reduction) and the soil expands laterally, inducing tensions within the nails. The final average soil stress at location 'χ' is "$\bar{\sigma} + \sigma_{s\chi}$" and

$$\epsilon_{s\chi} = \frac{(\sigma_{s\chi} - \sigma_o)}{E_s} = \frac{du_{s\chi}}{d\chi}$$

$\epsilon_{s\chi}$ = soil strain
$u_{s\chi}$ = soil displacement
E_s = soil modulus
σ_o = equivalent total stress reduction at each section (i.e., released boundary traction/s^2)

$$\therefore\ \sigma_{s\chi} = \sigma_o + E_s\,\frac{du_{s\chi}}{d\chi}$$

3. Nail load "$T_{n\chi}$" is

$T_{n\chi} = E_n s^2 \frac{du_{n\chi}}{d\chi}$ $u_{n\chi}$ = nail displacement; E_n = effective nail modulus $= \frac{E_G A_G + E_{st} A_{st}}{s^2}$

Where E_G = nail grout modulus; A_G = nail grout area; E_{st} = nail steel modulus; A_{st} = nail steel area

4. From equilibrium

$$(\bar{\sigma} + \sigma_{s\chi})s^2 + T_{n\chi} = \bar{\sigma} s^2$$

$$T_{nx} = -s^2 \sigma_{sx} \qquad \therefore \frac{du_{sx}}{dx} = -\frac{E_n}{E_s}\frac{du_{nx}}{dx} - \frac{\sigma_o}{E_s}$$

and integrating

$$u_{s\chi} = -\frac{E_n}{E_s} u_{n\chi} - \frac{\sigma_o \chi}{E_s} + \alpha \qquad \alpha = \text{constant}$$

5. Interface shear force per unit length of nail between the soil and the nail is

$$S_\chi = -\frac{dT_{n\chi}}{d\chi} = -E_n s^2 \frac{d^2 u_{n\chi}}{d\chi^2}$$

$$= K_s (u_{s\chi} - u_{n\chi}) \qquad K_s = \text{interface shear stiffness}$$

$$\therefore \frac{d^2 u_{nx}}{dx^2} = \frac{K_s}{E_n s^2}\left[(1+E_n/E_s)u_{nx} + \frac{\sigma_o}{E_s}x - \alpha\right]$$

6. Solving for $u_{n\chi}$, noting that $T_{n\chi} = E_n s^2 \frac{du_{n\chi}}{d\chi}$, and introducing the boundary conditions:

At $\chi = 0$: $T_n(\chi = o) = \bar{K}(u_n(\chi = o) - u_s(\chi = o))$ Where $\bar{K}$ = face-nail stiffness

At $\chi = L$: $T_n(\chi = L) = 0$.

This yields the general solution

$$T_{nx} = \frac{\sigma_o s^2}{(1+E_s/E_n)}\left[\frac{\sinh\lambda x + \frac{\bar{K}\lambda}{K_S}\cosh\lambda x + \sinh\lambda(L-x)}{\sinh\lambda L + \frac{\bar{K}\lambda}{K_s}\cosh\lambda L} - 1\right] \quad \textit{where } \lambda = \frac{1}{s}\sqrt{\frac{K_s}{E_s}\left(1 + \frac{E_s}{E_n}\right)}$$

7. Maximum nail load occurs at $\chi = \bar{L}$ where $\frac{dT_{n\chi}}{d\chi} = 0$ and $\bar{L}$ is determined by solving:

$$\cosh\lambda(L-\bar{L}) = \cosh\lambda\bar{L} + \frac{\bar{K}\lambda}{K_s}\sinh\lambda\bar{L}$$

$$\text{and } T_{max} = \frac{\sigma_o s^2}{(1 + E_s/E_n)}\left[\frac{1}{\cosh\lambda(L-\bar{L})} - 1\right]$$

T_{max} depends on the equivalent total stress reduction "σ_o", the nail spacing "s", together with three dimensionless parameters

$\frac{E_n}{E_s}$ = ratio of effective nail modulus to soil modulus

$\frac{K_s}{E_s}$ = ratio of interface shear stiffness to soil modulus

$\frac{(L-\bar{L})}{s}$ = ratio of nail length factor to nail spacing

APPENDIX 2

SIMPLIFIED NAIL PULLOUT/VALIDATION TEST ANALYSIS

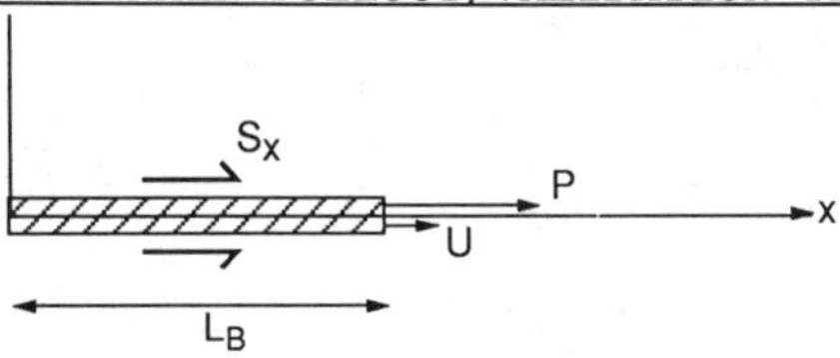

1. Consider a pullout test where load "P" is applied and displacement "u" is measured at the interface between the bonded and unbonded sections of the test nail (i.e., displacement "u" is the measured displacement at the face less the calculated extension of the unbonded section of the bar).

2. Interface shear force per unit length of nail between the soil and the nail is:

$$S_\chi = -\frac{dT_{n\chi}}{d\chi} = -(E_G A_G + E_{st}A_{st})\frac{d^2u_{n\chi}}{d\chi^2} = -K_s u_{n\chi}^*$$

where
$T_{n\chi}$ = nail load
$u_{n\chi}$ = nail displacement
E_G = nail grout modulus
A_G = nail grout area
E_{st} = nail steel modulus
A_{st} = nail steel area
K_s = interface shear stiffness

$$\therefore \quad \frac{d^2u_{n\chi}}{d\chi^2} = \frac{K_s}{(E_G A_G + E_{st}A_{st})} u_{n\chi}$$

3. Solving for $u_{n\chi}$, noting that $T_{n\chi} = (E_G A_G + E_{st}A_{st}) \dfrac{du_{n\chi}}{d\chi}$ and introducing the boundary conditions:

At $\chi = 0$: $T_n\ (\chi=0) = 0$
At $\chi = L_B$: $T_n\ (\chi=L_B) = P$

This yields the general solutions

$$T_{nx} = \frac{P \sinh \lambda\chi}{\sinh \lambda L_B} \quad \text{and} \quad u_{n\chi} = \frac{P}{(E_G A_G + E_{st}A_{st})} \quad \frac{\cosh\lambda\chi}{\lambda \sinh\lambda L_B}$$

$$\textit{where } \lambda = \sqrt{\frac{K_s}{(E_G A_G + E_{st}A_{st})}}$$

4. For small λL_B (i.e., stiff nail) then $K_s \approx P/L_B u$

At $\chi = L_B$: $u_n(\chi=L_B) = u$

$$\therefore P/u = (E_G A_G + E_{st}A_{st})\ \lambda \frac{\sinh\lambda L_B}{\cosh\lambda L_B} = \frac{K_s}{\lambda} \quad \frac{\sinh \lambda L_B}{\cosh \lambda L_B}$$

which can be solved iteratively for K_s

---

*For the pullout test, the average soil displacement can be considered as zero. Soil displacements occurring in the near vicinity of the test nail are incorporated into the stiffness of the "interface" between the nail and the soil, consistent with the assumptions incorproated into the simplified kinematic model.

The following symbols are used in this paper.

A_G = nail grout area
A_{st} = nail steel area
E_G = nail grout modulus
En = effective nail modulus
E_{st} = nail steel modulus
$\epsilon_{s\chi}$ = soil strain
Es = soil modulus
γ = soil specific weight
H = wall height
Ks = nail-soil interface stiffness
$\underline{L}$ = nail length
L = distance from face to point of maximum nail load
s = nail spacing
σ = stress
S_H = horizontal nail spacing
S_V = vertical nail spacing
$T_{n\chi}$ = nail load
$u_{n\chi}$ = nail displacement
$u_{s\chi}$ = soil displacement
z = nail depth

Recent Progress In American Pinpile Technology

by Donald A. Bruce, PhD, MASCE*

Synopsis

The benefits of pinpiling have been exploited in the USA for the last two decades. The demands and skills of American practice have led to developments of sufficient importance to now distinguish it sharply from aspects of the original European technology. This paper provides details from 20 recent projects and reviews in greater depth three major case histories. These particular projects illustrate major features about new developments in design, construction and performance in unique applications.

Introduction

A number of papers in the late eighties (3-6) described the state of practice in pinpile technology in the United States. These papers demonstrated that pinpiles - known elsewhere in the world as minipiles or micropiles, but defined generically as small diameter cast-in-place bored piles - had arrived relatively late in the United States, probably some 20 years or so after their European genesis in the early fifties. However, the growth in their use had been dramatic, as the specialist geotechnical construction industry responded to the demands of renovation, remediation and redevelopment, largely in urban and industrial locales. These reviews also detected that the pinpile market in the United States was generating a special identity for itself as a result of its ongoing pursuit of progressively higher unit capacities in a wide and complex range of subsurface conditions.

This paper presents brief reviews of three selected case histories of pinpile projects conducted by the author's company alone, in the few years since the first survey. To begin with, it should be noted that 20 pinpile projects can be cited in the two years from early 1989 (7), compared to the total of 25 projects listed in the previous 11 years (3). This is clear evidence of the rapidly increasing use of the

* Technical Director, Nicholson Construction of America, P.O. Box 308, Bridgeville, PA, 15017

technique of pinpiling, especially in the older cities of the East Coast (8). Equally noteworthy is the routine use of individual pile working loads appreciably higher than in earlier years, or elsewhere in the world. The test loads of 220 tons in fine sands in Brooklyn or 350 tons in poor rock in Pittsburgh (7) are, admittedly, exceptional: normal working loads of 75-110 tons in soils, and higher in rock are now fairly routine.

This expansion in application and growth in working loads has been supported by evidence from numerous exhaustive test programs, executed in a consistently routine and scientific manner as an integral part of each commercial project. In most cases the test piles have been loaded in incremental, cyclic fashion, so permitting movements to be partitioned into elastic and permanent components. Analysis of the former provides indications of load transfer mechanisms within the pile, whereas examination of the latter essentially permits the behavior of the founding medium to be investigated. This approach has significantly aided the understanding of how the ancillary ground treatment technique of post grouting (9) works to enhance pile performance - although space prevents a detailed account of the fundamental work at Augusta, Brooklyn and Seattle. More experience is also being obtained with the concept of preloaded piles wherein the service load is preapplied and locked into the pile before the pile is connected to the structure. In this way the possibility of later structural movement is eliminated. The potential of this particular pinpile variant, the NCA-Pile (Nicholson Compressed Anchor-Pile) in seismic areas is outlined.

In addition, it should be recorded that much recent attention has been focused on the performance of pinpiles when subjected to lateral loading. The impressive results from laboratory testing (2) have been supported by recent field test programs at Brooklyn and Augusta (7): provided casing is left in place through the upper, typically "soft" horizons, pinpile performance is excellent.

Review of Three Illustrative Case Histories

The three case histories selected for more detailed review to illustrate new developments are:

- Postal Square, Washington, D.C. - A classic application of pinpiles as an advantageous construction alternate to hand underpinning or large diameter caissons during the redevelopment of an old, massive structure, and which demonstrates the benefits of cyclic loading.
- Presbyterian Hospital, Pittsburgh, PA - High capacity pinpiles installed in very difficult access conditions, subsequently partially exposed, to operate without lateral restraint.
- Pocomoke River Bridge, MD - High capacity, temporarily preloaded pinpiles installed to underpin an existing bridge unable to accept further structural movements.

Postal Square, Washington, D.C.

Background - The original portion of the massive Old Postal Building, Postal Square, was completed in 1911. A major extension followed in

1931. For many years it served as the main Post Office for Washington, D.C., being located adjacent to the Union Station on Massachusetts Avenue, a few blocks north of the Capitol Building. The Developer (Gerald D. Hines Interests), acting for the Federal Government, planned to remodel the existing structure by adding new office floors in the center court area and constructing mechanical space below the existing lowest basement elevation of +23'. This meant that existing foundations had to be upgraded and new columns added to support new interior framing. The existing supports are steel and concrete columns on large concrete footings, and 14" square caissons end bearing on dense sands.

Originally a cumbersome underpinning scheme was envisaged, involving hand dug support, massive spread footings and large diameter caissons, both hand excavated and mechanically drilled. However, the hand work would probably have caused significant undermining of the existing footings, leading to settlement, whereas drilled caisson work would have been inhibited by the very restrictive access, and low headroom. Both techniques would have been unattractively time consuming and costly. The pinpile alternate resolved both concerns.

<u>Site and Ground Conditions</u> - The work was conducted underground in three main areas in the basement of the existing structure:

- B2 Level (EL 6'): large level area with about 12' headroom. Piles reached El -45'.
- B2 Level (El 11'): most restricted area, headroom 8'. Piles reached El -45'.
- B1 Level (El 23'): Open access with 15-20' headroom. Piles reached El -35'.

Under the concrete footings and a few feet of fill, the natural soils comprised recent alluvials, ranging from coarse to fine sands, laterally and vertically variable. Some gravel and mica were found sporadically, together with thin layers of cobbles or stiff clayey silt and silty sand in lower reaches. Typically the sands were dense to very dense. Natural groundwater level was at about El -5'.

<u>Design and Construction</u> - Past experience and standard texts (12) were used to design 390 vertical piles in the B2 levels and 310 piles in the B1 level, each with a nominal working load of 75 tons. About 25% of the piles were installed in groups of 4 or 6 through 15 existing B2 (El 6) footings comprising 7-14' of concrete. Pile centers were within 20" of existing columns.

Totals of 21 new reinforced concrete caps were created in B2 (El 6), 17 in B2 (El 11) and 53 in B1. These featured standard (and several non-standard, specially designed) plan geometries from 5'4" x 4'8" (3 piles) to 7'6" square (9 piles). The minimum pile separation was 26" center to center, but was typically 30".

Custom built, short mast diesel hydraulic track rigs were used to rotate 7" dia. 0.5" wall N80 casing with water flush, to target depth. Type I grout of w/c = 0.45 was injected under excess pressures of 80-

110 psi during progressive extraction of the casing over the lower 25'. The casing was then reinserted 5' into this pressure grouted zone as permanent support. The lowermost 25' of pile was reinforced by Grade 60, 1-3/8" dia. rebar in 3 coupled lengths. For those holes through existing footings, a 8-7/8" down-the-hole hammer was used to penetrate until significant steel was encountered. Thereupon, the hole would be completed with a 8" core bit. Load transfer between casing of the pile and the footing was ensured by the use of a special non-shrink, high strength grout. For the new pile caps, the pinpile casing was extended 4" up into the subsequent concrete, the horizontal reinforcing of which was fixed 2" above the top of casing.

Testing and Performance - Four special test piles (TP) were installed prior to constructing the production piles (Table 1).

	TP1	TP2	TP3	TP4
Area	B2-6'	B2-6'	B2-11'	B1-23'
Length (ft)	36	51	36	58
Bond (ft)	25	25	25	25
Max. Test Load (kips)	187.5 (failed)	300	300	300
Elastic Extension at Max Load (inches)	0.173	0.461	0.383	0.374
Permanent Deflection at End of Test (inches)	0.313	0.291	0.187	0.113
Total Cumulative Creep (inches) During Test	0.096	0.174	0.059	0.061

Table 1. Test Pile Data, Postal Square, D.C.

TP1 and 2 were tested cyclically, yielding the analysis provided in Figure 1. TP3 and 4 were also tested incrementally but progressively to maximum load in accordance with ASTM-1143 (1). TP1 failed at the grout-soil interface, the founding horizons being on average finer and less dense than those for the other piles. Figure 1 also shows that the elastic compressions of TP1 and 2 were similar at the failure load of the former. This shows that load must have been transferred to similar depths in both piles, despite the nominal difference in "free length" upon construction. The elastic performance of TP3 and 4 was likewise similar, supporting the observation.

Table 1 also highlights higher total creep amounts in TP1 and 2: simply a reflection that there were far more creep monitoring periods in the cyclic loading that in the progressive loading. This clearly impacts overall permanent displacement and is an important point to bear in mind when judging pile performance as judged by this criterion.

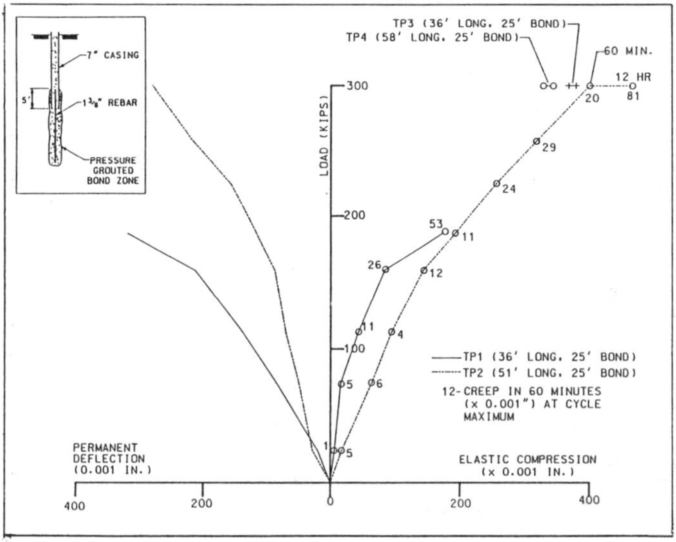

<u>Figure 1.</u> Elastic/Permanent St Performance of TP1, TP2, Postal Square, D.C.

A separate pullout test was conducted in an existing column footing in the B2 (El 6) level to explore concrete/grout bond. A special element was grouted 4'6" into an 8-7/8" hole drilled through the concrete. A special high strength, no shrink grout was used. After repeated cyclic loading to 525 kips (79% GUTS), and equivalent to 350% design load), the maximum uplift recorded was 0.005", reducing to 0.001" upon destressing. Assuming uniform bond distribution, an average grout-concrete bond of almost 350 psi had therefore been safely resisted.

Following installation of the pinpiles, the structural renovation has progressed, and the foundations have performed perfectly.

<u>Presbyterian University Hospital, Pittsburgh, PA</u>

<u>Background</u> - The Presbyterian University Hospital complex already occupies two extremely congested city blocks, and so when the need for more facilities became apparent the decision was made to vertically extend and laterally link several existing operational structures. Overall, 1.6 million square feet of new facilities are being built in four major additions.

This highly delicate operation, conducted within a fully functional facility, has necessitated some equally complex and innovative foundation engineering solutions involving excavation support and structural underpinning. One of the most dramatic operations has been associated with the completion of a new Magnetic Resonance Imaging Center. The construction of a new elevator pit called for a 30' deep excavation directly underneath three exterior column footings of the adjacent 13-story hospital structure. The pit, 60' x 32' in plan was further bounded on two sides by five additional footings, and so these sides required anchored lateral support.

Historically, the support of columns in such circumstances has been achieved by conventional underpinning pits and needle beams. However, in this instance, the difficult access conditions, and the specified requirement to limit downwards movement of the columns to less than 1/8" demanded a special solution, featuring high capacity pinpiles in rock.

<u>Site and Ground Conditions</u> - The access was very restricted laterally and vertically (as low as 12'), and the work had to be conducted within the confines of a fully operational medical facility. The piles were installed through 3'6" of existing nearby reinforced concrete footings cast directly on fractured, fissile medium hard-hard siltstone, occasionally calcareous or limey.

<u>Design and Construction</u> - At each existing footing, six pinpiles (4 working, plus 2 redundant) were installed in 8-1/2" holes drilled vertically by rotary percussive methods and air flush to the target depth (43' below the footing). Each had a design working load of 250 kips. The reinforcing element consisted of a 7" dia., 1/2" wall N-80 casing placed full depth in each hole which was then tremied full of neat cement grout of w/c = 0.45. The upper 23' of each pipe was greased on the outside to debond it from the surrounding grout in that region and so permit load transfer into the 20' long bond zone. The suitability and security of this design had been proved in the earlier test program, described below. A structural steel jacking frame was then erected over the top of the piles and fastened to the existing steel column. Each of the steel columns - supporting an occupied hospital building - were then sequentially lifted off their existing spread footing by a distance of 1/16 - 1/8". This effectively preloaded the piles to prevent any later settlement of the building, and transferred the column loading into the bedrock, but 23' below. Excavation then proceeded, supported laterally by beams, shotcrete lagging and prestressed rock anchors. As the excavation deepened, cross frames were welded to the pinpiles to limit the unbraced lengths of these piles, now exposed as steel columns.

<u>Testing and Performance</u> - By the end of excavation, the foundations of the existing structure could be seen resting on the pinpile groups, 22' off the bottom of the excavation. During and after excavation, no movement of the structure was measured.

One of the most common problems foreseen for pinpiles (10) is their potential for buckling or bending, as an inferred consequence of their high slenderness ratio. This unique project - featuring pinpiles with no surrounding ground to offer any lateral restraint - is proof irrefutable that correctly designed and constructed pinpiles can operate with surprising efficiency not only in the axial sense.

Clearly, testing of production piles was not possible in this instance, and so a full scale test pile was installed beforehand at an adjacent location. Using identical construction methods, a pinpile with 20' bond was formed in the same geological stratum. The total length was 50', including, therefore, 30' of debonded "free length". The casing was preassembled in the workshop and consisted of 5 separate lengths, hand tightened together. Two "telltales" were incorporated - one each at the top and bottom of the bond zone. A thick, soft wood plug was attached to the bottom of the reinforcing pipe to eliminate any possible end bearing contribution and to so allow only side shear to be mobilized. As part of the contract requirement, the pile was then tested to twice design working load (500 kips), according to ASTM-D1143-81 (1) (modified to allow cycles at 25%, 50%, and 75%). Results are summarized in Table 2. At 250 kips, the elastic compression of 0.227" was exactly that predicted, while the permanent displacement of 0.052" was proved by the telltales to be due to some inelastic compression of the steel casing itself. While loading from 400 to 425 kips, a "bump" was recorded and the load dropped to 300 kips. Load was then increased to 500 kips when a further "bump" occurred. However, when the data from the cyclic loading and the telltales were analyzed it became clear that:

(i) the pile elastic deflection at 500 kips was exactly as predicted, and

(ii) the apparently large permanent movement (Table 2) was due to an irreversible "one off" shortening of the steel pipe. The assembly records of the pile were then reviewed. It transpired that there had been several "unshouldered" hand tightened joints between adjacent casing sections. It was suspected that each joint was unshouldered about 1/4 - 1/8". Thus, the sudden 0.471" permanent compression of the pile material was readily explained, and when subtracted from the permanent set of 0.503", gave a true movement of the pile tip into the rock of 0.032" at 500 kips - an outstanding performance. There was negligible creep at all load increments.

Load Cycle Maximum (kips)	Total Butt Movement at Maximum (A) (inches)	Permanent Butt Movement at Subsequent Zero (B) (inches)	Therefore, Elastic Deflection at Maximum (A) - (B) (inches)	Apparent Bottom Telltale Movement (Relative to Butt) (inches)
125	0.127	0.042	0.085	0.037
250	0.279	0.052	0.227	0.044
375	0.448	0.077	0.371	0.063
300	0.663	0.329	0.334	0.295
500	1.020	0.517	0.503	0.471

Table 2. Summary of test pile performance, Presbyterian University Hospital, Pittsburgh

Thereafter, the pile was tested to a maximum load of 675 kips before it became clear that material failure of the steel casing was occurring. At this load, the steel had compressed 3.084 (from telltales), compared with the measured butt permanent displacement of 3.224". Thus, at 675 kips, a true permanent movement of the pile of 0.140" had been recorded, while analysis proved the perfect elastic performance of the pile with a calculated debonded length a few feet into the bond zone.

This project was therefore highly significant from several viewpoints:

- the excellent lateral and vertical performance of pinpiles was proved again;
- the value of telltales in aiding understanding of internal pile performance was demonstrated, and
- the warming fact that the boundaries of pinpile design are now those of the constituent materials - i.e., independent of the surrounding ground properties - was massively underlined.

Pocomoke River Bridge, MD (11)

Background - This 275' long movable bascule pier drawbridge (Figure 2) was built over the Pocomoke River in 1921. Bascule Piers 3 and 4 were originally supported on wooden piles driven through the soft riverbed muds into the underlying compact sand. The support offered by these piles had been compromised by river scour which had exposed them in several places. The pinpiles designed to stabilize the structure were remarkable on three counts:

(i) they had to be installed through the structure and through the scour zone,
(ii) they had to provide support without allowing any additional structural settlement so necessitating the use of preloading techniques, and
(iii) a very intensive test program was required on special test piles to verify the concept and the performance.

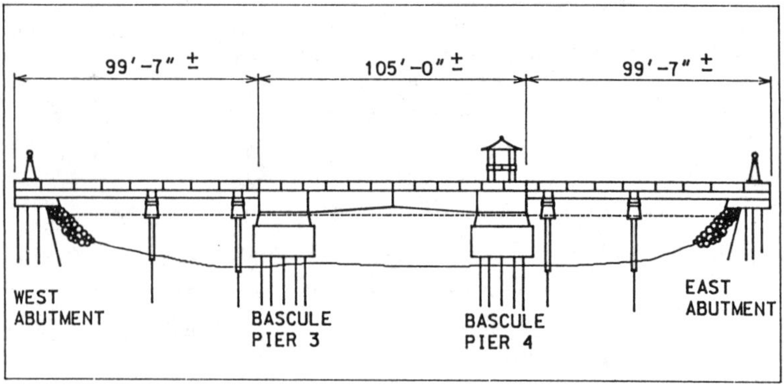

Figure 2. General Configuration of Pocomoke River Bridge, MD

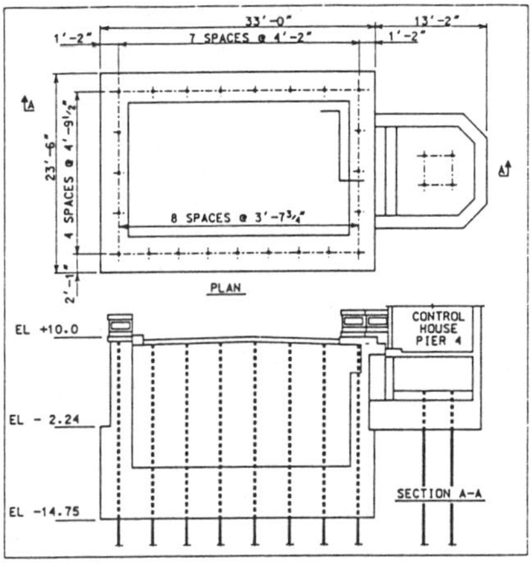

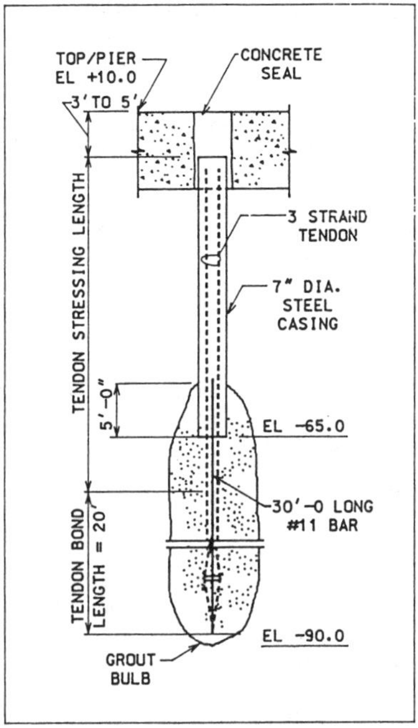

Figure 3. (above) Plan and Section of Bascule Pier #4, Pocomoke River Bridge, MD, Showing Pinpile Locations

Figure 4. (right) Typical Detail of Pinpile, Pocomoke River Bridge, MD

Site and Ground Conditions - In each of Piers 3 and 4 a total of 24 piles were drilled from the bridge deck. In addition, a further 4 piles were installed from the restricted access of the Control House of Pier 4 (8x8' plan, 14' headroom) (Figure 3). The riverbed materials into which the underpinning was installed comprised a Recent sequence of alluvial sediments. The founding horizon was dense medium to coarse sand, beginning about 60' below river surface level.

Design and Construction - Each pile was installed as shown in Figure 4. The allowable stresses used in the design were 30% U.C.S grout and 40% of the yield strength in both the casing and the epoxy coated rebar. To permit preloading of the pile a tendon comprising 3 ea. 0.6" dia seven wire strands was also installed through each hole, its 20' bond zone extending to 25' below the toe of the pinpile casing. Prior to drilling, grout filled bags had to be placed around the bases of the piers as framework for void filling grouting.

After the neat cement grout had reached 3500 psi, the tendon was stressed against the top of the steel casing, to the design load of 82 kips. The annulus between casing and structure was then grouted with special high strength grout. About 5-7 days later, the prestress was released at the tendon head, thereby allowing full structural load transfer to the pile but without obviously causing further pile compression. Slightly amended procedures had to be adopted in the restricted access of the Control House, but the same basic principles were followed, resulting in the perfect installation and functioning of all 52 pinpiles on the project.

Testing and Performance - Two special preproduction test piles were installed for intensive testing, 16' apart on the adjacent west bank, 145' north and 55' west of the west abutment. Each pile had an outer casing 8-5/8" outside diameter, predrilled from the surface to -30'. The 7" casing of the pile was then installed in standard fashion through this large casing, but without being bonded to it in any way. This arrangement was intended to simulate in the test the lack of resistance afforded by the river and the very soft soils on its bed as well as the portion of the pile which was within the confines of the bascule pier. Each of the identical piles had 25' of pressure grouted bond zone (maximum grouting pressure 100 psi), 30' of #11 rebar, and 70' of 7" casing (from surface to 5' into the bond zone). Soil anchors were installed to provide reaction to the test loads. This is a common and attractive feature of testing pin piles in restricted access locations.

The test had three phases:
Phase I: a "Preload-Unloading" test designed to verify the performance efficiency of the preloading system.
Phase II: a conventional pile load test to establish load-deflection performance within the scope of the specification (i.e., progressively to twice working load).
Phase III: on one pile, loading to failure.
The test was heavily instrumented, with load being measured independently by load cell and by hydraulic jack gauge, and deflection monitored by dial gauges supported from an independent reference beam and by piano wire and mirror scale. Dial gauges were also used to indicate movements of fixed end extensometers (telltales) located at elevation -70' (i.e. at top of bond length) and at elevation -90' (bottom of bond length). These telltales were intended to directly indicate displacement at these two points in the bond zone, relative to the pile cap.

- Phase I Tests (Preload - Unloading Tests) - The anchor tendon in each pile was loaded to 82 kips, creating an elastic shortening of each pile by 0.123" and 0.137" respectively. Upon unloading to zero (i.e. releasing the prestressing load), the pile cap rebounded totally elastically, indicating no measurable permanent shortening. As the procedure was demonstrated to work, and since the performance was elastic, this Phase of testing was accepted as being successful.

- Phase II Test (Load/Deflection Test to Twice Design Working Load) - Each pile was loaded progressively to 200 kips in 20 kip intervals, each with a 5-minute hold period. Details are summarized in Table 3. Major points are as follows:

- Performance of the piles was very similar, being virtually elastic, linear, and with minimal creep at intermediate holds.
- The total pile deflections (anticipated and observed) at 200 kips were less than one half inch, and the permanent deflections upon unloading were around 0.04" at 2 hours after final load release.

	Deflection at 200 Kips (inches)	Creep in 24 hrs at 200 Kips (inches)	Permanent Displacement Upon Unloading from 200 Kips Instantaneous/After 2 hrs (inches)
Pile 1			
Pile Cap	0.442	0.038	0.044/0.020
Upper Telltale	0.344	0.028	0.024/0.021
Lower Telltale	0.374	0.031	0.095/0.093
Pile 2			
Pile Cap	0.437	0.059	0.047/0.027
Upper Telltale	0.385	0.033	0.041/0.037
Lower Telltale	0.420	0.021	0.067/0.063

Table 3. Highlights of Load/Deflection Data, Test Piles 1 and 2, Pocomoke River Bridge, MD

After 12 more hours, the piles had returned to full extension (i.e. no measurable permanent shortening).

- The performance of the telltales was wholly consistent. They reflected the internal elastic performance of the piles, and so provided movements less than the total pile displacement (i.e. elastic plus permanent). Predictably, the upper telltale, monitoring a shorter length, provided the smaller movements. These data compare closely with the net elastic deflection obtained by subtracting total cap movement (at 200 kips) from the residual (at zero), as shown in Table 4.
- Total creep at 200 kips ranged from 0.038" to 0.059" over 24 hours. However, the amount of "internal" creep was smaller and more uniform (0.021" to 0.033", Avg. = 0.028").
- There was a time related "rebound" evident in all points of measurement after unloading. Overall, this was 0.020" to 0.027" at the pile cap including 0.002" to 0.004" of "internal" pile rebound.

• Phase III Test (Load/Deflection Test to Failure, Test Pile 2) - Once the required test to two times design load (200 kips) was satisfied, an attempt was made to determine the ultimate skin friction. The hold down reaction system was sized for about 360 kips which was initially felt to be sufficient to plunge the pile. Surprisingly, after four successive cycles to about 360 kips, the pile had not yet failed, despite a cumulative permanent displacement of 0.567". Thereafter, the test set up was overhauled, and the test rerun: a maximum load of 390 kips was reached before plunging of the pile was recorded.

Pile Number	Net Elastic Deflection* at 200 kips (inches)	Measured Elastic Deflection+ at 200 kips (inches)
1	0.398	0.374
2	0.390	0.420
Average	0.394	0.397

Table 4. Comparison of net and measured elastic pile performance, Pocomoke River Bridge, MD

* total deflection at 200 kips less permanent deflection at subsequent zero.
\+ from Lower Telltale.

Again the evidence of the telltales was that of virtually perfect elastic performance within the pile structure. The difference, at maximum load, between overall elastic performance (Lower telltale) and total deflection was 2.874" - 1.128" = 1.746": very close to the measured permanent set at zero load, of 1.712". The difference is probably due to the fact that the telltale was not exactly at the pile tip. Creep values were only significant from about 340 kips onwards.

As a final point, this project represented the second time that this particular preloaded pinpile concept had been used. Some years before a structure at the Mid Orange Correctional Facility, NY, had been saved by preloaded 55 kip piles (3). In both instances, the subsequent structural movements in service have been of the order of a few thousands of inches. These successes have recently encouraged the possible use of permanently preloaded piles in California to underpin transmission lower footings threatened by "rocking" actions produced by seismic events. The basic requirements of the problem can be readily satisfied with this technique:

- resistance to uplift forces in the range of 100-400 kips
- no additional compressive loads are imposed upon a new or existing footing
- no structural movement within the design capacity of the system in service
- each pile installed is tested (during application of the prestress).

Final Remarks

The recent case histories described in this paper clearly underline the advances being made in pinpile technology in the United States. One can cite the unusually high load capacities, the use of preloading and postgrouting, the quality of the testing, and the growing understanding of lateral loading behavior. It is questioned if the same rate of advance is being sustained anywhere else in the world. It is queried if there is the same growth in market potential anywhere else in the world.

References

1. American Society for Testing and Materials (1981). "Testing of Piles Under Static Axial Compressive Load". D1143-81.
2. Brown, D.A., Morrison, C. and Reese, L.C. (1988). "Lateral Load Behavior of Pile Groups in Sand". Jour. Geot. Div., ASCE, 114 (11), November.
3. Bruce, D.A. (1988, 1989). "Aspects of Minipiling Practice in the United States". Ground Engineering, 21 (8) pp 20-33, and 22, (1) pp. 35-39.
4. Bruce, D.A. and Nicholson, P.J. (1988). "Minipiles Mature in America". Civil Engineering, 58 (12) December, pp. 57-59.
5. Bruce, D.A. and Nicholson, P.J. (1989). "The Practice and Application of Pinpiling". ASCE Foundation Engineering Conference, Evanston, IL, June 25-29, 19 p.
6. Bruce, D.A. (1989). "American Developments in the Use of Small Diameter Inserts as Piles and In Situ Reinforcement". DFI International Conference on Piling and Deep Foundation, London, May 15-18, pp. 11-22.
7. Bruce, D.A. (1992). "Recent Advances in American Minipiling Practice". Intl. Conf. "Piling Europe", London, 7-9 April. 9 pp.
8. Bruce, D.A. (1988). "Developments in Geotechnical Construction Processes for Urban Engineering". Civil Engineering Practice, 3 (1) Spring, pp. 49-97.
9. Bruce, D.A. (1989). "Postgrouted Anchorages: Current Practice of Nicholson Construction Company". August, 70 pp.
10. Bruce, D.A. and Yeung, C.K. (1984). "A Review of Minipiling with Particular Regard to Hong Kong Applications". Hong Kong Engineer. June, pp. 31-54.
11. Bruce, D.A., Pearlman, S.L. and Clark, J.H. (1990). "Foundation Rehabilitation of the Pocomoke River Bridge, MD, Using High Capacity Preloaded Pinpiles". Proc. 7th Intl. Bridge Conf., Pittsburgh, PA, June 18-20, Paper IBC-90-42, 9 pp.
12. Post Tensioning Institute (1986). "Recommendations for Prestressed Rock and Soil Anchors". Post Tensioning Manual, Fourth Edition, pp. 236-276.

Permanent Excavation Support and Underpinning in Sands: A Case History

Russell J. Morgan, P.E.[1], M. ASCE, Lawrence F. Johnsen, P.E.[2], M. ASCE, and Franklin M. Grynkewicz[3]

Abstract

The expansion of a shopping center called for the transformation of an underground access tunnel and storage basement into a lower level of retail stores. This required the lowering of foundations and slabs to increase the ceiling height, the underpinning of basement foundation walls and the construction of three major earth support systems. Underpinning was accomplished with friction minipiles in sand. Permanent excavation support methods included soil nailing and tied back H-piles with shotcrete lagging.

Introduction

Expansion of the Connecticut Post Mall in Milford, Connecticut called for the transformation of an underground access tunnel and storage basement into a lower level of retail stores. This required the lowering of foundations and slabs to increase the ceiling height in the basement, the underpinning of 1500 linear feet of basement foundation walls and the construction of three major earth support systems to provide walk-in access to the new retail level. An outline of the mall with the locations of the major underpinning and shoring activities is shown on Figure 1.

---

1 Senior Project Manager, GZA GeoEnvironmental, Providence, RI

2 Associate Principal, GZA GeoEnvironmental, Trumbull, CT

3 Senior Project Manager, GZA GeoEnvironmental, Newton Upper Falls, MA

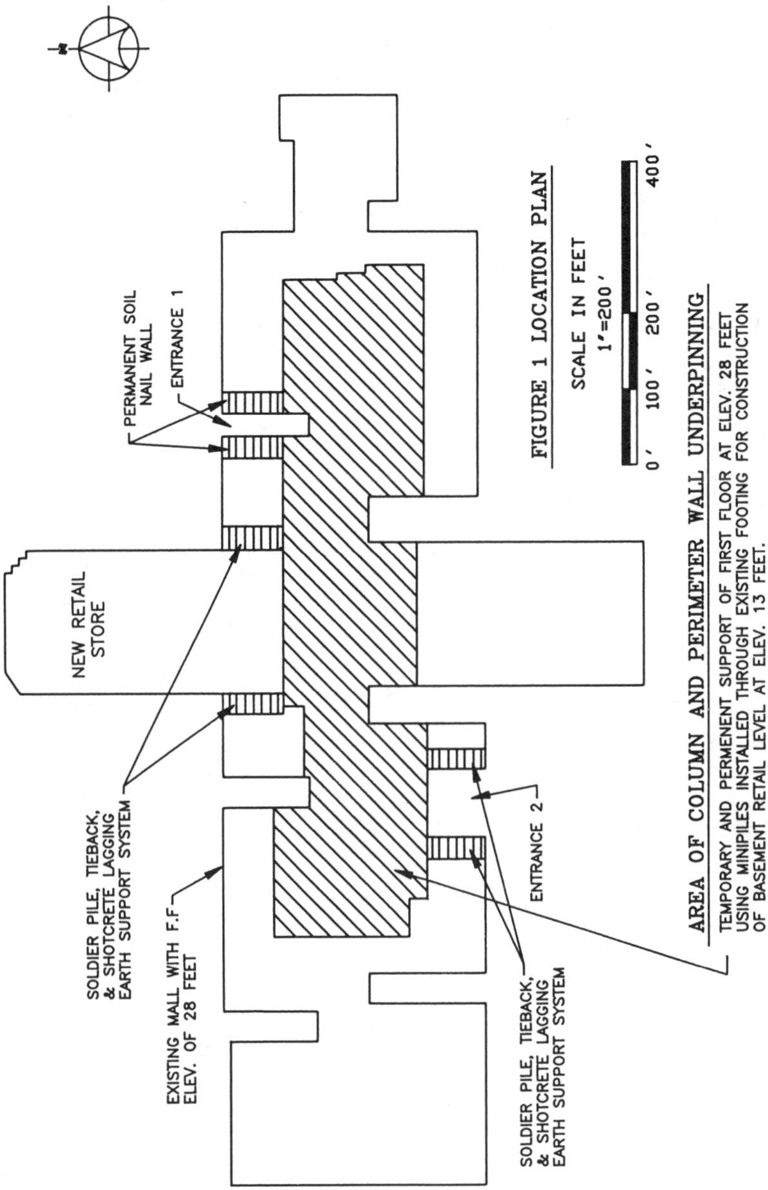

FIGURE 1 LOCATION PLAN

Underpinning was accomplished with friction minipiles in sand carrying up to 620 kips (2.75 MN) per four pile group. Permanent excavation support methods included soil nailing at one entrance and tieback walls at two entrances. Additional minipile foundations were installed to carry increased loads from a redesigned roof and future upper level. A permanent dewatering system was also installed, but is beyond the scope of this paper. Similarly, the conventional underpinning of slabs-on-grade and a short length of a bearing wall are not discussed.

The subsurface conditions consist of sands and silts extending to a silty glacial till overlying schist bedrock. The overlying sands and silts graded finer from east to west with a peat layer at the extreme west end of the construction area. Bedrock was generally encountered at depths greater than 40 feet (12.2 m) but several narrow ridges oriented northeast to southwest rose to within a few feet of ground surface.

Underpinning Isolated Columns

Drilled minipiles were used to underpin existing footings in the mall basement to allow the lowering of the floor below the elevation of the existing footings. Refer to Figure 2. A total of 600 drilled minipiles were installed as permanent support for existing columns. The minipiles were limited to 24 or 26 inch (61 to 66 cm) square pile caps due to architectural considerations. Minipiles were also installed at several new column locations.

Design. The construction sequence necessitated temporary support for the existing columns as the area around and beneath the column was excavated from an existing slab elevation of 17 feet (5.2 m) to a drainage system subgrade at elevation 12 feet (3.7 m), mean sea level (MSL). Following excavation, permanent column support was installed.

The design issues included the contractor's desire to utilize a single support system for the temporary and permanent conditions, the transfer of load from columns to minipiles, buckling of the free standing minipiles during excavation and dimensional constraints imposed on the permanent support system.

Minipiles were chosen by the contractor as the load bearing element for both the temporary and permanent condition. Minipiles consist of uncased reinforced grouted piles installed with temporary casing. The minipiles consisted of 6-inch (15.2 cm) diameter piles of

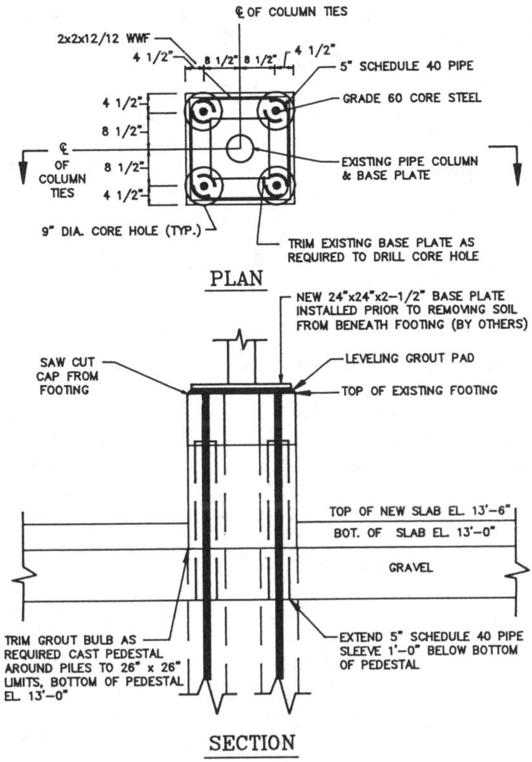

FIGURE 2 COLUMN UNDERPINNING

75 kip (0.33 MN) capacity (low capacity) and 9-inch (22.8 cm) diameter piles of 155 kip (0.69 MN) capacity (high capacity). Local codes did not address minipiles. Therefore, the design criteria was based on the addendum to Section 742.0 of the Massachusetts State Building Code (Ref 1). The production minipiles were provided with 4.8 ksi (33 MPa) grout and grade 60 core steel.

The bond developed between the minipile and existing column footing was utilized for the transfer of load from column footing to minipile. Since ACI 318 (Ref 2) does not address this condition, full scale bond load tests were performed. Based on the results of bond tests an allowable value of 190 psi (1.31 MPa) was adopted for design.

A 3 inch diameter, .216 inch wall thickness or 5 inch diameter, .258 inch wall thickness (schedule 40) steel pipe was installed through the unsupported length of the minipile to provide the required buckling resistance for the temporary condition during excavation. The dimensional constraints on the pile caps resulted in edge distances below ACI 318 requirements. The pile caps were laterally reinforced with welded wire fabric and welded strapping to provide the required restraint.

Pile Load Tests. Pile load tests were performed on a three-pile cluster of low capacity minipiles and a four-pile cluster of high capacity minipiles. The low capacity cluster was intended to support design loads of up to 225 kips (1.0 MN) and the high capacity cluster was intended to support design loads of up to 620 kips (2.76 MN).

The low capacity test was performed near the west end of the construction area. The soils consisted of 5 feet (1.5m)of medium dense, medium to fine sand underlain by stratified layers of loose, silty fine sand and loose, fine sandy silt extending to glacial till at 51 feet (15.5 m). The piles were drilled to 30 ft (9.14 m) depths with 5.5 inch diameter casing, grouted with a 4.8 ksi grout mix and reinforced with a #10 bar (3.12 cm diam.) extending the full length of the pile.

In the low capacity testing, a plunging failure occurred at approximately 230 kips (1.02 MN) which did not provide the required factor of safety of 1.5. The piles were regrouted and retested. The regrouted pile group sustained a load of 342 kips (1.52 MN). A settlement of slightly over 0.25 inch (0.64 cm) occurred during the final load test.

The high capacity test was run in Entrance I as shown on Figure 1. The soils consisted of medium dense, fine to medium sand from 0 to 20 feet (6.1 m), medium dense, silty sand from 20 to 55 feet (6.1 to 16.7 m), medium dense, sandy silt from 55 to 81 feet (16.7 m to 24.7 m) and glacial till at 81 feet (24.7 m). The piles were drilled to 71 ft. (21.6 m) depths with an 8 inch (20.3 cm) diameter casing, grouted with a 5000 psi (34.5 MPa) grout mix and reinforced with two #10 (3.12m diameter) bars extending the full length of the pile. The top ten feet (3.0m) of the piles were sleeved resulting in effective lengths of 61 feet (18.6m). Regrouting was performed with little or no additional grout take.

The high capacity pile group was loaded in stages to a maximum applied load of 900 kips (4 MN) with a resulting net settlement of 1/8 inch (0.32 cm). The test was terminated at 900 kips (4 MN) due to concerns regarding the reaction frame.

Bond Tests. Three bond tests were run to substantiate a design bond stress to be used for the transfer of load from the the existing footings to the minipiles. The first bond test was run with a small bearing plate on a pile cap constructed for the low capacity pile load test cluster. The plate was sized and positioned so as not to extend over the piles. Loads were applied in stages to a maximum load of 190 kips (0.85 MN).

Two additional bond tests were run on existing footings to determine bond strength of the existing footing concrete. Six inch diameter cores were made through two footings of 14 (35.6 cm) and 22 inch (55.9 cm) thickness. A 1-inch (2.54 cm) diameter Grade 150 threaded bar restrained by a coupler at the base of the footing was then placed and grouted. Both tests were run to 100 kips (0.44 MN), 90% of the ultimate strength of the bar. Bond stresses of 240 and 380 psi (1.65 MPa - 2.62 MPa) were achieved with no indication of failure. Based on these results an allowable bond stress of 190 psi (1.31 MPa) was used for design.

Construction Procedures. Drilled mini-piles were installed using specialized low head room drilling rigs. Drilling methods consisted of wet rotary cased boreholes with external casing flush. Depending on pile capacity, casing diameter was either 6-inch (15.2 cm) or 9-inch (22.9 cm). Existing column footings were cored prior to drilled pile installation.

Once the piles were installed, an excavation was made from the top of footing to elevation 12. At this time the load transfer from the footing to the drilled minipiles occurred entirely through the bond between the footing and minipiles. Then the 24 or 26-inch (61 to 66 cm) square pedestal/pile caps were cast below the existing footing, and the existing footings saw cut to the required dimensions. Finally, the top of the poured pedestal was dry packed to the underside of the existing footing.

Field Observations. Since the construction sequence relied on the pile to footing interface for complete load transfer, the effects of the drilling on the shape of the pile and the loss of grout into the surrounding sands was of importance. It was initially expected that the drilling operation would remove the sands and silts sufficiently beyond the 5.5 inch (14 cm) or 8 inch (20.3 cm) diameter drill casing to create a rim of grout beyond the 6 inch (15.2 cm) or 9 inch (22.9 cm) core hole through the footing. Bent reinforcing bars were used to probe for the overdrill at the completion of the drilling. Results were inconsistent and the design modified so as not to rely on an overdrill.

It was found that the soils were sufficiently coarse so that grout from the pile would seep into the surrounding soils for several hours following the drilling of the pile. All piles were periodically refilled following drilling to restore grout to the proper level. The concern was that low pile tops would collect soil which could later be covered with grout without being noticed.

Quality control of the 4.8 ksi (33 MPa) psi grout mix consisted of specific gravity tests on all grout batches and cubes being taken on the samples with the lowest specific gravity. Cores taken at the completion of the project provided compressive strengths of 6000 to 9000 psi (41 to 62 MPa).

Underpinning Basement Foundation Walls

The lowering of the mall's basement floor by five feet necessitated an underpinning system capable of providing temporary and permanent support to both the vertical and lateral loads acting on the foundation walls. Also, the sequencing of the underpinning, shoring and excavating operations became critical. The system consisted of minipiles to provide vertical support to the existing basement walls when excavated beneath, shotcrete underpinning to further support the existing walls and to retain soil, and tiebacks and wales to provide lateral support for the basement walls. Refer to Figure 3.

Design. Design issues included load transfer from the column footings to the minipiles, potential buckling of the minipiles, balancing the eccentric loads on the minipiles to reduce bending stresses, estimating the lateral earth pressure acting on the wall and the wall's moment capacity between installed tie backs, in a horizontal direction.

The load transfer and buckling issues were dealt with similarly to the isolated columns: bond stress between pile and footing was utilized for load transfer, and steel pipe was inserted for buckling resistance. Tiebacks were installed to both resist lateral earth pressures acting on the wall and to prevent the development of bending stresses in the battered minipiles.

The design of the tieback system was based on a triangular distribution using an at rest earth pressure coefficient of 0.4, and no consideration for the bending capacity of the existing foundation wall. The final

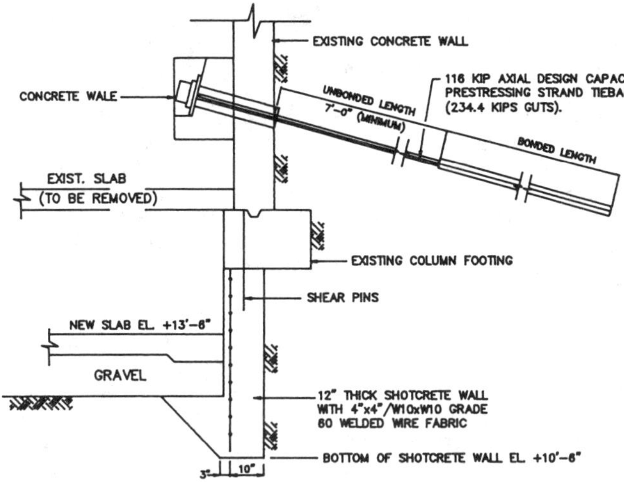

FIGURE 3 FOUNDATION WALL UNDERPINNING

design consists of 116 kip (0.51 MN) capacity, single corrosion protected, tiebacks and a reinforced concrete wale.

A shotcrete wall was installed beneath and tied to the wall and column footings. It provided permanent support for the wall and earth pressures developed below the base of the wall.

<u>Construction Procedures</u>. The five foot excavation and resulting loss of passive pressure required a careful sequencing of operations.

First, two 75 kip (0.33 MN) capacity minipiles were installed at each pilaster to provide temporary support during excavation and underpinning. Then, concrete wales were constructed along the foundation wall to distribute wall loads to the tiebacks. The tiebacks consisted of permanent single corrosion protected Grade 270 strand anchors installed by conventional rotary drilling equipment. Then, existing slabs and underlying soils were excavated in 30 foot (9.1 m) sections. Finally, shear dowels were installed into the underside of the existing wall, required reinforcing and welded wire fabric were set and the shotcrete was applied.

<u>Field Observations</u>. The design relied on the minipiles to support the walls temporarily until shotcreting was complete. At one minipile location the footing cracked between the minipile and the pilaster resulting in a

settlement of less than one inch to the pilaster. Shoring was installed to support the wall until the load could be carried by the shotcrete underpinning.

Significant loss of soil occurred at five locations along the 1500 lf (457 m) of foundation wall which required underpinning. At these locations, voids formed behind the foundation wall. Following the installation of shotcrete, the voids were filled by coring through the wall and pumping from 4 to 30 cubic yards (3 to 22.9 m^3) of concrete. No damage was observed or reported in the stores above.

All tiebacks were proof tested to 150% of design load. The 116 kip (0.51 MN) capacity tiebacks were typically drilled to lengths of 35 feet (10.7 m) with an unbonded zone of 10 feet (3 m). All tiebacks met the Post Tensioning Institute criteria (Ref 3). Maximum creep rates were less than 0.04 inch (0.1 cm) per log cycle of time.

Tied Back H-Pile and Shotcrete Walls

Tied back H-pile walls with reinforced shotcrete lagging were installed for permanent earth support at the new retail store entrance and Entrance 2. Refer to Figure 1 for the locations of these entrances. At each entrance the existing mall was demolished. The contractor was provided a 13 inch (33 cm) width from the demolished structure to install the earth support systems. Each support system had to provide for an excavation extending 16 feet (4.88 m) below the existing mall floor.

Design. Space limitations necessitated that a single system provide both temporary and permanent support at each entrance. Walls were subjected to vertical loads of the adjoining mall roof and the inclined tiebacks, and lateral earth pressures including building and seismic surcharges. To minimize wall movements, the design lateral earth pressures were based on "at rest" conditions with a triangular distribution to the tieback and rectangular from the tieback to the base of the excavation. The tiebacks were designed for capacities of 77 kips (0.34 MN).

The lagging installed between the H-piles consisted of a 7-inch thick working slab of reinforced shotcrete. Reinforcing consists of horizontal #5 bars (1.59 cm diam.) at 6 inches (15.2 cm) center to center and vertical #4 (1.27 cm diam.) bars at 10 inches (25.4 cm) center to center. The earth pressures on the shotcrete were based on "at rest" earth pressures reduced by 50% due to soil arching effects between the piles.

Construction Procedures. The permanent tied back H-pile walls were installed using conventional driven piles.

Reinforced shotcrete "lagging" was provided in lieu of conventional wood lagging due the permanent construction requirements for the walls. Excavation was typically performed in 4-foot (1.22 m) increments with drainage board, reinforcing and primary shotcrete placed between the H-pile flanges.

Tiebacks were installed using rotary drilling methods. Soldier piles were equipped with prefabricated "through-holes" at the required tieback design elevations. This feature precludes the use of walers. Tiebacks consisted of Grade 150 steel bars with simple corrosion protection. Tieback testing was performed in accordance with Post Tensioning Institute recommendations for testing of permanent ground anchors.

Field Observations. Vertical and lateral movements were monitored at Entrance II with maximum recorded movements of 1/2 inch (1.27 cm) settlement and 3/8 inch (0.95 cm) outward deflection. No damage to the adjoining stores was either observed or reported.

Sloughing of soil occurred at many locations prior to shotcreting. The sloughing ranged up to 18 inches (45.7 cm) in thickness and up to 30 feet (9.14 m) in length. The shotcrete wall was thickened in these areas.

The permanent tiebacks were proof tested to 150% of design load. The 77 kip (0.34 MN) capacity tiebacks were typically drilled to lengths of 35 feet (10.7 m) with an unbonded zone of 10 feet (3 m). All tiebacks met the Post Tensioning Institute criteria.

Soil Nail Walls

An earth support system was required at Entrance I to allow direct access from the front parking lot to the lower level of the mall. Refer to Figure 4. Excavations of sixteen feet depth were required adjacent to mall areas which remained in use. The existing mall construction consisted of shallow spread footings and slabs-on-grade.

Design. The project criteria required that the underpinning and earth support system at Entrance I be constructed within 13 inches (33 cm) of the existing foundation walls. A soil nail wall was chosen for this application. It allowed top down construction and when complete provided permanent walls for the entrance.

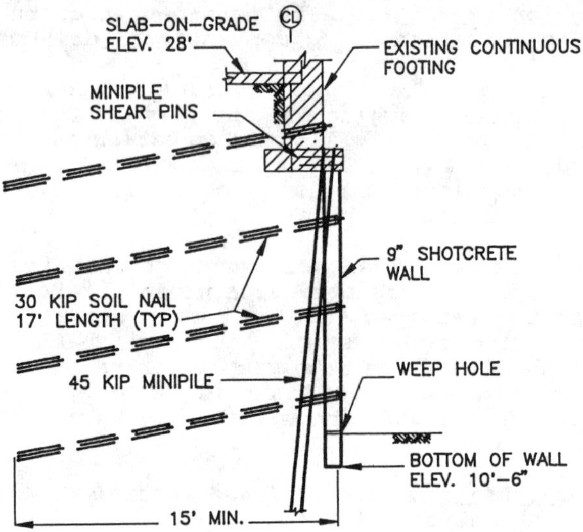

FIGURE 4 TYPICAL SOIL NAIL WALL WITH MINIPILES

The design considerations for this earth support system included settlement of existing structures, the installation of an earth support system beneath the existing shallow foundation, design pressures for the shotcrete facial slab, lateral earth pressure distribution and method of internal analysis of wall system.

Using a rectangular "at rest" lateral earth pressure distribution, and an analysis of the external stability of the reinforced soil block, a soil nail length of 12 feet (3.66 m) (width of reinforced soil block) was required to achieve factors of safety against sliding and overturning of 1.5 and 2.0, respectively. An analysis of the internal stability of soil nails was based on an ultimate bond stress between soil nail and granular soil of 5 kips/L.F. (0.07 MN/m), an angle of internal soil friction of 32° and a reinforced soil block width of 11 feet (3.35 m) and a soil nail spacing of 4 feet (1.22 m). The internal stability analysis indicated a factor of safety of greater than 2.0 and the required soil nail capacity was 30 kips. The final design configuration of the soil nail wall consisted of 17.5 feet (5.33 m) long drilled and grouted soil nails battered 10° from horizontal in a 4 foot (1.22 m) on-center grid spacing. Soil nails consisted of 1-#9

(2.86 cm. diam.) epoxy coated reinforcing bar, A-615 steel.

A quadralateral pressure distribution of "active" earth pressure (Juran, 1987) was used to determine the pressure distribution on the shotcrete. The final facing consisted of a 7-inch (17.8 cm) thick, 4000 psi (27 MPa), shotcrete working slab reinforced with 2 vertical and 2 horizontal #6 (1.91 cm diam.) reinforcing "waler" bars at each soil nail head and 2-inch square,10/10 welded wire fabric throughout the slab.

Minipiles, of 45 kip (0.2 MN) capacity, were designed to support the column footings above the soil nail wall. The continuous footing extended only 6 inches (15.2 cm) beyond the edge of the foundation wall resulting in insufficient area for the load transfer to occur by bond between the footing and minipile. Load transfer was achieved by installing a steel angle over the top and side of the minipile and then bolting the angle into the concrete footing.

A major construction issue was the underpinning of the existing structures to the east and west of the proposed entrance. The existing foundations were underpinned with drilled minipiles prior to excavation for the soil nail structures.

All soil nails were 17.5 feet (5.33 m) long, and were installed with wet rotary drilling methods on a grid spacing of 4 feet (1.22 m) and at 10 degree batters to horizontal. Following drilling to the design length, the drill holes were filled with grout and epoxy coated Grade 60 deformed steel bars equipped with centralizers were installed through the drill string. During casing removal grout was injected into the drill hole through the drill head.

Field Observations. Settlements and lateral movements on the order of one-half inch occurred during the construction of the earth support system at Entrance I. The movements resulted in a sloping and cracking of a tile floor in the retail store located on the east side of the excavation.

During excavation for the second lift of shotcrete, loss of soil occurred at the south end of the east wall where it abutted a basement foundation wall. The void was later filled by pumping grout through the top lift of shotcrete.

Selected soil nails were proof tested to 150% of design load of 30 kips (0.13 MN). All tested soil nails met the Post Tensioning Institute criteria.

Conclusions

All aspects of the construction proceeded without interruption to the existing mall operation. Two factors were critical to the success of the chosen methods. The on-site soils were found to have sufficient silt binder and apparent cohesion from partial saturation to hold a vertical cut which allowed the use of soil nailing and shotcrete lagging. The bond testing showed that a considerable load can be carried by bond between grout and concrete. This allowed the minipiles to support columns in both temporary and permanent conditions. However, the permeation of grout into surrounding soils required releveling of grout to prevent loss of bonding area.

Acknowledgements

The authors wish to acknowledge the specialty foundation contractor Terra Drilling Company, Inc. who had prime responsibility for execution of the design. In addition, the Owner, Westfield Design and Construction, Inc. must be praised for the relatively unique method of contracting the work; the performance specification concept was particularly well suited to the operational constraints imposed by this development.

References

1. Massachusetts State Building Code addendum to Section 742.0

2. ACI 318, "Building Code Requirements for Reinforced Concrete", 1988

3. Post Tensioning Institute, "Recommendations for Prestressed Rock and Soil Anchors", 1986

4. Juran, I., and Elias, V., "Soil Nailed Retaining Structures: Analysis of Case Histories", Soil Improvement - A Ten Year Update, Geotechnical Special Publication No. 12, ASCE, 1987

Retention System Using Compaction Grouting In Clay Soils

Gary T. Brill[1]
Kenneth E. Darnell[2]

Abstract

At a large industrial mill, the owner desired to install a prefabricated steel tunnel in a heavy machinery and high traffic equipment area. The necessary work involved making a 13 feet vertical by 116 feet long excavation in highly variable, clayey soils paralleling and immediately adjacent to a series of heavy columns supported on shallow spread footings. Typical methods of providing excavation support and underpinning the column footings would have required a rather lengthy shutdown of the mill support lanes, creating an unacceptable situation. To minimize the shutdown time and the associated loss of revenue, an in-situ ground support system consisting of compaction-grouted piles was developed. The technique enabled the necessary combination earth retention and underpinning system to be entirely installed prior to any excavation while the mill continued to operate.

This paper presents the results of the exploration program and a detailed site description. In addition, the design and analyses of the compaction-grouted pile system are described along with the specialty drilling and grouting techniques involved. Finally, the results of the grouting program and correlations developed from these results are presented.

----

[1] Senior Project Manager, ERC Environmental and Energy Services Company, Inc., 725 Pellissippi Parkway, Knoxville, Tennessee 37932

[2] Manager, Eastern Operations, ERC Environmental and Energy Services Company, Inc., 725 Pellissippi Parkway, Knoxville, Tennessee 37932

Introduction

In the past 30 or so years, compaction grouting methods have become increasingly popular for improving soils in-situ and for remediating foundation structures.

New applications for the compaction grouting technique have been introduced in recent years using columns or "piles" created by pulling the installation casing in increments. It is clear that compaction grouting will play an important role in repairing America's aging infrastructure.

In the majority of the published case histories on compaction grouting, very little, if any experience is presented on compaction grouting in clayey soils. In fact, many of the recognized authorities on compaction grouting consider the method impractical in clayey soils. Admittedly, compaction grouting methods in saturated, plastic, homogeneous clay soils would typically be difficult. However, expansion of grout in a five-inch diameter injection hole to ten or more inches in diameter was accomplished at shallow depths and at relatively low pressures using a low pumping rate in the soft, non-homogenous clayey soils at a karst site in East Tennessee. Pressure grouting in a karst setting can be futile at times, especially when using neat cement grout because of the tendency of the grout to flow through open cavities in the soil and rock mass. Using a stiff, low slump grout mix, the injected grout was successfully contained providing a combination underpinning and excavation support system. This paper presents a case history of the design and construction of the compaction-grouted pile system.

Subsurface Conditions

In general, the subsurface conditions at the site are highly variable and complex. The site is underlain by soluble bedrock of the Knox Group. This bedrock is steeply dipping, interbedded limestone and dolomite, containing various percentages of chert and cherty sandstone. The bedrock has been severely fractured and deformed, and subsequently solution-weathered to form a highly irregular mass containing numerous enlarged fractures and both open and soil-filled cavities. The general bedrock surface is typically broken by deep soil-filled slots and steep-sided pinnacles. The overlying residual soil was developed from weathering of the bedrock and its characteristics are also highly irregular and somewhat erratic. Typically, in this geologic setting, the upper soil interval is reasonably competent and the soil generally becomes less competent with depth, (the extreme case being extremely soft soils within deeply weathered slots in the bedrock). Ground water was generally encountered near or below the bedrock surface.

Although the residual soil was erratic in grain size and consistency, the general soil type was firm, moderately plastic, reddish-brown, clayey silt and silty clay containing varying amounts of sand, as well as chert fragments ranging in size from sand to cobbles. Contrary to the general soil type described above, there were periodic layers and lenses of interbedded fine, silty sand and sandy silt.

It was extremely difficult to establish soil shear strength parameters which were representative of the general soil matrix. The subsoils exhibited consistencies ranging from very stiff to very soft, and the "N" values from Standard Penetration tests were often inflated by the chert inclusions. In addition, the soil system contained relict weaknesses and fractures (slickensided soils), which sometimes resulted in lower shear strengths when subjected to unconfined compressive strength testing than the soil had in a confined or massive state. Therefore, a compaction-grouted system that could not only provide a known amount of support but could also improve random zones of less competent soils became an attractive option.

Design Considerations

Because one of the primary concerns of the owner was to minimize the shutdown time of the mill, it was necessary for the earth reinforcement and retention system to be completely installed within a limited space before essentially <u>any</u> excavation was made. In addition, excavation for installation of the prefabricated tunnel was scheduled to occur subsequently during a shutdown period of no more than seven days, and no significant deflection of the support system was permitted during the excavation. Furthermore, the work was to be performed in a high-traffic, active work area of the plant.

The earth reinforcement and retention system requirements included preventing the transfer of any significant lateral load from the soil to the walls and bottom of the tunnel. The system had to be designed to provide adequate temporary stabilization of the excavation and the adjacent column footings, which were to be significantly loaded by overhead cranes for installation of the tunnel. The permanent loading conditions were to occur after the tunnel was in place and not only included supporting the heavily loaded column footings, but also supporting proposed pile caps. The system requirements were further complicated by other restraints which included: (1) limited headroom, (2) a confined work area, and (3) the requirements for limiting airborne particles or other waste release from the confined area.

In summary, the selection of the earth reinforcement and retention system was ultimately based upon the following criteria:

(1) Time required for completion;
(2) Potential for the disruption of the mill operations during installation;
(3) Cost;
(4) The quality of the end product; and
(5) The qualifications of the contractor.

Support System Description

The selected system consisted of constructing over 117 grout piles installed using compaction grouting techniques. Compaction grouting consisted of injecting a very stiff mortar-like grout material under pressure in order to displace and compact soils in-place.

The system was made up of vertical and battered soil retention piles for excavation support, and vertical bearing piles for supporting the soils beneath the heavy column footing loads and the eventual pile cap. The soil retention grout piles were installed to the design depth below the proposed excavation, while the bearing grout piles for underpinning support were socketed into the lower, variable bedrock surface. A single reinforcing bar was used in each grout pile to provide continuity and reinforcement. An unconventional theory of underpinning was developed whereby the grout piles were used to prevent a mass soil shear failure, but they were structurally separated from vertical bearing on the footings to reduce the potential for creating "hard spots" or point bearing contacts. The separation was achieved by sleeving the piles through the footings with PVC casing. In addition, because of the potential for collapse at the face of the excavation due to the "blocky" nature of the soils, piles were located along the face of the proposed excavation.

Design and Analyses

Because the required system had to provide excavation support for the project as well as support new pile caps, both short-term and long-term analyses were performed. The construction loads (short-term analyses) were to occur during the seven-day period for installation of the tunnel, and the operating loads (long-term analyses) would occur any time after installation of the tunnel was complete. Total stress shearing strength parameters were estimated for the short-term analyses while effective stress shearing strength parameters were estimated for the long-term analyses. The soil considered in the analyses was that which is contained in the theoretical active soil pressure wedge, as shown in Figure 1. Loading applied to this wedge of soil consisted of the weight of the soil in the wedge and any loads acting on the surface of the wedge. These forces were combined into a resultant vertical pressure. The vertical pressure was then divided into components acting tangent to the active wedge shear plane and acting normal to the active soil wedge. In general, the pile analyses were performed assuming that the pressure acting tangent to the base of the active wedge would be resisted by a combination of the shearing strength of the soil and the shearing resistance of the piles. The piles were assumed to have a minimum diameter of six inches.

The construction loads included bearing pressures from the footings of up to 3200 psf, and were primarily due to the weight of overhead cranes that would be used to move and place the prefabricated steel tunnel. In the short-term analyses, the piles in the front row were included to maintain the face of the excavation, but were not considered as structural elements providing resistance to the loads on the active soil wedge. The back row of piles under the footings, and the piles installed on a batter, were analyzed as structural elements resisting the imposed loads. For the short-term analyses, it was assumed that the piles acted as continuous vertical footings or "straps" and the bearing capacity of the soil behind the piles dictated the lateral capacity of the piles.

The operating loads included bearing pressures from the footings of up to 5300 psf, and the bearing piles were to be incorporated into pile caps and axially loaded to 30 tons each. The bearing piles were assumed to have fixed ends for the long-term analyses. The capacity of the system for these analyses was therefore based upon the shearing strength of the individual pile members. The piles were designed to derive their axial load-carrying capacity through a combination of soil-grout shear transfer, rock-grout shear transfer, and end-bearing on rock, as shown on Figure 1.

Conditions Encountered During Drilling of Piles

The pile system was installed using a rotary-percussion drilling rig with a water flush. The grout holes were approximately five and three-quarters inches in diameter in the soil. The drilling equipment advanced the

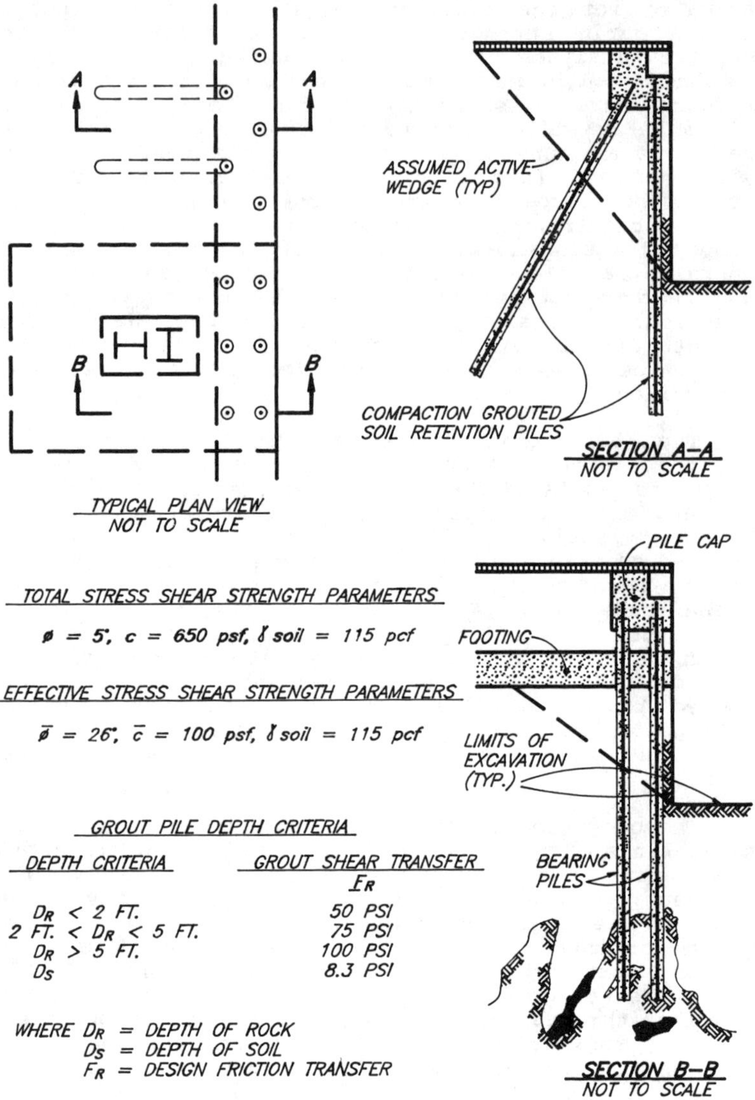

TOTAL STRESS SHEAR STRENGTH PARAMETERS

$\phi = 5°$, $c = 650$ psf, γ soil $= 115$ pcf

EFFECTIVE STRESS SHEAR STRENGTH PARAMETERS

$\bar{\phi} = 26°$, $\bar{c} = 100$ psf, γ soil $= 115$ pcf

GROUT PILE DEPTH CRITERIA

DEPTH CRITERIA	GROUT SHEAR TRANSFER F_R
$D_R < 2$ FT.	50 PSI
2 FT. $< D_R <$ 5 FT.	75 PSI
$D_R > 5$ FT.	100 PSI
D_S	8.3 PSI

WHERE D_R = DEPTH OF ROCK
D_S = DEPTH OF SOIL
F_R = DESIGN FRICTION TRANSFER

FIGURE 1 TYPICAL SECTIONS USED FOR STABILITY ANALYSES

drill hole using an outer steel casing and an inner drill bit. For the bearing piles, the outer casing (5¾-inch diameter) was extended one foot into rock, and an inner drill bit (4¼-inch diameter) was extended to the required embedment depth in rock.

One advantage of the selected support system was that during the drilling for installation of the grouted piles, additional exploration of the subsurface could occur. Numerous undisturbed soil samples (Shelby tubes) were obtained and tested for unconfined compressive strength throughout the project. In this manner, the system could be adjusted in the field if the soils encountered did not meet the assumed design parameters. Adjustments consisted of installing additional compaction grouted piles where necessary. The unconfined compressive strengths of the tested soil samples ranged from as low as 200 psf to as high as 5000 psf. The average unconfined compressive strength was approximately 1900 psf.

As expected, the top of rock encountered during the drilling for installation of the piles was very irregular. The top of rock varied from two feet above the proposed excavation level to approximately 20 feet below, sometimes between adjacent columns (spaced five feet apart). In addition, floating boulders, voids, and intervals of extremely soft soil conditions were common. The voids typically occurred near the soil-bedrock interface with their vertical extent approximating 20 feet in some cases. Isolated voids up to ten feet in height were also encountered in the underlying limestone.

Grouting Program/Procedures

The grout mix consisted of a low slump (two inches nominal) mortar-like material combining Type I Portland cement with aggregate sand. For ease of pumping the low slump mix, rounded, naturally silicious sand is typically recommended. However, because this type of sand was not available locally, angular, manufactured sand was used. To adjust the mix for difficulties in pumping which occurred because of the coarseness of the sand, fly ash and powdered bentonite were added to make the mix "slicker" and enable material with a slump as low as one inch to be pumped through the grout lines. The specified minimum strength for the grout was 3500 psi. Numerous grout cube sets were made and tested throughout construction. The majority of the cubes exceeded 5000 psi in strength. The reinforcement for the system was epoxy-coated, threaded No. 8 bars for the soil retention piles and No. 11 bars for the bearing piles.

The primary controlling influence in compaction grouting is the grout pressure behavior. The effective radius of a grout hole will vary according to the rate and resulting pressure of grout injection, as well as soil type, moisture, and relative density. In terms of soil type and relative density, an advantage of using compaction grouting is its tendency to find and compress the weakest soil zones. Because of this, we expected that the radius of individual piles would vary along their length depending on the densities encountered. The nominal pumping rate used for compaction grouting was 1.8 cubic feet per minute. A given grout quantity was pressure injected through the drill casing and the casing was raised in one-foot intervals according to the return pressure reading included in Table I.

TABLE I
GROUTING PROCEDURES

Return Pressure at Hole (psi)	Volume of Grout Injected per One-foot Lift of Casing	Estimated Equivalent Pile Diameter (in.)
0 - 50	1.8 CF	18
50 - 200	1.2 CF	15
> 200	0.6 CF	6 - 10

The casing was removed simultaneously with grouting if surface upheaval or grout show around the casing occurred.

Projected Results/Observed Results

To estimate the success of the in-situ ground modification system at improving the subsoils, we staggered the installation of the piles in a given group and observed the pressure behavior. Figures 2 and 3 are the results of pressure monitoring during compaction grouting operations at various locations along the excavation. As shown thereon, there is definitely a marked trend towards increased pressures at subsequent grouting sessions. We interpret these increases as densification of the soils in-place during compaction grouting. One of the reasons that the compaction-grouted pile support system was selected was that the technique was expected to fill the numerous voids and slots in the soil and to displace extremely soft soils. During the second phase of drilling for installation of the grout piles, we encountered grout from the first phase of compaction grouting at 15 different locations. Some of the locations where we encountered grout were three or more feet away from the closest grout pile, indicating that the compaction grouting process was successfully filling the slots and voids.

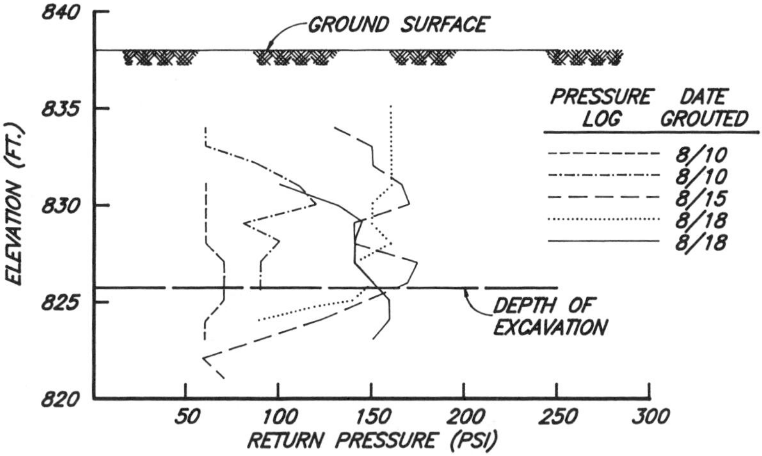

FIGURE 2(a) PILES BETWEEN COLUMN FOOTINGS

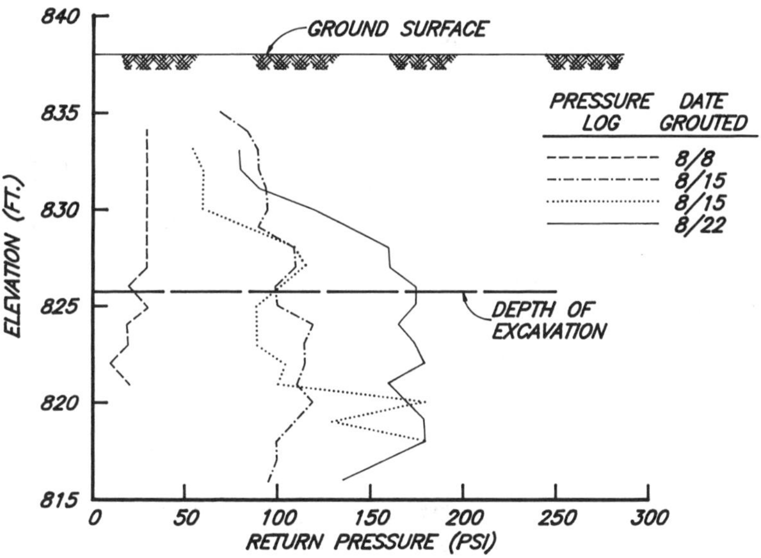

FIGURE 2(b) PILES BENEATH COLUMN FOOTING

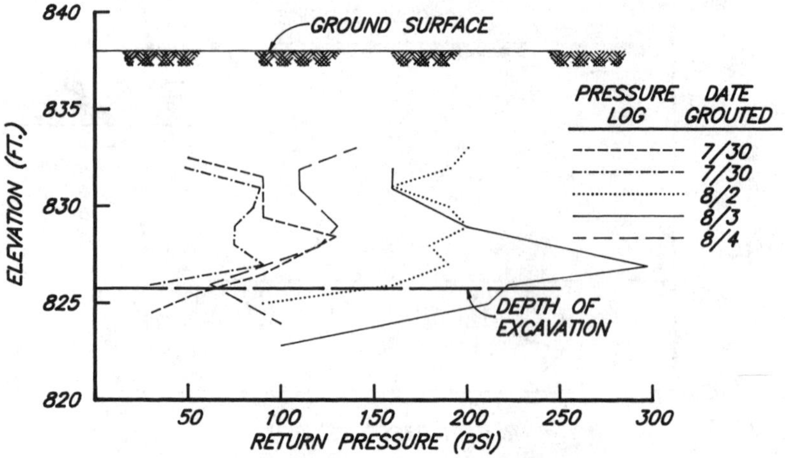

FIGURE 3(a) PILES BENEATH COLUMN FOOTING

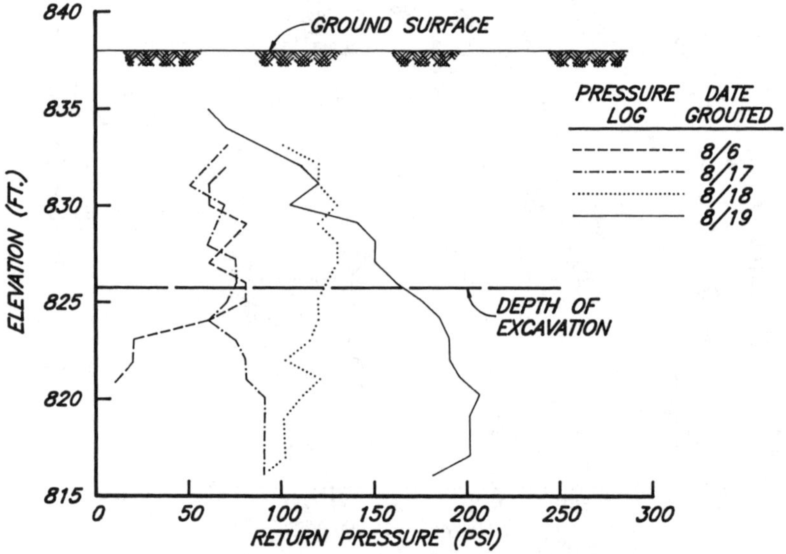

FIGURE 3(b) PILES AT CORNER OF EXCAVATION

FIGURE 4 FINAL EXPOSED EXCAVATION

FIGURE 5 INSTALLATION OF PREFABRICATED TUNNEL

The excavation for installation of the prefabricated tunnel was made five months after completion of the grout pile support system. The excavation support system performed extremely well. No noticeable deflection of the system or sloughing along the sides of the excavation was evident during the tunnel installation. The grout piles that were exposed along the face of the excavation varied in diameter from 9 inches to 12 inches. Based on the diameters of the exposed portions of the piles in the upper interval of soil, we expect that the portions of the piles below the excavation grade have even larger diameters. In addition, numerous bulbs of grout were encountered during the excavation which extended out into the excavation from the grout piles. We expect that these bulbs indicate voids or soft soil zones which were present in the soil prior to grouting. Figures 4 and 5 are photographs showing the exposed excavation and the final tunnel installation, respectively.

Conclusions

Compaction grouting methods are generally not considered effective at shallow depths or in clayey soils. However, we found that expansion of grout in a five-inch diameter injection hole to ten or more inches in diameter could be accomplished at relatively low pressures (0-250 psi) using a low pumping rate in the softer, clayey soils at the site. In addition, we found that compaction grouting was effective in filling the numerous voids and slots in the soil and rock without excessive flow beyond the subject area.

The use of the compaction grouted pile system provided the necessary soil retention and underpinning support and allowed the excavation to be made with no discernable soil or footing movements. The grout piles that were installed for bearing capacity performed so well that an increase in their allowable design capacity to 40 tons was used on subsequent projects at the same site.

Conversions

1 inch = 2.54 centimeters
1 foot = 0.3048 meters
1 pound = 0.4536 kilograms
1 ton = 2000 pounds

ACKNOWLEDGEMENTS

The specialty contractor for the project was Hayward Baker, Inc. of Tampa, Florida.

Use of Cement-Bentonite for Cutoff Wall Construction

B.L. Kilpatrick[1] and S.J. Garner[2]

Abstract

B.C. Hydro's John Hart Dam on Vancouver Island in British Columbia consists of a concrete gravity dam, a concrete power intake structure and earthfill sections. The dam is in a high seismic risk area. Early studies in 1985 and 1986 concluded that the dam was susceptible to failure due to liquefaction during a relatively minor earthquake. To minimize the liquefaction potential of the foundation soils and to facilitate reconstruction, slurry trench and jet grout cutoffs were constructed through the intake structure and adjacent earthfill dams under full reservoir and normal operating conditions. To minimize the potential for cracking of the cutoffs under large earthquakes, a flexible cement-bentonite mix was used. A series of strength and permeability laboratory testing were carried out and the results are presented. Field quality control tests were also carried out to ensure that the selected mix had the required design properties. Field test results for the same cement-bentonite mix placed by the two different methods are also presented.

Introduction

John Hart Dam is situated near the east coast of central Vancouver Island in British Columbia, about 8 km from Campbell River. The dam, built in 1947, consists of a 30 m high concrete gravity dam, earthfill sections and an 11.7 m high power intake dam. The earthfill sections and power intake dam are founded on a deep sequence of post-glacial deltaic sands, interlayered sands and silts and pro-deltaic silts. Beneath the deltaic deposits is a

[1]Section Head, Soil Mechanics Section, Geotechnical Dept., B.C. Hydro, 970 Burrard St., Vancouver, B.C., Canada V6Z 1Y3

[2]Sr. Geotechnical Engineer, Geotechnical Dept., B.C. Hydro, 970 Burrard St., Vancouver, B.C., Canada V6Z 1Y3

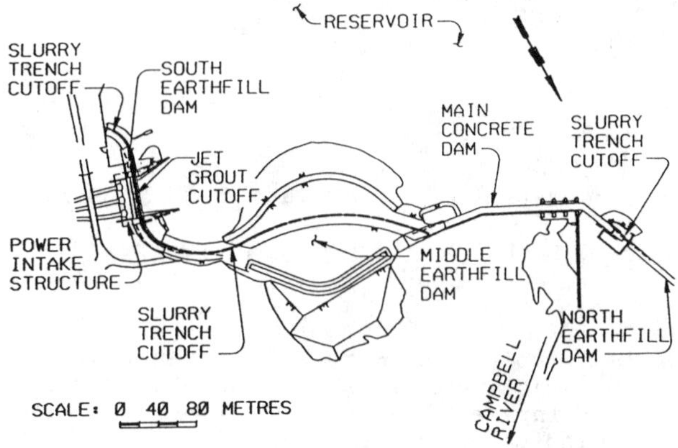

Figure 1 John Hart Dam

blanket of dense till which overlies a buried bedrock channel. A partial sheetpile cutoff was installed through and under the earthfill sections to a depth of about 8 m and under the power intake structure to a depth of about 9 m.

As part of B.C. Hydro's Dam Safety Program a Comprehensive Inspection and Review was undertaken in 1985 in order to detect any deficiencies which could affect the safety of the dam. Of particular interest was the seismic resistance of the foundations beneath the earthfill sections and power intake structure. Previous seismic studies indicated that the dam site could experience a maximum credible earthquake (MCE) of 0.6g. As well, slopes downstream of the earthfill sections and the power intake structure had experienced a history of seeps, boils and shallow slides which required numerous repairs, including reverse filter blankets and attempts at grouting.

Investigation

In 1985 and 1986 continuous standard penetration tests (SPT) and cone penetration tests (CPT) were performed within and in the foundations of the earthfill embankments and in the foundation of the intake structure. Results of the tests indicated that most areas were very loose and could be susceptible to liquefaction at accelerations much less than the MCE. The upper 5 m immediately below the power intake structure, however, was determined to be dense uniform fine sands with average depth and energy corrected blows $((N_1)_{60})$[3] of about 40. The underlying interlayered sands and silts consisted of layers up to 1.5m thick with $(N_1)_{60}$ varying between 15 and 35.

Design

As a result of the field investigation programs, it was determined that an improved cutoff barrier was required under the power intake structure and in the earthfill dams in addition to densifying and replacing sections of the embankments and abutments of the concrete structures.

The cutoff selected for the earth structures was a slurry trench with jet grouting chosen for beneath the operating power intake concrete gravity structure. Panel construction was the selected method for the slurry trench because the restricted size of the working area and the curvilinear alignment of the cutoff were not conducive to the conventional method of slurry trench construction.

The cutoff barrier would be installed through the sands and interlayered sands and silts into the underlying silts or keyed into till or bedrock. This represented a cutoff depth of about 14 m below the power intake structure and adjacent earthfill sections and up to 30 m in the earthfill structures. The cutoff would consist of cement-bentonite slurry panels in the earthfill sections connected to a double row of flexible jet grout installed through the power intake structure. This would form a continuous cutoff barrier under all the structures founded on soils.

In order to reduce the possibility of windows in the jet grout wall under the intake structure, the design called for a double row of jet grout columns at a spacing sufficient to allow overlapping of individual columns in each row and between adjacent rows thereby reducing the possibility of voids or unconnected zones of grout.

Detailed seismic analyses based on the procedure by H.B. Seed[3] indicated that the cutoff barrier would be subject to large cyclic stresses induced by the MCE. This would cause some softening or modulus reduction in the interlayered sands and silts leading to post-earthquake movements of up to 20 cm. A flexible membrane, therefore, was essential to maintain the integrity of the cutoff barrier during and after a major earthquake event.

The selected design was a grouted soil cutoff wall which contained a cement-bentonite mix proportioned to

---

[3] $(N_1)_{60}$ is the Standard Penetration Resistance corrected for depth and energy.

create a wall which would meet the following criteria:

a) Coefficient of permeability $\leq 10^{-6}$ cm/sec.
b) A compressive strength at 7 days of between 30 and 100 kPa as determined by consolidated undrained triaxial tests.
c) Stress-strain characteristics which would enable the wall to undergo a minimum of 10% strain without cracking as determined by consolidated undrained triaxial compression tests with a confining stress of 100 kPa carried out on minimum 7 day old samples.

Laboratory Testing

The first phase of the laboratory testing program included mix design and material properties analysis of the cement-bentonite. The viscosity, setting time, strength, permeability and stress-strain characteristics were measured for a number of trial mix designs. The second phase included testing of field samples for quality control and insitu properties.

The percentage of bentonite was selected based on

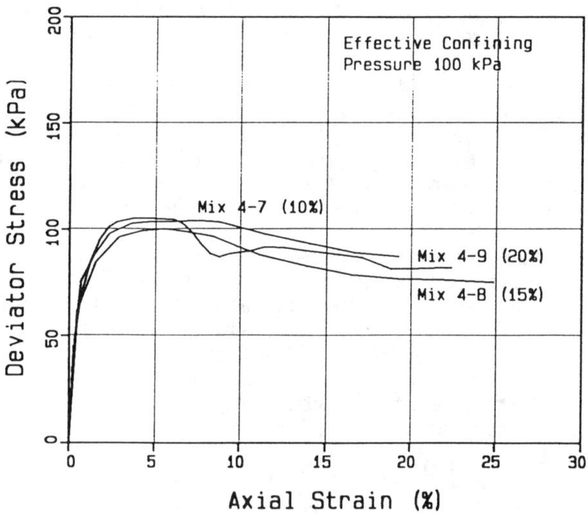

Figure 2 CU Tests on Cement-Bentonite

the requirement that the fresh grout mix should have a Marsh cone flow time of about 45-55 seconds. This value of viscosity is used to prevent settlement of the cement suspension and at the same time enable the jet grout operation to be carried out smoothly. Initial trials using COREGEL, a high yield bentonite slurry, indicated that 4 to 5 per cent by weight of bentonite would produce a flow time of about 50 seconds. The addition of cement increased the viscosity slightly so the lower bentonite content of 4 per cent was selected for production.

Trial mixes were prepared using 4 per cent bentonite and 10, 15 and 20 per cent cement. Typical stress-strain curves obtained from consolidated-undrained triaxial tests of samples from these mixes are shown on Figure 2. Increasing the cement content had the beneficial effect of increasing the strength and decreasing the permeability. However, with increasing cement content, the stress-strain characteristics showed that the material became more brittle.

Measured permeability values from these trial mix samples were generally in the order of 10^{-6} cm/sec. Pinhole erosion tests using maximum gradients of 4 were also carried out on these trial mixes. No evidence of erosion or dispersion of materials was detected.

In the field, grout samples were obtained from the mixer as well as from the return lines. These were cast into 76 mm diameter specimens, field cured for at least 48 hours and shipped to the laboratory for testing. Laboratory testing of these specimens consisted of consolidated-undrained triaxial tests and permeability tests. Unconfined compression and falling cone tests were also carried out, however, the results were quite scattered.

Peak compressive strength values measured from these specimens were generally in the range of 100-150 kPa. A large scatter of strength values was observed, however, none of these were less than 85 kPa. These values, obtained for both the slurry trench and jet grout cutoff, are shown on Figure 3. An apparent lack of strength gain with age may be due to consolidating the samples after they had set, thereby causing breakage of some of the cement bonds.

The average coefficient of permeability for the jet grout samples was 2.5×10^{-6} cm/sec. This compares to a value of 9×10^{-6} cm/sec for the cement-bentonite samples obtained from the slurry trench. Although these values are moderately higher than the 1×10^{-6} cm/sec specified, they were accepted because it was considered that, as in the strength samples, the permeability samples were not consolidated until testing commenced and the strength of

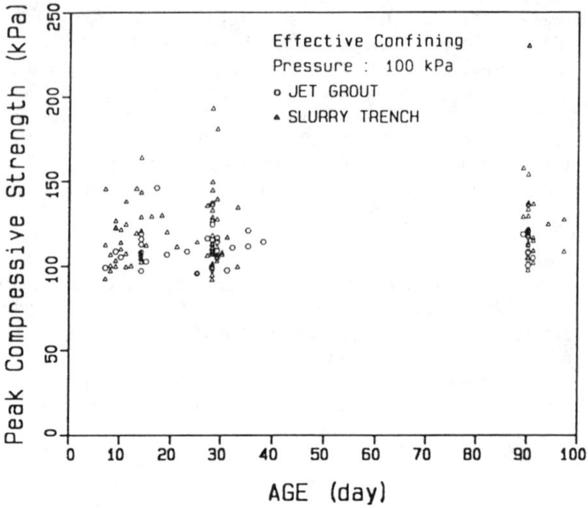

Figure 3 Peak Strength of Field Samples

the cement prevented the voids from completely compressing.

An accelerated percolation test was also carried out to ensure stability of the cement-bentonite against water percolation. A test specimen of about 75 mm diameter by 80 mm high trimmed from a Shelby tube sample taken from a slurry panel at 7 m depth was used for this test. The test was carried out in a triaxial cell using a confining pressure of 100 kPa and water was percolated through the specimen from bottom to top at a gradient of about 100. The test was continued for more than 90 days without any sign of material degradation. The resulting coefficient of permeability, calculated on the rate of water percolation through the sample is shown on Figure 4.

Field Tests

Control tests performed during the cutoff construction included viscosity and specific gravity tests of the grout with samples taken from both the mixer and the return line. Samples were also taken and cast into 76 mm diameter cylinders for records testing as discussed previously.

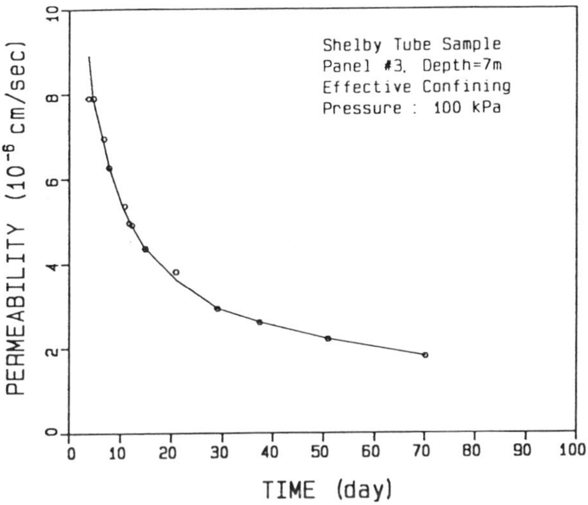

Figure 4 Permeability Values from Percolation Tests

Field tests of the completed jet grout columns included a series of 21 CPT tests and 118 pore pressure dissipation tests taken at about 10 to 35 days age. A further 49 CPT profiles and 29 electric vane tests were performed in the slurry panels. CPT testing proved to be a quick and effective method for determining in situ properties of the cement-bentonite.

Undrained shear strengths were determined from the CPT data based on correlations with laboratory and in situ electric vane tests. Typical CPT results for jet grout columns are shown on Figure 5. A summary of the undrained shear strengths determined from CPT compares very well with the strengths determined in the laboratory. These are shown on Figure 6.

For comparison, typical undrained shear strength results for the identical mix used in the adjacent slurry trench panels and about the same age are shown on Figure 7. A summary of the undrained strengths determined from CPT are shown on Figure 8.

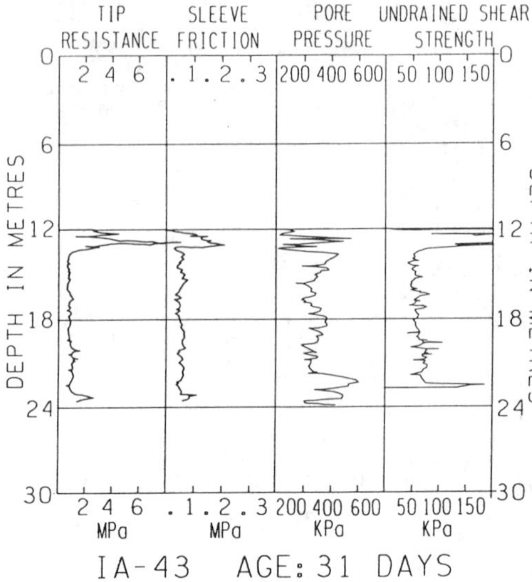

Figure 5 Typical Jet Grout Column CPT and Undrained Strength Results

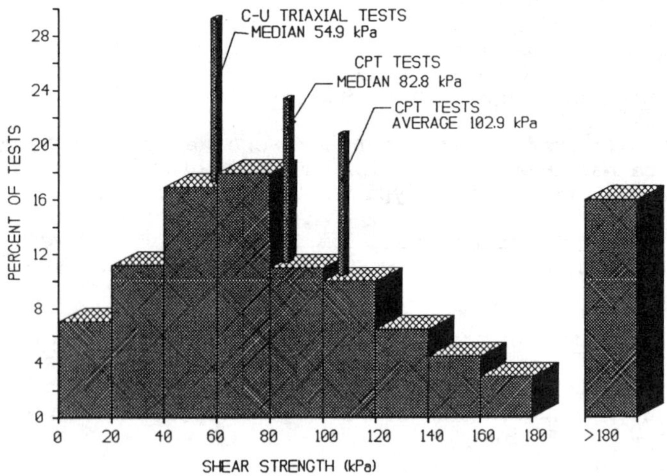

Figure 6 Jet grout Shear Strength Summary of Results From CPT

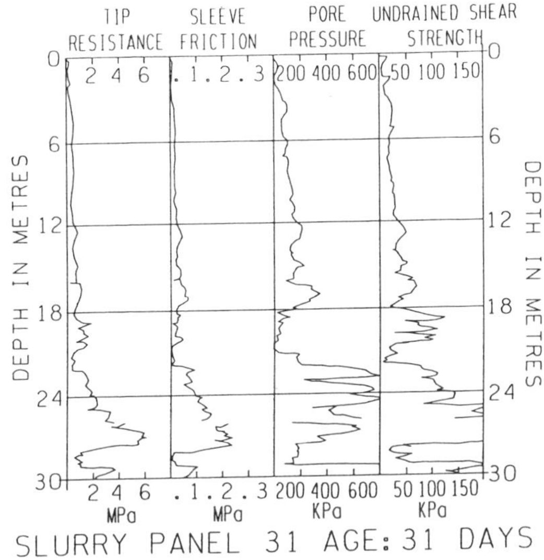

Figure 7 Typical Slurry Panel
CPT and Undrained Strength Results

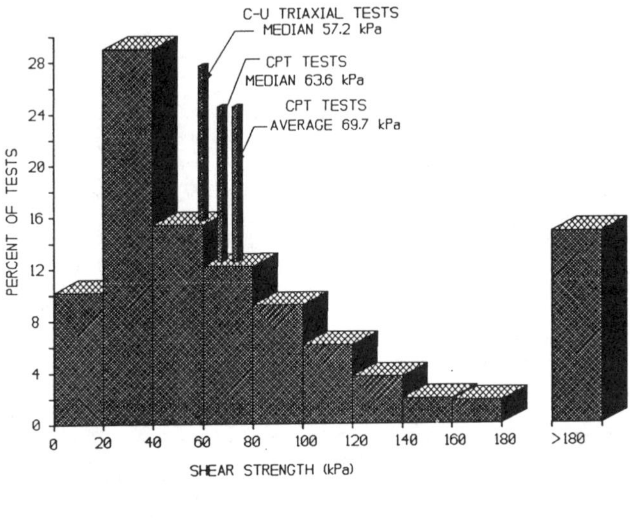

Figure 8 Slurry Panel Shear Strength
Summary of Results From CPT

The results indicate that the jetted columns achieved about 30 percent higher strengths than the slurry panels.

The CPT results also indicated that the upper metre of both slurry panels and jet grout columns were unacceptably soft. This is attributed to an increased water cement ratio created during settlement of the cement-bentonite. The upper metre, therefore, was excavated and replaced in all panels and rejetted in alternate jet grout columns. Settlement of the jet grout columns also created a risk of separation between the top of the columns and the base of the intake slab. Steel drums filled with fresh cement-bentonite therefore were applied to the tops of the jet grout columns immediately after jetting to ensure that a bond was maintained. CPT and pore pressure dissipation tests indicated that a tight bond was made.

Discrete high strength layers were encountered in the jet grout columns, particularly immediately below the concrete slab. These are interpreted to be cement-rich layers possibly caused by hydraulic fracturing or "claquage". This is also verified by the occasional cement nodule which appeared in the refuse grout.

CPT tests were performed at various ages of the jet grout columns and slurry panels. The results, shown on Figures 9 and 10 indicate that the rate of strength gain in the field is slower than that indicated in the laboratory. However, very little strength gain occurs after about 28 days.

Performance

A total of seven 300 mm diameter wells were installed downstream of the cutoff. A series of piezometers were installed in arrays around each well to determine the drawdown and transmissivity characteristics of the foundation soils. Pump tests were performed before and after the cutoff wall installation. By comparing the directional transmissivities, the effectiveness of the cutoff was determined.

As shown on the transmissivity ratio rosettes in Figure 11, results of the tests indicate up to 74 percent reduction in transmissivity in the direction of the cutoff. As anticipated, no significant changes in transmissivity occurred from the downstream direction or from the open end of the slurry panel cutoff or in zones of low flow. In the vicinity of the intake structure, recharge flows were reduced by up to 50 percent in the upper layer and 20 percent in the lower layer.

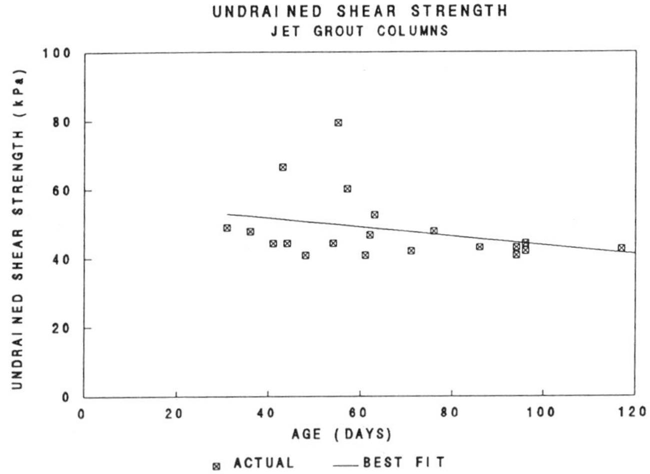

Figure 9 Jet Grout
CPT Undrained Strength vs Age

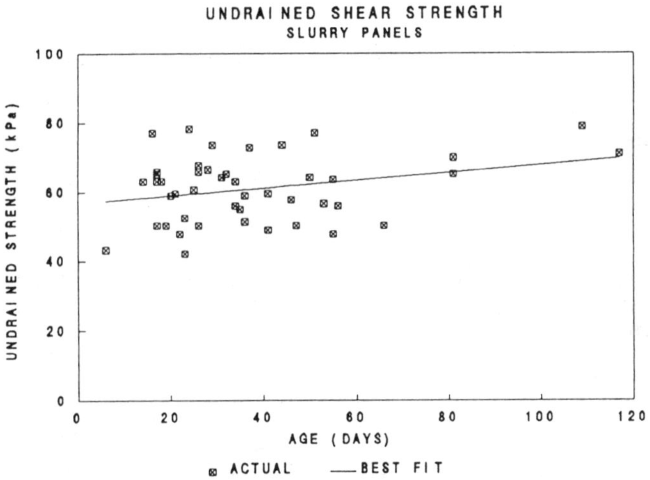

Figure 10 Slurry Panels
CPT Undrained Strength vs Age

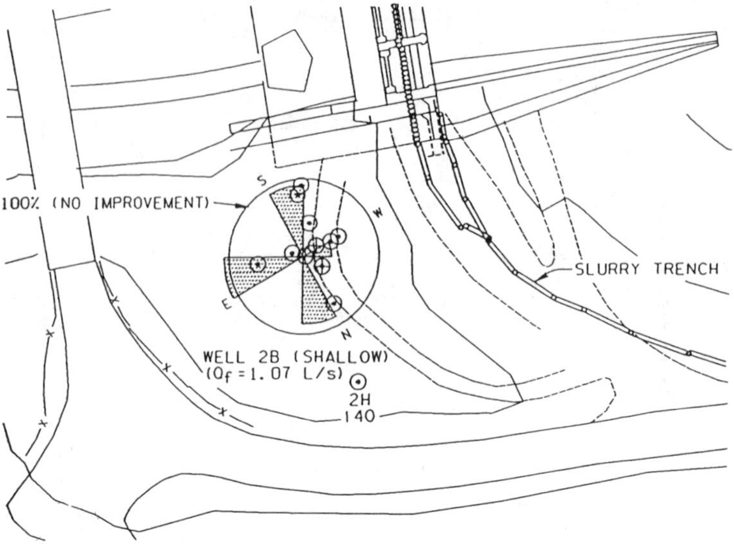

Figure 11 Transmissivity Ratio Rosette
$\frac{\text{After Cutoff}}{\text{Before Cutoff}} \times 100\%$

These tests indicate that the groundwater table downstream of the intake structure can be readily lowered if required by the use of two pumps installed in the existing test wells.

Conclusions

For the given site conditions, cement-bentonite jet grout, as measured in situ, has about 30 percent higher strength than the same mix used in slurry panels.

The jet grout method has been used to create a flexible, impermeable cutoff beneath an existing concrete power intake structure while it remained fully operational.

Low cement content cement-bentonite grout will experience significant settlement due to consolidation and bleeding. Replacing the upper metre of fully set cement-bentonite with fresh mix may be required due to a high water-cement ratio created from the bleeding. The method of topping up each jet grout column using fresh grout at the collar is successful in maintaining a tight contact at the underside of the concrete slab.

Consideration should be given to curing laboratory

test samples under the confining pressures at which the test will be performed. This would be applicable to both laboratory strength and permeability tests.

References

1. Chapuis, R.P., Pare, J.J., and Loiselle, A.A. (1984). "Laboratory test results on self-hardening grouts for flexible cutoffs." Can. Geotech. J., 21, 185-191.

2. Imrie, A.S., Marcuson, W.F., Byrne, P.B., (1988). "Seismic Cutoff." Civil Engineering, 12, pp. 50-53.

3. Seed, H.B., Idriss, I.M., and Arango, I. (1983). "Evaluation of liquefaction potential using field performance data." J. Geotech. Engrg., ASCE, 101(9), pp. 889-912.

SITE IMPROVEMENT FOR A STEEL MILL COMPLEX

Eun C. Shin[1], A.M. ASCE, Bang W. Shin[2], and
Braja M. Das[3], M. ASCE

Abstract

Sand compaction piles can be used in marginal sites for improving stability, controlling liquefaction, and reducing settlement of various structures. These piles can significantly accelerate the pore water pressure dissipation process and, hence, the time of consolidation. This paper relates to the case history of a steel mill complex construction project in Korea in which the beneficial effects of sand compaction piles, sand drains, and preloading were fully utilized. Construction methods of the sand compaction piles and sand drains are described along with the reasons for some unsuccessful cases of sand pile construction. Preloading techniques with staged loading and the resulting settlement and quality control have been outlined. Finally the methods for quality control with automatic monitoring recording graphs, field tests, and the grain-size distribution of the backfill material have been presented.

Introduction

In industrialized countries, the intensity of development activities has resulted in considerable increase in land value for the dwindling supply of suitable land needed for construction. When marginal land is used for development purposes, improvement of poor soil and subsoil conditions for supporting structures will generally be

---

[1] Graduate Research Assistant, Department of Civil Engineering and Mechanics, Southern Illinois University, Carbondale, Illinois 62901-6603, U.S.A.

[2] Professor, Department of Civil Engineering, Chungbuk National University, Chong-Ju, Korea

[3] Professor and Chair, Department of Civil Engineering and Mechanics, Southern Illinois University, Carbondale, Illinois 62901-6603, U.S.A.

required. Construction of sand compaction piles is one potential method for improving marginal sites to increase the load-bearing capacity of the soil and to reduce the probable settlement and liquefaction potential. They also accelerate the time required for primary consolidation in saturated clayey soils.

Sand piles were first constructed in Japan between 1830 and 1850 (Ichimoto, 1981). Large-diameter compacted sand columns were constructed in 1955 using the Compozer technique (Aboshi et al., 1979). The Vibro-Compozer method of sand piles construction was developed by Murayama in Japan in 1958 (Murayama, 1962).

Sand compaction piles were first used in Korea in 1984 for the construction of the Kwang Yang Steel Mill Complex. The project site, Kwang-Yang, is located on the South Sea about 300 km (187 miles) south of Seoul, where a delta is formed at the convergence of the Sum Jin and Su Oh rivers (Figure 1a). This site was selected for the steel mill complex because it is directly connected to the sea which facilitates the import of raw material and export of finished products.

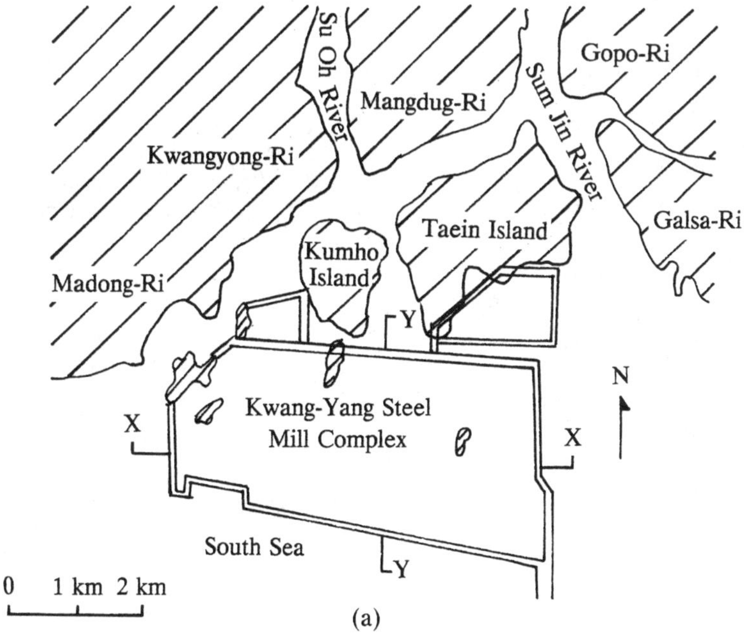

(a)

Figure 1. (Continued)

Site improvement techniques used in this project include sand compaction piles as well as some sand drains with preloading. The sand compaction piles and the sand drains had diameters of 0.7 m (2.3 ft) and 0.4 m (1.3 ft), respectively, with an average length of about 25 m (82 ft) each. The center-to-center spacing of the sand drains and sand compaction piles ranged from 1.75 m (5.8 ft) to 2.5 m (8.2 ft). The reclaimed area was about 1,450,000 m^2 (15,602,000 ft^2). The volume of sand used for construction of sand piles and preloading, including the backfill sand layer, was about 11,600,000 m^3 (15.2 × 10^6 yd^3). The improved site presently supports a stockyard of heavy materials, a slab yard, oil tanks, embankments for roads and railways, and steel mill factories.

Soil Conditions

Figure 1b shows the general nature of the *in situ* soil along with the backfill which was about 5 m (16.4 ft)

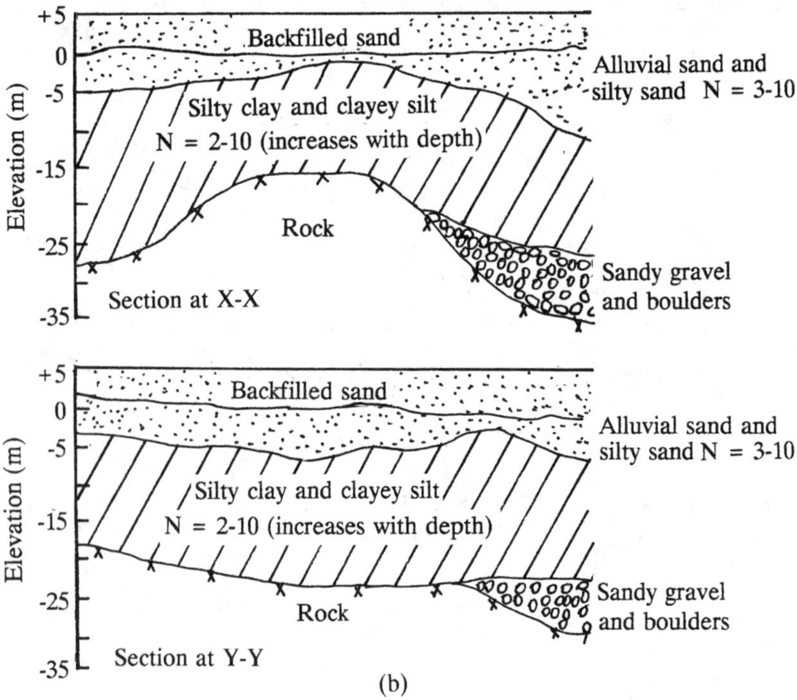

Figure 1. (a) Site location of the Kwang-Yang Steel Mill Complex; (b) general nature of soil profiles (Note: N = standard penetration number)

thick. In general the subsoil conditions can be described as about 5 m (16.4 ft) of alluvial loose sand and silty sand at the top, followed by soft silty clay and clayey silt up to a depth of 15 to 20 m (49.2 to 65.5 ft). Below the clay layer there was, in general, bedrock and in some places sandy gravel and boulders. The field standard penetration resistance values (N) of the top sand layer varied from about 3 to 10. The clayey soil underlying the top sand layer was normally consolidated. The sensitivity of this clay varied from 3 to 6. The physical properties of this clayey soil can be summarized as follows: natural water content, 44%-90%; liquid limit, 40%-80%; plastic limit, 20%-40%; plasticity index, 20%-40%.

Construction of Sand Compaction Piles and Sand Drains

The construction procedures for the sand compaction piles and sand drains are schematically shown, respectively, in Figures 2a and 2b. For the construction of the sand compaction piles, the following procedure was used:

1. Installing a casing pipe having an outside diameter of 0.4 m (15.75 in.) and a wall thickness of 16 mm (0.63 in.) at a designated position on the ground with an allowable maximum horizontal deviation of 0.3 m (1 ft).

2. Driving the casing pipe into the ground by the vibro-hammer with an allowable deviation of two degrees with respect to the vertical. The vibrator used has the following characteristics: weight, 54 kN (6.1 ton); frequency, 580 cpm; acceleration, 8.2 g. The vibrator was driven by a 90-kW (120 hp) motor and had a vibrating force of 443 kN (50 ton). When a localized hard spot was encountered (usually SPT N-values > 20), water jetting was used to aid the penetration. Due to water jetting, if the possibility existed for mud to enter the casing, air pressure in the range of 294-392 kN/m^2 (42.6-56.8 lb/in^2) was applied.

3. When the casing pipe reached the desired depth (that is, penetrated the full depth of the soft clayey soil which varied from 20 m to 25 m), about 2 m^3 (2.6 yd^3) of sand was filled through the upper hopper of the pipe. The casing pipe was then drawn up at the speed of 9 m/min (29.5 ft/min) through a distance of 3 m (9.84 ft).

4. An air pressure of 588 kN/m^2 (55 lb/in^2) was applied to the top of the sand for extrusion from the tip of the casing.

5. The casing pipe was redriven through a distance of 2 m (6.6 ft) under vibration to compact the sand pile.

6. Steps 3, 4 and 5 were continued until the sand pile was built up to the ground level.

Construction of sand drains consisted of Steps 1, 2, 3, 4 and 6 outlined above.

The progress of the sand compaction pile construction was observed from a graph drawn on an automatic recording system for the sand level and the casing level. A total of

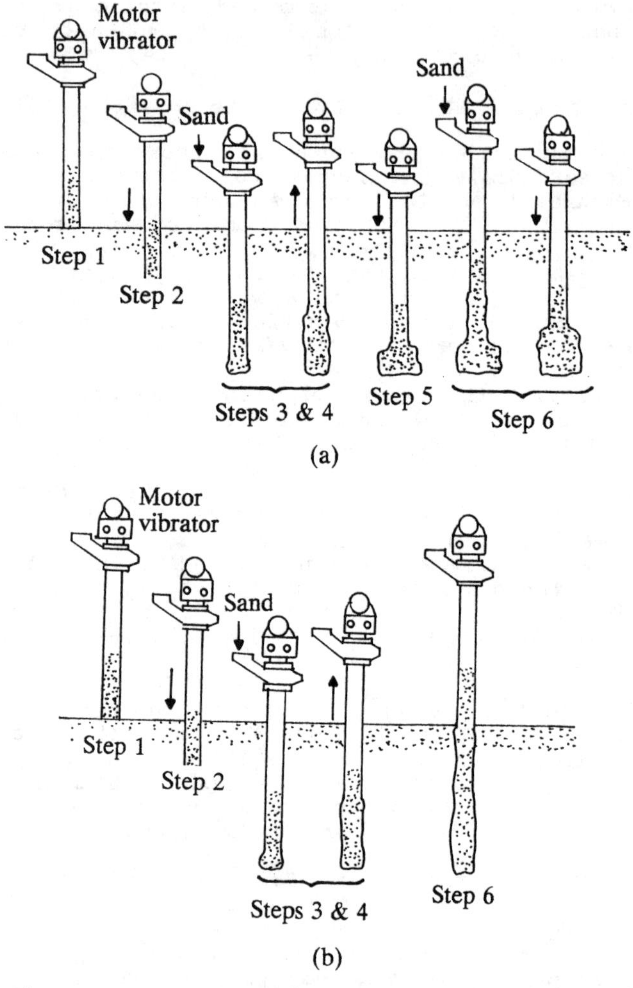

Figure 2. Construction sequence for (a) sand compaction piles and (b) sand drains

23,217 sand compaction piles and 27,224 sand drains were attempted, out of which only 0.62% of the sand compaction piles and 0.36% of the sand drains were not successful. The cases for which sand pile construction was not successful were due to one or more of the following reasons:

1. The slope of the sand pile deviated more than two degrees from the vertical.

2. The pre-estimated length of the sand pile was not achieved due to localized hard spots.

3. The casing pipe did not reach the required depth past localized hard spots due to the malfunctioning of water jet equipment.

4. Horizontal deviation of the actual sand pile location was more than 0.3 m (1 ft) from the designated location.

5. The sand pile was broken during the construction process. The reasons for breakage are (a) malfunctioning of the lower end of the casing (shoe) and (b) careless operation of the construction equipment.

6. The level of sand inside the casing drawn on the automatic recording graph was more than 0.3 m (1 ft) due to the incomplete extrusion of sand from the bottom end of the casing. This was due to insufficient air pressure.

7. Automatic recording graphs were not drawn well due to malfunctioning of the recording system.

8. The gradation of the sand was not in agreement with the design requirements.

The total volume of loose sand brought to the site for the sand pile construction was about 140% of the actual volume of the piles. This was based on the assumption that there would be a 20% reduction due to the densification during the placement process and 20% loss during construction.

The area replacement ratio a_s is an important parameter in the design of sand compaction piles. It is defined as

$$a_s = \frac{A_s}{A} \tag{1}$$

where A_s = cross-sectional area of the completed sand compaction pile, and A = total area tributary to the sand compaction pile. The area replacement ratio can also be expressed as (Barksdale and Bachus, 1983; Barksdale, 1987)

$$a_s = \frac{\pi}{4}\left(\frac{D}{S}\right)^2 \quad (\textit{for square pattern}) \qquad (2)$$

and

$$a_s = \frac{\pi}{2\sqrt{3}}\left(\frac{D}{S}\right)^2 \quad (\textit{for equilateral triangular pattern}) \qquad (3)$$

where D = diameter of sand compaction piles and S = center-to-center spacing. For the project under consideration, the magnitude of a_s was about 0.09.

Preloading

Construction of staged load for preloading purposes needed to be done carefully to avoid base failure. Based on careful calculations, it was decided to construct the staged load in such a manner that the minimum factor of safety against base failure remained about 1.5 or more at all times. This translated to about 5 m (16.4 ft) maximum height of sand for each stage of loading having a unit weight of about 17.6 kN/m^3 (112 lb/ft^3). The period of preloading and the magnitude of the maximum preload varied respectively from 4-12 months and 20-265 kN/m^2 (2.9-38.4 lb/in^2) depending upon the usage of improved site. The period of preloading and the magnitude of the maximum preload were designed by considering the rate of pore water pressure dissipation, expected settlement of soil, and stability of the sand pile reinforced foundation. The detailed design procedures are given in Reference 4. Significant modification in the construction procedure was not required based on initial field experiences.

The stability against base failure due to the staged loading was carefully monitored by observation of the lateral displacement and the vertical settlement of the surface layer supporting the preload (Figure 3). A surface settlement plate, layer settlement plate, slope indicator, pore water pressuremeter, and water level indicator were installed to monitor the settlement and lateral displacement of various subsurface layers, pore water pressure in the clay layer and water level. A detailed layout of the instrumentation program is given in Reference 4.

Field Performance

Stability Control Against Base Failure. The largest preload was applied at the proposed site for raw material (coal, ore, lime and blend) stockyard; therefore, the stability control monitoring technique at the stockyard site only will be described in this section.

The stability control monitoring technique used in this project was in accordance with the procedure described by Matsuo and Kawamura (1977) and is shown in Figure 4. In this figure, q is the preload per unit area causing displacements s and δ for any loading stage (Figure 3) and q_f is the failure load per unit area. The observed plot of s and δ/s is also shown in Figure 4. Practically at all times it fell below the control limit line.

At the end of the preloading period, the total settlement at various locations of the stockyard varied from 0.51

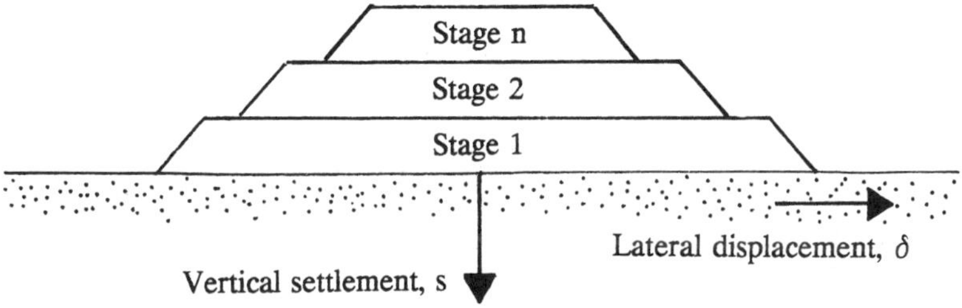

Figure 3. Staged construction process for preloading

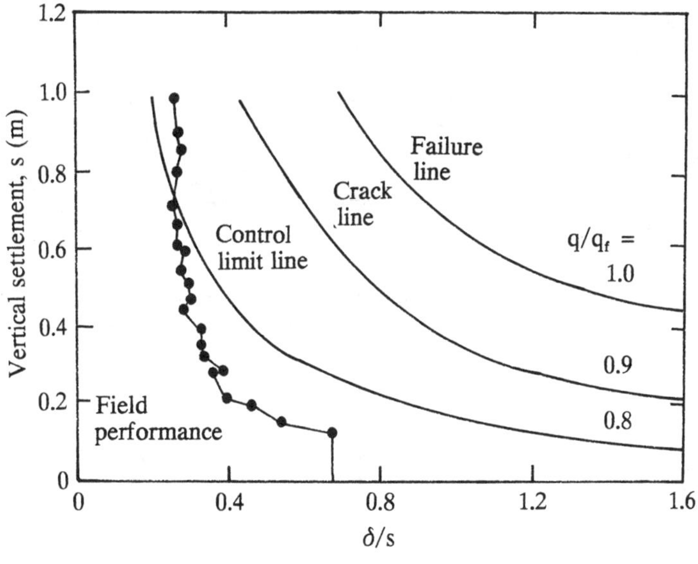

Figure 4. Stability control by Matsuo and Kaomura method (1977) for the stockyard site

to 1.03 m (20.1 to 40.6 in.). Figure 5 shows a plot of the variation of the excess pore water pressure at the middle of the clay layer [which was about 16 m (52.5 ft) thick] with time during the preloading period. Also shown in this figure is the variation of settlement with time at elevations of -7 m (23 ft), -10 m (32.8 ft), -13 m (42.7 ft) and -16 m (52.5 ft) (see Figure 1b).

Undisturbed Shear Strength of the Clay Layer. Unconfined compression tests on undisturbed specimens obtained from various depths were conducted before the application of the preload and also a month before the removal of the preload. The preload was removed after six months. Table 1 shows

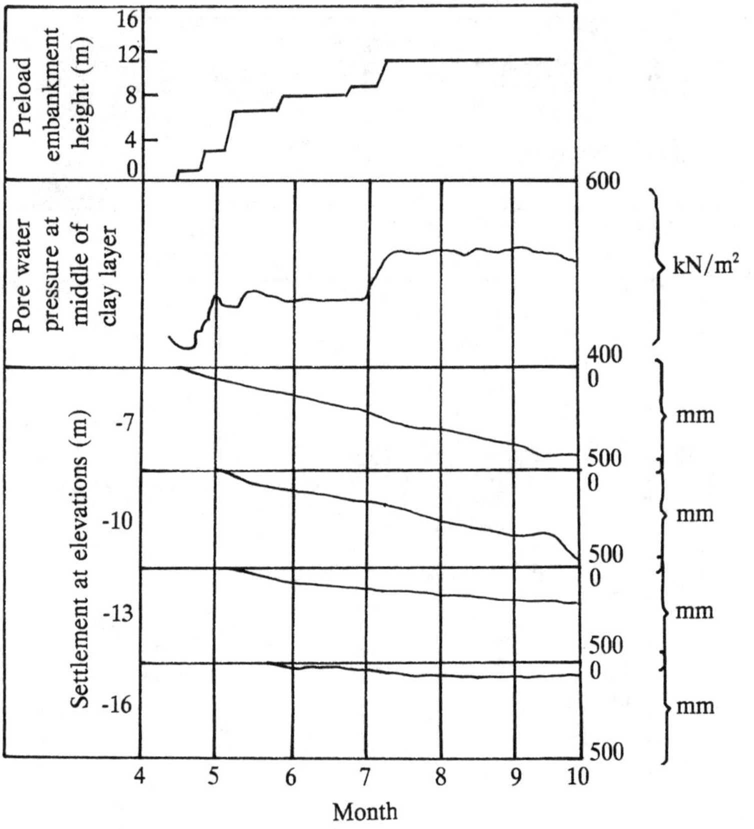

Figure 5. Pore water and settlement measurement with staged loading condition at the stockyard site

the effect of preload on the average undrained cohesion (ϕ = 0 condition) of clay at the stockyard site. It can be seen from the table that the average magnitude of the undrained cohesion increased from 60% to 195% at z = 0 to about 60% to 90% at z = 15 m, depending on the weight of the surcharge and the period of loading.

Table 1 Preload Effect on the Average Undrained Cohesion of Clay at Stockyard Site

Location	$c_{u(i)}$[a] (kN/m^2)	$c_{u(f)}$[b] (kN/m^2)
B1 coal	30.4 + 1.72 z	83.4 + 1.716 z
B2 ore	33.3 + 1.72 z	68.6 + 1.716 z
B3 lime	49	78.5
Blending	23.1 + 1.72 z	51.5 + 1.716 z

$c_{u(i)}$[a] = initial undrained cohesion
$c_{u(f)}$[b] = undrained cohesion a month before the removal of the preloading
z = depth measured from the top of the clay layer

<u>Standard Penetration Test</u>. At the beginning of the project, standard penetration tests were carried out at depth intervals of 2 m (6 ft) throughout the site. The uncorrected N-values ranged between 3 to 7 in the area of the construction of sand drains and 5 to 12 in the area of the construction of sand compaction piles. It was intended that the SPT N-value in the granular soil after improvement should be at least between 10 and 20 in order to reduce the liquefaction potential and increase the stability of the foundation. After removal of the preload, the range of field standard penetration resistance (N) values were as follows:

1. Between sand drains (in sand layer) . . . . . 6 - 15
2. Between sand compaction piles
   (in sand layer) . . . . 12 - 27
3. Center of sand compaction piles in the zone of
   top sand layer . . . . 28 - 43
   clay layer . . . . 25 - 27

The plot of some of the N-values after sand pile construction versus N-values after removal of preloading in sand is shown in Figure 6. It appears that the N-values below 14 were somewhat increased, whereas N-values above 14 were not affected due to the preloading.

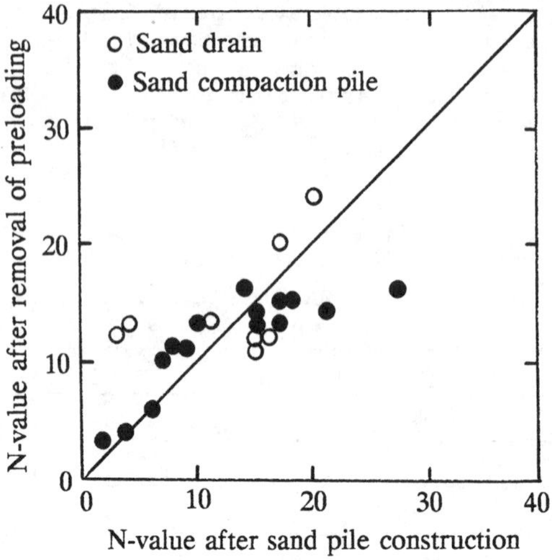

Figure 6. Effect of preloading on SPT-values after sand pile construction

Quality Control

During the construction of the sand compaction piles, an automatic recording system made a great contribution to quality control. It continuously recorded the casing level and the sand level inside the casing, the electric current of the vibro-hammer, the volume of the sand used per unit length of sand pile, and the sand pile depth. Elapse time was also recorded on an oscillograph during penetration and extrusion of a casing.

The undrained shear strength of the cohesive soil was evaluated before construction of the sand pile to make certain that the surrounding soil could provide enough confining pressure to the sand pile. The SPT tests were performed to ascertain the quality of the sand pile construction.

The average grain-size distribution curve of the backfill material is shown in Figure 7. The permeability of backfill material for the sand piles was about 1×10^{-2} cm/sec at a dry unit weight of 15.20 kN/m^3 (96.7 lb/ft^3). The silt content of the backfill material was less than 5%. The backfill material was dredged around the site and stored for three months before use to maintain a moisture content of less than 6% which facilitated its feeding to the hopper and extrusion from the lower end of the casing pipe.

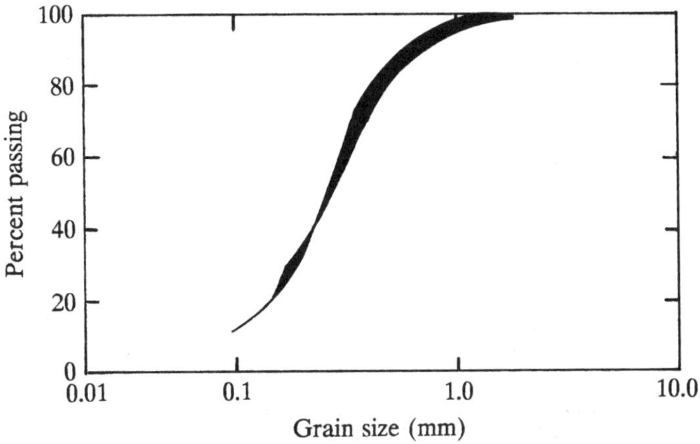

Figure 7. Average grain-size distribution curve of improvement sand for sand pile

Conclusions

The following conclusions can be drawn from the results of field tests, engineering analyses, and construction performance of the project.

1. The sand compaction pile technique is an efficient method to improve weak soil.

2. Matsuo and Kawamura's (1977) stability control method proved effective in monitoring the stability of the staged preloading.

3. Average undrained shear strength increase of the cohesive soil was about 60% to 90% due to the preloading.

4. Large increases of N-values in sand were observed due to the construction of the sand compaction piles rather than preloading.

5. An automatic recording system made a great contribution to the quality control during construction by monitoring the process of sand compaction pile construction.

References

Aboshi, H., Ichimoto, E., and Harada, K. (1979). "The compozer--a method to improve characteristics of soft clay by inclusion of large diameter sand column."

Proc, Int Conf on Soil Reinforcement, Reinforced Earth and Other Techniques, I, Paris, 211-216.

Barksdale, R.D. (1987). "State-of-the-art for design and construction of sand compaction piles." *Tech Report REMR-GT-4*, U.S. Army Engineers Waterways Experiment Station.

Barksdale, R.D., and Bachus, R.C. (1983). "Design and construction of stone columns." *Fed Hwy Admin Report FHWA/RD-83/026*, I.

Hyundae Construction Co., Ltd. (1985). "Construction report of kwang yang steel complex." (in Korean).

Ichimoto, A. (1981). "Construction and design of sand compaction piles." *Soil Improvement, Gen Civ Engg Lab*, 5 (in Japanese).

Matsuo, M., and Kawamura, K. (1977). "Diagram for construction control of embankment on soft ground." *Soils and Found*, JSSMFE, 17(3), 37-52.

Murayama, S. (1962). "An analysis on vibro-compozer method on cohesive soils." *Mechanization of Construction*, 150 (in Japanese).

Design Of Floating Stone Columns In Hydraulic Fill

Raymond A. DeStephen[1], M. ASCE, David W. Kozera[2], M. ASCE and Frank J. Swekosky[3], M. ASCE

Abstract

Floating stone columns were used to support a Processing Center at the Salem/Hope Creek Nuclear Power Station, located along the Delaware River in southern New Jersey. This foundation system was proposed as an alternative to more costly deep driven pile foundations typically used for other structures at the station. The stone column foundation was achieved by installing short vertical columns of compacted stone into the deep hydraulic fill that was placed to reclaim the power station area known as Artificial Island. The stone columns act as a soil reinforcement, forming a composite with the soft hydraulic fill to stiffen it and reduce deformation. The system's purpose was to reduce settlements of the foundations and floor slab and provide more uniform support within the upper portions of the hydraulic fill. The total foundation settlements were predicted to be about 6.0 inches (15.2 cm) considering both area loadings by floor slabs/new fill and the column loads. The actual settlements were recorded at about 2.5 inches (6.2 cm) over a period of 2.5 years. Half the settlement occurred under a preload prior to construction.

Introduction

An economic foundation alternative to driven piles was provided for the one story Processing Center con-

---

[1]Principal, Schnabel Engineering Associates, Inc., One West Cary St., Richmond, VA 23220.
[2]Principal, Schnabel Engineering Associates, Inc., 1600 N. Calvert St., Baltimore, MD 21202.
[3]Regional Manager, Hayward Baker, Inc., 675 North Court, Palatine, IL 60067.

sisting of shallow spread footings supported on stone columns. The stone columns were designed for lengths of only 18 ft (6m) to improve soils within this zone, which included sandy end-dumped fill near the surface and about 12 ft (3.6m) of soft hydraulic fill. Since the hydraulic fill thickness is about 35 ft (11.5m), the stone columns were considered as "floating" stone columns with no firm support by underlying strata. Although recognition has been given to the concept of floating stone columns in the geotechnical literature (2), constructed projects cannot be cited except for an eight story dormitory in Konstanz, Germany (3), built in 1965.

The Processing Center site was graded with the addition of up to three ft (one m) of compacted fill to provide floor support, as well as to provide a slight preload to account for live and dead slab loadings. Stone columns were placed on five ft (1.5m) spacings under foundation elements. Nine foot (2.7m) centers were used throughout the floor slab area to affect more uniform settlements. Stone column installation criteria were set based upon control stone columns installed prior to installation of production stone columns. To confirm design, an area load test was performed to simulate a footing improved by two stone columns. Presented herein are the geologic conditions, soil parameters for the hydraulic fill, stone column design considerations and construction procedures.

Site Geology

The Processing Center site is flat with ground surface at about El 12 ft (3.7m) above mean sea level and lies within the Atlantic Coastal Plain Physiographic Province. Only the Tertiary age Vincentown formation and overlying Miocene and Pleistocene age formations were previously of interest with regard to foundation support for structures at the Power Station.

The Vincentown sands consist of gray-green fine shell and quartz sand and form the bearing stratum for virtually all pile supported structures at the Power Station as well as the 70 ft (21m) thick concrete mats for the nuclear reactors. The overlying Miocene age Kirkwood clays lie above the Vincentown sands and are about 50 ft (15m) thick as indicated by the geologic profile, Figure 1. The top of the Kirkwood clays is covered with recent to Pleistocene age sand and gravel which formerly comprised the bed of the Delaware River estuary. The Pleistocene, where present, is only about five ft (1.5m) thick and contains layers and lenses of peat. Hydraulic fill was subsequently placed over these deposits to a thickness of about 35 ft (11.5m). The hydraulic fill was placed sometime prior to the early 1970s. In order to provide a trafficable surface, the hydraulic fill was covered with a variable thickness of poorly graded sand which averaged about six ft (two m).

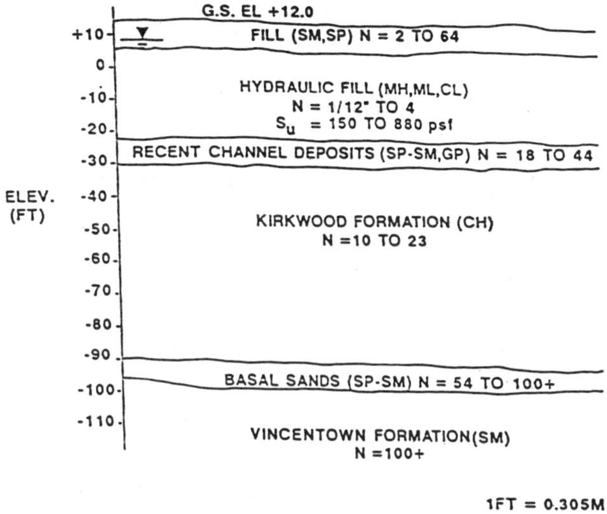

Figure 1 - Geologic Profile

Properties of Hydraulic Fill

The hydraulic fill primarily consists of fine-grained material classifying as elastic silt (MH) per ASTM D-2487 with occasional traces of organic matter. This material exhibits a high plasticity with typical Liquid Limits of 55 to 90 and Plasticity Index values of 30 to 48.

Unconsolidated undrained triaxial test data indicate very low to moderate undrained shear strength values of 150 to 881 psf (7.2 to 42.2 kN/m^2). Cone penetrometer (CPT) tests also indicated low undrained shear strength values of 200 to 1000 psf (9.6 to 47.9 kN/m^2). One dimensional constrained moduli for the elastic silts obtained from correlations with CPT data varied from 21 to 70 tsf (2016 to 6720 kN/m^2). Consolidation test data revealed an average Compression Index of 0.532. A coefficient of consolidation, C_v, of 2.4 x

10^{-4} cm sq/sec was obtained from the consolidation tests. Due to dewatering previously performed at the nearby nuclear reactors, some slight preconsolidation effects within the hydraulic fill may be present. However, this could not be determined by conventional one-dimensional consolidation tests. The hydraulic fill was therefore assumed to be normally consolidated for the purpose of analysis.

Foundation Alternatives

The Processing Center is a 33,000 sf (3070 m^2) one story, high bay, steel frame building with masonry exte rior. Column loads vary from 19 to 75 kips (84 to 333 kN). Column spacing is typically 33 by 37 ft (10.8 by 12.1m). Foundation walls between columns are grade beam supported.

Several foundation types were considered for sup port of the building including shallow spread footings, deep foundations (driven piles), and ground modification of the hydraulic fill. Shallow spread footings could not be supported directly on the fill without producing excessive differential settlement.

Deep foundations for the majority of permanent structures at the Power Station consist of various driven piles, most notably concrete-filled pipe piles founded in the dense Vincentown sands. The piles cannot be efficiently used since they must be designed to support potential down-drag of up to 30 to 45 tons (267 to 400 kN) per pile resulting from the compressible hydraulic fill. Thus, for relatively lightly loaded buildings, deep foundations represent a significant portion of the construction costs.

Several ground modification methods were initially considered including Deep Dynamic Compaction, Vibro-Compaction, and Vibro-Replacement stone columns. Only stone columns, which involve the partial replacement and displacement of unsuitable cohesive hydraulic fill with compacted vertical columns of stone, was considered a viable foundation alternative.

Stone columns are typically advanced by jetting a hole using either water or air. The weight of the fol lower tube attached above the probe, and the vibration of the probe, aid in advancing the hole. The hole is backfilled in lifts with open graded crushed stone and the stone is densified by the vibratory action of the probe. Each lift is penetrated several times with the vibrating probe to densify the stone and force it into the surrounding soil. Incremental feeding and compacting produces a dense granular column within the native cohesive soil. Final stone column diameters will vary depending on the consistency of the soil, with softer soils allowing for larger diameters to be formed. In this regard the stone column system is self-compensating.

Ultimate Load Analyses

Three failure modes were considered in designing the floating stone columns: (1) bulging failure of a single stone column, (2) shear failure of a stone column group, and (3) punching failure of the short stone columns.

There is a potential for bulging failure to occur for stone columns having a length to diameter ratio equal to or greater than four to six, and bulging failure can develop whether the tip of the column is floating in soft soil or resting on a firm bearing layer (2). A stone column's length to diameter ratio of six was estimated for this project, and thus this failure mode was evaluated. The bulging type failure of a single column was considered by Hughes and Withers (5) to be similar to the cavity expansion developed when conducting a pressuremeter test. Their cavity expansion approach considers elastic-plastic theory of a frictionless material and an infinitely long expanding cylindrical cavity for predicting the ultimate stress σ_3, surrounding the stone column. Assuming q_{ult} is the ultimate stress σ_1, that can be applied to the stone column, it can be written in terms of the elastic modulus of the soil, the initial in situ lateral stress, and friction angle of the stone as follows:

$$q_{ult} = \{\sigma_{ro} + c_u\, [1 + \log_e \frac{E_c}{2c_u(1+v)}]\}(\frac{1+\sin\phi_s}{1-\sin\phi_s}) \ldots\ldots (1)$$

where: σ_{ro} = initial total in situ lateral stress
E_c = elastic modulus of soil
ϕ_s = angle of internal friction of stone column
c_u = undrained shear strength of cohesive soil
v = Poisson's ratio

Using an elastic modulus of soil E_c=20 tsf (1916 kN/m^2), from correlations with SPT and CPT tests, ϕ_s = 40° from published values for compacted crushed stone, c_u = 500 psf (24 kN/m^2), v= 0.3, and σ_{ro} = 400 psf (19.2 kN/m^2) a factor of safety against bulging failure of 5.9 resulted for the stone columns at the site.

The second mode of failure evaluated was the bearing capacity of a stone column group. For this analysis the concept of stress concentration must be considered. An important aspect of stone column reinforced ground is the ability of stress concentration to occur in the stone column, with an accompanying reduction in stress in the surrounding soft soil. This transfer of a stress from the soil to stone column is considered within a "unit cell", defined as a stone column and the tributary area of soft soil surrounding it (4). The stress concentration will occur since the vertical settlement of the stone column and surrounding soil is approximately the same.

The stress concentration factor, n, is defined as the stress in the stone column, σ_3, divided by the stress in the surrounding cohesive soil, σ_c. The stress concentration factor, n, can be obtained by measurement of stresses in full scale installations, or estimated as the ratio of the constrained modulus of the stone divided by the constrained modulus of the soil. The latter method generally gives high values of n. Published values of n (2) range from two to five. For equilibrium of forces to exist within the unit cell the stress in the cohesive soil and stone using the stress concentration factor n gives (1)

$$\sigma_c = \sigma/[1 + (n-1)a_s] = \mu_c\sigma \quad (2)$$

and

$$\sigma_s = n\sigma/[1 + (n-1)a_s] = \mu_s\sigma \quad (3)$$

where: σ = average stress applied over the unit cell
a_s = area replacement ratio (ratio of the area of the stone column to the area of the unit cell)

The capacity of a group of stone columns can be approximated by assuming that the soil immediately beneath the foundation fails on a straight rupture surface, at some angle, β, with the ground surface, forming a wedge of soil.

Assuming the ultimate vertical stress q_{ult} (σ_1) and ultimate lateral stress σ_3 to be principal stresses, equilibrium of the wedge requires (2)

$$q_{ult} = \sigma_3 \tan^2 \beta + 2c_{avg} \tan \beta \quad (4)$$

where: $$\sigma_3 = \frac{\gamma_c B \tan \beta}{2} + 2c_u \quad (5)$$

wherein:

γ_c = wet unit weight of cohesive soil

B = foundation width

β = inclination of the failure surface with the horizontal = $45 + \frac{\phi_{avg}}{2}$

ϕ_{avg} = $\tan^{-1}(\mu_s a_s \tan\phi_s)$ with μ_s, a_s and ϕ_s as previously defined

c_u = undrained shear strength of unreinforced cohesive soil

c_{avg} = composite cohesion on the shear surface beneath the foundation = $(1-a_s)c_u$

For the Processing Center, the area replacement ratio under the footings was calculated as $a_s = 0.39$. A stress concentration factor of n = 4 was used, as lesser values were considered too conservative owing to the sand fill surrounding the upper portions of the stone columns. Values of $\phi_s = 40°$, c_u = 500 psf (24 kN/m^2), and a buoyant unit weight of 27.6 pcf (4.3 kN/m^3) were also used. The factor of safety against ultimate load was calculated at 2.3. Although this is slightly lower than desired, it does not accurately reflect conditions at the Processing Center site where the upper six ft (2m) of material surrounding the stone columns is granular fill that would be in a dense state following the Vibro-Replacement process. Therefore, for this project, this mode of failure was not considered critical as the shear resistance in this upper material surrounding the stone column would be greater than in the lower fill.

For floating stone columns without support on a firm base, the ultimate capacity against punching failure must also be determined. This can be accomplished by calculating the end bearing capacity of the stone columns using conventional bearing capacity theories and adding the skin friction load developed along the side of the column. The placement of stone columns is expected to cause remolding of the hydraulic fill adjacent to the stone columns. Such remolded clays are liable to consolidate further under their own weight. The installation of the free draining stone columns is expected to affect this consolidation adjacent to the columns during and shortly after installation. During loading of the stone column it is calculated that up to four inches (10 cm) of movement is required to cause a bearing capacity failure in the punching mode, whereas movements of less than 0.5 inch (1.2 cm) will develop positive skin friction between the stone and the soil. Therefore, prior to bearing capacity failure, positive side friction will be developed and can be considered in the analysis of punching failure.

For this analysis similar soil parameters as above were used resulting in a factor of safety of 2.8. A factor of safety of at least two was considered adequate. Again, the upper sands contributed significantly, as they accounted for much of the skin frictional load.

Settlement

Several methods of calculating stone column reinforced settlements are available including approximate methods with simplifying assumptions (equilibrium method) and more sophisticated methods based on elasticity and/or plasticity theory in which material and boundary conditions can be modeled by finite element methods. The equilibrium method assumes equilibrium is maintained within the unit cell, vertical displacement is equal within the stone and soil, and a uniform vertical stress due to external loading exists throughout the length of the stone column (2). With this method, the level of improvement is dependent on the stress concentration factor n, the initial effective stress in the clay, and the magnitude of applied stress. Considering consolidation theory, the settlement ratio, S_t/S, representing the ratio of settlement for treated versus untreated soil is as follows assuming:

σ_o = average initial effective stress in the clay
σ = the applied stress and μ_c is as previously defined

$$S_t/S = \frac{\log_{10}\left(\frac{\sigma_o + \mu_c\sigma}{\sigma_o}\right)}{\log_{10}\left(\frac{\sigma_o + \sigma}{\sigma_o}\right)} \quad \text{.................(7)}$$

Based on a stress concentration factor of four and a_s of 0.39 underneath the footings, μ_c is 0.46. Using the consolidation theory for the treated zone of hydraulic fill beneath the end dumped sand (average 10 ft), the calculated settlement was 2.2 in (5.6 cm) under a loading of 1500 psf (72 kn/m^2). Applying Eq 7 resulted in a settlement ratio of 49% and estimated settlement for the treated zone of 1.1 inches (2.8 cm). Based on consolidation theory, the untreated zone below the stone columns was estimated to settle 0.84 inches (2.1 cm) for a total predicted footing settlement from the column loads of 1.94 inches (4.9 cm). Secondary compression settlements in the untreated zone were estimated to be about one inch

(2.5 cm), occurring over a 20 year period.

Based on the stone column configuration shown in Figure 2, an area replacement ratio of 0.10 was calculated for the floor area with a settlement ratio S_t/S = 0.79 for the stone column improved soil. Total settlement under a load of 160 psf (7.7 kn/m^2) due to compacted fill and floor loads was estimated at 1.0 inch (2.5 cm) within the improved soil and 2.0 inches (5 cm) within the unimproved soil assuming the hydraulic fill to be normally consolidated. Secondary compression settlement of about one inch (2.5 cm) was anticipated to occur over a 20 year period. Settlement was considered negligible within the surficial granular fill densified by the stone columns. If the hydraulic fill was slightly preconsolidated due to previous dewatering, it would reduce the above settlement estimate considerably.

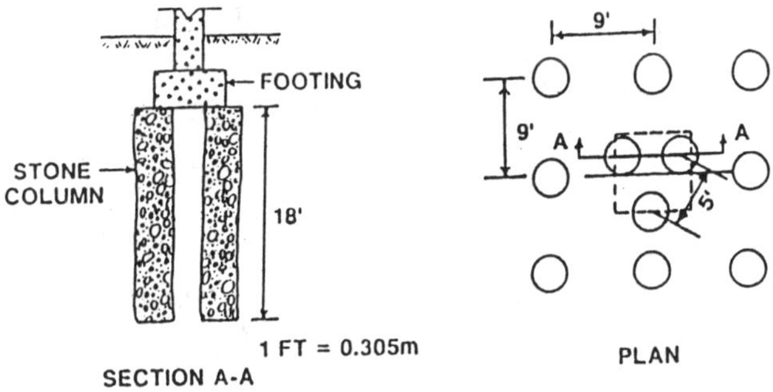

Figure 2 - Stone Column Configuration

Half of the total settlement was expected to occur under the surcharge load prior to construction. Compacted fill and surcharge load amounted to about 375 psf (18 kN/m^2) and were placed approximately three months prior to footing construction. Settlements of less than 1.25 inches, (3.2 cm) were recorded prior to stone column construction due to the fill and surcharge loading. An important design consideration was to improve the soil and provide for more uniform deformation characteristics. The installation of the stone columns created a composite matrix in the upper layer of soft silts which minimized the differential settlements of columns and floor slabs.

<u>Load Testing and Installation Criteria</u>

Prior to installing production stone columns, five control stone columns were put in throughout the site to compare stone volumes and amperage buildup at various locations. A location was chosen for load testing considering the stone columns with lesser amperage buildup. An area load test was performed using a five ft (1.5m) square plate centered between a two-stone column group. The plate was loaded at the proposed footing depth to 50 tons (445 kN), representing twice the design bearing pressure of 2000 psf (96 kN/m^2). Results of the load test are indicated on Figure 3, and indicate gross settlement at twice the design bearing pressure of 0.6 in (1.5 cm). This settlement represents both elastic and primary consolidation settlement. Movement of 0.18 in (0.5 cm) was recorded during the 13 hour holding period, with settlement rate to below 0.006 in/hr (0.015 cm/hr). Because this test represented a short term loading, the secondary compression did not contribute to the recorded settlements.

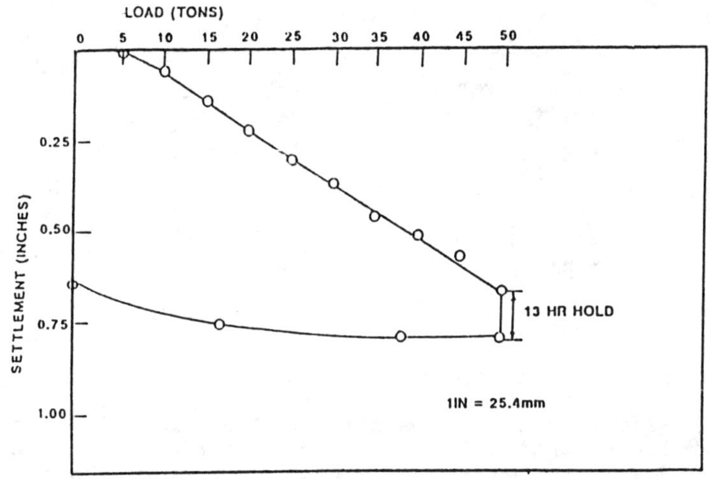

Figure 3 - Load vs Settlement For Footing Area Load Test

Installation criteria were set based on the load test results in that a minimum stone volume of seven cubic yards (5.4 m^3) per total length and a minimum amperage increase of 15 amps over the free standing draw of 130 amps of the vibrator was required, similar to the successfully load-tested stone columns. Since stone columns were all the same length, the volume requirement represented a minimum diameter criteria of 3.6 ft (1.1 m).

Where amperage buildup was not achieved for stone columns beneath footings, additional stone was added until the amperage criteria was met or a maximum additional stone volume of three cy (2.3 m^3) was installed per column, giving a maximum diameter of 4.4 ft (1.3 m). A total of only 14 of the 236 stone columns supporting footings required additional stone. A total of 638 stone columns were installed on the project.

Construction Procedures

Because of environmental considerations and because the site did not have a sufficient source for the volume of water required, the wet method of stone column installation (typically used below the water table to keep the hole open) could not be used. Thus, to ensure a continuous stone column, the dry method of stone column installation was used, but with a bottom feed system for delivering the stone to the bottom of the column. Following creation of the hole to planned depth using a 20 in (51 cm) diameter vibrating probe, stone was fed to the hole in 1.4 cy (1.1 m^3) charges by means of a slave crane carrying a small hopper. The crushed stone gradation for stone column construction is shown below.

Particle Size (in)	Particle Size (mm)	Percent Passing By Weight
1½	38	100
3/4	19	50-75
3/8	9.5	15-40
1/4	6	0-25

Equipment

Two rigs were used for installing production stone columns. An average production rate of 20 stone columns per day per rig was attained. Due to concrete obstructions in the surface fill, the majority of stone column locations were preaugered.

Monitoring

Three settlement plates were used to monitor the settlements produced by the compacted fill and surcharge loads. Settlements measured during stone column construction stabilized at 0.25 to 1.25 inches (0.64 to 3.2 cm) six weeks after fill placement, as indicated by Figure 4.

Settlement plate heave was noted as stone column construction proceeded near the plates. Six heave points were also installed to monitor heave. Several inches of ground heave were observed as indicated on Figure 4. Ground surface heave was monitored and generally dissipated in about two weeks for a given area of the building.

Post construction settlements were monitored at six locations on the floor slab, and at one footing. Settlement measurements taken up to 2.5 years after building completion indicate total settlements of 0.91 to 1.24 in (2.3 to 3.1 cm) for the floor slab, and 1.27 in (3.2 cm) for the footing. These values essentially represent primary consolidation settlement and secondary compression of the lower untreated hydraulic fill. Owing to the drainage afforded by the stone columns, primary consolidation settlements within the treated zone were estimated to have occurred under the preload before above-grade construction began. Data is not sufficient to determine when primary consolidation settlements in the lower fill had ceased.

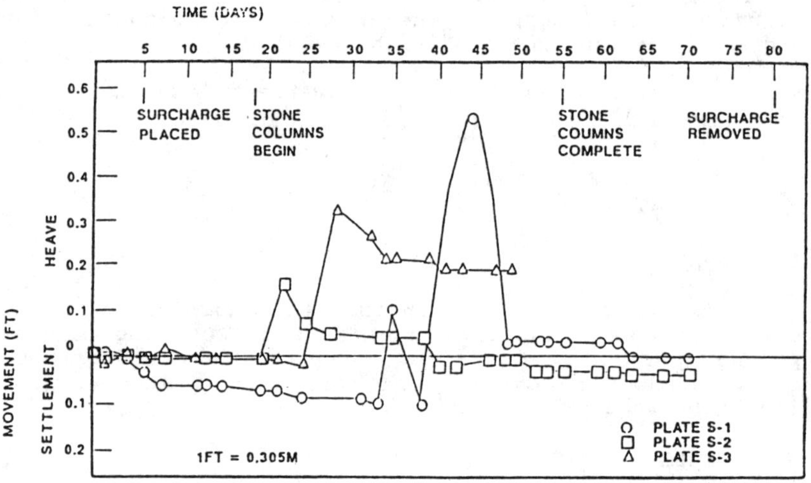

Figure 4 - Settlement/Heave Readings vs Time

Considering both preload settlements and post-construction settlements, the total foundation settlement was recorded as 2.5 inches (6.3 cm). The predicted total settlement of 6 inches (15.2 cm) overestimated this settlement. The overestimate is believed to be due to the assumption of normally consolidated hydraulic fill, whereas previous dewatering has likely provided a slight preconsolidation of the fill. A preconsolidation of only 100 to 150 psf (4.8 to 7.2 kN/m^2) in excess of previously existing overburden pressures would be enough to account for the differences in predicted and actual settlement.

Conclusions

Floating stone columns can be an acceptable foundation support method for moderately loaded buildings in deep compressible soils. Consideration must be given to punching failure at the base of the stone columns as well as evaluation of bulging failure and shear failure of stone column groups. Settlements of both the improved soil layer and underlying unimproved soils should be predicted.

The dry method of stone column construction has proven to be successful even in soft soils with a high water table by use of a bottom feed system. However, because the dry method displaces the soil, heave of the general ground surface will occur. Thus heave should be monitored during and following stone column construction.

Acknowledgements

The writers wish to express their appreciation and gratitude to the Public Service Electric and Gas Company; George Evans Associates, Structural Engineers; and Donkervoet, Tipton Architects for their confidence in the foundation system and their assistance in development of the project.

References

Aboshi, H. Ishimoto, Enoki, E. M., and Harada, K. (1979). "The Composer - a method to improve characteristics of soft clays by inclusion of large diameter sand columns." Volume I, Paris, 211-216.

Barksdale, R. D., and Bachus, R.C. (1983). "Design and construction of stone columns." Volume 1, Federal Highway Administration, FHWA/RD-83/026.

Baumann, V., and Bauer, G.E.A. (1974). "The performance of foundations on various soils stabilized by the vibro-compaction method." Canadian Geotechnical Journal, Volume II, 518-530.

Chambosse, G., and Dobson, T. (1982). "Stone columns I - estimation of bearing capacity and expected settlement in cohesive soils." GKN Keller, Inc., Tampa, Florida.

Hughes, J.M.O., and Withers, N.J. (1974). "Reinforcing of soft cohesive soils with stone columns." Ground Engineering, Volume 7, No. 3, 42-49.

Design and Performance of Two Port Silos on Improved Ground

M.U. Ergun[1], M. ASCE

Abstract

Two reinforced concrete grain silos were constructed on improved ground. The paper summarizes the description and profile of the foundation soils, the loading, the expected settlements, the type of improvement selected, the control tests and the assessment of the improvement done. Settlement records of the two silos are presented during their first loading. Static cone penetration test proves very useful both for control of the improvement works and for the estimation of settlements. Several liquefaction criteria are studied. Vibro-compaction and vibro-replacement methods are found efficient methods of ground improvement.

Introduction

Four reinforced concrete grain silos were constructed in the harbours between 1987-1990 in Turkey. Two of them were placed on improved ground due to better economy and suitability. The basic criteria for an acceptable foundation have been considered allowable amount of total and differential settlements and low liquefaction potential of the soils provided that there is not a bearing capacity problem. The natural strata did not satisfy these, and improvement methods were applied. Vibro-compaction was the method selected for the Bandìrma silo, both vibro-compaction and vibro-replacement methods were used at the Iskenderun site. The improved ground was evaluated against the above criteria. Static penetration test was the primary tool utilized to control the increased moduli.

---

1 Lecturer and consultant, Civil Engineering Department, Middle East Technical University, Ankara, Turkey.

A special program of settlement monitoring and controlled filling of the cells was conducted during their first loading. The purpose of the paper is to contribute to the knowledge on deep compaction by vibrofloatation and stone-columns through two case histories which present recent data concerning the pre and post-ground conditions, improvement patterns and settlements.

The Bandìrma Silo

The Bandìrma silo is located on the Southern coast of the Marmara sea. It has been designed for a 200 000 kN capacity at the first stage, and will be expanded to double capacity in the near future. There are two blocks on raft containing nine cells, each having dimensions of 27.8x27.15 m. Each block has a 100 000 kN design capacity, and the dead load of each block is 70 000 kN. Hydraulic fill was placed between perimeter quay walls during construction of the harbour. The typical soil profile is shown in Fig.1 where standard penetration numbers have also been plotted.

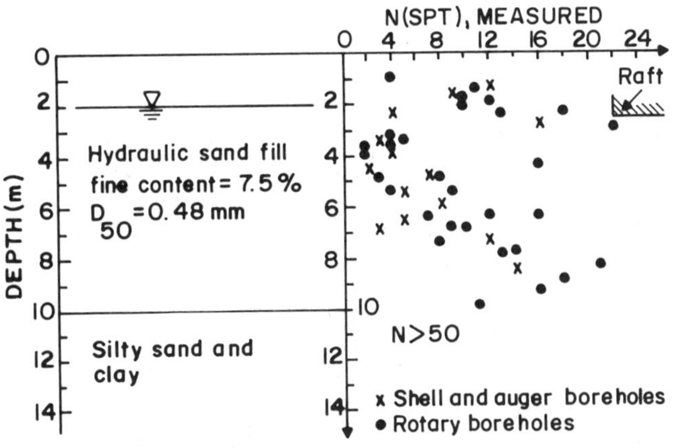

Figure I. Soil Profile and SPT Results, Bandırma Site

Static cone penetration tests were performed as a part of design soil investigation and before the improvement as reference tests. Test results are summarized in Fig.2. It can be seen that the hydraulic fill is thick (7-8 m) and loose. It is a silty medium to fine sand, and the silt content varies.

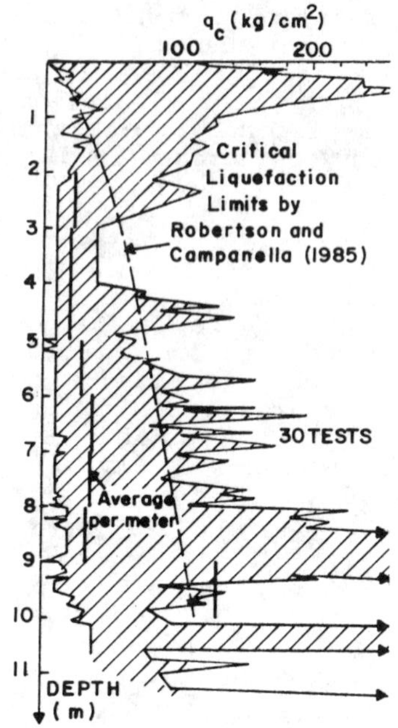

Figure 2. Envelope for the Cone Penetration Tests Before the Improvement, Bandırma Site

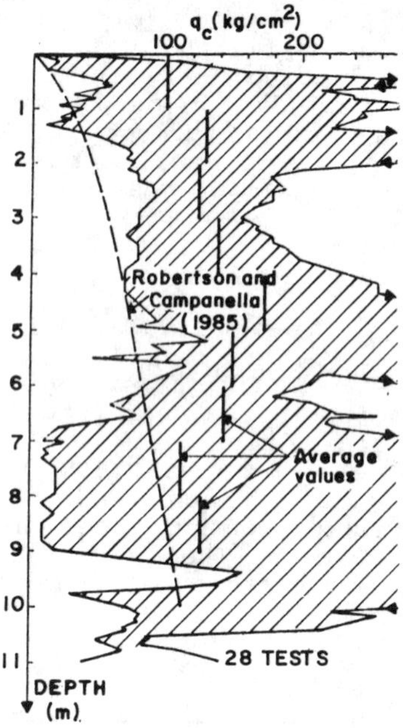

Figure 3. Envelope for the Cone Penetration Tests After the Treatment, Bandırma Site

Settlement

Settlement analysis based on the penetration profile in Fig.1 yields a maximum value on the order of 230 mm. Very silty pockets of hydraulic fill are expected to be compressible and to have low constrained moduli. The hydraulic fill has been assigned a constraint moduli between 5000-9000 kN/m². The stiff clays and compact sand and gravel layers starting at a 9 m depth do not significantly influence settlement results. Foundation gross pressures are 93 kPa and 223 kPa at empty and full load states. The adjacent headhouse unit has a deep basement and almost rests on the stiff parent soil layers, and exerts a gross foundation pressure of 175 kPa.

Improvement

Deep compaction of the hydraulic fill by vibro-compaction was the improvement method selected. Silty zones or pockets would necessiate use vibro-replacement technique due to the inefficiency of vibrations to densify silt or clayey silt. The extent of such zones was greater than was estimated during the site investigation. A cost comparison showed that deep compaction of 8-9 m thick hydraulic fill was considerably cheaper than the pile solution.

A triangular pattern and 1.8 m spacing were selected for the whole foundation area and there were 1100 application points. The diameter of the 120 kW vibrator was 0.42 m. The diameter of the completed columns was 0.80 m. A 4 m zone around the silo raft was also improved, and the width of this zone was increased to 8 m on the side of the rail tracks. A total of 9000 m^3 of added material was used in 8500 m of treatment. The backfill material used for stone-columns was 20-70 mm crushed rock. 0.2-30 mm sand-gravel mixtures have been used for vibro-compaction in the both projects.

Design of Vibro-compaction

Several methods are applied to densify the soils so that the relative density achieved decreases settlements under loading, and liquefaction risk is substantially reduced. Various criteria have been proposed for design against liquefaction, and these will be summarized in the following sections. Each requires a certain density profile. Settlement computations also require density-compressibility correlations. Improved densities will give higher moduli and will lead to lower settlements.

Vibro-compaction technique aims at the mass densification of loose to medium dense cohesionless soil layers. Compaction spacings to achieve predetermined values of relative density or bearing pressure are based on penetration tests or other in-situ tests. Greenwood and Thomson (1984) and Thorburn (1976) proposed design charts based on such observations. These are, however, related to a specific type of vibrator and soil, and it is possible to obtain different correlations with different vibrators in different type of soils. Therefore, such design charts must be regarded as general guides. Extensive field data are available to the specialist improvement companies. The spacings and the patterns of the vibrofloatation points have been selected on the basis of the general design recommendations above and the company's experience with the vibrator.

Quality Control

Static cone penetration tests have been performed at locations which are equidistant from three compaction points after treatment. This location gives a clear indication of the densification achieved throughout the depth. It has been observed that silty and clayey zones are less affected by the vibro-compaction process. Quantity of added backfill material and power consumption (ampere recording) at each compaction point are additional indicators about type of soil encountered and compaction achieved. Usually liquefaction criterion is tighter than that of static settlements. An 80% complience of the minimum requirement with the target profile is acceptable. That means at most two profiles may fail to satisfy the target profile in ten control tests. Fig.3 shows the envelope of 28 such tests. Some of the tests are almost at the same locations as the pre-treatment test locations. Critical liquefaction q_c limit is also plotted. It is observed that the selected pattern is effective in increasing the density of the fill.

Liquefaction

Liquefaction potential of the hydraulic fill was a significant issue. Initial cone penetration resistances can be seen in Fig.2. Several studies on liquefaction of sandy soils have been reviewed and both the standard penetration numbers and the static cone end resistances indicate the hydraulic fill is highly liquefiable.

The largest earthquakes in the vicinity of Bandìrma are Gönen-Yenice of 18 March 1953 and Manyas of 6 October 1964, of magnitude 7.3 and 7.0 respectively. These events occurred at minimum fault distances from Bandìrma of 30 and 27 km respectively. Ambraseys (1988) gives 0.16 g and 0.20 g maximum accelerations for magnitudes of 7.1 and 7.5 respectively for a fault distance of 30 km.

Since the performance control was monitored by cone penetration tests, these values were compared with limit values for liquefaction in various references given below. The influence of effective vertical overburden pressure was taken into account, and the end resistance values (q_c) were modified (q_{c1}). Cyclic stress ratios (Seed, 1979) were calculated for all levels in the hydraulic fill using the maximum acceleration of 0.2 g, and critical limit values of q_{c1} were obtained as in Seed et.al. (1983), Ishihara (1985), Robertson and Campanella (1985), Shibata (1987), Seed and Alba (1986)(Fig.4). A correction factor c_q was then used to convert q_{c1} to q_c values which represent measured values. Such q_c vs. depth profiles can be prepared for all criteria presented in Fig.4. The criterion by Robertson and Campanella (1985) is the most conservative, and this criterion has been used to prepare the limiting q_c-depth profiles for liquefaction. Similar curves by other criteria will be under this curve. The relationships

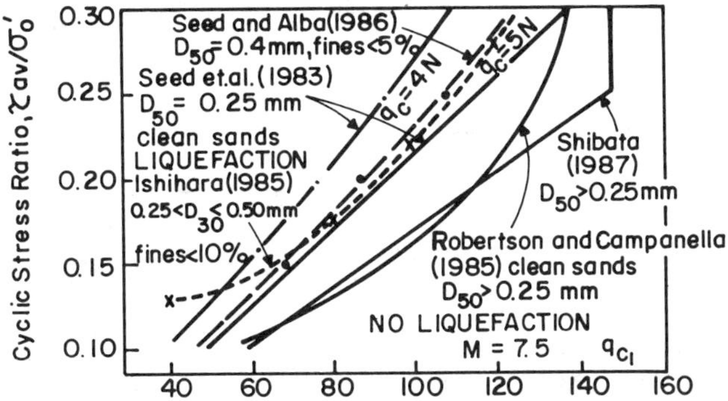

Figure 4. Various Liquefaction Criteria Versus Modified Cone Penetration Resistance (kg/cm²) q_{c_1}

presented in Fig.4 are for clean sands which have fine contents less than 5%. The hydraulic fill has an average fine content of 7.5%, and this adds further conservatism, and yet it is quite common to apply safety factors of about 1.5 to the cyclic stress ratios in using the relationships in Fig.4, and it has been also applied in this project.

Settlement Observations

The test loading with grain and the corresponding settlements are given in Fig.5. The loading was done as uniform as possible to prevent any differential settlements, but due to different types of grain there have been deviations from uniform loading. An estimate settlement was done prior to construction for the improved ground under full load of the silo and the dead weight of the silo making use of expected end resistances after improvement. An average value of q_c =100 kg/cm² was predicted for the hydraulic fill after treatment resulting in average total settlements of 26 mm and 57 mm for the empty and the full load cases respectively. A 20 m thick soil layer was considered and the end resistance values for the deeper compact layers were assumed at least 25 MPa. Unfortunately the settlement records for the dead load of the silo are not available but the observed average settlement of 21 mm for the grain loading can be compared with the previously calculated 31 mm (57-26). The actual cone resistance of the improved hydraulic fill is around 13.5 MPa (See Fig.3), and recalculation of the settlement for the grain loading making use of measured cone resistances after treatment would give the average measured settlement. The final settlements of silos are shown in Fig.5.

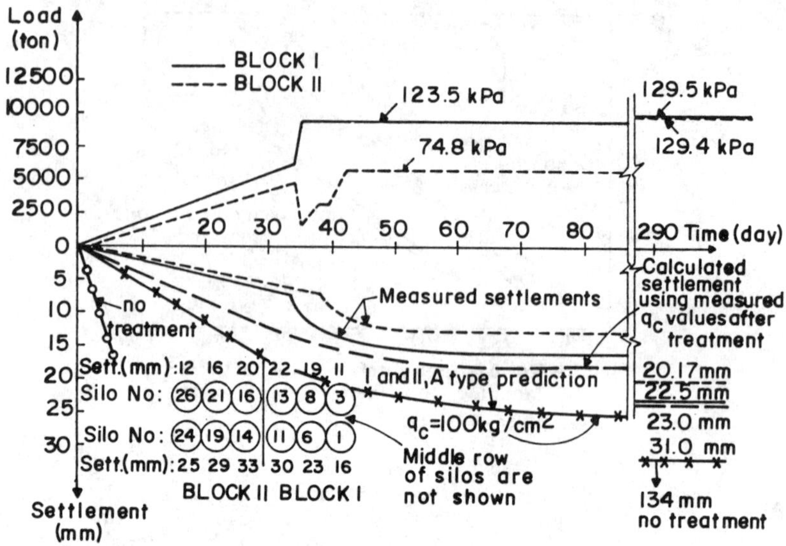

Figure 5. Load-Settlement Behavior of the Bandırma Silo, First Time Grain Loading Only (3 Months)

The Iskenderun Silo

General Description and Soils at the Site

The new Iskenderun silo has 400 000 kN capacity, and is located at the East Mediterrenian coast of Turkey. The Iskenderun harbour is one of busiest harbours in the Country. The new reinforced concrete silo has been constructed adjacent to the old silo which was constructed in 1954, and it has 200 000 kN capacity.

The typical soil profile and standard penetration test blow count profile are shown in Fig.6. Between 1 m and 4 m depth there are very compressible organic layers. Theg upper sands between 4-8 m are denser. Initial static penetration test profiles are shown in Fig.7 in form of end resistance envelopes. Such envelopes throughout the depth after the treatment are given in Fig.8. The liquefaction criterion based on Robertson and Campanella (1985) is plotted on both figures.

Settlement

The gross raft pressures are 93 kPa and 263 kPa for the empty and the full load cases. Average raft settlements may be either calculated assuming flexible loading and then converting to rigid raft settlements or directly calculating the settlements at characteristic points under flexible loading which correspond to average raft settlements. A 20 m depth of soil has been assumed compressible under the silo loads. The calculated average raft settlements for the empty and full load cases are 70 mm and 185 mm respectively. Since there are two expansion joints the three silo blocks will settle separately, and the side blocks will settle somewhat less than the central block, and there is a slight tilting action due to differences between the edges and towards the central block. This difference is expected at the order of 10 mm. Out of 185 mm total settlement 115 mm has been calculated due to a few meters thick silty organic clay at the upper levels. Constrained moduli have been assigned to various levels based on static cone penetration tests, and the compression has been assumed to occur in one dimension. The average idealized end resistance values throughout the depth have been multipled by 3.5 in all sand layers and by 7 in the silty clay layer to find constrained moduli Schmertmann (1978), Trofimenkov (1974).This modulus is a one-dimensional compression modulus.

Liquefaction

Study of the Iskenderun area and possible epicentral distances indicate a value of maximum acceleration of 0.15 g in the case of an earthquake having a magnitude of 7.5. The cyclic stress ratios have been calculated at every meter depth and the end resistance profile based on Robertson and Campanella (1985) has been plotted using the relationship in Fig.4 and making corrections for the overburden pressure. Other criteria in Fig.4 will end up with smaller q_c values. It is seen in Fig.7, that sands between 8-13 m depths have some potential to liquefy. On the other hand 12% fine content of this layer and the depth level considered may be regarded as advantages. The company responsible for the design recommended the improvement of this layer (8 to 13 m) reasoning that a factor of safety of 1.5 should be applied to the cyclic stress ratios. Considering the conservatism in the Robertson and Campanella (1985) approach and the fine content of the sand this decision may be open to discussion. The upper sands between 4-8 m are dense.

Improvement

A two-level improvement was planned. The upperlevel

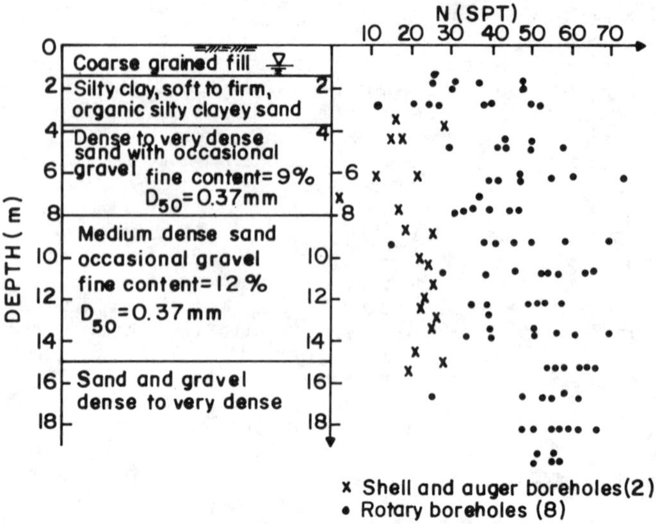

Figure 6. Soil Profile and SPT Results, İskenderun Site

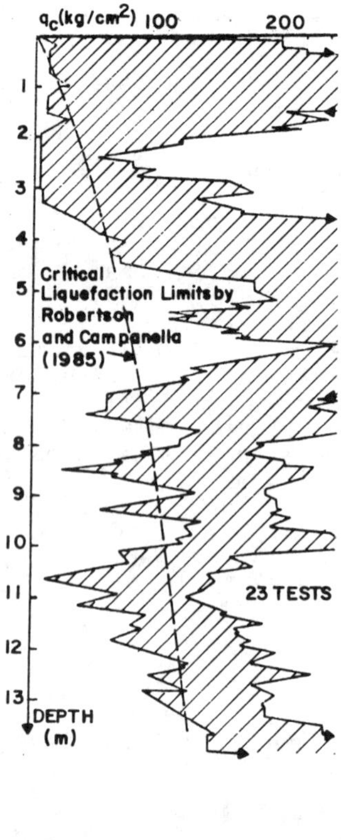

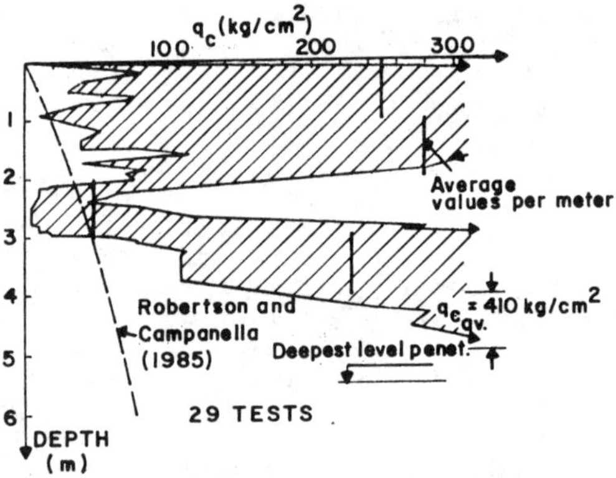

Figure 8. Envelope of the Cone Penetration Tests After the Treatment, İskenderun Site

Figure 7. Envelope of the Cone Penetration Tests Before the Treatment İskenderun Site treatment consists of 1.2x1.2 m square grid of vibro-replacement (stone-columns) 5 m deep resting on dense sands, and the lower treatment was designed as vibro-compaction extending the columns of the upper layout down to 13 m depth skipping one column in both directions i.e. 2.4mx2.4m square spacing. In other words the area was covered by 1.2x1.2m patterned stone columns from the ground surface to a depth of 5 m and 2.4x2.4 m spaced vibro-compaction probes between 8 to 13 m depth. Therefore half the stone-columns were based on the vibro-compaction probe locations. The upper treatment of stone-columns aims at the reduction of settlement in the soft, organic clay layer. The lower treatment down to 13m is vibro-compaction and, it is performed to eliminate the risk of liquefaction. There was also a peripherial treatment 4 m wide.

Design of Stone-Columns

There is no problem of liquefaction in cohesive soils, but on the other hand settlements may reach unacceptable levels. Vibro-replacement or stone-column design philosophy differs from that of vibrocompaction in that discrete sand, gravel or crushed stone columns are formed in the soil which is cohesive in most applications. The aim of the improvement is basically to reduce the settlement of the cohesive soil under foundation loading. The cohesive or fine soil between the columns is largely untreated unlike in the case of vibro-compaction. However, the fine soil gives lateral support. Settlement rate is also improved. Different approaches are available for the design of stone columns e.g. Priebe (1976), Balaam and Poulos (1983) and Van Impe and De Beer (1983).

Priebe's (1976) reduced stress approach has been employed in the Iskenderun case. It is basically an elasticity solution and the column is assumed rigid-plastic. The ratio between the tributary area corresponding to a single stone-column and the area of a single stone-column is called the area ratio. This ratio and the strength of the aggreate in the compacted state determine the settlement reduction ratio which is defined as the ratio between the settlement of the ground in the untreated and treated states.

Observed Settlements

The settlements during the initial loading of the silo were measured. Records up to 60% loading have been plotted in Fig.9 for the three blocks. The readings are average for each block. The actual values taken from the side silos are also given in the figure. The settlements corresponding to the dead load of the silo (93 kPa) were not available.

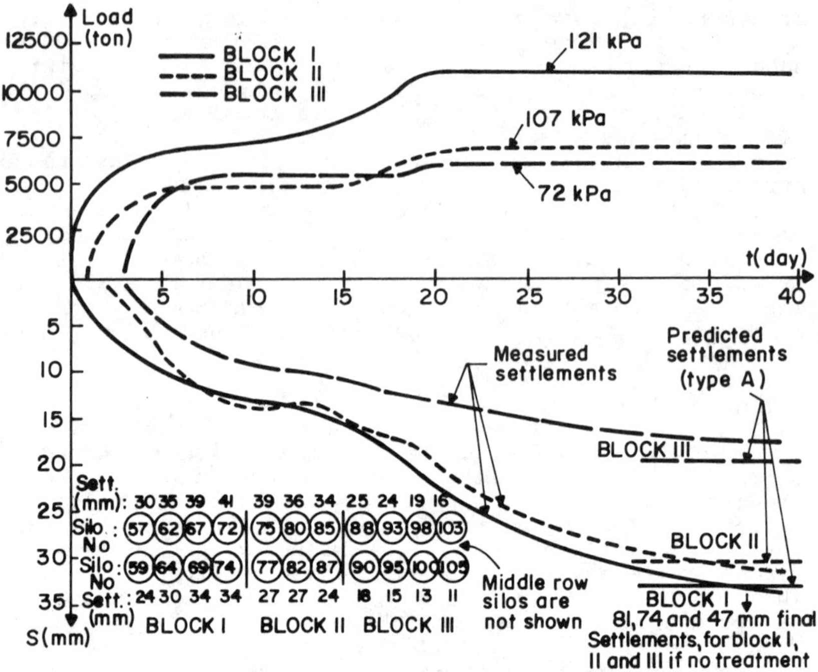

Figure 9. Load-Settlement Behaviour of the İskenderun Silo, First Time Grain Loading Only

To evaluate the observed settlements the following procedure was followed. The average raft pressures at 60% loading including dead loading (93 kPa) are 214, 200 and 165 kPa for blocks 1, 2 and 3 respectively. The average settlements of the blocks 1,2 and 3 have been calculated 150 mm, 145 mm and 116 mm respectively. The average settlements for the dead loading at each block is around 70 mm. These figures have been obtained for the untreated profile making use of the initial CPT profile. Therefore if the foundation soils were not treated the average settlements of the blocks 1,2 and 3 due to grain loading only at 60% capacity would be 81, 74 and 47 mm respectively. Settlements of structures on stone-columns in cohesive soils are usually predicted by settlement reduction factors. It is indeed very interesting to see the Priebe's (1976) correction factor of 4 for the upper layer (5m) where major part of the settlements originate explains the present observations, i.e. area ratio is 1.2x1.2/0.5=2.88 and the angle of shearing resistance for the crushed stone used is 43°, and the particle sizes are between 20-70 mm. The settlement of the lower layers (5-20 m) have been calculated by assigning 50% increase in CPT values due to the treatment. The final calculated settlements of the blocks 1,2 and 3 on the treated foundation soils are 33, 31 and 20 mm respectively, and such a picture may create doubts about this class A type of prediction !. 50% increase in q values may be regarded rather arbitrary but lower levels already settle quite less compared to the upper layers.

Conclusion

Two reinforced concrete port silos were constructed on improved foundation soils. The following conclusions have been reached in these experiences:

1. Static cone penetration test proved very useful both for control of the improvement works and for the estimation of the settlements. Standard penetration test is prone to various errors and modifications. A few numbers of extra blows may change the interpretation. For example, it was not possible to detect the dense sands through standard penetration tests under the Iskenderun silo between 4-8 m depth.

2. It may still not be possible to predict the q profile in a specific soil exactly after ground improvement works are completed following a specific improvement pattern. A range may be possibly given, and there is still need for field data from improvement sites.

3. Several liquefaction criteria have been studied. Robertson and Campanella (1985) has been shown in the figures because it represents a conservative upper limit for q values. Selection of higher criteria applying safety factors to cyclic stress ratio may lead to close improvement patterns and more conservative design.

4. Time dependent settlements occur in stone-column treatment of very soft and soft clays due to lateral yielding of the columns as clearly observed in the Iskenderun case, but stone-columns reduce the total settlement of the ground in comparison with non-treated soil.

5. Sophisticated design procedures may be required if more than a single soil layer is to be improved or the soil profile shows variability at a site.

6. Vibro-compaction and vibro-replacement methods of ground improvement are efficient ways of preparing foundation soils to support structures, and may be cheaper alternatives to deep foundation solutions in many cases.

References

1. Ambraseys, N. (1988) "Engineering Seismology" Mallet-Milne Lecture, Journal of Earthquake Eng. Struct. Dyn., Vol.16.

2. Balaam, N.P. and Poulos, H.G. (1983). "The Behaviour of Foundations Supported by Clay Stabilized by Stone-Columns", Proc. 8 th European Conf. on Soil Mech. and Found. Eng., Helsinki, Vol.1, pp. 199-204.

3. Greenwood, D.A. and Thomson, G.H. (1984) Ground Stabilization Deep Compaction and Grouting, ICE Works Construction Guides, Thomas Telford Ltd., 47p, London.

4. Ishihara, K. (1985). "Stability of Natural Deposits during Earthquakes", Proc., 11 th International Conf. on S.M.F.E., Vol.1, pp.321-376.

5. Priebe, H. (1976) "Abschatzung des. Setzungsverhaltens eines durch stopverdichtung verbesserte Baugrundes", Die Bautechnik Vol.53, No.5, pp.160-162.

6. Robertson, P.K. and Campanella, R.G. (1985) "Liquefaction Potential of Sands using the CPT", Jour. of Geotech. Eng. A.S.C.E. Vol.111, No.3.

7. Shibata, T. (1987) "Discussion", Jour. of Geotech. Eng., A.S.C.E., Vol.113, No.6, pp.676-678.

8. Schmertmann, J.H. (1978). "Improved Strain Influence Factor Diagrams" Jour. of the Geot. Eng. Div., A.S.C.E., Vol.104, No.GT8.

9. Seed. H.B. (1979) "Soil Liquefaction and Cyclic Mobility Evaluation for Level Ground During Earthquakes, Jour of Geotech. Eng., A.S.C.E., Vol.109, No.3.

10. Seed, H.B. and DeAlba, P. (1986) "Use of SPT and CPT Tests for Evaluating the Liquefaction Resistance of Sands," Proc. of In Situ 86, A.S.C.E., pp.281-302.

11. Seed, H.B., Idriss, I.M. and Arango, I. (1983) "Evaluation of Liquefaction Potential using Field Performance Data", Jour. of Geotech. Eng., A.S.C.E., Vol.109, No.3.

12. Thorburn, S. (1976). "Building Structures Supported by Stabilized Ground", in Ground Treatment by Deep Compaction, The Inst. of Civil Eng., London, pp.83-94.

13. Trofimenkov, J.G. (1974). "Penetration Testing in Eastern Europes", Proc. European Symposium on Penetration Resistance, Stockholm, Vol.2.1, National Swedish Building Research, 1975, pp.24-28.

14. Van Impe, W.F. and De Beer, E.E. (1983). "Improvment of Settlement Behaviour of Soft Layers by means of Stone-Columns", Proc. 8 th European Conf. on Soil Mech. and Found. Eng., Helsinki, Vol.1, pp.309-312.

Effect of Soil Treatment on Dynamic Response of Foundations

M. H. Maher[1], A.M. ASCE and J. P. Welsh[2], M. ASCE

Abstract: This paper describes the results of an analytical study on the response of rigid and flexible foundations (single degree of freedom), resting on a treated segment of a half-space, subjected to steady-state source vibration. The study uses both analytical and finite-element methods to obtain the foundation response, and examines the effect of treatment type and geometry (i.e., depth and width) on the response amplitude. Treatment of soil significantly reduced the amplitude of vibration of both rigid and flexible foundations subjected to source excitation. The amount of reduction was influenced by the treatment geometry (i.e., width and depth of the treated zone) and treatment type.

Key Words: Soil Trteatment, Soil Dynamics, Soil-Structure Interaction, Foundation Vibration.

1.Introduction

Soil stabilization using admixtures and chemical grouts is an effective method for improving soil's engineering properties. In-situ stabilization techniques such as chemical grouting (Baker, 1982), compaction grouting (Mitchell, 1981), jet grouting, vibro-compaction and vibro-replacement (Welsh, 1986) are highly practiced in Europe and Japan, and are being increasingly used in the United States for a wide range of geotechnical applications. Compaction and chemical grouting techniques have been particularly successful in controlling settlement and enhancing the bearing capacity of foundations.

Although treatment of soils significantly enhances the bearing capacity under static loads, its effect on the response of foundations undergoing dynamic loads is yet to be investigated. This paper describes the results of an analytical study on the response of rigid and flexible foundations (single degree of freedom), resting on a treated segment of a half-space, subjected to steady-state source vibration. The dynamic response of the foundations were determined using both numerical (finite-element) and analytical procedures.

1- Assistant Prof., Dept. of Civil and Environmental Engineering, Rutgers University, Piscataway, NJ 08855-0909

2- Vice President, Hayward Baker Inc., Odenton, Maryland 21113

2. Effect of Treatment on the behavior of soils subjected to low strain, high frequency dynamic loads.

The effect of treatment on the low strain, high frequency dynamic moduli and damping of soils has been investigated by a number of investigators in recent years (Chae and Chang, 1978; Chang and Woods, 1987 and Li and Woods, 1987). Treatment types in these investigations have included admixtures such as cement, lime and fly ash, and chemical grouts such as sodium silicate. Based on the results from these works, it can be concluded that treatment significantly improves soil's dynamic shear modulus and damping under low strain, high frequency dynamic loads such as those encountered in machine foundation problems. A special observation is that the contribution of treatment to shear modulus decreases with increasing depth (Chang and Woods, 1987), with the reason being that the dynamic shear modulus at large depths is already high due to high confining pressures and the increase due to treatment is not as significant as that observed in shallower depths. This will help to specify optimum depths of treatment for a given treatment type. A relationship describing the effect of increasing depth on the increased shear moduli of soil, as a result of treatment by a number of different additives (Table.1), as obtained by Chae and Chiang, and Chang and Woods is represented in Fig.1.

Table.1 Treatment Types and Their Composition (Chang and Woods, 1987; and Chae and Chiang, 1978)

Type	Mixture Description	Mixture Proportion
I	Fly ash / Cement/ H2O	45:5:50
II	Lime / Cement / H2O	45:5:50
III	Sodium Silicate / H2O / Ethylacetate & Foramide	50:42:8
IV	Lime / Fly Ash / H2O	----

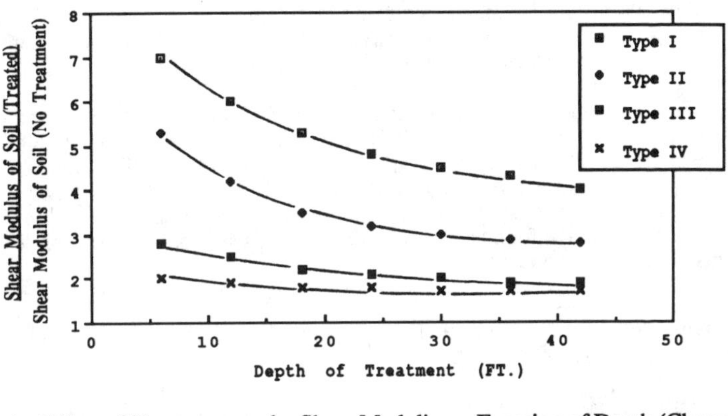

Fig.1.- Effect of Treatment on the Shear Moduli as a Function of Depth (Chang and Woods, 1987; and Chae and Chiang, 1978). (1 ft = 0.305 m)

3. Effect of Treatment on the Amplitude of Elastic Response (Rigid Foundation).

3.1. Method of Analysis: The effect of treatment on the response of a rigid strip footing is analyzed using a "complete" soil-structure interaction procedure. The rigid strip footing, which has an embedment depth of 5', is considered to be 10' wide and 10' high and resting on a 100' layer of soil overlaying a rigid bedrock. A schematic diagram of the footing and a corresponding treatment zone is shown in Fig.2. A finite element model was used to measure the amplitude of the response of the foundation undergoing steady-state machine type loading. The soil-structure interaction problem was thus classified as a "source problem."
The computer program DYNAFLOW (Prevost, 1985), a general purpose dynamic finite element program, was used to perform the finite element analysis. A two dimensional plain strain element was used and linear elastic material response was assumed for the low strain, high frequency loading condition. The finite element mesh used to model the foundation and underlying soil consisted of 258 nodes and two hundred twenty four 4-node isoparametric elements. Four rigid elements model the concrete footing (Young's modulus $E = 4.32 \times 10^{12}$ psf, Poisson ratio $\upsilon = 0.45$, mass density $r = 4.6$ lb.s^2.ft^{-4} (2.4 kN.s^2.m^{-4}) The remaining 220 elements model the underlying soil. Each layer of the soil elementwas defined with varying shear modulus which reflects the stiffening of the soil with depth. A Poisson ratio of $\upsilon = 0.35$ and mass density of $\rho = 3.31$ lb.s^2.ft^{-4} (1.73 kN.s^2.m^{-4}) was used throughout.

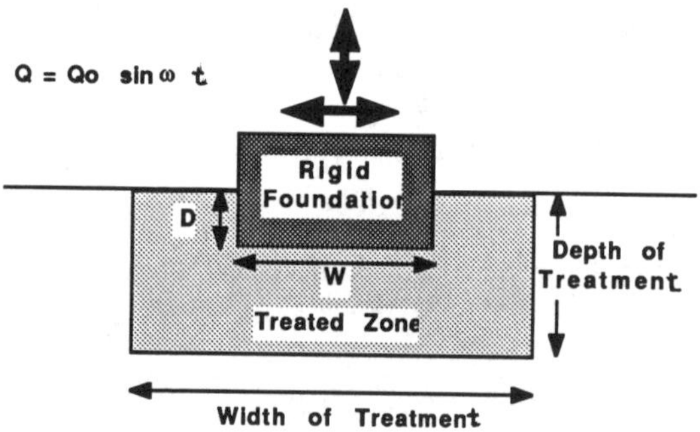

Fig.2 Schematic Diagram of a Rigid Foundation Resting on a Treated Zone

The magnitude of the shear modulus of the untreated soil was estimated using Hardin's empirical relationships (Hardin and Richart, 1963). The nodes along the base of the mesh were constrained from movement to model the rigid bedrock underlying the soil stratum. To model infinite domain in the horizontal direction, a standard viscous boundary developed by Lysmer and Kuhlemeyer (1969) was used. The viscous boundaries were placed at about one shear wavelength from the source of the excitation. The shear modulus of the treated segment of the foundation soil was obtained from Fig.1.
The harmonic loading for both cases of vertical and coupled rocking-sliding vibration was applied for one second, at 10 cycles / sec. with a maximum amplitude of 1600 lbs. A time step of 0.005 seconds or 20 steps per cycle of loading, was used. Time integration of the semi-discrete finite element equations was performed using an implicit Newmark method (Newmark, 1959) with integration parameters $\alpha = 0.55$ and $\beta = 0.28$.

<u>3.2. Effect of Treatment Geometry on the Response Amplitude.</u> The reduction in the amplitude of vibration in a rigid foundation due to treatment is greatly influence by the extent and the geometry of the treatment zone . In all the cases studied, increasing depth and width of treatment reduced the amplitude of response on the foundation. A typical frequency response diagram showing the effect of uniformly increasing width and depth of treatment on the amplitude of vertical vibration is presented in Fig.3. This effect, observed at the resonant frequency, is shown in Figs. 4 and 5 for vertical, and coupled sliding and rocking oscillations. Similar trends in these relationships were observed for the other treatment types which were studied.

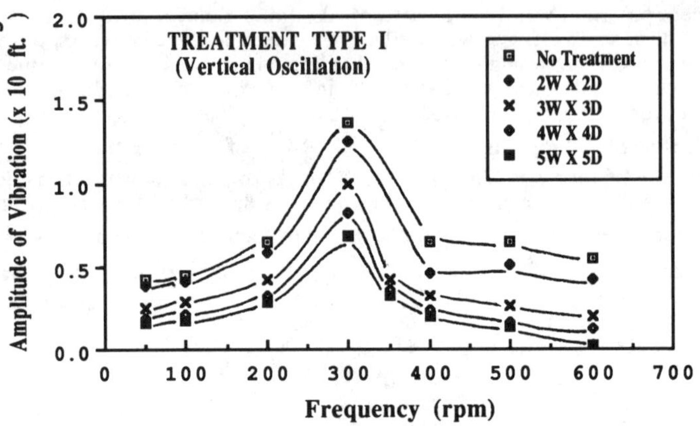

Fig.3.- Effect of Treatment on the Amplitude of Vertical Vibration (Treatment TypeI) (1 ft = 0.305 m)

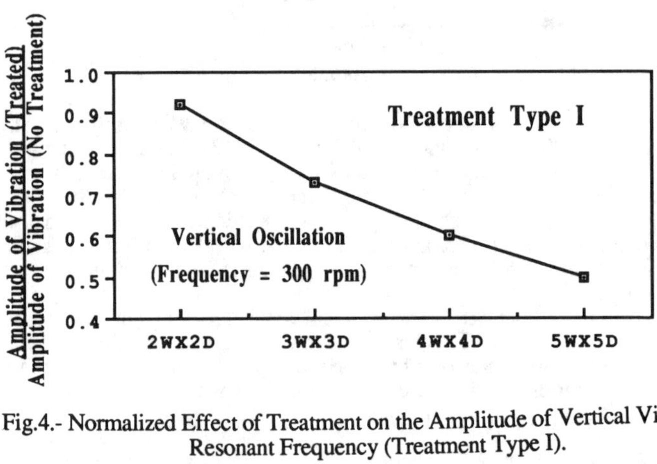

Fig.4.- Normalized Effect of Treatment on the Amplitude of Vertical Vibration, at Resonant Frequency (Treatment Type I).

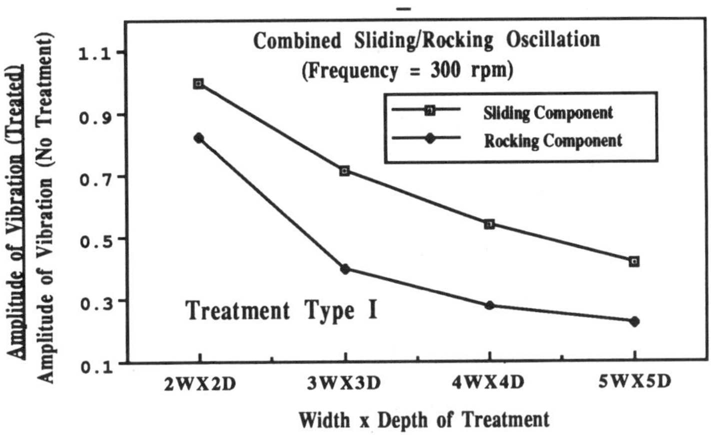

Fig.5.- Normalized Effect of Treatment on the Amplitude of Coupled Sliding/Rocking Vibration, at Resonant Frequency (Treatment Type I).

The effect of treatment was also studied as a function of increasing width or depth of treatment separately. The Influence of increasing width, with constant depth of treatment, is presented in Figs.6,7 and 8 for both cases of vertical and coupled sliding and rocking oscillation. As seen in these figures, increasing the width of treatment significantly reduces the amplitude of vibration of the foundation. This reduction is, however, more pronounced for rocking component of the coupled sliding/rocking type of vibration (up to 60% for type I treatment).

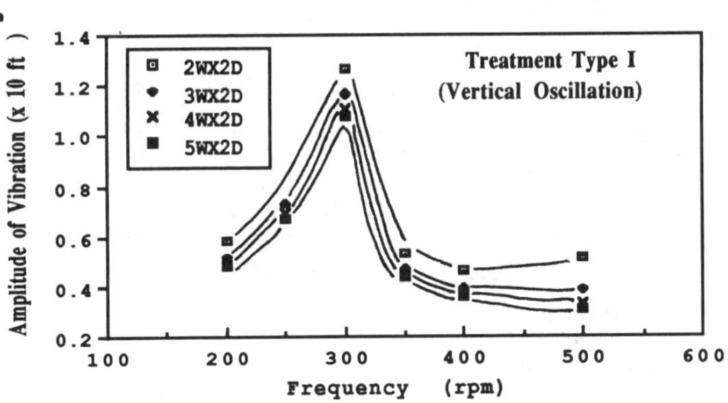

Fig.6.- Effect of Increase in Width of Treatment on the Reduction of Amplitude (Vertical Oscillation)
(1 ft = 0.305 m)

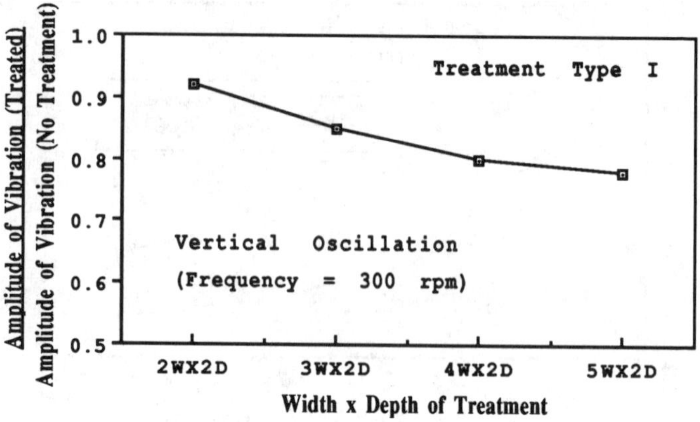

Fig.7.- Normalized Effect of Increase in Width of Treatment for Vertical Oscillation (at Resonant Frequency)

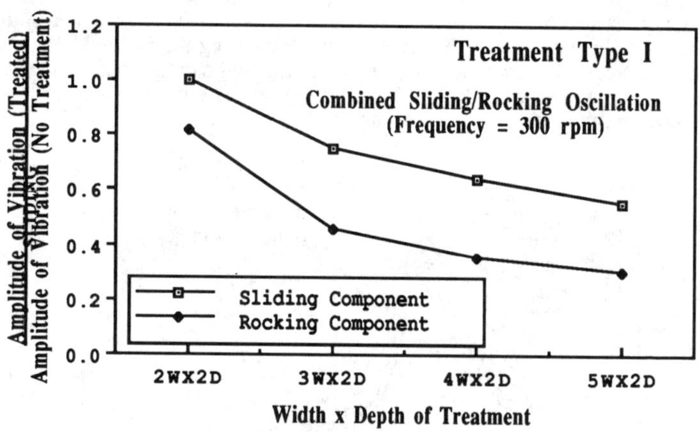

Fig.8.- Normalized Effect of Increase in Width of Treatment for Coupled Sliding/Rocking Oscillation (at Resonant Frequency)

The influence of increasing depth while keeping the width of treatment constant is shown in Figs. 9, 10 and 11. For the type of treatment shown in these figures, the increasing depth reduced the amplitude of vibration for both vertical and coupled sliding/rocking vibration. The magnitude of reduction, although significant, was less than that observed when increasing the width of the treatment.

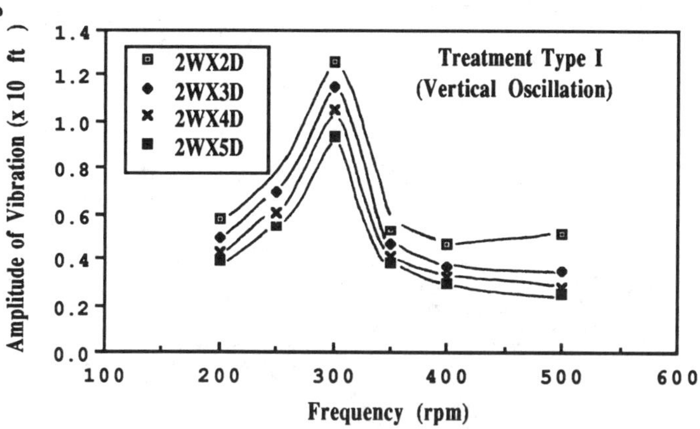

Fig.9.- Effect of Increase in Depth of Treatment on the Reduction of Amplitude (Vertical Oscillation) (1 ft = 0.305 m)

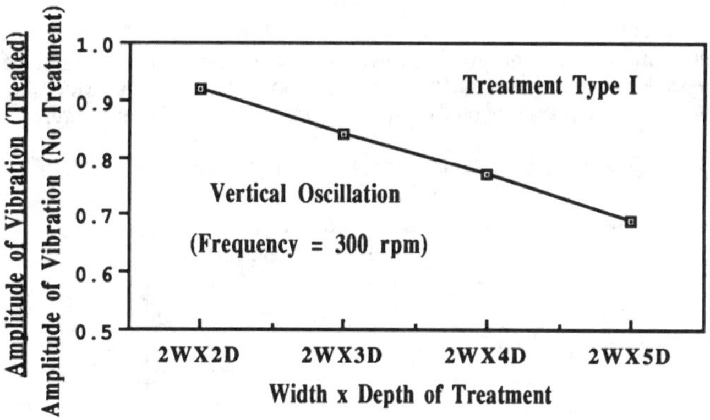

Fig.10.- Normalized Effect of Increase in Depth of Treatment for Vertical Oscillation (at Resonant Frequency)

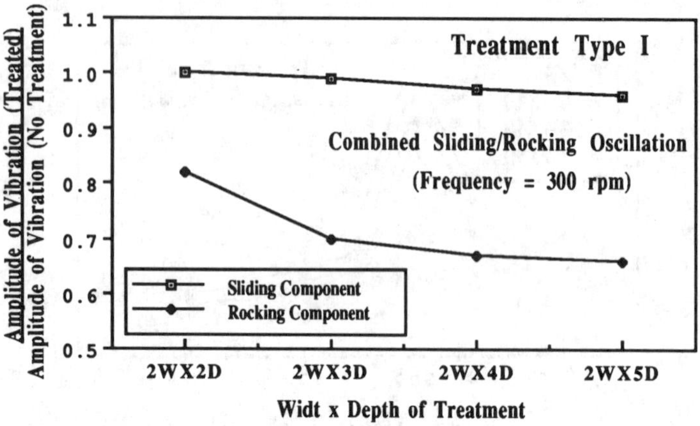

Fig.11.- Normalized Effect of Increase in Depth of Treatment for Coupled Sliding/Rocking Oscillation (at Resonant Frequency)

In all the cases studied, it was observed that the resonant frequencies of the treated and untreated systems were almost the same. To further collaborate this, an eigenvalue analysis of both a treated and an untreated system was performed. In this analysis the first ten fundamental frequencies of the untreated system were compared to those of 2W x 2D, 3W x 3D and 5W x 5D treated systems. The comparison showed no significant change in the magnitude of the fundamental frequencies.

3.3. Effect of Treatment Type on the Response Amplitude: The effect of treatment as a function of treatment type is shown in Tables 2 and 3. Although all of the treatment types studied significantly reduce the amplitude of vibration, the reduction is, however, more pronounced for type I which is a strong bond mixture.

Table 2. Effect of Treatment Type (at resonant Frequency) on the Reduction of response Amplitude (3W x 3D treatment)

Treatment Type	% Reduction in Amplitude		
	Vertical Oscillation	Coupled Sliding and Rocking Oscillation	
		Sliding	Rocking
I	27	29	60
II	27	27	57
III	18	22	47
IV	13	14	30

Table 3. Effect of Treatment Type (at resonant Frequency) on the Reduction of response Amplitude (5W x 5D treatment)

Treatment Type	% Reduction in Amplitude		
	Vertical Oscillation	Coupled Sliding and Rocking Oscillation	
		Sliding	Rocking
I	50	58	78
II	48	54	74
III	39	46	62
IV	25	34	43

4. Effect of Treatment on the Amplitude of Elastic Response (Flexible Foundation).

4.1. Method of Analysis: The effect of treatment on the response amplitude of a flexible foundation was determined using an analytical soil-structure interaction procedure. The computer program FLEXFOUN (Gucunski, 1991), which evaluates the elastic response of a circular flexible plate under vertical harmonic loading, was used to evaluates the impedance function, soil reaction and displacement distribution below the foundation. In this program, following the determination of the stiffness and the mass matrix of the flexible foundation, the influence coefficient matrix of the soil system is evaluated. Then both solutions are combined, using the compatibility of displacements at the plate interface, to obtain the response.

4.2 Effect of Treatment on the Response Amplitude: The schematic diagram of the flexible foundation resting on a treated segment of a half-space is shown in Fig.12. The foundation is a circular concrete plate with a radius, R, of 5.6 ft. and a thickness of R/20. The treatment type used was type I (Table 1.). The depth of treatment was varied from 0 (no treatment) to 40'(12.2 m) and the effect of treatment on the amplitude of response was observed. Similar to rigid foundations, treatment significantly reduces the amplitude of vibration due to low strain, harmonic loads. A relationship showing the reduction of amplitude at radial distance from the center of the foundation is shown in Fig.13. It is seen from this figure that 40' of treatment results in approximately 90% reduction in the amplitude of vibration.

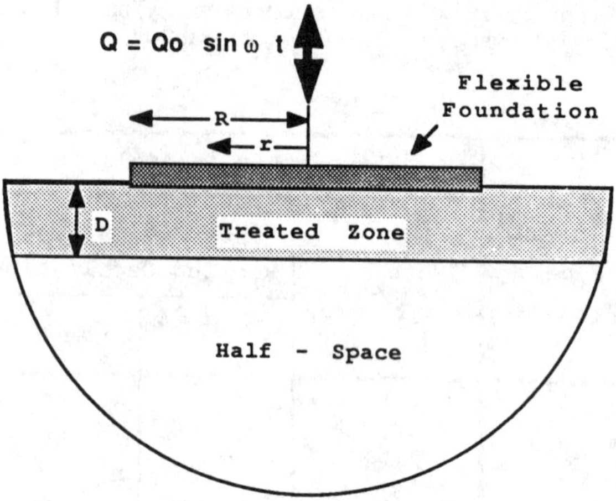

Fig.12.- Schematic Diagram of a Circular Flexible Foundation Resting on a Treated Segment over a Half-Space

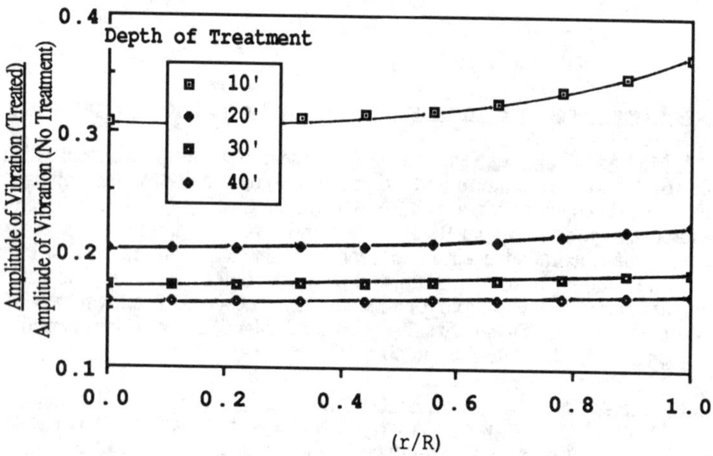

Fig.13.- Effect of Treatment on the reduction of Response Amplitude of a Circular Flexible Foundation.
(1 ft = 0.305 m)

5. Conclusions

Results from this analytical study show the potential of foundation soil treatment for control and improvement of a foundation's response to low strain dynamic loads. Important conclusions are the following:

1. Treatment of foundation soil significantly reduced the elastic response amplitude of both rigid and flexible foundations undergoing low strain, high frequency dynamic loads.

2. Response amplitude decreased with increasing both width and depth of treatment. These increases reached a limiting value near 5W (width) x 5D (depth) for all of the cases studied.

3. Increase in either the width or the depth of treatment significantly reduced the amplitude of vibration for both vertical and coupled sliding/rocking oscillations. For the case of coupled sliding/rocking, increasing the width of treatment was more effective.

4. Treatment had little effect on the fundamental frequencies of the soil-structure systems studied.

5. Treatment type had significant influence on the reduction of response amplitude. Strong bond treatments had more effect on the reduction of amplitude than those of lesser strengths.

References

Baker, W. H. (1982). "Planning and Performing Structural Chemical Grouting." *Proc., Specialty Conf. on Grouting in Geotechnical Engineering*, ASCE, New Orleans, Louisiana, 515-539.

Chae, Y. S. and Chiang, A. M. (1978). "Dynamic Properties of Lime and LFA Treated Soils," Proc.,*Specialty Conf. on Earthquake Engineering and Soil Dynamics*, ASCE, Pasadena, California, 308-324.

Chang, T. S. and Woods, R. D. (1987). "Effect of Confining Stress on Shear Modulus of cemented Sands." *Soil - Structure Interaction*, A. S. Cakmak ed., Elsevier.

Chang, T. S. (1986). "Dynamic Behavior of Cemented Sand," thesis submitted to the University of Michigan, at Ann Arbor, Mich., in partial fulfillment of the requirements for the degree of Doctor of Philosophy.

Chiang, Y. S. and Chae, Y. S. (1972). "Dynamic Properties of Cement Treated Soils." *Highway Research Record,* 379, 39-51.

Gucunski, N. (1991). "Generation of Low Frequency Rayleigh Waves for Spectral-Analysis-of-Surface-Waves Method," thesis submitted to the University of Michigan, at Ann Arbor, Mich., in partial fulfillment of the requirements for the degree of Doctor of Philosophy.

Hardin, B. O. and Richart, F. E., Jr. (1963). "Elastic Wave Velocities in Granular Soils." *J. Soil Mech. and Found. Div.*, ASCE, Vol.89, No. SM 1, 33-65.

Li, N. and Woods, R. D. (1987). "Dynamic Behavior of Grouted Sand," *Soil - Structure Interaction*, Cakmak A. S. ed., Elsevier.

Lysmer, J. and Kuhlemeyer, R. L. (1969). "Finite Dynamic Model for Infinite Media." *J. of the Engrg. Mechanics Div.*, A.S.C.E., Vol.95, No. EM4, 859-877.

Mitchell, J. K. (1981). "Soil Improvement State-of-the-Art Report." *Proc. Xth Int. Conf. on Soil Mech.and Found. Engrg*, Stockholm, 509-565.

Newmark, N. M. (1959). "A Method of Computation for Structural Dynamics." *J. of the Engrg. Mechanics Div.*, A.S.C.E., 67-94.

Prevost, J. H., (1981). "DYNAFLOW: A Nonlinear Transient Finite Element Analysis Program." Dept. of Civil Engineering, Princeton University, 1981, last update 1987.

Welsh, J. P. (1986). "Construction Considerations for Ground Modification Projects." *Proc. Int. Conf. on Deep Foundations*, Beijing, China.

Notations

Q = Harmonic force
Q_o = Amplitude of the Harmonic force

ω = loading Frequency (rpm)
t = duration of load,
R = radius of the circular flexible foundation
r = radial distance
W = width of the foundation,
D = embedment depth of the foundation,

υ = Poisson ratio

ρ = mass density of soil

α = integration constant

β = integration constant

Seismic Repair at Seventh Street Marine Terminal

John A. Egan[1], M.ASCE, Robert F. Hayden[2], M.ASCE,
Larry L. Scheibel[1], A.M.ASCE, Mahmut Otus[1], M.ASCE,
and Gerald M. Serventi[3], M.ASCE

Abstract

The 17 October 1989 Loma Prieta earthquake heavily damaged the Seventh Street Marine Terminal at the Port of Oakland, California. Distress to the wharf, which included bayward dike translation, inboard crane rail settlement, and damaged or broken deck-support batter piles, was due primarily to liquefaction-related lateral spreading and settlement of hydraulically-placed sand fill and native silty sands beneath and behind the perimeter rockfill dike. Ground improvement alternatives were evaluated for their potential to mitigate liquefaction hazard to the wharf structure during future earthquakes. The Port's desired dike performance and cost/benefit objectives were achieved by a stone column solution in which the stone columns were installed using the vibro-replacement process to improve a 12-m (40-ft.) wide zone of soil in the perimeter dike along the wharf alignment. Primary ground improvement/strengthening objectives were densification of loose liquefiable soils and replacement and reinforcement with compacted stone. Extensive post-construction testing verified that the vibro treatment attained the specified soil densification, as reflected by measured $(N_1)_{60}$ blowcount values, to effectively mitigate the liquefaction-related hazard of lateral spreading.

Introduction

The Seventh Street Marine Terminal at the Port of Oakland, California, is situated at the margin of San Francisco Bay, near the eastern end of the Oakland-San

---

[1] Geomatrix Consultants, 100 Pine Street, Suite 1000, San Francisco, California 94111

[2] Hayward Baker, Inc., Western Region, P.O. Box 7690, Ventura, California 93006

[3] Port of Oakland, 530 Water Street, Oakland, California 94607

Francisco Bay Bridge, and is constructed on land reclaimed from the Bay. The San Francisco Bay Area is considered to be one of the more seismically active regions of the world. Several major faults within the area are potential sources of future major earthquakes, including the San Andreas, Hayward, San Gregorio, Calaveras, and Healdsburg-Rodgers Creek faults (Figure 1). During the historical period (approximately 160 years), these faults have produced ten moderate to large earthquakes (magnitude, M, >6) affecting the Bay Area.

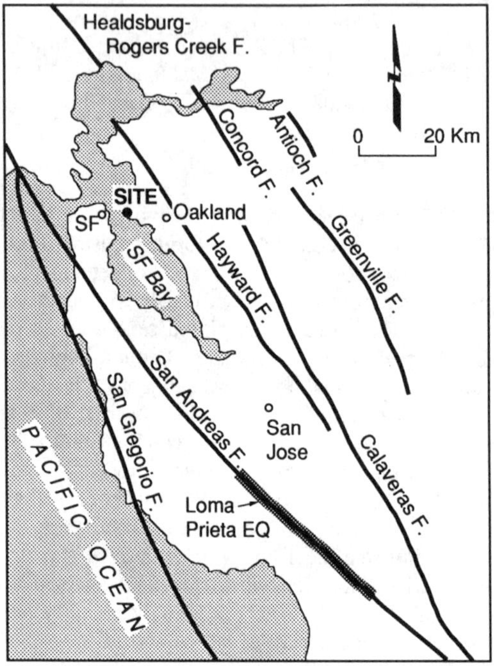

Figure 1. Location of Seventh Street Marine Terminal and Known Active Faults in the San Francisco Bay Area.

On 17 October 1989, the M7 Loma Prieta earthquake ruptured along a portion of the San Andreas fault in the southern Santa Cruz Mountains, approximately 80 km (50 mi.) south of the Terminal site (Figure 1). Ground shaking recorded at the Port's Berth 25 Wharf, located approximately 2 km (1-1/4 mi.) northeast of the Terminal, is characterized by a horizontal peak ground acceleration of 0.29g (CDMG,1989). The duration of strong shaking for this record was about 8 seconds and was typical of ground shaking duration produced by the earthquake. It is expected that ground motion levels experienced at the Terminal were similar to those recorded at the Berth 25 Wharf. Following the earthquake, effects of strong ground shaking were evident at the Terminal in the form of liquefaction sand boils, lateral spreading fissures, and differential settlement. Significant damage was sustained by the Terminal wharf structure, due primarily to liquefaction-related lateral spreading and settlement of hydraulically-placed sand fill materials and native silty sands beneath and behind the perimeter rockfill dike. The perimeter dike moved bayward 0.3 m (1 ft.) or more and many of the 405-mm (16-in.) square prestressed concrete batter piles supporting the rear of the wharf deck were damaged or broken. Some isolated vertical piles in other rows also suffered minor damage. The inboard crane rail, which was supported on a 1.5-m (5-ft.) wide reinforced concrete footing bearing on fill, settled 225 to 350 mm (9 to 14 in.) and there reportedly was as much as a 75-mm (3-in.) narrowing of the gauge of the crane rails. The Terminal was fully out of operation about six months and had limited crane capability for twenty-two months after the earthquake while seismic repair work proceeded.

Future-earthquake design considerations for Terminal repairs are dominated by the Hayward fault, situated approximately 10 km (6 mi.) to the east, and the San Andreas fault, situated approximately 21 km (13 mi.) to the west. The Hayward fault has been the source of two M6.8 earthquakes during the historical period, in 1836 and 1868 (Toppozada and Parke, 1982). The Working Group on California Earthquake Probabilities (1990) has estimated a 45 percent probability for a M7 earthquake occurrence on the Hayward fault during the 30-year period between 1990 and 2020. The San Andreas fault was the source of the largest Bay Area event during the historical period, the M8 1906 San Francisco earthquake. In addition to the M7 Loma Prieta earthquake in 1989, a M7 earthquake in 1838 and a M6.3 earthquake in 1865 have been attributed to the San Andreas fault. The Working Group on California Earthquake Probabilities (1990) has estimated a less than 5 percent probability for a M8 earthquake occurrence on the San Andreas, but a 23 percent probability for a M7 earthquake occurrence on the San Francisco Peninsula segment of the fault during the 30-year period between 1990 and 2020. Given the historical record and the estimated likelihood of future earthquakes in the Bay Area, future occurrence of strong ground shaking at the Terminal site is expected.

Earthquake ground motion characteristics utilized for evaluations of Terminal repair alternatives were developed using the results of a probabilistic ground-motion assessment. The probabilistic assessment was based upon a geologic model of the spatial and temporal occurrence of earthquakes on all potential seismic sources in the San Francisco Bay Area, and appropriate attenuation relationships, to define peak ground acceleration and response spectral acceleration hazard levels at the Terminal associated with various potential earthquake occurrences. Ground motions corresponding to a 20 percent probability of exceedance in 50 years was established as the design criterion for Terminal repairs. This level is characterized by a horizontal peak ground acceleration of 0.35g.

Terminal Construction and Site Conditions

The Terminal, which is located bayward of the former San Francisco Bay shoreline, is constructed on land reclaimed from the Bay by filling performed in the mid-1960s. Water depths at the site prior to reclamation varied between about 4.5 and 9 m (15 and 30 ft.) below Mean Lower Low Water (MLLW). The site was reclaimed by first constructing a perimeter dike around the area and then filling behind the dike to raise the grade to near its present elevation of approximately +4.3 m (14 ft.). Before the perimeter dike was constructed, approximately 1.5 to 3 m (5 to 10 ft.) of soft recent Bay sediments (known locally as Bay Mud) were removed by dredging within the perimeter dike footprint.

The generalized stratigraphic conditions and perimeter dike configuration at the Terminal are illustrated in Figure 2. The dike base was created by hydraulically placing dredged-sand fill from bottom-dump barges up to the elevation of the original mudline. A small rock toe dike then was constructed atop the sand base, and a second lift of sand was hydraulically placed up to the crest of the toe dike. The thickness of the sand-fill base increases westward along the dike alignment from about 4.3 to 8.2 m (14 to 27 ft.). The hydraulic sand dike fills are predominantly clean, fine-to-coarse-grained sands as indicated by the grain-size characteristics shown in Figure 3. These dike fill sands were loose to medium dense, having $(N_1)_{60}$ values in the range between 9 and 20, with an average value of 14. Assessment of liquefaction potential, using the simplified procedures described in Seed and others (1985) and the National Research Council (1985), indicated that these fill sands were expected to experience liquefaction during ground shaking characteristic of both the Loma Prieta earthquake and design level ground motions from future earthquakes.

A rock-fill dike extending up to approximately elevation +2.4 m (+8 ft.) was then constructed on top of the second sand lift. Construction records indicate that this portion of the dike was constructed with sound rock that generally was a mixture of 380-mm (15-in.) minus and 150-mm (6-in.) minus materials obtained from two borrow sources, although some sections were constructed with rock from only one source. These rockfill materials were not considered to be susceptible to earthquake-induced liquefaction.

Fill materials placed in the Terminal area behind the perimeter dike included hydraulic sand fill, hydraulic clay-ball and sand fill, submerged fill placed by end-dumping of dry soil, and engineered fill placed above water. The materials placed immediately adjacent to the perimeter dike consisted of relatively clean sand fill. These various fill materials experienced liquefaction during the Loma Prieta earthquake and, given this recent occurrence, are expected to also liquefy during design level ground shaking from future earthquakes.

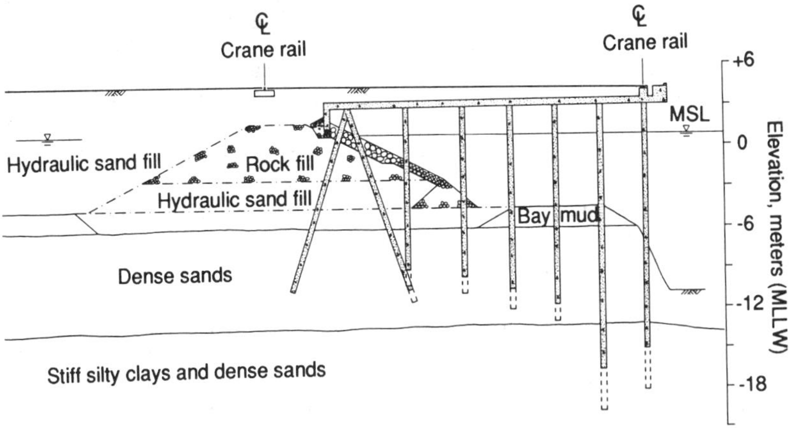

Figure 2. Generalized Terminal Stratigraphy and Wharf Cross-Section at the Time of the 1989 Loma Prieta Earthquake.

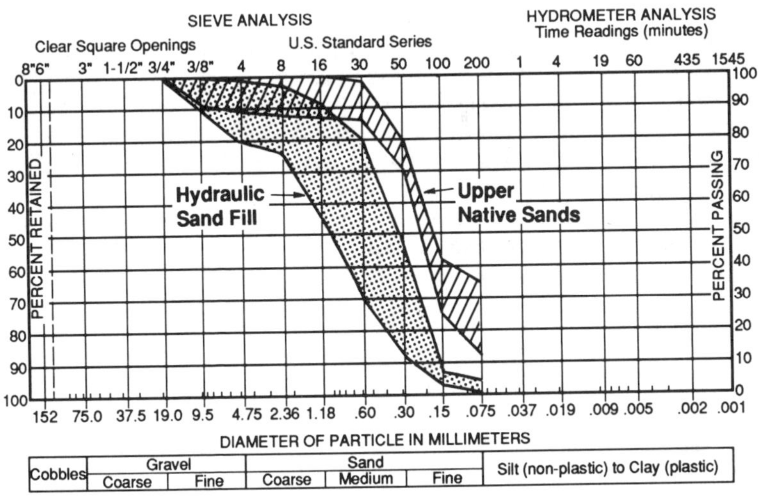

Figure 3. Grain-size Characteristics of Hydraulically-placed Dike Sand Fill and Upper Native Sands.

The dike fill materials and soft Bay Mud deposits, where present, are underlain by an approximately 11 to 15-m (37 to 50-ft.) thick native sand stratum consisting primarily of medium dense to dense silty or clayey sands, with lenses of stiff silty clays. The upper 1.5 m (5 ft.) of these native sands are fine-grained silty or clayey sands, with silt/clay contents ranging between 12 and 35 percent. Grain-size characteristics for these sands are also shown in Figure 3. These upper native sands were medium dense, having $(N_1)_{60}$ values in the range between 13 and 18, with an average value of 15. Based upon assessments of liquefaction potential as previously discussed, it is considered marginal whether these upper native sands could have experienced liquefaction during the Loma Prieta earthquake ground shaking, but would be expected to liquefy during design level ground motions from future earthquakes. The deeper native sands, which are fine-grained and dense (occasionally very dense), have $(N_1)_{60}$ values typically in excess of 30 and are not expected to experience earthquake-induced liquefaction. Deposits of stiff to very stiff silty clays, locally known as Old Bay Mud, underlie the native sand stratum. These clays which lightly to moderately overconsolidated, extend at least to depths of Elevation -36.5 m (-120 ft.).

After the filling operation, wharf structures were constructed along the north and west sides of the Terminal. The total length of the wharf alignment is approximately 915 m (3000 ft.). The wharf structures consist of a reinforced concrete deck that varies in thickness from 460 to 510 mm (18 to 20 in.) and is supported on 405-mm (16-in.) square prestressed concrete piles. The deck is overlain by 0.6 to 0.9 m (2 to 3 ft.) of ballast fill. Piles on the back row (landside) were battered at an inclination of 1 horizontal to 3 vertical, with alternating inboard and outboard batters. All other piles that support the wharf deck are vertical or nearly vertical. The longitudinal spacing of the piles that support the outboard crane rail girder varies between approximately 1.8 and 2.4 m (6 and 8 ft.) center-to-center. The remaining piles have a longitudinal center-to-center spacing of 3.7 m (12 ft.). The container cranes that operate at the site have a gauge of 29.3 m (96 ft); the outboard crane rail is supported on the wharf structure and the inboard crane rail supported on a 1.5-m (5-ft.) wide by 0.5-m (1.6-ft.) thick reinforced concrete footing bearing on dike fill materials. A typical cross-section of the Terminal wharf structure at the time of the earthquake is illustrated in Figure 2.

Dike Performance Evaluations

As a result of the Loma Prieta earthquake, the perimeter dike at the Terminal had deformed bayward 0.3 m (1 ft.) or more. During future large earthquakes, significant further lateral spreading and/or failure of the perimeter dike was anticipated. Based on the observed dike performance, ground-shaking from future earthquakes is expected to affect the stability of the perimeter dike primarily by seismic inertia forces within the dike and by shaking-induced strength reductions in the hydraulic sand fills and upper native sands. Changes of stability would be manifested by lateral movements of the dike toward the Bay. Such movements would involve portions of the rock and sand dike fill, the hydraulic sand fill behind the dike, and the upper native sands. During ground shaking, transient increases of seismic force could exceed the dike's yield resistance, creating temporary instability that would enable bayward dike movement. The duration of ground shaking characteristic of earthquakes that dominate the design ground-motion level ($7<M<8$) could create many such occurrences, producing accumulation of bayward movement. The methodology used to evaluate dike performance for earthquake loading is described in Updike and others (1988). This methodology incorporates simplified procedures, modified from Newmark (1965) and Makdisi and Seed (1978), to estimate slope movements.

Existing Perimeter Dike

The performance of the wharf and perimeter dike during the Loma Prieta earthquake afforded a site-specific case history with which to examine the appropriateness of the strength characteristics that were interpreted for liquefied-sand from the field exploration and laboratory testing programs (Seed and Harder, 1990); to calibrate those characteristics with known performance; and to examine the sensitivity of dike stability to variations in selected soil characteristics. In the analyses, soil/fill liquefaction behavior was made consistent with that observed and/or expected for the Loma Prieta earthquake, as previously described. The hydraulic sand fill was assumed to have liquefied quickly during the Loma Prieta ground shaking and to have remain liquefied throughout the approximately 8-seconds of strong ground shaking. Liquefaction of the upper native sands, considered marginal for the Loma Prieta earthquake, was assumed to have occurred near the end of ground shaking, thus primarily affecting post-earthquake stability. For undrained residual liquefied-sand strengths of about 20 kPa (400 psf), stability analyses and dike movement evaluations for these conditions estimated bayward movements of 350 to 400 mm (14 to 16 in.), which are consistent with the movements of 0.3 m (1 ft.) or more observed for the perimeter dike following the Loma Prieta earthquake. Stability analyses conducted for gravity loading conditions immediately following the earthquake indicated factors-of-safety in excess of unity, implying no additional post-earthquake movement, which also was consistent with observations at the Terminal.

Stabilized Perimeter Dike

To mitigate the liquefaction hazard and minimize the risk of major damage to the wharf due to future lateral spreading or dike failures, alternatives were examined for stabilizing the perimeter dike by strengthening a zone of soil within the dike fill and upper native sand deposits below the dike. Stabilization alternatives examined included grouting, stone columns, and compaction piles. Stability analyses and dike movement evaluations were conducted for several dike sections, various widths and depths of stabilization, and a range of soil strengthening anticipated for the stabilized zone. Based on the degree of soil densification/strengthening that was considered practical to obtain within the dike fill materials, it was reasonably assumed in the evaluations of expected performance that the strength and integrity of the stabilized zone would be maintained throughout and following the earthquake ground shaking. The dike performance evaluations indicated that a zone of stabilization approximately 12 m (40 ft.) wide and extending in depth to at least the bottom of the upper native sands would be required to achieve the Port's performance criterion that dike movements be limited to a few inches during future design-level earthquake ground shaking. At the east end of the Terminal, where the thickness of the dike sand fill is the least, evaluations indicated that for a uniform degree of soil improvement, a somewhat lesser width of improvement would be satisfactory to achieve the same desired performance as at the west end of the Terminal. Similarly, if a constant 12-m (40-ft.) wide zone of soil stabilization was assumed, a lesser degree of soil improvement at the east end would produce the same performance as at the west end. The specifications were written to allow for this variation of required soil strength and density, such that average $(N_1)_{60}$ values for the soil within the improved zone could range from 25 at the Terminal's eastern end to 30 at the western end. These blowcount values were identified, by assessments of liquefaction potential (Seed and others, 1985), as levels that would effectively mitigate liquefaction hazard and maintain necessary soil strength within the stabilized zone. Note that the simplified procedures of Seed and others (1985) used in this study to assess liquefaction potential indicate that for $(N_1)_{60}$ values greater than 30, a soil is generally not considered to be liquefiable.

Ground Improvement Measures

Dike movement evaluations indicated that it was technically feasible to achieve the performance criterion by each of the stabilization alternatives examined. However, based upon comparisons of estimated costs for implementation of each stabilization alternative examined, ground improvement made by installing stone columns constructed using the vibro-replacement process was recommended to achieve the Port's cost/benefit goals. The stabilizing effects on the dike performance would be accomplished by increasing the density of the loose to medium dense sand materials, replacing some of those sands with compacted stone columns, and providing a drainage path through the stone columns to dissipate excess pore water pressures generated during earthquake shaking. As described above, a 12-m (40-ft.) wide zone of soil in the dike slopes along the wharf alignment would be improved to depths ranging between approximately Elevation -7.5 and -11.5 m (-25 and -38 ft.). Because it was assumed that the integrity and strength of the stabilized zone would be maintained throughout and following earthquake, analyses were conducted using methods presented by Seed and Booker (1977) to identify stone column spacings necessary to minimize accumulation of excess pore water pressure, and thus minimize strength reduction in the improved zone during the expected future earthquake ground shaking conditions. Recommended stone column spacings were developed for various column diameters, with a maximum spacing of 2.4 m (8 ft.) specified for 0.9-m (3-ft.) diameter stone columns. The column spacing required to attain the specified soil density was assessed by the contractor based upon experience with similar conditions at other sites.

As part of the wharf repair work, two rows of new 610-mm (24-in.) octagonal piles were also driven. The vibrations and soil displacement caused by driving these piles were expected to help densify the sands in the vicinity of the new piles. This zone of densification around the piles was incorporated into the required 12-m (40-ft.) width of soil stabilization to reduce the number of stone columns required at the site. The stone column pattern was developed to minimize the densification effects on pile drivability and to take advantage of the densification that would result from the pile driving. The ground improvement scheme developed for the wharf repair, which incorporated stone columns and new piles, is illustrated in Figure 4.

Construction Aspects

Soil densification and stone column installation was achieved using the vibro-replacement process. Electrically driven depth vibrators were utilized to densify potentially liquefiable soils in the perimeter dike, while constructing columns of dense stone to reinforce the soils. The vibrators used at the Terminal were driven with 160-hp motors operating at a frequency of 1800 rpm. Approximately 280 kN (28 tons) of lateral force were developed having amplitudes of 16 mm (0.63 in.) and frequencies of 30 Hz. The vibrator had a diameter of up to 0.55 m (22 in.) and a length of 2.9 m (9.5 ft.). Extension tubes attached to the top of the vibrator enabled the vibrator to penetrate to the maximum treatment depth of approximately 13.5 m (44 ft.). Water pipes built into the vibrator provided jetting to assist penetration of the vibrator into the ground.

The concrete wharf deck was cut approximately 2 m (7 ft.) outboard of the location of the batter pile connections to the deck, then a trench was excavated in the approximately 18- (60-ft.) wide gap created between the wharf deck and the Terminal yard. The stone columns were installed from the bottom of the excavation, which corresponded to approximately the top of the rock dike. Field procedures involved controlling probe locations by using flags to mark a grid pattern with dimensions of 2.1 by 2.4 m (7 by 8 ft.). Maximum center-to-center spacings were specified as 2.4 m (8 ft.), contingent upon attaining the densification criteria. The vibrator was suspended from a crane above each marked location. Then the water jets were turned on and the vibrator penetrated to full treatment depth under its own power. When it reached the final base elevation, the vibrator was lifted a short distance from the bottom of the hole, and a front-end loader dropped stone backfill through the water to the vibrator. The vibrator was withdrawn approximately one meter and then re-penetrated backfill to densify the stone. Re-penetration process was repeated until the ammeter reading on the crane increased to approximately 200 to 230 amps, indicating a dense volume of stone and surrounding soil at the bottom. The vibrator then was raised about one meter (3 ft.) and the process repeated. Construction operations were sequenced so that the vibro-treatment preceded installation of new vertical prestressed-concrete piling that replaced the damaged batter piling.

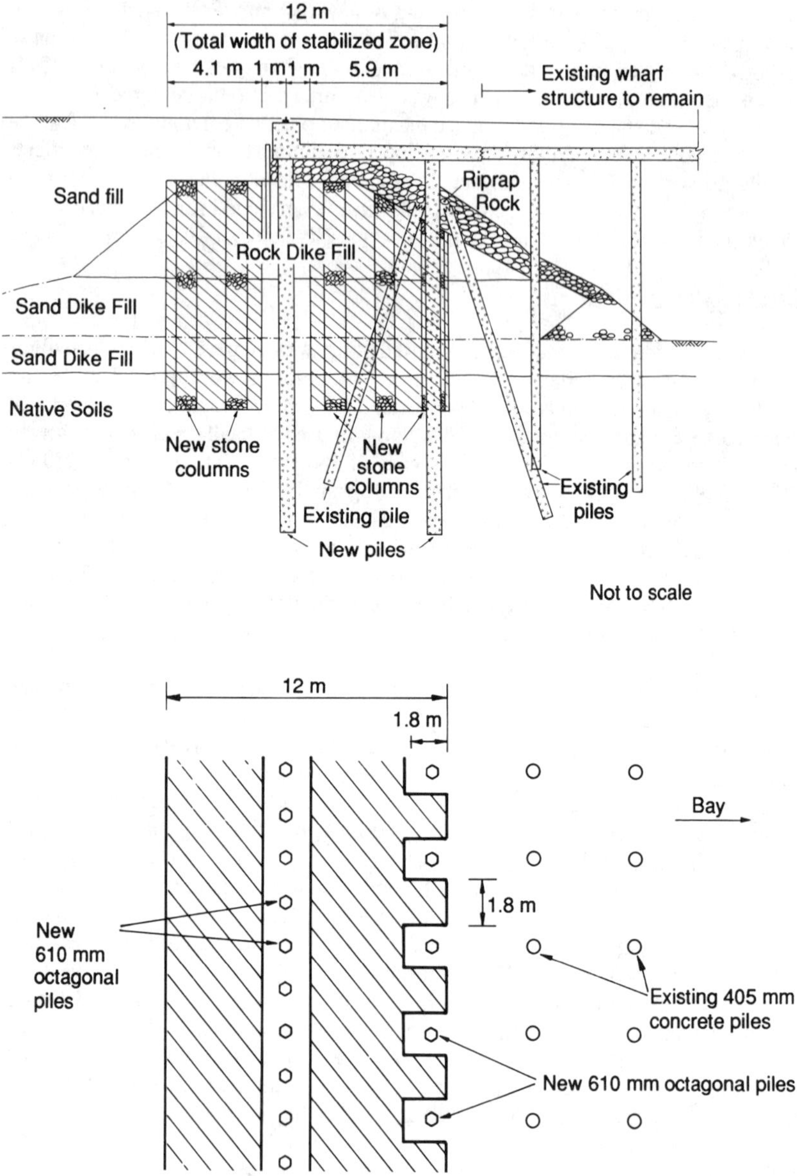

Figure 4. Stone Column and New Piling Configuration for Seismic Repair of Wharf.

The backfill consisted of hard, clean crushed stone that was delivered by barge from a local rock quarry. To provide desired drainage characteristics in the columns, the gradation of the stone backfill was specified to have a maximum size of 38mm and to less than 5 percent passing the 10mm sieve. The minimum durability index of the stone was 40 as measured by California Test Method 229. Jetting water was pumped from the bay adjacent to the site.

The vibro-replacement process very effectively densified subsurface materials along the entire reach of the wharf. Before construction began, the key questions about construction concerned the vibrator's ability to penetrate the rock dike and the need to work in the tidal zone at the base of the excavation. The specifications included a contingency plan for the contractor to predrill the rock dike, but the vibrator readily penetrated the matrix of rock without drilling. To accommodate the tidal influence on ground stability in the excavated trench, a tracked crane was mobilized that was capable of reaching out from the asphalt-surfaced Terminal yard to above the trench. Because of difficulty pushing a cone penetrometer (CPT) through the rock dike, post-construction testing of the degree of ground improvement was conducted using the SPT. The extensive post-construction testing verified that the vibro-treatment improved the penetration resistance of the loose sand, as measured by $(N_1)_{60}$ blowcount values, by as much as 400 percent, as illustrated in Figure 5.

Appendix. References

California Division of Mines and Geology (1989). "CSMIP strong-motion records from the Santa Cruz Mountains (Loma Prieta), California earthquake of 17 October 1989." *Report OSMS 89-06*, Office of Strong Motion Studies.

Earthquake Engineering Research Institute (1990). "Loma Prieta earthquake reconnaissance report." *Earthquake Spectra*, 6(Supplement), Report No. 90-01.

Makdisi, F.I., and Seed, H.B. (1978). "Simplified procedure for estimating dam and embankment earthquake-induced deformations." *J. Geotech. Engrg. Div.*, ASCE, 104(GT7), 849-867.

National Research Council (1985). *Liquefaction of soils during earthquakes*: Committee on Earthquake Engineering, Commission on Engineering and Technical Systems, National Academy Press, Washington, D.C..

Newmark, N.M. (1965). "Effect of earthquakes on dams and embankments." *Geotechnique*, 15(2), 139-159.

Seed, H.B., and Booker, J.R. (1977). "Stabilization of potentially liquefiable sand deposits using gravel drains." *J. Geotech. Engrg. Div.*, ASCE, 103(GT7), 757-768.

Seed, H.B., Tokimatsu, K., Harder, L.F., and Chung, R.M. (1985). "Influence of SPT procedures in soil liquefaction resistance evaluations." *J. Geotech. Engrg. Div.*, ASCE, 111(12), 1425-1445.

Seed, R.B., and Harder, L.F. (1990). "SPT-based analysis of cyclic pore pressure generation and undrained residual strength." *Proc. of the H.B. Seed Memorial Symposium*, 2, 351-376.

Toppozada, T.R., and Parke, D.L. (1982). "Area damaged by the 1868 Hayward earthquake and recurrence of damaging earthquakes near Hayward." *Proc., Conf. on Earthquake Hazards in the Eastern San Francisco Bay Area*, 321-328.

Updike, R.G., Egan, J.A., Moriwaki, Y., Idriss, I.M., and Moses, T.L. (1988). "A model for earthquake-induced translatory landslides in Quaternary sediments." *Geological Society of America Bulletin*, 100, 783-792.

Working Group on California Earthquake Probabilities (1990). "Probabilities of large earthquakes in the San Francisco Bay Region, California." U.S. Geological Survey, *Circular 1053* .

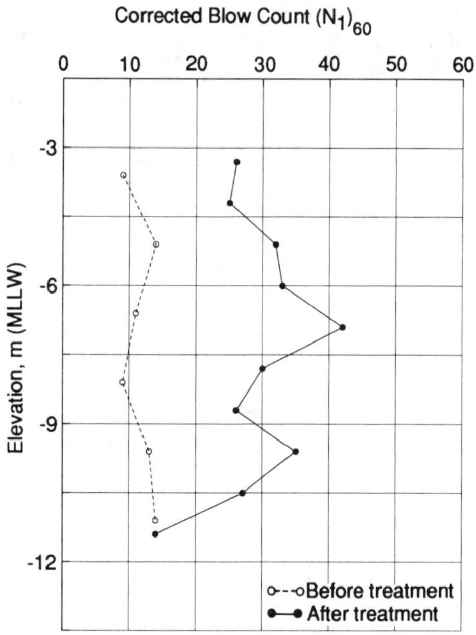

Figure 5. Representative Pre- and Post-Improvement Standard Penetration Test Resistance, $(N_1)_{60}$, Values.

Earthquake Support Grouting In Sands

Edward D. Graf[1], Member ASCE

Abstract

Four case histories illustrate some of the uses of chemical grout and compaction grout in sands for the support of structures under earthquake loading. Case histories demonstrate applications for overloaded footings, shock densification, liquefaction, and vertical uplift. Rehabilitation and new construction projects are described.

Introduction

Stringent regulations have been placed on structural and foundation design in California because of strong motion earthquake potential. State and local building codes require investigation, analysis, and design for larger structures, predicated on the potential seismic risk of the geographic area in which the project is located. School and hospital analyses and designs, together with construction inspection reports and tests, are carefully reviewed by structural engineers from the Office of the State Architect (OSA). Many cities like San Francisco, a Zone 4 earthquake area, require earthquake structural rehabilitation to meet current building codes if there is remodeling that affects more than 30% of a structure. With the high rents in the downtown area, it has become economically attractive to update older buildings. San Francisco has many masonry, concrete, and steel structures founded on spread footings built before current earthquake design standards were established, resulting in a need for a sizable amount of structural rehabilitation to meet modern earthquake design requirements.

Clean dune sands, both natural and fill, cover much of the downtown area of San Francisco. These uniform fine sands usually contain only 3% to 6% passing a No. 200 sieve. Older spread footings are commonly subject to earthquake overload under the new design codes and to shock densification settlement when founded near the surface in loose sands. In the flatland area of downtown San Francisco, the high water table adds the potential of liquefaction of loose sands.

---

[1]Grouting Consultant, 715 San Miguel Lane, Foster City, CA 94404.

Following are four brief case histories outlining the conditions and illustrating different solutions using chemical grout to form "weak sandstone" in clean sands and compaction grouting to densify liquefiable sands. Only outline case histories are presented; each of them could be developed into a full paper.

Case I - New Footing By Chemical Grouting.
1049 Market Street, San Francisco, California.

An existing six story and basement unreinforced brick office building built in 1910 was completely remodeled. The relatively narrow structure, 50 x 165 feet (15 x 50 m), was sandwiched between two similar structures of the same vintage and construction that had not received structural rehabilitation. The original design for the seismic strengthening of the foundations required an exceptionally large continuous concrete footing extending down into the dense sands (Fig. 1). Both adjacent structures had basements with their footings at the same or slightly higher levels. To install the new footings, this design required the support of the entire structure, the underpinning of the immediately adjacent building footings, and dewatering.

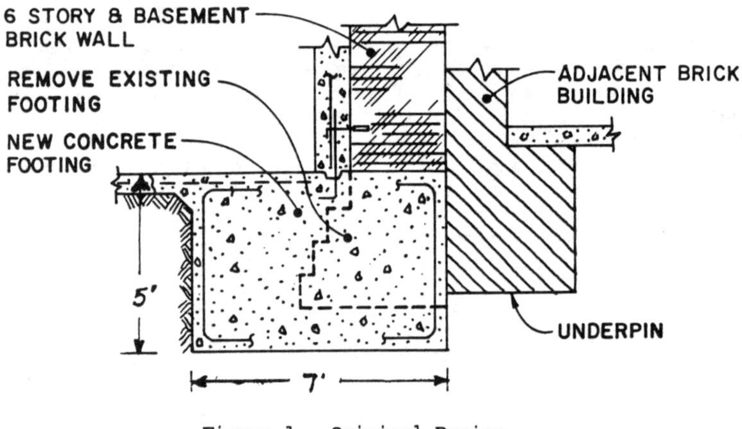

Figure 1 - Original Design

A redesign used chemical grouted sand footings, injected in-situ, in lieu of the original design of new concrete footings (Fig. 2) and saved:

a. Dewatering for the deep footings excavation.
b. Underpinning support of the long seven story masonry walls for demolition of the existing footings.
c. Excavation for the new concrete footings.
d. Underpinning support for both of the adjacent structures which would have been necessary to allow excavation for the new concrete footings.
e. Two to four months of construction time.
f. Potential risks and legal fees involved with underpinning

adjacent structures 75 years old.

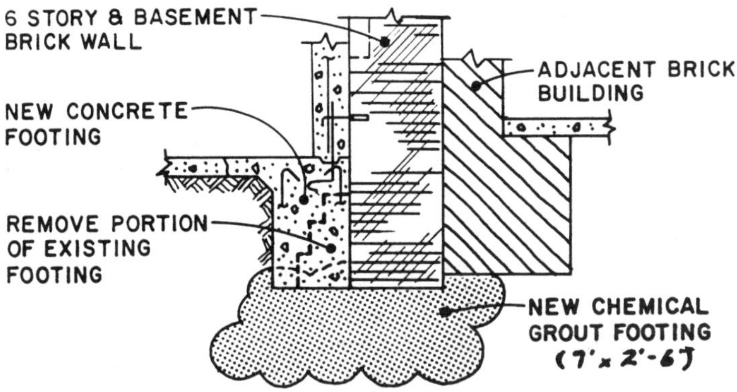

Figure 2 - Redesign.

In one area, a new elevator pit required the removal of the existing brick footing. At that area, the adjacent building was underpinned utilizing a higher strength chemical grout than was used for the continuous footings (Fig. 3).

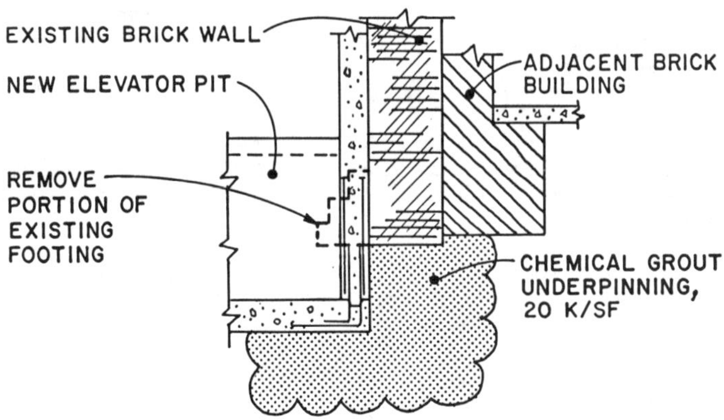

Figure 3 - New Elevator Pit

The 7 day minimum unconfined compressive strength of the grouted sand was specified to be 10 ksf (478 kPa) for the continuous footings (Fig. 2) and 20 ksf (957 kPa) for the elevator pit excavation underpinning (Fig. 3). A sodium silicate base grout with an organic reagent (ROC) was used at different concentrations for the injections. Carved samples from several locations showed unconfined compressive strengths from the lowest strength of 10.3

ksf (493 kPa) for the 10 ksf (478 kPa) specification and ranging from 20.9 to 25.1 ksf (1.0 to 1.2 MPa) for the 20 ksf (957 kPa) specification. From previous field and laboratory testing, it is estimated that the chemical system used should almost double in strength in about a year under the project conditions.

In these uniform fine sands, it is known from previous observations that the permeation of the chemical grout into the sand is radial from each point of injection. Each injection point was preplanned and detailed with a calculated injection quantity based on an overlapping spherical pattern to assure permeation to the minimum design sections. Because of the necessary spherical injections and the minimum design sections, detailing of each injection is critical to both the success and cost of the grouting. [Each cubic foot of grouted sand, whether within the structural design sections or wasted outside due to the necessary spherical injections, adds to the cost of the work. To reduce waste, engineers are encouraged to detail well rounded corners in their basic structural design instead of the usual square or angular corners.]

Field execution of the grouting was through one and three-quarters inch (44 mm) holes drilled through the old basement slab and brick footing. Locations and angles of drilling were carefully monitored. Injection was staged down through the tips of one and one-half inch (38 mm) hand held rotary wet-head drills. The one and three-eights inch (35 mm) drill rod was scored at one foot (30 cm) increments to facilitate accurate tip location. Advances of the drill tip were made while injecting the chemical grout through the wet-head water swivel. No packers were used except for the first stage injections in holes drilled through the brick footings.

During the grouting, an area about six feet (2 m) long on one wall did not "take" the grout injection. Subsequent excavation revealed a clayey fill which was removed and replaced with lean concrete.

The grouted zones were probed with drills at numerous locations to verify the extent of the grout permeation. No "windows" were found and no remedial grouting was required.

The use of chemical grouting in lieu of conventional concrete design resulted in savings of $400,000 in direct construction costs and several months of construction time in addition to reducing potential problems with adjacent property owners.

Case II - Loose Sands And Overloaded Footings, Chemical Grouting
4041 Geary Street, San Francisco, California.

An existing concrete frame and concrete block structure, without a basement, was being remodeled into a supermarket. The building had been constructed circa 1940, before current earthquake building codes. Foundation soils were a clean dune sand well above the water table. The existing footings were subject to earthquake overload, shock densification of near-surface loose sand, and additional loading at additions and new wall openings.

The design called for a low minimum unconfined compressive strength grouted material of 3 ksf (144 kPa) to enlarge the footing bearing areas, prevent shock densification of the near-surface loose sands, and to transfer the loads to dense sands (Fig. 4). At overloaded footings and at footings with new loads added, the footings were to be enlarged with a minimum unconfined compressive strength grouted sand specified to be 8 ksf (383 kPa).

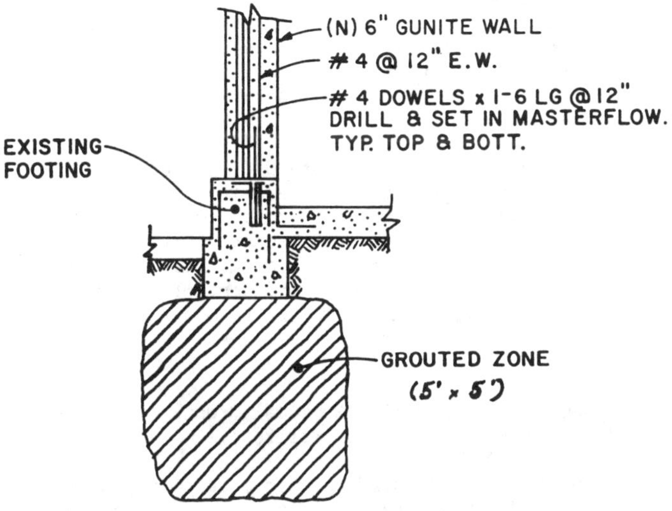

Figure 4 - Chemical Grouting for Overloaded Footings, etc.

Grout injection proceeded as planned using a sodium silicate base grout with an inorganic reactant (T-57) for the low strength areas and a sodium silicate base grout with an organic reactant (ROC) for the higher strength requirements. The viscosity of the grouts was in the range of 2 to 4 cps.

Each injection point was preplanned, detailed, and grout quantity calculated prior to injection work on the site, all as described in Case I above. Where necessary, one and one-half inch (38 mm) holes were drilled through the existing footings to allow injection directly below the footings. Chemical grout was then injected in downward stages through one-half inch (13 mm) open end pipe jetted into place with the chemical grout. The injection pipes were scored at six inch (15 cm) intervals to assure proper pipe tip location.

Based on previous experience with similar dune sands under near-surface concrete footings, each footing soffitt was pressured with a cement slurry to fill any open voids prior to the chemical grout injection. [Previous work developing this technique had almost always shown areas under footings where the sand had settled away, leaving passageways for the water-like chemical grout to advance uncontrolled, which resulted in ungrouted areas.]

Extensive drill probing, together with excavations for new footings, showed the design sections to be properly permeated with grout. Unconfined compressive strengths of all carved samples were well above the specified minimum. No regrouting was necessary.

During the engineering investigation of the structure, it was found that a large percentage of the old concrete block shear wall cells had not been grouted although the reinforcing bars were in place. These empty cells were drilled and grouted using a structural portland cement grout to establish the required structural strength.

This chemical grout approach, instead of the conventional concrete approach, saved extensive excavation, underpinning, construction time and other construction costs. [This structure was in the same vicinity as a large brick school that had extensive earthquake rehabilitation utilizing chemical grout foundation stabilization under the requirements and control of structural engineers from the California Office of the State Architect (Ref. 3).]

Case III - Tiedown Anchored With Chemical Grout.
1200 Market Street, San Francisco, California.

A two story steel frame building with a basement, built in 1917, was being remodeled for modern retail store and office usage. The underlying soils were dune sands with the water table about 3 feet (0.9 m) below the basement floor. Earthquake analysis showed strong foundation uplift forces that required restraint by tiedowns. Economical drilling was complicated by the factors of the clean fine sand below the water table and the need to drill within the basement of an existing structure.

The problem was solved by designing three feet (0.9 m) nominal diameter chemically grouted "weak sandstone" piers, 18.5 feet (5.6 m) deep. Tie-down bars were to be epoxy grouted into six inch (15 cm) holes drilled down through the center of each pier. A 20 ksf (957 kPa) minimum unconfined compressive strength was specified for the grouted sand.

Grouting proceeded as planned without incident using a sodium silicate base grout with an organic reactant (ROC) in a formulation that had a viscosity of about 3 cps. Grout was injected in downward stages of 18 inches (46 cm) through the tips of one and one-half inch hand-held rotary wet-head drills. A calculated quantity of grout was injected at each stage to form a minimum three feet (0.9 m) diameter injection.

During the drilling of the six inch (15 cm) anchor bar holes, caving occurred at about 15 to 16 feet (4.6 to 4.9 m) depth in all six of the grouted anchor piers. Solidly grouted sand was found above and below the caving zone. The drilling indicated a layer of clayey organic material at that level which was not previously known from the soils investigation. After re-grouting the caving zone twice with a relatively high strength organic polymer chemical grout (REG), the tiedowns were successfully installed and tested. Tension tests at each of tiedowns were taken to 54 kips (240 kN), a design factor of safety of 2.0.

In this case, there were substantial savings over the drilling of much deeper conventional anchors within the limited confines of the building basement, particularly with the problems of drilling in clean fine sands below the water table.

Case IV - Liquefiable Sands Densified By Compaction Grouting.
1200 El Camino Real, South San Francisco, California

An addition to an existing pile-supported hospital underlain by liquefiable sands was complicated by the necessity to keep the hospital in full operation during construction.

The site soil conditions consisted of about 7 to 10 feet (2 to 3 m) of silty sand fill overlying liquefiable silty sands, most generally 20% to 25% passing the No. 200 sieve, to a depth 25 to 35 feet (7.5 to 11 m). The soils investigation showed Standard Penetration Test (SPT) and converted continuous Cone Penetration Test (CPT) "N" values in the range of 10 to 25. Below the sands were silts and clays not subject to liquefaction.

Although compaction grouting had been used successfully on hundreds of remedial projects in the San Francisco Bay Area in the preceding 25 years, it had never been used for the defined and tested purpose of densifying liquefiable sands to meet the strict California hospital foundation requirements in a Zone 4 earthquake area. The foundation design engineers decided to run a test grouting program to determine the feasibility of the method. The test results showed that compaction grouting did densify the sand (Fig. 5) and the work proceeded.

The final grouting pattern consisted of a horizontal equilateral spacing of ten feet (3 m) for the full depth, up to 35 feet (11 m). For the seven to seventeen foot (2 to 5 m) depth, the horizontal spacing was split-spaced to 5 feet (1.5 m) equilateral spacing. The grout was staged up in 3 feet (0.9 m) stages. The grout mix was a lean cement:silty sand with water to yield a slump of not more than two inches (50 mm), usually about one inch (25 mm) at the collar of the injection pipes. The top seven feet (2 m) of fill, just above the water table, was excavated and recompacted after the grouting operation.

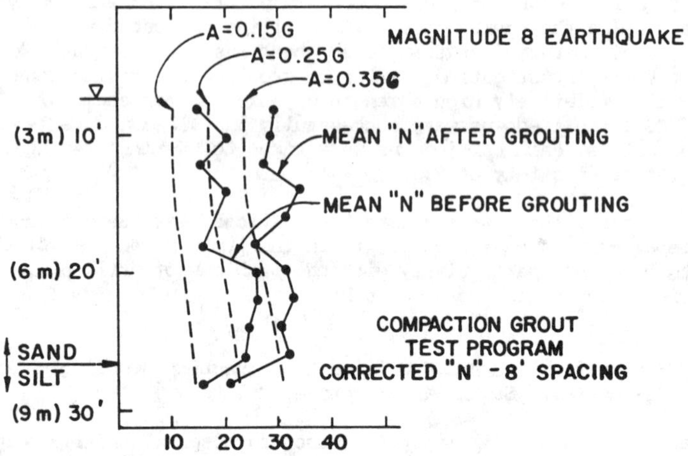

Figure 5 - Compaction Grout Test Results.

Results of the extensive continuous CPT testing during the test grouting, production grouting, and after completion, showed the converted CPT "N" values to be in the range of 30 to 35 with an infrequent drop to as low as 25 and an occasional high of over 40.

Due to noise and vibration reflecting into an operating hospital, the use of piles, vibroflotation, or dynamic compaction were automatically eliminated. Alternative methods considered included: (1) install a diaphragm wall, dewater, excavate, and then recompact; (2) chemical grouting; (3) drilled piers; and (4) compaction grouting. Cost analyses showed that compaction grouting saved two to three million dollars for this 0.8 acre (0.8 ha) site.

Author's Comments

The four case histories cited are only a sampling of the large number of projects successfully completed in the field of grouting for earthquake support and structural rehabilitation in sands. They are not large projects, but they do illustrate some of the ways that grouting may be used to achieve effective and economic engineering goals. An innovative engineer will find many other applications for grouting in his endeavor to provide a low cost and well engineered product for his clients.

Higher strengths of chemical grouted sands than specified in the cited cases can be readily achieved but usually at greater expense. Chemical grouted sands are relatively massive and the arbitrary selection of a "safe" strength, without an analysis of what is actually necessary, will result in higher costs to the client.

There are no established standards for these grouting methods or materials yet. An experienced grouting professional working closely with the design engineer through design and construction is necessary. Severe problems may arise with a specialty contractor that is not experienced in the particular method and materials and/or with inexperienced field supervisory personnel.

Quality of the work may suffer as a result of untrained field inspectors or engineers. Both quality problems and claims can be expected if they impose their personal inexperienced concepts on the contractor.

If the work is to be bid, the design engineer should have other possible bidders review the specifications before they go out to bid and/or write performance specifications; there is no single method, special material, or special equipment to provide quality work on most projects, and the cost of the work can double because of special-interest specifications.

Conclusions

These case histories demonstrate the practical use of grouting methods that are not in standard engineering textbooks but have been used successfully to solve difficult problems and save the client substantial amounts in construction costs and construction time.

Verification of these methods during and after completion is readily accomplished although the inspector does not yet have the controls established for concrete, steel, and masonry construction. It must be noted that conventional sampling methods cannot be used to test most grouted sands because of the brittle and relatively weak nature of the final product.

The case histories presented were completed prior to the 1989 Loma Prieta earthquake. There was no damage to these structures from the Loma Prieta earthquake.

Acknowledgments.

The author is indebted to the very many engineers and field personnel who have assisted in developing the concepts and practices used in the cases cited which were accomplished by Pressure Grout Company, South San Francisco, during the 31 years he owned the company.

Special appreciation is expressed to the engineers responsible for designing and specifying the use of the grouting methods used on the cited cases:

Case I - Henry J. Degenkolb Associates, Engineers, San Francisco.

Case II - Robert S. Gefken, Structural Engineer, San Francisco.

Case III - Peter Culley & Associates, Structural Engineers, San Francisco.

Case IV - Dames & Moore, San Francisco.

References

1. Baker, W.B., "Embankment Foundation Densification by Compaction Grouting", ASCE Proc. Issues in Dam Grouting, Denver CO, April 1985.
2. Graf, E.D., "Compaction Grouting Technique and Observations", ASCE Journal of the Soil Mechanics and Foundation Division, Vol. 95, No. SM5, Proc. Paper 6766, September 1969.
3. Graf, E.D. and Zacher, E.G., "Sand to Sandstone: Foundation Strengthening with Chemical Grout", ASCE "Civil Engineering", January 1979.
4. Warner, J., "Compaction Grouting - The First Thirty Years", ASCE Proc. Grouting in Geotechnical Engineering Conference, New Orleans, LA, February 1982.

Deep Compaction by Vibro Wing Technique and Dynamic Compaction

Kaare Senneset[1]
Jarle Nestvold[2]

Abstract

A ground improvement project was carried out at an offshore platform yard near Trondheim, Norway. The soil conditions over an area of 50.000 m² were improved by deep compaction and surface compaction. Two deep compaction methods were used, Vibro Wing and dynamic compaction (heavy tamping). The deposit consists of sand, from coarse to fine, partly with silt and gravel.

The obtained compaction results were investigated in the field by cone penetration tests and dilatometer tests, before and after compaction, together with plate load tests at the surface. Depth of influence for the two methods were found and the changes in soil properties evaluated.

The project provided a unique opportunity to compare two effective methods of deep compaction.

Introduction

The improvement of soil properties may be obtained by a number of methods, depending on type of soils and conditions at the site in question.

A relatively new technique for deep compaction of cohesionless soils has been introduced in the 1970's, the Vibro Wing method. This method is successfully used to densify sandy soils to a depth of 10-15 meters.

----

[1]Professor, Geotechnical Division, Norwegian Institute of Technology, University of Trondheim, 7034 Trondheim, Norway
[2]Director, Geotechnical Consultants O.Kummeneje A/S, 7034 Trondheim, Norway

The Vibro Wing method, together with dynamic compaction or heavy tamping was used for the improvement of soil conditions at an offshore platform yard near Trondheim, Norway. The compacted sand area served as a construction and assembly site for structural offshore jackets. In particular the requirement for only minor differential settlements during the launching of jacket to barge made an improvement of deformation properties necessary.

The compaction work occured under the supervision of Ottar Kummeneje a.s. acting as geotechnical consultant to Aker Verdal a.s. The Geotechnical Division at The Norwegian Institute of Technology was engaged to assist in the compaction control by performing cone penetration tests. An experimental program was carried out by the Norwegian Geotechnical Institute to evaluate use of dilatometer testing as an alternative and supplement to other methods for compaction control.

Methods for soil compaction used

Vibro Wing method

For the first part of the compaction project a relatively new technique for deep compaction, the Vibro Wing method, was chosen. The equipment used consists of a heavy crane mounted vibrator with a mass of 7 tons attached to the top of a 15 meter long steelrod which is provided with wings 0.8 m long and 0.5 - 1.0 m apart. The rod is driven by the vibrator to the decided depth of compaction, facilitated, if necessary by jetting at the bottom of the steel rod.

The time for penetration and retrieval of the Vibro Wing and the frequency of vibration can be varied according to the soil conditions. A typical frequency used is 20 Hz.

The method is principally based on the liquefaction phenomena occuring in saturated sandy soils when exposed to vibrations. The method is best utilized for deposits beneath ground water level. Fig.1 shows in principle the Vibro Wing technique. As the probe is retracted with the vibrator working, the soil is liquefied in a volume corresponding to the wing diameter. The liquefied soil beneath the retracting probe is then consolidated by the weight of the overlying soil mass.

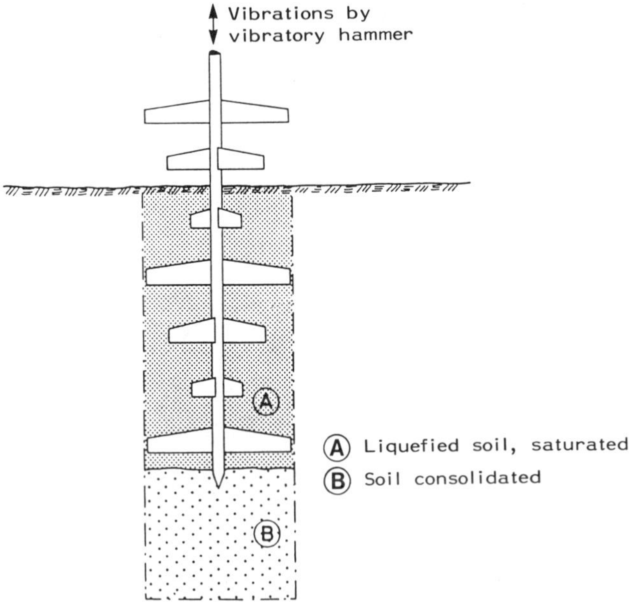

Fig.1 The Vibro Wing system and technique.
(Jendeby 1990)

Only minor shear stresses occur between the liquefied and the nonliquefied soil, causing an effective consolidation process beneath the probe being retrieved. Both the liquefaction process and the consolidation process are highly dependent on soil permeability. If a coarse sand is present, the liquefaction effect may be substantially less than in a finer material. In order not to have the probe locked or stuck during the retraction process, shorter times should be used. On the other hand, if a relatively high content of finer sand or silt is present, the consolidation process is slow due to slow excess pore pressure dissipation. For a silty sand, with silt content up to 30%, the VibroWing compaction capacity was approximately 500 m^3 a day according to experiences in Sweden (Jendeby, 1990). In the literature (Mitchell, 1981) a rule of thumb is mentioned saying that vibrocompaction is ineffective in soils containing more than 20 percent fines. This rule may also be valid for the Vibro Wing method.

Field and laboratory test results concerning the Vibro Wing method have been reported by Massarch and Broms (1983), Jendeby (1990) and others.

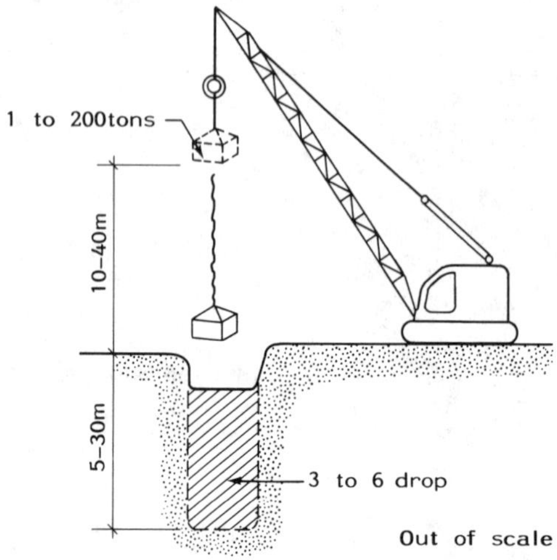

Fig.2 Dynamic compaction by heavy tamping

Dynamic compaction

Soil compaction by repeated dropping of heavy weights on the ground surface is termed dynamic compaction, dynamic consolidation, heavy tamping or pounding. The method in its present form for improvement of large areas was pioneered by Menard.

Fig.2 describes the method principally. Weights up to 172 tons have been used in the tamping process (Broms, 1991). Drop heights up to 40 m have been used. The pounders have dimensions and weights depending on the dynamic bearing capacity at the surface of the ground to be treated. When applied to partly saturated soils, the densification process may be compared to compaction tests made in the laboratory. Dynamic compaction or heavy tamping is considered to be a simple, rapid and low cost soil improvement method. The depth of influence is of particular influence when soil improvement methods are considered.

Mitchell (1981) presents a chart showing the depth of influence as a function of impact energy for heavy tamping, based on case records. In the chart is shown the proposed relation by Menard and Broise (1975).

$$D = \sqrt{W \cdot H}$$

where D = maximum depth of influence, meters
W = falling weight, metric tons
H = height of drop, meters

Likewise is shown the conclusion by Leonards et al. (1980) based on seven cases indicating half the depth of influence given above. The majority of cases presented by Mitchell lie between the values representing Menard's and Leonard's proposals. The depth of influence experienced in the case described in this paper is $D = 0.6 \sqrt{WH}$.

Site description

The construction site is located on one of the larger sand deposits along the Trondheim fjord (Mid-Norway), at Verdal.

The deposit is of at least 40 meters thickness, the upper 20 meters consist of sand, from coarse to fine and silty. Layers of silt are recorded at greater depths than 20 meters. A hydraulic sand fill of 4 meters thickness was placed over the natural deposit at the site. The compacted sand area serves as construction and assembly site for structural jackets to be used offshore. Following assemblys the jackets are shipped by barge. The loading process provides critical concerns for bearing capacity and especially settlements.

The first part of the ground improvement was carried out in 1984, over an area of 25000 m², by the Vibro Wing technique. In 1988 dynamic compaction was used at a neighbouring area. This method was chosen mainly because most of the area was covered by a layer of 50 cm surface compacted rockfill.

The use of the two different deep compaction techniques in the same deposit made an interesting comparison possible. For compaction control cone penetration tests, dilatometer tests, plate load tests, and ground surface measurements were made.

Soil conditions

The soil investigation before the compaction work included sampling to 20-25 m depth in 6 borings and piezocone tests at 6 locations to 20-25 meters depth.

The laboratory investigations and sample descriptions gave soil profiles showing sand layers varying from coarse to fine with irregular and infrequent layers of silty sand and silt. Accordingly, the water content varied from 10 to 20% in the sand and up to 40 percent in silty layers. Some of this silt layers had a silt content of 50-70%.

The piezocone tests delimit some highly resistant layers to a depth of 4 to 5 meters. Below those layers the cone resistance increases approximately linearly with depth down to 20-22 meters which borders a silt layer. The pore pressure recorded by piezocone tests are approximately hydrostatic in the mainly sandy upper 20-22 meters. In the underlying silt, the penetration resistance decreases sharply and the pore pressure recorded increases. The range of point resistance recorded by the 5 piezocone tests to a depth of 20 meters is shown in Fig.3. The ground water table is 3 meters below ground surface. The laboratory investigations included 6 triaxial and 8 oedometers tests.

Laboratory test results:

- angel of internal friction: $\varphi = 35° - 42°$
- compression (constrained) modulus (Janbu, 1967)
  $M = 120 - 150\ (p' \cdot p_a)^{1/2}$

where

p' = stress at which modulus is obtained
p_a = reference stress (~ 100 kPa)

Field work

Vibro Wing compaction

The Vibro Wing equipment used was 15 meters long, with 0.8 meter long "wings" in pairs 0.5 to 1.0 meter apart along the rod, Fig.1. The vibrator had a mass of 7 tons and a frequency of vibration of 20 Hz. The penetration was done in a triangular pattern with 2.5 meters spacing. As penetration difficulties were encountered, some of the wings were cut along the upper two-third length. It was also necessary to adjust the planned penetration time from 15 minutes to 30 to 45 minutes. This included in most compaction points a repeated cycle of penetration.

The unexpected difficulties for the Vibro Wing compaction at some parts of the Verdal site, may be explained by the high penetration resistance found by some CPT-recordings. As may be seen from Fig.3 the point resistance varied from 2 to 10 MPa measured before compaction. An additional explanation is an unfavourable grain size distribution (coarse sand) for the liquefaction of respective layers.

Dynamic compaction

The combination of weight of pounder and drop height was evaluated with respect to depth of influence by the Menard and Broise relation: $D = \sqrt{W \cdot H}$. Chosen weight of pounder was 12 tons and drop height 20 meters. Tests were carried out at the site, in order to find an optimum relation between weight of pounder and drop height. According to the results, 6 drops for each point were chosen. The distance between drop-points was 2.4 meters, obtained by 3 coverages.

Compaction control and results

The levelling of the ground surface after compaction work, gave an average settlement of 0.7 meters for the Vibro Wing compacted area (including surface compaction). At an area compacted by heavy tamping, the average ground surface settlement was 0.5 - 0.55 meters. This corresponds well with the observed depth of influence for the two methods as both methods give approximately 5 percent of ground surface settlement relative to depth of influence. This should also indicate that the densification effect is close to the same for the two compaction methods used.

The depth of influence is evaluated from Fig.3a presenting the cone penetration resistance before compaction and after compaction with the two methods. After compaction 3 cone penetration tests are performed, and the figure shows clearly the inhomogenities at the site both before and after compaction. The resistance measured indicate a poor compaction at former mudline (4 meters depth) and at a depth of 10 to 11 meters where a silt layer is observed by neighbouring sampling. The depth of influence for Vibro Wing compaction is about 14 meters and for heavy tamping about 10 meters. There is a significant increase in point resistance for Vibro Wing compacted soil between 5 and 10 meters compared to heavy tamping. Above that level the average resistance is close to the same for both methods.

In Fig.3b the ground improvement effect is illustrated by the two deep compaction techniques used. This is shown by the range of cone penetration resistance measured before and after compaction (including both Vibro Wing and dynamic compaction). Such an overall picture shows that deep compaction densifies the soil with a factor which may vary in the order of magnitude from 2 to 4 or more. The highest value corresponds to the layers with low point resistance before compaction. The increase in internal friction may be calculated as described by Senneset et al (1982).

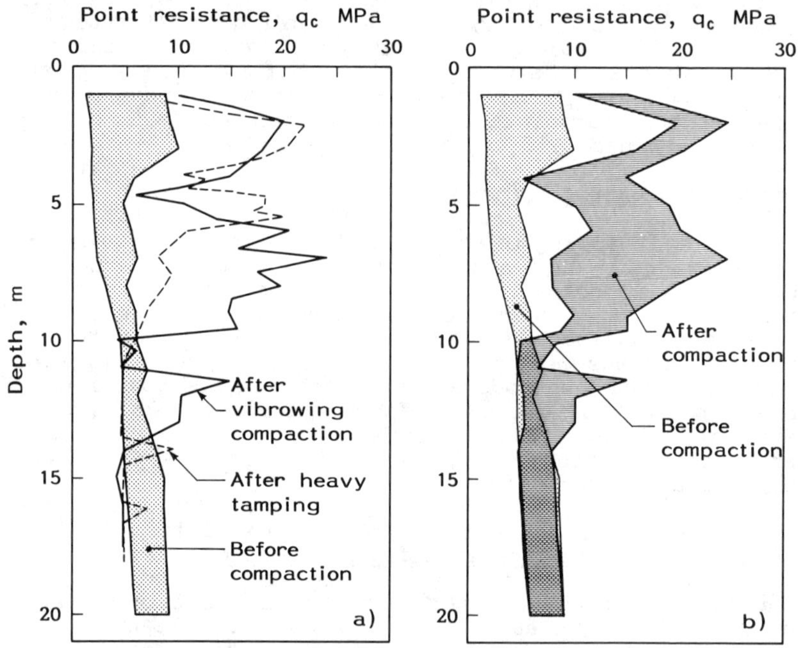

Fig.3 Cone penetration resistance before and after deep compaction.
a) separate results for the two methods
b) overall picture for deep compaction at the site

Dilatometer tests were carried out by the Norwegian Geotechnical Institute at the Vibro Wing compacted area, Lacasse and Lunne (1986), Lacasse (1986). The dilatometer test results suggest that the compaction was effective over a 10 to 15 m depth.

Furthermore a conclusion is that dilatometer results should not be used alone to evaluate soil design parameters. The reliability of the different soil characteristics could be improved by comparison with other available information for this sand.

Plate load tests were carried out after deep compaction. A total of 10 tests were run with plate sizes varying from 0.9 m² to 3.0 m². The plate load tests and applied loads during construction of jackets, confirm the high shear strength parameters interpreted from in situ tests. Likewise, it is confirmed that the deformation parameters are of a realistic order of magnitude.

Compressibility - settlements

Strict requirements were set for differential settlements during the construction period and load-out of the structural jackets. After compaction and compaction control, an evaluation of settlements was made.

For the calculations a set of compression moduli could be obtained, based on cone penetration tests and dilatometer tests. Also a number of plate load tests were available for the confirmation of calculations, but of a limited size and corresponding limited depth of influence.

According to Senneset et al (1982, 1989) it is possible to obtain a bulk modulus of (vertical) compression from cone penetration testing by semiempirical rules. For sand and coarse silt, the compression modulus may be expressed by

$$M_s = m_s \sqrt{q_n \cdot p_a}$$

Herein

$q_n = q_c - p$ = net cone resistance

p_a = 100 kPa = reference stress, ~ 100 kPa

An average value of m_s = 40 is suggested, but the likely variation may be up to ±10 or, or more. The largest values correspond to the densest state. Using the CPT-recordings together with this semiempirical formula, a profile with compression moduli might be established, Fig.4.

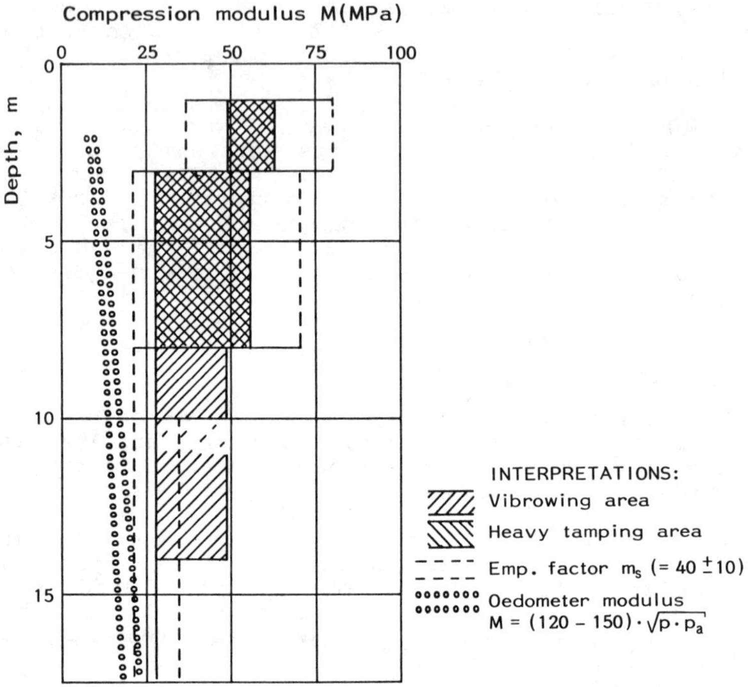

Fig.4 Compression moduli derives from cone penetration tests and oedometer tests

In this figure, the shadowed areas indicate the best estimate for compression moduli based on the CPT-recordings from Vibro Wing and dynamic compacted areas. The wide range of moduli for a variation of empirical factor m_s = 40 ± 10 is also indicated. In this figure is also shown a range of values of the 9 oedometer moduli obtained through laboratory investigations before compaction. The overall picture as given in Fig.4 shows that the compressibility of the deposit has been reduced to 1/3 - 1/4 of its original value in the upper layers.

The constrained modulus from dilatometer tests is difficult to evaluate. But an average modulus of magnitude M = 60 MPa to a depth of 6 meters and M = 40 MPA from 6-10 meters may be reasonable (after compaction).

The change in compressibility by deep compaction, may furthermore be illustrated as given in Fig.5. Settlements are approximately calculated for a foundation of 3 m² based on the compression moduli given above, based on CPT and dilatometer tests (averaged range of values). The findings are compared to results from a plate load test with plate area 3m².

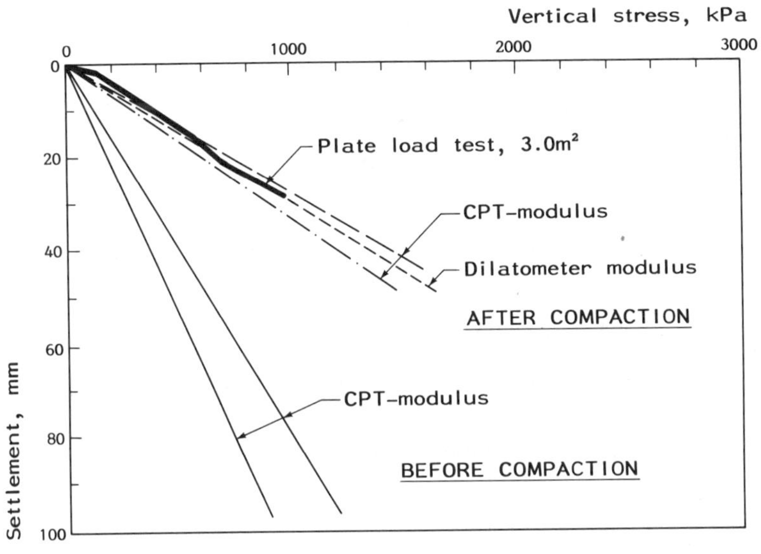

Fig.5 Calculated settlements based on in situ tests compared to plate load tests (Vibro Wing compacted area)

There is a good correlation between plate load test results and calculated settlements based on compression moduli from cone penetration tests and dilatometer tests after compaction. Based on this we may also have a certain confidence in moduli obtained from

CPT tests before compaction. Also the corresponding settlements calculated, indicate that the deep compaction work reduced the settlements to 1/3 for a foundation of 3 m² and external loads reaching 1000 kPa.

Conclusions

- Both the Vibro Wing and the dynamic compaction used in this project are technically effective deep compaction methods for a sandy- silty deposit. Both methods have compacted the volume of soil influenced by 5 percent.

- Coarse materials may increase the time, and thereby costs, of the Vibro Wing method.

- The depth of influence for Vibro Wing compaction corresponds to the depth to which the equipment is brought.

- The depth of influence observed by heavy tamping corresponds to a relation between depth and energy per drop given by $D = 0{,}6\sqrt{W \cdot H}$.

- The cone penetration test is a very applicable compaction control method, even when very dense and coarse sand materials are present. The dilatometer test require finer and more homogenious soils to give consistent values without problems.

- Heavy tamping effect is not influenced by layers of rock and other coarse layers. Vibro Wing compaction is a method more applicable in homogeneous deposits, medium to fine sands, and is a more gentle method for use close to existing structures.

Acknowledgement

The authors are grateful to the Norwegian Geotechnical Institute for the opportunity to use some of the dilatometer test results from their experimental program at the Verdal site.

We also appreciate very much the assistance given by dr. Leif Jendeby, Chalmers University of Technology, by providing us with test and research results from Sweden.

References

Broms, B.B. (1991). "Deep compaction of granular soils. Foundation Engineering Handbook. Second Edition, Vn Nostrand Reinhold, New York. pp.814-832

Janbu, N. (1967). "Settlement Calculations based on the Tangent Modulus Concept". Bulletin No.2, Geotechnical Division, The Norwegian Institute of Technology, Trondheim

Jendeby, L. (1990). " Djuppackning, ett teknisk/ekonomisk alternativ till påling". (Deep compaction, a technical /economical alternative to piling). NCC Bygg AB, Chalmers University of Technology

Lacasse, S. and Lunne, T. (1986). " Dilatometer tests in sand". ASCE Geotechnical Special Publication. Speciality conference on use of in situ tests in geotechnical engineering, Blacksburg. pp.686-699

Lacasse, S. (1986). " Dilatometer tests before and after compaction of sand foundation in Verdal". Norwegian Geotechnical Institute, Internal Report

Leonards, G.A. Cutter, W.A. and Holtz, R.D. (1980). "Dynamic compaction of granular soils, Journal of the Geotechnical Engineering Division, ASCE 106, No.GT-1, pp.35-44.

Massarch, K.R., Broms, B. (1983). " Soil Compaction by Vibro Wing method". Eight European Conf. on Soil Mechanics and Foundation Engineering, Helsinki

Mellerud, T. (1988). "Compaction of soil". Master thesis. Geotechnical Division, The Norwegian Institute of Technology, Trondheim

Menard, L. and Broise, Y. (1975). "Theoretical and practical aspects of dynamic consolidation". Geotechnique 25, No.1, pp.3.18.

Mitchell, J.K. (1981). "Soil Improvement - State-of-the-art Report". 10th International Conference on Soil Mechanics and Foundation Engineering, Stockholm. 4, pp.509-565

Senneset, K., Janbu, N., Svanø, G. (1982). "Strength and Deformation Parameters from Cone Penetration Tests". Proceedings of the Second European Symposium on Penetration Testing. Amsterdam. pp.863-870

Senneset, K., Sandven, R., Janbu, N. (1989). " Evaluation of Soil Parameters from Piezocone Tests". Transportation Research Record 1235. National Research Council, Washington D.C. pp.24-37.

Compaction of Granular Soils - Remarks on Quality Control

by Michele Jamiolkowski[1] and Erio Pasqualini[2]

Abstract

Based on recent experimental data, it is pointed out that a quality control program based only on the evaluation of D_R can be inadequate; a better estimation of the densification of sands is possible if the effects of the stress and strain history induced on the improved soil by compaction are considered. It is also stressed the importance of taking into account the mineralogy of the improved soil when evaluating its geotechnical parameters from in situ tests.

Introduction

The densification of granular soil is generally required to avoid liquefaction and/or to limit settlements. The quality control of sand deposits improved by vibroflotation, vibrocompaction, tamping and blasting is usually based on a relative density acceptance criterion. Since the direct determination at different depths of the in-place void ratio is uncertain due to difficulty in obtaining good undisturbed samples in granular deposits, relative density is often evaluated from the results of Standard Penetration Test (SPT) and the Cone Penetration Test (CPT), for which empirical correlations with relative density are available. However, a large amount of experimental data show that both blow count, N, from SPT and cone resistance, q_c, from CPT are strongly influenced by effective horizontal stresses, σ'_{ho}, existing in the soil prior to

1 Prof., Technical University of Torino, Italy.
2 Prof., Fac. of Engrg., University of Ancona, Italy.

penetration; therefore the correct use of the correlations for evaluating D_R from N and/or q_c requires the knowledge of σ'_{ho}. This knowledge, however, represents one of the most complex geotechnical problems; moreover, when improved soils are examined, it is fundamental to know if compaction produces not only the densification of sands but also a change in effective horizontal stresses. This subject, already investigated by other researchers, [Saito (1977), Leycure and Schroeder (1987)], is discussed, considering recent experimental data, to point out that quality control based only on values of D_R can be insufficient to evaluate the ground modifications achieved by compaction. A better understanding of the effects of the densification is possible if one considers the influence of the stress and strain history (SSH) induced by the compaction on the resistance against liquefaction and the deformability of the improved soils. The present study is mainly based on CPT results, but its conclusions also apply, at least qualitatively, to SPT.

Evaluation of relative density from CPT

As penetration resistance, q_c, is influenced by several factors, its correct use to estimate D_R requires the consideration of the following aspects.

a) The dominant influence of initial horizontal stress (σ'_{ho}) on penetration resistance has long been recognized from the results of large calibration chamber, (CC), tests (Schmertmann, 1978; Baldi et al., 1985; Jamiolkowski et al., 1985). This has important implications for the evaluation of D_R. Based on extensive CC studies (Baldi et al., 1986) the following relationship, shown in Fig. 1, has been obtained for medium uniform Ticino sand (TS) to evaluate D_R:

$$D_R = \frac{1}{2.38} \ln \left[\frac{q_c}{248\ (\sigma'_{ho})^{0.55}} \right] \tag{1}$$

where q_c and σ'_{ho} are in kPa (1 ton/ft^2 = 107 kPa = 0.107 MPa). This formula reflects the fact that the penetration resistance is controlled by in situ effective horizontal stresses (Houlsby and Hitchman, 1988). The analogous expression related to effective overburden stresses (Schmertmann, 1978) is applicable to NC deposits only, having K_o constant with depth, where K_o = coefficient of earth pressure at rest.

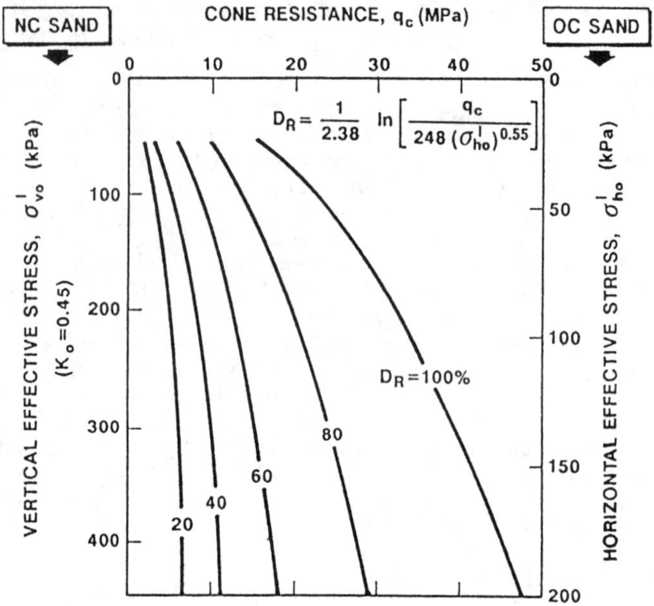

Figure 1. Relative density correlation for uncemented, unaged silica sand (Ticino sand), after Baldi et al. (1986). (q_c, σ'_{ho} in kPa).

b) Skempton (1986) has shown the importance of aging on the interpretation of penetration tests in cohesionless soils. Since the CC correlations were developed on pluvially deposited, unaged, uncemented sands it is likely that the D_R vs q_c correlations (Fig. 1) will lead to an overestimation of D_R when applied to natural sand deposits. However, the same correlations will underestimate D_R if applied to more crushable and compressible sands or sands containing more than 5 to 10% fines.

c) Bearing in mind the importance of factors like compressibility, aging, and the presence of fines, the relationship illustrated in Fig.1 can be used as a guide for evaluating in-situ density of clean, predominantly silica sands of recent, uncemented deposits. In the evaluation of normally consolidated (NC) sands the vertical effective stresses (σ'_{vo}) can be applied: in Fig. 1, K_o is assumed equal to 0.45. Of course in NC soils, K_o decreases with increasing D_R; for the sake of simplicity (here and in the following) K_o is assumed to be constant. The evaluation of D_R in OC sands becomes less reliable because of the inherent difficulties in proper assessment of σ'_{ho} (the evaluation of σ'_{ho} for improved soils is difficult too).

Modulus of sands from CPT results

The evaluation of deformation characteristics (moduli) of cohesionless soils from penetration tests is a very complex problem that has been examined by Bellotti et al. (1989) and Baldi et al. (1989). Therefore only the main practical aspects pertaining to the subject are here considered.

The deformation characteristics of a given soil depend on many factors, among which the stress and strain history of the deposit is very important. On the other hand this history has little influence on penetration test results. This was initially shown by the results of small scale laboratory tests (Lambrechts and Leonards, 1978) and fully confirmed by the results of CC tests (Baldi et al., 1985, 1986; Jamiolkowski et al., 1988).

Fig. 2a shows correlations between the CC results of CPT's and deformation characteristics (E'_s) of the predominantly silica, Ticino sand (TS); where E'_s corresponds to the secant drained Young's modulus inferred from CK_oD triaxial compression tests performed on pluvially deposited TS. The E'_s values refer to a given effective mean consolidation stress (σ'_m) and to a level of axial strain (ε_a) of 0.1% in OC sand and 0.1 and 0.25% in NC sand. The value of ε_a = 0.1% corresponds to the average vertical strain of practical interest when designing foundations in cohesionless soils (Jardine et al., 1986; Battaglio and Jamiolkowski, 1987). An analysis of Fig. 2a reveals that even for the same sand the ratio of the reference modulus (E'_s), to the penetration test result (q_c) is substantially higher for mechanically OC specimens than for NC specimens. This trend, which is observed in all silica sands and for a variety of penetration tests validated in CC's around the world, reflects very high sensitivity of the reference moduli and low sensitivity of the penetration test results to the strain and stress history of the sand. The above results from CC studies performed on freshly deposited silica sands. Further research is necessary to evaluate to what extent these findings are applicable to natural and not necessarily silica sands. However the writers believe that, at least qualitatively, similar overall trends should be also expected in natural aged sands. For natural aged NC sand deposits the writers believe that correlations between penetration resistance and moduli, for example E'_s and q_c, may lay somewhere between the NC and OC sand correlations developed for unaged sands in the CC studies, as shown in Fig. 2b (For details see Bellotti et al., 1989 and Baldi et al., 1989). Fig. 2 emphasizes that a correct use in practice of empirical correlations between penetration resistance and moduli must take into account, at least qualitatively, the influence of stress and strain history.

Influence of compaction on stress and strain history of the soil

The previous discussion shows that a correct evaluation of D_R and moduli of improved soils from penetration test results requires knowledge of the stresses and strains induced in the soil by compaction. Unfortunately, at present only very few experimental data on this topic exist, all of which concern the influence of vibrocompaction on K_o.

Saito (1977) shows that the change in magnitude of

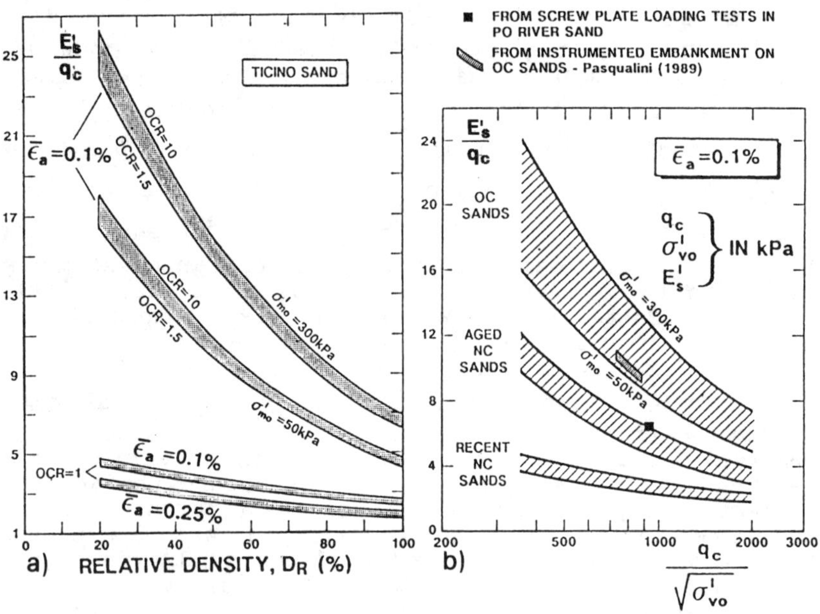

Figure 2. Evaluation of drained Young's modulus from CPT for silica sands, after Baldi et al. (1989) and Bellotti et al. (1989).

total lateral stress, σ_{ho}, obtained by pressuremeter tests in a borehole, can be very high as a result of probing. The computed K_o values of the improved soil are from three to six times the K_o values of the virgin soil. The results of Schroeder et al. (1977) and those of Sorota et al. (1981), both deduced from circumferential strain measurements on steel sheet piles, show that only small increases of the coefficient of the lateral earth pressure were produced by probe compaction. Because of limitations due to the lack of experimental data and to the differences among them, it is not possible, at present, to draw firm conclusions.

Based on presently available data the writers believe that the compaction of granular soils can generally cause both an increase in D_R and generate effects similar to those produced by cyclic prestressing and/or prestraining; an increase in σ'_h is probably also present.

Influence of stress and strain history on D_R, settlements and liquefaction potential

In this section it is shown, by means of a simple example, that for a given penetration resistance the increase in horizontal effective stress influences D_R and moduli values of improved soils in an opposite manner. Control of densification of sands usually requires that a pre-established value of D_R be obtained. In this connection, it is important that the engineer and contractor agree beforehand upon the in-situ test and correlation to be used for evaluating D_R. This evaluation can be considered correct only if it is possible to confidently @@estimate the K_o value of the improved soil. For example, if we assume D_R = 80%, Fig. 3 shows the q_c values versus depth for different values of K_o. The cone resistance values in Fig. 3 were obtained from equation 1, assuming the water level at the ground surface and the total unit weight, γ_t = 20 kN/m^3. In the same way, for example, it is easy to see that the q_c-versus depth curve A of Fig. 3 gives different values of D_R according to the assumed K_o value; for example we obtain: D_R = 80% when K_o = 0.45; D_R = 75% when K_o = 0.55; D_R = 55% when K_o = 1.42 and so on. Therefore to assume that compaction increases K_o is a burden for the Contractor, on the other hand, assuming that K_o doesn't increase can lead to an unsafe

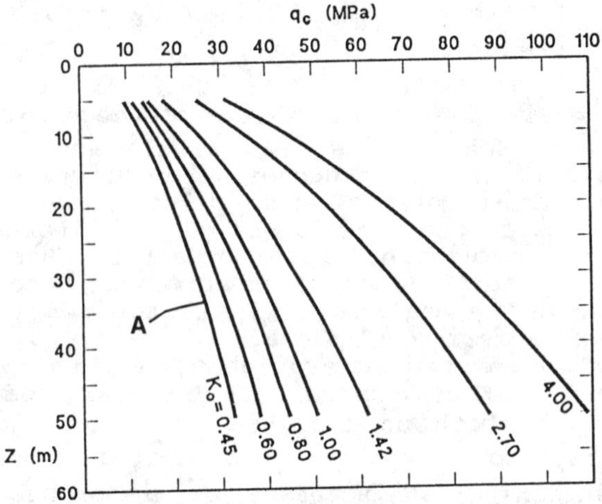

Figure 3. Example of the influence of K_o on q_c versus depth relationship, for a given D_R = 80%.

prediction of D_R.

The above considerations apply only when limiting quality control to the evaluation of relative density; in fact, the consequences of an increase of K_o are quite different when quality control is based on the evaluation of the compressibility of the improved soil or on the prediction of settlements of foundations. This is pointed out in Fig. 4, where the settlements of shallow foundations founded on an improved soil, are plotted using the q_c values of curve A shown in Fig. 3. The settlements were calculated using the method by Schertmann (1970) and Schmertmann et al. (1978) and the values of E'_s/q_c ratio from Fig. 2. The E'_s/q_c ratios were selected at the corresponding D_R values at σ'_m = 50 kPa and K_o has been related to OCR using the following relation (Mayne and Kulhawy, 1982)

$$K_o^{OC} = K_o^{NC} (OCR)^{0.5} \leq K_p \qquad (2)$$

OCR = overconsolidation ratio
K_o^{OC} = coefficient of earth pressure at rest of the OC soil
K_o^{NC} = coefficient of earth pressure at rest of the NC soil, assumed equal to 0.45
K_p = coefficient of passive earth pressure.

The example of Fig. 4 shows that the largest

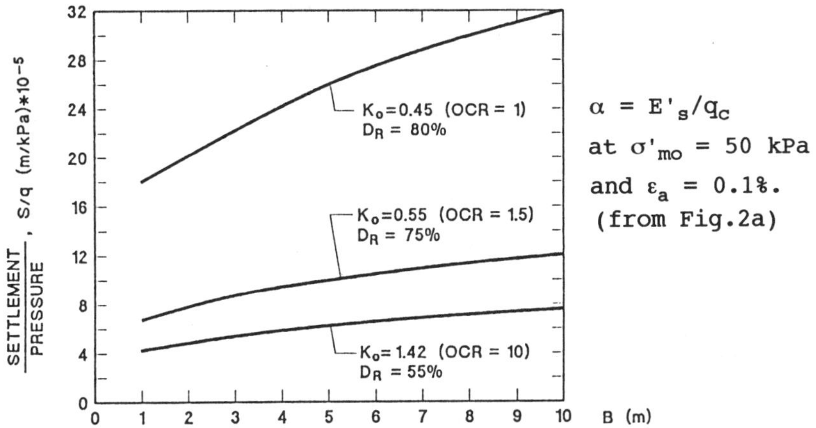

Figure 4. Settlement versus foundation sizes for the different K_o-D_R combinations from curve A of Fig. 3.

settlements for the deposit occur for K_o = 0.45 although it has the highest D_R (= 80%). On the other hand the predicted settlements are least for the deposit with the lowest D_R and greatest value of K_o. This example points out that the mechanical overconsolidation of the soil has an important influence on the deformability of the soil, greater than the one produced by a variation in D_R. The results of Fig. 4 are only valid for cohesionless soil improved by pre-loading (and, maybe, by tamping), where σ'_{ho} and K_o are strictly interrelated to OCR. The increment of σ'_{ho} produced by vibrocompaction, vibroflotation and blasting has probably less effect on the deformability of the improved soil; in these cases the increment of soil-stiffness is correlable, in first approximation, to the square root of the increment of the mean effective stresses produced by compaction. Moreover, even if we want to cautiously assume that σ'_{ho} does not increase, we should consider that densification may induce, in the improved young soil deposit, effects similar to those produced by cyclic prestressing and prestraining.

It is interesting to note that Schmertmann et al. (1986), in order to check the soil at a power plant site improved by dynamic compaction and compaction grouting, initially used an equivalent CPT-based relative density acceptance criterion. However, during the work they modified criteria to include a dilatometer test-based modulus criterion. This indirectly proves the need to overcome the limits of a quality control specification based only on the evaluation of D_R.

Finally, the possible effects of compaction on the cyclic stress and strain history of the improved soil must also be considered when densification of granular soils is performed to avoid liquefaction; in fact, the liquefaction potential is usually correlated with the relative density of the saturated sand deposit (or directly with the penetration resistance), but the experimental data by Seed and Peacock (1971) and Seed et al. (1975) clearly show that sands with same D_R greatly increase their resistance to liquefaction both when they have higher K_o values and when they are subject to prestressing and/or prestraining.

Influence of mineralogy

At present all the correlations for the evaluation of geotechnical parameters from in situ tests results have involved clean, predominantly silica sands; therefore, it is necessary to be very cautious when dealing with silty sands or sands having different mineralogical composition.

Particularly in the case of sands with a high carbonate content the use of the correlations like those shown in Figs. 1 and 2 can lead to glaring predictions. This is proved by the preliminary experimental data obtained in CC on the Quiou sand and shown in Figs. 5 and 6 (Bellotti and Jamiolkowski, 1991). Based on these preliminary data, we observed that when correlations obtained for silica sands are used to predict the behaviour of calcareous sands D_R was underestimated and moduli greatly overestimated. For this reason, and because correlations applicable to calcareous and crushable sands are not yet available, placement control of these sands must not solely depend on the results of penetration tests; for these sands it seems appropriate to perform load tests.

Closing remarks

a) Recent CC research has clearly shown that correlations between any kind of penetration resistance and D_R require the knowledge of σ'_{ho}.

b) All correlations between penetration resistance and D_R through σ'_{vo} are applicable only to uncemented NC soils having K_o more or less constant with depth; when using these correlations in soil deposits that do not meet the

Sand	e_{max}	e_{min}	n
QS-calcareous	1.281	0.831	0.50*
TS-silica	0.923	0.575	0.38
HS-silica	0.894	0.549	0.42

(*) Assumed as first approximation

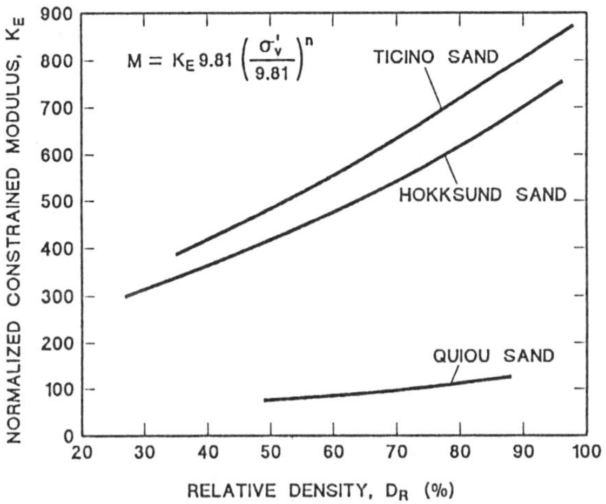

Figure 5. Normalized constrained modulus of two silica and Quiou NC sands versus relative density, adapted from Bellotti and Jamiolkowski (1991).

above conditions the tendency is to overestimate D_R.

c) In addition it must be pointed out that all existing correlations between penetration resistance and D_R through effective stresses are valid only in moderately crushable and uncrushable sands. Their use in more crushable sands leads to underestimated values of D_R.

d) The above statements demonstrate that an acceptance criterion for a densification treatment in terms of minimum value of D_R to be achieved, is correct if we are able to know or to estimate the post densification σ'_h and the mineralogy of soil to be densified.

e) Unfortunately very little is know about σ'_h after densification and the information reported in literature results to be partially contradictory. Therefore, it seems necessary to carry out experiments similar to those reported by Saito (1977), Schoeder et al. (1977) and Sorota et al. (1981). Such experimental researches can

Test n.	356 (D)	357 (D)	358 (S)	359 (S)	360 (S)	361 (S)	362 (S)	363 (D)
D_R (%)	86.0	49.1	88.2	51.1	51.8	62.0	96.9	69.8
q_c (QS) MPa	7.97	5.18	7.88	3.97	5.00	6.25	11.37	8.00
q_c (TS) MPa	17.06	6.90	16.96	6.64	6.82	9.15	21.17	11.25

(D) = dry specimen, (S) = saturated specimen

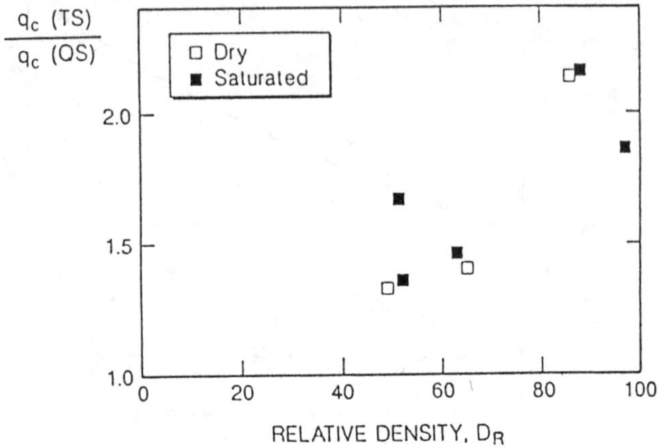

Figure 6. Comparison of cone resistance of calcareous QS and silica TS at equal relative density, after Bellotti and Jamiolkowski (1991).

produce significant practical effects in the evaluation of the compaction results.

f) It is important to note that the assumption that compaction does not produce any increment in the horizontal effective stresses is unconservative if penetration resistence is used to determine relative density while is conservative if it is used to evaluate settlements. In these circumstances, it appears more rational to fix the acceptance criterion for a densification treatment in terms of a minimum value of penetration resistance to be achieved. Such a value can be determined on the basis of the allowable settlement evaluated on the conservative hypothesis that no increment in the horizontal stress occurs.

g) Finally, at present, the execution of load tests seems necessary to check the improvement of site soil where significant structures have to be built, and to predict the behavior of crushable sand deposits.

References

1) Baldi, G. et al. (1985) "Laboratory Validation of In-Situ Tests". AGI Jubilee Volume, 11th ICSMFE, San Francisco.
2) Baldi, G., Bellotti, R., Ghionna, V.N., Jamiolkowski, M. and Pasqualini, E. (1986) "Interpretation of CPT's and CPTU's. Second part: Drained Penetration on Sands", Proceed. IVth Int. Geotechnical Seminar on Field Instrumentation and In Situ Measurements, Nan Yang Tech. Inst., Singapore.
3) Baldi, G., Bellotti, R., Ghionna, V.N., Jamiolkowski, M. and Lo Presti, D.C.F. (1989) "Modulus of sands from CPT's and DMT's", XII ICSMFE Rio de Janeiro, Brasil.
4) Battaglio, M. and Jamiolkowski, M. (1987) "Analisi delle Deformazioni", XIII Ciclo di Conferenze di Torino, in Italian.
5) Bellotti, R., Ghionna, V.N., Jamiolkowski, M. and Robertson, P.K. (1989) "Design Parameters of Cohesionless Soils from in-situ tests", Speciality Session on "In Situ Testing of Soil Properties for Transportation Facilities". Committee.
6) Bellotti, R. and Jamiolkowski, M. (1991) "Evaluation of CPT and DMT in Crushable and Silty Sands", Third Interin Report. ENEL C.R.I.S., Milan.
7) Houlsby, G.T. and Hitchman, R. (1988) "Calibration chamber tests of a cone penetrometer in sand" Géotechnique, 38, No. 1, 39-44.
8) Jamiolkowski, M., Ladd, C.C., Germaine, J.T. and Lancellotta, R. (1985) "New Developments in Field and Laboratory Testing of Soils", Theme Lecture, Proc. XI ICSMFE, San Francisco, Vol. 1, pp. 57-153.
9) Jamiolkowski, M., Ghionna, V.N., Lancellotta, R. and Pasqualini, E. (1988) "New Correlations of Penetration Tests for Design Practice, Proc. ISOPT-I, Orlando, Fla.
10) Jardine, R.J., Potts, D.M., Fourie, A.B. and Burland, J.B. (1986) "Studies of the Influence of Non-Linear Stress-Strain Characteristics in Soil Structure Interaction", Geotechnique, 36.
11) Lambrechts, J.R. and Leonards, G.A. (1978) "Effects of Stress History on Deformation of Sand", Journal of Geotechnical Engineering Division, ASCE, No. GT 11, pp. 1371-1387.
12) Leycure, P. and Schroeder, W.L. (1987) "Slope Effects on Probe Densification of Sands", Soil Improvement - A Ten Year Update, ASCE Geotechnical Special Publication No. 12, pp. 197-214.
13) Mayne, P.W. and Kulhawy, F.H. (1982) "K_o-OCR Relationships in Soils", Journal of the Geotechnical

Engineering Division, ASCE, Vol. 108, No. GT6, pp. 851-872.

14) Pasqualini, E. (1989) "Analisi dei cedimenti di un rilevato di precarico", CNR, Roma, in Italian.

15) Saito, A. (1977) "Characteristics of Penetration Resistance of a Reclaimed Sandy Deposit and Their Change Through Vibratory Compaction", Japanese Society of Soil Mechanics and Foundation Engineering, Vol. 17, No. 4, pp. 31-43.

16) Schmertmann, J.H. (1970) "Static Cone to Compute Static Settlement over Sand", Journal of the Soil Mechanics and Foundation Engineering, Division, ASCE, Vol. 96, No. SM3, pp. 1011-1043.

17) Schmertmann, J.H. (1978) "Guidelines for cone Penetration Test Performance and Design Report N. FHWA-TS-78-209, U.S. Department of Transportation, Federal Highway Administration, Washington, D.C.

18) Schmertmann, J.H., Baker, W., Gupta, R. and Kessler, K. (1986) "CPT/DMT QC of Ground Modification at a Power Plant", in Use of In Situ Tests in Geotechnical Engineering, Geotechnical Special Publication No. 6, ASCE, pp. 985-1001.

19) Schmertmann, J.H., Hartman, J.P. & Brown, P.R. (1978) "Improved Strain. Influence Factor Diagrams", Journal of the Geotechnical Engineering Division, ASCE, Vol. 104, No. GT8, pp. 1131-1135.

20) Schroeder, W.L., Marker, D.K., and Khuayjarernpanishk, T. (1977) "Performance of a Cellular Wharf", Journal of the Geotechnical Engineering Division, ASCE, Vol. 103, No. GT3, pp. 153-168.

21) Seed, H.B., Mori, K. and Chan, K. (1975) "Influence of Seismic History on the Liquefaction Characteristics of Sands", Report No. EERC 75-25, Berkeley, California.

22) Seed, H.B. and Peacock, H. (1971) "Test Procedure for Measuring Soil Liquefaction Characteristics", Journal of the Soil Mechanics and Foundation Division, ASCE, Vol. 97, No. SM8, pp. 1099-1119.

23) Skempton, A.W. (1986) "Standard Penetration Test Procedures and the Effects in Sand of Overburden Pressure, Relative Density, Particle Size, Aging and Overconsolidation", Geotechnique, 36, No. 3, pp. 425-447.

24) Sorota, M.D., Kinner, E.B. and Haley, M.X. (1981) "Cellular Cofferdam for Trident Drydock: Performance", Journal of the Geotechnical Engineering Division, ASCE, Vol. 107, No. GT12, pp. 1657-1676.

A New Technique for Quality Control of Dynamic Compaction

Chaim J. Poran[1], King-Sen Heh[2], and Jorge A. Rodriguez[3]

Abstract

In recent years dynamic compaction (DC) has become a popular method for deep improvement of loose cohesionless soils and man made fills of disposed waste and rubble. Many DC sites have irregular subsurface conditions where it is difficult to interpret verification test results and assess soil properties after DC improvement. The paper describes a new technique for quality control of DC that is based, primarily, on results from an extensive experimental model study of impact response in dry sand. A method is presented for the evaluation of the dynamic stiffness of the soil mass affected by the impact by computing a dynamic settlement (DSM) from integration of the impact acceleration record with respect to time using measured integration constants. DSM values obtained in the model tests show good correlation to soil densities and corresponding elastic moduli obtained from laboratory tests. The proposed method may have immediate construction application as it offers a reliable and cost effective alternative to quality control of dynamic compaction.

Introduction

Dynamic compaction (DC) is a soil improvement method which is rapidly

---

[1] Asst. Prof., Civil Engineering Dept., Univ. of North Carolina at Charlotte, Charlotte, NC 28223

[2] Former PhD student, Dept. of Civil and Environmental Engineering., Polytechnic University, 333 Jay St., Brooklyn, NY 11201

[3] Graduate Student, Civil Engineering Dept., Univ. of North Carolina at Charlotte, Charlotte, NC 28223

gaining acceptance in the United States and abroad. The method involves the repeated application of high energy impacts on the soil surface using steel or concrete tampers weighing 100 to 250 KN, dropped from heights of 10 to 30 m, compacting the soil strata to considerable depth. The tampers are dropped repeatedly at each impact point on a predetermined pattern over the treatment area. Depending on soil conditions, two or more phases (or "passes") of DC are carried out on a progressively closer grid spacing pattern until the required ground improvement is confirmed by verification testing or by predetermined number of drops for each pass. The DC method is particularly effective in granular soils, deposits of rubble, and landfills. Generally, site investigation and verification testing may include one or more of the following tests: SPT, CPT, pressuremeter, dilatometer and plate load tests. However, in many such sites where soil conditions are irregular (such as in boulder and rubble deposits) verification testing is difficult and the results are often confusing and misleading. Although DC is widely used in such soils, the dynamic soil-tamper interaction at impact is not well understood and, to date, there is no comprehensive model to quantify the impact induced deep densification and its relationship to impact characteristics.

Ellis (1986), Orrje and Broms (1970) and other researchers have conducted analytical and experimental investigations of soil-tamper interaction and impact response of granular soil. However, none of these investigations directly correlated soil-tamper interaction during impact to the permanent effect of that impact on soil density and deformation characteristics.

The new quality control (QC) technique presented herein is based, primarily, on results from an extensive laboratory model study of DC of dry sand. In that study (Heh, 1990) measurements included tamper acceleration and soil pressures during impact, and soil densities and strains before and after impact. A procedure is presented to analyze the dynamic results and a dynamic settlement modulus (DSM) is calculated as subsequently shown. Based on this procedure, a telemetry system has been designed for on-site, instantaneous quality control and verification of DC projects.

Analysis of Experimental Data

As reported by Poran et al. (1991) and Heh (1990) the data analysis procedure developed is based on simple dynamics concepts. The response of soil strata to an impact of a large tamper is influenced by global properties of a large soil mass under the impact point. Therefore, this response realistically represents these properties. The tamper-soil interaction is defined by the boundary conditions and soil properties. This global response is believed to be much better suited to represent the variable soil conditions encountered in many DC sites. The dynamic soil-tamper interpretation of such interaction during impact is therefore the key to successful interpretation of such global soil properties. The analysis procedure, based on the dynamic settlement modulus (DSM) concept, is described as follows.

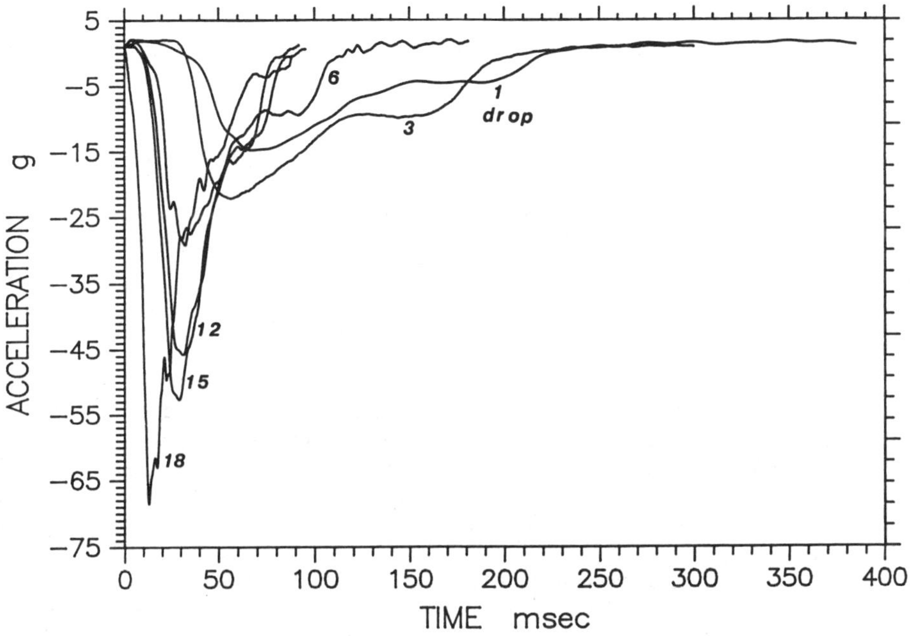

Figure 1 - Impact Acceleration vs. Time

The DSM Concept.

Typical accelerations recorded on the tamper during impact are shown in Figure 1 (Heh, 1990). Based on Lucas (1986), these records are also typical to what was observed in field monitoring of DC projects. It may be noted that as the number of drops increase so does the global dynamic stiffness of the compacted sandy soil mass, as reflected by the increase in the acceleration

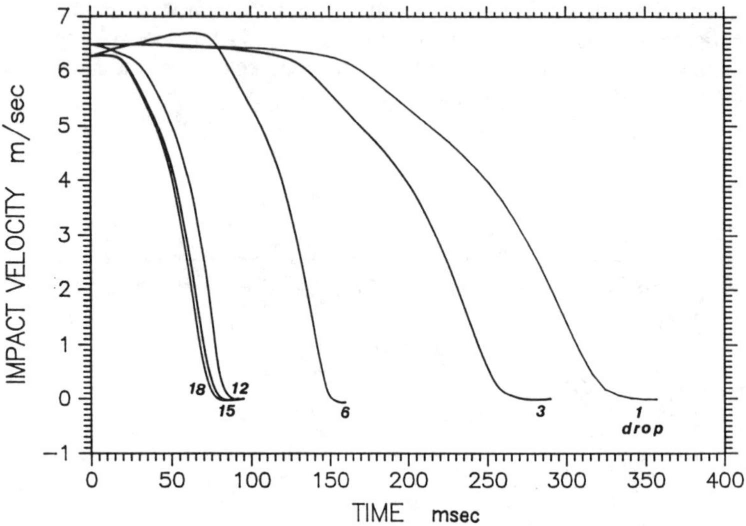

Figure 2 - Impact Velocities vs. Time

peak and the shortening of the response duration as was found in the test results. In the first processing step, each acceleration record, a_t, was integrated once with respect to impact time, t, to obtain the tamper velocity V_t, as:

$$V_t = \int a_t dt + C_v \tag{1}$$

where C_v was determined at the end of the acceleration record when $V_t = 0$. C_v could also be determined at t=0, when the velocity at impact is known. A typical adjusted velocity record is shown in Figure 2. This and the following integrations were performed at test time by a microcomputer with a high speed special software, which enabled rapid visual inspections and modifications of the digitalized records as appropriate at any time during the signal processing.

The next step was to obtain tamper displacement record, d_t, by integrating the digitized velocity record as:

$$d_t = \int V_t dt + C_d \tag{2}$$

Adjusted d_t curves are shown in Figure 3, where C_d was determined based on the actual "at rest" incremental tamper displacement for each drop, i.e., C_d was calculated by equating the computed displacement value, d_t, at the end of each impact to the measured, net incremental tamper displacement for that drop.

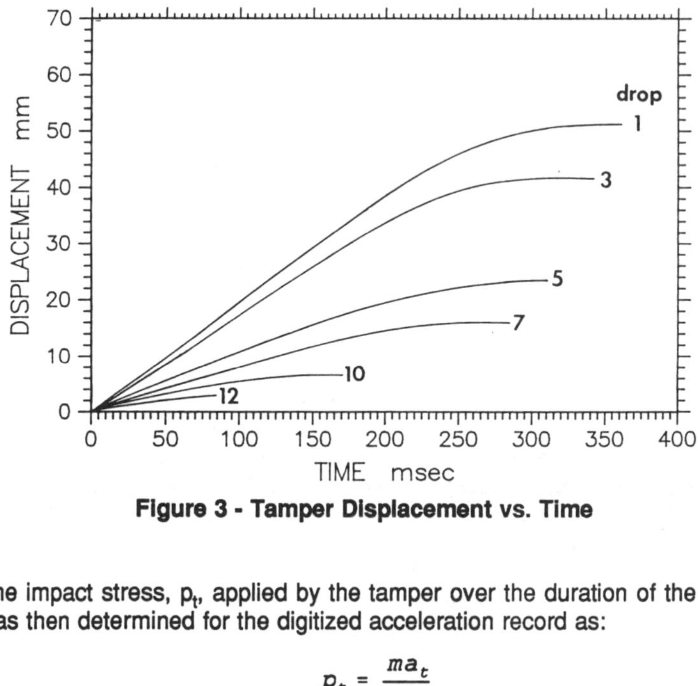

Figure 3 - Tamper Displacement vs. Time

The impact stress, p_t, applied by the tamper over the duration of the impact was then determined for the digitized acceleration record as:

$$p_t = \frac{ma_t}{A} \tag{3}$$

where m and A where the mass and contact area of the tamper, respectively. Then, the adjusted dynamic displacement record, d_t, was plotted versus the corresponding impact stress, p_t. The DSM was defined as the slope of the tangent of the loading portion of the impact stress-relative settlement curve, as: where the relative tamper settlement, ε_i, is defined as the tamper displacement,

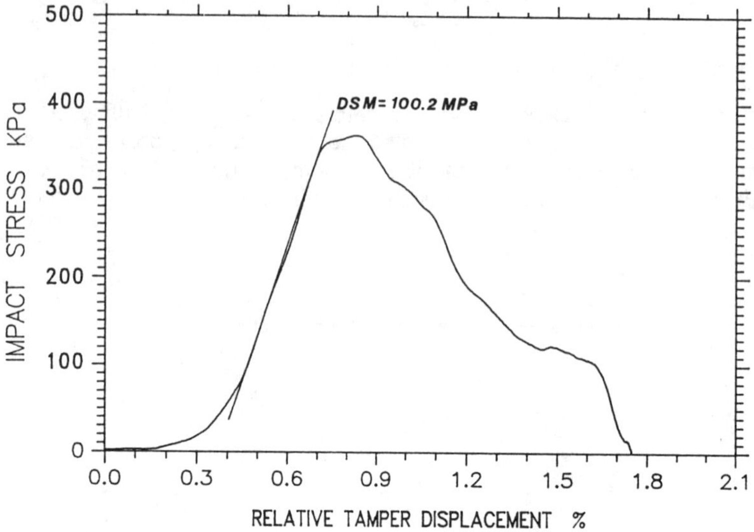

Figure 4 -Impact Stress vs. Relative Tamper Settlement

$$DSM = \frac{\Delta p_t}{\Delta\left(\frac{d_t}{D}\right)} \tag{4}$$

d_t, divided by its diameter, D. The DSM value was determined by the processing of the dynamic loading slope data (dDSM/dε_i), as graphically shown in Figure 4. This data processing was automated with the computer software and carried out very rapidly.

The dynamic stiffness of the soil at impact, K' (in force/displacement units), may be expressed in terms of DSM as:

$$K' = \frac{\pi D}{4} DSM \tag{5}$$

Finally, the DSM values were plotted as a function of tamper drops, as shown in Figure 5. For most of the tests, the rate of DSM increase was significantly reduced from the 12th drop on. This relative change in DSM values was found to be proportional to the rate of densification (density increase as a function of number of drops) as shown in the Figure, where results of accumulated, net crater depth are also plotted versus number of tamper drops. As shown in the Figure, the DC was more effective in the first seven drops, whereas after 12 drops there is relatively small effect.

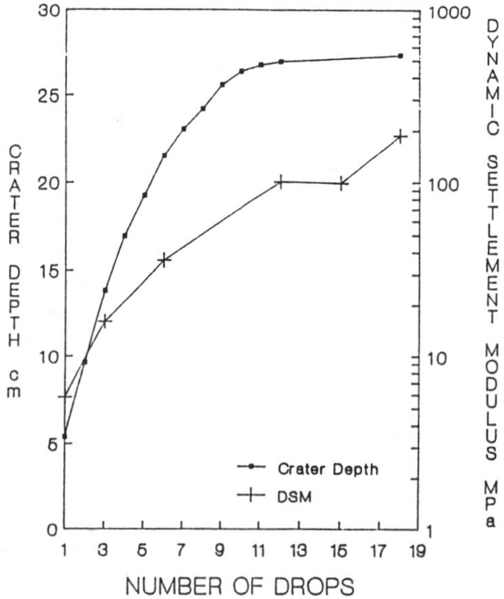

Figure 5 - Crater Depth and DSM vs. Number of Drops

Correlations with Elastic Properties

Poran et al. (1991) reported that DSM values may be correlated with the elastic moduli for this sand as it was compacted from loose to dense state. Heh (1990) computed the range of elastic properties for this sand from static triaxial and oedometer test results as indicated in table 1. In order to compare the DSM to the elastic modulus is first necessary to look at an equivalent dynamic system. Gazetas (1991) expressed the stiffness of a circularly loaded area with a diameter D on an elastic half space as:

$$K' = \frac{1.135\ E\ D}{(1+\mu)\ (1-\mu)}\ k(w) \qquad (6)$$

where K' is the dynamic system stiffness as previously defined, k(w) is a dynamic stiffness coefficient, μ is the Poisson's ratio and E is the elastic modulus. Based on the frequency content of the DC impact pulses found in the laboratory a $k(w) \approx 1.0$ may be assumed for loose and medium-dense sandy soils with $\mu < 0.4$. Therefore, the dynamic stiffness is practically equal to the static value. The dynamic stiffness computed from experimental data in Equation 5 should be the same as in Equation 6. Subsequently, E may be expressed in terms of DSM as:

$$E = 0.692\ (1-\mu)\ (1+\mu)\ DSM \tag{7}$$

As shown in Table 1 for loose and medium-dense sand E values correlate well with the range of elastic modulus values from static test results and with values reported in the literature. E values computed for medium dense sand do not agree as well. These computed results are significantly higher than E values obtained from static laboratory tests and from the literature. This discrepancy, which becomes even larger for dense sand, may be due to the assumption that $k(w)=1.0$, which may not be valid for denser sand, where the frequency content of the impact pulse becomes wider. This issue is presently being investigated further.

Field Monitoring System

Lucas (1986) reported on results of field measurements of tamper deceleration at impact. These few measurements were not used for DC monitoring but rather for research aimed at understanding the soil-tamper interaction. Holeyman (1986) and the Port Authority of New York and New Jersey (1987) reported on DC projects were an accelerometer was mounted on the tamper and wired to a data acquisition system. These results were later analyzed and used for QC in these sites where penetration testing was difficult. The main problem with such a system is durability. It has to withstand repeated high acceleration shocks and the harsh environment of DC field operation. To date, these measurement systems were rendered useless following several impacts due to damages to the accelerometers and their wiring.

In order to provide QC and verification data simultaneously with the DC construction project and without unnecessary interruptions, such a system must have an instantaneous, interactive data processing capabilities. This has not been available in the past. A new system was developed based on the experience acquired during the laboratory testing. A block diagram of the system is shown in Figure 6. The system includes several components, as follows:

a) Three rugged accelerometers are placed with special mounts at 120^0 around the circumference of the tamper. These accelerometers are

TABLE 1 · Correlation Between Elastic Properties from Static and DC Tests.

Sand Density	Estimated Modulus of Elasticity (MPa) (1)	Estimated from Results of DC Laboratory Tests		Range From Literature (3)	
		DSM (MPa)	E (MPa) (2)	Poisson's Ratio	Elastic Modulus (MPa)
Loose Dr≤35%	14-21	25	15-17	0.2-0.35	10 - 30
Medium Dense 35%≤Dr≤65 %	28-37	70	43-45	0.25-0.35	30 - 50
Dense Dr>65%	51-82	215	125-135	0.3-0.4	50 - 80

(1) Heh (1990) estimated the range of tangent modulus of elasticity from triaxial compression tests and results computed from constrained modulus obtained from oedometer tests (based on the range μ from Hunt, 1986).

(2) Computed from Equation 7 based on DSM values.

(3) Based on Hunt (1986).

connected to a central unit (CU) also mounted on the tamper. The connecting wires are highly protected in special impact resistant encasement tubes.

b) The CU consists of several components. Each of the three accelerometers

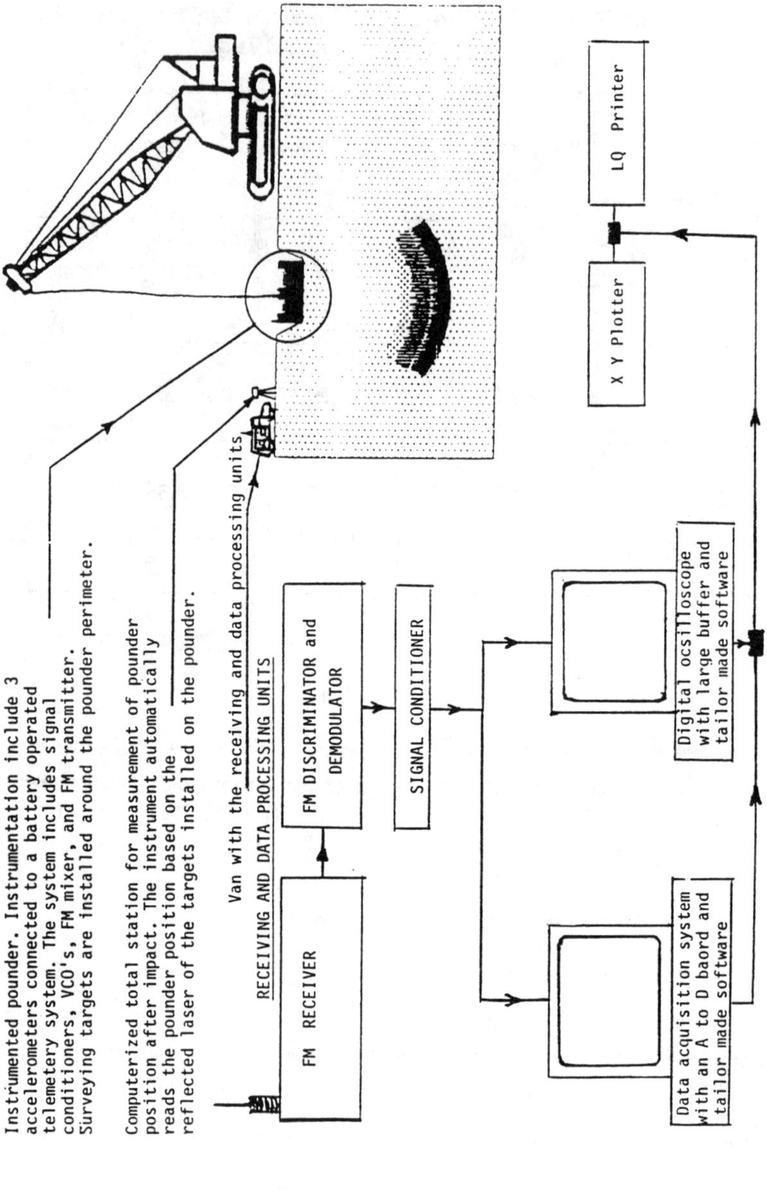

Figure 6 - Schematic Layout of the New QC System for Dynamic Compaction.

has a preamplifier and a voltage controlled oscillator (VCO). The signals from the VCOs are connected to a FM mixer (multiplexer) which produces a single mixed FM signal connected to a short range FM transmitter with a rugged antenna. The transmitter may be operated in the L or S frequency bands, commonly used for telemetry. The internal power supply is a battery providing DC voltage for all the components and the accelerometers. The miniature, shock resistant CU components are mounted on a specially designed chassis housed in a rouged steel box which is mounted on the tamper with special shock absorbing padding. In that way the CU is well protected and suited for large number of repeated impacts.

c) The remote monitoring station (RMS) is operated off a suitable vehicle on site. All RMS components are battery operated. The RMS includes a FM receiver, a demultiplexer playback unit, and a computerized data acquisition system (DAS).

d) The DAS consists of a multi channel analog to digital converter and a four channel digital oscilloscope connected to a laptop computer with special interactive software for signal processing, integration, DSM computation and instant display of DSM and settlement data versus number of drops.

e) The surveying system includes a total station which is placed near the monitoring equipment. Several reflected targets are attached to rugged steel bars mounted around the tamper circumference. This allows an average reading of position even if the tamper is partly hidden from the sight line in the depression created from the impact. The total surveying station operates with an infrared laser source aimed at the reflecting targets. It provides instantaneous data on target position. The average vertical offset of the tamper with respect to its position in the previous impact is used as an integration constant in the DSM computation, as previously indicated.

At present, the detailed design of this telemetry system has been completed and the DAS (hardware and software) is operative. However, due to lack of resources, the CU and the receiver units have not been assembled yet. The cost of items (a) through (d) (excluding the total surveying station) is presently estimated at $69,000.

Conclusions and Recommendations

The DSM concept which is based on the global dynamic response of the affected soil mass has been successfully applied to monitor dynamic compaction of dry sand in laboratory model tests. DSM values may be correlated to the density and global elastic moduli of the sand and may be used to estimate elastic settlements. Relative change in DSM values together with tamper settlement data for successive impacts provide useful information on the DC process. Based on the DSM concept, a field monitoring system is being developed for DC projects. Eventually, this system may be used for verification and quality control to determine the tamper drop which results in sufficient compaction, without interrupting the compaction process at any grid point. During construction when the QC criteria is met and the drop number is equal or larger than a minimum, initially specified number of drops, the DC process is moved to the next grid point on the site.

At present, additional resources are required to complete the telemetry components of this system so it can be successfully demonstrated in a full scale field study.

Acknowledgement

The results presented in this paper are from a study on Dynamic Compaction of Granular Soils funded by the National Science Foundation Awards No. MSM8847092 and MSM8709280.

References

Ellis, G.W.(1986). "Dynamic Consolidation of Fly Ash," Proc., Intl. Symposium on Environmental Geotechnology, Pennsylvania, 564-543.

Gazetas, G. (1991). "Foundation Vibrations," Chapter 15 in Foundation Engineering Handbook, H.Y. Fang, Editor. Van Nostrand Reinhold, 571-575.

Heh, K.S. (1990). "Dynamic Compaction of Sand," PhD Dissertation, Polytechnic University, Brooklyn, NY, December.

Holeyman, A. (1986). "Dynamic Testing of Soil Compaction," lecture presented at the May 28 Workshop in Delft, Holland, FRANKI International Technology Report, Liege, Belgium.

Hunt, R. (1986). Geotechnical Engineering Techniques and Practices, McGraw Hill, NY, 127-138.

Lucas, R.G. (1986). "Dynamic Compaction For Highway Construction, Volume I: Design and Construction Guidelines," FHWA Report No. RD-86/133, July, 204-219.

Orrje O., and Broms, B. (1970). "Strength and Deformations of Soils as Determined by a Free Falling Weight," Swedish Geotechnical Institute, Proceeding No. 23.

Poran C.J., Heh, K.S., and Rodriguez, J.A. (1991). "Impact Response of Granular Soils," Proc., Second Intl. Conf. on Recent Advances in Geotechnical Earthquake Engineering and Soil Dynamics, Vol. II, March, 1387-1398.

Port Authority of New York and New Jersey (1987). "Dynamic Compaction at Bathgate Industrial Park," video tape and report, February.

Dynamic Compaction : Predicting Depth of Improvement

Vince LUONGO[1]

Abstract

Predicting depth of improvement has long been estimated by the empirical relationship first proposed by Menard and Broise (1975). The present paper discusses briefly factors which can influence depth of improvement. A new empirical method for predicting depth of improvement is presented and the effects of applied energy, soil type, water level and tamper contact pressure is considered and evaluated.

Introduction

The empirical relationship for predicting depth of improvement proposed by Menard and Broise (1975), found that the energy per blow (i.e. WH) could be related to the depth of improvement, D. For preliminary depth estimate the relationship was given by :

$$WH > D^2 \qquad (1)$$
$$D < (WH)^{1/2} \qquad (2)$$

where : D = depth of improvement, meters
W = weight of tamper, tonnes
H = height of drop, meters

---

[1] Project Engineer, FRANKI CANADA LIMITED, 4795, St-Catherine West, Westmount, Quebec, H3Z 1S8 CANADA

Several investigators have recognized that some empirical factor, n, should be considered to account for reduction in depth of influence (Leonards et. al., 1980; Lukas, 1980; and Mayne et. al., 1984). The depth reduction factor, n, results mainly from the effects of soil type and characteristics; applied energy; tamper contact pressure; influence of cable drag; presence of energy absorbing layers; and the ground water level (Lukas 1986). Thus, a more appropriate relationship for estimating depth of improvement is given by :

$$D = n\ (WH)^{1/2} \qquad (3)$$

where : n = depth reduction factor less than 1.0

Influence of soil type

Dynamic Compaction is best suited for densification of deposits grouped as pervious and semi-pervious, as defined by Lukas, 1986. Furthermore, the position of the water table and the amounts of fine contents generally influence the effectiveness of dynamic compaction. Experience has indicated that the method is generally less effective when the percentage of clay content is greater than 15 percent finer by weight. This is because the permeability of soils containing larger amounts of fine content is too low to allow rapid dissipation of excess pore water pressures. In addition, the soil structure may be more difficult to break down owing to cohesion contributed by the water-particle interaction (Mitchell 1960). It is also not recommended to use dynamic compaction on impervious fills with high water tables since these deposits have shown marginal improvement (Lukas, 1986).

Equipment limitations

The type of equipment used will also have an effect on depth of influence. Conventional crawler cranes with a rated capacity of 136 tonnes are commonly used. This equipment can perform properly when lifting a maximum weight of 16 tonnes with drops of 20 meters. Therefore, if using equation (1), with a depth reduction factor of 0.5, it would result in a depth of influence of approximately 9 meters. Obviously, to achieve greater depths of influence would require a higher energy per blow and consequently special lifting equipment.

Influence of tamper contact pressure

Contact pressure is defined as the weight of tamper divided by the contact area. Commonly used tampers have contact pressures ranging from 30 to 75 kN/m^2. Low contact pressures could develop a crust of soil and prohibit any soil improvement below this crust. Conversely, contact pressures higher than those indicated above could result in the tamper punching into the ground upon impact, which reduces energy efficiency.

Influence of cable drag

Reduction in depth of influence is directly associated with repeated operations of lifting and dropping of the tamper, producing some friction which results in energy losses. Variables contributing to energy losses include : 1) friction of the cable unwinding over the spool drum and 2) reduction in tamper velocity due to air resistance.

Influence of hard layers

The presence of hard layers underlying the compressible layers can enhance densification. This is produced by compression and shear waves manifested by tamping which are reflected back up from the hard layer causing additional densification.

The presence of soft layers such as cohesive deposits, underlying compressible layers can reduce the effectiveness of compaction. These soft deposits appear to absorb energy and greatly attenuate the effective depth of compaction achieved. Leonards et.al., 1980 discuss a site where energy was doubled to achieve adequate depth of influence due to the presence of clay deposits underlying a sand deposit.

Degree of saturation and water level location

Depth of influence is greater in partially saturated, cohesionless soils, including some fines and old refuse fills. The mechanism of compaction is similar to that which occurs from Proctor compaction. The depth of improvement in saturated soils is less since the sudden large energy release gives rise to immediate build-up in pore water pressure which greatly attenuates the dynamic loadings with depth.

Observed depth of improvement

Of all factors affecting depth of improvement, soil type appears to have the greatest effect. Therefore the 30 field cases studied in this paper and presented in Tables 1,2 and 3 are classified as pervious, semi-pervious and partially saturated impervious soils, as defined by Lukas, 1986. The field cases presented in Tables 1, 2 and 3 list parameters such as soil type, tamper weight, drop height, observed depth of improvement, tamper base area and water table level. Rockfill deposits were not considered since their compaction effectiveness is not associated with in-situ soil stresses but rather volume decrease.

Method of analyses and results

There exists a trend between observed depth of improvement D, and energy delivered per blow WH, (Mayne et. al., 1984). This trend is used in the present paper to evaluate the effects of parameters such as water level, soil type and tamper contact pressure on observed depth of improvement.

For predicting depth of improvement, sample data from Tables 1, 2 and 3 are used and the method of least squares curve fitting is adapted to correlate the field data into a linear relationship between observed depth of improvement, D, and variation of energy per blow, WH, for which empirical equations with different field parameters can be formulated.

This analysis gives a general formula for predicting depth of improvement which is given by :

$$D = k_1 + k_2 \ (WH) \qquad (4)$$

where : D = depth of improvement, meters.
W = weight of tamper, tonnes.
H = height of drop, meters.
k_1 = depth constant.
k_2 = energy constant.

From equation (4), empirical equations are derived from sample data with common site characteristics such as, high water table, low water table, tamper base areas of 4 and 5 m^2 and soil type.

TABLE 1 Field data for pervious deposits

SAMPLE NO.	SOIL DESCRIPTION	REFERENCE	TAMPER W (tonnes)	DROP HEIGHT H (meters)	OBSERVED DEPTH OF IMPROVEMENT (meters)	TAMPER BASE AREA (sq.m)	DEPTH TO WATER TABLE (meters)
1	Sand fill	Smoltczyk, U., (1983)	15.0	15.0 and 25.0	12.0	----	----
2	Sand fill	Koponen, H.,(1983)	12.3	10.0	10.0	2.8	3.5
3	Sand fill with trace of gravel and silt	Franki (Aqua Park)	13.6	20.0	9.5	5.0	7.0
4	Compact sand with trace of silt	Franki (Quebec City)	16.0	20.0	8.0	5.0	5.0
5	Granular backfill	Franki (Lauzon)	13.6	16.0	6.0	5.0	1.0 to 3.0
6	Sand fill	Franki (Lauzon)	16.0	20.0	9.5	5.0	3.0 to 5.0
7	Clean sand	Gambin *	16.4	23.0	15.0	----	3.0
8	Silty sand	Gambin *	16.4	23.0	12.0	----	3.0
9	Sand and silt	Hansbo *	12.0	12.0	12.0	----	2.5
10	Pervious mine spoil	Guyot, et. al.,*	15.0	20.0	10.0 to 12.0	----	13.0
11	Fine to medium sand fill	Lukas*	15.0	18.3	9.1	----	3.0
12	Sand and silty sand	Cognon, et.al.,(1983)	20.0	20.0	12.0	----	2.0
13	Cinders, decomposed refuse and misc. fill	Stienberg and Lukas *	13.6	18.0	9.0	----	2.0

* After Lukas, R.G., (1986)

TABLE 2 Field data for semi–pervious deposits

SAMPLE NO.	SOIL DESCRIPTION	REFERENCE	TAMPER W (tonnes)	DROP HEIGHT H (meters)	OBSERVED DEPTH OF IMPROVEMENT (meters)	TAMPER BASE AREA (sq.m)	DEPTH TO WATER TABLE (meters)
1	Loess	Minkov, M., and Donchev,P., (1983)	8.3 and 15.0	10.0	4.0 and 4.5	5.1	––––
2	Silty sand to silt	Lovelace, et. al.,*	11.0 to 18.0	22.0	6.1 to 7.6	––––	12.0
3	Silty sand with peat and clayey silt seams	Gambin*	16.4	24.4	8.0	––––	––––
4	Clay, sand and ash with gravel, brick fragments,etc	Charles, et. al., (1981)	14.0	14.0	6.5	4.0	>6.5
5	Mine spoil consisting of shale and siltstone fragments	Mayne, et. al., (1984)	20.9	19.0	10.7	––––	>12.0
6	Sandy silt	Franki (Vancouver)	13.6	20.0	8.0	5.0	5.0
7	Heterogeneous silty fill	Franki (Montreal)	16.0	20.0	9.0	5.0	9.0
8	Heterogeneous silty fill	Franki (Montreal)	16.0	20.0	9.0	5.0	7.0
9	Sandy silt	Hartkainen, J., and Valtonen, M., (1983)	12.0	12.0	7.0	6.25	––––
10	Sandy silt	Franki (Laprairie)	16.0	20.0	9.0	5.0	7.0
11	Sandy silt	Cleaud, J.J., et. al., (1983)	15.0	20.0	12.0	––––	––––
12	Black ash with fragments of brick, wood and glass	Charles et. al., (1981)	15.0	20.0	8.0	4.0	––––

* After Lukas, R.G., (1986)

TABLE 3 Field data for partially saturated impervious fill deposits

SAMPLE NO.	SOIL DESCRIPTION	REFERENCE	TAMPER W (tonnes)	DROP HEIGHT H (meters)	OBSERVED DEPTH OF IMPROVEMENT (meters)	TAMPER BASE AREA (sq.m)	DEPTH TO WATER TABLE (meters)
1	Heterogeneous cohesive fill with gravel and rubble	Charles, et. al., (1981)	15.0	20.0	6.5	4.0	>24.0
2	Cohesive strip mine spoil	Thompson, G.H., and Herbert, A., (1978)	15.0	24.0	6.0 to 7.0	----	>24.0
3	Cohesive strip mine spoil with shale and sandstone fragments	Lukas, R.G.,*	16.4	23.0	7.6	----	9.8
4	Cohesive fill	Franki (Chicoutimi)	13.6	20.0	6.0	5.0	6.0
5	Black ash and clay with fragments of wood, glass and brick	Charles, et. al., (1981)	15.0	20.0	6.0	4.0	----

* After Lukas, R.G., (1986)

The results of the analysis is presented in Table 4 which illustrates the mean depth of improvement with the corresponding standard deviation (i.e. upper and lower bound values). The general category in Table 4 considers all projects studied for the respective soil deposit, regardless of their specific site characteristics.

The equations presented in Table 4 were analysed based on the following criteria :

1) Rockfill deposits were not considered for the analysis.

2) The sample data for the analysis ranged between 100 T.M. and 400 T.M. of energy per blow, WH.

3) The reduction in energy per blow influenced by cable drag and absorbing layers are included in the observed depth of improvement presented in Tables 1, 2 and 3.

4) Position of water table is defined as follows :
   High water table $\leq$ 1/2 D
   Low water table $>$ 1/2 D

   where : D = observed depth of improvement.

5) Depth of improvement equations for the Tamper Base Area Category include sample data for both high and low water tables.

To illustrate generally how well the sample data correlates with equation 4, sample data from the different soil deposits studied are plotted and the corresponding general category equations are presented in Figures 2 through 4. Other categories with the respective plotted data are illustratd in the report by Luongo, 1990.

Values in Table 4 should be used with great caution and only as a guideline since depth of improvement observed at a specific site will depend not only on the variables discussed, but also on specific site characteristics and the method of measurements used to quantify depth of improvement achieved. Some in-situ test procedures used may indicate greater or lesser depth of improvement than others. It is therefore essential to consider a field testing program with more than one method to assess and confirm results. It must be pointed out that what constitutes improvement may differ from one investigator to another.

Table 4 Anticipated depth of improvement for different soil deposits.

CATEGORY		DEPOSIT		
		PERVIOUS	SEMI−PERVIOUS	PARTIALLY SATURATED IMPERVIOUS FILL
GENERAL	L.B AVG U.B	D= 5.6 + 0.009WH D= 7.9 + 0.009WH D=10.1 + 0.009WH	D= 1.4 + 0.016WH D= 3.4 + 0.016WH D= 5.3 + 0.016WH	D= 2.0 + 0.012WH D= 2.6 + 0.012WH D= 3.3 + 0.012WH
SAMPLE RANGE		125<WH<400	150<WH<400	250<WH<400
HIGH WATER TABLE	L.B AVG U.B	D= 4.6 + 0.012HW D= 7.4 + 0.012HW D=10.1 + 0.012HW	Insufficient Data	Dynamic Compaction Not Recommended
SAMPLE RANGE		125<WH<400		
LOW WATER TABLE	L.B AVG U.B	Insufficient Data	D= 1.7 + 0.016WH D= 3.3 + 0.016WH D= 4.9 + 0.016WH	Same as for General Category
SAMPLE RANGE			200<WH<400	
TAMPER BASE AREA 4 & 5 sq.m ONLY	L.B AVG U.B	*D=−0.2 + 0.024WH *D= 1.5 + 0.024WH *D= 3.2 + 0.024WH	D=−0.4 + 0.024WH D= 1.3 + 0.024WH D= 3.0 + 0.024WH	*D= 4.9 + 0.004WH *D= 5.1 + 0.004WH *D= 5.4 + 0.004WH
SAMPLE RANGE		200<WH<325	150<WH<400	250<WH<350

* Because of the small number of samples these equations should be interpreted with caution.

D= Anticipated depth of improvement, meters
W= Tamper weight, tonnes
H= Drop height, meters
L.B= Lower bound values
AVG= Average values
U.B= Upper bound values

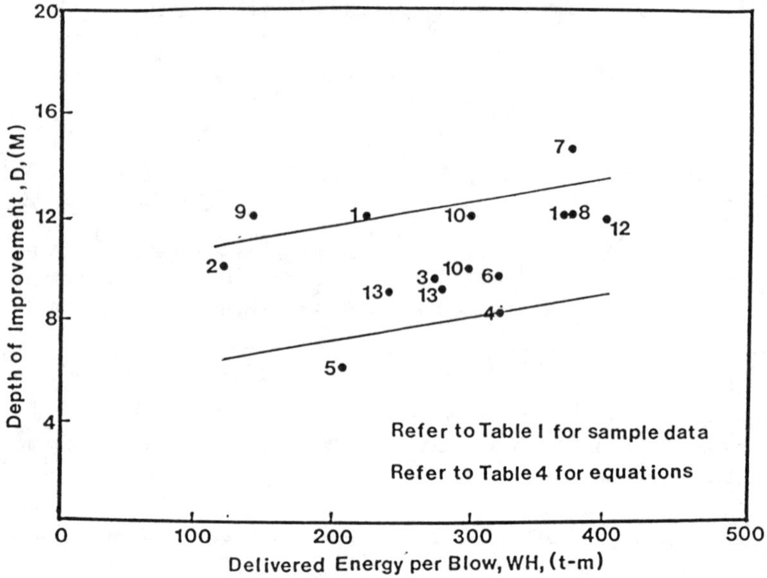

Figure 2. Observed depth of improvement with variation of energy per blow for all field cases having pervious deposits.

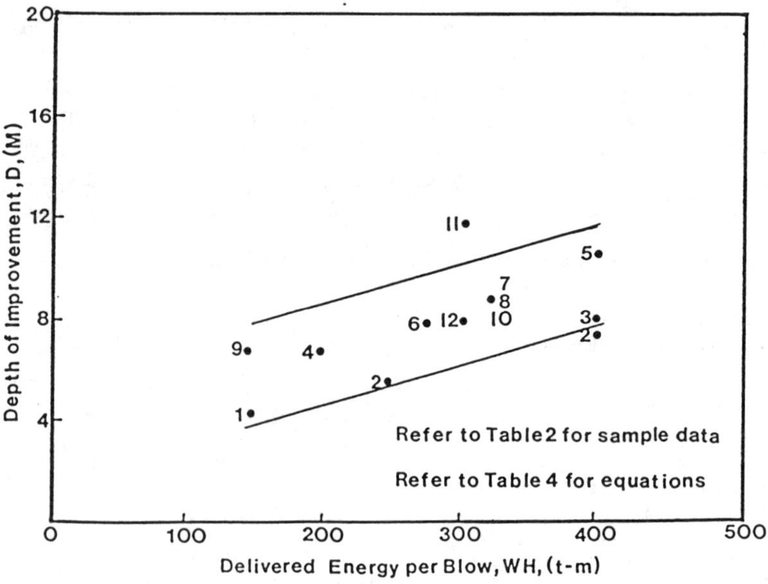

Figure 3. Observed depth of improvement with variation of energy per blow for all field cases having semi-pervious deposits.

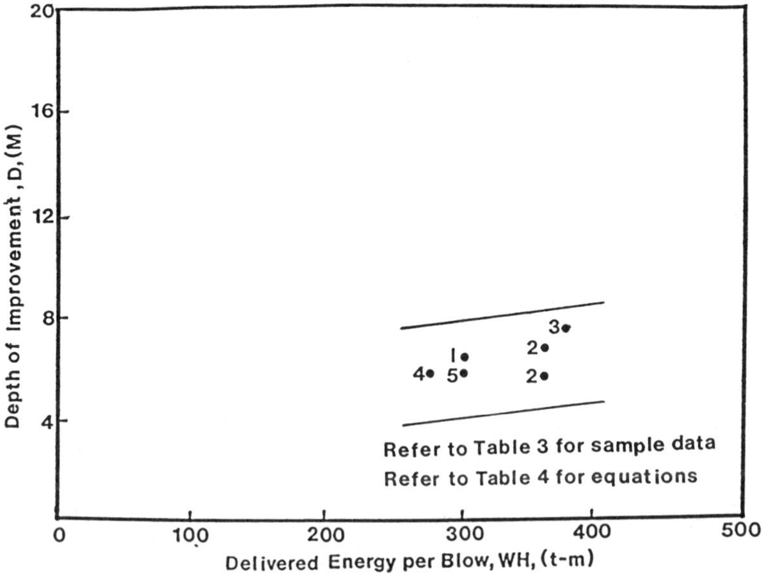

Figure 4. Observed depth of improvement with variation of energy per blow for all field cases having partially saturated impervious deposits.

Conclusions

The study has presented a new method for predicting depth of improvement with the influence of factors such as, water level, soil type and contact pressure.

From the study presented in the present paper, the following conclusions are observed.

1. From the established analysis, it is observed that the empirical equations formulated correlate well with the plotted data and that the data generally correspond with the assumed linear relationship between variation of energy per blow, WH and depth of improvement, D.

2. Depth of improvement is influenced by several factors, mainly soil type, ground water conditions and contact pressure. From the analysis it is observed that in general the greatest depth of improvement is associated with pervious soils and the least with partially saturated impervious fill deposits.

3. Insufficient sample data have been observed to properly evaluate and quantify the effects ground water conditions and tamper base area have on depth of improvement. Therefore, more emphasis and sample data are required for these parameters especially the influence of tamper size. Varying sizes of tampers can give significant differences in results for the same site conditions.

Acknowledgment

The author wishes to thank the reviewers for their constructive criticism.

Appendix - References

Charles, J.A., Burford, D. and Watts, K.S.(1981). "Field Studies of the Effectiveness of Dynamic Consolidation", *Proceedings of the Tenth International Conference on Soil Mechanics and Foundation Engineering,* Stockholm, Volume 3, pp. 17-62.

Cleaud, J.J., Bourdon, L., and Karaki, P. (1983). "Analysis of Results Obtained on a Dynamic Compaction Site" *Proceedings of the Eighth European Conference on Soil Mechanics and Foundation Engineering,* Helsinki, Volume 1, pp. 19-22.

Cognon, J.M., Liausu, P., and Vialard, R. (1983). "Combination of the Drain and Surcharge Method with Dynamic Compaction", *Proceedings of the Eighth European Conference on Soil Mechanics and Foundation Engineering,* Helsinki, Volume 1, pp. 219-222.

Hartikainen, J., and Valtonen, M. (1983). "Heavy Tamping of Ground Aimarautio Bridge", *Proceedings of the Eighth European Conference on Soil Mechanics and Foundation Engineering,* Helsinki, Volume 1, pp. 249-252.

Koponen, H. (1983). "Soil Improvement by Deep Compaction at the Site of a Harbour Storage", *Proceedings of the Eighth European Conference on Soil Mechanics and Foundation Engineering,* Helskinki, Volume 1, pp. 267-269.

Leonards, G.A., Cutter, W.A., Holtz, R.D. (1980). "Dynamic Compaction of Granular Soils", *Journal of the Geotechnical Engineering Division,* ASCE, Vol. 106, No. GTI, Jan. pp 35-44.

Lukas, R.G. (1986). "Dynamic Compaction for Highway Construction, Volume 1 : Design and Construction Guidelines", *U.S. Department of Commerce, National Information Service,* Springfield, Va 22161, July pp 1-130.

Lukas, R.G. (1980). "Densification of Loose Deposits by Pounding", *Journal of the Geotechnical Engineering Division,* ASCE, Vol. 106, GT4, April pp. 435-446.

Luongo V. (1990), "Dynamic Compaction : Predicting Depth of Improvement for Different Soil Deposits" *M. Eng. report,* McGill University, Montréal, Canada.

Mayne, P.W., Jones, Jr, J.S., and Dumas, J.C. (June 1984) "Ground Response to Dynamic Compaction", *Journal of the Geotechnical Engineering Division,* ASCE, Vol.110, No 6 June pp 757-771.

Menard, L., and Broise, Y. (1975). "Theoretical and Practical Aspects of Dynamic Compaction", *Geotechnique 25,* No.1, March pp 3-18.

Minkov, M., and Donchev, P. (1983). "Development of Heavy Tamping of Loess", *Proceedings of the Eighth European Conference on Soil Mechanics and Foundation Engineering,* Helsinki, Volume 2, pp. 797-800.

Mitchell, J.K. (1960). "Fundamental Aspects of Thixotropy in Soils", *Journal of the Soil Mechanics and Foundation Division,* ASCE, Vol. 86, No. SM3, June pp 19-52

Smoltczyk, U. (1983). "Deep Compaction", *Proceedings of the Eighth European Conference on Soil Mechanics and Foundation Engineering,* Helsinki, Volume 3, pp. 1105-1114.

Thompson, G.H., and Herbert, A. (1978). "Compaction of Clay Fills In-Situ by Dynamic Consolidation", *Proceedings of the Clay Fills Conference, The Institution of Civil Engineers,* London, pp. 197-204.

Dynamic Compaction Engineering Considerations

Robert G. Lukas[1], Member ASCE

Abstract

Dynamic compaction has been used to densify a wide variety of deposits thereby rendering them capable of supporting structural loading with acceptable settlements. The engineering factors that have to be considered before using dynamic compaction include the suitability of the deposit for improvement by dynamic compaction, the predicted depth of improvement, the anticipated degree of improvement, and the amount of energy to apply to achieve this improvement.

The Federal Highway Administration funded a study to develop guidelines for the use of dynamic compaction for highway construction. This paper summarizes the portion of this study that deals with engineering considerations. Construction related factors such as grid spacing, offsite vibration control and monitoring of the improvements are also important, but are beyond the scope of this paper.

Introduction

Since its introduction into the United States in 1971, dynamic compaction has been used to densify a wide variety of deposits, including natural soils and miscellaneous fill materials. The process is relatively simple. A crane is used to raise and drop a steel tamper with a single cable from varying heights to impact onto the soil, thereby causing the densification. Because this process does not require highly sophisticated equipment, it has been relatively economical, thereby leading to its popularity. The tampers generally range in size from 4 to 20 tons, although tampers as high as 35 tons have been used on a regular basis. The drop heights generally range from 25 to 75 feet, but drop heights as high as 120 feet have been used.

---

[1] Senior Principal Engineer, STS Consultants, 111 Pfingsten Road, Northbrook, IL 60062

Dynamic compaction has been used for a variety of purposes, including:

- an increase in load supporting capacity and reduction in compressibility of various types of soil deposits to support structural loading such as buildings, slabs-on-grades, or embankments[8,11,17,18]
- a reduction in liquefaction potential of sands or silty sand deposits located beneath earth dams[6,23]
- densification of surface mine spoil or former landfills to allow for construction of highways or buildings over these deposits[1,7,21,22,24]
- densification of sinkhole and Karst deposits to reduce the risk of having a void present below foundation bearing level[5,12]
- densification of collapsible soils beneath highway pavements to reduce pavement roughness[9,20]
- densification of nuclear or hazardous waste deposits to reduce the risk of future subsidence of the surface cap[14]

Over the last 20 years, it is estimated that approximately 400 to 600 projects have been completed in the United States and Canada using dynamic compaction. Because of its popularity and promise for economically densifying marginal sites, the Federal Highway Administration funded a study for developing guidelines for dynamic compaction for highway construction[13]. This paper summarizes the portion of the FHWA report that deals with engineering considerations.

Deposits Suitable for Dynamic Compaction

The most important parameter relating to the suitability of a deposit to be improved by dynamic compaction is the ability for excess pore water pressures to dissipate. During dynamic compaction, the soil particles are displaced into a tighter state of packing and if water is present in the voids of the soil, it is necessary for this excess pore water pressure to dissipate before additional densification under repeated blows can take place. If pore water dissipation does not occur, then repeated blows from the tamper impacting the ground will only cause displacement of the ground and not densification.

Using this parameter as a guide, the deposits most suitable for dynamic compaction would include coarse grained deposits such as sands, gravels, cobbles, building rubble, some types of mine spoil, and decomposed refuse deposits. Even if these deposits are located below the water table, pore water dissipation is rapid and densification occurs relatively quickly.

By a similar line of reasoning, the soil deposits that are unfavorable for dynamic compaction would include the relatively impervious soils such as clays, a mixture of silts and clays and organic deposits that are saturated or nearly saturated. When these deposits are impacted, excess pore water pressures develop quickly, but because of the low permeability, long periods of time are required for dissipation which make dynamic compaction impractical. Frequently, these deposits when saturated have a very low strength, so when the tamper strikes the ground, large shear displacements take place, and even if excess pore water pressure were to dissipate, the improvement would be minor [2]. Some fine grained soil deposits have been successfully densified by dynamic compaction, but these are generally deposits that are elevated well above the water table and are only partially saturated or artificial drainage paths such as wick drains were installed. Included in this category are the collapsible soils of the western United States which are frequently classified as clayey silts or clayey silty sands. Organic soils have also been stabilized by driving granular soils into these deposits. This process is called dynamic mixing and replacement [10,19] and although similar, is distinct from dynamic compaction and not discussed in this paper.

Between the two broad limits of coarse grained and impervious deposits, there is an intermediate zone of soils which can be classified as semi-pervious deposits. Dynamic compaction has been successful in these deposits, but the construction procedure has to be carefully planned so as to allow excess pore water pressures to dissipate between impacts. Deposits that would be included in this category would generally consist of silty soils, possibly containing minor amounts of clay. These deposits could be natural silty deposits or fill such as some mine spoils or fly ash deposits. In general, the plasticity index of the soils should be less than 8. Excess pore water pressures resulting from impacting requires minutes to days to dissipate, so the field densification procedure has to be undertaken with multiple passes, and applying the energy on widely spaced grid points during one pass, followed by intermediate grid point locations on subsequent passes. This could lengthen the time required for the densification process, and thereby increase the cost. However, by careful planning of the construction process, the cost will not increase significantly. On some projects where semi-pervious deposits are located between impervious soils and drainage is difficult to accomplish, wick drains have been installed to allow the excess pore water pressures to dissipate more quickly [3,23].

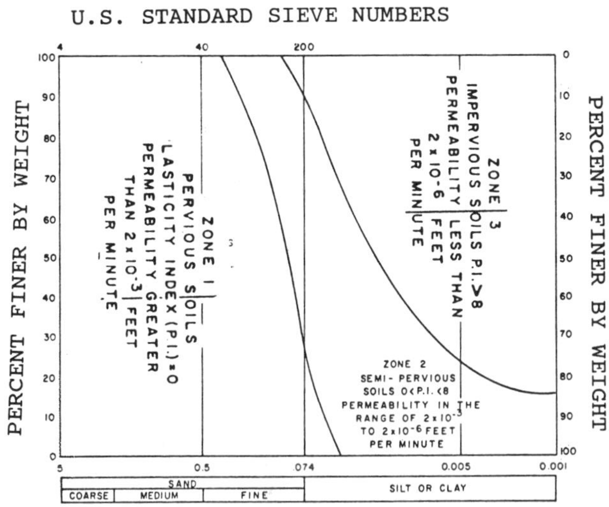

FIGURE 1: GROUPING OF SOILS FOR DYNAMIC COMPACTION

Table 1. - Recommended n Value for Different Soil Types

Soil Type	Degree of Saturation	Recommended n Value*
Pervious Soil Deposits - Granular Soils	High	0.5
	Low	0.5 to 0.6
Semi-Pervious Soil Deposits, Primarily Silts with Plasticity Index <8	High	0.35 to 0.4
	Low	0.4 to 0.5
Impervious Deposits Primarily Clayey Soils with Plasticity Index >8	High	Not recommended
	Low	0.35 to 0.40 Soils should be at a water content less than the plastic limit

*For an applied energy of 34 to 100 ton-ft/ft^2 and for a weight dropped using a single cable with a free spool drum.

As a guide to categorize the soil deposits into one of the three categories described above, either field permeability tests or laboratory grain size gradation tests could be undertaken. The preferable method would be to perform field permeability tests. If this is not possible, then grain size gradation tests should be performed on representative portions of the deposits. Figure 1 shows the approximate grain size gradation and permeability ranges for each of the three categories of deposits previously described.

While most deposits consist of relatively stable and inert particles, certain deposits may continue to degrade or soften with time following densification, so this should be kept in mind when considering dynamic compaction. Deposits that would fall into this category would include recent landfills where organic decay could result in future subsidence [24]. Also included would be soft rock particles that are flaky in shape left over from the mining of oil shales. These materials have been found to exfoliate in the presence of the water and disintegrate over a period of time. Other waste products may experience similar types of difficulties.

Depth of Improvement

The depth of improvement can be predicted in advance by the use of equation 1.[15]

$$D = n\ (WH)^{1/2} \qquad (1)$$

where:
- D = depth of improvement in meters
- W = weight of tamper in metric tons
- H = drop height in meters
- n = an empirical coefficient

The coefficient, n, accounts for factors which can affect the depth of improvement other than the weight of the tamper and the drop height. These other factors can include:

- the influence of the cable drag which reduces the impact energy
- the type and characteristics of the deposit being densified
- the contact pressure of the tamper
- the presence of energy absorbing layers

During the FHWA study [13], the influence of the first two factors was evaluated. The third factor, contact pressure of the tamper, was typically on the order of 800 to 1400 psf for most projects, so there was insufficient range of pressure to evaluate this factor. It was recognized that the presence of a clay within an otherwise granular deposit can act as an energy absorbing layer which reduces the depth of improvement, but there was insufficient data to evaluate the extent of the reduction.

The influence of cable drag was determined by measuring the velocity of a falling weight immediately prior to impact for different rigs and different tamper weights used at 5 separate dynamic compaction projects. This velocity was compared to the maximum theoretical velocity for a freely falling object without resistance.

Wherever the tampers were raised and dropped with a single cable with a free spool, the measured velocity was found to range from 0.88 to 0.93 of the theoretical velocity.[13] These results are encouraging in that while there is an influence of cable drag on the energy applied, it is a relatively constant influence, so it does not have to be measured for each particular piece of equipment or tamper weight. For comparison purposes, the velocity of a free fall tamper was measured at one site and ratio of measured to theoretical velocity was found to range from 0.97 to 0.98. The air resistance below the tamper is believed to be the reason for the ratio to be less than 1.0.

The influence of applied energy on the depth of improvement was investigated by different methods, one of which consisted of examining inclinometer deflections in three soil types, with the inclinometer situated at a distance of 10 feet from the center of the point of impact. Figure 2 shows the results of the inclinometer movements for numerous drops at one site. At any depth, if the inclinometer showed movements in excess of 0.1 inch which is considered the accuracy of the inclinometer measurement, this would be an indication of soil strain from stresses transmitted into the formation and hence, some soil improvement at that level. Using this criteria, the data obtained from inclinometer measurements at three sites are replotted in Figure 3 to illustrate depth of improvement with consecutive drops. The greatest depth of improvement occurs during the initial 2 to 5 drops with a continuing but minor increase in depth of improvement with the application of additional energy. This implies that after the initial few drops, the depth of improvement is not greatly influenced by the applied energy. Hence, equation 1 does not have an energy quantity input.

On typical projects, approximately 7 to 12 drops are imparted at each drop point. Within this range of drops, the depth of improvement for each soil type shown on Figure 3 falls within a narrow range. For the micaceous silty sand, the fine to medium sand and the clayey mine spoil the depth of improvement ranges from 34 to 35.5 feet, 32.5 to 34 feet and 24 to 26 feet, respectively. Using these depths of improvement and

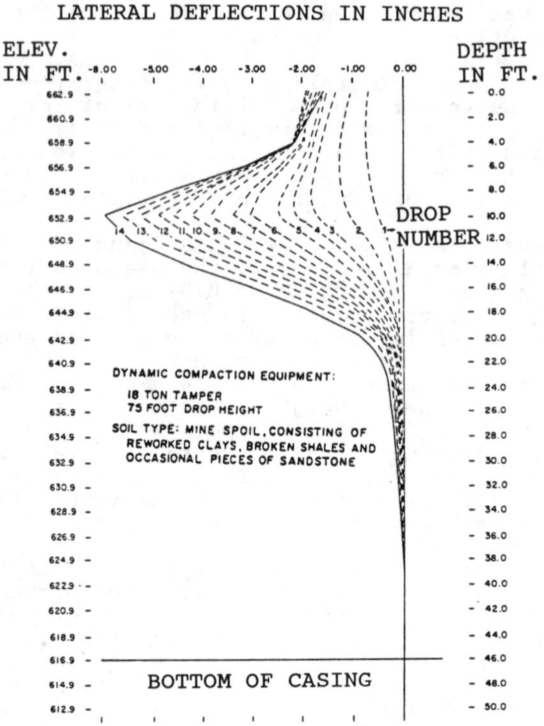

FIGURE 2: LATERAL DEFLECTIONS AT DEPTH MEASURED WITH AN INCLINOMETER 10 FEET FROM DROP POINT.

Table 2 - Upper Bound Test Values After Dynamic Compaction

Soil Type	Standard Penetration Resistance blows/foot	Static Cone Tip Resistance tons/foot^2	Pressuremeter Limit Pressure tons/foot^2
Pervious Coarse Grained Soil - Sands & Gravels	40 - 50	200 - 300	20 - 25
Semi-Pervious:			
Sandy Silts	34 - 45	140 - 180	15 - 20
Silts and Clayey Silts	25 - 35	100 - 140	10 - 15
Partially Saturated Impervious Deposits - Clay Fill - Mine Spoil	30 - 40*	N/A	15 - 20
Landfills	20 - 40*	N/A	5 - 10

*Higher test values will occur when sampling on large particles present in the soil mass.

equation 1, the calculated value of n is approximately 0.36 for the micaceous silty sand, 0.63 for the fine to medium sand, and 0.42 for the clayey mine spoil.

By comparing the depth of improvement at various sites and correlating these depths with different soil types, it appears that the greatest influence on the n value appears to be related to the type of deposit being densified, as well the position of the water table. Table 1 lists the recommended n value for the three general broad categories of deposits that are densified by means of dynamic compaction. For the pervious soils, the suggested n value is on the order of 0.5 to 0.6. This value drops to 0.35 to 0.4 for the partially saturated impervious deposits and intermediate values for the semi-pervious soils.

Degree of Improvement

The primary factor affecting the degree of improvement on a dynamic compaction project is the average energy applied at the ground surface. The greater the amount of energy applied, the greater the improvement in the soil properties, although there is usually a diminishing effect for additional increments of energy applied beyond some reasonable amount. The ground improvement can be likened to the improvement that occurs in a Proctor mold when different energy is applied. The unit weight that is achieved by compacting the soil in a Proctor mold would be higher for the Modified Proctor energy than the Standard Proctor energy. The increase in unit weight is not directly proportional to the energy, but does result in an increase in strength and reduction in compressibility. A similar phenomena occurs in the ground during dynamic compaction. Figure 2 demonstrates the amount of lateral ground compression and hence overall site improvement increases with repeated drops. Numerous investigators [4,8,16], have measured the increase in soil properties with the application of additional energy. The improvement was measured with pressuremeter tests, as well as cone penetrometer tests, and the sites investigated ranged from pervious soils to impervious deposits.

Based upon a review of the testing performed after completion of dynamic compaction, upper bound test values that can be reached are shown in Table 2. As stated, these are upper bound values, and it is quite likely that values lower than shown in the table will be attained, depending upon the amount of energy applied.

The values shown in the table are also the upper bound with respect to depth below grade. Normally, the maximum improvement in the soil deposit occurs at a

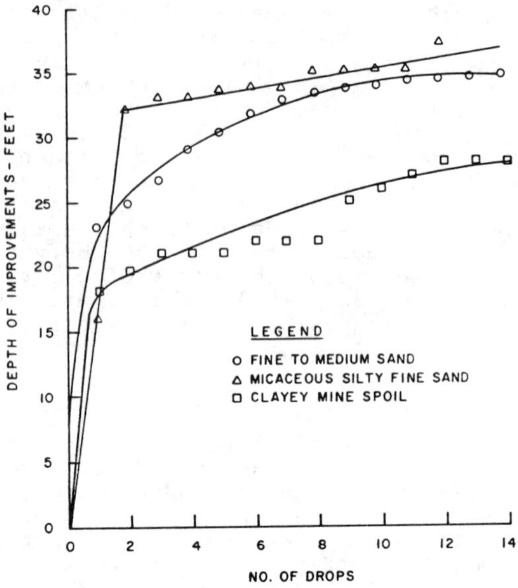

FIGURE 3: DEPTH OF IMPROVEMENTS INFERRED FROM LATERAL DEFLECTION OBTAINED AT INCLINOMETER LOCATED 10 FEET FROM CENTER OF DROP POINT.

(1FT = 0.305m)

Table 3. - Applied Energy Guidelines

Type of Deposit	Applied Energy tonne meters/meter^3
Pervious Coarse Grained Soil - Zone 1 of Figure 1	20 - 25
Semi-Pervious Fine Grained Soils - Zone 2 and Clay Fills Above the Water Table - Zone 3 of Figure 1	25 - 35
Landfills	60 - 110

NOTE: Standard Proctor energy equals 60.5 ton meter/meter^3

depth approximately equal to 1/3 to 1/2 of the depth of improvement as predicted by equation 1. Below this point of maximum improvement, the degree of improvement becomes less with increasing depth until at the depth predicted by equation 1, there is no significant increase in properties over the initial conditions.

For any particular project, soil borings with property measurements should be made both before and after dynamic compaction to confirm that the depth and degree of improvement that is desired has been attained.

Applied Unit Energy Requirements

The applied unit energy is calculated on the basis of the number of drops times the weight of the tamper, times the drop height, divided by the area encompassed within the drop point spacing. This unit energy is generally specified in tonne/meters per meter^2, although U.S. units could also be used.

On most projects, two levels of energy are applied. The first level is called the primary energy or high level energy, and represents the energy used to obtain the depth and degree of improvement that is desired. Following dynamic compaction, the surface craters left from the impact of the tamper with the ground, result in a loosened surface zone of soil, so a second level of energy that is called an ironing pass is used to densify the upper portion of the soil mass. The unit energy applied during the ironing pass is generally only a small portion of the primary energy.

Determination of the precise amount of energy to apply at any given project site is difficult, because many factors enter into this decision. These factors include:

- the type of deposit being densified
- the initial relative density of the deposit
- the thickness of the deposit being densified
- the required degree of improvement

Generally, the greatest amount of applied energy is used at landfill sites, because of the extreme loose nature of the deposit prior to densification. Deposits where the least amount of applied energy is necessary are the pervious deposits.

Depth also enters into the applied energy requirements. Some deposits are of limited thickness, and less energy is required for densification than for an otherwise equal quality deposit of greater thickness.

The required degree of improvement for any particular site may be less than the maximum test parameters shown on Table 2, in order for the deposit to behave satisfactorily. As an example, a shallow embankment over a loose deposit may only have to be improved a slight amount to still perform satisfactorily, whereas, the same deposit would have to be improved more if spread footings for a bridge were to be placed upon the deposit.

Suggested values of applied energy requirements are shown in Table 3. Soil types are grouped into three broad categories. Within each category, the range of applied energy is to compensate for the initial condition of the deposit. More energy is required for those deposits that are initially looser, than for those that are initially denser. The depth factor is taken into account by specifying the required energy in tonne meters per meter^3. To convert this energy into the more conventional tonne meters per meter^2 would require multiplying the suggested values by either the thickness of the compressible zone that is to be densified, or the maximum depth of improvement that can be reached by equation 1, whichever is least. If the suggested amount of energy is applied, then the improvements listed in Table 2 should be reached.

For comparison purposes, Standard Proctor energy equals 60.5 tonne meters per meter^3. The recommendations for applied energy in Table 3 are less than this amount, except for landfills. In the case of coarse grained soils, approximately 1/3 of Standard Proctor energy is suggested, while for the semi-pervious soils or the partially saturated cohesive soils, approximately half of this energy is suggested. This may seem like a low amount of energy for these two soil categories. However, it must be kept in mind that the deposits that are being densified are initially at least partially compact prior to densification, whereas, in the case of landfills, the deposits are in an extremely loose condition, and it is necessary to apply the full Standard Proctor energy to densify these deposits.

Naturally, the applied energy requirement suggested by Table 3 is to be used merely as a guide, since all soil deposits will respond differently to dynamic compaction, and adjustments are necessary in the field at specific project sites. Soil borings with measurements of property values at different depths below grade are suggested to confirm that the desired depth and degree of improvement are reached.

Conclusions

Some of the important engineering considerations that need to be considered and addressed prior to using dynamic compaction at a site have been discussed in this paper. Guidelines have been presented for the following factors.

1. Methods for establishing the suitability of various deposits for dynamic compaction are discussed. The importance of the rate at which excess pore water pressures can dissipate is suggested as a prime consideration. However, since this is sometimes difficult to estimate in advance of construction, ranges in field permeabilities and soil gradations are also given to assist in rating the suitability of deposits for densification.
2. Factors that influence the depth of improvement have been discussed and suggestions have been made for values of the empirical coefficient, n, for use in equation 1 to predict the depth of improvement.
3. Upper bound values of properties of soils as measured by the standard penetration resistance, cone penetration testing, or pressuremeter test values have been given based upon measurements from actual project sites.
4. Guidelines have been presented for calculating the amount of energy to apply at a site to attain the desired depth and degree of improvement.

APPENDIX I. ACKNOWLEDGMENTS

The writer wishes to thank the Federal Highway Administration for funding the dynamic compaction research effort and in particular, Mr. Al DiMillio for his guidance during the study.

APPENDIX II. REFERENCES

(1) Arkansas State Highway Department, (1985) "Dynamic Compaction of a Sanitary Landfill", Final Report, **Contract DTFH61-81-00117.**

(2) Charles, J.A. and Watts, K.S., (1982) "A Field Study of the Uses of the Dynamic Consolidation Ground Treatment Technique on Soft Alluvial Soil", **Ground Engineering**, pp. 17-23.

(3) Choa, V., et al, (1979) "Consolidation of Changi Marine Clay of Singapore Using Flexible Drains", **Seventh European Conference on Soil Mechanics and Foundation Engineering**, Brighton, England, pp. 29-36.

(4) Gambin, M.P., (1983) "The Menard Dynamic Consolidation Method at Nice Airport", **Eighth European Conference on Soil Mechanics and Foundation Engineering**, Helsinki, Volume I, pp. 231-234.

(5) Jones, J.S., (1987) "Report on Dynamic Compaction to Collapse Sinkholes in Knoxville, Tennessee", **Prepared by Geosystems, Inc.**

(6) Keller, T.O., Castro G., Rogers, J.H., (1987) "Steel Creek Dam Foundation Densification", **ASCE Geotechnical Special Publication 12, Soil Improvement - A Ten Year Update.**

(7) Koutsoftas, D.C., Kiefer, M.L., (1990) "Improvement of Mine Spoils in Southern Illinois", **ASTM STP 1070, Geotechnics of Waste Fills**, pp 153-167.

(8) Leonards, G.A., Cutter, W.A., and Holtz, R.D., (1980) "Dynamic Compaction of Granular Soils", **Journal of the Geotechnical Engineering Division**, ASCE, Vol. 106, GT4, pp. 35-44.

(9) Lovelace, A.D., Bennett, W.T., and Lueck, R.D., (1982) "A Test Section for the Stabilization of Collapsible Soils on Interstate 25", New Mexico State Highway Department, **Report MB-RR-83-1**, Algonones, New Mexico.

(10) Lo, K.W., Ooi, P.L, and Lee, S.L. (1990) "Dynamic Replacement and Mixing of Organic Soil with Sand Charges", **ASCE Journal of Geotechnical Engineering**, Vol. 116, No. 10, p. 1463.

(11) Lukas, R.G., (1980) "Densification of Loose Deposits by Pounding", **Journal of the Geotechnical Engineering Division**, Vol. 106, American Society of Civil Engineers, GT4, pp. 435-446.

(12) Lukas, R.G., (1984) "Report on Dynamic Compaction of the Karst Deposits for the Glenhardie Corporate Center Building in Valley Forge, Pennsylvania", **Prepared by STS Consultants, Ltd.**

(13) Lukas, R.G. (1986) "Dynamic Compaction for Highway Construction, Volume 1: Design and Construction Guidelines", **FHWA Report RD-86/133.**

(14) Lukas, R.G., Schexnayder, C., (1990) "Dynamic Compaction of Nuclear Waste Deposits", **ASTM 2nd International Symposium on Stabilization/ Solidification of Hazardous, Radioactive and Mixed Waste.**

(15) Menard, L., Broise, Y., (1975) "Theoretical and Practical Aspects of Dynamic Consolidation, **Geotechnique.**

(16) Mayne, P.W., Jones, J.S. and Dumas, J.C., (1984) "Ground Response to Dynamic Compaction", **Journal of the Geotechnical Engineering Division,** American Society of Civil Engineers, Volume 110, No. 6, pp. 757-774.

(17) Mayne, P.W., Jones, J.S., and Boudra, L.H., (1983) "Report of Field Testing Program, I-65 Impact Densification Study, Jefferson County, Alabama", for State of Alabama Highway Department by **Law Engineering & Testing Co.**

(18) O'Brien, J.F., and Gupton, C.P., (1984) "In-Situ Stabilization of Two Industrial Sites by Dynamic Compaction", **International Conference on Case Histories in Geotechnical Engineering**, University of Missouri-Rolla, Volume 3, pp. 1259-1264.

(19) Ramaswamy, S.D., Aziz, M.A., Subrahamanyam, M.H., Abduhl Khader, M.H., and Lee, S.L., (1979) "Treatment of Peaty Clay by High Energy Impact", **ASCE Journal of Geotechnical Engineering Division**, Vol. 116, No. 10, p. 957.

(20) Rollins, K.M., Rogers, G.W., (1990) "Mitigation Measures for Small Structures on Collapsible Alluvial Soils", Submitted for publication to **ASCE Geotechnical Division.**

(21) Snethen, D.R., Homan, M.H., (1986) "Evaluation of Dynamic Compaction in Tulsa, Oklahoma", **FHWA Report OK86(3).**

(22) Steinberg, S.B., and Lukas, R.G., (1984) "Densifying a Landfill for Commercial Development", **International Conference on Case Histories in Geotechnical Engineering**, University of Missouri-Rolla, Volume 3, pp. 1195-1200.

(23) VonThun, J.L., (1989) "As Built Report on Stage 1 Dynamic Compaction - Jackson Lake Dam", **Bureau of Reclamation Report.**

(24) Welsh, J.P., (1983) "Dynamic Deep Compaction of Sanitary Landfill to Support Superhighway", **VIII European Conference on Soil Mechanics and Foundation Engineering**, Helsinki, Volume 1, pp. 319-321.

APPENDIX III. METRIC EQUIVALENTS

1 tonne = 1.1 U.S. ton
1 meter = 3.28 feet
1×10^{-3} centimeters/sec = 2×10^{-3} feet/min.
.305 meter/sec. = 1 foot/sec.
95.8 kN/m^2 = 1 ton/square foot

Densification of Loose Sands by Deep Blasting

By Ulrich La Fosse,[1] P.E., and Theodore von Rosenvinge IV,[2] P.E., Member ASCE

Abstract: An unconventional method, deep blasting with explosives, was used to improve the foundation soils for construction of industrial park buildings on a 90 acre site in Chicopee, Massachusetts. Ninety feet of alluvial sands of varying density blanket the site. Although the uppermost 20 feet was sufficiently dense to support shallow foundations, a very loose to loose zone extended from depths of 20 to 50 feet. It was determined that this zone would require densification to provide an adequate factor of safety against liquefaction in the event of an earthquake. Although densification methods such as vibro-compaction and deep dynamic compaction were considered, deep blasting proved to be faster and more economical, in part because the energy could be delivered directly to the loose zone. The design process included a detailed seismic hazard evaluation and an instrumented test blasting program. Production blasting included post-densification in-situ tests which demonstrated the success of the method as compared to design criteria. Vibration data was collected and proved useful in reducing vibrations and identifying limitations of the method for more developed areas.

Introduction

Deep densification was used to improve foundation soils for new commercial/industrial park buildings (low rise) in Chicopee, Massachusetts. The site location is depicted on Figure 1. The site consisted of a relatively flat 90-acre parcel which was once used for single-story wood frame housing at the former Westover Strategic Air Command Base.

Below approximately 17 acres of the site, loose to very loose saturated alluvial sand deposits were encountered at depths of 20 to 50 feet below the ground surface. Based on building code design criterion, these deposits were determined to have an inadequate factor of safety against soil liquefaction in the event of an earthquake. A subsequent site-specific seismic hazard assessment confirmed that the loose sand layer required improvement to increase its density and achieve an adequate factor of safety against liquefaction for support of shallow foundations. Several soil improvement methods were evaluated, including vibro-compaction and deep dynamic compaction. Cost and schedule considerations led to the evaluation of a third less common alternative, deep blasting. Design studies indicated, and construction demonstrated this to be a very efficient method for soil densification at this site. This paper describes the project including the geologic setting, a summary of the site specific seismic hazard assessment, and an evaluation of the site's

[1]Sr. Project Manager, GZA GeoEnvironmental, Inc., 27 Naek Road, Vernon, Connecticut 06066
[2]Sr. Vice President, GZA GeoEnvironmental, Inc., 27 Naek Rd., Vernon, CT 06066

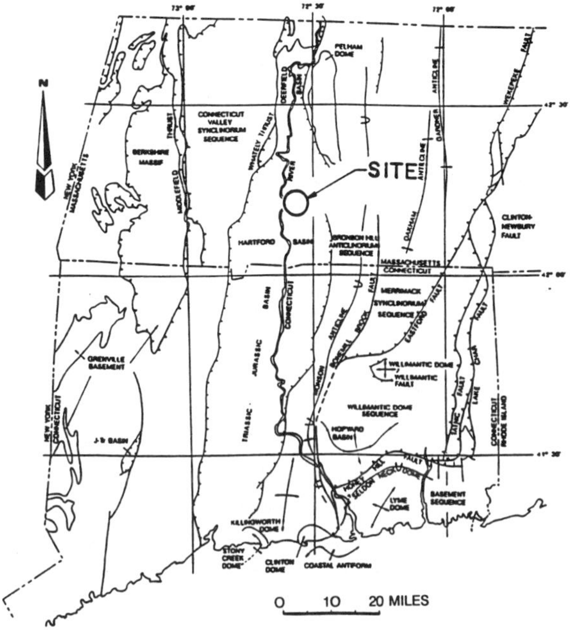

FIG. 1. Site Locus

liquefaction potential. Ground improvement alternatives, deep blasting precedence, and details of the design and the performance of the blast densification program are presented.

Geologic Conditions

A generalized subsurface profile is shown as Figure 2. The upper 90 feet consists of Chicopee-Delta sands deposited as part of prehistoric Glacial Lake Hitchcock (pleistocene). The sands overlie a 15 to 20 foot thick layer of varved silt and clay, likely deposited in the deeper waters of the lake. The varved silt and clay overlies a thin Glacial Till deposit, itself overlying sedimentary Shale and Siltstone bedrock. Groundwater was encountered at depths ranging from about 7 to 9 feet. The delta sand deposits are relatively uniform and clean in gradation. A summary of laboratory grain size analyses is provided as Figure 3. Initial test borings obtained at the site indicated sand densities, determined by the N-values, as follows:

Depth	Average N-value	Density Description
0 to 10 feet	8	Loose to medium dense
10 to 20 feet	22	Medium dense to dense
20 to 50 feet	6	Very to loose to loose (locally medium dense)
50 to 90 feet	22	Medium dense to dense

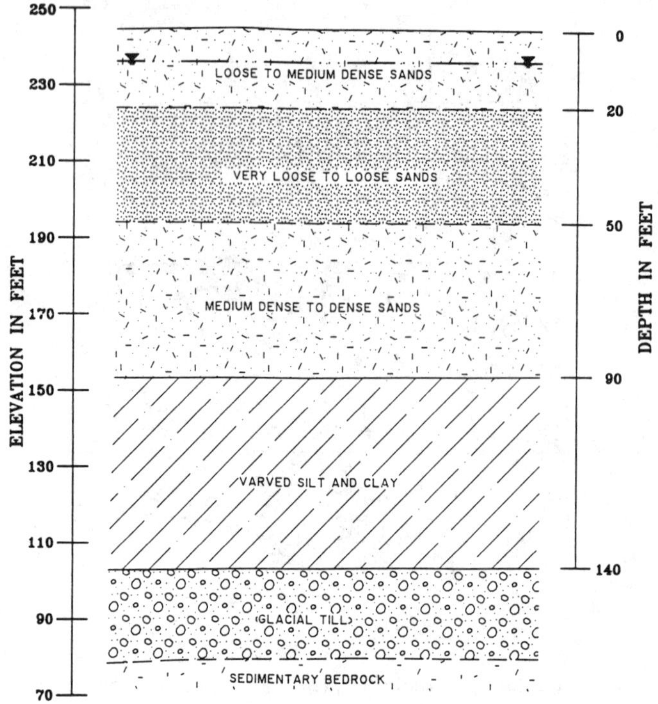

FIG. 2. Typical Subsurface Profile

Seismic Hazard Assessment

Comparison of the N-values to the Massachusetts State Building Code (MSBC) soil liquefaction guidelines (2), indicated that the sands were liquefaction-susceptible because of their saturated state, loose consistency, and uniform gradation. Thus, further study was required. To characterize the site's liquefaction potential, a site-specific seismic hazard assessment was performed to quantify the seismicity of the site. This was done with the objective of documenting and justifying the use of more specific, and lower, design peak ground acceleration (pga) and earthquake magnitude (M) than those used by the MSBC. The assessment indicated that the site is in a less seismically active area than other parts of Massachusetts, particularly the Boston area. Using an earthquake event return period of 500 years (10% probability of exceedence in 50 years) inherently assumed in the MSBC, the assessment justified a design of 8.2% of gravity (0.082g) and a range of magnitudes of 5.0 to 6.5 M. This pga is approximately 50% less than that assumed in the preliminary evaluation (0.12g) based solely on the MSBC. The magnitude range is also lower than the magnitude range inherently assumed in the MSBC (6.25 to 6.5).

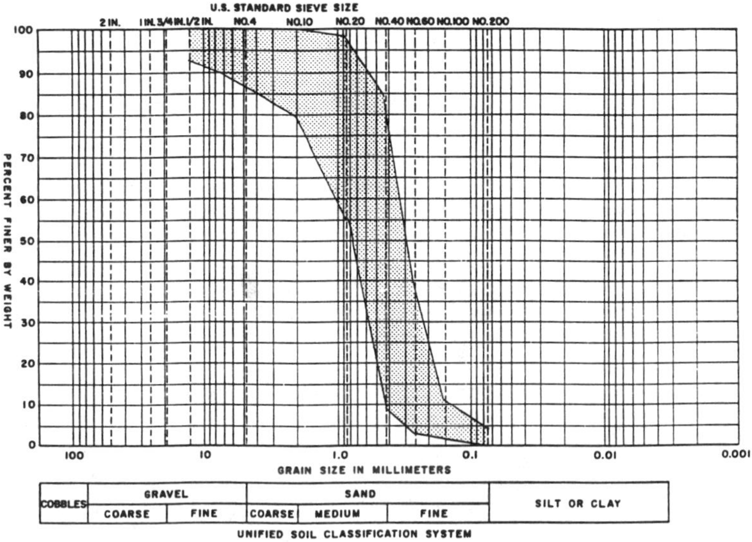

FIG. 3. Summary of Grain Size Analyses

Liquefaction Evaluation

Liquefaction evaluations were performed using methods developed by Seed and Idriss (12) with peak ground acceleration and magnitudes determined from the seismic assessment. The minimum N-values required to achieve a factor of safety against liquefaction equal to 1.0 were computed. These N-values are plotted as a solid line on Figure 4. N-values to the left of the plotted line indicate soil samples deemed to be liquefaction susceptible, and N-values to the right of the line indicate soils that are not liquefaction susceptible.

The analyses indicated an inadequate factor of safety against liquefaction between depths of 20 to 50 feet below the ground surface. This depth was designated as the target zone to be improved. The shaded area to the right of the solid line on Figure 4 indicates the design range of N-values required to achieve an adequate factor of safety against liquefaction for this project. The recommended design degree of densification is that which will result in N-values of at least 6 to 8 blows per foot. This N-value range was subsequently correlated to tip resistances (Qc) obtained with electronic cone penetration tests (CPTs) ranging from approximately 400 to 500 psi.

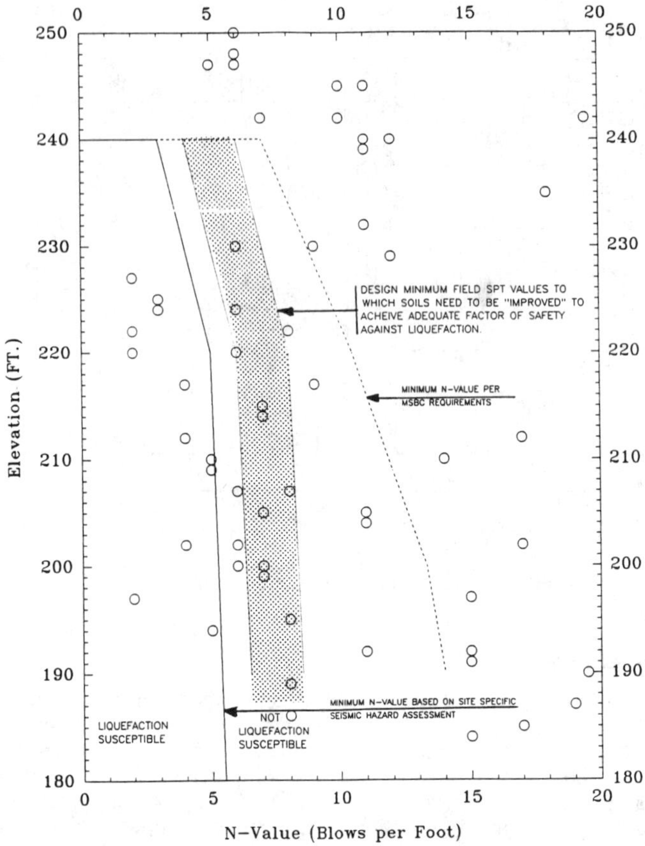

FIG. 4. N-Values vs. Elevation (with Liquefaction Criteria)

Ground Improvement Alternatives

Figure 5 depicts three ground improvement methods which were considered to increase the density of the sands in the target zone to the minimum levels described above. Deep dynamic compaction (DDC) and vibro-compaction equipment must penetrate the denser upper 20 feet in order to deliver densification energy to the underlying target zone. In contrast, deep blasting is efficient in that the energy of the dynamite charges can be placed directly within the loose layer. Additionally, despite the modest degree of densification sought, both DDC and vibrocompaction typically have a small effective densification radius (typically 4 to 6 feet), while blasting has a large radius of influence (typically 15 to 20 feet). This results in few blast locations as compared to the large number of locations required by the other two methods.

Both cost and time savings could therefore be realized by using deep blasting vs. DDC or vibrocompaction.

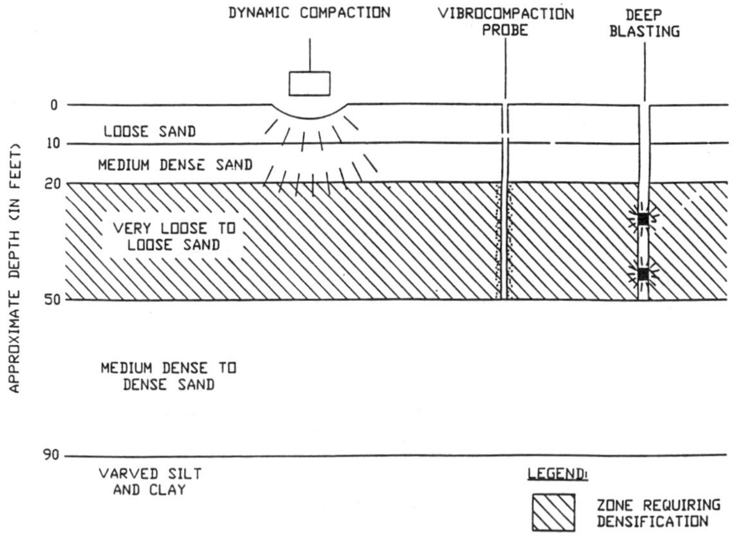

FIG. 5. Ground Improvement Methods

Design Precedence

Densification of soils by blasting has not been used frequently in the past, but there are several recorded effective uses of this method in the literature. The method has been used to improve foundations for such structures as dams and electric transmission line support towers. The most notable recorded uses in the United States were in Franklin, New Hampshire, and Pittsfield, Massachusetts. Blast densification was used in 1940 at Franklin Falls Dam, (10) to reduce foundation settlement and seepage. The method was also used in the Housatonic River flood plain near Pittsfield (9) to densify foundation soils for three power line towers in 1960. Blast densification was also used to improve foundation soils at the Karnafulli dam in Pakistan (5), in 1960, and at the Jebba Dam near the town of Jebba, Nigeria (14,15,16) in 1979-1980. The method has been used in the Soviet Union (6), and in the Netherlands for harbor construction (1, 7). The depth of the target zone (the loose saturated granular soils) for these projects ranged from 10 to 120 feet below the ground surface.

No use of the method to densify soils for building sites specifically to improve their resistance to earthquake induced liquefaction was found in the literature. However, the literature confirms that the method had been successfully employed to achieve a moderate degree of densification in loose saturated granular soils - precisely the conditions present at the site, and precisely the objective of the soil improvement program.

Test Program

After selection of blast densification, design of a blasting program was initiated. It became apparent that a large scale testing program would be beneficial to verify soil density increase, determine optimum blasting parameters, correlate densification with ground surface and deep settlements, and obtain ground vibration data. The verification of the method was also used to demonstrate to the project team that this unconventional method (as compared to more commonly used alternatives) would indeed achieve the intended purpose.

The test program served to better define the numerous parameters which comprise the blasting design. These include: charge weight, diameter, depth, spacing, and pattern (e.g., square or triangular); number of charge levels (decks); casing material (e.g., PVC or steel); explosive type (e.g., dynamite or TNT); number of coverages for a given number of charges per unit area (e.g., single coverage at a tight spacing or multiple overlapping coverages at a wider spacing); horizontal and vertical detonation sequencing; number of simultaneous charges detonated; charge delays; and delay between coverages. These factors affect the performance and cost of blast densification, and are often interrelated.

The multiple variables combined with the scarcity of experience with this method led to the performance of a large scale test to achieve an economical design. Three extensively instrumented square test areas, termed A, B, and C, were tested in February and March, 1990. Charges of high-velocity ammonia gelatin dynamite were set in two levels (decks), installed in two overlapping square grids, and detonated successively (staggered). Geometry, charge spacing, and charge size for each area were as follows:

Test Area	A	B	C
Area Width (ft.)	120	150	180
Square Grid Spacing (ft.)	40	50	60
Effective Spacing (ft.)	28	35	42
Charge Weight per Deck (lb.)	5.6	7.5	9.4
Powder Factor (lb./C.Y.)	0.0126	0.0108	0.0094

Nine first coverage and four second coverage locations were detonated in each test area, resulting in a total of 13 blast locations per test area. Figure 7 depicts this arrangement for test Area B spacing. The upper and lower decks were detonated with 25 milliseconds (ms) and 500 ms electric detonation caps, respectively. This design resulted in the upper decks being detonated almost one-half second before the lower decks. This delay was selected to allow the intersticial pore water pressure to increase in the lower portion of the target zone, due to the detonation of the upper charge, prior to detonating the lower charge. The intent was to reduce the intergranular stress in the lower zone thereby increasing its liquefaction susceptibility and increasing the effect of the blasting in this zone (increased densification) without requiring a larger charge.

In each test area, the degree of densification achieved within the target zone due to blasting was measured or inferred from in-situ tests, from settlement measurements at the ground surface and with "Borros" anchors installed at a depth of 20 feet, from water level measurements in shallow observation wells, and from piezometric measurements at depths of 30 and 40 feet. These tests and measurements were made prior to, during, and following blasting. Each test area was instrumented with one pre- and one post- blasting boring, one pre- and one post-blasting CPT, eighteen settlement platforms, four "Borros" anchors, one shallow observation well, and two piezometers.

The test program allowed the correlation of soil density increases as measured by pre-and post-blasting in-situ strength tests with soil volumetric strain as measured by ground settlements. Measurement of settlements allows more timely and economical verification of soil density increases than in-situ strength testing. Hence, demonstrating the validity of settlement results and calibrating them against in-situ tests allowed their extensive use during the production blasting phase. Ground vibration data, also obtained during the test was critical, due to the necessity of controlling vibrations at buildings on and surrounding the site.

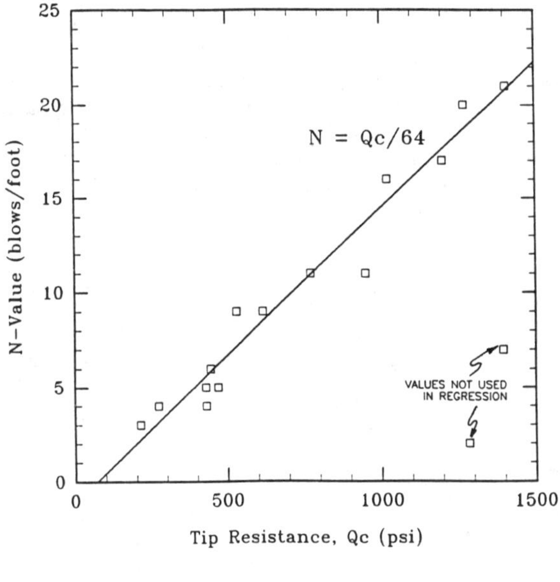

FIG. 6. N-Value vs. Qc Correlation

During the test program N-values were correlated to tip resistance (Qc) obtained with electronic cone penetration tests (CPTs). Figure 6 depicts Qc (in psi) vs. N-values obtained at the same elevation in adjacent CPTs and borings. The relationship between the N-values and Qc is quite good resulting in the N-value being approximately equal to Qc divided by 64 (N = Qc/64). This relationship agrees quite closely with published relationships by Robertson & Campanella (N = Qc/58 to Qc/72) for similar material (11). CPTs were used exclusively to locate zones susceptible to liquefaction, and to verify and document the degree of soil densification achieved.

The pre- and post- densification test borings generally indicated a substantial increase in N-values. In almost all samples the minimum design post-densification N-values were achieved at all three test areas. However, a few of samples taken in post-blasting borings indicated a reduction in N-values, in some cases, slightly below the target minimum N-values established in the liquefaction evaluations. Test program results indicated average ground surface settlements ranging from approximately 2 to 5 inches at all three test areas. Comparison of settlement measurements made at the ground surface and at a depth via the "borros" anchors confirmed that approximately three-fourths of the settlement, and hence the majority of the densification, had occurred below the 20 foot depth as desired.

Although sufficient densification was determined to have been generally achieved at all three test areas, the variation in the ground surface settlement and the variation in the N-value changes was greatest in test Area C. In this test area, the average surface settlement at points furthest from the charge locations was approximately 50% less than at points nearest the charge locations. This compared to average settlement reductions between near and far monitoring locations of approximately 10 and 20% at Areas A, and B, respectively. Similarly, while the changes in N-values at Areas A and B ranged from about -50% to +200%, the changes in Area C ranged from about -250% to +330 %. The data suggested that at Area C, the ground densification was less uniform than at Areas A or B. In addition, the larger charge size at Area C, resulted in more ground vibrations than at the other two areas. The designs for test Areas A and B were both deemed to be adequate to densify the site. The charge spacing of Area B was selected for production blasting due to the time and cost savings which it would provide over the test Area A spacing.

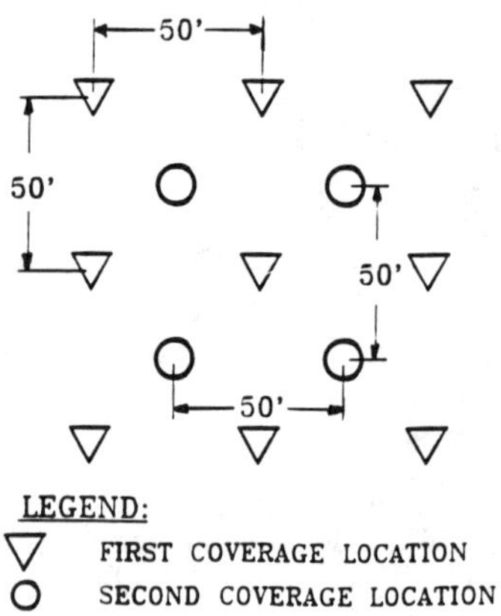

FIG. 7. Typical Blast Hole Layout Detail

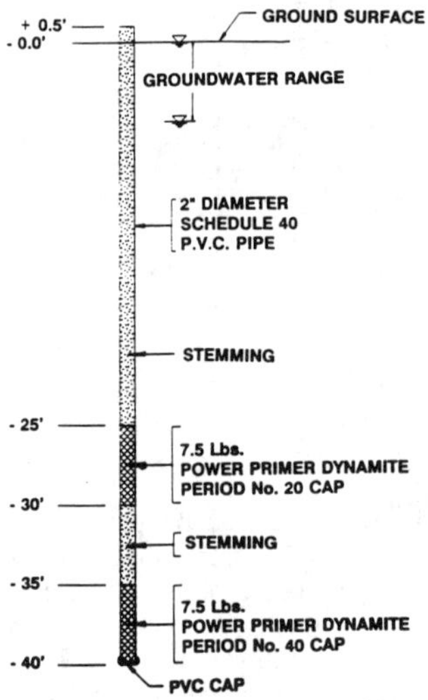

FIG. 8. Typical Loading Detail

Production Blasting

Production blasting was performed in April and August, 1990. In areas where the distance between the blasting and the buildings adjacent to the site was sufficient to allow ground surface vibrations to attenuate to acceptable levels, two decks of 7.5 lbs. of dynamite per deck were used in two 50 foot-on-center square grid overlapping coverages. This charge layout, which results in an effective square pattern spacing of about 35 feet on center and a powder factor of 0.011 lb./C.Y., (the same as Test Area B) is depicted in Figure 7.

In areas where vibrations were reduced due to the proximity of buildings, charges were reduced to two decks of 5.6 lbs. each, and the layout was reduced to two overlapping 30 foot-on-center square grid overlapping coverages. This layout resulted in an effective square pattern spacing of approximately 21 feet on center and a powder factor of 0.022 lb./C.Y.

As shown in Figure 8, the charges were loaded in two 5-foot decks, each with its own detonation cap. The decks were set at depths of 25 to 30 feet, and 35 to 40 feet inside previously installed 2-inch inside diameter Schedule 40 PVC water-filled casings (pipes). Three eighths inch stone stemming was used between the decks and above the upper deck. The upper and lower decks were detonated with electric caps with delays of 100 ms or 500 ms, and 600 ms or 1,000 ms, respectively. This resulted in a one-half second (500 ms) delay between the detonation of the upper and lower decks and allowed the use of longer delays between blast locations to reduce ground vibrations.

Upon detonation, little or no visible evidence of the blast was apparent. However, significant ground vibrations (discussed below) were experienced. Approximately 10 to 12 seconds following the blast, a spectacular gushing of groundwater in sand boils and from the blast casings was observed, often to a height of up to approximately 30 feet. The water was mixed with soil carrying with it sand and gravel size particles to the ground surface. While the bulk of the gushing usually receded within approximately one minute, evidence of water rising to the ground surface and of soil liquefaction in the form of sand boils continued for up to one-half hour. The boils created crater topped sand cones usually centered at the locations of blast casings, while some cones appeared between casings. The cones reached heights of up to approximately 4 feet and diameters of up to 30 feet.

Including the test program, 680 locations were blasted. These locations encompassed a total area of approximately 710,000 square feet (17 acres). The total construction cost for the blast densification project, including the test program, was approximately $250,000. Thus, the approximate construction cost per square foot of improved area was $0.35. This cost is estimated at approximately three to six times less than that of deep dynamic compaction or vibrocompaction at this site.

Vibration Control

While the test area locations and the early stages of the production blasting were at least 650 feet from existing structures, subsequent blasting proceeded as close as 175 feet from existing wood frame structures. In addition, the project schedule required beginning the construction of an industrial building before the remainder of the site could be densified. This resulted in blasting near newly poured concrete. Hence, devising and implementing methods to reduce vibrations became necessary.

During the test program up to nine locations were detonated simultaneously. During the early stages of the production blasting, up to three rows were detonated simultaneously, and 8 millisecond (ms) delays were used between adjacent charge locations within each row. As the blasting progressed closer to existing buildings adjacent to the site, fewer and fewer rows were exploded simultaneously until only one row, hence only one location, per delay was detonated. Thus, the charge per delay was reduced to a minimum of 7.5 lbs./delay for the 50 foot spacing, and to 5.6 lbs./delay for the 30 foot spacing. Additionally, the detonation delays between charge locations were increased from 8 ms to up to 125 ms to further reduce vibration levels.

These techniques successfully reduced blast induced ground vibrations (8). However, as noted in the literature (3), vibrations levels created by blasting in saturated soils are substantially higher than vibrations commonly created by rock blasting. In addition, ground vibration frequencies measured during the soil blasting (8) were significantly lower than those commonly measured in rock blasting (13). Because low frequencies are generally more detrimental to structures than high ones, additional precautions and vibration control had to be undertaken for this project as compared to precautions usually taken when blasting rock.

Evaluation of Densification

The performance of the blast densification was evaluated by two methods; measuring ground settlement caused by the blasting, and performing in-situ tests before and after blasting. Ground settlement was determined by surveying settlement platforms placed approximately two feet below the ground surface. The settlement platforms were used as the primary method to verify the performance of the program. More than 350 of these platforms were used.

Ground settlement was generally measured several hours, one day, and several days after blasting. At several locations ground settlement measurements, made weekly for a period of one month after blasting, confirmed that the settlement was complete within several days after blasting. Total ground surface settlement ranged from 1 to 14 inches and averaged approximately 5 inches within the areas which were densified by blasting.

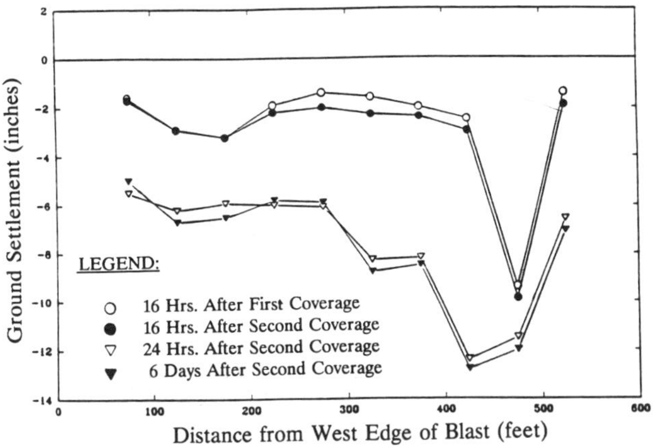

FIG. 9. Typical Post Blast Settlement

Figure 9 depicts a typical post-blast ground settlement profile across the width of an area which was improved. The average settlement at the site corresponds to approximately 2% strain within the zone which was densified. This is in good agreement with the strain anticipated during the design phase.

In situ electronic cone penetration tests (CPTs) were generally made prior to, and two to six weeks after blasting. In addition, at several locations CPTs were made five months post-blasting. The delay in the performance of the post-blasting in-situ tests was provided to investigate time dependent increase in strength of the sands. This was done in consideration of observations made by other researchers (15, 16, 17) that a reduction in tip resistance has been noted in tests made immediately post-blasting, and that a strength increase not accompanied by density changes was observed several weeks to several months following blasting as the soil "ages".

The post-blasting CPTs indicated a significantly higher strength between approximate depth of 20 to 50 feet and a slight decrease in strength in the upper 10 feet after blasting as compared to CPTs made prior to blasting. Figure 10 depicts CPT tip resistances pre-blasting, two weeks post blasting, and five months post-blasting and is typical of the strength improvements measured at the site. While no reduction in the tip resistance was noted in CPT's made only two weeks post-blasting as observed by other researchers (15,16), there was a significant tip resistance gain in the CPTs made five months post-blasting as compared to the CPTs made two weeks post-blasting. This observation confirms that the "aging" effect of sands densified by blasting is significant and needs to be considered in determining the timing of post-blasting in-situ tests.

The strain measurements and the post-blasting CPTs indicate that the soil strength, in the previously loose zones, was sufficiently increased to resist the design earthquake forces at the site.

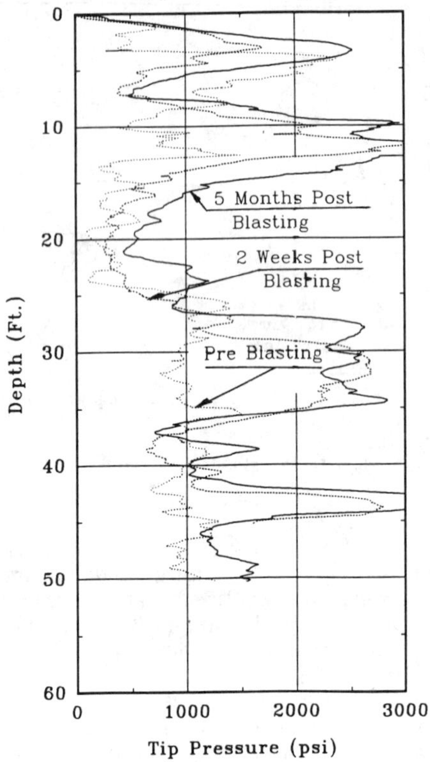

FIG. 10. Typical Pre- and Post-Blasting CPT Results

Conclusions

The following conclusions may be drawn from the results of the liquefaction evaluation and the deep blasting densification at the Westover site:

1. Some sites have such loose and clean granular soil conditions that they are unsuitable for support of shallow foundations without soil improvement due to an inadequate factor of safety against liquefaction in the event of a moderate design earthquake. This is particularly important in consideration of the recent increased awareness of seismic risk in parts of the United States other than the most commonly known seismically active zones (e.g., the west coast). A site specific seismic hazard assessment can be useful to help assess the degree of soil improvement needed.

2. Of the available soil improvement methods, deep blasting proved to be the most rapid and economical on this project where a modest degree of improvement was required over a large area.

3. Test program and production phase results conclusively show that deep blasting for soil densification is a viable and cost effective soil improvement method where a limited degree of densification of loose soils is sought. This is usually the case when foundation soils need to be improved to increase their resistance to liquefaction when subjected to moderate earthquake forces.

4. Comparison of in-situ tests made immediately after densification by deep blasting with tests made several months later indicate apparent in-situ strength increases with "aging" of the soil (e.g., tests made after the soils have had an opportunity to "age" indicate a strength increase not accompanied by density changes). Thus, post-densification in-situ tests made to confirm results of the deep blasting soil improvement method should include testing at least four to eight weeks post-blasting to avoid misleading conclusions regarding the efficacy of this method.

5. While, significant ground vibrations are associated with this method ruling out its use in densely developed areas, this project conclusively shows that, with proper precautions, the vibration levels can be controlled sufficiently to allow use in moderately developed areas (in this case deep blasting was accomplished within less than two hundred feet of occupied wood frame houses). Modern vibration monitoring methods combined with sophisticated blasting control techniques allowed the successful revival of the deep blasting method of soil improvement, and this project demonstrates that this method is indeed viable and should be considered when evaluating soil improvement methods.

6. The ground vibration data obtained shows that vibration attenuation and frequency characteristics associated with blasting in saturated soils differ markedly from those associated with rock blasting. The attenuation is much lower for soil than for rock blasting, especially for the lower frequencies which are usually the most detrimental to structures. Hence, particular care must be exercised when using this soil improvement method.

Acknowledgements

We would like to acknowledge the following persons who assisted the authors in the research and implementation of this project by providing consultation and assistance in their respective areas of expertise: George C. Klimkiewitcz, Mgr., Seismology Dept., Weston Geophysical, for his input regarding the site's seismic hazard assessment; James K. Mitchell, Ph. D., P.E., Professor, Dept. of Civil Engineering, University of California at Berkeley, for his review of the test blasting program.; and Mishac K. Yegian, Ph. D., P.E., Chairman Dept. of Civil Engineering, Northeastern University, and Earthquake Engineer, for his input regarding the evaluation of the site's seismology and liquefaction potential.

Appendix I. --References

1. Carpentier, R. et. al. (1985). "Compaction By Blasting in Offshore Harbour Construction"*Proceedings of the Eleventh International Conference on Soil Mechanics and Foundation Engineering,* San Francisco 1985, Vol. 3, 1687-1692.

2. *Commonwealth of Massachusetts State Building Code (1988),* Fourth Edition, 780 CMR, July 1, 1988, Sections 716.7 and 720.4, 285 & 292-294.

3. Dowding, C.H. (1985). *Blast Vibration Monitoring and Control.* Prentice-Hall, Inc., New Jersey, 30-34.

4. Dowding, C.H. (1986). "A Laboratory Study of Blast Densification of Saturated Sands" *Journal of Geotechnical Engineering,* ASCE, Vol. 112, No. 2, Feb. 1986, 187-199.
5. Hall, C.E. (1962). "Compacting A Dam Foundation By Blasting" *Journal of the Soil Mechanics and Foundation Division,* ASCE, (88), SM3, June 1962, 33-52.
6. Ivanov, P.L., (1967). *Compaction of Noncohesive Soils by Explosions.* Translated from Russian in 1972, U.S. Department of Commerce, TT70-57221.
7. Jaworski, W.E. (1978). "Methods and Control For Deep Densification" *International Symposium on Foundation Aspects of Coastal Structures,* Delft, Netherlands, Oct. 1978, V.3, 1-9.
8. La Fosse, U. and Gelormino, T.A., (1991). "Soil Improvement By Deep Blasting - A Case Study" *Proceedings of the Society of Explosive's Engineers 17th Annual Conference on Explosives and Blasting Technique,* Las Vegas, Feb. 3-7, 1991, 205-213.
10. Lyman, A.K.B., (1941). "Compaction of Cohesionless Foundation Soils by Explosives" *Proceedings of the ASCE,* May 1941, Paper No. 2160, 1330-1348.
11. Robertson, P.K., and Campanella, R.G., (1984). *Guidelines for Use and Interpretation of the Electronic Cone Penetration Test.* Second Edition, Sept. 1984 Hogentogler & Company, Inc., Maryland, 140.
12. Seed, H.B. and Idriss, I.M., (1971). "Simplified Procedure for Evaluating Soil Liquefaction Potential" *Journal of the Soil Mechanics and Foundations Division,* ASCE, Vol. 97, No. SM9, 1249-1273.
13. Siskind, D.E., Stagg, M.S., Kopp, J.W., and Dowding, C.H. (1980). "Measurements of Blast Induced Ground Vibrations and Seismograph Calibrations" *U.S. Bureau of Mines,* Report of Investigations No. 8506.
14. Solymar, Z.V. (1984). "Compaction of Alluvial Sands by Deep Blasting" *Canadian Geotechnical Journal,* Vol. 21, No. 2, May 1984, 305-321.
15. Solymar, Z.V., (1984). "Earth Foundation Treatment At Jebba Dam Site" *Journal of Geotechnical Engineering,* Vol. 110, No. 10, Oct. 1984, 1415-1430.
16. Solymar, Z.V. (1986). "A Comparison of Foundation Compaction Techniques" *Canadian Geotechnical Journal,* Vol. 23, Aug. 1986 No. 3, 271-280.
17. Solymar, Z.V. and Mitchell, J.K. (1986). "Blasting Densifies Sand" *Civil Engineering,* ASCE, Vol. 56, No. 3, March 1986, 46-48.

Appendix II. --Notation

The following symbols are used in this paper:

CPT = electronic cone penetration test
C.Y. = cubic yard
DDC = deep dynamic compaction
lbs = pounds
M = earthquake magnitude (Richter Scale)
Ms = millisecond
MSBC = Massachusetts State Building Code
N = blow count per foot (standard penetration test)
pga = peak ground acceleration
psi = pounds per square inch
PVC = polyvinyl chloride
Qc = electronic cone tip penetration resistance
TNT = trinitrotoluene

Appendix III. --Conversion to SI Units

To Convert	To	Multiply by
ft.	m	0.3048
ft.^{2}	m^2	0.0929
mile	km.	1.6094
acre	hectare	0.4047
lb.	kg	0.4545
psi	kN/m^2	6.9000

Investigations on Influence of Vibration Parameters on Compacting of Cohesionless Soils

Jerzy Sękowski *

Abstract

In this paper there are presented results of laboratory tests concerning the evaluation of the influence that basic parameters have on the compactibility of cohesionless soils by means of vibration. For the sample's moisture content (w), static load (σ), frequency and amplitude of vibrations(f, A) and for the process duration(T) such values were determined at which the compacting process is best. Moreover results of proof tests in situ have been given.

Introduction

Most frequently used for the construction of the so-called sand cushions are cohesionless mineral soils. These soils are also used for the construction of banks, embankments and dykes. The effectiveness of these ground mass structures depends on compacting processes. Compacting is also used to improve the existing weak subgrade. In practice are used static, percussive and vibrational methods of soil compacting [5](Wiłun, 1987), making use of various compacting equipment, described e.g., in paper [1](Arqui'e, 1972). It should be noted that in Polish practice more and more frequently used is the vibrational system of soil compacting, both during the formation of long embankments, and volumetrically small sand cushions for cubage housing objects.

Laboratory tests aim at the determination of the value of those parameters at which the volume density of the compacted soil skeleton reaches the value $P_d = P_{ds}$.

* Dr.Sc. Silesian Technical University. Institut of Road Construction. 44-100 Gliwice, ul. W. Pstrowskiego 5/214, Poland

As it is known, quality inspection of an embankment consists in a comparison of the value of this soil compactness in an embankment already made, and the value obtained in laboratory tests. A measure of compactness quality is the so-called soil Relative compaction I_S ($I_S = P_d / P_{ds}$). In the now-used Polish norm [3], parametr P_{ds} is determined by the method of soil compacting in Proctor's apparatus. The number of layers, the number of ram strokes, its mass and height of descend, as well as cylinder volume is so assumed that the compacting energy is equal 0.589 J/cm^3 (the so-called normal method), or 2.649 J/cm^3 (the so-called modified method). The parameters obtained in these tests for the particular kind of soil is the already mentioned P_{ds} and the corresponding moisture content called optimum moisture content (w_{opt}).

It seems that in the context of executive practice and higher and higher popularity of vibrational methods of compaction of soils, mainly cohesionless ones, it is justified to use in laboratory tests the vibrational methods of determining the values of those parameters at which the efficiency of compacting, measured by the value P_d is the highest. Such methods are used in many western countries (e. g. [4]). An accurate definition of the initial conditions of the laboratory method is rendered difficult on account of the not yet fully clarified influence of many factors on the final effect of soil compaction. Among the many variables may be mentioned: the frequency and amplitude of vibrations, vibration time and the load applied to the compacted soil.

In this paper will be presented the results of laboratory tests, the effect of which is the determination of the influence of the selected vibration parameters on the compacting ability of two different cohesionless soils. Analyzed were the moisture content of the material (w), frequency and amplitude of vibrations (f, A), vibration time (T), and the static load (σ). The tests were carried out in a specially adapted for this purpose testing apparatus. The test results have been compared with the results obtained by the recommended standard method [3] . Also the comparative results for the particular actual example have been given.

The materials used for the testing

Subject to tests were two soils which in accordance with the Polish standards were classified as uniformly graded medium-size grain sand and non-uniformly graded gravel. The basic grading characteristics of these materials have been denoted in Fig. 1 by no (1) and (2), respectively. In the same figure is shown the grain size distribution curve of the material used for the sand cushion of the building for which inspection testing described in one of the successive parts of the paper was made, whereas in Table 1 have been given the statistical characteristics of grain size distribution and some of the geotechnical parameters for all the materials.

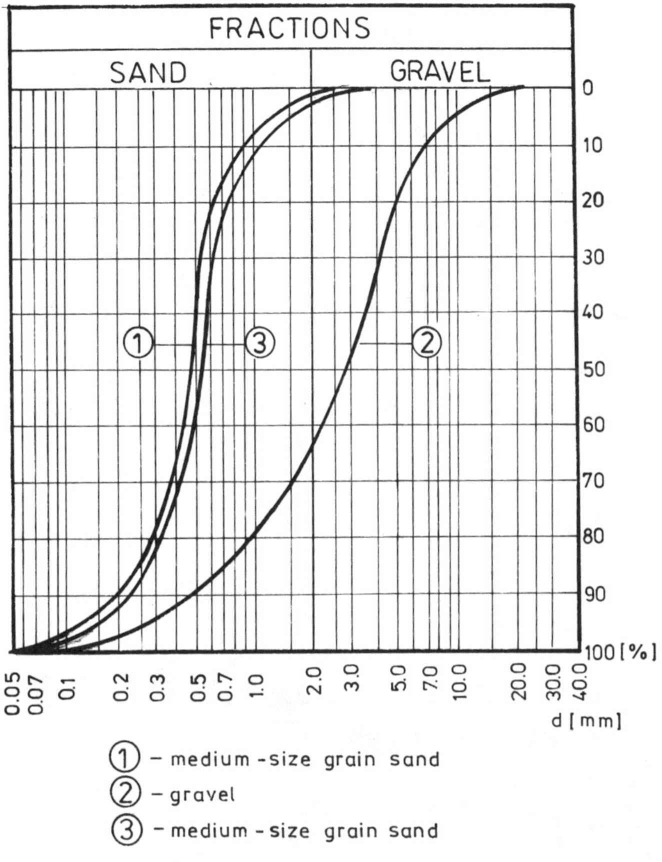

Fig. 1. Grain-size curves for the tested soils

Table 1. Basic physical parameters of materials

Parameters	MATERIAL		
	①	②	③
Soil [PN-86/B-02480]	SAND	GRAVEL	SAND
d_{60}, mm	0.47	0.4	0.50
d_{50}, mm	0.44	2.9	0.47
d_{30}, mm	0.35	1.6	0.33
d_{10}, mm	0.20	0.5	0.21
$U=d_{60}/d_{10}$, 1	2.35	6.8	2.6
ϱ_s, $g \cdot cm^{-3}$	2.650	2.649	2.650
e_{max}, 1	0.695	0.573	0.675
e_{min}, 1	0.468	0.354	0.380
W_{opt}, %	11.9	7.25	13.0
ϱ_{ds}, $g \cdot cm^{-3}$	1.69	1.86	1.85

The testing apparatus and laboratory test methods

The basic laboratory tests were made on the vibrating table of type St 500/300 installation, manufactured by the Italian firm Reuenstein, also equipped with a control desk, vibration meter and fans. The parameters of the vibrating table (used i.a., in the testing of concrete mixtures), amouting to:

frequency $f \in \langle 10 \div 5000$ Hz $\rangle$,

amplitude $A \in \langle 0 \div 6$ mm $\rangle$,

in these investigations were practically limited to 10 ÷ 100 Hz and 0.1 ÷ 1.2 mm. The soils were compacted in a cylinder [3] of the height h=120.6 mm and diameter Ø=152.4 mm ($V=2.2$ dm^3), fixed to the vibrator plate. Thanks to a simple manually operated screw mechanism a constant unit thrust on the compacted soil sample could be achieved. A general view of the test stand has been presented in Fig. 2a, and the vibrating table itself with the fixed cylinder and vibration meter in the background in Fig. 2b.

Fig. 2a. Testing stand - general view

Fig. 2b. Details of the testing stand (vibration plate and cylinder)

In the preliminary tests use was made of a vibrating table type VT. 355/580 CY manufactured by Tonindustrie Pruftechnik GMBH Berlin with parameters:

frequency $f=50$ Hz,
amplitude $A \in \langle 0 \div 1 \rangle$ mm.

The cylinder with a screw mechanism described above was also used in these tests.

The preliminary tests were concentrated on the determination of the moisture content level (w) and vibration time (T) as well as the static load quantity (σ), at which the compacting effect is the best. Since there is a conformance concerning these paramaters (e.g., Pisarczyk, 1977 - [2]), the tests made were of control character only. The tests were made for several A amplitude levels ($A=0.1$; 0.3 ; 0.5 mm), with different values of the static load σ ($\sigma=0$; 10 ; 15 ; 20 ; 25 and 30 kPa) and for different moistures content of the material (w). The compaction measured by parameter P_d was determined after 1, 2, 4, 6, 8 etc. minutes until the moment of repeated value P_d. The elaborated preliminary tests results made possible the assumption of the variable values w, σ, T for the basic tests.

The basic tests were made seperately for each soil assuming the following levels of amplitude A ($A=0.1$; 0.3 ; 0.5 ; 0.8 ; 1.0 ; 1.2 mm) and for each of them - several levels of frequency f ($f=10$; 25 ; 30 ; 40 ; 50 ; 60 : 70 : 80 : 90 : 100 Hz - depending on the results of tests from three to ten). In each of the tests the volume density of the ground skeleton P_d was determined.

The laboratory test results

The preliminary tests conformed that the course of the dependence $P_d = F(w)$ is almost identical in the testing of the compacting ability by the method of soil compacting in Proctor's apparatus and by the vibration method. The tightest compacting of the grains in the case of both soils took place with the moisture content close to zero, independently of the amplitude and the load applied. The obtained values of density P_d corresponded to those obtained with the moisture content called optimal (w_{opt}), i.e., in the curve bending point $P_d=F(w)$. As in the tests by Proctor's method, the lowest values P_d, corresponding to the most loose grain distribution were obtained with the moisture content of $0.5w_{opt}$.

The preliminary tests also confirmed the fact that compacting is not promoted either by a lack of load or excessive static load applied to the soil and derived from the vibrating installation. The best compactness has been obtained for the load of the range $10 \div 15$ kPa, which is illustrated in Fig. 3.

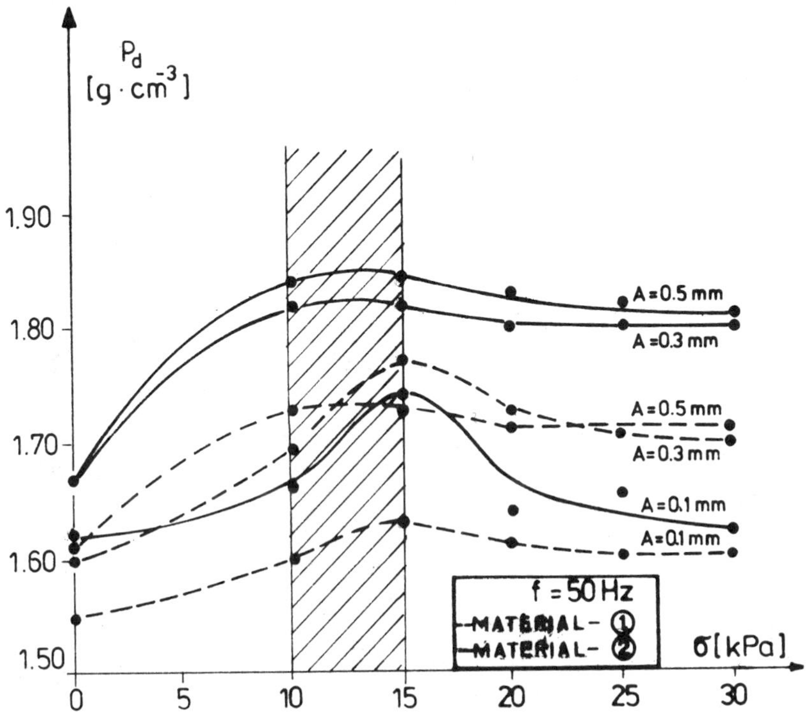

Fig. 3. Compacting as functions of the static load

The tests described above also confirmed the observation about the influence of vibration time on compacting efficiency, this time being also dependent on the kind of material, its moisture content and amplitude, as well as the frequency of vibrations. This is illustrated by the way of an example in Fig. 4. Assuming, on the basis of the results of the described earlier preliminary test, the moisture content level w=0%, and the static load σ=15 kPa, it has been found that a complete stabilization of compacting took place after the time T=8 min for sand and T=10 min for gravel with the greatest part of compacting (70÷90%) being practically realized after the time of 4 minutes.

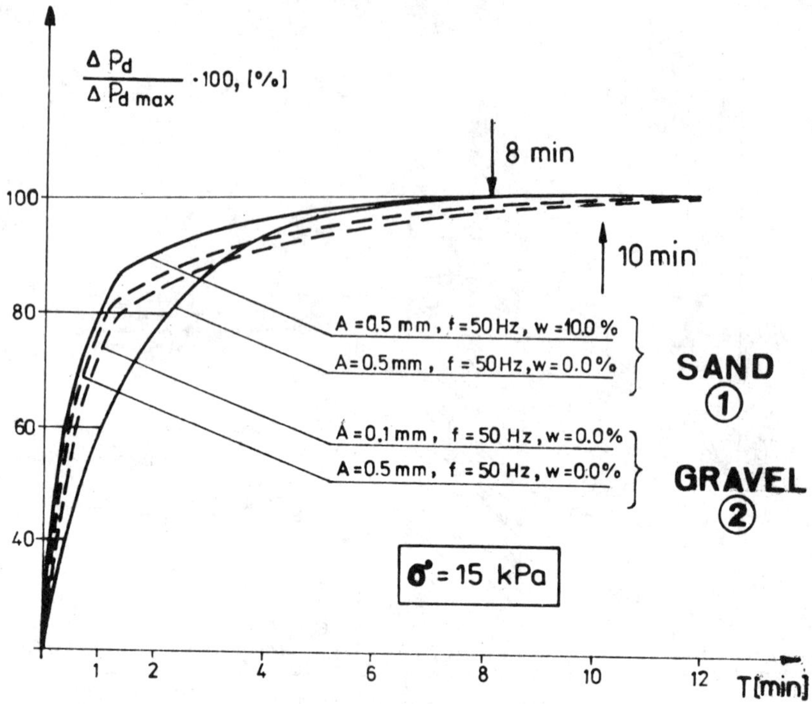

Fig. 4. Compacting as a function of vibration duration

Having determined the values of the variables w, σ, T, the effect of the frequency and amplitude on the efficiency of compacting of both soils was found in the basic tests. The test results have been given in Fig. 5 and 6. Each of the curves corresponds to a constant value of the amplitude A. The envelopes of the particular curves picture a more general dependence $P_d = F_1(f, A)$. From these envelopes it results that the medium-size grain sand achieved the best compactness with the frequency f=50 Hz and amplitude A=0.3 mm, whereas gravel-with frequency f=25 Hz and amplitude A=1.0 mm.

From the inspection tests made it results that a similar influence of the frequency and amplitude has been observed in the case when soil moisture content corresponded to the optimum moisture content determined at the point of the curve of the curve bending $P_d = F(w)$, and the vibration time (T) and the static load (σ) was assumed on the levels discussed above. However, the value of the volume density of the skeleton, obtained during the tests was somewhat lower than when soil moisture content was close to zero.

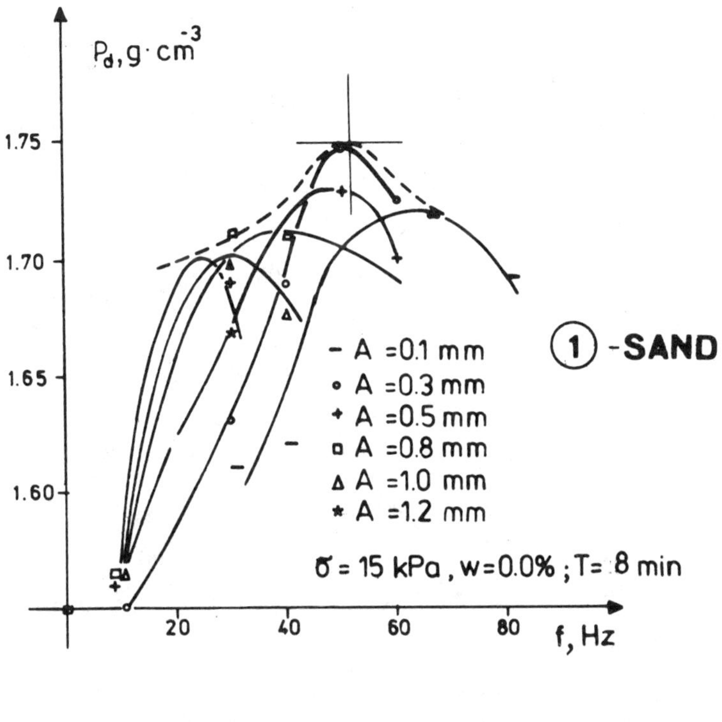

Fig. 5. Compacting as a function of amplitude and frequency of vibrations (sand)

From a comparison of the maximum values of the compactness P_{ds} obtained for both soils by the normal Proctor's method (acc. to the Polish standard - Tabl. 1) and by vibrations method (Fig. 5 and 6) it results that the latter ones about 4÷7% higher.

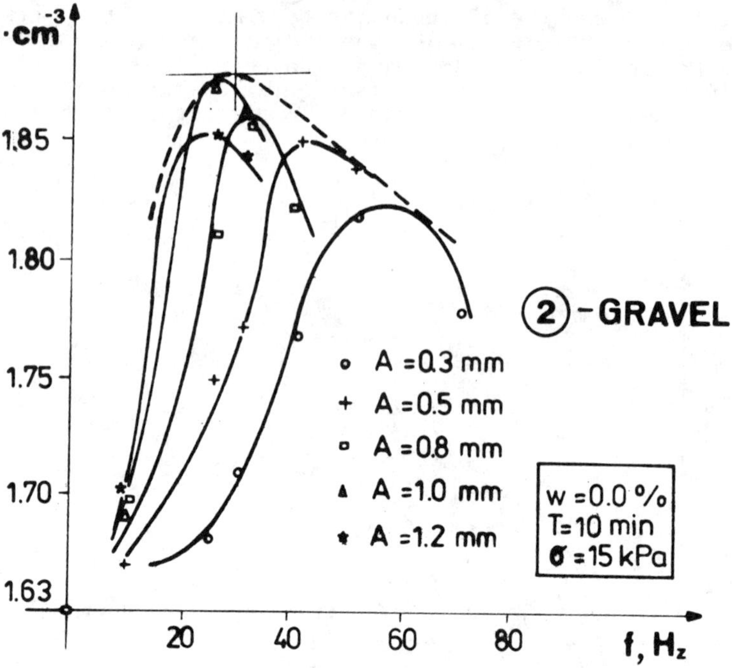

Fig. 6. Compacting as a function of amplitude and frequency of vibrations (gravel)

An example of the field control tests

In the town of Zabrze, in place of a house destroyed during the war, a three storey, no-cellar, social building was designed. The designer decided to change the uncontrolled embankment located three and characterized by a differentiated compaction and composition to a sand cushion, demanding at the same time that relative compaction of the cushion material should not be less than 0.9. This relatively small relative compaction resulted on one hand from the small loads transferred from the construction designed, and on one the other-from the hazards of the so-called mining effects. The building of the embankment was started after an execution of a wide-space excavation of the projected dimensions of about 28.0x15.0 m (with the projected dimensions of the object 21.61x9.52 m) and depth of about 3.0 m. A material whose grain size distribution curve has been shown in Fig.1 was used for the construction of the sand cushion with some of the geotechnical parameters being given in Table 1 (the material has been denoted by no ③). This was a medium-size sand (very similar to the material denoted by no ①). The sand supplied to the spot was manually spread at the bottom of the excavation in 0.3 m layers and next compacted by means of a walking vibrating compactor owned by the investor, of the symbol WZK-10. The amplitude and frequency of vibrations were 0.5 mm and 15 Hz, respectively, and the exciting force was 50 kN (σ = 10 kPa). Before the compacting of the successive layer, the material was moisturized with water. As a result, the cushion made had a total thickness 1.8 m. In the course of determinations of the mass density were made twice at the level 0.9 and 1.4 m counting from the cushion floor. An average from 50 determinations was $P_d = 1.7\ g/cm^3$ with standard deviation s=1.97%. For the material used for the construction of the sand cushion, determined were by the compacting method recommended by the Polish standards [3], and the vibrating method (for the assumed w=0.0% , σ = 15 kPa , T = 8 min , A = 0.3 mm, f = 50 Hz), the maximum values of the volume densities of the skeleton. These were respectivelly: P_{ds}=1.85 g/cm^3 and 1.89 g/cm^3. The values of the compaction relative I_S for both methods were: I_S=0.918 and 0.900, and although they meet the design requirements, they are a little different. Of course, in order to compare field results and laboratory results both tests need to be performed by the same method, in this case using vibrations.

Final remarks

The laboratory tests made indicate an essential effect of many variables on the of compacting effectiveness of cohesionless soils by vibration with the frequency and amplitude of vibrations being essential here. Detailed quantitative conclusions are included in the paragraph "The laboratory test results" above. The effectiveness of compacting of cohesionless soils by vibration methods is somewhat higher than by other methods. However, the problem requires further studies.

References

[1] ARQUIÉ G.: Le compactage. Routes et Pistes d'envol. Edition Eyrolles 61, boulevard Saint-Germain-Paris-V^e, 1972

[2] PISARCZYK S.: Compactibility of Coarse-Grained and Stony Soils (in Polish). Politechnika Warszawska,

Warszawa, 1977
[3] PN-88/B-04481. Building soils. Laboratory tests (in Polish), Warszawa, 1988
[4] United States Department of the Interior Bureau of Reclamation: Development of a maximum density test for cohesionless soil by a vibratory method. Denver, Colorado, May 8, 1961
[5] WIŁUN Z.: Zarys geotechniki. Wyd. Kom. i Łączności, Warszawa, 1987

Cement-Stabilized Soil for Coal Retaining Berms

Gary J. Van Riessen[1] M.ASCE and Kenneth D. Hansen[2] F.ASCE

Abstract

The use of cement-stabilized soil (also called soil-cement) to construct steep, stable slopes for coal-retaining berms satisfies the needs of state-of-the-art coal storage and mixing reclaim systems at power plant sites. A review of five projects constructed between 1974 and 1982 indicates this method can be used to solve a number of technical requirements in an economical manner. Readily available construction equipment can be used to produce slopes up to 55° with little modification. Design and construction methods using cement-stabilized soil for coal retaining berms has improved over the years and the resulting structures have shown excellent performance.

Introduction

As electrical energy demand in the United States continues to increase, the need for coal as a primary energy source will increase proportionately. At coal-fired power generating facilities throughout the country, the requirement for innovative, low-maintenance coal-handling facilities is a primary design consideration.

One solution for storing then transporting coal to power generating units involves the use of steeply-sloped coal retaining berms adjacent to a gravity-fed conveyor located in a reclaim tunnel. A large piece of moving equipment called a stacker-reclaimer feeds coal to the trough which is formed by berms on one or both sides of the tunnel. The stable slopes, which are as steep as 60° to

---

[1]Geotechnical Section Supervisor, Black & Veatch, 8400 Ward Parkway, P.O. Box 8405, Kansas City, MO 64114.

[2]Senior Water Resources Engineer, Portland Cement Association affiliate, 241 West 56th Avenue, No. 200, Denver, CO 80216.

the horizontal, are made possible by facing the berms with cement-stabilized soil (also called soil-cement), which has a higher angle of internal friction than untreated soil.

History and Development of Cement-Stabilized Soil or Soil-Cement

Soil-cement is a highly compacted mixture of soil, (usually sand), portland cement, and water. It was developed in the 1920s as a method for stabilizing local soils with cement to produce an economical road building base material. During the 1940s, the American Society for Testing and Materials (ASTM) adopted several test standards for soil-cement, including a method to determine adequate cement content for varying soil conditions based on research by the Portland Cement Association (PCA).

The use of soil-cement to protect embankments from reservoir wave action got its start in 1951 when the US Bureau of Reclamation (USBR) constructed a full-scale field test section on the south shore of Bonny Reservoir in eastern Colorado. Following 10 years of successful exposure to wave action and an average of 140 freeze-thaw cycles per year, the USBR and private consulting firms started to specify soil-cement for upstream slope protection for earth dams. Ute Dam in New Mexico became the first embankment dam constructed with a soil-cement face in 1963 and Cochiti Dam, also in New Mexico, was the site of the first use of cement-stabilized gravel for a mass structural use in 1968. Here, 58,500 yd^3 (44,700 m^3) of cement-stabilized fill replaced a weak shaly clay and was placed as the foundation for a triple barrel concrete outlet conduit to produce physical properties similar to the adjacent sandstone bedrock.

With these early developments as a background, the initial use of soil-cement for a coal storage facility occurred in 1973 at the Sarpy Creek Coal Mine near Hardin, Montana. Mass soil-cement was used to produce a coal storage slot 750 ft (230 m) long and 20 ft (6.1 m) deep. The soil-cement walls for the trough, which were sloped at a 50° angle, were faced with more wear-resistant shotcrete. The mass soil-cement buttress was also used to support loads from the coal handling equipment and roof structure. A coal storage slot only differs from the coal retaining berms described in this paper in that it is constructed below the ground surface rather than above.

Engineering Properties of Cement-Stabilized Soil

The physical properties of cement-stabilized soil of greatest interest to designers of coal-retaining berms are durability, compressibility, and shear strength. All of these properties increase with increased cement content and with time, similar to concrete.

Adequate durability is determined by standard freeze-thaw and wet-dry tests to determine the minimum cement content required to produce a structural material that will resist volume changes produced by changes in moisture and temperature. For durability testing, specimens with varying cement content and optimum moisture content as determined by the standard Proctor test (ASTM D558) are compacted, cured for 7 days, then subjected to 12 cycles of freezing and thawing or wetting and drying. The test procedure and criteria for determining suitable cement content are detailed in PCA's Laboratory Handbook.

Compression strength is directly proportional to durability, except for coarse sand-gravel mixtures with little or no fines passing the US No. 200 (0.075 mm) sieve. These cement-treated materials may have high compressive strength and relatively poor durability. Nearly all cement-stabilized soils exceeding 750 psi (5.2 MPa) at 7 days and 85% of those exceeding 600 psi (4.1 MPa) at 7 days will have adequate freeze-thaw durability.

To produce the steep slopes required for coal-retaining berms, an increase in shear strength over typical untreated soils is required. The total shear strength of both untreated and cement-stabilized soils can be determined using Coulomb's equation:

$$s = c + p \tan \phi$$

where s = unit shear stress
c = unit cohesion
p = unit normal stress
ϕ = the angle of internal friction

Even when small amounts of cement are added to soils, an increase in shear strength is appreciable after 28 days curing, as shown in Table 1. Cohesion (c) continues to increase with time for well-bonded layers. However, the angle of internal friction (ϕ) may in some cases decrease slightly at later ages.

History of Coal Retaining Berms

The past twenty years have seen many new trends and developments in the coal handling and storage industry. With many coal-fired units exceeding 600 MW capacity, adequate coal handling and storage facilities at power generation facilities have become critical factors in power plant design. Continuing trends also include combining production from different mines by mixing coals at the plant site. Most of these developments have been driven by the

Soil (AASHTO)	Percent cement by weight	Cohesion, psi	Slope angle, degrees, Ø
A-2-4	0	20	29
	2	50	41
	3	58	44
	4	70	44
	6	90	48
	8	100	49
A-1-b	0	10	38
	1	27	45
	2	37	49
	3	50	51
	4	72	52
	5	95	55
A-4	0	5	37
	2.5	30	46
	5.5	65	45
	7.5	85	45
	9.5	125	45

Table 1. Triaxial Strength and Cement Content After 28 Days Curing (Nusbaum and Colley 1971)

use of the unit train, which has required the construction of innovative storage, high-speed loading and unloading, and high-capacity reclaim systems, such as rotary plow tunnels and stacker-reclaimers.

Current design and power production philosophies require not only that the flow of coal proceed unabated from unit train unloading to pulverizing, but also that adequate coal be available to act as a buffer between the coal supply source and its burning in the boilers. For this reason, the coal stockpile has become more important. Along with these stockpiles, the stacker-reclaimer has become necessary as coal handling facilities adapt to meet high-capacity requirements. The stacker-reclaimer system may also include a tunnel reclaim system to facilitate blending of different coals. A key component of the stacker-reclaimer system is the coal retaining berm.

The coal retaining berm consists of a steeply-sloped embankment either founded directly upon a prepared subgrade or immediately above the structural concrete portion of the reclaim tunnel. The steep slope is necessary to limit the adherence of coal to the side walls, thereby minimizing the chances of spontaneous combustion of exposed coal. The coal-retaining berm also reduces the volume of dead storage in a stacker-reclaimer system, and decreases the required number of mobile pieces of equipment, such as dozers and scrapers, in the coal stockpile area.

Some coal retaining berms have been constructed by compacting select soil materials at steep temporary angles and covering these slopes with a reinforced concrete overlay. Other berms have used compacted soil reinforced by metal strips connected to precast concrete panels inclined at the appropriate angle. However, some coal retaining berms have been constructed using cement-stabilized soil for steep, stable slopes. These innovative designs provide technical, economical solutions which satisfy the requirements for high-capacity coal storage and reclaim systems.

The use of cement-stabilized soils for coal retaining berms has changed considerably over the past fifteen years. Initial applications had limited height, typified by a compacted earth embankment enclosed by a shell of cement-stabilized soil. Early construction methods included in-place mixed soil-cement and ranged from overbuilding and cutting back at appropriate slopes to shaping the completed material after compaction. More recent applications have involved placement of cement-stabilized soils mixed in central plants called pugmills and placed on berms to heights approaching 40 ft (12.2 m) at slopes of 0.7 H to 1.0 V (55°) using laser guided equipment.

Case Histories

The following five projects are representative of the gradual transition and technical improvements in the use of cement-stabilized soil for coal retaining berms, and clearly demonstrate the transformations in their design and construction. In all cases, the shear strength properties of the cement-stabilized soil were used to ensure stability for a wide variety of design loads, including seismic events. The projects were typically designed using an effective angle of internal friction (Ø) equal to 35° and an effective cohesion equal to 50 psi (350 KPa). All the projects were designed in the Kansas City offices of Black & Veatch.

Case History 1. As part of the construction of this northern plant, in 1974 the utility began development of a new coal storage area. After a detailed study of alternative design sections for coal retainage dikes, they selected an embankment design consisting of a composite section of cement-stabilized soil and on-site compacted sands. Figure 1 shows a typical section for the coal retaining system. Onsite silty sands were pugmill-mixed with approximately 8 to 10% cement by dry weight of soil, compacted to densities greater than 95% of standard Proctor density, and placed with conventional equipment to lines and grades slightly greater than plan dimensions. After the cement-stabilized soil had initially hardened (approximately one day), the freshly set material was mechanically trimmed to the required embankment and slope configurations. The final slope of the retaining berms was 0.7 H to 1.0 V (55°). The approximate length of the bermed retaining system is 900 ft (274 m). The completed berm as it appeared in 1983 is shown in Figure 2.

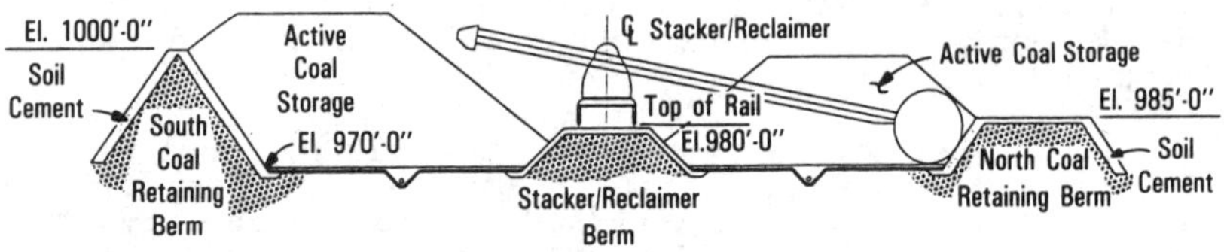

Figure 1. Typical Section, Coal Retaining System

Figure 2. Completed Coal Berm

<u>Case History 2</u>. This midwestern utility made use of an 800-ft (244 m) long solid cement-stabilized soil retaining berm. Constructed in 1976, the berm was approximately 35 ft (11 m) high, at a slope of 0.7 H to 1.0 V. The entire berm consisted of cement-stabilized soil which was inplace mixed and compacted to a minimum of 95% of standard Proctor density. Cement content for the completed mixture ranged from 6 to 8%. No special forming or shaping of the sloping surface was performed other than hand compaction and shaping after placement. Figure 3 shows the nature of the berm several years after completion.

Figure 3. Completed Solid Soil-Cement Coal Berm

Case History 3. A second midwestern utility constructed a berm using a sand embankment fill and an outer shell of cement-stabilized soil to retain coal in a stacker-reclaimer operation. This berm, constructed in late 1980, is 715 ft (218 m) long, 34.5 ft (10.5 m) high, and has steep 55° angle side slopes. The cement-stabilized soil facing provides stability and erosion protection. A typical cross-section is shown in Figure 4.

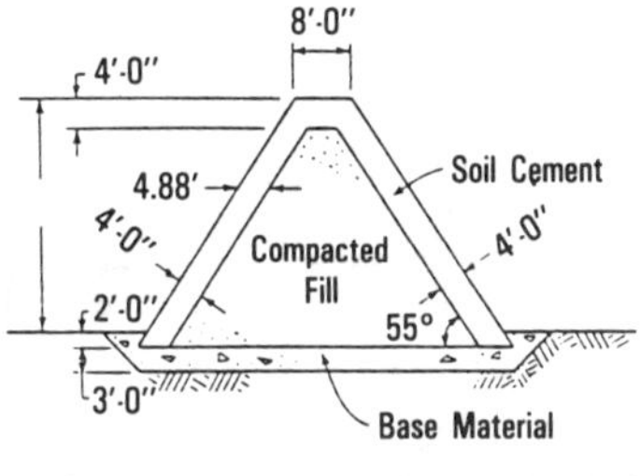

Figure 4. Typical Section of Soil-Cement Coal-Retaining Berm

The cement-stabilized soil faces and the interior uncemented sand fill were placed and compacted together in 9-inch (23 mm) thick vertical lifts. The 8,500 cubic yards (6,500 m^3) of cement-stabilized soil, containing 10% cement by weight, was central-plant mixed. A modified asphalt paving machine was used to place and initially compact the cement-stabilized soil to proper width and thickness, and to form the 55° outer slope (Figure 5). Much of the cement-stabilized soil was compacted using hand compactors on a horizontal surface. Following final compaction, the cement-stabilized soil was sprayed with an asphalt emulsion curing compound to prevent premature drying. Shear strength properties of the cement-stabilized soil were used in a slope stability analysis to minimize thickness of the soil-cement cap to 4 ft 0 in. measured perpendicular to the slope. This design effort resulted in an economical structure with a strong, durable exterior surface (Figure 6).

Figure 5. Modified Asphalt Paver Forming Edge and Placing Cement-Stabilized Soil for Berm

Figure 6. Completed Cement Stabilized Soil Shell for Coal Retaining Berm

Case History 4. Cement-stabilized soil was used extensively in 1982 for construction of the new coal mixing facility at this southwestern generating station. The coal mixing system consists of a coal-retaining and support embankment 1,100 ft (335 m), a stacker-reclaimer, and a reclaim tunnel as shown in Figure 7. The design of this system made use of the shear strength of the cement-stabilized soil to ensure adequate factors of safety for both static and seismic design conditions. Cement-treated soil was also used to provide a support mat beneath the stacker-reclaimer.

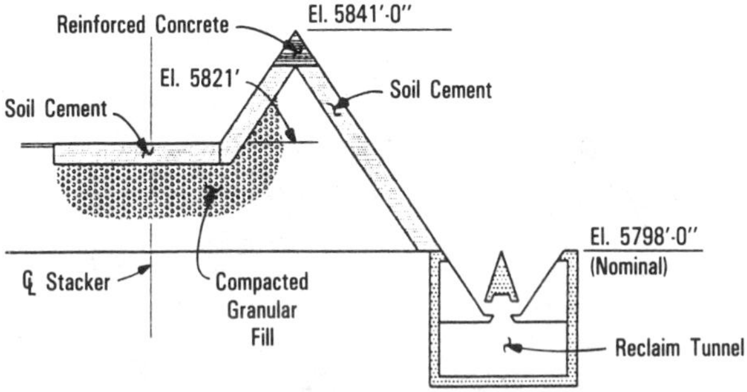

Figure 7. Typical Berm Section

Figure 8. Modified Asphalt Paver Forming Outer Edge of Coal Retaining Berm

The cement-stabilized soil was placed similar to that for the midwestern utility at which a trailing plate inclined at 55° was attached to a modified asphalt paver as shown in Figure 8. All cement-stabilized fill was mixed with approximately 10% cement by weight at a centrally located pugmill and hauled to the project by trucks. Compaction was accomplished using vibratory plates near the edge of the slope and a steel-wheeled compactor on the remaining cement-stabilized material. This particular approach also used pneumatically placed concrete (shotcrete) as the finished surface for the cement-stabilized soil on the 55° slope. This surface had relatively low adherence, and when coupled with the steeply inclined berm surface, resulted in minimal "hang-up" of coal above the reclaim tunnel. The completed structure is shown in Figure 9.

Figure 9. Completed Coal Berm

Case History 5. This western utility used a massive cement-stabilized soil block to retain coal in a stacker-tunnel reclaim system, as well as provide foundation support for the stacker. A section of the system is shown in Figure 10. The massive block minimizes potentially large horizontal seismic-induced deformations that might occur with more conventional, less rigid retaining wall systems. The cement-stabilized soil contained about 10% cement by weight and was mixed in a pugmill.

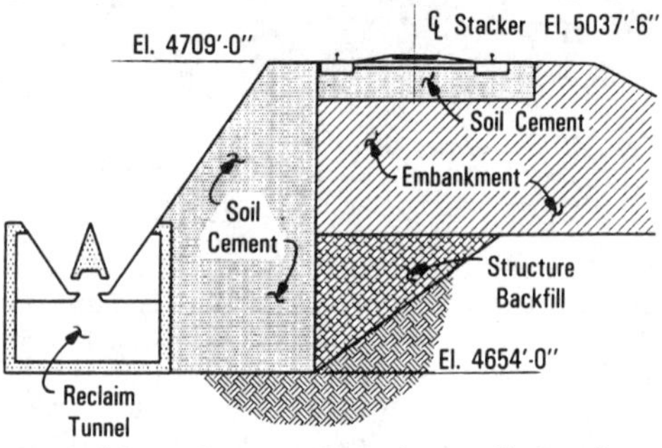

Figure 10. Cross Section of Coal Berm

The 55° slope was also constructed using a modified asphalt paver and a fabricated trailing edge, but with two innovative modifications. One modification attached a concrete form-vibrator to the trailing edge, compacting the outside slope surface at the time of placement. Additional compaction away from the edge was provided by plate compactors and a steel-wheeled vibratory compactor. The second modification, as shown in Figure 11, attached a laser receiver on the paver and used a laser system to guide the paver to maintain desired line and grade. This system was more efficient and accurate than a typical string line guiding system. A pneumatically-placed concrete (shotcrete) liner was applied to the finished surface to reduce adherence of coal to the 55° slope. The shotcrete was anchored by bars embedded in the cement-stabilized soil at a grid spacing of 5 ft by 5 ft. The completed cement-stabilized soil is shown in Figure 12.

Figure 11. Laser Guidance System (Note Target to Right of Machine)

Conclusions

After service ranging from 10 to 18 years, the coal retaining berms described in this paper are performing well. During this time, several valuable lessons were learned which improved both the design and construction of cement-stabilized coal retaining berms.

- Slopes of up to 0.7 H to 1.0 V (55°) can be constructed using a modified asphalt paver with a steel trailing edge fabricated to the required slope.
- Cement contents between 8 and 10% by weight provide the desired physical properties and durability for these applications.

Figure 12. Coal Berm with Finish Surface

- Pugmill mixing ensures mixture uniformity.
- Coal retaining berms described here can be cost competitive with reinforced soil or cast-in-place concrete and can be constructed using readily available construction equipment with little modification.
- The high shear strength and bearing capacity of cement-stabilized soil provide for designs with innovative, unique berm geometries.

Appendix - References

1. Nusbaum, P.J. and B.E. Colley. "Dam Construction and Facing with Soil-Cement," Portland Cement Association, RD010, 1971.
2. Portland Cement Association. "Soil-Cement Laboratory Handbook," EB052, 1971.
3. Portland Cement Association. "Coal-Handling and Storage Facilities," PL152, no date.

Flow Capacity Effect on Vertical Drain Performance

R. Robert Goughnour[1], A.M.ASCE

Abstract

The influence of discharge capacity on vertical drain performance is studied by computer simulation. Effects of variation in soil compressibility with depth, nonuniform stress increase with depth, and discharge capacity variation with depth and with strain are studied and reported. In general, results indicate that limited drain discharge capacity has less detrimental effect on performance than indicated by the results of analytical techniques using average parameter values.

Introduction

The potentially detrimental effect of inadequate discharge capacity on vertical drain performance has long been recognized. Analytical solutions (Barron, 1948; Yoshikuni & Nakanodo, 1974; Hansbo, 1981) indicate that drain flow capacity can be important for small-diameter sand drains and prefabricated drains with great length, particularly if they are closed at the bottom.

Derivations of these equations were based on the assumption that the soil parameters m_v and k_h, the coefficient of volume compressibility and the horizontal permeability respectively, and q_w, the drain discharge capacity, remain constant in time and position. Average, or representative constant values must be used. In reality, all three of these parameters are dependent on overburden pressure, and vary with depth. In addition, the discharge capacity of prefabricated vertical drains can change if soil compression becomes great enough to cause buckling of the drain. It would be of interest from a practical point of view to investigate the theoretical effect of m_v values and k_h values that change with depth, as well as the effect of q_w values that change with both depth and vertical strain.

A numerical procedure which accommodates the required parameter behavior has been developed. Typical discharge capacity behavior is described in terms of confining pressure and deformation. Representa-

---

1 - Vice Pres. and Senior Design Engr., Geotechnics America, Inc., P.O. Box 2324, Peachtree City, GA 30269.

tive situations are studied using this behavior and the numerical procedure. In general the results indicate that limited drain discharge capacity has less detrimental effect than indicated by the analytical solutions using average parameter values. Only cases where the drains are closed at the bottom are considered.

Analytical Solution

Hansbo (1981) presented a closed-form solution for equal strain conditions which includes the effects of drain resistance. To provide more versatility, this solution will be modified to include the effect of a drain extension through an overlying incompressible material, as illustrated on Figure 1. The influence of the effective overburden, O_b, will be considered later.

The flow, Q, is assumed to be confined completely within the drain through the incompressible material, and the pressure in excess of hydrostatic within the drain at z' = E is

$$(u)_{z'=E} = QE\gamma_w/q_e \tag{1}$$

where E is the thickness of the incompressible layer, γ_w is the unit weight of water, and q_e, the discharge capacity of the drain through the incompressible material, equals $k_w A_w$, the permeability in the longitudinal direction of the drain times the cross-sectional area of the drain. In terms of flow within the drain through the compressible material, Q is given by

$$Q = \frac{q_w}{\gamma_w}\frac{\partial u}{\partial z} \tag{2}$$

where q_w is the flow capacity of the drain within the compressible material. Substituting equation (2) into (1) we obtain

$$(u)_{z'=E} = (u)_{z=0} = \frac{q_w E}{q_e}\frac{\partial u}{\partial z} \tag{3}$$

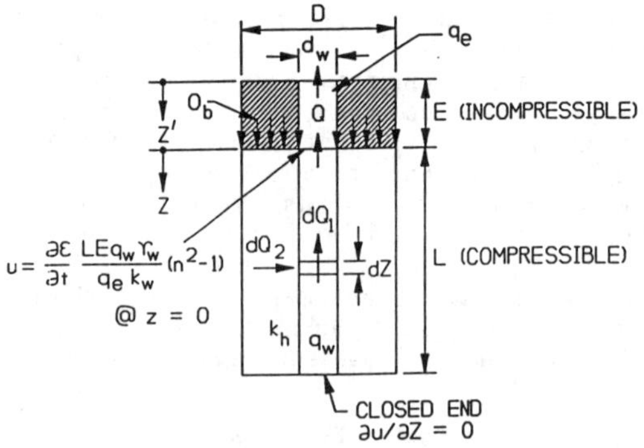

Figure 1. Schematic picture of soil cylinder dewatered by vertical drain.

Equation (11) from Hansbo (1981) relating the pore pressure at the soil-drain interface to the vertical strain in the soil was given as

$$\frac{\partial^2 u}{\partial z^2} = -\frac{\gamma_w}{k_w}(n^2-1)\frac{\partial \varepsilon}{\partial t} \tag{4}$$

where $n = D/d_w$, the diameter of the cylinder of influence of the drain divided by the equivalent diameter of the drain, and $\partial\varepsilon/\partial t$ is the rate of vertical strain. Integrating and inserting the boundary condition, $\partial u/\partial z = 0$ at $z = L$

$$\frac{\partial u}{\partial z} = \frac{\gamma_w}{k_w}(n^2-1)\frac{\partial \varepsilon}{\partial t}(L-z) \tag{5}$$

Now set $z = 0$ and substitute equation (5) into equation (3)

$$(u)_{z=o} = \frac{q_w E}{q_e}\frac{\gamma_w}{k_w}(n^2-1)\frac{\partial \varepsilon}{\partial t}L \tag{6}$$

Substituting this boundary condition instead of $u = 0$ at $z = 0$, as presented by Hansbo, and proceeding with the derivation yields

$$\bar{U}_h = 1 - \exp[-8tC_h/(D^2F)] \tag{7}$$

where $\bar{U}_h$ is the average degree of consolidation at the depth under consideration, t is the time, C_h is the coefficient of consolidation, and

$$F = F(n) + F(r) + F(e) \tag{8}$$

$$F(n) = \frac{n^2}{n^2-1}\left(\ln n - \frac{3}{4} + \frac{1}{n^2} - \frac{1}{4n^2}\right) \approx \ln n - \frac{3}{4} \tag{9}$$

$$F(r) = \frac{4k_h}{D^2k_w}(n^2-1)(2Lz-z^2) \approx \pi z(2L-z)\frac{k_h}{q_w} \tag{10}$$

$$F(e) = \frac{8q_wk_hEL}{D^2k_wq_e}(n^2-1) \approx \frac{2\pi k_hEL}{q_e} \tag{11}$$

The term $\bar{\bar{U}}_h$ is defined as the overall average degree of consolidation, and is given by

$$\bar{\bar{U}}_h = 1/L\int_0^L \bar{U}_h\, dz \tag{12}$$

When required, this integration was accomplished by numerical methods. Equation (7) describes the average pore pressure versus depth as a function of the various parameters. The term $C_h = k_h/m_v\gamma_w$ is assumed to be unvarying in both time and position. The horizontal permeability, k_h, also appears in the equation, and is assumed to be unvarying in both time and position. Therefore, m_v must be unvarying in time and position. Since the initial excess pore pressure distribution was taken to be uniform throughout the problem region, the vertical stress increase must be constant with depth. Thus, this solution applies only to conditions where the ultimate vertical strain is uniform throughout the problem region. The same reasoning applies to solutions by Barron (1948) and Yoshikuni and Nakanodo (1974).

It is convenient to express equation (7) in terms of the dimensionless time factor, $T_h = C_h t/D^2$, as follows

$$T_h = \frac{F(n) + F(r) + F(e)}{8} \ln \frac{1}{1-\bar{U}_h} \tag{13}$$

$$T_h = T_h(n) + T_h(r) + T_h(e)$$

$$T_h(n) = (1/8)\ F(n)\ \ln(1/1-\bar{U}_h)$$

$$T_h(r) = (1/8)\ F(r)\ \ln(1/1-\bar{U}_h)$$

$$T_h(e) = (1/8)\ F(e)\ \ln(1/1-\bar{U}_h)$$

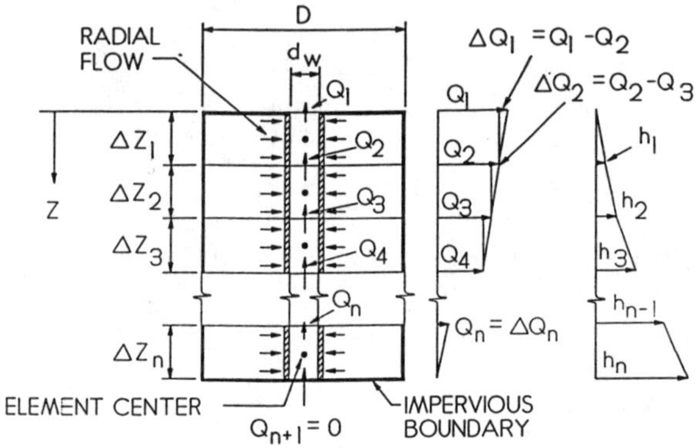

Figure 2. Element layout through vertical drain and soil cylinder.

Numerical Solution

The cylinder of influence is divided into disc-shaped elements, as shown on Figure 2. Solution is achieved by considering a series of relatively small time increments, during which steady-state conditions prevail. The total surface load, L_t, applied at time zero, is assumed to remain constant. The magnitude of this load distributed to the depth of the center of element i, $(L_t)_i$, also remains constant. The equal strain assumption has been adopted, and $(L_t)_i$ is equal to the vertical stress integrated over the horizontal projection of the element.

Pore water, expelled from the soil into the drain, is forced upward through the drain with attendant head loss. The head is zero (or considered known) at the ground surface and increases to a maximum at the bottom of the drain. The quantities $W_{i,1}$, $W_{i,2}$, etc. represent the excess water pressures within drain element i, during time increments 1, 2, etc. At time zero the actual net load applied to element i, $L_{i,1}$, is assumed to be $(L_t)_i - AW_{i,1}$, where A is the cross sectional area of the cylinder of influence given by $\pi D^2/4$. At time Δt the load

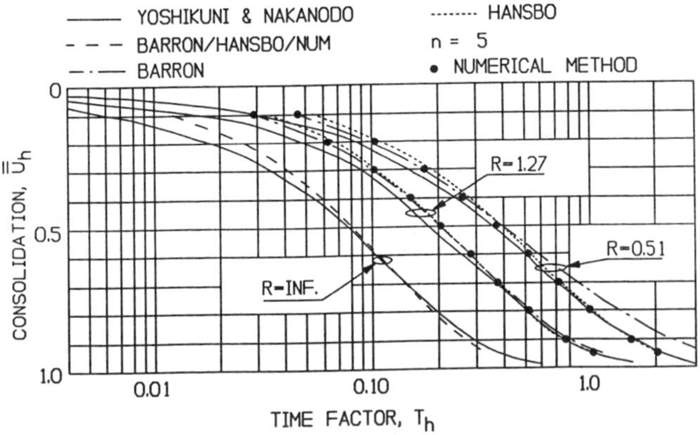

Figure 3. Comparison of results from various methods for n=5.

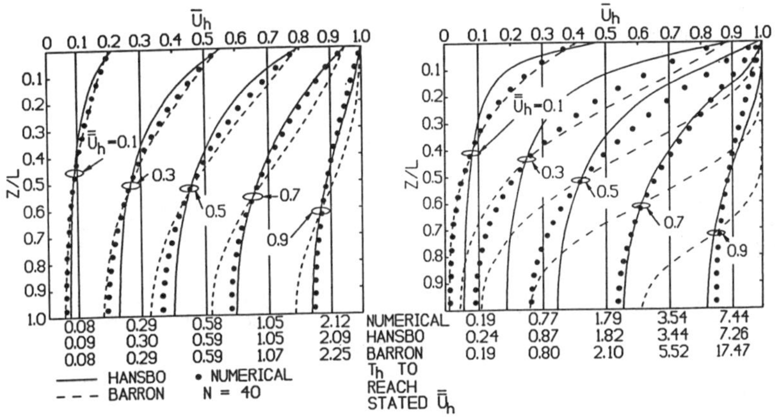

a) $R = q_w/k_hL^2 = 0.5$ b) $R = q_w/k_hL^2 = 0.1$

Figure 4. Comparison of consolidation results with depth.

applied to element i abruptly changes to $L_{i,1} + L_{i,2} = (L_t)_i - AW_{i,2}$. The first subscript indicates the element number, and the second indicates the time increment. Consolidation of the soil under the loads $L_{i,1}$, $L_{i,2}$, etc. proceeds simultaneously, with the excess pore water pressures, $u_{i,1}$, $u_{i,2}$, etc. for each time increment transferring to effective stresses, $\bar{\sigma}_{i,1}$, $\bar{\sigma}_{i,2}$, respectively. Note that the consolidation times for the loads $L_{i,j}$ are not all the same, i.e. $L_{i,1}$ starts at time zero, $L_{i,2}$ starts at time Δt, $L_{i,3}$ starts at $2\Delta t$, etc.

Although m_v and k_h are allowed to change from one element to the next, their ratio is assumed to remain constant, i.e. C_h remains constant. Details of this procedure are given elsewhere (Goughnour, 1991).

Comparison With Analytical Solutions

To compare results with those of the analytical methods, only cases where m_v, k_h, and q_w remain constant with both depth and time, and the stress increase is constant with depth can be considered. This comparison is presented on Figure 3 for an n value of 5, as a plot of overall average consolidation, $\bar{\bar{U}}_h$, versus time factor, T_h. The data are grouped according to various values of the dimensionless quantity $R = q_w/(k_h L^2)$ ranging from 0.51 to infinity. Since the results of Hansbo's, Barron's, and the numerical methods converge to the same solution when this factor approaches infinity, these results have been plotted as one curve for $R = \infty$. Data for Yoshikuni and Nakanodo's method were taken from Figure 2 of their paper.

Correlation of results by the numerical method and Hansbo's method is excellent. It appears that results from Yoshikuni and Nakanodo's method also converge to essentially the same results as consolidation increases to 60% to 70% or more. Barron's solution appears to follow this pattern for values of R greater than about 1 to 1.5, but tends to produce much larger time factor values for small R values and consolidation percentages greater than about 50% to 60%.

Figure 4 presents average degree consolidation as a function of depth for n equals 40, and R equals 0.5 and 0.1. Barron's, Hansbo's, and the numerical methods are represented. The numbers at the bottom of the plots are the time factor values when the overall average consolidation reaches the indicated values. A wide disparity between Hansbo's and Barron's methods for large consolidation percentages and low R values (Figure 5b) is apparent in the pore pressure distribution. Although the correlation between the numerical method and Hansbo's method is still very good, the numerical method appears to correlate more closely with Barron's solution for low degree consolidation where the difference between Hansbo's and Barron's solutions is also small. This trend can also be observed on Figure 4.

Effect of Nonuniform Soil Compression with Depth

In normally consolidated soil, compressibility decreases with depth. The result is that vertical compression is greatest near the surface, even if the stress increase is constant with depth. This tends to mitigate head losses due to well resistance because water flowing into the drain flows a shorter distance to reach the ground surface. The coefficient of volume compressibility, m_v, for normally consolidated soil can be defined by

$$m_v = \frac{C_c}{\Delta p(1+e_o)} \log_{10}\left(\frac{p_o+\Delta p}{p_o}\right) \quad (14)$$

where C_c is the compression index, Δp is the change in vertical stress at the depth under consideration, p_o is the effective overburden stress, and e_o is the initial void ratio. Normal practice for settlement calculations is to break the soil profile into sublayers, assigning values of m_v to each sublayer based on Δp and p_o. The compression in each sublayer is then calculated, and the sum of the effects of all sublayers is the computed ultimate settlement. This scheme is used to account for m_v variation with depth. The average value of m_v is taken as that value which produces equivalent settlement when applied to the entire soft layer. Since C_h is considered invariant, the average value of k_h is also determined.

The change in magnitude of m_v across the soft layer, is affected by both p_o and Δp. The effect of m_v variation with depth has been quantified in terms of two dimensionless factors, $O_b/(L\Gamma')$ and $\Delta p/(L\Gamma')$, where O_b is equivalent to p_o at the top of the soft soil, L is the thickness of the soft layer, and Γ' is the submerged unit weight of the soft material.

Figure 5 illustrates the effect of m_v variation on consolidation behavior for conditions where R = 0.3, $\Delta p/(L\Gamma')$ = 0.1, and $O_b/(L\Gamma')$ = 0.2. R was computed from the average value of k_h. Data presented on Figure 5a are for the case where Δp is constant with depth, while that presented on 5b is for the condition that Δp varies linearly from a maximum at the top of the soft layer to zero at the bottom. The head loss in the section of drain through the crust, was taken as zero. The curved lines plot percent consolidation as a function of depth for times when the overall average consolidation, $\bar{\bar{U}}_h$, reaches 0.1, 0.3, 0.5, 0.7, and 0.9, as indicated by the vertical lines. The solid lines are for the case where m_v varies with depth according to equation (14), and the dashed lines represent the case where m_v is constant with depth. The numbers at the bottom of the plot are the time factor values when the indicated overall average consolidation percentages have been reached. It can be seen that when m_v varies, consolidation proceeds at a faster rate than when m_v is constant with depth; equivalent consolidation percentages are reached in about 0.73

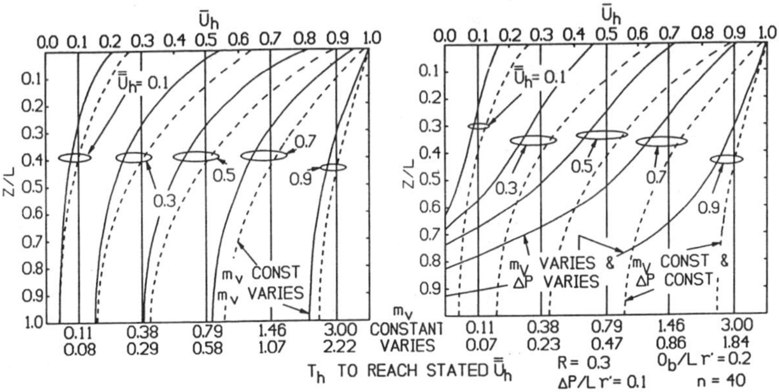

a) Δp constant with depth

b) Δp varies linearly to zero at bottom of soft layer

Figure 5. Effect of varying m_v on pore pressure dissipation.

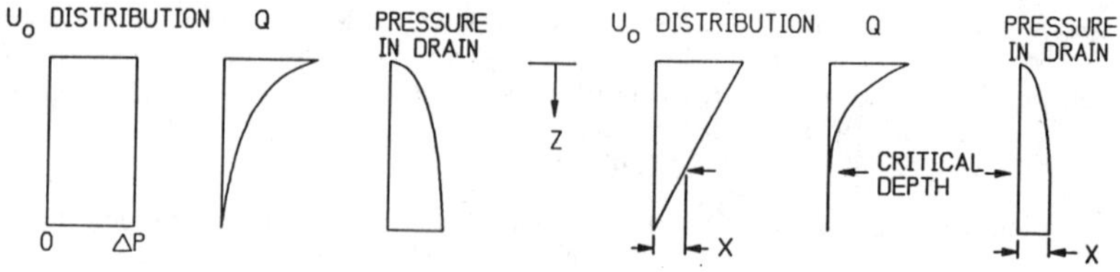

a) Δp uniform   b) Δp diminishes with depth

Figure 6. Effect of nonuniform pore pressure distribution.

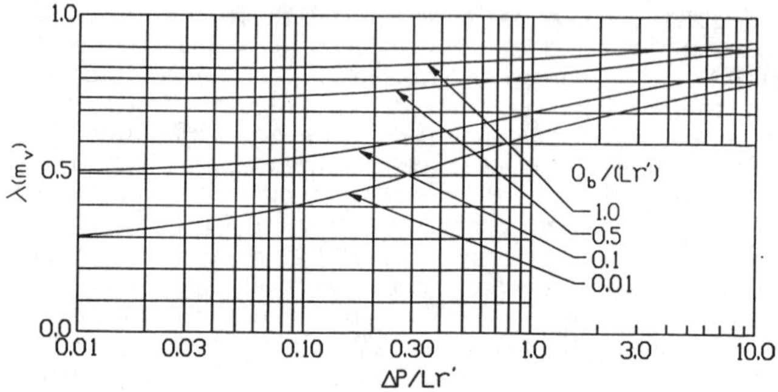

Figure 7. $\lambda(m_v)$ as a function of $O_b/L\Gamma'$ and $\Delta p/L\Gamma'$.

to 0.76 as much time when Δp is constant with depth, and about 0.59 to 0.62 as much time when Δp varies.

The broken lines in both Figures 5a and 5b represent conditions where both m_v and Δp are constant with depth, and so the ultimate soil compression must also be uniform with depth. Under this condition the area enclosed to the right of the overall average consolidation line, at the top, must equal the area enclosed to the left of this line, at the bottom. The solid lines in Figure 5a represent data for varying m_v with Δp constant, and for this condition the areas no longer balance because the ultimate vertical compression is greater near the top of the soft layer. When, in addition, Δp diminishes with depth, as shown in Figure 5b, the variation of compression of the soft layer from top to bottom is even greater, and the unbalance of these areas is even more pronounced.

Another effect of varying Δp results from the nonuniform distribution with depth of the initial excess pore water pressure. Referring to Figure 6a, illustrating the condition where Δp is uniform with depth, the initial excess pore pressure distribution is also uniform with depth and equal to the applied load. The flow in the drain, Q, varies with time but is always a maximum at the top and diminishes to zero at the bottom of the drain. The excess water pressure in the drain varies from zero at the top to a maximum at the bottom. This maximum varies with time, but is always less than Δp.

When Δp diminishes with depth the initial excess pore pressure also diminishes, as illustrated on Figure 6b. Now the excess water pressure within the drain increases with depth only to some critical level where it is equal to the excess pore pressure. Below this depth consolidation is prevented, and since no water can enter the drain, Q is zero from this point downward. Below this critical depth the excess water pressure in the drain is greater than the excess pore pressure, and swelling would actually take place. In the analysis, however swell ing was not considered.

As the top portion of the soil consolidates, Q decreases with attendant decrease of excess water pressure in the drain. The critical depth then gradually moves downward, until it eventually reaches the bottom of the drain. This downward movement can be observed in the results presented on Figure 5b as the depth at which the degree of consolidation goes to zero.

The effect of m_v variation with depth for the case of Δp constant with depth is summarized on Figure 7 in terms of $\lambda(m_v)$. To apply this correction factor simply multiply either $T_h(r)$ or $F(r)$ by $\lambda(m_v)$. The behavior of $\lambda(m_v)$ is defined in terms of $O_b/L\Gamma'$ and $\Delta p/L\Gamma'$.

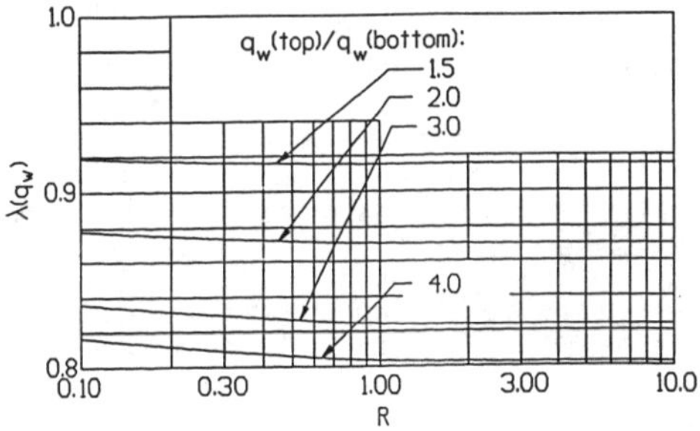

Figure 8. $\lambda(q_w)$ as a function of R and q_w(top)/q_w(bottom).

Effect of Changing q_w With Depth

Compression resulting from lateral soil pressure tends to inhibit water flow within prefabricated vertical drains, and their flow capacity at depth may diminish to only a fraction of that near the surface. Measured values (Jamiolkowski, Lancelotta, & Wolski, 1983; Lawrence & Koerner, 1988; Faisal Ali & Yong Kong Why, 1989; Oostveen & Troost, 1990) indicate that the decrease in flow capacity, q_w, is roughly proportional to lateral pressure, and thus roughly proportional to depth. A linear variation has been assumed, and it is convenient to present the results in terms of the ratio of the flow capacity of the drain at the top of the soft layer to the flow capacity of the drain at the bottom of the soft layer, q_w(top)/q_w(bottom).

Figure 8 illustrates the effect of varying q_w with depth in terms of the factor $\lambda(q_w)$. $\lambda(q_w)$ is a function of R, and is applied in the same manner as $\lambda(m_v)$.

When m_v and q_w variations are considered simultaneously, the combined effect is slightly greater than applying $\lambda(m_v)$ and $\lambda(q_w)$ in combination. This effect can be approximated very closely as follows

$$\lambda(\text{comb}) = [\lambda(m_v)]^{1.2}\ \lambda(q_w) \quad \text{if } \lambda(m_v) \le \lambda(q_w) \qquad (15)$$

$$\lambda(\text{comb}) = \lambda(m_v)\ [\lambda(q_w)]^{1.2} \quad \text{if } \lambda(m_v) > \lambda(q_w)$$

The added time caused by a drain extension, $T_h(E)$, applies unchanged in all cases, as long as the average value of k_h is used in computing F(e). If the overlying incompressible material is very permeable, water may escape from the drain into the compressible material, and $T_h(e)$ may only partially apply.

Effect of Changing q_w With Strain (Buckling)

As the soil surrounding a prefabricated vertical drain is compressed during consolidation the drain is forced to compress vertically by an equal amount. If such deformation becomes great enough the drain will accommodate the forced compression by folding or buckling. This folding or buckling can diminish the flow capacity of the drain.

Attempts to quantify this phenomenon have so far taken one of two forms; mechanically "kinking" the drain while measuring the loss of flow capacity (Lawrence and Koerner, 1988; Oostveen and Troost, 1990), or inducing drain deformation by consolidation of a large soil sample containing a vertical drain (Oostveen and Troost, 1990; Ali and Why, 1990). The latter method more closely simulates field conditions, and provides insight concerning flow capacity dependence on overall soil and drain compression.

The results of these tests have been somewhat surprising in that the decrease in discharge capacity due to drain deformation appears to be only partially related to the rigidity of the drain (Lawrence and Koerner, 1988; Ali and Why, 1990). Drains which were classed as rigid in some cases performed better than those classed as flexible and vice-versa.

Although there is considerable variation in measured values presented in the literature, it appears that the behavior can be roughly represented by a linear relationship between flow capacity and relative compression. Such a relationship was assumed as illustrated on Figure 9. This assumed relationship still permits a linear variation with depth.

A number of different problem conditions were simulated by use of the numerical procedure. Typical results are presented on Figure 10 for the conditions indicated. The ratio of t_{90} to t_{90}(ideal) is given as a function of the ratio of the ultimate soil relative compression to the relative compression at which the drain flow cuts off, i.e. the intercept with the horizontal axis on Figure 9. The ultimate relative

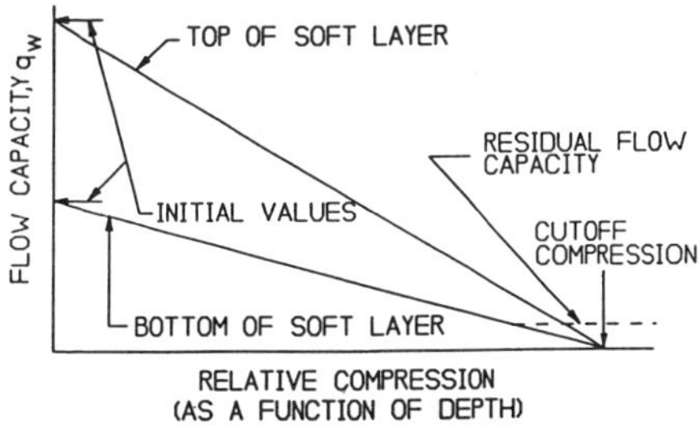

Figure 9. Assumed dependence of well capacity on deformation.

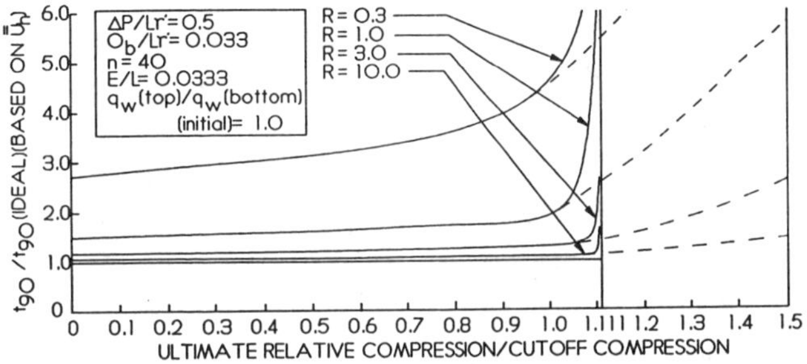

Figure 10. Effect of drain buckling on t_{90}.

compression of the soil is defined as that at the top of the soft layer computed using m_v obtained by means of equation (14). R values indicated apply to the initial average q_w. The solid lines represent the condition where the discharge capacity diminishes totally to zero, while the dashed lines represent the condition where the discharge capacity is allowed to diminish only to 1/100th of its initial value (see Figure 9).

The largest flows through the drain occur at the beginning of the consolidation process and diminish thereafter. Since folding occurs only after considerable compression takes place, the effects of folding should be minimal. This reasoning has been postulated by a number of authors, and its validity appears to be confirmed here. The results displayed on Figure 10 indicate that the time for 90% consolidation increases only nominally for ratios of ultimate compression to drain cutoff compression smaller than 1. For ratios greater than 1 behavior depends on the extent to which the discharge capacity is allowed to diminish. If the discharge capacity diminishes totally to zero, the time for 90% consolidation increases sharply, approaching infinity at a ratio of 1/.9 = 1.111. This increase in t_{90} depends also on the initial R value. For R initially greater than 1 there is little change in t_{90} when the compression ratio is less than 1.

Measuements indicate that most strip drains retain a small residual flow capacity, even under severe folding (Lawrence and Koerner, 1988; Oostveen and Troost, 1990). The dashed lines on Figure 10 show the dramatic effect on the time for 90% consolidation to occur by considering a small residual flow capacity of 1/100th of the initial value.

Summary

A numerical procedure has been developed which will accommodate parameters that change with depth, as well as changing q_w values with strain or time. Results by this method show excellent agreement with analytical methods. A number of different problem conditions have been investigated using the numerical procedure. The following conclusions are based on these results.

1. As compared with using overall average parameter values, consideration of the following tend to mitigate the detrimental effect of drain well resistance:

   a) soil compression that increases near the surface, because the water must travel a shorter distance to reach the surface. Nonuniform compression can be caused by either increasing m_v values near the surface, or decreasing Δp with depth.

   b) drain flow capacity that increases near the surface, because flow capacity is greatest where the flow is greatest.

2. Drain extensions at the top of the soft layer tend to increase consolidation times.

3. The largest water flow through the drain occurs at the beginning of consolidation and decreases thereafter. Since buckling occurs only after significant deformation, the decreased flow capacity has small effect until relative compression approaches the drain flow cutoff value. Small residual flow capacity dramatically affects consolidation time at large strain levels.

References

1. RA Barron, "Consolidation of Fine-Grained Soils by Drain Wells", Trans. ASCE, Vol. 113, Paper No. 2346, 1948.
2. Faisal Ali & YK Why, "Effects of Consolidation Settlement on the Performance of Vertical Drains", Symp. on the Application of Geosynthetic and Geofibre in Southeast Asia, 1st-2nd August, 1989, Petaling, Jaya, Malaysia.
3. RR Goughnour, "Discharge Capacity and Vertical Drain Performance", The Ninth Asian Regional Conf. on Soil Mech. and Found. Engr., 9-13 Dec., 1991, Bangkok.
4. S Hansbo, "Consolidation of Fine-Grained Soils by Prefabricated Drains", Proc. 10th Int. Conf. on Soil Mech. & Found. Engr., Stockholm, 1981.
5. M Jamiolkowski, R Lancellotta, & W Wolski, "Precompression and Speeding Up Consolidation", Proc. VIII ECSMFE, Helsinki, 1983.
6. W Kjellman, Discussion of "Consolidation of Fine-Grained Soil by Drain Wells", Trans. ASCE, Vol. 113, Paper No. 2346, 1948.

7. CA Lawrence, & RM Koerner, "Flow Behavior of Kinked Strip Drains", Proc. Symp. on Geosynthetics for Soil Improvement, Geot. Engr. Div., ASCE, 1988, pp. 22-39.
8. JP Oostveen & GH Troost, "Discharge Index Tests on Vertical Drains", Proc. 4th Int. Conf on Geotextiles and Geomembranes, The Hague, 1990.
9. H Yoshikuni & H Nakanodo, "Consolidation of Soils by Vertical Drain Wells with Finite Permeability", Jap. Soc. Soil Mech. Found. Engr., Vol. 14, No. 2, 1974.

Appendix - Small Strain Assumption

In the derivation of Terzaghi's consolidation theory the assumption of small strains occurs when the vertical drainage distance is assumed to remain constant during consolidation. In the case of vertical consolidation with radial drainage only, the drainage distance, D/2 (Figure 1), does indeed remain constant during consolidation, and this assumption is not invoked.

When the drain capacity is assumed to be infinite the solution is independent of vertical position. Although C_h is required to remain constant with respect to radial position and time, this restriction is not the result of an assumption of small strains.

When finite drain resistance is considered, the solution is no longer independent of vertical position, and the assumption of small strains is involved. First, the vertical drain length is assumed to remain constant during consolidation. Although sand drains probably shorten during consolidation, it can be argued that since prefabricated drains accommodate vertical soil strain by folding or buckling, their drainage length remains essentially constant, and this assumption is proper. The second assumption of small strains occurs when the cylindrical area through which pore fluid flows from the soil into the drain is assumed to remain constant. These assumptions are common to the methods considered here.

Selection and Laboratory Evaluation of Modifying Additives for Soil-Cement-Bentonite

T.S. McFarlane[1], A.M., ASCE, and R.D. Holtz[2], F. ASCE

Abstract

Additives were selected for the purpose of increasing the flexibility of soil-cement-bentonite mixtures commonly used in cutoff walls. Based on a review of the literature, additives selected for further investigation included SB latex, acrylic latex, epoxy resin, steel fibers, carbon fibers, Kevlar fibers, polypropylene fibers, and asphalt emulsion.

Samples of soil-cement-bentonite mixtures plus additives were prepared in the laboratory. A fine sand and a silt were used. Two sets of samples, one with each type of soil, were cast and left unmodified as control samples. The samples were tested for compressive, tensile, and flexural strength; modulus of elasticity; and permeability. In addition to soil and additive type, test variables included curing time and, for the most promising additives, mix ratios.

Generally, all the additives caused a decrease in the modulus of elasticity and compressive strength relative to the corresponding unmodified control. Tensile strengths of sand samples modified with SB latex, steel fibers, and Kevlar fibers were increased over the unmodified samples. Asphalt emulsion provided a low modulus, together with compressive, tensile and flexural strengths near those of the control mix. The strength of the silt samples was generally lower than for the sand samples for all additives. Permeability was reduced by the polymer latexes and asphalt emulsion. As commonly reported for cementitious materials, the modulus and strength increased with increasing cement content and with increased curing time.

Based on the test results, SB latex and asphalt emulsion appear to be the most promising additives.

---

[1]Staff Engineer, Applied Geotechnology, Inc., 300 120 Ave. NE, Bellevue, WA 98009

[2]Professor, Department of Civil Engineering, University of Washington, Seattle, WA 98195

Introduction

The technology of soil-cement cutoff walls is quite well established. For example, Taki and Yang (1989, 1991) describe how one such cutoff wall is constructed using a three axis hollow-stem auger. During penetration and withdrawal at the location of the proposed wall, cement-bentonite grout is pumped out through the tips of the auger shafts. The auger flights and paddles mix the soil and grout in place to form overlapping columns, a line of which forms a continuous wall of low permeability. Because the cured soil-cement-bentonite mixture is relatively brittle, there is some concern as to whether such a cutoff wall has sufficient flexibility to enable it to remain intact as well as impervious during and after seismic activity or large settlements and deformations. The objective of our research was to identify additives to the soil-cement mix that could potentially provide increased flexural strength combined with relatively low modulus of elasticity. Rather than conducting an analysis of in situ cutoff wall behavior from which to design special laboratory testing, the thrust of this study was to provide a preliminary comparison of the effects of different additives and mix designs on the strength, modulus of elasticity, and permeability of soil-cement mixes using standard laboratory tests.

Additives

It was assumed that the effect of suitable additives on the properties of a soil-cement would be similar to their effect on concretes or mortars. With this in mind, a search was directed toward additives that improve the strength and flexibility of concretes and mortars. Other criteria included the effect of additives on the permeability and workability of the soil-cement mix, and the cost of the additive. As a result of the literature review, several potentially suitable additives were identified, including different polymers, various types of reinforcing fibers, and asphalt emulsion. All have been used to some extent in concrete, mortar, or slurries, and each appeared to have distinct advantages and disadvantages. Polymeric additives selected for this project included styrene-butadiene (SB) latex, acrylic latex, and epoxy resin. Selected fibers included steel, carbon, Kevlar, and polypropylene. The asphalt emulsion used was a slow setting cationic type (CSS). McFarlane (1990) presents a detailed description of these additives and their primary properties.

Other Materials

The cement grout used for the soil-cement mix was of type I-II Portland cement and tap water. In the control samples and in those samples with fiber additives, a powdered bentonite clay was also used; no bentonite was used in other samples.

The two soils, a slightly silty fine sand and a very fine sandy silt, were used for the soil-cement, and their classification and engineering properties are given in Table 1.

Table 1. Classification and Engineering Properties of Sand and Silt Soils (without additives).

Soil	Unified Symbol	Percent Fines	LL	PI	Effective Friction Angle (deg)	Cohesion (kPa)	Permeability (cm/s)
Sand	SP	≈1	--	NP	37	0.0	2×10^{-4}
Silt	ML	71	22	2	36	0.0	4×10^{-6}

Laboratory Testing Program

The purpose of laboratory testing program was to obtain a measure of the effect the various additives on the flexibility, strength, and permeability of the soil-cement-bentonite mixture. Testing variables included: additive type, soil type, curing time, and mix ratios (soil, cement, bentonite, water, and additive). Strength and modulus tests were conducted to determine the flexibility of each mix, and permeability was obtained directly from flexible wall permeability tests. Triaxial compression test results enabled a comparison of the strength properties of the soil-cement-bentonite mix with those of unstabilized soils. Specifically, laboratory tests included compressive strength (ASTM C-39), flexural strength (ASTM D-1635), splitting tensile strength (ASTM C-496), modulus of elasticity (ASTM C-469), permeability, and CU triaxial compression (ASTM D-4767). Testing details are given by McFarlane (1990).

In preparing the samples, an attempt was made to simulate in the laboratory as closely as possible the conditions under which the soil-cement design would be mixed in the field. The components included soil (sand or silt as described above), type I-II Portland cement, potable water, bentonite clay (for the control samples and those with fibers), and the additive (except for the control samples). The samples were prepared in the laboratory by hand using proportions in the following ranges in percent of total weight: soil, 63 to 70%; cement, 12 to 14%; bentonite, 0 to 0.2%; water, 15 to 18%; and additive, 0 to 4%.

The cement and bentonite were mixed dry before any water was added. The water and cement (and bentonite when used) were thoroughly mixed before the additive was included, except in the samples in which polymer and asphalt emulsions were used. In these cases the water and additive were added simultaneously. The soil component was added to the cement-bentonite-water-additive mixture and stirred thoroughly. The mixture was placed in the appropriate molds and each mold was vibrated for 30 sec on a vibrating table to densify the sample and drive out air bubbles. The samples were placed in a fog room at 23 C and 100% humidity to cure until tested. Samples tested for compressive strength, splitting tensile strength, and modulus of elasticity were cast in 102mm diameter × 203mm high (4.0 in. × 8.0 in.) cylindrical molds. Permeability and triaxial samples were 53mm diameter × 102mm high (2.1 in. × 4.0 in.) cylinders, and flexural strength testing was performed on 76mm × 76mm × 292mm (3.0 in. × 3.0 in. × 11.5 in.) beams.

Based on observations made during mixing and testing, the mix designs were changed for some of the samples prepared subsequently. This was done on the more promising additives in order to explore more desirable mix designs. The properties sought were greater strength, lower modulus, and improved workability.

The samples were identified by the type of soil (S for sand and M for silt), the type of additive, and the particular mix design that was being used. Each additive was given a symbol as listed in Table 2. For example, the designation S-Acr-1 represents the first mix design of a stabilized sand soil with acrylic latex as an admixture.

Table 2. List of Symbols Used for Additives

Additive	Symbol
None (control)	C
Steel fibers	SF
Carbon fibers	CF
Kevlar fibers	KF
Polypropylene fibers	PPF
Styrene-butadiene latex	SB
Acrylic latex	Acr
Epoxy resin	ER
Asphalt emulsion	AE

Summary of Results

Initial testing was conducted on a stabilized sand soil at an age of 7 days with no additives. After that, each test variable was changed in turn to determine the effect of that variable on the parameters of interest. Some additives during early testing were found to be clearly unsuitable for use in a flexible cutoff wall, so they were eliminated to reduce the total number of tests. The effect of curing time was considered only in the sand samples, since it was assumed that age would have the same effect on the properties of the stabilized silt samples as it did on the properties of the stabilized sand samples.

Unconfined Compressive Strength -- Although an installed flexible cutoff wall may not need great strength, it is useful to know the compressive strength of the stabilized soil samples for comparison between samples. Also, a stress-strain curve and modulus of elasticity are easily obtained from these tests. Unconfined compression test results with various additives are given in Fig. 1 for the stabilized sand soil and in Fig. 2 for the stabilized silt soil. In all cases, compressive strengths of the stabilized soils were less than the control. Of the modified samples of stabilized sand, the greatest strengths (or least strength losses) were provided by steel fibers, polypropylene fibers, SB latex, and asphalt emulsion. For all these additives, 7 and 28 day compressive strengths were greater than 60% of the control. In the modified stabilized silt soil, steel and Kevlar fibers provided compressive strengths greater than 70% of the control after 7 days.

<u>Tensile Strength</u> -- Tensile strength was measured in the stabilized soil samples by the flexural and splitting tensile tests (ASTM D-1635 and C-496, respectively). These tests are a convenient way to measure, albeit indirectly, the tensile strength of these materials. Tensile and flexural test results are summarized in Figs. 3 and 4 for stabilized sand soils with 7 and 28 day strengths. The tensile strengths of the stabilized soils were not increased significantly by any additive. In the stabilized sand soil, SB latex provided strengths consistently greater, on the order of 9 to 14%, than those of the control samples under both the tensile and flexural tests at 7 and 28 days. Steel fibers, polypropylene fibers, and asphalt emulsion additives provided strengths near those of the control samples. Steel fibers and Kevlar fibers in the stabilized silt soil improved the tensile strength by 20 to 40% over the control sample (McFarlane, 1990).

<u>Modulus of Elasticity</u> -- For this project, samples were tested in unconfined and confined compression from which stress strain curves and thereby moduli were obtained. For the unconfined compression tests, ASTM test method C-469 defines the modulus of elasticity as the slope of the line passing through the points on the curve at strain of 5×10^{-5} and at 40% of maximum stress. This modulus is a drained modulus (E_d), since the strain rate was slow enough to allow excess pore pressures to dissipate. From the confined (CU triaxial) compression tests, a secant modulus was determined. It is defined as the slope of the line passing through the origin and a point on the curve at 50% of maximum stress. This modulus is an undrained modulus (E_u) because the triaxial specimens were tested under undrained conditions.

The drained modulus of elasticity values for the stabilized soil samples with various modifying additives at 7 days (and 28 days for the stabilized sand samples) are given in Figs. 5 and 6.

The results indicate that each of the additives reduces the drained and undrained modulus of elasticity in the stabilized soils to some extent. The greatest reduction in the modulus was produced by the polymer latex additives and asphalt emulsion. Of the

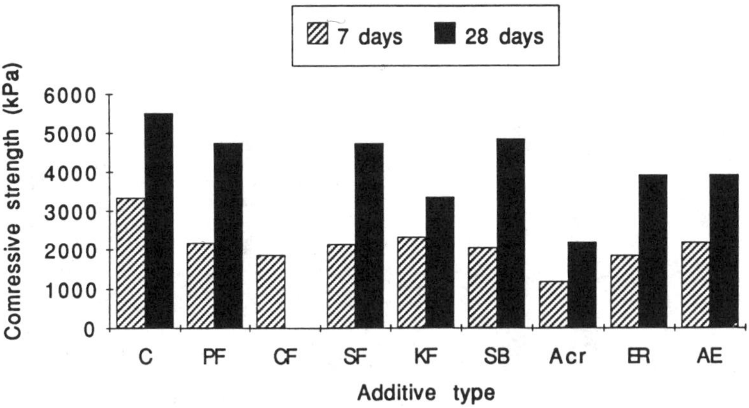

Fig. 1. Unconfined Compressive Strength for Stabilized Sand Soil.

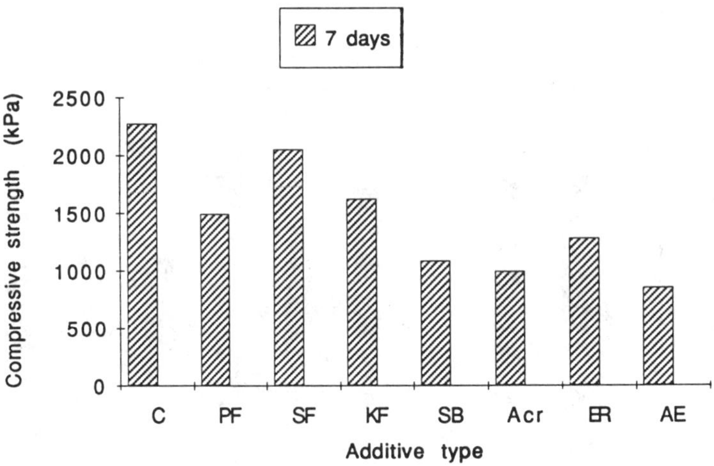

Fig. 2. Unconfined Compressive Strength for Stabilized Silt Soil.

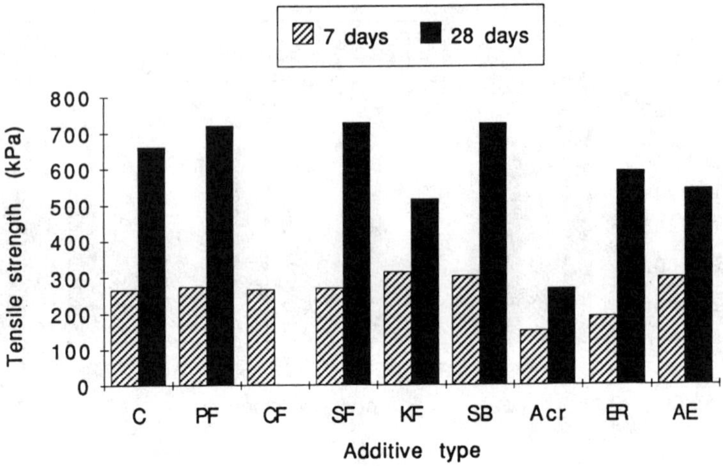

Fig. 3. Tensile Strength for Stabilized Sand Soil.

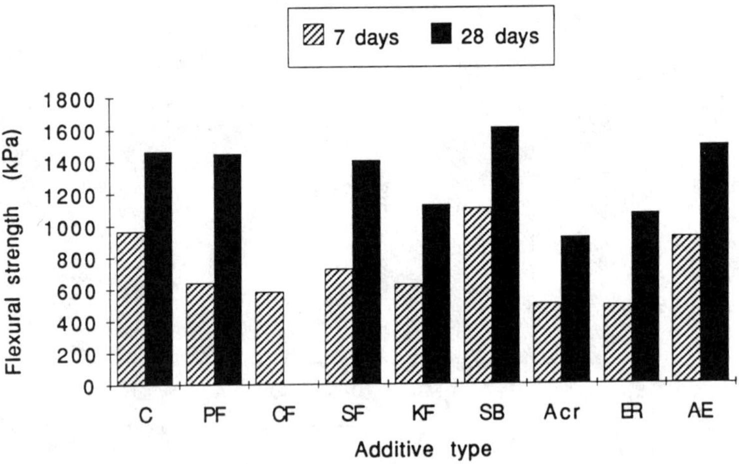

Fig. 4. Flexural Strength for Stabilized Sand Soil.

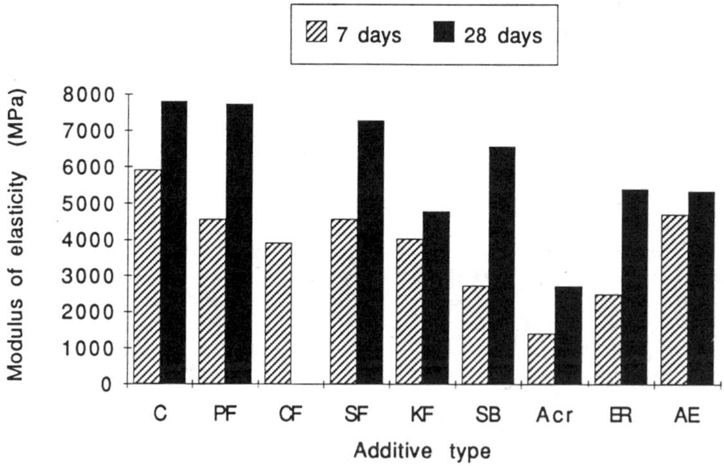

Fig. 5. Drained Modulus of Elasticity for Stabilized Sand Soil.

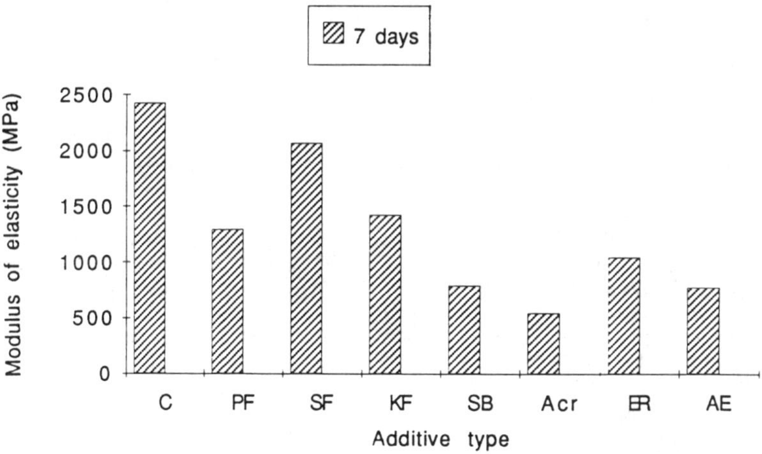

Fig. 6. Drained Modulus of Elasticity for Stabilized Silt Soil.

polymers, acrylic latex was most effective in reducing the modulus from that of the control. Increases in modulus of from 20 to 125% were observed for the stabilized sand samples between 7 and 28 days.

Permeability -- As the purpose of cutoff walls is to provide a hydraulic barrier within a soil formation, the permeability of the soil-cement-bentonite mix is of particular interest.

The results of permeability testing are given in Figs. 7 and 8 for the stabilized sand and stabilized silt soils respectively. They indicate that the permeability of the sand was reduced by about two orders of magnitude when stabilized with cement, while the polymer latexes decreased the permeability of the sand soil-cement mix only slightly. The asphalt emulsion additive reduced the permeability of the sand soil-cement mix from about 2×10^{-6} cm/sec to about 5×10^{-10} cm/sec.

The permeability of the silt was reduced from about 4×10^{-6} cm/sec to about 1×10^{-6} cm/sec when stabilized with cement. The permeabilities of the silt soil-cement samples modified with SB latex and acrylic latex were both about 2×10^{-8} cm/sec. The sample modified with asphalt emulsion provided even greater resistance to water flow with a permeability of 2×10^{-9} cm/sec.

Confined (Triaxial) Compression -- From consolidated undrained (CU) triaxial tests, the friction angle of the stabilized sand was altered very little by any additive. A friction angle of about 45 deg was fairly consistently obtained from the samples tested. Cohesion is related to tensile strength, and those additives that provided the greatest tensile strength also had the greatest cohesion. For the stabilized sand, these included SB latex and asphalt emulsion.

The modifying additives had more effect on the cohesion and friction angle of the stabilized silt. These parameters for the modified samples were less than those of the control. The Kevlar fiber modified sample was an exception in that its cohesion was greater than that of the control, but it had a much lower effective angle of internal friction. The fibers apparently reduced the friction within the soil, although when tensioned during cracking of the sample, they probably helped hold the sample together which increased the cohesion intercept.

Mix Design -- In terms of improving the flexibility of soil-cement mixes, the most promising additives were SB latex, acrylic latex, and asphalt emulsion. Additional samples containing these additives were cast with different ratios of the components in order to optimize the mix designs, that is, to obtain the lowest possible modulus of elasticity in conjunction with the highest possible strength. In the case of the latexes, the cement-additive ratio was kept constant and the proportions of the other components were changed.

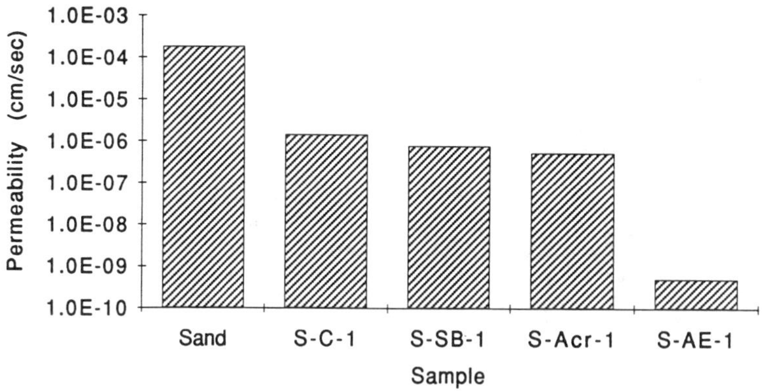

Fig. 7. Permeability of Stabilized Sand Soil.

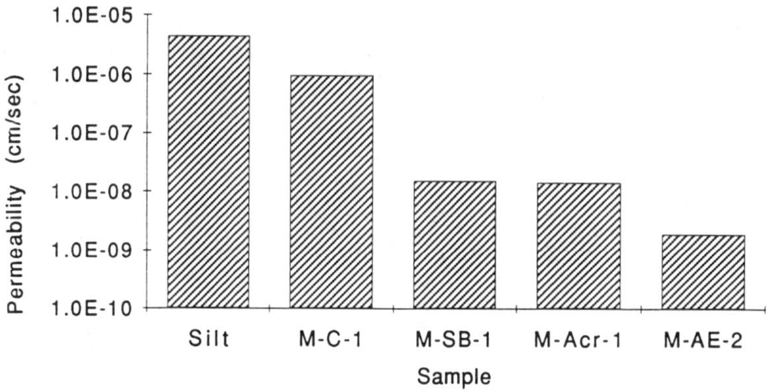

Fig. 8. Permeability of Stabilized Silt Soil.

The test results indicated that by increasing the content of polymer and cement in the stabilized sand and silt soils, the strength and drained modulus increased. In the asphalt modified samples, an increase in the amount of asphalt was not accompanied by an increase in cement. Thus the strength and drained modulus decreased with increasing asphalt content. Increasing the sand content by 1.2% and lowering the water content by 1.2% resulted in a slightly higher strength and a slightly lower drained modulus.

Discussion of Results

Polymer additives -- Of the polymer additives, epoxy resin was eliminated from consideration early in the experimental part of the project. During an initial attempt to test an epoxy modified sample under triaxial compression, the epoxy was observed to wash out of the sample during the saturation phase of the test. Another contributing factor to its elimination was its relatively high cost.

The polymer latexes (SB and acrylic) decreased the drained and undrained modulus of the modified samples as predicted in the literature. The SB latex improved the tensile and flexural strengths in the stabilized sand samples, but no increase in compressive strength was observed. Acrylic latex and epoxy resin did not provide the anticipated flexural and tensile strengths, and strength increases were not observed in the stabilized silt soil with the polymer additives.

Permeability of stabilized sand was essentially unaffected by polymer latexes, but an order of magnitude decrease in permeability was observed in the stabilized silt. This is probably due the small size of the polymer particles which apparently are effective in plugging the pores in the stabilized silt soil. The pores of the stabilized sand soil are larger, and the polymer particles probably do not fill them completely. Ohama (1987) had reported reductions in permeability of about an order of magnitude in cement or mortar by the addition of polymer latexes.

The latexes were surprisingly easy to work with and mixed well with the cement and soil. The polymer latexes would probably not be feasible for use in a flexible cutoff wall due to their relatively high cost. It is possible that in some small but critical applications, the cost of using the polymer latexes could be justified.

Fiber additives -- From information gained in the literature review, the fiber additives were expected to improve the tensile and flexural strengths of the stabilized soils along with a slight improvement in the compressive strength. The test results show no improvement in compressive strength, and little or no improvement in the tensile or flexural strengths.

A particular problem with the carbon and Kevlar fibers was that since they are very fine and light, they were difficult to mix uniformly in the soil-cement mix, remaining instead in small bundles. Most of the fibers in each bundle were not in contact the soil-cement matrix and, therefore, provided no reinforcement. In effect, each bundle of fibers created a small void in the sample that contributed to a decrease in compressive and tensile strength. This effect was less pronounced in the stabilized silt samples.

When fibers are placed in a stabilized soil sample, they provide little reinforcing until the sample has cracked. At that point the fibers are loaded in tension and provide post-crack load capacity. This effect produced increased tensile strength in the stabilized silt with steel and Kevlar fibers.

The test results do not indicate that fibers contribute much improvement to the flexibility of the stabilized soil samples; therefore, they do not appear promising as additives for a flexible cutoff wall.

Asphalt Emulsion -- In a mix design of 56% sand, 6% cement, 19% water, and 19% asphalt, emulsion, the mix was nearly unworkable. During mixing and casting, the sand separated from the liquid portion of the mix and the mix could not be made homogeneous. The result was that a large portion of the asphalt and cement washed away with the liquid as the sand settled to the bottom of the mold during casting. The cured samples had a low drained modulus and very little strength. The mix design was changed to 70% sand, 13% cement, 12% water and 5% asphalt emulsion, and this resulted in a very workable mix with higher modulus and strength.

Of the additives studied in this project, asphalt emulsion is one of the more promising. Using the proper mix design, it should be possible to create a mix having an efficient combination of permeability, modulus and strength.

Conclusions

The best additive to make a cutoff wall more flexible is one that would provide increased flexural strength and low modulus at a reasonable cost while not adversely affecting the permeability of the wall. From the test results discussed above, none of the additives tested provided dramatic improvement in the flexibility of the soil-cement mix. A limited amount of improvement, however, was provided by SB latex and asphalt emulsion.

SB latex has the potential to increase the tensile strength of the stabilized soil while decreasing the modulus and the permeability. The main drawback to using it for a flexible cutoff wall is its relatively high cost. Asphalt emulsion provides limited tensile strength, but potentially very low modulus and permeability. The cost of asphalt emulsion is much lower than the polymer emulsions, so it would not be prohibitive for most applications.

SB latex and asphalt emulsion are readily obtained and are fairly easy to work with. They mix well with cement and soil and are considered good candidates for use as modifying additives in a flexible cutoff wall.

Acknowledgements

This research was supported by S.M.W. Seiko Inc. The encouragement and assistance provided by O. Taki and D. Yang are especially appreciated.

References

American Society for Testing and Materials (1988) "Concrete and Aggregates," Vol. 04.02, and "Soil and Rock, Building Stones; Geotextiles," Vol. 04.08, Annual Book of ASTM Standards.

McFarlane, T.S. (1990) "Selection and Laboratory Evaluation of Modifying Additives for the Flexible SMW Wall", M.S. Thesis, University of Washington, 156 pp.

Ohama, Y. (1987) "Principle of Latex Modification and Some Typical Properties of Latex Modified Mortars and Concretes," American Concrete Institute Materials Journal, Vol. 84, Nov.-Dec., pp. 511-519.

Taki, O. and YANG, D.S. (1989) "Excavation Support and Groundwater Control Using Soil-Cement Mixing Wall for Subway Projects," Proceedings of the Rapid Excavation and Tunneling Conference, Los Angeles, pp. 156-175.

Taki, O. and YANG, D.S. (1991) "Soil-Cement Mixed Wall Technique," Proceedings of the ASCE Geotechnical Engineering Congress '91, Boulder, Geotechnical Special Publication No. 27, Vol. I, pp. 298-309.

The Design of A Reclamation Scheme by Preloading

S. Ossama Mazen[1], B. Sc., M.Sc., Ph.D.

Abstract

The natural ground at the site of Port Said development plan consists of the original bed of Lake Manzala. It contains a surface stratum of soft silty clay overlying fine to medium sand. Work carried out indicated that the most appropriate solution is to use hydraulic fill over large areas of land for preloading of surcharging the soft clay layer. This would create land suitable for development using stiff shallow foundations at a production rate appropriate for the project goals.

Following the results of the investigations, a Pilot Housing Project was designed for implementation within the context of work. The project consisted of the construction of 84 units on a prototype development area. It was anticipated that the performance of the Pilot Housing Project would confirm the results of the study.

Introduction

A master plan program for the city of Port Said was designed to provide an economic and physical framework for development until the end of the century. In order that the city of Port Said develops its economic potential within the programed period, some 3000 hectare of reclaimed land would be required together with 125000 new dwellings and all integrated facilities such as schools, hospitals, shops, factories, public buildings, etc. Therefore, a new construction method was needed to achieve the building production rates necessary to achieve this goal.

----

1 Swnior Researcher, general Organization for Housing Building and Planning Research, El-Tahreer Street, Dokki, P.O.Box 1770, Cairo, Egypt.

Furthermore, the need for a new approach to housing construction was identified from the observation of existing structures in Port Said by independent consulting engineers. Buildings were observed to have suffered large settlements,both total and differential. Considering these conditions,the present investigation was planned. The main objective of the work was to locate suitable land for development, carry out a detailed soil investigation,and to design a reclamation scheme.

Geological Background

Port Said is located in the north eastern margin of the Nile Delta overlooking the Mediterranean sea.It is bounded on the south and west by Lake Manzala and on the east by Suez Canal. Lake Manzala is a large shallow lake separated from the sea by a narrow sand bar. Prior to the construction of Suez Canal, lake Manzala also extended to the east of Port Said into what is now known as Lake Mallaha (Fig. 1).

The geology of Port Said has been studied by a large number of research workers within their study of the geology of Egypt. Among them are: Attia (1954), Said (1962), Abu Al-Ezz (1977),and Said (1981). Their studies can be summarized as follows:

In the Middle Miocene, the northern part of the present Nile Delta was an embayment bordered on the west and south by cliffs of cretaceous and Eocene rocks rising more than 1000 meters above the Miocene Sea.During the upper Miocene the sea regressed,causing the northern part of the present Nile Delta area to be dried up. Thus causing the sediments to become hard, consolidated and possibly eroded. In Early Pliocene time, the straight of Gibraltar was opened, thus the Mediterranean became flooded as a result of the connection with the Atlantic and the sea transgressed rapidly.

During Pleistocene and Recent times, sea level fluctuated, flooding parts of the delta at certain times, retreating from it at other times. As a result, the land became subjected to subsidence and intermittent transgression by the sea, causing the area to be infilled with alternate marine and Nile sediments.

The sand bar separationg Lake Manzala from the sea formed in its present position under the combined effects of sedimentation, littoral drift and erosion.

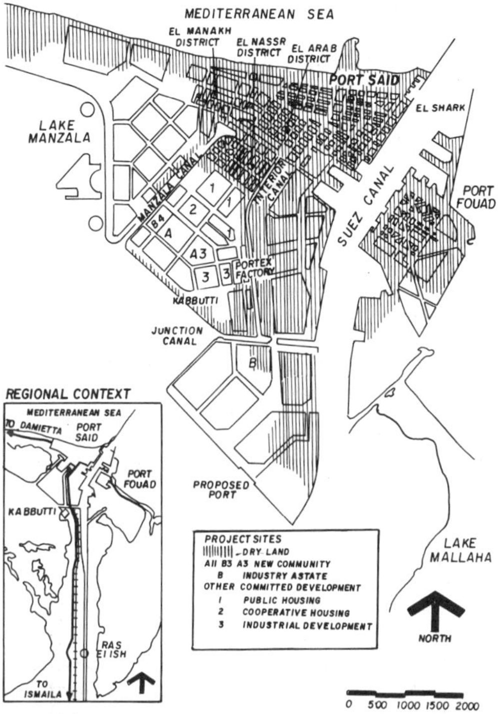

Figure 1 Site Location

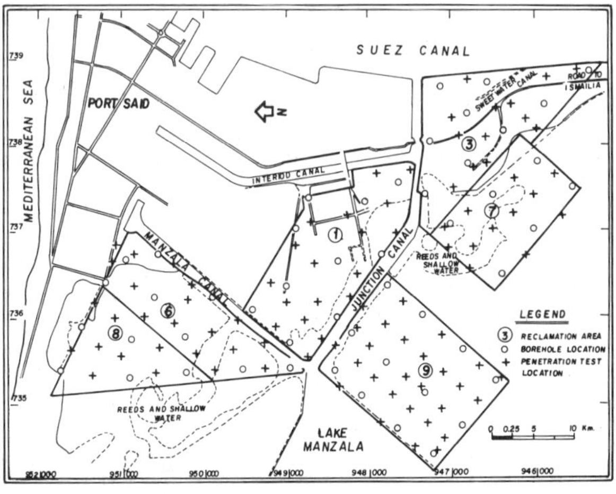

Figure 2 Reclamation Areas and Boundaries

After the bar was formed and separated Lake Manzala from the sea, a thin layer of soft lacustrine clay was deposited in the lake on top of the sand. It increases in thickness from a very thin layer in the north part to some two meters in the south.

Location and Description of Site

As natural sites became exhausted, the possible alternative was to use the marginal lands. The location of the chosen areas of these marginal lands in its Port Said context is illustrated in Figure 1 and their boundaries are shown in Figure 2.

The site in question originally consisted of mudflats with saline deposits, the majority of which have been inundated with water from Lake Manzala. Fill material dredged from Suez Canal was used to raise the elevation of the swampy or even the submerged land. However, at present, most of the site is at or below the level of Lake Manzala.

Planning and Layout of Investigation

Subsurface investigation program in the studied areas consisted of 50 boreholes to depths ranging between 20 and 55 meters, disturbed and undisturbed sampling, standard penetration resistance tests, in situ vane shear test, dynamic cone penetration tests and grounwater level measurements. Laboratory testing included tests for physical properties, index properties shear strength and consolidation characteristics.

Subsurface geotechnical data were, then, gathered to determine the magnitude and rate of settlements and the stability under the proposed fill and building loads. Particular emphasis was given to primary geotechnical parameters influencing the design of the reclamation scheme such as shear strength and compressibility. The shear strength controls stability during construction, while compressibility controls the post-filling settlement.

Design aspects included selection of method of treatment, predication of field performance, development of planning concept, structural and constructional parameters influencing design. A pilot Housing Project consisting of the construction of 84 units on a prototype development area was implemented.

Soil Conditions and Geotechnical Properties

Typical soil profile at the site in question is shown in Figure 3. It comprises in succession: 1 m or less of top soil, 2-3 m of soft surface clay,4-10 m of compact to dense fine sand, 2-5 m of a layered transition zone, 30-40 m of firm to stiff lower clay, basal deposit of very dense sand and hard clay.

The surface clay is grey, very soft, highly plastic with high organic content. The clay was so soft that difficulty was found in extruding suitable samples for consolidation test. Measured in situ vane test values ranged between 6 and 15 k N/m2. In the dewatered poldered area the clay is stronger because it is slightly dessicated with undrained shear strength near the surface of about 20 kN/m2. Liquid limits varied between 50 and 130 with plasticity indices varying between 30 and 90. The natural water content has been generally close to the liquid limit. The compression index, Cc, varied between 0.38 and 1.66.

The sand stratum is fine to medium grained and uniformly graded.The relative density is medium to dense.The density increases with depth to a maximum at the center of the stratum then decreases towards the bottom of the stratum. Results of standard penetration tests indicate values of N ranging between 10 and 40 blows/300 mm.

The transition zone occurs at an average depth of about 8 meters as thin layers of silt and clay in the sand stratum. This marks the start of a gradational change from the sand stratum to the lower clay. The layers of clay increase in frequency until the soil is predominantly a clay with some sand layers. Both boundaries of the zone are indistinct but it can be recognized as a layer in which the standard penetration resistance reduces sharply with depth. The shear strength of the transition zone is governed primarily by the clay layers which are difficult to measure accurately in site because field tests invariably predict a mixture of sand and clay layers. However, because the clay layers are thin, consolidation under load and the consequent gain in shear strength will be rapid.

The lower clay extends up to 50m below the surface. It is dark grey organic highly plastic silty clay. Its consistency increases with depth from firm to stiff. The undrained shear strength increases from about 50 kN/m2 near the top of the stratum to about 110 kN/m2 at the base. The clay mineral content of the stratum increases from 65 percent near the top of the deposit to about 85 percent near the base. Measured organic contents are between 1.3 and 6 percent. The plasticity increases with depth as the clay content increases. Liquid limits of about 75 percent were measured near to the top of the stratum. At the bottom, the liquid was in the order of 150 percent. The plastic limit ranged between 25 and 45 and the moisture content varied slightly with depth from about 50-60 percent near the top of the deposit to 70-75 percent at the bottom. The clay has been preconsolidated and the amount of over-consolidation averages about 100 kN/m2. The compression and recompression indices shown that the soil is quite compressible and that the compressibility increases with depth. Average Cc ranges from 0.6 at the top of the stratum to about 1.0 at bottom.

Basal deposits consist of very dense sand and hard clay. The basal deposits are so named because they underlie the relatively soft deposits which affect the performance of foundations in Port Said.

Ground Water Conditions

There are two water regimes in the reclamation area. One is perched on top of the impervious lower

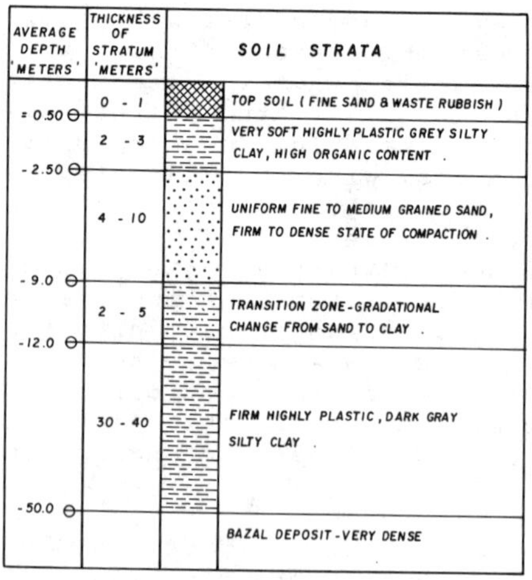

Figure 3 Typical Soil Profile

clay and reflects the lake level. The other originates under pressure at depth in the granular basal deposits beneath the lower clay.

On the land,the upper water level is found to be slightly above lake level.Lake level is about elevation +0.5m.The lower level stabilizes at elevations between +2m to +3m.

The upper ground water which could come into contact with the structural foundations of buildings has a very high sulfate and chloride content.

Field Problem and Selection of Method of Treatment

There are many alternative methods cited in the literature for ground improvement to serve different purposes.They are reviewed by Mitchel and Katti(1981), Hartikainen (1983) and others. The use of soil improvement method depends on many factors. However, the basic aspect in the choice of the method is the possible soil conditions.

Soil conditions of the site under investigation indicate that there are a number of problems which necessitate special study to assess the most appropriate solution. The poor condition of the surface clay layer is the most significant feature.

The two or three meters thick top layer of soft clay is weak and highly compressible and would be unable to accept the imposed loads without excessive settlement. It is far too compressible to build on directly. Therefore this layer would either be penetrated by pilling, removed completely or strenthened by in-situ consolidation.

The purpose of piling is to transfer loads from a weak surface strata to a stronger strata at depth. On these sites the sand stratum below the surface clay, when loaded by piling,would itself impose unacceptable stresses into the lower clay layer leading to high settlements. From an operational point of view, piling would allow rapid mobilization of the site. However, high costs are incurred and few benefits are to be gained. The site loses its flexibility if piling is used and the untreated ground supporting roads and services would continue to settle.Therefore piling was abandoned.

The possibility of excavating this clay and backfilling with a suitable material was found unacceptable because of its excessive cost and long duration of work to be executed. It was estimated that removal of the surface clay layer and replacement by better structured,well graded sand would be twice as expensive as placing fill directly on the surface clay layer.

The third possibility of preloading by a surcharge of sand fill laid over the area of the site was found to be the most economical and appropriate method. The surface clay would then be consolidated and strengthened and high structures would be built on top of fill material. It was anticipated that the combined layers of fill, clay and sand will,after consolidation be capable of supporting the imposed loads.

The beneficial effects of this method is evidenced by the low lying nature of the ground as well as by the presence of ground water at or near original ground surface. Cosequently any excavations for foundations below water table would present problems. Furthermore, the upper ground water which could come into contact with the structural foundations of building has a very high sulfate and chloride content. In the adopted method, foundations are located on top of the proposed fill material,i.e. far above the water table.

Characteristics of Borrow Meterial

Previous work (Aldrich, 1965; Whitman, 1970 and others) show that the most appropriate and economical method for reclamation of marginal areas for building construction is by hydraulic filling.

Envelope of grading curves of samples of dredged fill mateial obtained from Lake Manzala is given in Figure 4. It suggests that this layer consist of uniform fine grained sand with average diameter of 0.2 mm. This would indicate that this material have the free draining properties required for a hydraulic fill material and that it should also be readily transported by pumping.

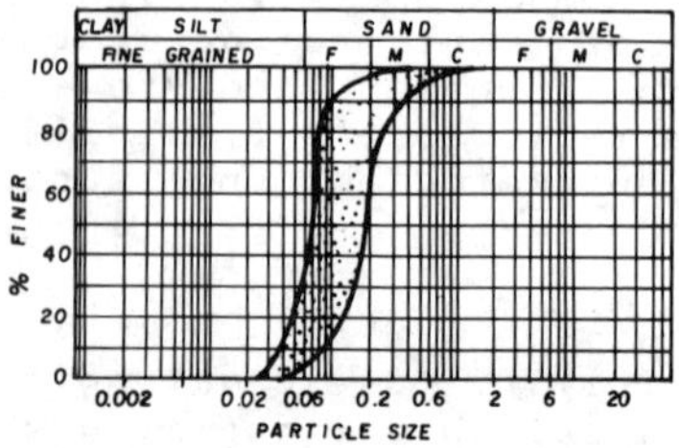

Figure 4 Envelope of Grading Curves of Borrow Material

Prediction of Field Performance

The main parameters affecting field perfomance are the consolidation characteristics of both clay deposits as well as the shear strength of the surface clay. In the reclamation areas geotechnical data has been gathered to evaluate the stability under the proposed fill and building loads. A summary of the in situ vane shear tests as well as predicated times for 50 % and 90 % consolidation settlements of the surface clay are given in the following talbe 1.

Table 1. Characteristics of Surface Clay

* Area	Average Fill Thickness(m)	Average undrained shear strength (kN/m2)	Time for 50% consolidation	Time for 90% consolidation
8	1	13	2 weeks	2 months
6	1	14	3 weeks	3 months
1	1.2	16	4 months	1.25 years
9	1.7	7	6 months	2 years
3	1.9	13	5 months	1.75 years
7	1.9	9	7 months	2.5 years

* Indicated in Figure 2.

It was found that immediate settlement and primary consolidation would form the greatest part of the total settlement and the secondary settlement could only be significant for relatively heavy structures. Therefore, the construction of relatively light structures with foundations above the surface clay in fill material is very appropriate in order to distribute bearing pressure through the fill.

The stresses imposed by the fill on the underlying deposits will extend virtually undiminished throughout the whole profile. Therefore, the consolidation characteristics of the lower clay is as well important, especially the preconsolidation pressure in relation to the imposed stresses.

Assuming an average depth of fill of 1.8 meters the allowable bearing capacity of 80 kN/m2 was found to be appropriate. Estimates of settlement have been made for various reclamation areas. It was found that for the 24x15m building block, settlements on the order of 40mm can be expected and differential settlements of the order of 20mm could, therefore, be realized.

Planning Considerations to Accommodate Field Conditions

In order to accommodate field conditions, it was found that the construction should not involve more than three story high buildings to provide stresses at foundation level within the predicted alowable bearing capacity. This made the previously proposed five story high building unsuitable. Therefore, a new planning concept was introduced which led to a form of three story development and that gives the same number of housing unit per hectare as has been achieved by the traditional five story development (Fig. 5).

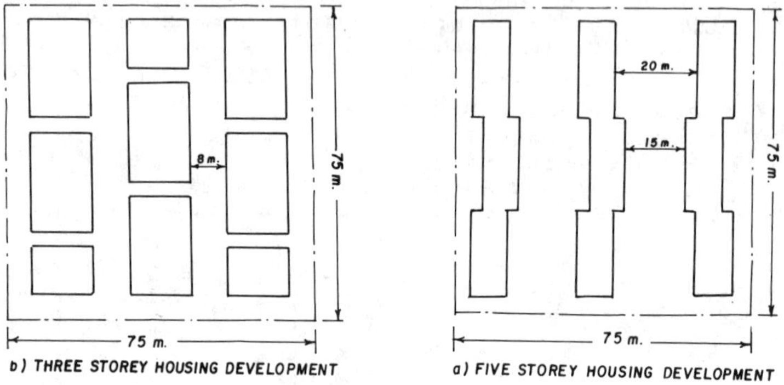

Figure 5. The Proposed and the New Planning Concepts

Implementation and Justification of The Adopted Solution

As a result of the previously menthioned investigation and to try out the adopted solution, a pilot Housing project consisting of the construction of 84 unites on a prototype development area was decided for implementetion. It was anticipated that performance of the constructed building would confirm the results of the study.

Before starting construction, more detailed soil investigations were made on the Pilot Project Site.The results confirmed the findings of the previous more general site investigations. This indicated that it would be feasible and practical to proceed with the project.

At the time of pilot project commencement, no hydraulic fill was available in the quantity and in the location required. Therefore, it was decided to import sand having as near as possible similar characteristics as the sand dredged from Lake Manzala.

The specification for the fill and its compaction was written for imported fill in a similar way as that for the hydraulic fill from Lake Manzala which will be used in future to maintain the same high standard of uniform density as at present. Sand imported from Salhiya, an area to the south of Port Said, was found conforming with the required specifications. Figure 6 shows the gradation curves of two types of the imported material and the results of laboratory tests to determine their maximum dry density.

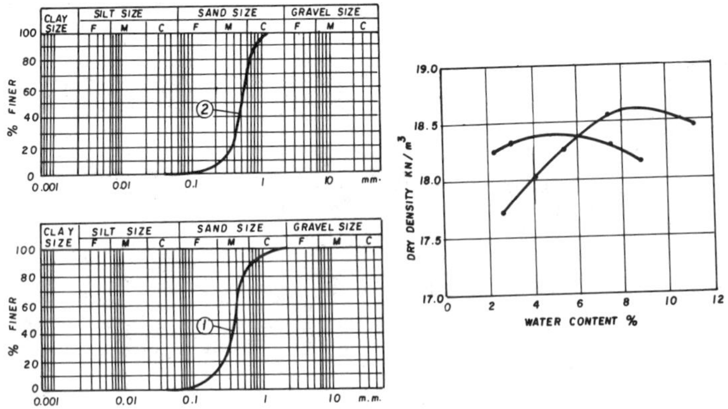

Figure 6. Grading Curves and Compaction Curves (Modified Proctor) of Imported Sand.

The area proposed for the project has been located immediately adjacent to area 6. Table 1 indicates that the corresponding 50% consolidation would occur within 3 weeks and the 90% consolidation would occur within about 3 months.

The ultimate bearing capacity of the compacted fill was found to be on the order of 350 kN/m2. As the applied load at foundation level is within 80 kN/m2, this gives a factor of safety of 4.38.

Design Criteria

1) As previously indicated, settlements of about 40 mm can be expected and differential settlements of the order of 20mm could be realized. In order to minimize the effect of these differential settlement, it was decided to provide strip footings in two directions having upstand beams to stiffen the foundation. For the same reasons,columns were provided at close regular intervals, thus minimizing the point load effect on the foundations and ensuring a reasonalble spread of loading under all parts of the foundations.

2) The stress imposed by the fill on the underlying deposits extends virtually undiminished throughout the whole profile. Furthermore, due to the proximity of adjacent buildings, stresses at foundation level will extend considerably into the lower clays (Fig. 7). Therefore, the level of stresses due to fill and foundations at these lower levels has been checked to ensure that they do not exceed 50% of the preconsolidation pressure of 100 kN/m2.

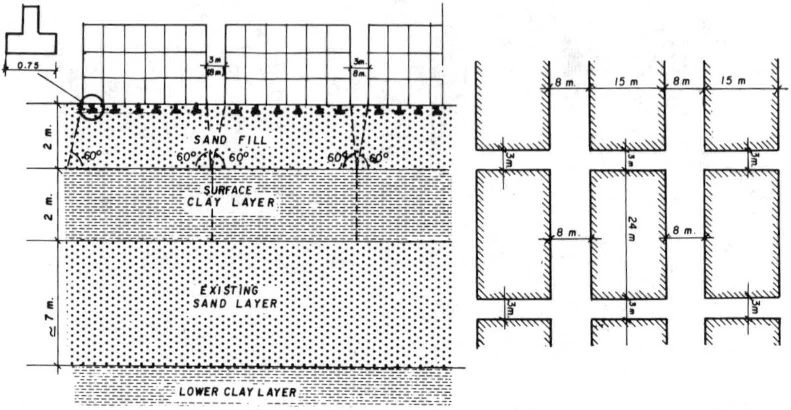

Figure 7. Elevation and Layout of Blocks

Appendix-References

Abu Al-Ezz, M.S.,(1971) "Land forms of Egypt". The American University in Cairo, Cairo, Egypt.

Aldrich, H.P. (1965) "Precompression for support of shallow foundations", J. of the Soil Mechanics and Foundations Division, 91(2)5-20.

Attia,M.I.,(1954) "Deposits in the Nile Valley and the Delta" The Geological Survey of Egypt.,Cairo, Egypt.

Hartikainen, J. (1983) "On the Geotechnical Design of Foundation in Improved Subsoil", Proceedings, 8th European Conf. on Soil Mechanics and Foundation Engrg. Helsinki, Finland, 3, 1319-1332.

Mitchel, J.K. and Katti, R.K.,(1981)"Soil Improvement. State of the Art Report", 10th Int. Conf. on soil Mechanics and Foundation Engrg,Stockholm, Sweden, 4, 567-579.

Said, R., Geology of Egypt (1962), Elsevier Publishing Co., New York, N.Y.

Said,R., (1981) "The Geological Evolution of the River Nile",Springer. Verlag Publishing Co.,New York, N.Y.

Whitman, R. V. (1970) " Hydraulic Fills to Support Structural Loads", J. of the Soil Mechanics and Foundations Division, ASCE,96(1) 23-47.

Influence of Particle Structure on Properties of Fly Ash and Sand

Krzysztof Parylak[1]

Abstract

This paper presents the compaction and triaxial test results obtained on silty sand and coal fly ash. The tested soils have a very similar sphericity and particle roudness, and similar grain size distribution. A new device for the determination of surface rougness is also presented. The great differences of compaction effects obtained in Proctor tests. The triaxial tests were carried out on the compacted samples in five comparable relative compactions at constant degree of saturation. The fly ash at all different initial porosity shows constant value of the cohesion intercept caused by especially high roughness of particles. The curvature of the Mohr's envelope for the fly ash was higher than for the sand.

Introduction

Improvement of cohesionless soils in engineering practice may be carried out in many ways. For shallow layers, improvement methods include rolling, vibrating as well as compacting with tamping rollers, vibrating rollers, impact rollers or vibro-tampers. For deeper layers the densification process can be realized, among others, by preloading, vibrocompaction, vibrodisplacement or compaction by blasting (Bell et al. 1987; Chang and Chae 1987).

The effect of this process is affected by the size of particles, size-distribution,shape and roughness of particles, density,impact velocities of the particles during compaction, presence or lack of lubrication, vibratory characteristics, temperature and humidity.

----

[1]Parylak, D.Sc. Assistant Professor,Institute of Applied Hydraulics and Earth Structures, Agricultural University of Wrocław, P o l a n d

The influence of these parameters have been described by many authors (Bell et al. 1987; Chang and Chae 1987; Chaplin 1961; Kézdi 1979; and all). The densest packing condition of particles occurs when large voids get filled with smaller particles and so on. This process is reduced by the phenomenon of bulking (low densities obtained at low water content) and by the effect of grain crushing. Well-graded soils are subject to less breakdown than uniform soils composed of particles of the same size because of more interparticle contacts in the former ones (Lambe and Whitman 1969). Angularity of particles and their roughness get reduced at high relative porosities by greater number of contacts per grain and smaller space (solid + void) per particle (Chaplin 1961). Experiments by Kolbuszewski (1967) showed, that changes in sand particle roundness from 0.15 to 0.48 induced by vibration brought about a decrease in porosity from 0.48 to 0.36. The influence of roughness particles didn't explained yet. Chaplin (1961) and Kolbuszewski (1967) summarizing the influence of particle shape and surface roundness on the fundamental stress-strain properties of sand showed, that further urgently and detailed study are necessary.

In this paper the comparative compaction and shearing resistance of two very similary graded materials but with different roughness were performed. The influence of these properties on compactibility effect, and on the angle of internal friction were determined. The continuity of existence of structural cohesion in the fly ash was analized. This phenomenon in cohesionless soils is very controversial.

Materials Tested

This study was performed on crushed sand,obtained from the crushed gravel from a plant in Lower Silesia. The gravel used for crushing was obtained from the Nysa Kłodzka - River valley, near Ząbkowice. A mineralogical composition indicated that above 80% of the tested sand consisted of quartz grains. The remaining particles consisted of feldspar, mica and other minerals. Several microphotographs of these particles show mainly irregular grains (Fig. 1). The tested fly ash was obtained from deposited lagooned fly ash of Power Coal Station "Blachownia" near Opole. The chemical composition was as follows: SiO_2 about 45%, Al_2O_3 about 20%, Fe_2O_3 about 4%, CaO about 3% and unburnt carbon about 10%. The petrographic analises of fly ash from this station indidated that about 80% crystalline form consisted of glass. Typical glass contents crystal structures such as quartz, mullite, magnetite and hematite. The rest, about 20% consisted of amorphous components. As a result of

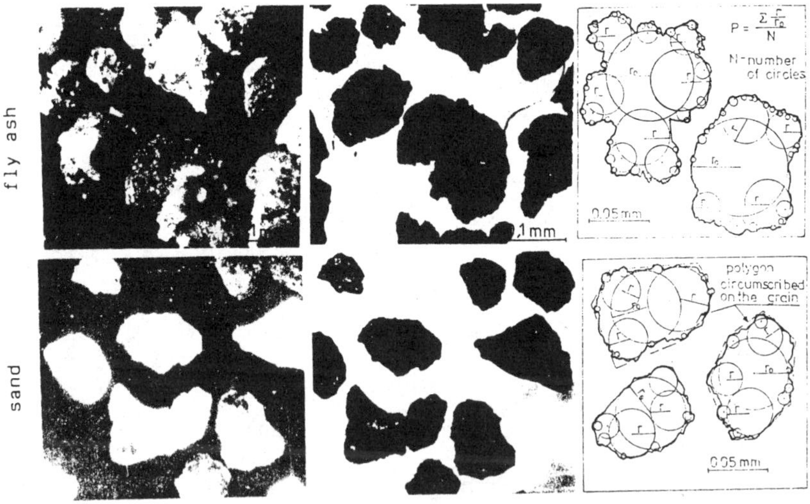

Figure 1. The Microphotography of Fly Ash and Sand Particles in Reflected and Transmited Light. The Procedure of Determination of Shape Particles Angularity

the combustion process some irregular and rough surface structures of ash grain occured. Very small spherical silty fraction particles were fusioned on these surfaces. The process of speedy cooling of fly ash caused the formation of round edges and round peak particles (Fig. 1). The aggregated structure of grain with the presence of numerous holes and cavings was made. In order to compare grains shape and their roughness effect on compactibility and shear strength, the analises of the particles area were made. They were realized on the particles photographies under 320 x magnification. The grains were measured using a cartographer apparatus connected on computer system with high accurency. Apart from general particle shape definitions (sphericity and roundness of particles), the ratio of cross section area to the perimeter of the grains was also determined. The ratio of the grain perimeter to the perimeter polygon circumscribed on this particle was also analized. It was assumed, that the described polygon area was the same as the cross section area of the particles. These dependencies and some other relevant properties are shown in Table 1. The differences of grain shape parameters between tested soils were also calculated. The mean results evidenced slightly higher values of the ash particles sphericity and higher values of perimeters, which were about 14.5% higher than for the sand. These

Table 1. Index of Properties of Investigated Soils

Properties	Sand (A)	Fly ash (B)	B-A	$\frac{B-A}{A}\cdot 100\%$
Specific gravity kN/m³	26.7	23.7	-3.0	
Specific surface m²/g |Stępkowska 1977|	0.66	3.35	2.67	407
Sphericity	0.792	0.813	0.021	2.6
Roundness |Herdan 1953|	0.334	0.392	-0.05	-1.5
Cross section area of grains mm²	0.175	0.181	0.035	3.4
Perimeter of grains mm	0.334	0.382	0.048	14.5
Ratio of cross section area to perimeter	0.52	0.47	0.05	-10
Difference of perimeter grains and of perimeter polygon circumscribed on grains (about the same area) mm	0.02	0.11	0.09	450

differences were not proportional to the degree of roundness of grains P, determined as in Fig. 1. The differences of the roughness of particles were demonstrated by the difference of the grain perimeter, and the grain perimeter of the polygon circumscribed on the particle. They were confirmed by the differences of specific surfaces obtained for the fly ash and for the sand (Table 1).

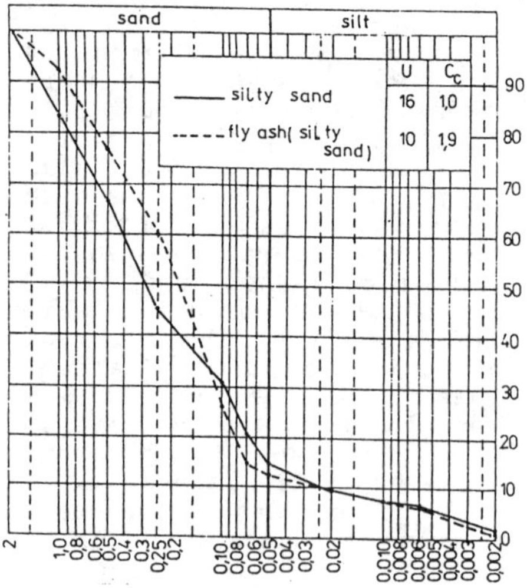

Figure 2.Grain Size Distribution

The results obtained from sieving and hydrometer analises indicated very similar curves for the grain size distribution bellow D_{30} diameter (Fig. 2). The differences of the curves above D_{40} diameter should not have important influences on the compactibility and shear strength (Kézdi 1979). As it was shown by Koerner (1970) the changes of the coefficient uniformity from 1 to 5 in quartz sand, caused the increase of angle shearing resistance Φ' in about 2°. The differences of grain size distribution above the D_{40} diameter had similar small influence on the compactibility of fine sands (Kézdi 1979). From these and other data we can assume that the differences of grain size distribution (Fig. 2) do not significantly influence the compactibility and shear strength.

Compactibility Tests

The compactibility tests of these soils were carried out according to the Standard Proctor Compaction Test. The results are presented in fig. 3.

For sands the optimum moisture contents usually corresponds to a degree of saturation S_r between 0.6 to 0.7. This was also observed in these tests. For fly ashes the optimum moisture content usually appears at higher degrees of saturation. In this test it was at S_r about 0.8. The existing differences between these materials were caused by water contained in voids and

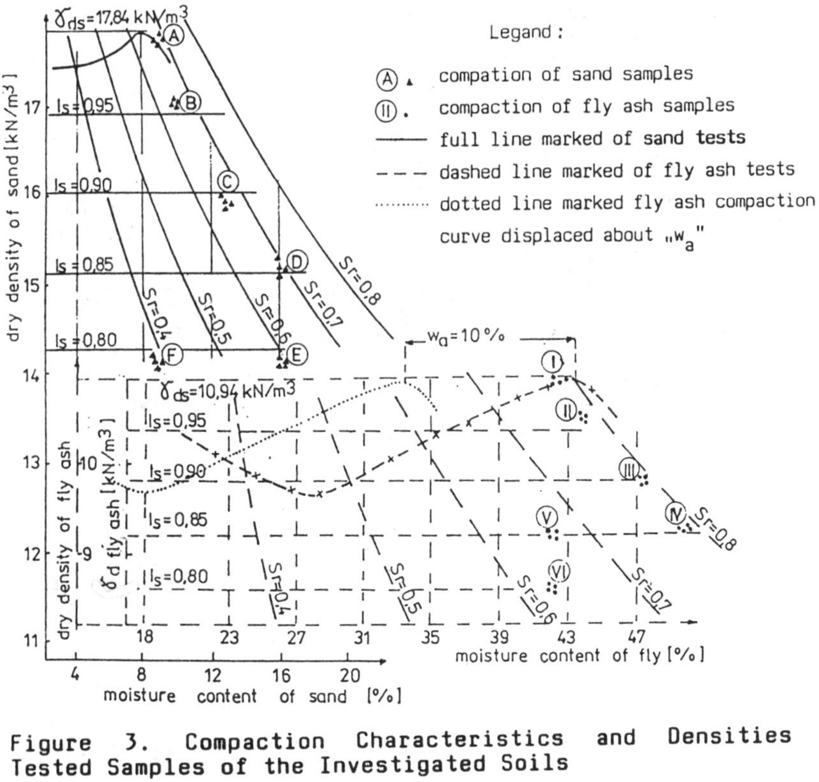

Figure 3. Compaction Characteristics and Densities Tested Samples of the Investigated Soils

holes of the fly ash aggregates. This moisture content called aggregate moisture content w_a was not included in the compactibility process. This quantity was determined by the "blotting paper test" and calculated by equations presented in(Parylak 1984, 1987a) giving a value w_a = 10%. When w_a is substracted, the interparticles degree of saturation is S_{ri} = 0.68. This result was comparable with S_r of the tested sand. For closer analysis of the compaction effects of the moisture contents, the ash axis was shifted in relation to the sand axis about 3 kN/m³. This value is caused by the difference between specific gravity of sand and fly ash (Table 1). As it is known, a smaller specific gravity of fly ash is caused by air bubbles, entrapped inside the ash particles. As shown by Jarrige (Bastian 1970) after four hour milling of fly ash, the specific gravity increased from 19 kN/m³ to 26.7 kN/m³. The weight of compacted fly ash were also reduced by the air bubbles entrapped in the holes of the particles by capilary water filling these holes. This volume was determined by the tests and the solutions described in (Parylak 1987a) and was about 4% of the sample volume. It is about 0.4 kN/m³ of the ash dry density. Interpreting the influence of difference of uniformity index U on the compactibility of the tested soils by Kézdi's nomogram (Kézdi 1979) the difference of dry density between sand and fly ash equals 0.7 kN/m³. The influence of the interparticle capilary water effect at a degree of saturation of about 0.6 was not significant (Kézdi 1979). Very small differences of sphericity and of roundness index (Table 1) did not significantly influence on compactibility. Hence, it could be estimated that the differences of the maximum dry densities of sand and of fly ash were caused in 43% by the specific gravity differences. In 10% they were caused by the differences of the uniformity index of particles, and in 4% by the volume of the air bubbles closed in the hollows of the particles. The remaining 43% was caused mainly by the roughness of the grains surface.

Shear strength tests

Unconsolidated, undrained triaxial tests with pore pressure measurement were performed on 38 mm diameter samples. The samples were formed in a new apparatus for compaction of samples designed for triaxial tests (Parylak 1987). This procedure is comparative to Standard Proctor compactive effort. The tested series were performed at constant degree of saturation S_r = = 0.8. The porosities of the samples (both soils) was between 1.0 and 0.8 of the relative dry densities - I_s.

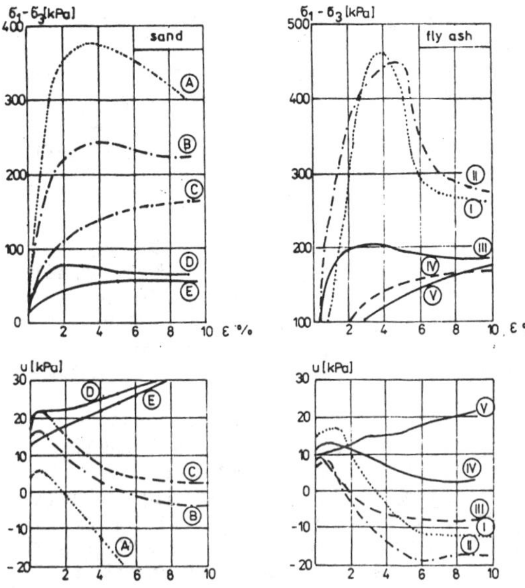

Figure 4. Variation of the Deviator Stress and Pore Water Pressure with Axial Strain for σ_3 = 50 kPa

The compression triaxial tests were performed 6%/h axial strain rate. The minor principal stresses were 25 kPa, 50 kPa, 100 kPa and 200 kPa. The results were analized in effective stress. The changes of the interparticle angle friction Φ' as function of confining pressure, porosity and particle shape were analized.

The results (in Fig. 5) showed curved Mohr's envelopes. Generally, the curvatures were obtained from drained tests on dense sands. But as shown by Bora (1984), similar curvature Mohr's rupture envelopes were obtained in

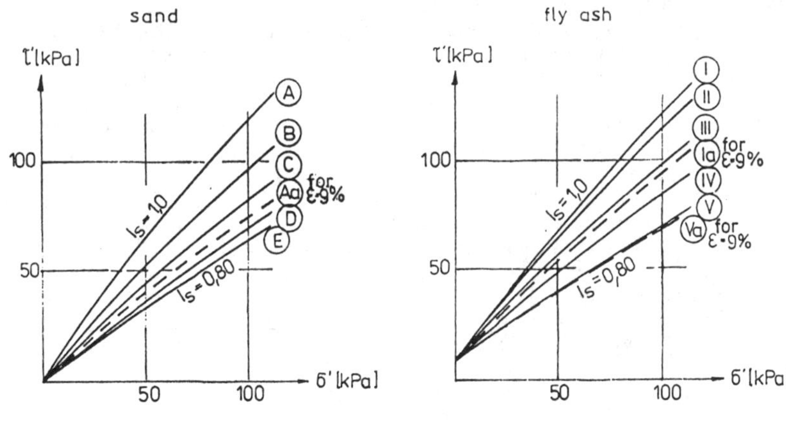

Figure 5. Comparison Between the Mohr Envelopes for Both Soils

the case of drained and undrained tests results. In the case of tests the tested samples had about 25 cm^3 of air spaces in which air could be tightened (Parylak 1987a). Hence, inside those samples the migration of water was possible, as is illustrated by the pore pressure distribution in Fig. 4. A new interpretation of the Mohr's envelope curvature demonstrated that the curvature mainly depends on effective confining contact stress $n \cdot \sigma_3$ (where n is the porosity) (Bora 1984). To a smaller degree it depended on breakdown of grains described also by Berezantzev and Kovalev (1968). The tendency of the first hypothesis was confirmed by Vesić and Clough (1968). Their tests showed that above the normal stress 10 MPa the continuity of crushing of grains was observed, when the obtained secant angle of shearing resistance was constant. Mohr's envelopes plotted in Fig. 5 have small curvatures.

Curvature was higher for ash than for sand. For both soils they were the greatest for I_S = 0.90, and smaller for I_S = 0.96 and 1.0. The smallest curvature occurred for I_S = 0.85, and for I_S = 0.80 the envelopes were straight lines. In the case when the porosities of samples were smaller than critical porosity, compressed during the test samples by great σ_3 were compacted at a greater degree, than they would be compressed upon smaller σ_3. The additional compaction causes the formation of greater quantity of interparticles contacts ($\sigma_3 \cdot n$). As a result the coefficient of friction decreased in relation to the value, which would occur at smaller σ_3.

In the second case, when the initial porosity of samples was higher from critical porosity (Fig. 4, 5, lines D, E, IV, V), at the shearing process the relocation of grains was continued. Hence, short lasting interparticles contacts were continuously unstable. Probably, as a result, the influence of higher minor effective stress σ_3 was not important. Thus, the envelopes for relative dry density I_S = 0.80 were straight lines.

The comparative analyses of the influence of particles roughness on changes of the friction angle and the cohesion intercept were made by tangengt lines for normal stress σ_n' = 50 kPa (Fig. 6). The same values were obtained for Φ' at failure and at great strains (ε = 9%), and also at I_S = 0.80. The increase of densities involved greater increase of Φ' for the fly ash than for the sand. The values of Φ' for fly ash are reduced by smaller uniformity index of ash (Fig. 2). It cause the additionally increase of Φ' for fly ash about 6° (Koerner 1970). Hence, we can assume that differences of Φ' between fly ash and sand are caused mainly by particle roughness.

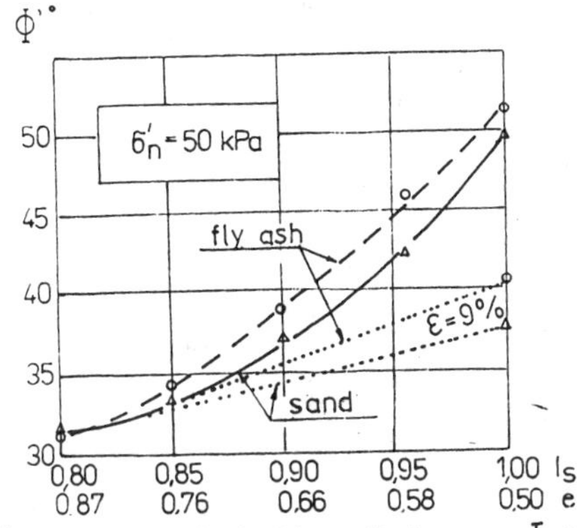

Figure 6. Relation Between Φ' and Relative Dry Densities (values e calculated for sand)

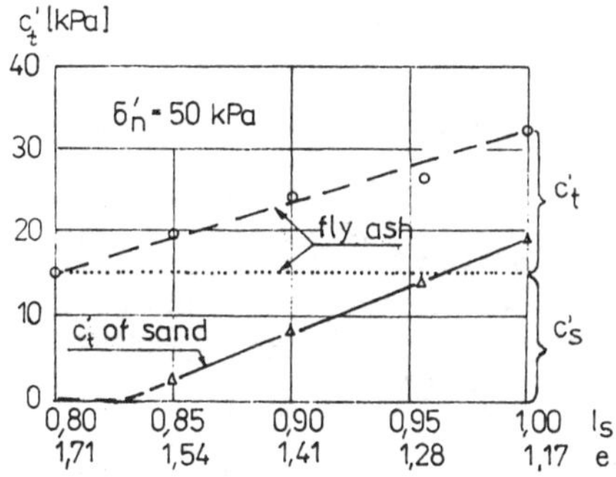

Figure 7. Relation Between Cohesion Intercept and Relative Dry Densities (values e calculated for ash)

At the lossest states (I_s = 0.80) of the fly ash, the values of structural cohesion intercept c_i = 14 kPa were obtained. It appeared at failure as well as at a strain of $\mathcal{E}$ = 9% (Fig. 7). It was caused mainly by the roughness of grains and aggregates, as the main parameter differentiating both soils. This phenomenon is the consequence of the structural cohesion c_s produced by succesive interlocking of aggregates (dotted line Fig. 7). These results indicate that c_s' is constant for tested fly ash. The second component of cohesion intercept - the tangent cohesion c_t, probably was caused by the factor resulting from the curvature of Mohr's envelope (Hansen 1979), which was described earlier.

Similar values of cohesion intercept were also obtained in unconfined compression tests (Parylak 1984). Great values of cohesion intercept for ash were found by Gray and Lin (1972), as well as by Leonards and Bailey (1982) who suggested that the cohesion intercept was caused only by the capilary forces. According to Kézdi (1979), the maximum values of capilary forces in fine dense sands are smaller than 0.1 kPa.

Main Conclusions

Differences in roughness can be satisfactority determined by the differences of grain perimeters of the cross section of particles as well as by the differences of grain perimeters and perimeters of polygons circumscribed on these grains. The obtained results are comparable with differences of the specific surfaces of the tested soils.

The particle roughness significantly effects the compaction process. The differences of maximum dry densities of tested sand and fly ash were caused at least in 43% by the higher roughness of the ash particles.

Roughness has also a significant effect on the angle of internal friction and on the curvature of Mohr's envelope. It also influences the cohesion intercept value. In very loose fly ash, notable values of cohesion intercept occurred probably produced by the succesive interlocking of aggregates.

Acknowledgements

The writer wish to thank of the Organizing Committee of Conference and Reviewers for comments and suggestions.

References

Bastian,S., (1980). "Structural Concretes with Fly Ash" Arkady (in Polish).

Bell,F.G., Gambin,M.P., and Parsons,A.W., (1987). In: "Ground Engineer's Reference Book", Butterwoths, London.

Berezantsev,V.G., and Kovalev,I.V., (1968). "Consideration of the Curvilinearity of the Shear Graph when Conducting Tests on Model Foundations", Osnovaniya, Fundamenty i Mekhanika Gruntov, 1, 1-4 (in Russian).

Bora,P.K., (1984). "Shear Failure in Granular Media", Journal of Geotechnical Engineering, 110 (5), 582-598.

Chang,P.W., and Chae,Y.S., (1987). "A Parametric Study of Effect of Vibration on Granular Soils",Proc.Intern. Conf. Soil Dynamic and Earthquake Eng., 3rd, Princetown, vol. 42, 137-151.

Chaplin,T.K., (1961). "Compressibility of Sands and Settlements of Model Footings and Piles in Sand",Proc. Intern. Conf. Soil Mech. Found. Eng., 5th, Paris, 33--40.

Herdan,G., (1953). "Small Particle Statistics",Elsevier, Amsterdam.

Gray,D.H., and Lin,Y.K., (1972). "Engineering Properties of Compacted Fly Ash", Journal of the Soil Mechanics and Foundation Division, 4, 361-380.

Hansen,B., (1979). "Definition and Use of Friction Angles", Proceedings European Conference on Soil Mechanics and Foundation Engineering, Brighton, vol. 1, 173-176.

Kézdi,A., (1979). "Soil Physics", Akadémiai Kiadó, Budapest.

Koerner,R.M., (1970)."Effect of Particle Characteristics on Soil Strength", Journal of the Soil Mechanics and Foundation Division, (4), 1221-1234.

Kolbuszewski,J., (1967). Lecture given to SGGW in Warsaw.

Lambe,T.W., Whitman,R.V., (1969). "Soil Mechanics", John Willey a. Sons, New York.

Leonards,G.A., and Bailey,B., (1982). "Pulverized Coal Ash as Structural Fill", Journal of Geotechnical Engineering, (4), 517-531.

Parylak,K., (1984). "Shear Strength of Natural and Lime Treated Fly Ash as Engineering Material for Embankments Construction", Archiwum Hydrotechniki, (1-2), 137-152, these of fulfilment of the requirements for the degree of Doctor of Science; (in Polish).

Parylak,K., (1987). "Apparatus for Compaction Fine Grained Soils", Archiwum Hydrotechniki, (3-4),349-361, (in Polish).

Parylak,K., (1987a). "The Influence of the Structure of Bitumions Coal Fly Ash Particles on Pore Pressure in the Triaxial Test", VIII Polish National Conference on Soil Mechanics and Foundation Engineering, vol. 1, 99-104, (in Polish).

Stępkowska,E.T., (1977). "Water Sorption Test and Its Application in Different Tests", Archiwum Hydrotechniki, (3), 411-420, (in Polish).

Vesić,A.S., and Clough,G.W., (1968). "Behaviour of Granular Materials Under High Stresses", Journal of the Soil Mechanics and Foundation Division, (3), 661-689.

Stabilized Active Clay by Sand Admixture

Pat T. Leelani,[1] M.ASCE,
Maen M. Shaar[2] and Phil V. Compton[3]

Abstract

To improve the infrastructure of highway pavements, properties of an active clay with an admixture of a coarse sand were evaluated for earthwork construction. The active clay is a poor soil existing at a marginal site and the sand can be obtained from a nearby borrow pit. The borrow sand is treated as an admixture to improve the properties of the existing active clay at various ratios of mixtures. Experimental results indicate that a 20% sand admixture improved the properties of the existing active clay adequately for highway embankment construction.

Introduction

Improvement of poor soil conditions existing at marginal sites using various methods of stabilization was recently updated by Welsh (1987). Improved properties of an existing active clay by admixture of a fine sand were introduced by Leelani (1989, 1990). This paper describes the improved properties of an active clay with an admixture of a coarse sand at various ratios of mixtures. A method was first developed for obtaining consistency in the Harvard Miniature compaction. This method was then used to conduct compaction tests and prepare specimens for unconfined compression tests. The influence of coarse sand as an admixture at various ratios of mixtures was evaluated for its effects in improving the workability and properties of the active clay. The minimum admixture of

---

[1] Prof., Dept. of Civil Engr., Texas A&I University.
[2] Engineer, Dept. of Transportation, California, 92402.
[3] Prof. & Dean, Coll. of Engr., Texas A&I University, Kingsville, TX 78363.

borrow sand was determined by comparing the strength of the clay-sand mixture to that of a typical clay used for highway embankment construction.

Earthwork Construction using Sand Admixtures

An alternative chosen to improve the properties of an existing active clay by using a sand admixture should be economical when a near borrow pit of sand is available. In an urban area, this alternative is extremely attractive since the sand admixture increases the existing soil profile to the required elevation without using other backfill materials. The alternative should be selected with the consideration that the existing surface stratum of active clay is rather shallow and located above the ground-water table. The active clay should be adequately dried for scarifying and loosening to mix with the borrow sand. The thoroughly mixed soil should be compacted in layers, approximately 100-mm thick per layer. The clay-sand mixture should be compacted, layer by layer, until the required elevation is attained. The compaction should be made by sheepsfoot rollers to help mix sand particles with the clay as normally the clay lumps are not entirely broken down during the mixing process.

Scope of Present Study

A borrow sand consisting of coarse particles was evaluated as an admixture to improve the properties of an existing active clay. The minimum amount of sand admixture was determined for use in earthwork construction to improve the infrastructure of highway pavements. The influence of sand admixture in improving the workability and properties of active clay was evaluated at 1:0.5, 1:1, 1:2, 1:3, 1:4, and 1:5 ratios of sand:clay. At these ratios of clay-sand mixtures, compaction tests were conducted using a Harvard Miniature mold. The optimum moisture content obtained was then used to prepare test specimens for unconfined compression tests. Finally, the strength of clay-sand mixtures were assessed by comparing to that of a typical clay that has been successfully used for highway embankment construction.

Active Clay; Source of Samples and Index Properties

Samples of active clay were obtained from a site, approximately ten kilometers due West from the Gulf of Mexico. The active clay was geologically characterized as a Coastal Plain sediment deposited during the interglacial stages of the Pleistocene Epoch.

The active clay was a part of the younger formation unconformably overlaid above the older formation owing to erosion during the Pleistocene glacial stages. The active clay is recognized as a poor soil and has caused damages to structural buildings in this area.

The active clay is black in color and shows no trace of organic matter. The gradation of the active clay obtained by hydrometer analysis is shown in Figure 1. The active clay shows a wide range of gradation distribution and contains a clay fraction of approximately 35.00%. The liquid and plastic limits of active clay were determined to be 49.40% and 15.00%, respectively. The soil activity; i.e., a ratio of the plasticity index to the clay fraction, is approximately equal to 1.00. The active clay was therefore identified to be normally active based on the rating of soil activity given by Skempton (1953).

Sand Admixture

Samples of sand were obtained from a borrow pit located near the marginal site of the active clay. The sand is a clean beach sand with round quartz particles. The gradation distribution of the sand is shown in Figure 1. The coefficient of uniformity was determined to be less than 4.0 and the sand was rated as a poorly graded soil. The sand contains no silty material and was identified to be a coarse sand. The large particles of the sand have a rough surface texture, a darker brown color and are not quite round in shape. The smaller particles of sand are semi-angular and lighter in color.

Index Properties of Clay-Sand Admixtures

Samples of clay-sand mixtures were prepared at 1:0.5, 1:1, 1:2, 1:3, 1:4, and 1:5 ratios of sand:clay on a dry weight basis. All samples were prepared from air-dried soils and the existing moisture content was taken into account in making specified ratios of clay-sand mixtures. After air-dried samples of the active clay and sand admixture were combined, the samples were passed through the number 40 sieve (0.425 mm) for conducting Atterberg limits. The results obtained are shown in Figure 2 and the plasticity chart in Figure 3.

Figure 2 shows the variation of liquid and plastic limits of clay-sand mixtures with various percentages of the sand admixture. Both liquid and plastic limits of clay-sand mixtures uniformly decrease when the content of sand admixture is increased. These

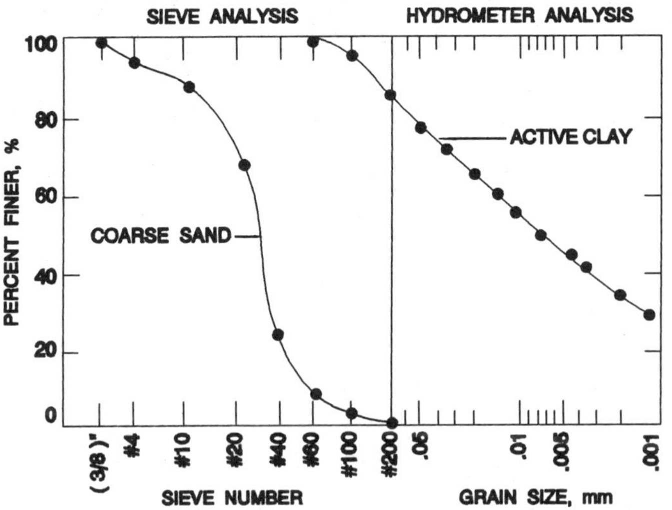

Figure 1. Grain Size Distribution of Active Clay and Sand Used as Admixture

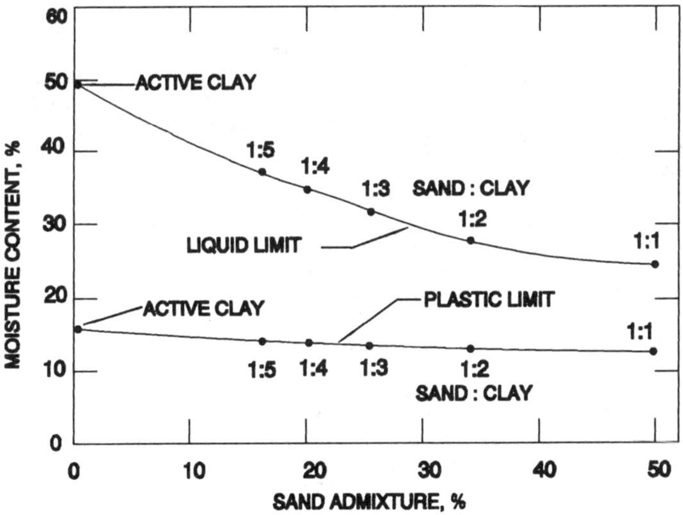

Figure 2. Variation of Liquid and Plastic Limits of Clay-Sand Mixtures at Various Ratios

results indicate that the sand admixture has, in general, a substantial influence in improving the workability and properties of active clay. This occurs because larger sand particles in the soil constituent adsorb less water, resulting in a decrease in values of the Atterberg limits at a higher content of the sand admixture. The clay-sand mixture becomes non-plastic at 66.67% sand admixture or a 2:1 ratio of sand:clay. At this ratio of mixture, the influence of sand admixture completely dominates the behavior of active clay.

Figure 3 is the plasticity chart showing the variation of liquid limits and plasticity indices of the clay-sand mixtures at various ratios. The soil index properties of active clay is also shown in Figure 3. All of the soil index properties are located in between the U line and the A line. The U line having an empirical expression of PI = 0.9(LL-8) is the upper limit of the soil index properties for all types of clay. The A line, empirically given by the equation PI = 0.73(LL-20), separates inorganic clays from inorganic silts and is used in the Unified Soil Classification.

With the increased content of sand admixture, the soil index properties of active clay conformably improves with the A line of the plasticity chart shown in Figure 3. This occurs because there is no silty material in the sand admixture. This result indicates that the presence of sand particles in the soil constituent effectively improves the properties of active clay. At a 1:5 ratio of sand:clay, the plasticity index of active clay decreases from 34.40% to 21.00% and the corresponding liquid limit decreases from 49.40% to 38.40%. The sand admixture improves the soil index properties of active clay from normally active to non-active. The clay-sand mixture becomes non-plastic when the properties of active clay are completely dominated by the sand admixture at 1:0.5 ratio of sand:clay or a 66.67% sand admixture.

Performance of Harvard Miniature Compaction

To simulate the kneading action of sheepsfoot rollers, the Harvard Miniature compaction was chosen to conduct compaction tests. Unlike the Standard Proctor compaction, the miniature compaction is achieved by kneading action rather than impact action. The miniature compaction consists of a cylindrical mold and a tamping rod. The cylindrical mold has the volume of 62.4 cucm or 1/454 cuft with the size of 1.3" (34 mm) in diameter and 2.8" (71 mm) high. The tamping rod is

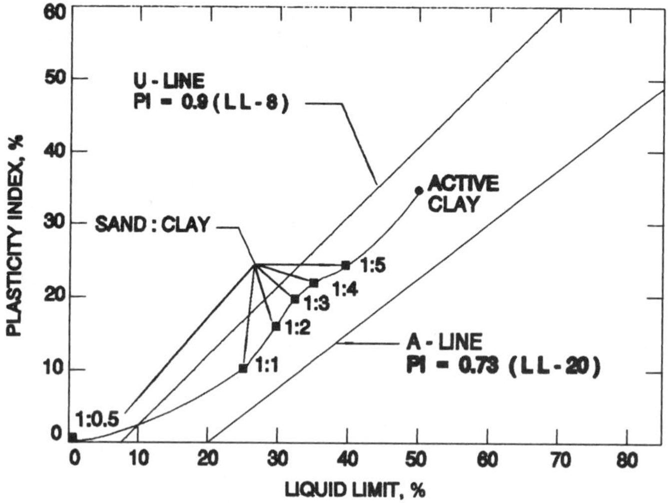

Figure 3. Plasticity Chart Showing Index Properties of Clay-Sand Mixtures at Various Ratios

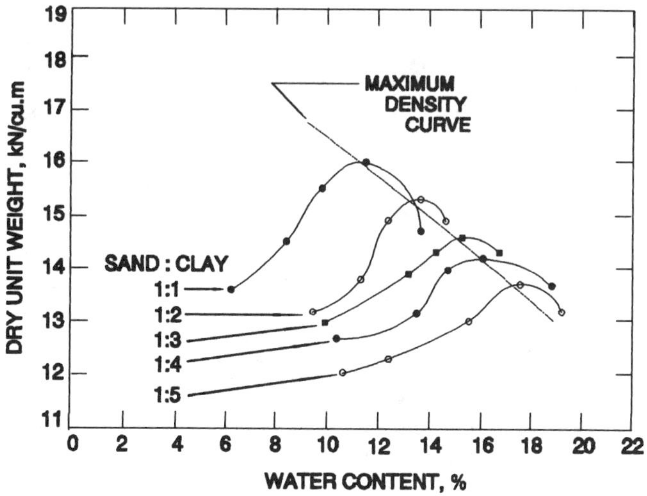

Figure 4. Dry Density versus Water Content of Clay-Sand Mixtures at Various Ratios

connected to a spring that delivers a static pressure of 20 lb (90 N). Kneading action of the compaction is achieved by tamping soil samples into the mold. The density of a soil specimen, obtained by tamping six layers of soil into the mold, with 12 tamps per layer is equivalent to the Standard Proctor compaction.

To obtain consistency in the Harvard Miniature compaction, two samples were consecutively compacted at the same moisture content. The unit weight of the two samples were then compared with each other and the difference between their unit weight was kept below 1.00 pcf (0.1572 kN/m^3) or approximately 1.00%. When the difference between the unit weight of the two specimens was larger than 1.00%, the third sample was immediately prepared. A comparison was then made among the three specimens to identify the sample to be discarded. The two compacted specimens having no more than a 1.00% difference in unit weight were taken as the final product. This method was applied to conduct compaction tests from five various ratios of clay-sand mixtures at 1:1, 1:2, 1:3, 1:4, and 1:5 ratios of sand:clay.

Figure 4. shows five density curves based on the results of compaction tests conducted from five ratios of the clay-sand mixtures. Each density curve in Figure 4 represents the variation of dry unit weight with moisture content at a specified ratio of the clay-sand mixture. The upper density curve is the result of the compaction test conducted from the clay-sand mixture at a 1:1 ratio or a 50% sand admixture. It indicates that the clay-sand mixture with 50% sand admixture exhibits higher values of dry unit weight at lower values of moisture content.

The bottom density curve in Figure 4 represents the variation of dry unit weight versus moisture content of the clay-sand mixture at a 1:5 ratio of sand:clay or a 16.67% sand admixture. As compared to other curves in Figure 4, the active clay with 16.67% sand admixture exhibits lower values of dry unit weight at higher values of moisture content. These results indicate that the clay-sand mixtures with a lower content of the sand admixture exhibit lower values of dry unit weight with higher values of moisture content. This effect is attributed to larger sand particles that adsorb less water to attain a higher density.

A dotted line shown in Figure 4 was drawn connecting the peaks of density curves. This dotted line represents the relationship between maximum dry density and optimum moisture content of the clay-sand mixtures at various ratios. When the content of a sand admixture is increased along the upward trend of the dotted line, the maximum dry density of clay-sand mixtures increases with a decrease in value of the optimum moisture content. These results are attributed to the presence of sand particles in the soil constituent. The larger sand particles decrease the surface area of the soil and adsorb less water. The clay-sand mixtures with higher content of the sand admixture exhibit higher values of the maximum dry density. This effect is attributed to the dense packing of soil particles since the workability in performing the miniature compaction is significantly improved for the clay-sand mixture with higher amounts of sand admixture.

Observation during the Harvard Miniature compaction indicates that the sand admixture improved the workability of active clay for compaction. Clay-sand mixtures with higher contents of sand admixtures were found to be easier to compact. Consistency in the compaction was also easier to obtain for the clay-sand mixtures with a higher content of the sand admixture. However, specimens of clay-sand mixture at 1:1 ratio of sand:clay were easily crumbled after ejection from the miniature mold due to the lack of soil cohesion.

Sample Preparation for Strength Tests

For the compaction of raw active clay using the Harvard Miniature compaction, the result obtained showed an erratic relationship between dry unit weight and moisture content. The workability of active clay was extremely poor for conducting the miniature compaction. The optimum moisture content of active clay was not known and miniature specimens of active clay were not prepared for conducting unconfined compression tests. Experimental results of the active clay were not included in this presentation.

Test specimens at various ratios of clay-sand mixtures were prepared from air-dried soils. With the existing moisture content of air-dried soils taken into account, specified amounts of distilled water were added to attain the optimum moisture content for compaction. The optimum moisture content was obtained from the results of compaction tests shown in Figure 4. The method of preparing miniature specimens was similar to that used in the compaction tests. Unconfined compression tests were conducted using various ratios of clay-sand mixtures at the optimum moisture content. Test

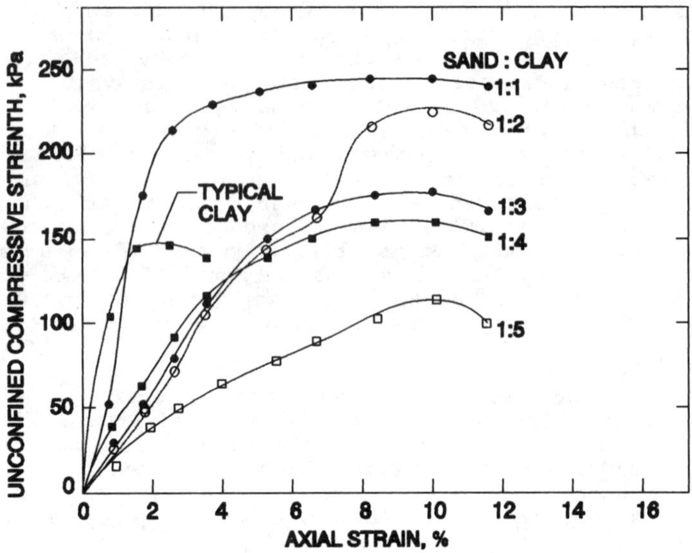

Figure 5. Stress-Strain Relation of Typical Clay and Active Clay Mixed with Sand at Various Ratios

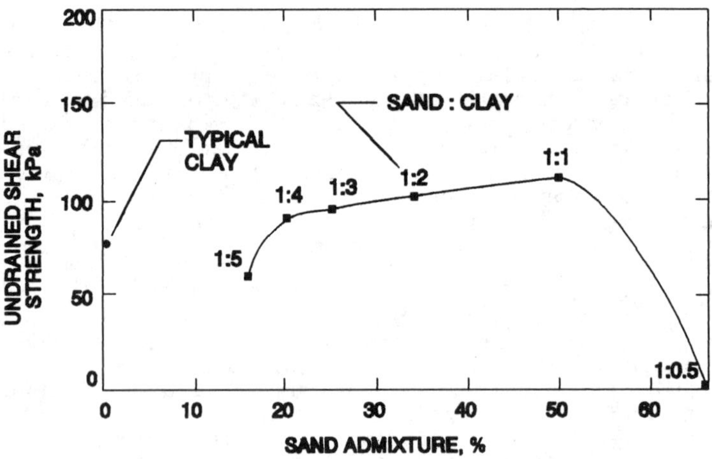

Figure 6. Undrained Shear Strength of Typical Clay and Active Clay Mixed with Sand at Various Ratios

specimens were also prepared for the strength test from a typical clay that has been successfully used in highway embankment construction.

Strength of Clay-Sand Admixtures

Figure 5 shows the results of unconfined compression tests of the clay-sand mixtures at various ratios. All of the stress-strain relationships obtained maximize at the axial strain of approximately 10.00%. The peak values of these stress-strain curves are the unconfined compressive strength of clay-sand mixtures at various ratios. The clay-sand mixture at a 1:1 ratio exhibits the maximum strength and that at a 1:5 ratio is minimum. It implies that the strength of clay-sand mixtures increases when the content of sand admixture is increased.

Figure 5 also shows the stress-strain relation of a typical clay that has been successfully used in highway embankment construction. The unconfined compressive strength of the typical clay is attained at approximately 2.50% axial strain as compared to 10.00% strain of the clay-sand mixtures. The clay-sand mixtures provide a much higher deformation at the peak strength and is therefore superior to the typical clay. The unconfined compressive strength of the typical clay was determined to be 146.18 kPa.

Based on Mohr-Coulomb's failure criteria, the undrained shear strength of soil can be taken to be one-half of the unconfined compressive strength. The results of the undrained shear strength shown in Figure 6 were obtained from one-half values of the unconfined compressive strength shown in Figure 5. Figure 6 shows the variation of the undrained shear strength of the active clay with various percentages of the sand admixture. It indicates that the undrained shear strength of the active clay improves with the increased content of sand admixture. This increasing trend steadily continues until it becomes maximum at 50% sand admixture. For more than 50% sand admixture, the strength rapidly decreases until it becomes zero at 66.67% sand admixture or a 1:0.5 ratio of sand:clay. At this ratio, the sand admixture completely dominates the behavior of active clay. Similar results are revealed by the soil index properties where the clay-sand mixture with 66.67% sand admixture becomes non-plastic.

Figure 6 also shows the undrained shear strength of the typical clay that has been successfully used in highway embankment construction. The undrained shear strength of the typical clay was determined to be 73.09 kPa. This value is lower than the 90.57 kPa of the clay-sand mixture at 20% sand admixture or at a 1:4 ratio of sand:clay. It is therefore recommended that the 20% sand admixture can be used to improve the properties of the existing active clay for highway embankment construction. It is also recommended that a sand admixture of at least 20%, but not more than 50%, can be used with the existing active clay as subgrade material to improve the infrastructure of highway pavements. These recommendations are made with the reservation that a nearby borrow sand is available at a low hauling cost.

Conclusions

Improved properties of an existing active clay with sand admixture were evaluated at the various ratios of clay-sand mixtures. Several conclusions were drawn from the results of this study.

1. The sand admixture improves soil index properties of active clay from normally active to non-active. The soil index properties conformably improve with the A line of the plasticity chart because there is no silt in the sand admixture. The active clay becomes non-plastic at a 66.67% sand admixture.
2. Consistency of the Harvard Miniature compaction was obtained from clay-sand mixtures by comparing the unit weight of two consecutively compacted specimens and the difference in the unit weight was kept below 1.00%.
3. The optimum moisture content of clay-sand mixtures decreases with the increasing content of the sand admixture. This occurs because the lower surface area of large sand particles require less water to attain the maximum density.
4. The undrained shear strength of clay-sand mixtures increases with the increasing content of sand admixture until it maximizes at a 1:1 ratio of sand:clay. It becomes zero at a 1:0.5 ratio of sand:clay because the clay-sand mixture is non-plastic.
5. By evaluating the strength of a typical clay used in highway embankment construction, it is recommended that a coarse-sand admixture of at least 20%, but not more than 50%, can be used to improve the properties of the existing active clay. Using this alternative is attractive when a borrow pit of sand is located near the job site and can be obtained at a low cost.

Acknowledgments

This investigation was supported in part by a Research Initiative Program of Texas A&M and A&I Universities. The authors express appreciation to Mr. Daqing Lu, a graduate research assistant, in preparing the figures of this paper.

References

Skempton, A.W. (1953), "The Colloidal Activity of Clays" Proc., 3rd Intl.Conf.SMFE, Zurich, Vol.1, p 57-61.

Leelani, P.T. (1989), "Improving Properties of Active Clay by Sand Admixtures," Proc. of the ASCE Foud. Eng. Congress: Current Principles and Practices, Edited by F.H. Kulhawy, Illinois, Vol.1, p 381-391.

Leelani, P.T. (1990), "Engineering Properties of Clay-Sand Admixtures," Paper Presented at the Technical Session of Geotechnical Engineering, ASCE Spring Convention, Austin, Texas, April 5.

Welsh, J.P. (1987), "Soil Improvement-A Ten Year Update" ASCE Geotechnical Spec. Pub. No. 12.

Seepage Control in Kaolinite Clay with Simulated Cracks

C. Vipulanandan[1], A.M.ASCE, and M. Leung[2]

Abstract

In this laboratory study the interactive nature of cracks (simulated) and methanol on the hydraulic conductivity of Kaolinite clay was investigated. Methods to decrease the hydraulic conductivity of cracked and contaminated Kaolinite clay by treating it with lean grouts were also studied. The hydraulic conductivity of clay ($<3 \times 10^{-8}$ cm/s) was increased above the U.S. EPA limit of 10^{-7} cm/s in a controlled manner using syringe needles to form simulated cracks (paths of least resistance for flow) in the compacted clay specimens. Relationships between hydraulic conductivity of clay and crack parameters (crack length and density) for both water and methanol have been developed. The hydraulic conductivity of cracked and/or contaminated clays ($>10^{-7}$ cm/s) were restored to satisfy the U.S.EPA limit using grouts such as sodium silicate and bentonite. Treatment of cracked and contaminated clay are also compared to the treatment of permeable clay (hydraulic conductivity $> 10^{-5}$ cm/s). The duration of treatment is very much dependent on the clay condition and the type of grout used.

Introduction

The failures of clay barriers in hazardous waste storage and disposal facilities has drawn much attention recently. But very little is known about repairing these failed clay barriers in situ to satisfy the performance

[1] Associate Professor, Department of Civil and Environmental Engineering, University of Houston, Houston, Texas 77204-4791.

[2] Engineer, Stone and Webster Environmental Services, Houston, Texas 77096.

criteria set by the U. S. EPA of 10^{-7} cm/s for hydraulic conductivity. In general the hydraulic conductivity of clay is affected by the interactive nature of local defects (desiccation cracks, fissures and fractures), chemical permeants and in situ stresses [1-7]. Improper modeling of clay, without flaws, in the laboratory tests are totally or partly responsible for the discrepancies in the reported field and laboratory test results. Since field tests are expensive, time-consuming and in many instances impractical, it is essential to develop reliable laboratory permeability tests, where the field conditions are closely modelled, to evaluate the effect of cracks and various repairing techniques to reduce the hydraulic conductivity of permeable clay.

After studying 39 hazardous waste sites, Freeze [5] found that they leaked after an average of 14 years. When they were repaired, only 16% of the time was the effort totally successful, and 43% of the repairs failed totally [5]. It is also important to realize that there are ways to reduce and improve land disposal, but there is currently no way to practically eliminate land disposal. Even if a policy of waste destruction and recycling is rigorously followed, there will always be waste that must be perpetually stored [2]. Deep-well-injected wastes should be totally isolated from deep aquifers by shale or clay aquicludes (confining layers). Since wastes are commonly injected under high pressure, the injection process could cause the confining layers to fracture and hence reduce their effectiveness and performance. How long clay liners for surface facilities or clay aquicludes can remain resistant to chemicals is presently unknown and the information in the literature is sometimes confusing and often contradictory [1,3-5,7]. These observations stress the need for improving the repair methods and the prediction of their life expectancy under hazardous waste environments for existing clay barriers both near-surface land disposal and deep injection.

Desiccation cracks can be as wide as 1 mm and could penetrate deep into clay liners, increasing their hydraulic conductivity at least by 10 times [8]. Laboratory permeability tests conducted on recompacted samples are unlikely to contain representative patterns of cracks and fissures and significantly under predict the hydraulic conductivity of clays in natural deposits or compacted liners [3]. Although large scale in situ permeability tests may provide correct permeability information in accessible soils (near-surface), cost and the time periods required to conduct these tests often make them impractical. Hence, it is essential to develop laboratory test procedures to model closely the field soil conditions.

If the effect of cracks (size, length, distribution and density) on transmission of hazardous waste is understood, it will be possible to assess the impact of faults, desiccation cracks, fractures and fissures on the clay barriers and predict their performance in the field. Also, methods must be developed to repair cracked clays in-situ and evaluate the performance of repaired clay under various chemical environment.

Objectives

The objectives of this study are: (a) to develop a reliable laboratory test procedure to simulate cracked clay, (b) to investigate the interactive nature of cracks and permeants on the hydraulic conductivity of clay, and (c) to determine the effectiveness of a few selected treating materials (grouts) in reducing the hydraulic conductivity of cracked and/or contaminated clay to satisfy U.S.EPA requirement of less than 10^{-7} cm/s.

Experimental Program and Results

For the initial study, commercially available Kaolinite clay and methanol were selected. A testing facility with twelve double ring rigid-wall permeameters were used exclusively in this study. Several series of tests were performed to demonstrate the repeatability of the hydraulic conductivity-time history for the Kaolinite clay with water as permeant and the results are summarized in Fig. 1. The Kaolinite clay specimens were compacted at 34 and 36% moisture contents, slightly higher than the optimum moisture content, and the permeability tests were performed at a hydraulic gradient of 100. Results from flexible wall permeameter tests are also in good agreement as shown in Fig. 1 [8]. Kaolinite clay, methanol and stabilizing materials were characterized based on their physical, chemical and interactive properties. The organic permeant, methanol, is neutral and polar with dielectric constant of 34 and is totally miscible in water. Various methods to introduce cracks in the compacted clay specimens were investigated. Fractured and/or contaminated clays were repaired using a few selected stabilizing materials and the effectiveness of repair was evaluated by measuring the changes in the hydraulic conductivity of the clay within the inner ring of the permeameter.

Figure 2 compares the effect of pure methanol on the hydraulic conductivity of Kaolinite clay specimens compacted at two moisture contents. Methanol was selected since it is totally miscible in water with a low dielectric constant. The selection of high concentration of methanol in this study is not only to simulate situations

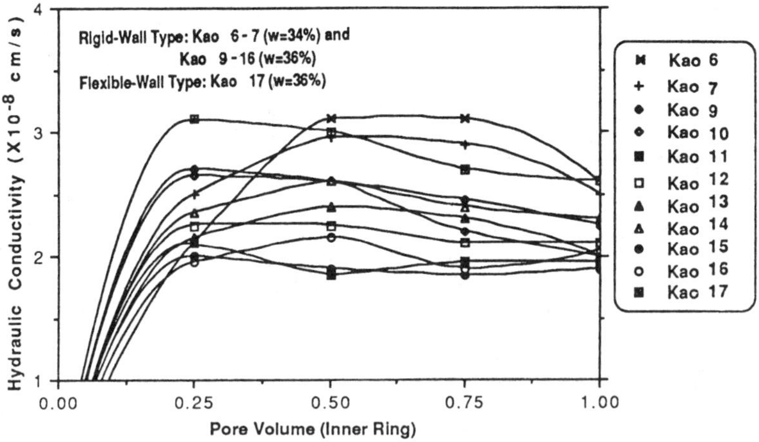

Figure 1. Reproducability of Hydraulic Conductivity-Time Relationship for Kaolinite Clay Using Double Ring Rigid-Wall Permeameter Under a Hydraulic Gradient of 100.

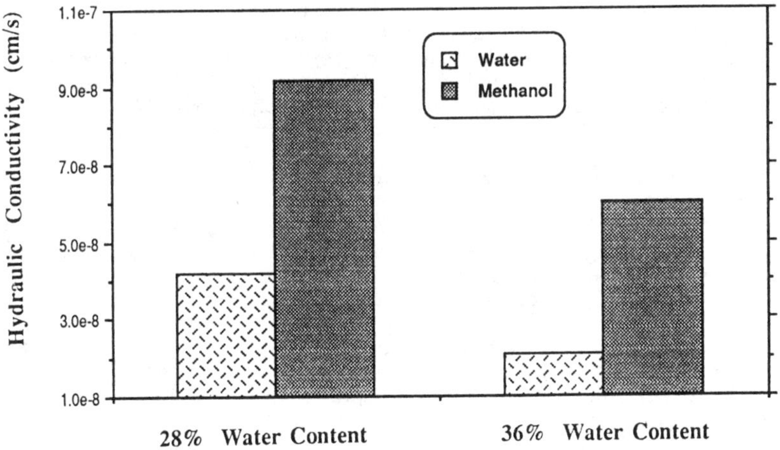

Figure 2. Effect of Methanol and Compaction Moisture Content on the Hydraulic Conductivity of Kaolite Clay.

such as spills and leakage from underground storage tanks but also to demonstrate the potential of the treatment under extreme conditions. The hydraulic conductivity of Kaolinite clay was more than doubled when permeated with methanol than with water. The hydraulic conductivity of the Kaolinite clay was doubled at 28% moisture content than at 36% moisture content, (<2 X 10^{-8} cm/s), which was the minimum measured for the Kaolinite clay [8]. Resistance to crack formation in clay is a function of the tensile strength of clay. Compacted clay samples were extruded and tested in splitting tension and the tensile strength reduced from 10 psi to 4 psi when the compaction moisture content was increased from 31% to 39%. The hydraulic conductivity and other properties reported herein are an average of at least three tests.

Methods of Crack Formation

In order to investigate the effect of cracks in clays, compacted samples (ASTM D698) were cracked (paths of preferential flow) before testing. Several methods of simulating cracks in the laboratory samples were investigated. The methods include (1) inserting permeable sand filled tubes, (2) introducing free standing cracks and (3) inserting syringe needles. Both thin walled glass and plastic tubes up to 5 mm diameter were filled with well graded sand (based on filter design) and buried in the clay to create paths of preferential flows (simulated cracks). But it was observed that the tubes were soon clogged with clay with no observable change in the hydraulic conductivity. In the second method, holes were set in a selected manner in the compacted clay by pushing rods of diameter less than 1 mm into the compacted clay to preselected depths and withdrawing them before starting the hydraulic conductivity test. It was observed that these holes closed as soon as the hydraulic conductivity test was started. In the third method, stainless steel syringe needles with 0.45 mm inside diameter were inserted (forming paths of least resistance to flow) into the the compacted clay in a selected pattern and were in place during the test. With this scheme it was possible to increase the hydraulic conductivity of the clay in a controlled manner. Inspection of the needles after the test showed only a few millimeters of clay had infiltrated into the tip of the needles. Of the methods investigated, inserting the syringe needles was the best. By inserting syringe needles, defects of various lengths and number (cracks/unit area) were introduced into the compacted clay samples as shown in Fig. 3. Using this technique several series of tests have been completed to investigate effect of crack length and number (cracks within the inner ring) on the hydraulic conductivity of Kaolinite clay. Typical crack configuration of 24 cracks, less than 0.1% of the cross

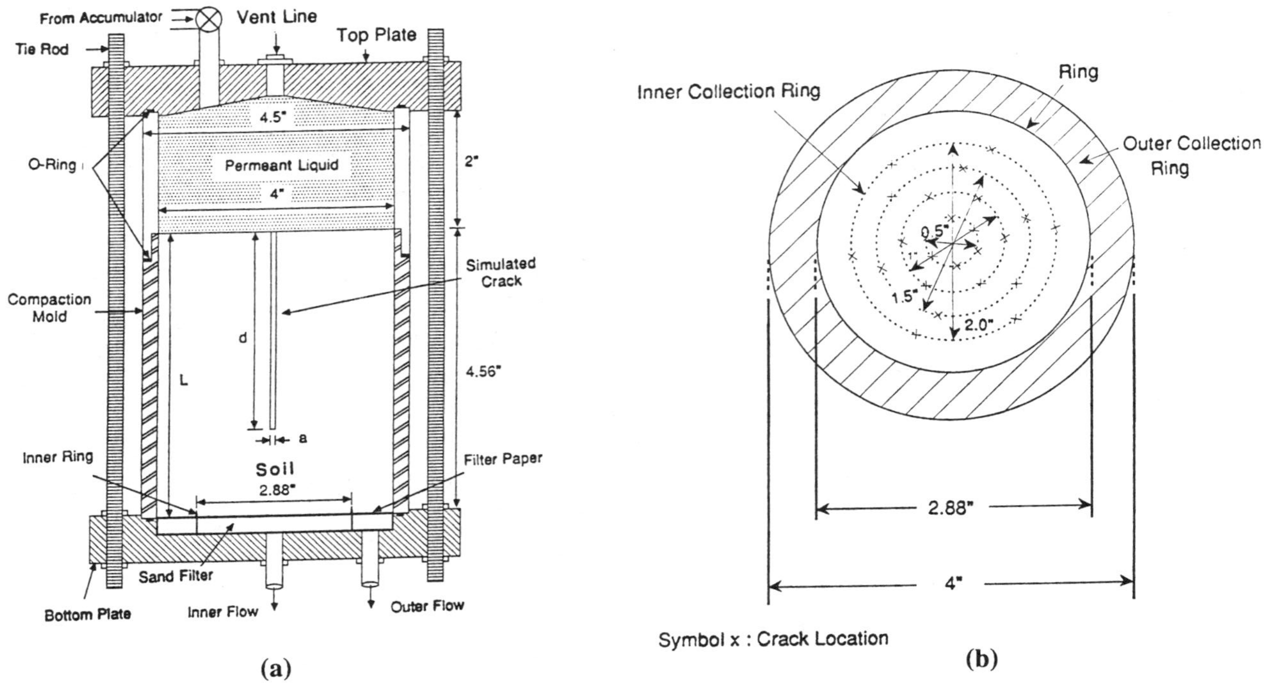

Figure 3. Schematic of the Double Ring Rigid Wall Permeameter with (a) a Simulated Crack (b) Typical Crack Pattern for 24 Cracks.

sectional area of the inner ring, are shown in Fig. 3b which affected the flow in the inner ring. Figure 4 summarizes the typical test results for relative crack length of 90% (where the relative crack length is defined a the ratio of crack length (d) to the thickness of the specimen (L)). For 24 cracks with d/l ratio of 0.9 the total cracked volume amounts to only 0.08% of the volume of clay within the inner ring. The normalized permeability is the ratio of hydraulic conductivity of cracked clay to uncracked clay. With cracks occupying less than 0.1% in area and 0.08% in volume the hydraulic conductivity of the Kaolinite clay was increased more than 5 times with water and 6.5 times with methanol after 1 pore volume (PV) of permeation. Figure 5 and 6 show how the hydraulic conductivity of clay ($<3 \times 10^{-8}$ cm/s) can be increased above the U.S.EPA limit ($>10^{-7}$ cm/s) using syringe needles. The cracked clay with hydraulic conductivity higher than the U.S.EPA limit was used for repairing studies.

Cracked Clay

Cracked clays with 24 cracks (hydraulic conductivity > 10^{-7} cm/s) were used for this study. Several grouts such as cement (0.1% solution), sodium silicate (8% solution) and bentonite (0.1% solution) with viscosity less than 2 centipoise were used for treating the cracked and contaminated clays. Viscosity of the grouts were limited in order to encourage rapid permeation through clay flaws. All repairing studies were done using the double ring rigid-wall permeameters using a hydraulic gradient of 100. The lean grouts were also introduced as permeants and are expected flow through the cracks (preferential paths of flow) and if possible seal them so as to reduce the hydraulic conductivity.

In Figs. 5 and 6, performance of various grouts in reducing the hydraulic conductivity of cracked clay have been compared. As shown in Fig. 5, 0.1% solution of cement showed only a temporary reduction in the hydraulic conductivity. Using 8% solution of sodium silicate, the hydraulic conductivity of clay ($>10^{-7}$ cm/s) was reduced to 3×10^{-8} cm/s within a 50 hour interval with 0.3 PV of grout thus restoring the hydraulic conductivity close to that of the uncracked clay. Figure 6 shows how 0.1% solution of bentonite could be used to reduce the hydraulic conductivity from 2×10^{-7} cm/s to 3×10^{-8} cm/s within 25 hours of treatment by permeating less than 0.4 PV of grout. Although 0.1% cement solution was unable to reduce the hydraulic conductivity of cracked clay, both 8% sodium silicate and 0.1% bentonite solutions were effective under the same condition.

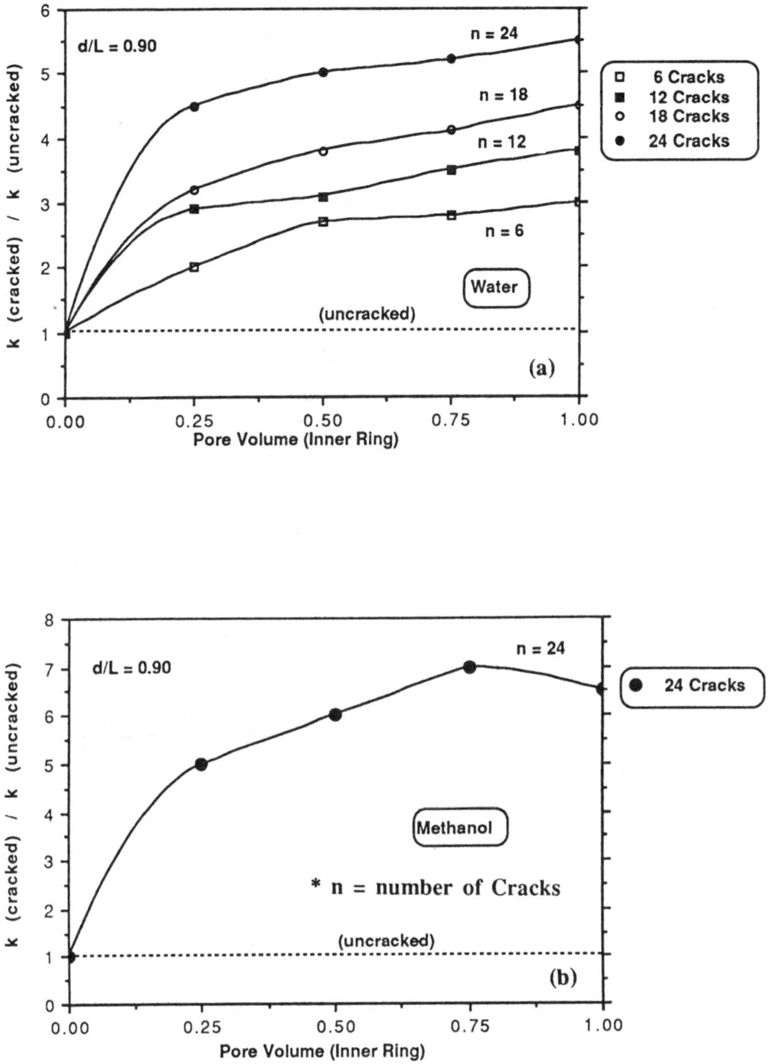

Figure 4. Normalized Hydraulic Conductivity of Cracked Clay With Crack-to-Depth Ratio of 0.90 (a) Water (b) Methanol.

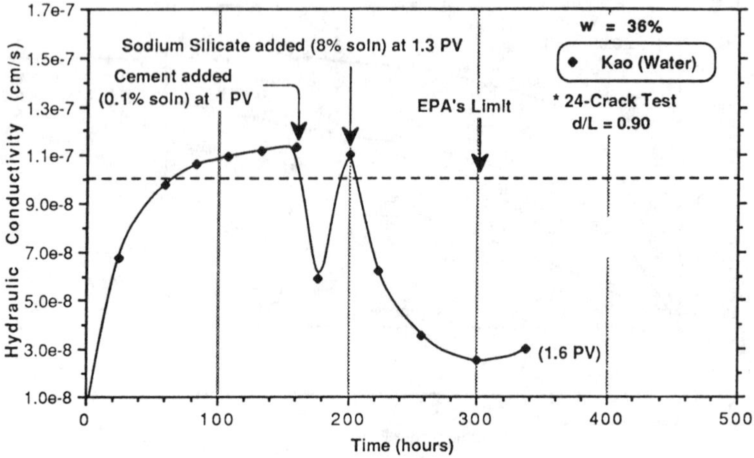

Figure 5. Repairing Cracked Clay With 0.1% Cement Grout and 8% Sodium Silicate Grout.

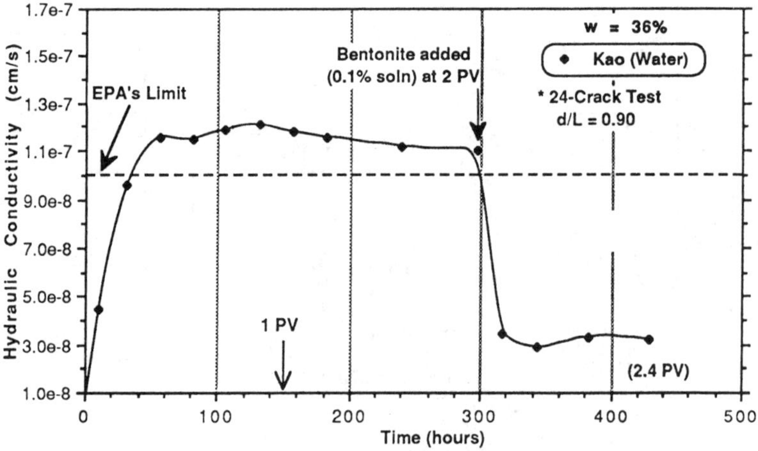

Figure 6. Repairing of Cracked Clay With 0.1% Bentonite Grout.

Cracked-Contaminated Soil

The hydraulic conductivity-time relationship for the cracked kaolinite clay with methanol permeant is shown in Fig 7a. The cracked clay was contaminated with methanol by permeating 1.9 pore volume (PV) of methanol and then treated with 8% sodium silicate solution. The hydraulic conductivity was reduced from 2.1 x 10^{-7} to 2 x 10^{-8} cm/s, an order of magnitude difference, in a period of 80 hours. Figure 7b shows the changes in the effluent pH during the treatment period. The effluent pH varied between 8.2 and 8.6 during this period, which is acceptable.

In order to demonstrate the potential of 8% sodium silicate solution, a Kaolinite clay-sand (20% clay and 80% sand) mixture with hydraulic conductivity of 10^{-5} cm/s was treated. As shown in Fig. 8, 0.1% cement solution was not effective but 8% sodium silicate solution reduced the hydraulic conductivity to 10^{-7} cm/s, a two fold reduction, in about 15 hours with 0.8 PV of grout solution.

Conclusions

Based on the laboratory study, the following could be concluded:

(1) Using syringe needles, the hydraulic conductivity of laboratory compacted Kaolinite clay can be increased in a controlled manner above the U.S.EPA limit to simulate cracked clays. The interactive effects of cracks (simulated) and permeants on the hydraulic conductivity of Kaolinite clay has been quantified.

(2) Bentonite grout (0.1% solution) and 8% sodium silicate grout are effective in reducing the hydraulic conductivity of cracked Kaolinite clay (> 10^{-7} cm/s) and to restore them to that of uncracked clay (<3 x 10^{-8} cm/s). Cement grout (0.1% solution) was not effective under the same condition.

(3) 8% sodium silicate solution is effective in treating cracked and contaminated Kaolinite clay (>2 x 10^{-7} cm/s) to reduce its hydraulic conductivity to less than 3 x 10^{-8} cm/s.

In this bench-scale study the potential of treating a cracked and contaminated clayey soil in situ to reduce the hydraulic conductivity was demonstrated. But further studies including pilot-scale studies are needed before using the treatment in the field.

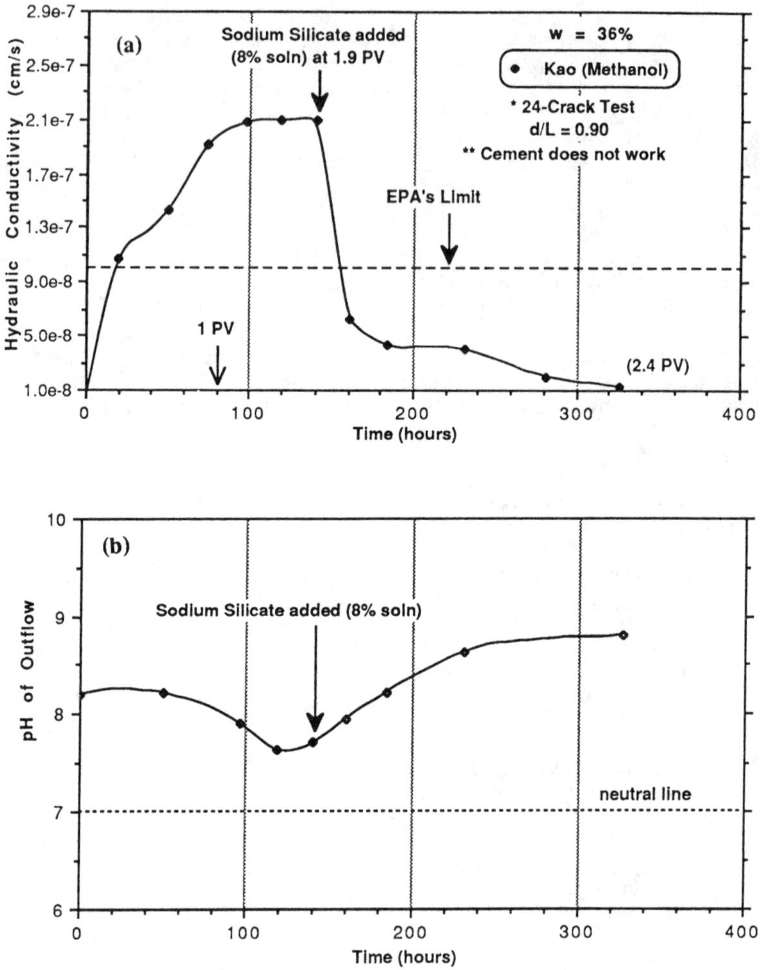

Figure 7. Cracked and Contaminated (with Methanol) Kaolinite Clay Before and After Treatment (a) Hydraulic Conductivity-Time Relationship (b) pH of Effluent

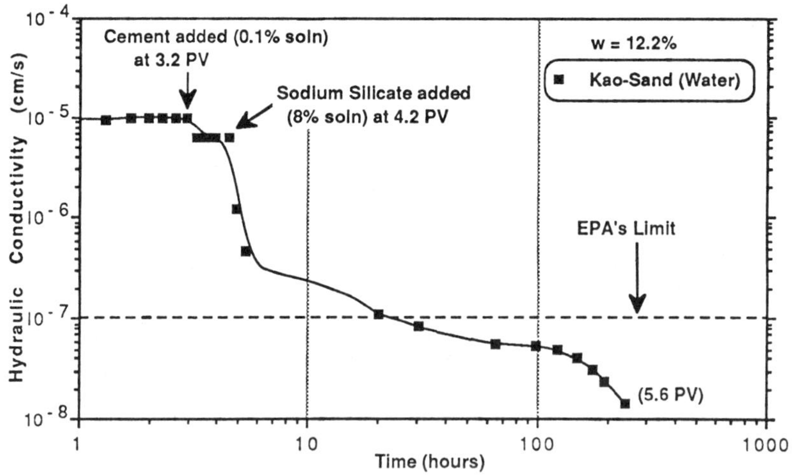

Figure 8. Hydraulic Conductivity-Time Relationship for Kaolinite Clay(20%)-Sand(80%) Mixture During Treatment with 0.1% Cement and 8% Sodium Silicate Solutions

Acknowledgement

This study was funded in part with Federal Funds as part of the program of the Gulf Coast Hazardous Waste Substance Research Center (GCHSRC) which is supported under cooperative aggreement R 815197 with the U.S. EPA and in part by funds from the State of Texas as part of a program of the Texas Hazardous Waste Research Center. The contents do not necessarily reflects the views and policies of the U. S. EPA nor the State of Texas.

References

[1] Acar, Y. B., Hamidon, A., Field, S.D, and Scott, L.,(1985)" The Effect of Organic Fluids on Hydraulic Conductivity of Compacted Kaolinite," Hydraulic Barriers in Soil and Rock, ASTM, STP 874, pp. 171-187.

[2] Anderson, D (1984). "Alternatives to Land Disposal of Hazardous and Radioactive Waste," Paper Presented at the Conference on Alternatives to Disposal of Hazardous Waste, Sponsored by Texas Agriculture Department, pp. 16.

[3] Boynton, S. S., and Daniel, D. E.(1985), "Hydraulic Conductivity Tests on Compacted Clay," Journal of Geotechnical Engineering, ASCE, Vol. 111, No. 4, pp. 465-478.

[4] Daniel, D. E.(1984), "Predicting Hydraulic Conductivity of Clay Liners," Journal of Geotechnical Engineering, Vol. 110, No. 2, 1984, pp. 285-300.

[5] Daniel, D. E., and Kurtovich, M.(1987), "Monitoring for Hazardous Waste Leaks," Civil Engineering, ASCE, Vol. 57, No. 2, pp. 48-51.

[6] Moore, C. A. and Ali, E. M.(1982), "Permeability of Cracked Clay Liners," Proceedings, Eighth Annual Reesarch Symposium on Land Disposal, Incineration and Treatment of Hazardous Waste, U. S. EPA, pp. 174-178.

[7] Tobin, W. R. and Wild, P. R.(1986), "Attapulgite: A Clay Liner Solution?" Civil Engineering, Vol. 58, No. 2., 1986, pp. 56-58.

[8] Vipulanandan, C. and Leung, M.(1990), "Performance and Repair of Cracked and Permeable Clays Under Hazardous Waste Environment," Department of Civil and Environmental Engineering, University of Houston, Houston, Texas, Report No. UHCEE 90-4, 273 pp.

[9] Vipulanandan, C. and Leung, M.(1991), "Effect of Methanol and Seepage Control in Permeable Kaolinite Clay," Journal of Hazardous Materials, Vol.27, pp. 149-167.

[10] Vipulanandan, C. and Leung, M.(1991), "Seepage Control in Contaminated and Permeable Houston Clay: A Laboratory Study," Journal of Hazardous Wastes and Hazardous Materials, Vol. 8, No.1, pp.17-32.

Pavement Improvement With Asphaltic Membrances

By

Ilan Ishai[1], Nathan Livnat[2], & Moshe Livneh[3]

Abstract

This paper describes a preliminary technological effort of using geotextile-reinforced asphaltic membranes for subgrade and pavement improvement. The membranes were applied either as a sealing and reinforcement membrane between the subgrade and the pavement or as a reinforcement membrane for the retardation of reflective cracking when overlaying existing cracked pavement surfaces. It is believed that this technology will provide a better solution to pavement damage due to swelling, shrinkage and collapse of active and sensitive soils, and to the premature reflective cracking in pavement overlaying. All projects described in the paper were designed and constructed during 1989-1991. A close control and follow-up system was applied for monitoring pavement performance with time under actual service and climatic conditions.

Background And Problem Definition

Among the many causes of structural pavement distresses, mainly expressed in cracking, two are related to subgrade volumetric changes, on the bottom side, and to asphaltic reflective cracking, on the surface side.

The first cracking distress is associated with the construction and maintenance of pavements on sensitive active soils, such as clays and silts. Here, severe structural problems often occur following the penetration of water into the active subgrade, causing swell, shrinkage and collapse due to substantial differential volume changes. As Israel and many areas of the U.S. are blessed with clay soils of high swell potential, and with silt and loess soils highly sensitive to collapse, the damage to road and airfield pavements on these soils are

[1] Associate Professor of Civil Engineering, Transportation Research Institute, Technion, Haifa, Israel.
[2] Manager, Marketing Division, Pazkar Ltd., Afula, Israel.
[3] Professor of Civil Engineering, Transportation Research Institute, Technion, Haifa, Israel

particularly frequent and severe. This constitutes one of the main reasons for the serious deterioration in the conditions of the pavement infrastructures in such areas.

One of the accepted solutions for countering the deterioration potential of these soils is to seal the subgrade surface with various types of membranes in order to prevent disruption of the moisture equilibrium attained during paving. The accepted techniques in this area include spraying the surface of the subgrade with bituminous materials, paving with an impervious layer of asphaltic concrete, or using plastic sheets impervious to water, on one side, or significantly increasing the thickness of the pavement structure in order to balance the swell pressures, on the other side. Decades of experience in this area, both in Israel and internationally, have indicated that the accepted solutions have, in many cases, failed to prevent the severe pavement damage which stemmed from volumetric subgrade changes. The main causes of this failure are well documented in the technical literature, and include the improper functioning of a number of types of membrane, (Kasiff et al., (1969); Steinberg (1980); McKeen (1980); and Weston (1980)).

The second distress is related to the reflective cracking which occurs in the asphaltic overlay of existing cracked pavements. This problem is quite common in many parts of the world. The numerous technical solutions and research works fail to achieve the desired prevention or significantly retard the phenomenon. Some improvement has been effected through the use of geotextile fabrics, but these too fall short of the desired levels and criteria by resulting in a very short period of crack retardation, if at all, (Button (1984); Colombier et al., (1986); Joseph and Haas (1989)).

Significant advances have recently taken place in the development and manufacture of special asphaltic membranes, mainly for the sealing of roofs and structures. These membranes are made of modified asphalt cement with improved flexibility, strength and durability, reinforced with polymeric geotextiles. Moreover, these membranes are manufactured by means of an advanced and innovative industrial process which allows maximum control of membrane characteristics and type, as well as various combinations of thickness, type of reinforcing material, etc. The result is an engineering product with excellent qualities of flexibility, durability and imperviousness.

The use of these membranes to seal roofs and structures has become very prevalent and has received great impetus from the perfection of the industrial process and product. On the other hand, the improved qualities of the industrial asphaltic membrane, and particularly its tensile strength, its resistance to tearing, its flexibility, its imperviousness, and the possibility of the controlled production of various thicknesses, all demonstrate promising potential for extensive use in construction and rehabilitation of pavements. Such usage can include the following possible applications:

1. A sealing and reinforcement membrane between the subgrade and the pavement;

2. Reinforcement membranes between the various pavement layers;

3. Reinforcement membranes for retardation of reflection cracking when overlaying existing pavement surfaces.

As this special technology is relatively new with regard to paving, it is extremely difficult at present to determine quantitatively the function of the asphaltic membranes in various pavement structure applications, and how well they perform in comparison to other options. There is an absence of information in all that concerns the mechanical and hydraulic qualities of the membrane, as well as pavement-membrane interaction in the various applications. In order to increase the efficiency of this technology, it was necessary to: (1) establish the optimal material properties for the specific functions of the geotextile in the pavement structure; (2) develop engineering criteria for the production of the membrane and its integration in the construction and/or rehabilitation of pavements; (3) establish reliable standardized testing procedure for relating performance and longevity of the reinforced pavement structure to the physical properties of the membrane; and (4) conduct a laboratory simulative test and demonstration field project for performance evaluation.

Preliminary technological effort in this respect has recently been made in Israel. Geotextile-reinforced asphaltic membranes have been practically applied in several highway and airport paving projects. These site applications are also accompanied by an experimental simulative laboratory research which is currently being carried out in the Transportation Research Institute of the Technion in Haifa (Kiff et al., 1991).

This paper describes the preliminary phase of this technology, namely, the practical in-situ application of geotextile-reinforced asphaltic membranes.

The Asphaltic Membrance

The asphaltic membrane in question is a prefabricated asphalt sheet consisting of upper and lower layers of polymer-modified asphalt cement mixed with a filler. These layers are separated and reinforced by a woven or non-woven geotextile fabric.

The membrane is fabricated by conveying and immersing a fabric strip through a heated asphalt bath in a continuous movement. The main matrix of the membrane is paving asphalt cement. Its viscosity or penetration grade is dependent on the specific role of the membrane and the climate. The asphalt cement is odified by a polymer amounting at least 2% by weight, and preferably in the range of 4% to 10%. The added polymer modifies the asphalt by imparting to its visco-elastic properties which enable it to undergo elongation of more than 100% and even up to 1500% without cracking. The polymer is usually selected from an elastomer and preferably from a thermoplastic rubber such as Styrene-Butadiene-Styrene (SBS) rubber, Styrene-Isoprene-Styrene rubber, a polyethylene such as polypropylene, polyethylene, or any other common polymer possessing elastomeric properties suitable for blending with asphalt.

Through the asphalt bath, an inert filler is added to the membrane in the manufacturing process. The filler can be of any mineral rock source, the most common of which are calcium carbonate (limestone or dolomite) or quartz. The desirable amount is between 5% to 25% of the asphalt weight.

The reinforcing fabric is usually a durable woven or non-woven geotextile fabric. The fabric should be elastic with an elongation capability between 30% - 70% and a proper tensile strength for its specific use. The fabric should be durable at high temperature to resist the asphalt temperature during production or during field application. Usually polyester, polypropylene, or fiberglass fabrics are useful for fabrication and use of the asphaltic membranes. The mechanical properties of the fabric are a major factor affecting the mechanical and elastic properties of the asphaltic membrane, as shown in Figure 1. It can be seen that membranes fabricated with polyester are more elastic compared to equivalent fiberglass membranes.

The thickness of the asphaltic membrane can be between 0.5 mm to 10 mm, depending on the specific intended use. The most desirable and useful thickness in the range of 3 to 5 mm being also most desirable from an economical point of view.

In Israel, geotextile-reinforced asphaltic membranes were originally developed and manufactured by a local asphaltic products company. Based on extensive practical applications in roofing, and preliminary feasibility applications in paving, a set of quality criteria have been adopted by several transportation agencies, such as the Public Works Department, Israel, Israeli Airforce, and Israel Airports Authority. As an example, a set of quality criteria for a 3/250 asphaltic-polyester membrane are given in Table 1. The table also contains results of actual quality control tests performed on the membrane in one of the field projects.

Case Study Projects

General

Geotextile-reinforced asphaltic membranes have been practically applied in several highway and airport paving projects in Israel. These projects cover two types of membrane application, namely: subgrade sealing and retardation of reflective cracking. At this stage, these projects serve as feasibility case studies under controlled paving conditions and systematic follow-up thereafter.

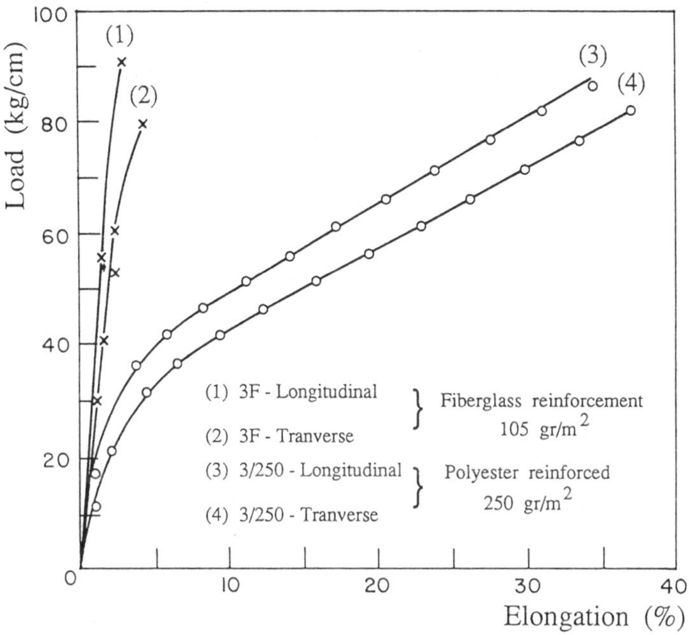

Figure 1: The influence of fabric type on the elastic behavior of the membrane

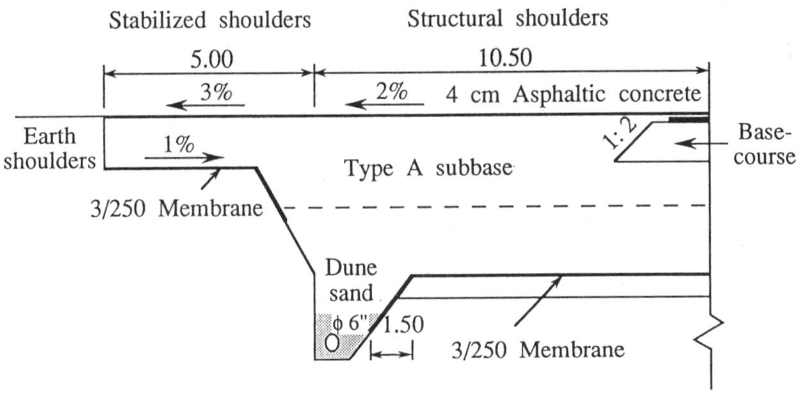

Figure 2: Application of asphaltic membranes on top of the subgrade in rapid exit-taxiway, Ben-Gurion International Airport - a typical cross-section

TABLE 1 - A SET OF QUALITY CRITERIA SPECIFICATIONS FOR A 3/250[1)] ASPHALTIC-POLYESTER MEMBRANE

PROPERTY	QUALITY CRITERIA	ACTUAL QUALITY CONTROL RESULTS
a. Dimensions		
1. Length (m)	10-15	
2. Width (m)	1.0	
3. Thickness (mm)	3.0±5%	3.0
b.Components		
4. Type of Reinforced Fabric	Polyester	
5. Weight per Area of Fabric (gr/m^2)	250	250
6. Penetration of the Asphalt Mass (1/10 mm)	25-35	
7. Ductility of the Asphalt Mass (cm)	>50	
8. Elongation of the Asphalt Mass (%)	>1300	
9. Type of Polymer in Asphalt Mass	S.B.S.[2)]	
c. Membrane		
10.Tearing Force in tension (N/5 cm) (According to DIN SL123)	Longitudinal >800 Transverse >750	1040 860
11.Elongation in Tension (%) (According to DIN SL123)	Longitudinal >40 Transverse >40	40 42.5
12.Water permeability when under pressure of 2 atm (according to DIN SL123)	No Cracking & No Leakage	
13.Stability in high temp. (^{o}c) (According to DIN SL123)	>100	
14.Flexibility at Low Temp. ($-20^{o}c$) (Acc.to DIN SL123)	No Cracks	No Cracks at 20°
15.Change in Rearing Force After 1000 hrs. of weatherly (UV) (%)	<5	3.4
16.Resistance to Static Punch at 25 kg. (Acc. to VEATC-MOAT 27/83)	No Leakage	
17.Welding Quality at Joints (Acc.to ASTM D 816)	Tear Outside The Welding Zone	

1) 3/250 - 3 mm thick membrane with 250 gr/m^2 polyester fabric reinforcement

2) S.B.S. - Styrene - Butadiene - Styrene

The following is a description of selected main projects:

Subgrade Sealing and Reinforcement:

1. Ben-Gurion Airport

A new parallel taxiway and a set of rapid exit taxiways were designed for the main 12-30 runway in Ben-Gurion International Airport. The active expansive clay (CH or A-7-6) and the traffic conditions called for a 110 cm pavement thickness.

The conventional pavement structure in this airport usually included a 5 cm thick dense bituminous concrete on top of the compacted subgrade. This layer is sealed with hot asphalt cement spraying. This sealing asphalt mat is protected by a 10 cm layer of fine uniform sand. The construction of such a system is costly and time-consuming. It is also almost impossible to pave the asphaltic layer in steep slopes adjacent to ditches or subsurface drains. The proper alternative to the asphaltic sealing layer is the geotextile-reinforced asphalt membrane. In this project, a 3/250 membrane (see Table 1) was laid directly on the top of the compacted clay subgrade extended to the entire lateral width of the runway down to the sub-surface drain. The specific membrane was "Polypaz" 3/250 as manufactured by Pazkar Company. An additional membrane was also laid underneath the stabilized shoulders and down to the drain on its other side, as can be seen in Figure 2 and in the photographs of Figure 3.

The laying procedure is very simple. The membranes are spread manually by rolling with a longitudinal overlap. The joints are later welded together by an open flame torch, as can be seen in Figure 4. The efficiency of welded joints can be tested by compressed air technique.

Because of the high resistance to tear and punching, no protective sand layer is needed. Practically all the bottom subbase course was laid and compacted exactly on top of the membrane. In general, the laying of the membranes and the construction of the entire pavement structure went along very quickly and smoothly. The fast covering of subgrade with the impermeable membrane ensured the preservation of the designed equilibrium moisture content achieved during the treatment and compaction of the subgrade.

The construction of two connecting exit-taxiways has been completed. At the time of writing, initial preparations are underway for the construction of the entire length of the parallel taxi-way, using the same pavement design and membrane application.

2. Kfar-Tabor Cross Highway

The new Kfar-Tabor cross highway was designated by the Israel Public Works Department (PWD) as a subgrade improvement experimental project. This highway section is located near the city of Afula in the middle of the 'Yizrael Valley' which is characterized by its heavy and expansive clay.

Figure 3: Application of asphaltic membrane beneath the pavement and adjacent the subsurface drain, rapid exit-taxiway, Ben-Gurion International Airport

Figure 4: Welding of membrane joints after laying

A similar 3/250 asphaltic membrane was placed directly on the treated subgrade right after compaction. It covers the entire cross section of the road, including the side drainage ditches, as can be seen in Figure 5. The photograph in Figure 6 shows the laid membrane while welding and paving the bottom subgrade layer.

Retardation of Reflective Cracking

1. General Procedure

As mentioned above, geotextile fabrics and geogrids are often used to control crack propagation when overlaying old cracked pavements. In this direction, the application of the geotextile-reinforced asphaltic membrane has the potential to provide a stress-relieving strong and flexible interlayer for the absorption, control and retardation of reflective cracking. Several pavement rehabilitation projects were also assigned as case studies for investigation of the effectiveness of these membranes for this pavement improvement use.

The application of the membrane can be made either on the entire cracked pavement surface area, or if the cracks are longitudinally delineated, it can be applied only in strips along the crack. Also, the membrane can be applied either directly on the top of the old pavement or alternatively, on the top of a pre-milled surface.

The following is a general procedure applied in several experimental projects:

1. Sealing wide and severe cracks by opening, cleaning and asphaltic sealing, cleaning the entire surface or the strip along a definite crack;

2. Applying a prime coat on the entire surface or in the crack strip (specifically, a CSS-1 emulsion at 300 gr/m^2 was used);

3. Covering the primed area or crack strips with 3/250 asphaltic membrane while heating and welding the membrane to the primed surface and to the overlapping joints;

4. Rolling the membrane by pneumatic rollers to enhance adhesion contact and to expel trapped air bubbles;

5. Spraying an asphaltic tack-coat on the entire surface to be overlaid, including membranes and exposed surface (specifically, a special tack-coat emulsion - STE - was used at a rate of 400 gr/m^2).

6. Asphaltic overlaying according to design.

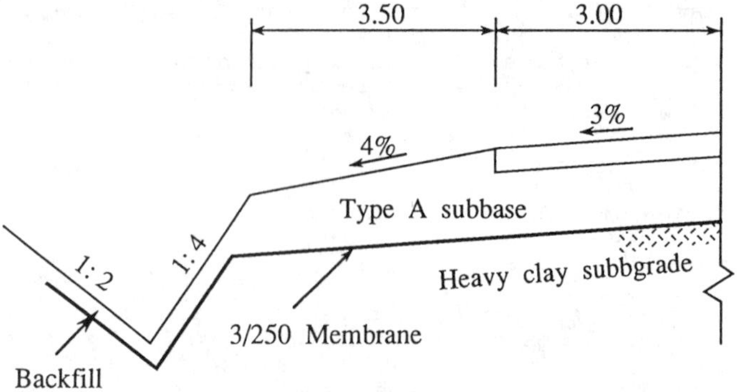

Figure 5: Application of asphaltic membrane on top of clay subgrade on Kfar-Tabor cross highway

Figure 6: Welding of membrane joints and application of first subgrade layer on Kfar-Tabor cross highway

2. Actual Projects

In all experimental projects using asphaltic membranes for retardation of reflective cracking, the membranes were applied only after milling the cracked and deteriorated pavement surface down to a depth of 2-5 cm. Then the outlined procedure was followed.

Both in the rehabilitation of a section in highway No. 2 (Tel-Aviv-Jerusalem Freeway), and a runway in Tel-Nof airport, the membranes were applied in strips along the well-delineated cracks (see Figure 7). On the other hand, in the rehabilitation of a runway in Ramat-David airport, the membranes were applied on the entire pavement surface area after milling the old weathered surface asphaltic course (see Figure 8).

Technically and logistically, the application of the asphaltic membranes and the paving of the asphaltic concrete on top, went along quite smoothly without any hindrance of pulling of the membrane by asphalt finishes.

Control And Follow-up

Most case study projects were carefully controlled during construction. Also, a follow-up system was applied for periodic pavement performance monitoring.

During the construction of the subgrade sealing projects, tests were performed just before the membrane coverage for monitoring subgrade density, moisture and strength (using Dynamic Cone Penetrometer -DCP). Similar tests were carried out on a specific control section without membranes. In the reflection cracking projects, a mapping of existing cracks was made prior to membrane coverage and overlaying. Control sections without membranes were also included. The specific locations of membrane strips were also recorded.

The follow-up system of the subgrade sealing projects will include periodic monitoring and comparison of pavement condition using the "PAVER" system, measuring pavement strength profiles using the DCP, and sampling of subgrade upper layer for variation in moisture and density. In the reflection cracking projects, monitoring and a comparison of pavement condition and crack mapping will be periodically performed.

Summary

This paper describes a preliminary technological effort of using geotextile-reinforced asphaltic membranes for subgrade and pavement improvement. The membranes were applied either as a sealing and reinforcement membrane between the subgrade and the pavement and also as a reinforcement membrane for the retardation of reflective cracking when overlaying existing cracked pavement surfaces.

Figure 7: Application of strips of asphaltic membrane along cracks on top of milled old asphaltic surface (Highway No. 1).

Figure 8: Application of asphaltic membrane on the entire area on top of milled old asphaltic surface (Ramat-David Airport)

It is believed that this technology will provide a better solution to pavement damages which occur due to swelling, shrinkage and collapse of active and sensitive soils, on one hand, and to the premature reflective cracking in pavement overlaying, on the other hand.

All projects described in the paper were designed and constructed during 1989-1991. A close control and follow-up system was applied for monitoring pavement performance with time under actual service and climatic conditions.

References

Button, J.W. (1984). "Evaluation of Fabric Interlayer -A Condition Survey." Research Report 261-2, Texas Transportation Institute, the Texas A&M University System.

Colombier, G., Lartant, M., & Machet, J.M. (1986). "Reflective Cracking: A New Test for SAMI." Proc., 3rd International Conference on Geotextiles, Vienna, Austria, June, 1986, pp. 105-108.

Joseph, P.E. & Haas, R. (1989). "Evaluating Alternative Solutions to Reflective Cracking Through Asphalt Overlays." Transportation Research Record No. 1215, Transportation Research Board, pp. 202-291.

Kasiff, G., Livneh, M., & Wiseman, G. (1969). "Pavement on Expansive Clays." Jerusalem Academic Press.

Kiff, O., Livneh, M., & Ishai, I. (1991). 'Use of Asphaltic Membranes in Flexible Pavements." Final Report (Draft), Research Project No. 115-090, Transportation Research Institute, Technion, Haifa, 1991.

McKeen, R. (1980). "Field Studies of Airport Pavements on Expansive Clays." Proc., 4th International Conference on Expansive Soils, Denver, Colorado, June 1980, pp. 241-261.

Steinberg, M.L. (1980). "Deep Vertical Fabric Moisture Seals." Proc., 4th International Conference on Expansive Clays, Denver, Colorado, June 1980, pp. 383-400.

Weston, D.J. (1980). "Expansive Roadbed Treatment for South Africa." Proc., 4th International Conference on Expansive Soils, Denver, Colorado, June 1980, pp. 339-360.

Experimental and Theoretical Study of Flexural Behavior of Polymer Fiber Reinforced, Cement-Treated Soils

Robert Liang[1], Member, ASCE

Abstract

The fracture resistance to cracking and post-peak flexural behavior of polymer fiber reinforced, cement-treated soils are studied both experimentally and theoretically. A series of 3-point beam tests provide quantitative information about the effects of variables such as fiber length and fiber content. Hillerborg's cohesive crack model is described and applied successfully to predict the load-load line deflection of a 3-point beam specimen. Determination of material constants in the stress-separation curve is also illustrated.

Introduction

The cement-treated soils have shown increased strength and stiffness over the untreated materials, and therefore can be used as an ideal construction materials for various applications such as highway base and subbase, slope protection for dams and embankments, liners for channels, reservoirs, and lagoons, and mass soil cement placement for dikes and foundation stabilization. However, one of the major problems with cement treatment soils is the more pronounced brittle behavior and the susceptibility to stress- and thermal-induced crackings. This brittle behavior has sometimes hindered applications of cement-treated soils or led to a more conservative design approach.

Incorporation of randomly distributed discrete fibers in an otherwise brittle material has been shown to improve the composite's mechanical properties. In particular, fracture resistance to cracking, post-peak load-

----

[1]Associate Professor, Department of Civil Engineering, University of Akron, Akron, OH 44325-3905

carrying capacity, resistance to weather-induced deterioration have been improved drastically by fiber reinforcing effect. From pavement structural analysis point of view, one of more significant improvements of fiber inclusions in cement-treated soils is the enhanced ductility and post-crack load-carrying ability. The main focus of this paper is therefore directed toward experimental and theoretical study of these characteristics.

The purpose of this paper is to present the findings pertaining to the following: (1) the results of a series of 3-point beam tests on fiber reinforced, cement-treated soil specimens; (2) the bond strength between fiber and cement-treated soil matrix as derived from the direct shear test techniques; (3) the applicability of Hillerborg's cohesive crack model to fiber reinforced, cement-treated soils. The characteristics of soils and polymeric fibers used in the present study are described, followed by a brief summary of sample preparation procedure. The succinct features of modified cohesive crack model for fiber-reinforced composites are provided. Furthermore, a procedure for determination of bond strength from direct shear test results is described.

Materials, Mixture Proportions

The materials used consisted of ASTM Type III cement, silty soil, and polyester fibers. The silty soil can be classified as AASHTO A-4 soil, with liquid limit = 27 and plastic limit = 25. Using the maximum density and optimum moisture test of the soil-cement mixture as outlined by ASTM D558 (maximum dry unit weight = 120 pcf, optimum moisture content = 14%), the minimum cement content was determined to be 12% by weight of dry soil.

The fibers used in the study are 11# gauge (nominal diameter = 0.078 in.) high-strength polyester wires. There was no surface treatment of fibers. The fiber has average initial Young's modulus of 2×10^6 psi and the breaking strength of 7×10^4 psi.

All the specimens were prepared using a small batch mechanical mixing method. The proportions of cement and water by weight of dry soil are 12% and 14%, respectively.

Cohesive Crack Model

A cohesive crack model originally proposed by Hillerborg, et al (1976) is adopted herein to study the crack propagation resistance of fiber reinforced, cement-modified soil composites. The model has been successfully used by Wecharatana and Shah (1983), Li and Liang (1986), among others, for studying the fracture resistance of fiber reinforced concrete. The succinct features of the cohesive crack model will be described in this section.

Under Mode I crack opening as shown in Fig. 1, the cohesive crack model assumes that the total crack length consists of a traction-free crack and a process zone. The process zone sustains two types of closing pressures: one due to fiber bridging and the other due to microcrackings in the matrix. For simplicity, however, these two effects can be combined and characterized effectively by a single stress-separation curve.

The stress-separation curve, or sometimes referred to as strain-softening curve (e.g., Li and Liang, 1986), constitutes one of the critical ingredients of the cohesive crack model in terms of its ability to capture the material's nonlinear fracture resistance behavior. The stress in the stress-separation curve represents the fiber bridging closing pressure; whereas, the separation indicates the crack-opening displacement. Typically, the stress-separation curve is established from uniaxial tension tests in a relatively stiff testing system. It has been shown that the post-peak displacements are essentially a result of a single crack. This means, during descending part of the stress-displacement curve of a uniaxial tension test, the total displacement of the specimen was the same as the width of a single crack. Therefore, the uniaxial stress-displacement relationship in the post-peak region can be taken as the closing pressure vs. the crack

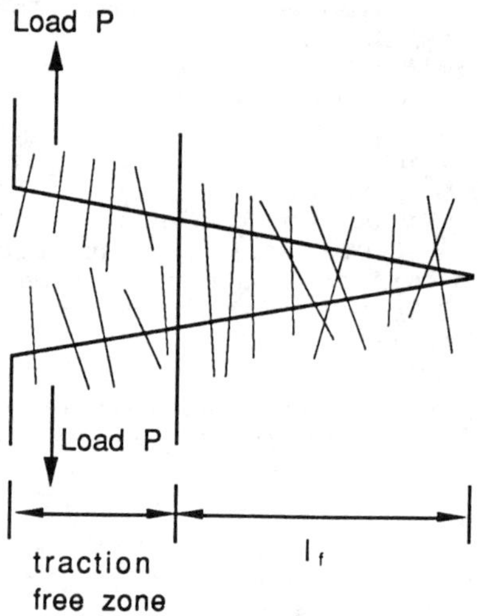

Figure 1. Schematics of Cohesive Crack Model

opening displacement needed for the cohesive crack model. Fig. 2 shows the procedure of deducing the stress-separation curve from the post-peak uniaxial stress-displacement curve.

In summary, the assumptions made in the cohesive crack model include the following: (1) the energy dissipation in the process zone contributes to the nonlinear fracture behavior, (2) the matrix cracking, fiber debonding and the eventual microcrack coalescing is a localized phenomenon in the process zone, thus the stress-separation type of constitutive law is applicable to describing the nonlinear behavior in the process zone. It should be pointed out that the cohesive crack model is more amenable to description of localized damage, whereas the J-integral approach is essentially a nonlinear deformation theory that cannot be applied to the fracturing involving substantial softening.

Although study on the stress-separation curve for the fiber-reinforced soil cement has not been reported before, a compilation of some available data for fiber reinforced concrete suggested that a normalized curve, as depicted in Fig. 3, can be obtained for various specimen sizes, fiber properties (aspect ratios, dimensions, fiber stiffness) and fiber volume content. In the normalized stress-separation curve, the stress is normalized with respect to the maximum post-cracking stress σ_{max} and the separation is normalized with respect to the maximum separation W_{max}. The general equation of the normalized stress-separation curve can be expressed as

$$\frac{\sigma}{\sigma_{max}} = \left[1 - \frac{W}{W_{max}}\right]^n \tag{1}$$

where n = parameter governing the nonlinearity of the curve. When n=1, the normalized stress-separation curve becomes a linear softening curve.

Regarding σ_{max}, there are many analytical and empirical expressions developed. In this paper, an approximation suggested by Swamy, et al (1983) is adopted

$$\sigma_{max} = \beta\,\tau\,V_f\,\frac{\ell}{d} \tag{2}$$

where β is a constant characterizing the statistical distribution of fibers in the specimen and therefore is dependent upon geometric and dimensional characteristics of fibers, τ is the average bond strength between fiber and cementitious matrix, V_f is the volume fraction of fibers, and ℓ/d is the length/diameter = aspect ratio of fibers.

Bond Strength from Direct Shear Test

The determination of bond strength is usually accomplished via a direct pull-out test. However, in our case the fiber possesses significant flexure rigidity to

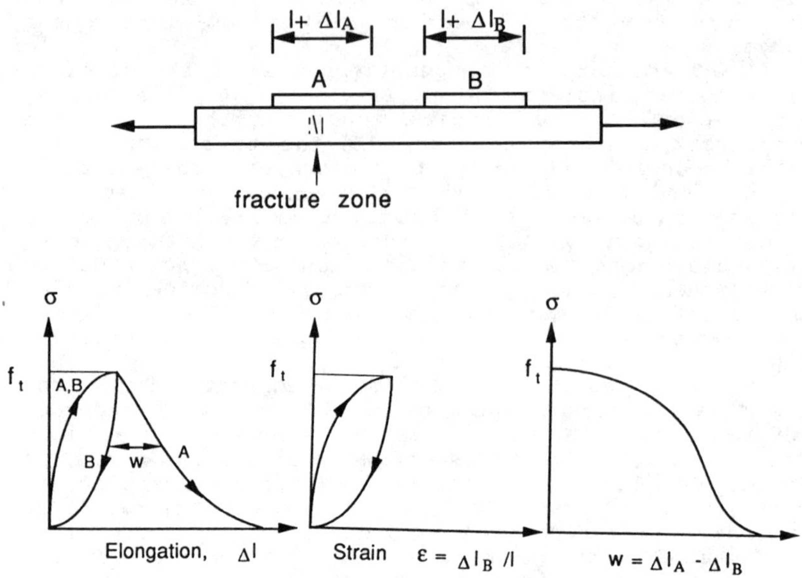

Figure 2. Procedure for Deducing Stress-Separation Curve

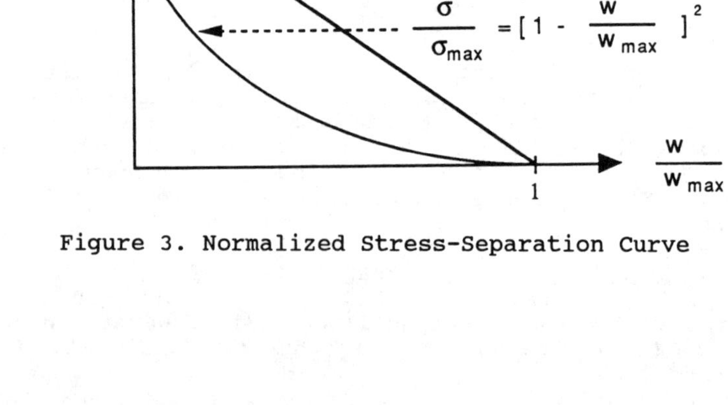

Figure 3. Normalized Stress-Separation Curve

warrant consideration of this additional energy absorption mechanism. In a beam specimen, where stable crack growth is present, it is certainly foreseeable that bending of fibers occur in the process zone. Therefore, a direct shear device is preferred over the direct pull-out test. This is because the fiber bends during debonding process in the direct shear test, thus representing more closely the energy consumption mechanisms in the process zone.

Fig. 4 shows a single fiber crossing the pre-defined shear plane in a direct shear box. Two separate plastic sheets were inserted in the pre-defined shear plane so that friction between the two halves of specimens during testing becomes negligibly small. When the upper half of the specimen is pushed horizontally tensile stresses are developed in the single fiber as a result of mobilization of bond stresses at the interface between fiber and soil cement. At the peak load P_{max} and the corresponding displacement Δ, it is assumed that the peak bond strength is fully mobilized and a limiting equilibrium condition is reached. Thus, the peak load P_{max} can be related to the maximum average bond strength τ_{max} as follows:

$$P_{max} = \tau_{max} \pi d \ell_o \sin\theta \tag{3}$$

where

$$\sin\theta = \Delta / \ell_o \tag{4}$$

The average maximum bond strength becomes

$$\tau_{max} = \left(\frac{P_{max}}{D}\right)\left(\frac{1}{\pi d}\right) \tag{5}$$

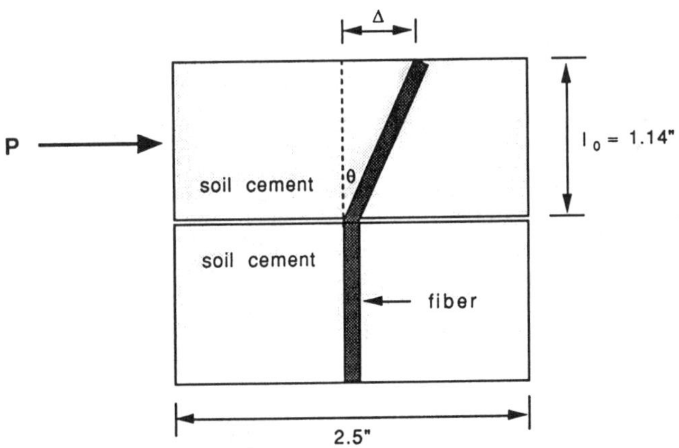

Figure 4. Direct Shear Test of Bond Strength

Experimental results of direct shear tests revealed that a large variation of the peak load value exists, as shown in Fig. 5. This large variation may be attributed to imperfect bonding between the fiber and the soil-cement matrix during sample preparation (i.e., compaction).

Further investigation on the reproducibility of direct shear test results is warranted. However, at present an average peak load of about 67 lbs at the corresponding displacement of 0.033 inch is adopted for calculation of the average bond strength in Equation 5. The average bond strength is determined to be about 647 psi.

<u>3-Point Beam Tests</u>

The variables included in the 3-point beam test program are: fiber length (0.5 in. vs. 1 in.), volume fraction of fibers (0%, 1%, 3%, and 5%). However, the cement content, water cement ratio, and compaction effort remain the same for all the samples prepared.

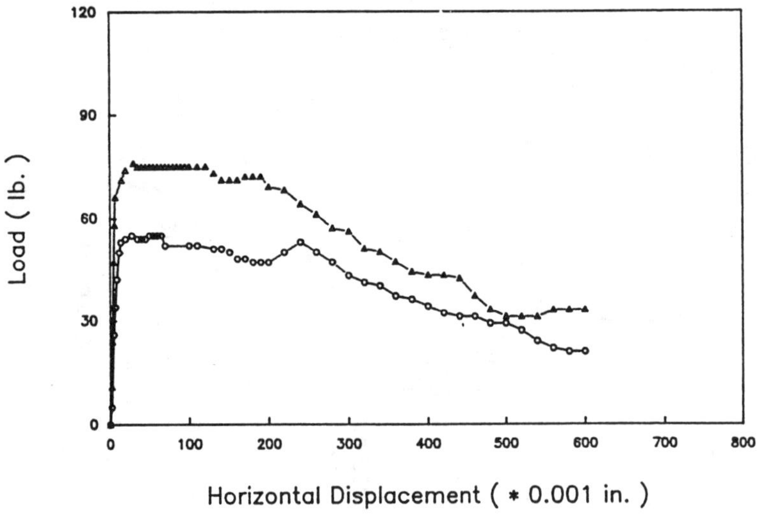

Figure 5. Typical Direct Shear Test Results

Figs. 6 and 7 show load-load line displacement curves for beam specimens mixed with 0.5 in. and 1.0 in. long fibers, respectively. The fiber length, within the range investigated, does not seem to affect significantly the peak load and the corresponding displacement. On the other hand, the volume content of fibers does seem to affect the load-carrying capacity of beams - the larger the volume content, the smaller the peak load. One apparent reason for this phenomenon can be attributed to the fact that as soon as fiber content exceeds a threshold value (about 1% in this case), the distribution of fibers tends to align themselves in a more or less preferred orientation along the axis of beam specimens, thus totally destroying the desired randomness of fiber distribution. Furthermore, this preferred fiber alignment tends to provide a weak plane in the beam specimen, thus allowing cracks to propagate along the interface between fiber and matrix. Further development of appropriate models is

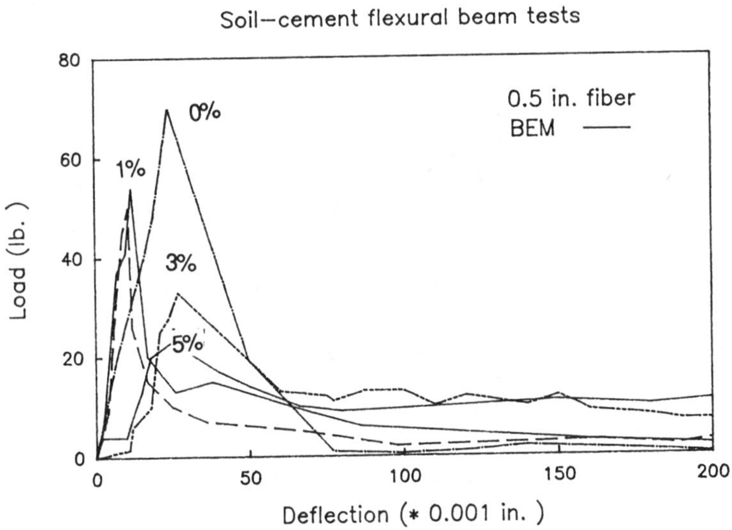

Figure 6. 3-Point Beam Test Results
(0.5 in. Long Fiber)

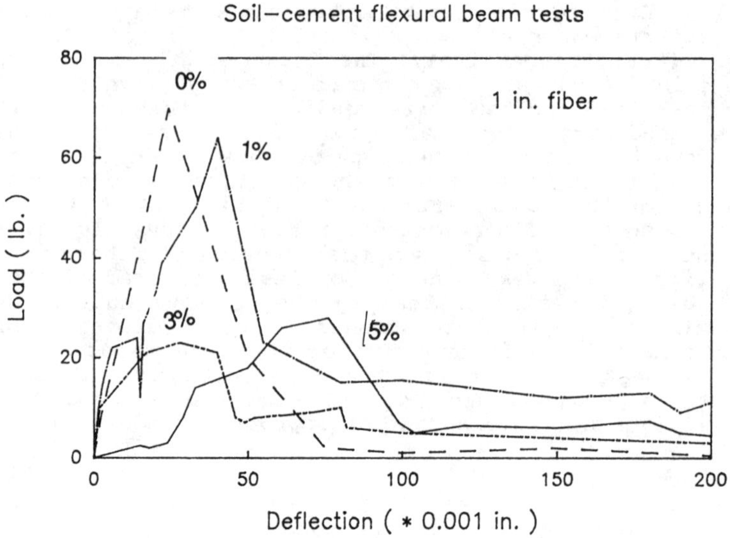

Figure 7. 3-Point Beam Test Results (1 in. Long Fiber)

needed to address the observed failure mode which is quite different from Hillerborg's cohesive crack model. Nevertheless, for beams with fiber content of 1%, it appears that fiber distribution is random and Hillerborg's approach seems to apply. In the section that follows, an exercise is made to examine the applicability of Hillerborg's cohesive crack model in predicting the fracture resistance behavior of fiber-reinforced soil cement in 3-point bending.

BEM Simulation of 3-Point Beams

In this section, a recently developed boundary element program (Liang and Li, 1991) is used to simulate load-displacement behavior of fiber reinforced soil cement specimens in 3-point bending conditions. The program adopts a linear softening stress-separation curve to characterize the behavior in the process zone. Numerical algorithms for both force-controlled (prescribed loading history) and crack-tip-controlled (prescribed crack tip position) have been implemented to allow for a whole range simulation, including strain-softening and snap-back. In the force-controlled approach, the propagation of crack is simulated via node release technique. Because crack opening displacement in the process zone depends on the traction force; therefore, an iterative process is involved until all nodes have reached their equilibrium position and the calculated tip node tensile stress is less than the tensile strength of the material. A nodal coordinate adjustment scheme is incorporated in the program to allow moving crack tip position such that the calculated tensile stress is exactly equal to the specified tensile strength.

The crack-tip-controlled approach is basically to address the issue of determining the corresponding external load for a given crack profile. This approach avoids numerical difficulties encountered in strain-softening material.

Basically, four material constants need to be prescribed in the program: Young's modulus, E, and Poisson's ratio, ν, of soil cement, threshold separation W_{max}, and maximum closing stress σ_{max}. The Young's modulus E is determined from a series of unconfined compression test and found to be 20,000 psi. The Poisson's ratio is simply assumed to be a typical value of 0.1. The threshold separation W_{max} is taken as half of fiber length. Finally, the maximum closing stress is calculated from Eq. (2), using bond strength τ = 647 psi, and β = 1. Based on these material constants, numerical simulation results are obtained for a 3-point beam specimen with fiber volume fraction V_f = 1% and fiber length ℓ = 0.5 in. The numerical simulation results are compared with experimental results in Fig. 7.

Conclusions

Presented in this paper is a preliminary study about the fracture resistance characteristics of polymer fiber reinforced soil cement composites. One of the unique features of the present study is that the polymer fibers possess high flexure rigidity due to large diameter of the fiber and special fiber manufacturing process (aligned extrusion process).

Due to the large size of fibers, it seems that a minimum of 1% of volume fraction is needed to ensure the random distribution of fibers in the matrix. Although more than 1% of fiber content can be compacted via conventional mechanical compaction technique, the fracture failure mode in those high fiber-content composite beams deviates from Mode I. Thus, it appears that the high denier, high flexure rigidity fibers may not be as useful as those monofilament, low denier fibers in terms of enhancing soil-cement's fracture resistance. Although clearly more experimental as well as theoretical work is needed, the following conclusions can be drawn from the results reported herein.

1. For the specific polymer fiber studied, the fiber content affects the failure modes of 3-point beam specimens. A fiber content of 1% (volume fraction) allows for process zone, to exhibit fiber bridging behavior. Beyond this threshold value, it seems that splitting along the interface between fiber and matrix is a dominate failure mode, thus invalidate the basic assumptions in the cohesive crack model.

2. Bond strength between high flexure rigidity fiber and cement-treated soil matrix can be derived from a direct shear test using a simple limiting equilibrium analysis. However, the sample preparation procedure seems prone to producing imperfect bonds due to non-uniform compaction. As a crude approximation, the bond strength for the conditions studied herein was found to be about 647 psi.

3. Experimental results of 3-point bending specimens showed that only the specimens with 1% fiber volume fraction satisfied the basic assumptions of cohesive crack model. A satisfactory comparison between model predictions and test results for this specific fiber content seems to support the applicability of the model. However, this is by no means an advocation of the model for all cases.

Acknowledgement

Suck-Hong Lee and Rhonda Boyd performed the tests described in this paper.

References

Hillerborg, A., Modeer, M., and Peterson, P.E. (1976), "Analysis of crack formation and crack growth in concrete by means of fracture mechanics and finite elements", Cement and Concrete Research, Vol. 6, 773-782.

Li, V.C., and Liang, E. (1986), "Fracture process in concrete and fiber reinforced cementitious composites", J. Engrg. Mech., ASCE, 112(6), 566-586.

Liang, R.Y. and Li, Y.N. (1991), "Simulation of nonlinear fracture process zone in cementitious material - a boundary element approach", Int. J. Computational Mechanics, 7(5).

Swamy, R.N., Mangat, P.S., and Rao, S.V.K. (1983), "Fiber reinforced concrete", ACI SP-44, 1-20.

Wecharatana, M. and Shak, S.P. (1983), "A model for predicting fracture resistance of fiber reinforced concrete", Cement and Concrete Research, 13, 819-829.

Soft Clay Subgrade Stabilisation Using Geocells

Mhaiskar S.Y.[1] , Mandal J.N.[2].

Abstract

The present study is aimed at investigating the efficacy of the geocell structure on a soft clay and studying three important parameters affecting its performance. They are the width of geocell [a] and the height of the geocell [b], effect of strength of the geocell material and relative density of the fill in the geocell.

Experimental and Finite Element [F.E.] procedures have been adopted to study the above parameters. Soft saturated marine clay was used as subgrade while sand was used as backfill material. Monotonic loading using a plate to represent on-highway loading conditions was used in plate load tests. The experimental results were simulated in a three dimensional F.E. procedure using ANSYS. [a general purpose F.E. package].

Considerable amount of improvement in the ultimate load and reduction in settlement was observed from the experimental results. Close agreement has been found between the experimental and F.E. results. The results of the F.E. analysis are used to study the improvement in stiffness.

Introduction

Geotextiles are being increasingly used in road construction and tank foundations. Numerous pavement design methods [Giroud and Noiray (1981), Sellmeijer et.al (1982), Milligan et.al. (1989)] have been suggested

---

1. Research Scholar, I.I.T., Bombay (Faculty Member of Sardar Patel College of Engineering, Bombay-400 058).

2. Associate Professor, Civil Engineering Dept., I.I.T., Powai, Bombay - 400 076.

for unpaved roads which are characterised by high allowable rut depth, low volume of traffic and no vehicle wander. Nagaraju et.al. (1990) have suggested the use of soil filled tubes for paved roads while Kazerani et.al. (1987), and several other investigators have suggested the use of geocell for paved roads. De Garidel et.al. (1990) have used geocells of width to height ratio of a/b = 0.5 and reported that the bearing capacity can be improved two to three times. Similar work has been reported by Shimizu et.al.(1990). However, according to them an a/b ratio of 0.66 is optimum. Guido et.al. (1990) have also reported the beneficial use of a texturized geocell with a a/b ratio of 1.0. Bush et.al. (1990) have reported the use of a geogrid geocell embankment on soft soil.

For paved road applications (characterised by low allowable rut depth and high volume of traffic) the geotextile should yield high resistance at low strains i.e. be of high modulus. High modulus geotextiles are not manufactured in India. Besides, the cost of low modulus nonwoven geotextiles is prohibitively high. Therefore, other alternatives need to be investigated.

Geocells which have axisymmetric cells have been proved effective for such applications. From the literature survey it can be observed that limited studies have been carried on a geocell structure on soft clay. Therefore, the present study aims at investigating three significant factors affecting the geocell design. These are, effect of the geometry of the geocell, effect of strength of the geocell material and the effect of relative density of the backfill. Experimental and Finite Element methods have been used to study the improvement in the load carrying capacity.

Experimental Study

An experimental model was fabricated to study the effect of cell width to height (a/b), strength of the geocell material and the relative density of the backfill on the geocell structure. It was designed in such a manner that full scale loading could be carried out. However, in the present study the geocell is scaled in such a way that the a/b ratio is varied while 'b' is maintained constant (10 cm.).

The dimensions of the mild steel tank used in the experiments were 85 cm. (length) by 75 cm. (width) by 60 cm. (height). The tank was placed under a loading frame and was filled with marine clay upto a depth of 48 cm. The geocell was placed on this clay subgrade. A loading plate 25x35 cm. was used to represent a dual wheel assembly subjected to a 80 KN load and 6.2 kg/cm^2 tyre pressure. Figure 1 shows the experimental setup. The

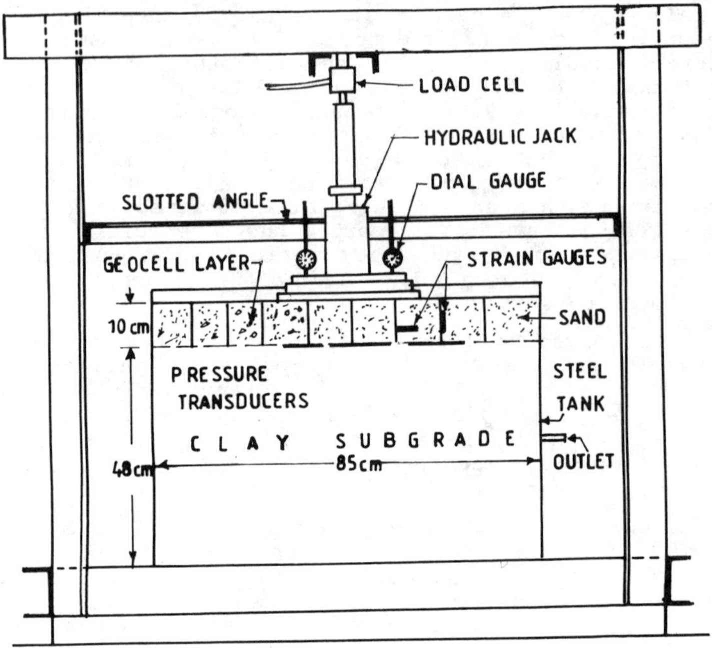

FIG. 1. Experimental Set-up

subgrade (marine clay) was placed in eight layers of six cm. thickness for the entire depth of 48 cm. Each layer was compacted by giving three blows of a Modified Proctor hammer (10 lbs. weight falling through a height of 18 inches) over an area of 225 cm^2 plate. Soil required for each layer was weighed to achieve a dry density of 10.4 KN/m^3 at a moisture content of two percent above the optimum. Wet soil was contained in a frame before compacting. The dimensions of this frame were 62 cm. by 75 cm. by 8 cm. The space between this frame and the tank was filled with sand to facilitate saturation. The tank was flooded for seven days to attain maximum saturation. This method of compaction and saturation yielded a degree of saturation of 93% and a cohesion of 11 kPa.

Two types of geotextiles were used for the geocell namely, needle punched nonwoven and woven slit film type to compare the effect of strength of a geocell wall on the load carrying capacity. Initial set of experiments (in the loose state) were carried out with needle punched nonwoven type only. This data was then used to optimize the cell geometry. Only the optimised sizes were used for the woven slit film type and nonwoven geocells in the dense state experiments. The cell sizes used were a/b = 0.5, 1.0, 1.5, 2.0 in the loose state while they were a/b = 0.5 and 1.0 in dense state. Photograph 1 a. 1 b. show typical type of geocell structure.

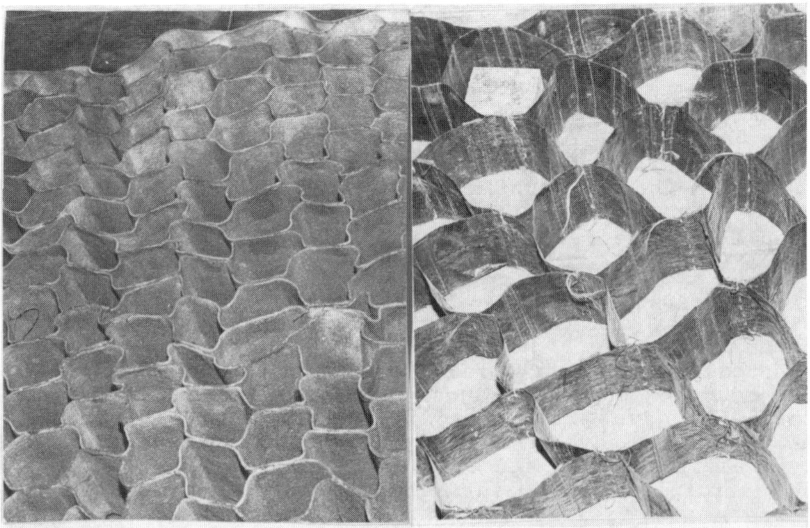

(a) (b)

Photograph 1. Typical type of geocells (a) Nonwoven and (b) Woven

The nonwoven type geocell was made by bonding (one inch length) with a heat bonded adhesive (AIKAMELT - 20), while the woven geocells were stitched by nylon threads. The handles of the frame for containing the soil prior to compaction were used to expand the geocell before placing the sand in the cells. The sand was filled at two different relative densities viz. 15% and 80%, to study the effect of relative density. The loose density (15%) was achieved by pouring the sand through a free fall of 15 cm., while the dense state (80%) was achieved by rodding the sand with a tamping rod 500 gm. in weight for 25 times for a geocell of a/b = 0.5 or 100 times for a geocell of a/b = 1.0. The loading plate was directly placed on the geocell. The load was applied by a hydraulic jack and recorded by a load cell of 20 kN capacity. The settlement was recorded by three dial gauges (0.01 mm.

least count) symmetrically placed on the plate. The load was applied in increments till the footing exhibited a punching failure. Settlement corresponding to each increment was recorded.

Pressure transducers were placed at the interface between the geocell and the clay subgrade to observe the interface stresses.

Horizontal and vertical strain gauges were mounted on the geocell wall below the loaded area to study the nature of geocell deformation. Paperback, 50 mm. active length, single wire type strain gauges were bonded by an elastic adhesive (FEVIBOND). The adhesive being elastic transferred strains to the gauge effectively. Initial experiments were carried out to compare actual strains in the geotextile and those measure by the strain gauges. A uniform method of mounting the strain gauges was adopted for all the experiments. Negligible variation between the actual and the measured strains was observed upto 2.5% approximately, beyond which a large variation was noted.

Adequate amount of adhesive was applied on the strain gauge and the geotextile for bonding and allowed to dry for about 10 minutes. The strain gauges were then mounted on the geotextiles. Same procedure was followed for mounting copper bridges. Lead wires were welded the following day. Similar studies carried out by Sluimer et.al. (1982) have indicated that strains upto about 10% can be effectively registered with strain gauges. Fixing the strain gauges, maintaining them while placing the geocell and filling it with sand caused many strain gauges to be inactive at the time of the test. However, useful observations were made in a few tests.

The Mumbra sand used as a backfill had a minimum density of 16.05 kN/m^3, a maximum density of 18.1 kN/m^3, a Uniformity coefficient C_u of 4.6, while the marine clay used as a subgrade had a Liquid Limit of 80%, Plastic Limit of 40%, an optimum moisture content of 32% and a maximum dry density of 13.85 kN/m^3.

The nonwoven geotextile used for the geocell was made up of polypropylene fibres and had a weight per square metre of 300 gm., a breaking load of 8.35 kN/m, a failure strain of 40%, a thickness of 2.35 mm., a bursting strength of 100 kPa and a seam strength of 0.35 kN. The woven slit film type geotextile had a weight per square metre of 136.5 gm., a breaking load of 6.5 kN/m, a failure strain of 10%, a thickness of 0.4 mm. and seam strength of 0.16 kN.

Numerical Study

ANSYS a general purpose Finite Element (F.E.) package was used for the present study. It is based on classical engineering concepts and documented F.E. and numerical techniques.

The typical three dimensional grid used for the F.E. analysis is shown in Figure 2. As the problem was symmetric only the symmetric quarter was used in this analysis. An eight-noded iso-parametric solid element was used to simulate the geocell layer and the clay subgrade in a three dimensional analysis. The element had anisotropic and plasticity capabilities.

The geocell layer filled with sand and clay (subgrade) were modelled as isotropic, elastic-perfectly plastic, non-dilatant material (associated flow rule was used). The Drucker Prager yield criteria which uses the outer cone approximation to the Mohr's Couloumb theory was adopted for the analysis. Due to the uncertainties in the boundary constraints and difficulties in modelling stress-strain relationship for the soil-geocell composite system, the geocell layer was considered as a layer with an equivalent stiffness. [Kazerani et.al. (1987)]. The unreinforced sand layer was 10 cm. thick. The modulus of elasticity of sand varies with confining pressure i.e. it increases with increasing depth, while it is zero on the surface. In the present study the average of the modulus of elasticity at the surface and that at a depth z = 10 cm. was used. The modulus of elasticity was calculated with the help of the equation [Duncan and Chang (1970)].

$$E_t = E_i \left[1 - \frac{R_f \ (\sigma_1 - \sigma_3}{(\sigma_1 - \sigma_3)_{ult.}} \right]^2$$

where $E_i = K. P_a \ (\sigma_3/P_a)^n$

and $(\sigma_1 - \sigma_3)_{ult.} = 2 \ \sigma_3 \ \text{Sin} \ \emptyset \ / \ (1\text{-Sin} \ \emptyset)$

(E_i = initial tangnt modulus, E_t = secant modulus, R_f = failure ratio, P_a = atmospheric pressure, K = modulus number, $\emptyset$ is the angle of internal friction, σ_1 = major principal stress and σ_3 = minor principal stress, n = constant). For the present case [K (assumed) = 250 since the sand is resting on a soft clay subgrade, P_a = 100 kN/m^2, γ = 16 kN/m^3, R_f = 0.85, $\emptyset$ = 35°, n = 0.5]. With the above assumptions $E_{avg.}$ works out to 520 kN/m^2 $\simeq$ 500 kN/m^2.

Kazerani et.al. (1987) have introduced a Clegg hammer to measure the in-situ modulus. Due to the confinement of the soil in the reinforced case, the poissons ratio is also reduced. In the present study the modulus of elasticity of reinforced sand was increased in steps in the F.E. analysis till a fairly good agreement was found in the load settlement curve obtained from the experimental results. The properties used in the F.E. analysis are listed in table 1.

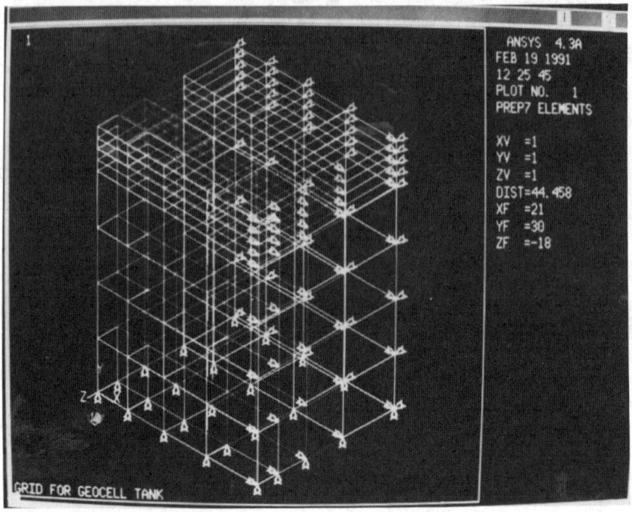

FIG. 2 Typical three dimensional grid used for F.E. Analysis.

Table 1. Summary of Properties used in F.E. Analysis

Modulus of Elasticity of reinforced sand = 1000 kPa
Poisson ratio of reinforced sand = 0.17
Modulus of Elasticity of unreinforced sand = 500 kPa
Poisson ratio of unreinforced sand = 0.35
Modulus of elasticity of clay (subgrade) = 500 kPa
Poisson ratio of clay subgrade = 0.48

Results and Discussions

Load settlement curves were drawn for all the experimental and Finite Element (F.E.) procedures both for unreinforced and reinforced cases. These curves were drawn on a semilog scale, in a dimensionless form ($Q/A \gamma B$ vesrus w/B, where Q is the applied load, A the area of the loaded plate, γ density of the backfill, B the width of loaded plate and w the settlement of the plate). The failure load was taken as the projection of the intersection of the initial tangent to the load settlement curve and the post yield curve on the x-axis.

Figure 3 shows load settlement curves for nonwoven type geocells in a loose state (relative density of 15%)

for a/b ratios of 0.5, 1.0, 1.5, 2.0 (a = width of the geocell, b = height of the geocell). It can be seen from Figure 3 that as the a/b ratio decreases the Ultimate Bearing Capacity (U.B.C.) and consequently the Bearing Capacity Ratio (B.C.R.) increases. However, the difference in the geocells of a/b 0.5 and 1.0 or that between geocells of a/b 1.5 and 2.0 in the elastic range is negligible. Therefore, for further studies on dense state conditions (relative density 80%) the geocell sizes were restricted to a/b of 0.5 and 1.0. It should be noted that the a/b was varied by changing 'a' the width of the geocell while 'b' the height of the geocell was kept constant (10 cm.).

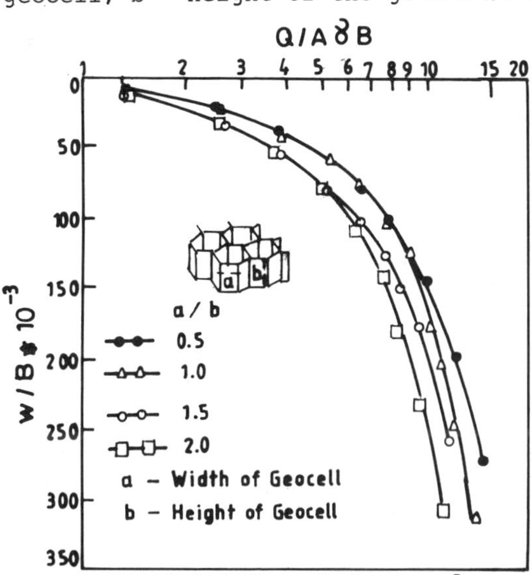

FIG. 3. Load Settlement Curves for Non-Woven Geocells in Loose State

Figure 4 illustrates load settlement curves for the non-woven geocell (a/b = 1.0) in loose (relative density 15%) and dense (relative density 80%) states. It can be inferred from Figure 4 that the BCR in a dense state is 1.35 while it is 1.33 in a loose state, indicating that only a marginal improvement in BCR is seen by compacting. However, compared to the unreinforced case (sub-grade plus loose state backfill) there is a substantial improvement of 47% (BCR = 1.47). Also w/B for a dense state (at Q/AγB of 7.3) is 0.060 while it is 0.080 for a loose state. It is therefore advantageous to compact the backfill in the geocell in a dense state.

Figure 5 shows load settlement curves for non-woven and woven slit film type geocells for an a/b ratio of 0.5. It should be noted that the dense state experiments were carried with a/b of 0.5 and 1.0 both for non-woven and woven slit film geocells. It is seen from Figure 5 that the woven slit film type which has a higher modulus compared to the non-woven type (though its seam strength is lower), shows a higher BCR of 2.41. As mentioned

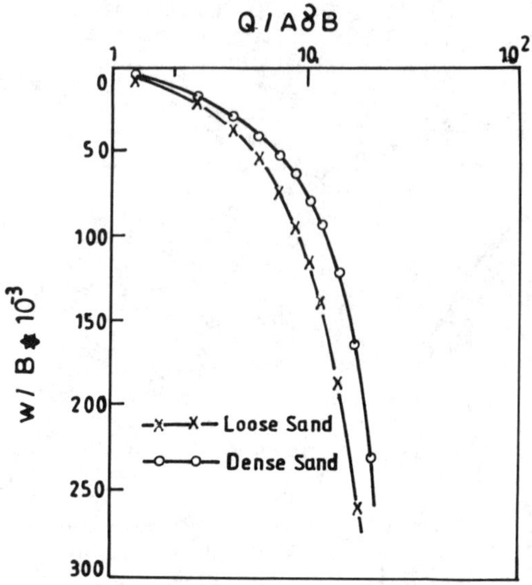

FIG. 4. Load Settlement Curves for Non-Woven Geocell (a/b = 1.0)

earlier, in the properties of geotextiles the failure strain for non-woven geotextile is 40% while that for the woven slit film type is 10%. It can be therefore inferred that a geotextile that has a higher modulus with less extensibility is more suitable for geocells.

Table 2 summarises the ultimate loads in a dimensionless from (Q/A γ B). From Table 2 it can be seen that as the cell size decreases the BCR increases.

Table 2 Summary of Experimental Results

Description of Test	a inches	a/b	Q/AγB					
			Non-Woven				Woven	
			Loose* State	BCR**	Dense@ State	BCR**	Dense@ State	BCR
Clay Subgrade	-	- -	4.5	-	-	-	-	-
Clay + Sand	-	-	5.5	-	5.8	-	5.8	-
Clay + Geocell filled with Sand	2	0.5	8.0	1.46	9.5	1.64	14.0	2.41
"	4	1.0	7.3	1.33	8.1	1.39	10.0	1.72
"	6	1.5	7.1	1.29	-	-	-	-
"	8	2.0	7.0	1.27	-	-	-	-

* Loose relative density = 15%, @ Dense relative density 85%

** BCR = Bearing Capacity Ratio i.e. the ratio of bearing capacity in reinforced to reinforced case.

However, from the view of making a geocell a/b = 0.5 seems to be a practical limit up to which geocell forming may be easy. Such a ratio would be easy to form especially for a 8 inch height geocell.

Figure 6 shows load settlement curves for unreinforced case in a dense state (relative density = 80%) as

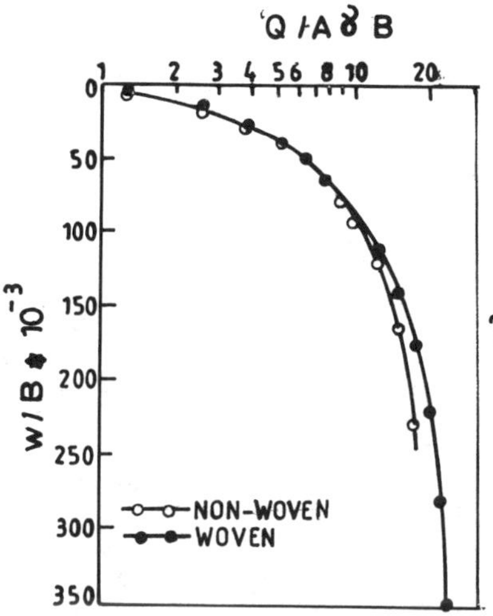

FIG.5. Load Settlement Curves for Geocell (a/b=0.5) in Dense State

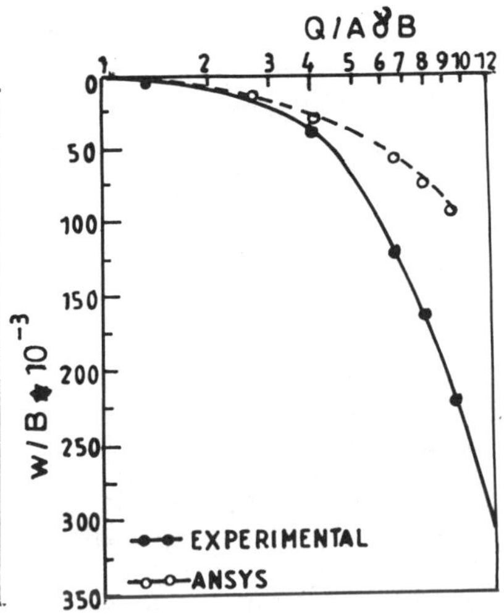

FIG.6. Load Settlement Curves for Unreinforced Case (Dense backfill)

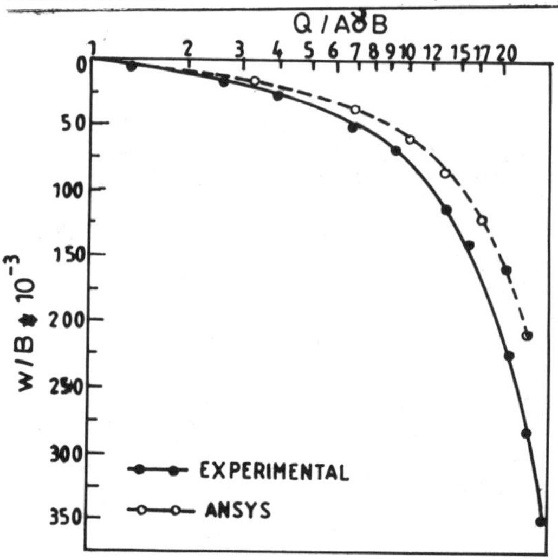

FIG.7. Load Settlement Curves for Reinforced Case with Woven slit Film Geocell a/b = 0.5 Dense Backfill

obtained from experimental and F.E. (ANSYS) procedures. It is seen from Figure 6 that there is a close fit in the elastic part of the load settlement curve. This indicates that the plasticity effects are not well accounted for in the F.E. procedure.

Figure 7 illustrates the load settlement curves for the woven slit film type geocell (a/b = 0.5) in dense state as obtained from experimental and F.E. procedures. The curve shows a good agreement in both elastic and plastic part.

The strain gauge readings mounted on the wall of the geocell below the loaded area have shown that the hoop strain (maximum 0.68%) is more than the vertical strain (maximum 0.13%). This shows that the hoop stress is a significant factor in the resistance to loads.

Table 3 shows the pressure transmitted to the clay layer as recorded by the pressure transducers in non-woven and woven slit film type geocells for a a/b ratio of 1.0.

Table 3 Summary of Pressures Transmitted to the clay at failure loads (observations from pressure transducers) in dense state (relative density = 80%).

Type of geocell	a/b	Applied Pressure Kg/cm^2	Pressures Recorded edge	centre kg/cm^2	edge
Woven	1.0	0.4526	0.312	0.227	0.2899
Nonwoven	1.0	0.4526	0.355	0.304	-

From the above table it is clear that the pressure distribution is as normally seen for a rigid footing on clay and woven type geocell disperses the load over a wider area compared to the nonwoven type, in view of the lower pressures transmitted.

Conclusions

The experimental and Finite Element (ANSYS) studied carried out on sand reinforced with geocell in loose and dense states over a soft clay subgrade at various width to height ratio (a/b) indicate that

(1) It is advantageous to compact the backfill to a dense state.

(2) As the a/b ratio decreases the U.B.C. and B.C.R. increase. However, from the point of making a geocell a/b = 0.5 seems to be a practical limit.

(3) A geotextile with a higher modulus and less extensibility is desirable as a geocell material.

(4) In the reinforced case F.E. and experimental results have shown good agreement in both the elastic and plastic part of the load settlement curve. However, in the unreinforced case plasticity effects are not well simulated.

(5) Experimental results show that the hoop stress in the geocell wall is the most significant contributing factor towards resistig loads. As the F.E. analysis considered the geocell reinforced sand layer as an equivalent layer, hoop stress in the geocell wall cannot be separately predicted and compared with the experimental results.

(6) The pressure distribution on the clay layer is as normally seen on a rigid footing resting on clay and the woven slit film geocell disperses the load over a wider area.

Acknowledgements

The authors are greatly indebted to Prof. B. Nag, Director, IIT, Bombay for his encouragement. The authors also wish to thank CAD Centre IIT, the Principal Sardar Patel College of Engg. and Soil Mechanics Laboratory staff for their help during the various stages of the investigations.

Appendix 1 - References

Bush, D.I., Jenner, C.G. and Basset, R.H. (1990) "The design and construction of a Geocell Foundation Mattress supporting embankments over soft ground", Geotextiles and Geomembranes, 9, (1990); 83-98.

De Garidel, R. and Morel, G. (1986), "New strengthening techniques by textile elements for low volume roads", Third International Conference on geotextiles, Vienna, 1027-1032.

Duncan, J.M. and Chang, C.Y. (1970), "Non-linear analysis stress and strain in soils", Journal of the Geotechnical Engineering Division, ASCE, 96, (SM5), 1629-1653.

Giroud J.P. and Noiray L., (1981), "Geotextile Reinforced unpaved Road design". Journal of the Geotechnical Engg. Division, ASCE, 107, GT9, 1233-1254.

Guido, V.A., Sobiech, J.P. and Christou, S.N. (1989), "A comparison of Texturised and Non-Texturized Geoweb Reinforced earth slabs" Geosynthetics 1989 conference,1, 215-230.

Kazerani, B. and Jamnejad, G.H., (1987), "Polymer grid cell Reinforcement in construction of pavement structures", Geosynthetics 1987 conference, 1, 58-68.

Khay, M., Morel G; and Perrier, H. (1986), "Use of geotextiles in construction of low cost highways: an experiment", Vienna,IV, 957-961.

Khay, M. and Perrier, H. (1990), "Sand confinement by geotextiles in Road construction", Fourth International Conference on Geotextiles, Geomembranes and Related Products; 255.

Milligan, G.W.E., Jewell R.A., Houlbsy G.T., Burd H.J., (1989), "A new approach to design of unpaved Roads", Ground Engineering; April 1989, 25-29.

Nagaraju S.S. and Mhaiskar S.Y. (1990), "Experiments on the use of Geotextiles and Geocomposites in Paved Roads, Indian Geotechnical Conference 1990; 89-92.

Sellmeijer J.B., Kenter C.J., Van den berg J. (1982), "Calculation method for a fabric reinforced road", Second International Conference on Geotextiles, Las Vegas, 393-398.

Shimizu, M., Inui, T. (1990), "Increase in the bearing capacity of ground with geotextile wall frame", Fourth International Conference on geotextiles, geomembranes and Related Products; 254.

Sluimer, G. and Risseeuw, P. (1982), "A strain technique for Measuring Deformations in Geotextiles", Proceedings of the Second International Conference on Geotextiles,Las Vegas, 835-838.

Reinforced Granular Soil For Track Support

G.P. Raymond[1], Member, ASCE, M.S.A. Abdel-Baki[2], R.G. Karpurapu[3], R.J. Bathurst[4]

Abstract

Tracks for gantry cranes are commonly subject to eccentric and inclined loads in conjunction with very heavy vertical loads. Recent practice has been to support these tracks on granular support rather than reinforced piled foundation beams. The paper investigates the effect of a single strong layer of reinforcement placed within a well compacted granular soil on the bearing capacity at collapse of eccentrically loaded footings. The practical significance of the investigation in terms of heavy track guided systems is presented.

Introduction

There are many cases of footings subject to eccentric and inclined loadings on both cohesive and granular soils. Most solutions do not require special measures such as soil reinforcement. As loads become heavier, and thus foundations costly, it becomes prudent to examine alternatives that may produce major economic advantages. A typical example is a heavy gantry crane foundation.

The guide rails for heavy gantry cranes are often placed on heavy reinforced beams that are supported on piled foundations. Alternative solutions of supporting these tracks on granular support, if successful, can reduce foundation costs by more than half.

In order to accommodate the large movements associated with granular supported track the gantry crane is constructed with one fixed upright and one hinged upright. The hinge is situated at the upright/cross beam

[1]Prof. and [2]Grad. Stud., Dept. of Civ. Engrg., Queen's University, Kingston, Ontario K7L 3N6, [3]Res. Assoc. and [4]Prof., Dept. of Civ. Engrg., Royal Military College, Kingston, Ontario K7K 5L0

intersection. The effect of the hinge is to cause any lateral loading from the lateral movements of the cross beam trolley and its load to be resisted in total by the track at the base of the fixed upright. Since the contact between the flange and the rail is made at the head of the rail the resultant load at the footing/soil contact area, as illustrated in Fig. 1, is both eccentric and inclined. Where loading is excessive or maintenance neglected, track support failure often occurs. One method of increasing the bearing capacity of the granular fill is to introduce reinforcement. In general these are well compacted and, if reinforced, only layer of reinforcement is placed well below the contact face. Herein only the problem of eccentric loading is investigated.

Formulation of Experimental Program

The initial objective was to investigate experimentally, by means of model tests, the effect of soil reinforcement on the bearing capacity of eccentrically loaded footings as related to track support. General practice is to use well compacted single sized ballast below track ties (footing). Thus a uniform sized aggregate placed at a high relative density (D_r = 90%) was a requirement for all tests. Footing size was selected by conducting tests to failure using a small (75 mm wide) footing resting on a foundation soil having embedded markers and observing the slip planes as described by Raymond and El Komos (1978). 3 mm aggregates, one crushed (ϕ_{tr} = 49) and one rounded aggregate (ϕ_{tr} = 34), were used in these initial tests. The failure was always observed to be symmetrical. This was probably due to good density control during aggregate placement. The pattern for the rounded material is shown in Fig. 2 and is mathematically

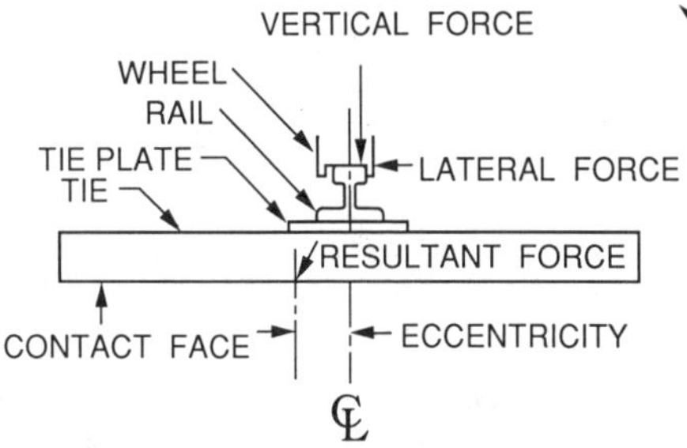

Fig. 1 Schematic of Forces Applied to Rail.

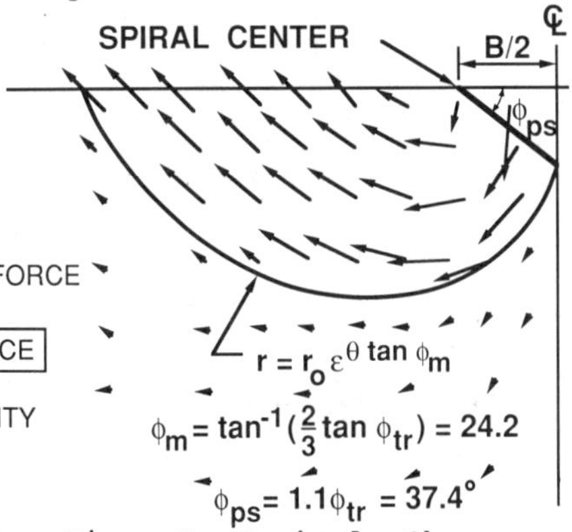

Fig. 2 Typical Observe Deformations Below Failed Footing.

the same as for the stronger material once the correct friction angles are inserted. A symmetrical wedge making an angle in degrees 10% greater than the triaxial friction angle was first constructed. This angle was assumed, based on Cornforth's (1964) results, to equal the plane strain friction angle (ϕ_{ps} = 37.4). Once drawn it established the starting length for a logarithmic spiral having a center at the edge of the footing. The initial radius is:

$$r = 0.5B/\cos(\phi_{ps}) \quad (1)$$

where B is the footing width. The failure plane then followed the logarithmic spiral pattern suggested by Terzaghi (1943) for a highly compressible soil, but using the triaxial friction angle rather than the plain strain angle.

$$\phi_m = \tan^{-1}(0.67 \tan\phi_{tr}) \quad (2)$$

Note that the ultimate bearing capacity measured corresponded to that obtained by using the estimated plane strain angle and the solution of Vesic (1977). Thus the constructions are purely empirical geometric fitted mathematical descriptions matching observations. They do not give ultimate bearing capacities. Terzaghi (1943) emphasized that his theoretical slip patterns, used to calculate his bearing capacity coefficients, did not even approximate his experimental observed patterns. This is confirmed by Fig. 2. The value of the fitted empirical procedure is that it permits the selection of a footing size that would fail without the slip plane intersecting the placement tank base or side. The critical dimension for the placement tank used is the depth of soil in relation to the footing width. This ratio is shown in Fig. 3 along with the slip plane depth ratios calculated for Terzaghi's (1943) rough and smooth cases and the Mandel and Salecon (1972) solution. If, as suggested by Hill (1950) a double wedge occurs at the footing base these minimum depths would be halved. These initial tests showed that the investigation could be performed using the rounded aggregate and a 200 mm wide footing in a 200 mm depth of soil. This gave a reasonable aggregate footing ratio for the track (prototype +20mm/2000mm; model 3 mm/200 mm).

Most track facilities selecting geosynthetics use geotextiles for separation and geogrids or confinement webs for reinforcement. Because ballast is large sized reinforcement mesh is generally selected for survivability during placement and is generally thick and robust. Initial tests were done with a polymeric micro mesh and a steel mesh. The micro mesh had a thickness of 0.3 mm or 10% of the aggregate size used. This compares with railroad type geogrid/ballast size ratios (i.e.+2mm grid and +20mm ballast particle). The results were within 5% of each other. The steel mesh, although thicker, was easier to handle and was selected for most tests.

The cost of undercutting track increases exponentially with depth. Most reinforcement is placed at considerable depth on new construction and at about 300 mm depth below the tie base when track is undercut. Tie length is typically 1800 mm long for a gantry crane and typically 2700 mm long for a railroad tie. A depth of 300 mm represents a placement depth to footing width of 0.167 or 0.111 respectively. Minimum depths to prevent tamper damage for a strong geogrid would be 150 mm or half these ratios. Thus tests are performed to ratios close to 0.06. For rehabilitation the use of more than one layer of reinforcement is clearly uneconomical. Thus testing is confined to the use of one reinforcement layer.

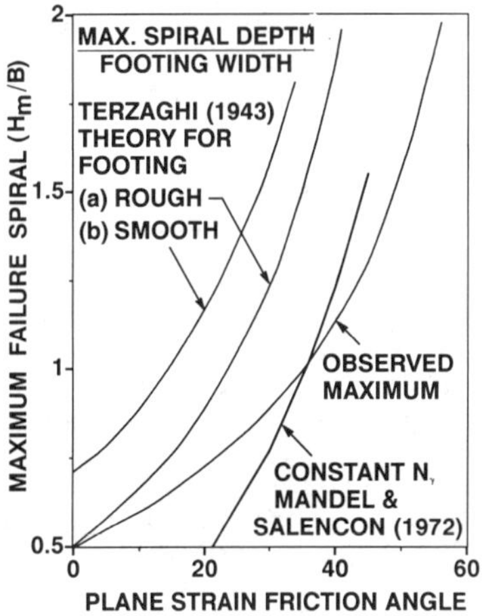

Fig. 3 Minimum Depth/Breadth Ratios to Prevent Base Interference with Slip Failure Plane.

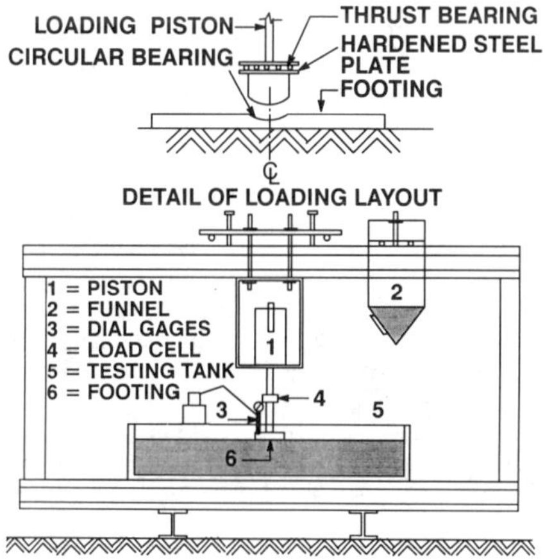

Fig. 4 Layout of Test Equipment.

Testing Equipment

The test equipment consisted of a material placement testing tank, loading frame, loading pistons, load measurement devices, rigid footing, air supply system, deformation dial gauge, tested soil, and reinforcement mesh. A layout is shown in Fig. 4.

The placement tank was 900 mm long, 200 mm wide, and 330 mm deep, and tightly connected to the loading frame.

The sides of the tank were made of 13 mm thick, Herculite transparent glass that fitted into grooves within the thick metal tank and was supported by a beam on its top edge. This glass minimized the friction between the soil and the tank. The rest of the tank was made of thick aluminum plates. Clamps across the open top of the tank were available to prevent lateral movement of the top edge but generally tank stiffness made them of little value. A moveable funnel simplified the placement of the soil in layers within the tank.

All loading tests used 19 mm thick aluminum model footings, whose length extended over the whole width of the tank (200 mm). Based on the initial justification tests a 200 mm wide footing was used for the eccentric loading tests. Four sets of thrust bearings, located on drilled seats in a rectangular plate, were used to ensure that the load always acted vertically on the loading plate. A 25.4 mm diameter semi-cylindrical rod was tightly connected to a second lower rectangular plate and fitted into longitudinal groves to ensure a constant eccentricity, even if the footing tilted.

Bellofram loading pistons were used to apply the load. A load cell was used to accurately monitor the load and cell response was recorded by a data acquisition system. The vertical deformation of the footing was measured by placing a dial gauge having a travel of 25 mm and sensitivity of 0.0025 mm near each corner of the footing. The average change of the two dial gauges on any side was taken as the settlement, or heave, on that side.

The main granular material used as the foundation soil was artificially manufactured ceramic known as denstone supplied by Norton Chemical Processing Co.. Crushed limestone and Ottawa sand were used to confirm the results. The ceramic used for the majority of tests was in the form of uniform rounded particles of an average diameter of 3 mm. The balls had a specific gravity of 2.4. The maximum and minimum test densities were 1.553 and 1.366 gm/cm^3 respectively. The placement density was 1.534 gm/cm^3 giving a relative density of 90%.

The soil reinforcement for the majority of tests was provided in the shape of steel mesh manufactured from welded galvanized wire with an aperture size of 19.0 mm x 19.0 mm. For the majority of tests the reinforcement was cut so that its length and width were 25.4 mm less than the length and the width of the tank. This clearance was provided to ensure that no friction was generated between the reinforcement and the walls of the tank. A few tests were done with different lengths of mesh. The shortest length used was equal to the footing width.

Unreinforced Foundation - Eccentrically Loaded Footing

The first series of tests was performed to determine the bearing capacity and strength properties of the soil. These tests permit a direct comparison with the eccentric footing failure theory of Meyerhof (1953). For a surface strip footing on a purely cohesionless soil, well above the water table, the failure line load per unit length is

$$Q_f = 0.5 \ \gamma \ (B-2e)^2 \ N_\gamma \qquad (3)$$

where γ is the unit weight of soil, B is the width of the footing, e is the eccentricity of the load and N_γ is the bearing capacity coefficient which, for a strip surface footing, is dependent only on the angle of friction of the soil ϕ. The back calculated value of N_γ = 78 agrees with the value from Vesic (1973) for $\phi = 38°$ (compared to ϕ_{ps} = 37.4° on Fig. 2). This comparison is shown in Fig. 5 where failure was defined by catastrophic collapse and as may be seen gives good agreement.

During the tests reported on Fig. 5 it was noted that for eccentricities of 37.5 mm and less (e/B < 0.1875) the footing remained in full contact with the soil. The value of 37.5 mm may be compared with the middle third rule, of an eccentricity of 33.3 mm. For eccentricities larger than 37.5 mm the unloaded end of the footing lifted off the soil during loading. For eccentricities of 75 mm and 87.5 mm this occurred during the first load increment.

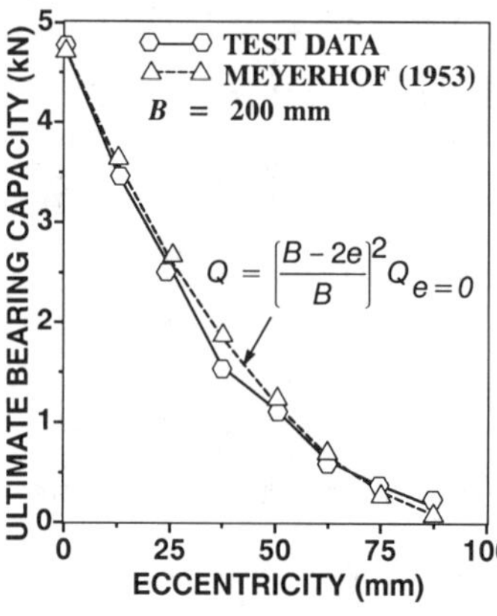

Fig. 5 Unreinforced Soil Eccentric Footing Failure Test Results.

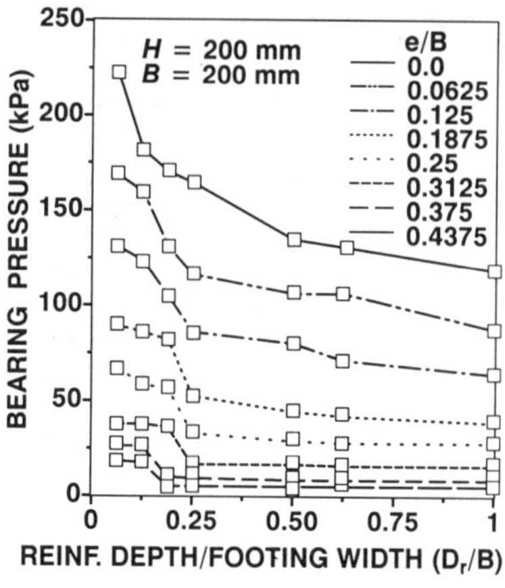

Fig. 6 Reinforced Soil Eccentric Footing Failure Test Results.

Eccentrically Loaded Footings on Reinforced Soil

The second series of eccentrically loaded tests used a 200 mm wide footing on a reinforced subgrade. The reinforcement was placed at difference depths; 12.5 mm, 25 mm, 37.5 mm, 50 mm, 100 mm, and 125 mm within a constant soil depth of 200 mm. In no case did the reinforcement fail. Every test resulted in a catastrophic and sudden soil failure. These failure loads are shown in Fig. 6. For shallow depth the slip plane extended well below the reinforcement and a well defined wedge existed below the footing, in much the same manner as for the unreinforced tests (Fig. 2).

The results in Fig. 6 clearly show that the bearing capacity increased as the reinforcement was placed closer to the contact surface. This is in agreement with the data obtained by Binquet and Lee (1975) and Guido et al (1986) for centrally loaded footings on reinforced granular soil. Akinmusuru and Akinbolade (1981), however found that the highest increase in bearing capacity occurred when the depth to the reinforcement was half the width of the footing. Both greater and lesser depths for the reinforcement resulted in a loss of bearing capacity according to their test data. One possibility was that Akinmusuta and Akinbolade used both an extensible and low interface friction angle. Some of the results shown in Fig. 6, were repeated using an extensible micro plastic mesh. The final results differed by less that 5%. No tests were conducted with a low coefficient of friction between the reinforcement and the soil. This suggests that the difference was due to slippage and grids rather than geotextiles or strips should be used for reinforcement.

In a similar manner to the footing tests on unreinforced soil, the footings stayed in contact with soil at small eccentricities and lifted off the soil when loaded at large eccentricities. When loaded at large eccentricities the reinforcement at 100 mm and 125 mm depth of soil gave similar results to the unreinforced soil. This would be expected from the data given in Fig. 3 since slip failure would occur above the reinforcement. As a practical matter it is recommended that where a single reinforcement layer is used it will have little benefit to the ultimate bearing capacity unless it intersects the contained soil wedge below the footing as defined in Fig. 2 by ϕ_{ps}.

Most of the load/settlement data was similar. At the beginning of the tests, the footing settled proportional to load. Near failure, the settlement rate increased until failure occurred. Some loading tests behaved differently.

These tests produce 's' shaped load settlement relationships. Examples of these few results are shown in Fig. 7. On loading, the footing settled in a similar manner to the normal tests until a first failure occurred. This first failure involved large settlements but not complete collapse. On further loading, the rate of settlement increased at a slightly faster rate than the rate of settlement observed at the beginning of the test. Finally, the footing failed catastrophically giving a second failure. This second higher failure is due to the benefit of large settlements resulting to the point where the reinforcement intersected the contained soil wedge below the footing. Without reinforcement, it is believed that catastrophic failure would have occurred at the first failure load.

For the concentrically loaded tests, a well-defined general failure pattern was observed. Approximately equal side berms were formed in all cases. For eccentric loading tests, unequal side berms were formed with the smaller on the loaded side. The slip plane formed under the footing pushing the major portion of soil to the unloaded side.

In order to appreciate the significance of the reinforcement Fig. 8 shows the increase in bearing capacity recorded from the reinforcement. From Fig. 8 it is obvious that the best place to position the reinforcement is within the contained soil wedge below the footing and close to the contact area. Reinforcement placed below D_r = 0.4B for

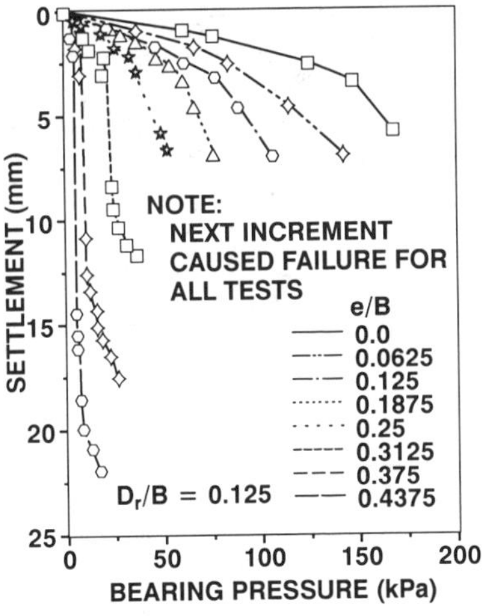

Fig. 7 Reinforced Soil Load Settlement Data for D_r/B = 0.125.

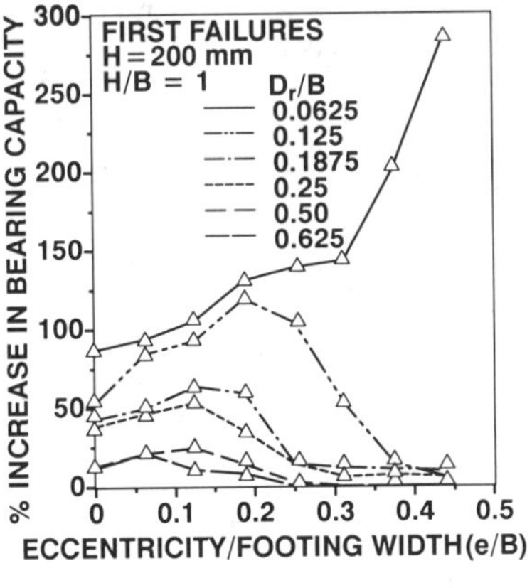

Fig. 8 Percent Increase in Failure Resulting From Reinforcement.

this soil with ϕ_{ps} = 37.4° may be seen to have little benefit to concentrically loaded footings. Similarly for eccentrically loaded footings the soil wedge may be based on B′ = B-2e. For large eccentricities and shallow depths the benefits are great. This is believed due to the fact that for eccentrically loaded footings a one sided rather than two sided slip plane occurs. Deep reinforcement has little benefit to footings with large eccentricities.

Effect of Reinforcement Length and Aperture Size

Reinforcements of different total lengths ℓ, from ℓ/B = 1.0 and over, and different aperture sizes (5.3 mm, 10.9 mm, and 23.4 mm) were tested using eccentrically loaded footings. Typical results are shown in Fig. 9. Contrary to the findings of others such as Akinmusuru and Akinbolade (1981) and Jewell et al. (1984), no significant change in bearing capacity was observed for values of ℓ/B = 1.25 and greater or changing aperture size. On the other hand, in agreement with the tests reported here Guido et al. (1986), using reinforcement of low stiffness in granular soil, observed that once the ratio ℓ/B reached 1.25, the length of the reinforcement had no measurable effect on the bearing capacity. The low value of ℓ/B agrees with the observation that the reinforcement needs to intersect the contained soil wedge below the footing. It also suggests that no seams or overlaps should be permitted in the reinforcement directly below the footing.

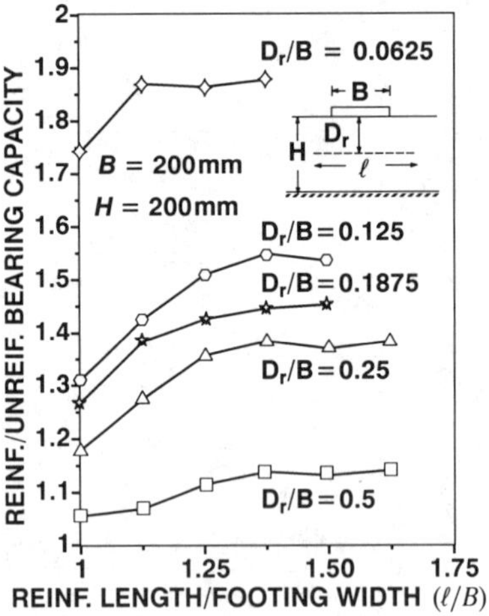

Fig. 9 Effect of Reinforcement Length on Centrically Loaded Footing.

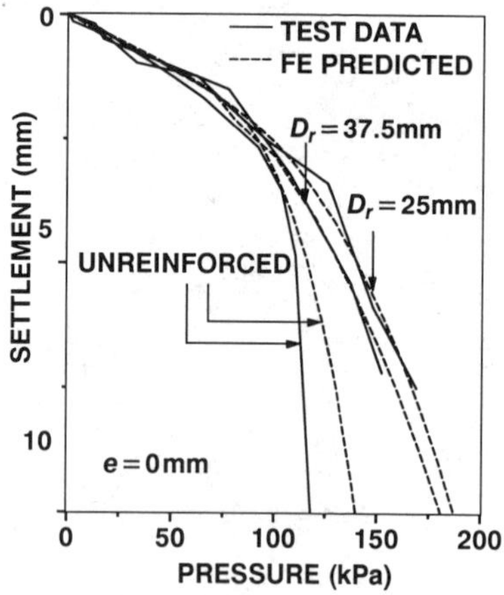

Fig. 10 Comparison Finite Element and Observed Data for Concentric Loading.

Finite Element Analysis

Finite element analysis of the experimental results was performed. The program used was developed at the Royal Military College of Canada. Bar elements modelled the reinforcement, beam elements the footing, joint (interface) elements the frictional behavior of the interfaces between two contacting faces, and three noded triangular elements the soil. The material behaviour of the soil was modelled using a hyperbolic constitute model formulated in terms of tangent Young's Modulus and tangent bulk modulus. When the stress state exceeds the yield surface the excess stresses are released along the flow direction allowing plastic strain to develop. The bar and beam elements are modelled as linear elastic elements. The joint elements are modelled as rigid perfectly plastic elements using the Mohr Coulomb yield surface.

Typical results of the modelling of the concentrically loaded footing tests are shown in Fig. 10. As seen the finite element does not give a catastrophic failure, however the initial load settlement data shows good agreement. The results showed that the forces in the reinforcement are greatest below the footing. This is in agreement with the experimental data that showed the reinforcement needed to intersect the contained soil wedge below the footing to be of benefit.

Estimated Bearing Capacity and Reinforcement Design

Based on the mechanisms observed, and assuming that when reinforced the containment wedge angle becomes 45° (conservative for ballast), a simple analysis may be undertaken that results in the following equations for the force in the reinforcement per unit length of footing:

$$T = \frac{(B - 2e - 2D_r)\ (\psi + \sin\psi)}{(B - 2e)\ \pi}\ \{Q_r - Q_u\} \tag{4}$$

$$\psi = 2\tan^{-1}\left[\ (B - 2e)/(2D_r)\right] \tag{5}$$

subject to pull out that gives

$$T_{max} \le (Q_r \tan\phi)/2 \tag{6}$$

$$\ell_{max} = 1.5\ B \tag{7}$$

where Q_r is the load at failure with reinforcement, and Q_u is the load at failure without reinforcement given by equation 3. Perfect interfriction has been assumed to obtain T_{max}. In practice the ℓ/B ratio is less than 1.5 for geogrids and geocells since particles lodge in the openings and produce a passive resistance, a factor not accounted for by assuming the reinforcement as a sheet with perfect interfriction only.

Practical Significance

The practical significance of the data presented is that reinforcement of granular fills subject to heavy loads approaching failure on stable subgrades are best reinforced by placing the reinforcement close to the footing contact area rather than deeply placed as often is the case. No seams or overlaps should be permitted below the loaded area. This is of importance in the design and rehabilitation of track support for heavy gantry cranes, heavy railroad tracks and the like. On rehabilitated track undercut to place reinforcement undercutting need only be performed to a minimum depth to prevent damage to the geogrid. This will result in considerable savings. The practice of using geogrids for reinforcement and geotextiles for separation is supported.

Conclusions

The research studied the effect of soil reinforcement on the bearing capacity of eccentrically loaded footings over granular soil. Tests done on a footing on unreinforced granular soil at different eccentricities verified the theory proposed by Meyerhof (1953) as seen in Fig. 5.

Reinforcement of the granular soil showed that placing the reinforcement near the foundation contact face resulted in the largest increase in bearing capacity independent of the eccentricity, see Fig. 6. The bearing capacity was not significantly affected by increasing the reinforcement length beyond 1.25 times the footing width, see Fig. 9. Reinforcement must intersect the contained soil wedge below the footing to be of value. A finite element program that showed good settlement agreements with the experimental data (Fig. 10) also confirmed the importance of intersecting the contained soil wedge. These mechanisms permitted the development of a mathematical model, expressed in the form of equations 4 to 7 for design purposes.

The percentage increase in bearing capacity of the eccentrically loaded footings was greatest at the largest eccentricities combined with the shallowest reinforcement. At large eccentricities and deep reinforcement little or no bearing capacity increase was measured. These conclusions are drawn from the results plotted on Fig. 8. For large eccentricities and shallow reinforcement two failures modes within the granular support were observed as seen in Fig.7.

Acknowledgements

Acknowledged is funding for the experimental work by the National Scientific and Engineering Research Council (Canada) grant awarded to G.P. Raymond and funding for the finite element work by the Department of National Defence (Canada) grant to R.J. Bathurst.

References

Akinmusuru, J.O., and Akinbolade,J.A. (1981). "Stability of Loaded Footing on Reinforced Soil". J. Geot. Div., ASCE, 107(6), 819-827.

Binquet, J., and Lee, K.L., (1975). "Bearing Capacity Analysis of Reinforced Earth Slabs". J. Geot. Div., ASCE, 101 (12), 1257-1276.

Cornforth, D.H. (1964). "Some Experiments on the Influence of Strain Conditions on the Strength of Sands". Geotechnique, 14, 143-167.

Guido, V.A., and Chang, D.K., and Sweeney, M.A., (1986). "Comparison of Geogrids and Geotextiles Reinforced Earth Slabs". Can. Geot. J., 23,435-440.

Hill, R. (1950). "The Mathematical Theory of Plasticity". Oxford University Press.

Jewell, R.A., (1984). "Interaction Between Soil and Geogrids". Proc. of the Symposium on Polymer Grid Reinforcement, London, 18-29.

Mandel, J., and Salecon, J., (1972). "Force Portante d'un Sol Sur un Assise Rigide (etude theorique)", Geotechnique, 22, 79-93.

Meyerhof, G.G., (1953). "The Bearing Capacity of Foundations Under Eccentric and Inclined Loads". Proc. 3rd. Int. Conf. on Soil Mech. and Found. Engrg., 1, 440-445.

Raymond, G.P., and El Komos, F., (1978). "Repeated Load Tests of a Model Plane Strain Footing". Can. Geot. J., 15, 190-201.

Terzaghi, K. (1943). "Theoretical Soil Mechanics". John Wiley and Sons.

Vesic, A.S. (1977). "Analysis of Ultimate Loads of Shallow Foundations". J. Soil Mech. Found. Div., ASCE, 99, 45-73.

Eccentrically Loaded Plates on Reinforced Subgrades

Vito A. Guido[1], M. ASCE
John J. Nocera[2], S.M. ASCE

Abstract

Presented herein are the results of 30 laboratory model eccentrically loaded plate tests on a geogrid reinforced granular subgrade. It was determined that as a result of the insertion of the geogrid reinforcement into the subgrade the bearing capacity of the eccentrically loaded plate was increased over the unreinforced case and the settlement/rotation of the plate was reduced. Several parameters were investigated, including, number of layers of the reinforcement, size of the reinforcement and the location and orientation of the reinforcement in the subgrade material.

Introduction

A foundation subjected to a bending moment, M, about one or both axes is common. This bending moment may be replaced by a vertical load, Q, placed at an eccentricity, e, from the centroid of the foundation, such that $M = Q \cdot e$. An investigation of the effect of eccentricity on the bearing capacity and settlement/rotation characteristics of a footing founded on unreinforced soil was performed by Meyerhof (1953). The authors have extended Meyerhof's research by including the effects of the inclusion of geogrid reinforcement in a granular subgrade. The results of these tests are presented herein.

The Testing Model

The 30 model eccentric plate loading tests were performed in a square stiffened wooden box, 48 in. (1.22 m) wide and 36 in. (0.92 m) deep. Based on the results

---

[1] Prof. Civil Engineering, The Cooper Union School of Engineering, New York, NY 10003

[2] Project Engineer, Malcolm Pirnie, Inc., White Plains, NY 10602

of over 250 prior model concentric plate loading tests, it was determined that this size test model minimizes edge effects. A poorly graded sand (SP) with a uniformity coefficient, C_u, of 1.90, a coefficient of gradation, C_c, of 1.23, an effective size, D_{10}, of 0.086 mm and an angle of internal friction, ϕ, of 37° at a relative density, D_r, of 55% or a dry unit weight, γ_d, of 91.6 pcf (14.4 kN/m³) was used as the subgrade soil. All of the tests were performed at this relative density to demonstrate the significant benefits of reinforcement in a potentially unsuitable subgrade soil.

Tensar BX1100 geogrid was used as reinforcement in the subgrade below the eccentrically loaded plate. It is a biaxially oriented polymer grid formed from polyolefine resine. Fig. 1 is a presentation of typical dimensions of the BX1100 geogrid.

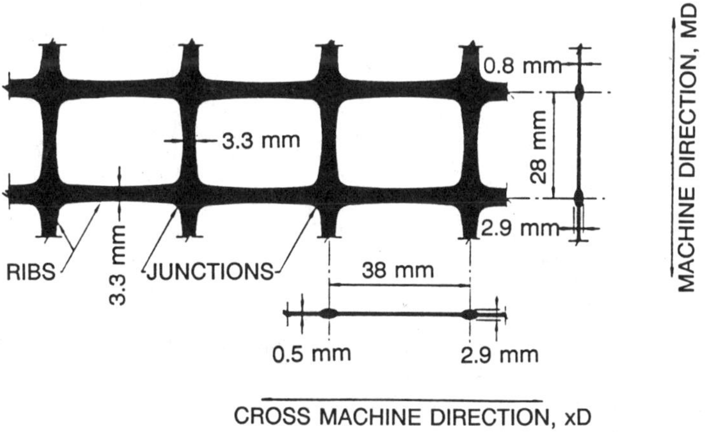

Fig. 1 Typical Dimensions of Tensar BX1100 Geogrid

A square aluminum plate 12 in. (305 mm) wide and 1 in. (25.4 mm) thick was loaded at the surface of the subgrade soil. A 25 ton (223 kN) capacity hydraulic cylinder and hand pump were used to apply a static load to the plate. The load was monitored on a 10 ton (89 kN) capacity load ring attached to the upper socket of a ball-and-socket joint. This load was transmitted through the ball to the lower socket and then to the plate. Four dial gages placed at the four corners of the plate were used to monitor the resulting displacements. The subgrade soil was placed in 2 in. (50.8 mm) lifts and new square sheets of BX1100 geogrid were used for each test performed. For the 91.6 pcf (14.4 kN/m³) dry unit weight indicated above, 244.3 lbs. (1.09 kN) of dry soil was required for a 2 in. (50.8 mm) lift. The pre-weighed soil was placed with a hydraulic drum dumper and compacted until the 2 in. (50.8 mm) lift height was achieved. The geometry of the test model is shown in Fig. 2.

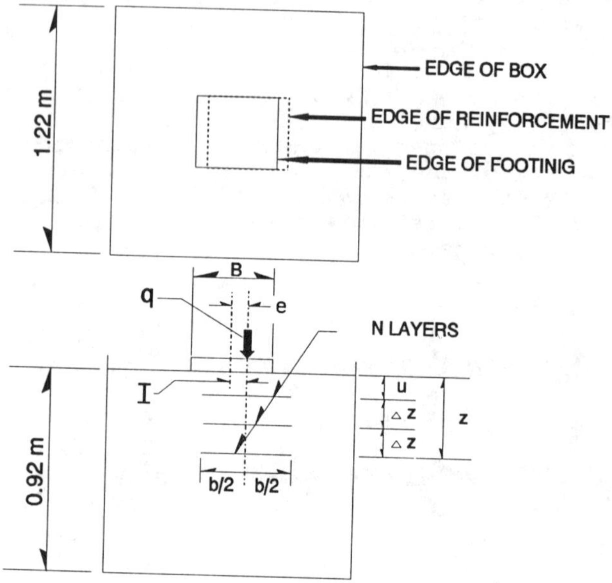

Fig. 2 Geometry of Test Model

Test Results

With the exception of 5 concentrically loaded tests, all of the remaining tests were performed at e = 2 in. (50.8 mm) or e/B of 1/6, where B is the width of the plate. To ensure that the entire plate was in compression the load was applied inside the kern. For a square plate the boundary of the kern is located at a distance B/6 from the centroid of the plate. In addition, for all the tests performed the depth below the plate to the first layer of reinforcement, u, was kept constant at 2 in. (50.8 mm) or u/B of 1/6 and the vertical distance between layers of reinforcement, Δz, was kept constant at 2 in. (50.8 mm) or $\Delta z/B$ of 1/6. The effect of the following four parameters on the bearing capacity and settlement/rotational characteristics of eccentrically loaded plates on geogrid reinforced subgrades was investigated:

1) Number of layers of geogrid reinforcement, N.
2) Width b of the square sheet of geogrid reinforcement, or in dimensionless form b/B.

3) Location of the centroid of the square sheet of geogrid reinforcement under the plate, I, or in dimensionless form I/B.
4) Orientation of the square sheet of geogrid reinforcement, either with the machine direction in the direction of the eccentricity (see Fig. 2) or the cross machine direction in the direction of the eccentricity.

To facilitate the expression and comparison of the test data the following definitions are used:

1) Average settlement, s, of the four settlement readings. This can be expressed in dimensionless form as the settlement ratio, s/B.
2) Rotation, θ, in radians (rads) is expressed as

$$\theta = (s_{(D)} - s_{(U)})/B \tag{1}$$

where $s_{(D)}$ and $s_{(U)}$ are the average downhill and uphill settlements, respectively.
3) In a private communication to the authors, Meyerhof (1990) concurred that for presentation of load-settlement/rotation curves the bearing pressure should be taken as the average applied bearing pressure, $\bar{q}$, or

$$\bar{q} = Q/A \tag{2}$$

where Q is the applied load and A is the area of the plate.

A comparison of curves b and d in Fig. 3 indicates that the bearing capacity curves based on average settlement and rotation do not agree. According to Meyerhof (1990), the ultimate bearing capacity for a given test for both the average settlement and rotation should be the same, when defined as the bearing pressure at which the curves pass into inclined straight lines. A series of tests performed by Bakkala (1990) and Nocera (1991) indicate that for small eccentricities (e/B = 1/12) and an unreinforced subgrade (N = 0), Meyerhof's statement appeared to be valid. However, for eccentricities approaching or exceeding the kern ($e/B \geq 1/6$) and/or a reinforced subgrade (N > 0), Meyerhof's statement is invalid. For large eccentricities, the rotation of the footing is large while the average settlement is approximately constant and the inclusion of reinforcement into the subgrade alters the failure mechanism of the subgrade soil; therefore Meyerhof's statement becomes invalid for $e/B \geq 1/6$ and/or N > 0. Following the above discussion it was decided to determine the ultimate bearing capacity based on load rotation curves and not on load settlement curves. For concentrically loaded plates the ultimate bearing capacity will be determined from a load-settlement curve. As can be seen in Fig. 3, there is approximately 0.007 rads of rotation for the concentrically loaded plate (curve c), which for all practical purposes is negligible.

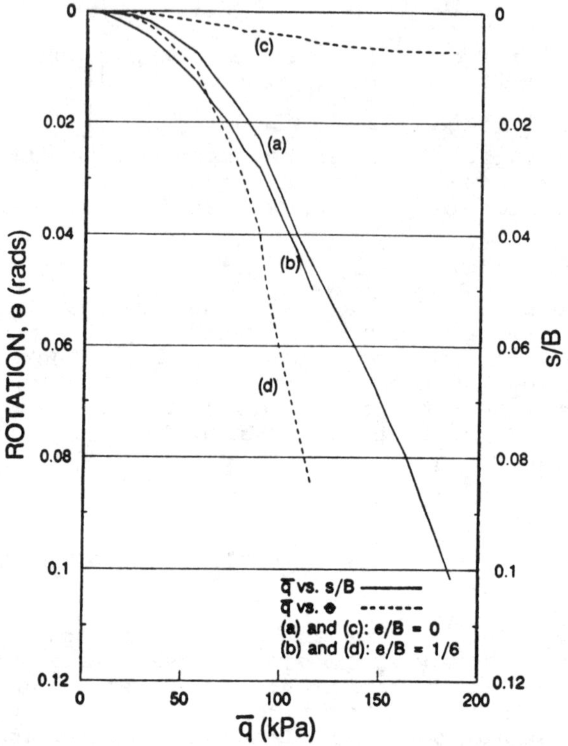

Fig. 3 Load-Settlement/Rotation Curves for Unreinforced Sand Subgrade

The theoretical ultimate bearing capacity of the concentrically loaded plate placed at the surface of the granular soil used in the test model was calculated to be 23.7 psi (163.5 kPa). This bearing capacity was calculated using Meyerhof (1963). From Fig. 3, this corresponds to s/B = 0.081 (see curve a).

The theoretical ultimate bearing capacity of an eccentrically loaded plate can be obtained in one of the following two ways:

(a) An eccentrically loaded plate of dimensions B x L with eccentricities e_B and e_L can be treated as a concentrically loaded plate of effective dimensions B' x L', where $B' = B - 2e_B$ and $L' = L - 2e_L$, Meyerhof (1953). The effective dimensions B' and L' are used in the bearing capacity equations, wherever the dimensions B and L are required.

(b) The ultimate bearing capacity of a concentrically loaded plate can be modified by a reduction factor R_e to account for the eccentrically applied load, Bowles (1988), where

$$R_e=1-(e/B)^{0.5} \text{ (cohesionless soil and } 0<e/B<0.3) \quad (3)$$

For the subgrade soil used in this series of tests and e/B = 1/6, method a yields an ultimate bearing capacity of 14.3 psi (98.6 kPa) and method b yields 14.0 psi (96.6 kPa). Therefore, the ultimate bearing capacity will be taken as 14.15 psi (97.6 kPa), the average of the two methods, which corresponds to a rotation of θ = 0.058 rads (see Fig. 3, curve d).

(4) Two bearing capacity ratios will be defined;

$$BCR = q_r/q_{uc} \quad (4)$$

and;

$$BCR' = q_{re}/q_{ue} \quad (5)$$

where q_{uc} is the ultimate bearing capacity of a concentrically loaded plate on an unreinforced subgrade soil given by 23.7 psi (163.5 kPa) for this series of tests; q_{ue} is the ultimate bearing capacity of an eccentrically loaded plate on an unreinforced subgrade soil given by the average of the two methods and for this series of tests is equal to 14.15 psi (97.6 kPa); q_r is the ultimate bearing capacity of any plate on a reinforced subgrade soil corresponding to θ = 0.058 rads for eccentrically applied loads or s/B = 0.081 for concentrically applied loads; and, q_{re} is the ultimate bearing capacity of an eccentrically loaded plate on a reinforced subgrade soil corresponding to θ = 0.058 rads. This is a settlement/rotation criteria, not bearing capacity failure since no catastrophic failure of the subgrade soil occurred. Settlement could no longer be determined as a result of the rotation of the plate. When the plate had rotated to such a degree that the dial gauges were no longer in contact with the plate, the test was stopped.

Number of Layers of Geogrid Reinforcement

The effect of the number of layers of geogrid reinforcement, N, on BCR and BCR' can be found in Fig. 4. A comparison of curves a, b and c shows a small improvement in the bearing capacity over the eccentric unreinforced case for one layer of reinforcement. This improvement is 25%, 13% and 3% for curves a, b and c respectively. With an increase in the number of layers of reinforcement above one, the bearing capacity increases sharply to values of 102%, 86% and 104% above the eccentric unreinforced case for curves a, b and c, respectively. Previous studies by Binquet and Lee (1975) and Guido et al. (1985, 1987, 1989) have shown that for concentrically loaded plates the optimum number of layers of reinforcement was approximately 3 with additional layers yielding no additional benefits. This was found to be within an effective depth B below the base of the plate or z/B = 1.0. For a plate having a single eccentricity, e_B, the effective dimension B' = B - $2e_B$ Meyerhof (1953). The reinforcement should lie within a depth z/B' = 1.0, which for this series of tests would indicate a maximum of 4 layers of reinforcement would yield optimum conditions. Curves a, b and c begin to turn downward for N greater than 2 but do not reach an ultimate value.

Curves d, e and f show similar trends to those of curves a, b and c. In addition, it can be seen that the bearing capacity of an eccentrically loaded plate on a geogrid reinforced subgrade can exceed that of a concentrically loaded plate on an unreinforced subgrade. Curves d, e and f cross BCR = 1.0 at N = 2. Therefore with the placement of approximately 2 layers of geogrid reinforcement the eccentrically loaded plate sustained a larger load than the concentrically loaded plate with an unreinforced subgrade.

Width of Geogrid Reinforcement

For a square sheet of geogrid reinforcement, the width of the sheet is designated by b. In Fig. 5, two sets of curves are presented: one for BCR and the other for BCR' as a function of b/B. Both sets of curves show similar trends, with an increase in the bearing capacity of the plate with increasing width ratio. In addition, optimal values of the bearing capacity are reached at values of b/B equal to approximately 1.3 and 2.5 for N equal to 2 and 3, respectively. Similar results have been observed for concentrically loaded plates by Fragaszy and Lawton (1984) and Guido et al. (1985, 1987, 1989). Curve d does not achieve BCR = 1.0, indicating that more than two layers of geogrid reinforcement are required to improve the bearing capacity above that of the concentrically loaded plate on an unreinforced subgrade. Curve c crosses BCR = 1.0 at approximately b/B = 0.8. Therefore, for N = 3, the width ratio must be greater than 0.8 to attain bearing capacities greater than the unreinforced concentrically loaded case.

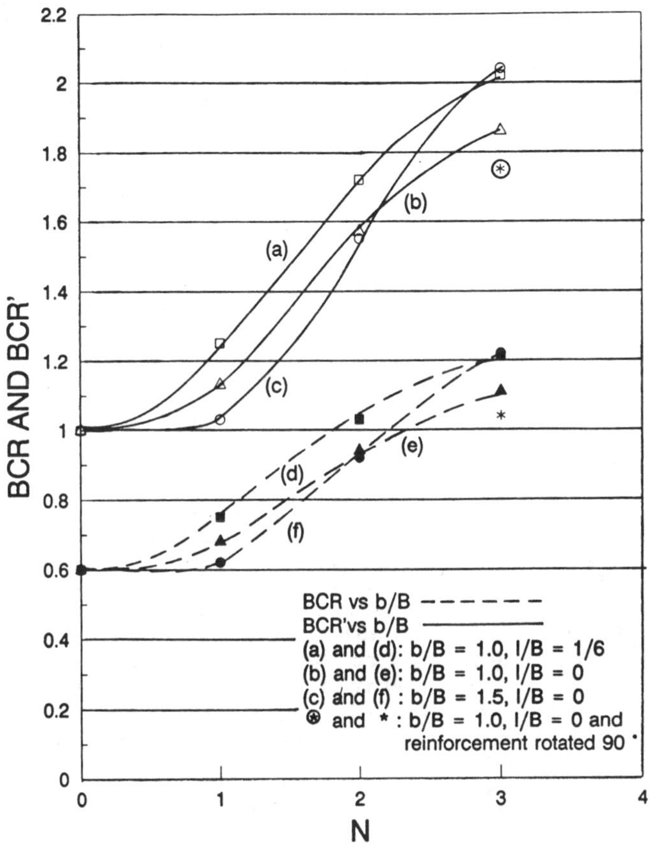

Fig. 4 Variation of BCR and BCR' with Number of Layers of Reinforcement (u/B=1/6, Δz/B=1/6, e/B=1/6, D_r=55%)

A comparison of curve b in Fig. 6 with curve b in Fig. 4 indicates that eccentrically loaded plates will benefit more from the placement of geogrid reinforcement in the subgrade than concentrically loaded plates. The placement of 3 layers of geogrid reinforcement below an eccentrically loaded plate increases the bearing capacity by 85% whereas 3 layers of geogrid reinforcement below a concentrically loaded plate increase the bearing capacity by 65%. This can be explained by Meyerhof's (1953) effective width theory. For the series of tests performed a value of b/B̄ of 1.0 represents an effective width ratio, b/B' of 1.5. Therefore a proper comparison between concentrically and eccentrically loaded tests would be for b/B = b/B'. The BCR values for b/B = 1.0 and e/B = 0 are compared with the BCR' values for b/B' = 1.0 and e/B = 1/6 in Fig 6. It can be seen that the two curves are almost identical, supporting Meyerhof's effective width concept and explaining why an eccentrically loaded plate benefits to a greater extent than a concentrically loaded plate from the placement of identically sized square sheets of geogrid reinforcement.

The width (surface area) of the reinforcement contributed to its anchorage in the subgrade and provides the mechanism for pull-out resistance. This is accomplished for geotextiles through the frictional resistance developed on the surface of the fabric, whereas, for geogrids it is a combination of frictional resistance and interlock between the subgrade soil and the grid. Therefore, when designing for the width of the reinforcement, an adequate factor of safety against pull-out must be provided.

Location of Geogrid Reinforcement

The distance from the centroid of the square sheet of geogrid reinforcement to the centroid of the plate was characterized by I or in dimensionless form by I/B. Two positions of the centroid of the sheet of reinforcement were investigated: directly under the centroid of the plate (I/B = 0) and directly under the applied load (I/B = e/B = 1/6). A comparison of either curves a and b or d and e in Fig. 4, indicates that those tests performed with I/B = 1/6 yielded higher bearing capacities than those tests performed for I/B = 0. By placing the geogrid reinforcement directly beneath the applied load, more of the reinforcement is placed in the zone of soil extension, allowing for a better utilization of its tensile and load spreading ability.

Orientation of Geogrid Reinforcement

Of the 30 tests performed, 28 had the grid oriented with the machine direction in the direction of the eccentricity and 2 had the grid oriented with the cross-machine direction in the direction of the eccentricity. These two tests are plotted as isolated points in Figs. 4 and 5 and should be compared to curves b or e in Fig. 4 and a or c in Fig. 5, respectively. It can be seen that when the grid is placed with the machine direction in the direction of the eccentricity the bearing capacity is greater than when it is rotated 90° from this orientation. Geogrids develop their anchorage strength from both frictional resistance between the soil and geogrid

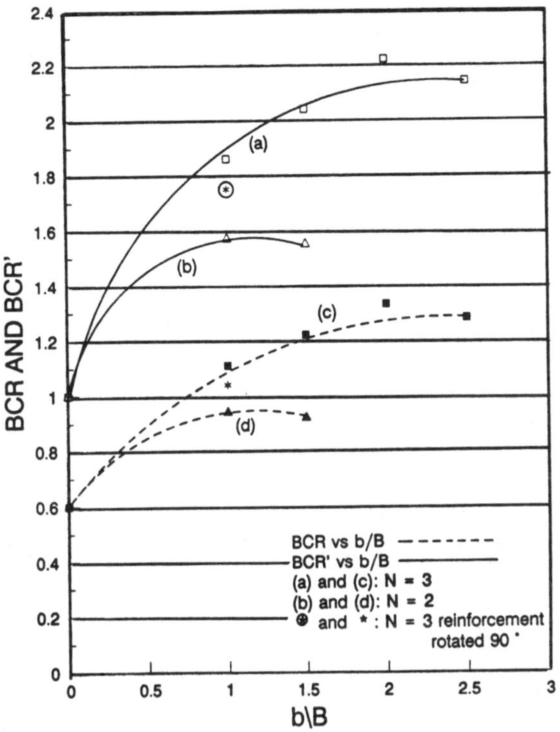

Fig. 5 Variation of BCR and BCR' with the Width Ratio ($u/B=1/6$, $\Delta z/B=1/6$, $e/B=1/6$, $I/B=0$ and $D_r=55\%$)

and transverse rib bearing strength, which is the major contributor. By placing the machine direction in the direction of the eccentricity, more ribs are parallel to the axis about which bending occurs than by placing the cross-machine direction in the direction of the eccentricity, and a better interlocking of the soil and grid occurs to resist bending. In addition, the larger rib length is placed perpendicular to the direction of the eccentricity indicating the bearing capacity would be influenced the greatest by this orientation of the geogrids since the largest transverse rib bearing strength is developed.

Conclusions

The results presented herein have shown that the insertion of geogrid reinforcement into a sand subgrade below an eccentrically loaded plate has substantially increased its bearing capacity over that of the unreinforced subgrade, and for some cases over that of the unreinforced concentrically loaded subgrade.

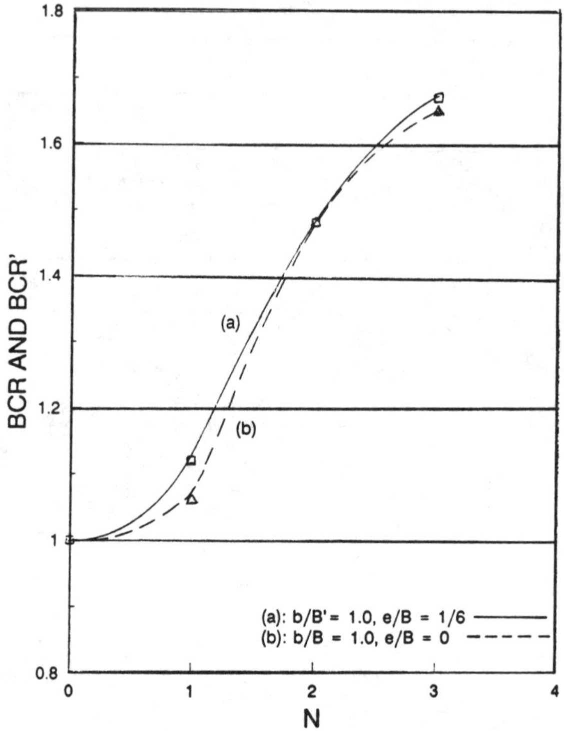

Fig. 6 Variation of BCR and BCR' with the Number of Layers of Reinforcement ($u/B=1/6$, $\Delta z/B=1/6$, $I/B=0$ and $D_r=55\%$)

The optimum number of layers of reinforcement to increase the bearing capacity is greater than 2 with the greatest increase per layer inserted occurring between 1 and 2 layers. In addition, an increase in the number of layers of reinforcement produces greater increases in bearing capacity than an equivalent increase in the width (or surface area) of a square sheet of reinforcement. For example, two layers of b/B = 1.0 reinforcement produce a 59% increase in the bearing capacity over the unreinforced case, whereas, one layer of b/B = 1.5 reinforcement produces a 3% increase of the bearing capacity. Therefore, for a specified safety factor with respect to pull-out it is more efficient to increase the number of layers of geogrid reinforcement, provided pull-out stability is ensured with an appropriate factor of safety.

A square sheet of geogrid reinforcement should be placed with its centroid directly below the point of application of the load, utilizing the full tensile capacity of the reinforcement. In addition, the geogrid should be oriented in the subgrade with the cross-machine direction parallel to axis of bending, insuring a maximum number of the largest rib lengths is placed parallel to the axis about which the plate is bending. This will increase the interlocking between the soil and the grid and increase its anchorage strength.

An optimization of the parameters will substantially improve the bearing capacity of the subgrade while reducing the rotation of eccentrically loaded foundations. Elimination of the rotation completely is not possible and the foundation must be designed with this in mind. In addition, for the series of tests performed the benefits associated with the inclusion of the geogrid reinforcement in the subgrade were not realized before large settlements/rotations occurred. The application of geosynthetics in subgrades underlying eccentrically loaded foundations is a viable solution to the resistance of the subgrade to withstand the stresses caused by the bending moment.

Appendix I - References

Bakkala, C.M. (1990). "Eccentrically loaded plate loading tests on geogrid-reinforced earth slabs", a thesis presented to The Cooper Union - Albert Nerken School of Engineering, New York, NY, in partial fulfillment of the requirements for the degree of Master of Engineering.

Binquet, J. and Lee, K.L. (1975). "Bearing capacity tests on reinforced earth slabs". J. Geotechnical Engrg, ASCE, 101(10), 1241-1255.

Bowles, J.E. (1988). Foundation analysis and design. McGraw-Hill Book Company.

Fragaszy, R.J. and Lawton, E. (1984). "Bearing capacity of reinforced sand subgrades". J. Geotechnical Engrg, ASCE, 110(10), 1500-1507.

Guido, V.A., Bieseiadecki, G.L. and Sullivan, M.J. (1985). "Bearing capacity of a geotextile-reinforced foundation". Proc. 11th I.C.S.M.F.E., San Francisco, Vol. 3, 1777-1780.

Guido, V.A., Knueppel, J.D. and Sweeney, M.A. (1987). "Plate loading tests on geogrid-reinforced earth slabs". Proc. Geosynthetics '87 Conference, New Orleans, Vol. 1, 216-225.

Guido, V.A., Christou, S.N. and Sobiech, J.P. (1989). "A comparison of texturized and non-texturized GEOWEB-reinforced earth slabs." Proc. Geosynthetics '89 Conference, San Diego, Vol. 1, 215-230.

Meyerhof, G.G. (1953). "Bearing capacity of foundations under eccentric and inclined loads". Proc 3rd I.C.S. M.F.E., Zurich, Vol. 1, 440-445.

Meyerhof, G.G. (1963). "Some recent research on the bearing capacity of foundations". CGJ, 1(1), 16-26.

Meyerhof, G.G. (1990). private communication to V.A. Guido.

Nocera, J.J. (1991). "Eccentrically loaded plate loading tests on geogrid-reinforced earth mattresses", a thesis presented to The Cooper Union - Albert Nerken School of Engineering, New York, NY, in partial fulfillment of the requirements for the degree of Master of Engineering.

The Evolution Of Geotextile Reinforced Embankments

C. Joel Sprague, M.ASCE[1] and Michael Koutsourais, A.M.ASCE[2]

Abstract

In 1978, the U.S. Army Corps of Engineers pioneered the use of high strength geotextiles in the United States at Pinto Pass in Mobile, Alabama with the objective of reinforcing an embankment over several feet of very soft silty clay. From this small, but technically significant demonstration project has grown an application which utilizes several hundred thousand square yards annually of high strength, high modulus geotextiles to reinforce embankments over soft foundations. Building upon experience, geotechnical engineers nationwide are designing larger and more critical embankments using stronger specially designed reinforcing geotextiles - saving owners millions of dollars over conventional embankment construction techniques.

This paper summarizes four especially significant projects which demonstrate the increasing acceptance of and major lessons learned from the use of high strength, high modulus geotextiles to reinforce embankments over soft foundations. Additionally, a detailed listing of major high strength geotextile installations during the last decade is included.

Introduction

Throughout history, soft soils have posed problems for engineers. Typically fine grained with high moisture contents, these soils are highly compressible and often saturated. The primary problems with soft soils result from low shear strengths and excessive consolidation settlements.

---

[1] **Manager, Engineering Services, Nicolon Corporation, 3500 Parkway Lane, Suite 500, Norcross, GA 30092**

[2] **Geosynthetics Engineer, Nicolon Corporation, 3500 Parkway Lane, Suite 500, Norcross, GA 30092**

Until recently, construction over soft soils was limited because of the high cost of stabilization. Today, however, engineers are taking advantage of a new cost effective improvement technique to utilize marginal sites. This new technique is based on a decade of experience, primarily involving projects designed by the U.S. Army Corps of Engineers. This paper presents four major projects and corresponding lessons learned that have been instrumental in the evolution of the use of high strength, high modulus geotextiles.

Overview

Cedergren and Weber (1962) summarized several methods of treatment available to reduce the problems associated with soft foundations. These methods include:

- Excavation and replacement of soft soil.
- Displacement of compressible material by end-loading.
- Staged construction - placing fill at controlled rates to allow for consolidation and strength gains.
- Installation of drains to facilitate consolidation.
- Pre-loading the site to reduce settlements of the structure and provide higher strength.
- Deposit improvement using admixtures or injections.
- Berms.
- Lightweight fill.
- Pile supported.
- Reinforcement of the soil matrix using a structural element.

While a wide variety of site improvement methods have been used during the past decade, soil reinforcement has emerged as an efficient, economical and effective solution to the problem of constructing embankments over soft soils.

Such embankments may be designed and constructed using a high strength, high modulus reinforcing geotextile at the base. The geotextile can reduce lateral displacement and improve the overall stability of the embankment.

This paper provides a comprehensive listing of the geotextiles and associated properties used in the most significant reinforced embankments to date. No attempt has been made to provide a geotechnical evaluation of geotextile suitability for each of the listed installations since little post-construction monitoring has been done. The authors are unaware of any failures associated with the listed projects and therefore offer these projects as evidence of an appropriate technique without drawing conclusions on each project.

General Design Approach

The following systematic six-step approach to the design of reinforced embankments has been presented by Fowler and Koerner (1987) and is shown graphically in Figure 1:

1) Conduct a bearing capacity analysis to determine the limits of the embankment geometry. (A)

2) Evaluate the embankment cross-section using a slope stability analysis. Locate the surface of maximum shear stress in both the embankment and the foundation. (B)

3) Conduct a deformation analysis to estimate the elongation of the reinforcement. (C)

4) Estimate the distribution of tensile force in the reinforcement along its length through an anchorage length evaluation. (D)

5) Gage the required interface shear strength between the embankment and the reinforcement using a base sliding evaluation. (E)

6) Establish the design tensile strength based on an evaluation of creep properties of the reinforcing geotextile.

This generally accepted design approach reflects both practical and theoretical advances which have resulted from building and monitoring significant projects since 1978.

Evolution of Projects

A detailed listing of major projects using high strength, high modulus geotextiles is given in Table 1. This table is an extension of an earlier compilation of projects by Fowler and Koerner (1987). Four of these projects have been thoroughly documented in the literature (U.S.ACOE, 1979; Fowler, 1985; Duarte and Satterlee, 1989; Schimelfenyg, Fowler, Leschinsky, 1990) and, as such, have been especially fruitful in advancing the technology associated with geotextile reinforced embankments over soft foundations.

These four projects were designed using comparable methodology within the Corps of Engineers and have in each case performed satisfactorily. No effort has been made to determine the relative performance of the other projects listed in Table 1.

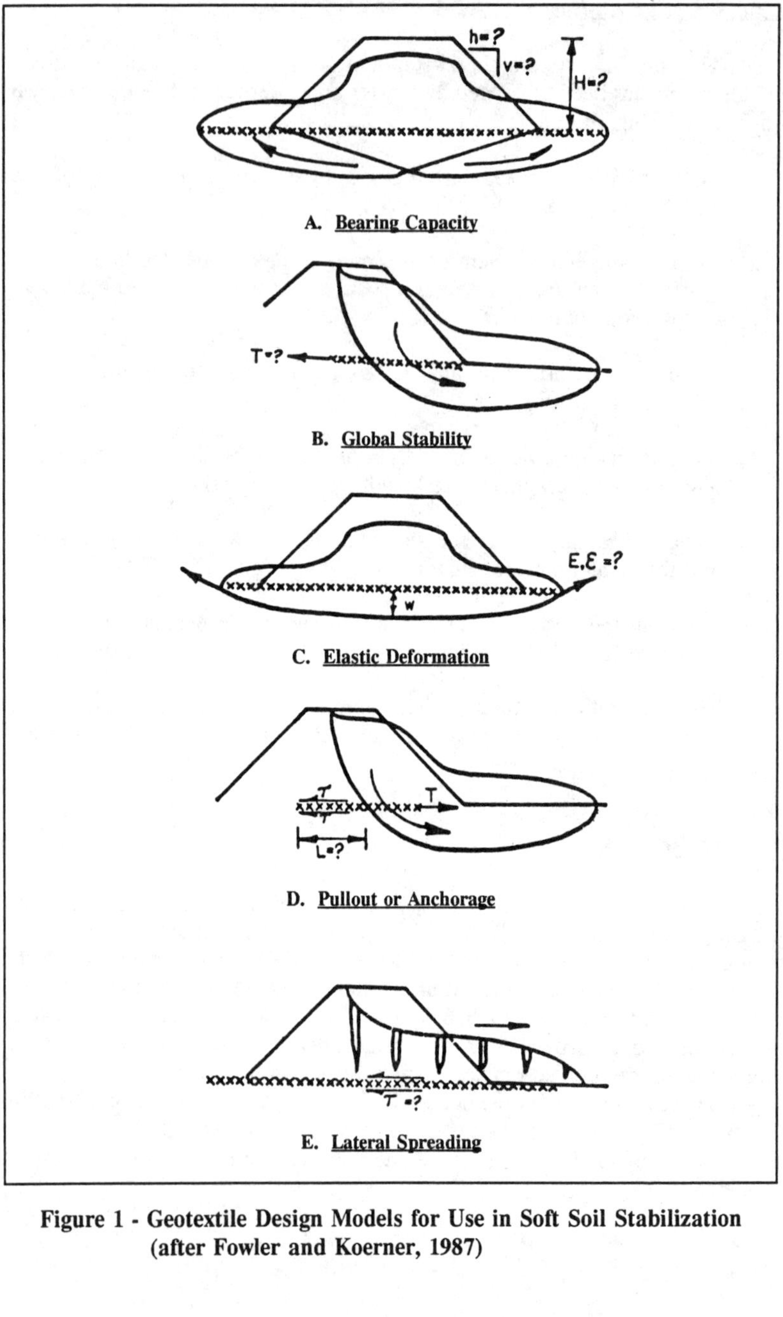

Figure 1 - Geotextile Design Models for Use in Soft Soil Stabilization (after Fowler and Koerner, 1987)

TABLE 1
HIGH STRENGTH REINFORCING GEOTEXTILE
MAJOR PROJECT HISTORY

FOUNDATION CONDITIONS				
	PROJECT NAME/ LOCATION	DATE	PROJECT DESCRIPTION	DESIGN ENGINEER
1.	PINTO PASS MOBILE, AL	1978	CONFINED DREDGED MATERIAL DISPOSAL AREA	C.O.E. MOBILE, AL
2.	INTERSTATE 494 ST. PAUL, MN	1981	ROADWAY EMBANKMENT	MINNESOTA D.O.T.
3.	CRANEY ISLAND NORFOLK, VA	1983	PERIMETER DIKE FOR DISPOSAL AREA	C.O.E. NORFOLK, VA
4.	WASHINGTON NATIONAL AIRPORT WASHINGTON D.C.	1983	RUNWAY EMBANKMENT EXTENSION	?
5.	BRUNSWICK PULP & PAPER BRUNSWICK, GA	1983	CONTAINMENT DIKE FOR BRINE STORAGE	BRUNSWICK PULP & PAPER
6.	SOUTH BLAKNEY ISLAND MOBILE, AL	1985	PERIMETER DIKE FOR DISPOSAL AREA	C.O.E. MOBILE, AL
7.	GREENWOOD ISLAND MOBILE, AL	1985	PERIMETER DIKE FOR DISPOSAL AREA	C.O.E. MOBILE, AL
8.	PARKS HIGHWAY NENANA, AK	1986	ROADWAY EMBANKMENT	ALASKA D.O.T.
9.	SEAGIRT BALTIMORE, MD	1986	AREA STABILIZATION OF DREDGED SPOIL MATERIAL	MARYLAND PORT ADMINISTRATION
10.	CRANEY ISLAND SOUTH NORFOLK, VA	1986	PERIMETER DIKE FOR DISPOSAL AREA	C.O.E. NORFOLK, VA
11.	NORTH BLAKNEY ISLAND MOBILE, AL	1986	PERIMETER DIKE FOR DISPOSAL AREA	C.O.E. MOBILE, AL
12.	WILMINGTON HARBOR SOUTH WILMINGTON, DE	1986	PERIMETER DIKE FOR DISPOSAL AREA	C.O.E. PHILADELPHIA, PA
13.	REACH A TEST SECTION PLAQUEMINES PARISH, LA	1986	HURRICANE PROTECTION LEVEE	C.O.E. NEW ORLEANS, LA
14.	E-W RUNWAY NEW ORLEANS, LA	1986	RUNWAY EMBANKMENT EXTENSION	LAWRENCE LAMBERT & ASSOC./EUSTIS ENGINEERING
15.	SOUTHWEST PASS PLAQUEMINES PARISH, LA	1987	FORESHORE DIKE	C.O.E. NEW ORLEANS, LA EUSTIC ENG.
16.	REACH A PLAQUEMINES PARISH, LA	1988	HURRICANE PROTECTION LEVEE	C.O.E. NEW ORLEANS, LA
17.	SOUTHWEST PASS PLAQUEMINES PARISH, LA	1988	FORESHORE DIKE	C.O.E. NEW ORLEANS, LA
18.	NEW BEDFORD NEW BEDFORD, MA	1988	CONTAMINATED SPOIL CONTAINMENT AREA	C.O.E. NEW ENGLAND
19.	CACHE SLOUGH SOLANO CO., CA	1988	SACRAMENTO RIVER FLOOD CONTROL	C.O.E. SACRAMENTO, CA
20.	BONNET CARRE NEW ORLEANS, LA	1988	HURRICANE PROTECTION LEVEE	C.O.E. NEW ORLEANS, LA
21.	WILDCAT RICHMOND, CA	1988	WILDCAT RIVER FLOOD CONTROL	C.O.E. SACRAMENTO, CA
22.	REACH A PLAQUEMINES PARISH, LA	1989	HURRICANE PROTECTION LEVEE	C.O.E. NEW ORLEANS. LA
23.	SOUTHWEST PASS PLAQUEMINES PARISH, LA	1989	FORESHORE DIKE	C.O.E. NEW ORLEANS, LA
24.	SEAGIRT BALTIMORE, MD	1989	AREA STABILIZATION OF DREDGED SPOILMATERIAL	MARYLAND PORT ADMINISTRATION
25.	PONTCHARTRAIN NEW ORLEANS, LA	1989	HURRICANE PROTECTION LEVEE	C.O.E. NEW ORLEANS, LA
26.	RECOVERY I NEW ORLEANS, LA	1990	CONTAINMENT LEVEE	WINK ENGINEERING
27.	REACH A PLAQUEMINES PARISH, LA	1990	HURRICANE PROTECTION LEVEE	C.O.E. NEW ORLEANS, LA
28.	US 90 TEST SECTION MORGAN CITY, LA	1990	ROADWAY STABILIZATION	LOUISIANA D.O.T.

TABLE 1 - CONTINUED

FOUNDATION CONDITIONS					DIKE GEOMETRY	
	SOIL	DEPTH (FEET)	SHEAR STRENGTH (PSF)	MOISTURE CONTENT (%)	FILL	HEIGTH (FEET)
1.	SOFT SILTY CLAY/PEAT	40	50-150	70-150	SAND	7
2.	PEAT	20	?	200-300	GLATIAL TILL	10
3.	MARINE CLAY	30-40	25-150	60-180	SAND	7-10
4.	CLAYEY SILT	16-20	40-100	75-100	SAND	16
5.	MARINE CLAY	30-40	20-40	50-100	SALT	25
6.	ORGANIC CLAY	16	40-100	70-150	SAND	10
7.	ORGANIC CLAY	15-30	40-100	70-150	SAND/CLAY	10
8.	GLACIAL TILL	13-26	20-60	50-100	VARIED	7
9.	ORGANIC CLAY	30	40-100	50-100	VARIED	3-16
10.	ORGANIC CLAY	16	100	70-180	SAND	10
11.	ORGANIC CLAY	16	80-100	70-150	SAND	7
12.	ORGANIC CLAY	30-40	40-100	70-160	SAND/GRAVEL	16-20
13.	ORGANIC CLAY	30-60	150-450	70-100	SAND/SILT	5-10
14.	MUCK/SOFT ORGANIC CLAY	70	20-400	20-730	SAND	5
15.	SOFT SILTY CLAY	50	160-420	70	SHELL	8
16.	PEAT/SOFT CLAY	30-60	150-450	25-285	SAND	10
17.	SOFT SILTY CLAY	50	160-420	70	SHELL	8
18.	ORGANIC CLAY	4-17	25-240	105	SILTY SAND	15
19.	ORGANIC SILTS & SILTY CLAYS	20	360-1440	24-130	SANDY SILTY CLAY	25
20.	SOFT CLAY	50	100-460	70	SHELL/CLAY	8-12
21.	SOFT CLAY	25	-	30	SANDY CLAY	8-15
22.	PEAT/SOFT CLAY	30-60	150-450	25-285	SAND	10
23.	SOFT SILTY CLAY	50	160-420	70	SHELL	8
24.	DREDGED SPOIL SILT	15-25	40-100	50-100	VARIED	7-12
25.	SOFT CLAY	100	25-500	20-100	SAND	5-12
26.	ORGANIC SOFT CLAY	15	50	300	SHELL/CLAY	6
27.	PEAT/SOFT CLAY	30-60	150-450	25-285	SAND	10
28.	MARSH/ORGANIC SILTY CLAY	150	100-200	53-110	SAND/SHELL	7-10

TABLE 1 - CONTINUED

	DIKE GEOMETRY				REINFORCING GEOTEXTILE (ACTUAL VALUES)			
	WIDTH (FEET)	SLOPE (H:V)	YARN TYPE	QUANTITY (SQ.YDS.)	5% WARP STRENGTH (LB/IN)	ULTIMATE WARP STRENGTH (LB/IN)	ULTIMATE FILL STRENGTH (LB/IN)	ULTIMATE SEAM STRENGTH (LB/IN)
1.	170	10:1	PP	3800 3300	250 50	900 225	170 245	260
2.	100	5:1	PP	-	-	1000	200	100
3.	200-235	(10-12):1	PP	93520	-	1025	455	300
4.	600	5:1	PET	-	-	1050	225	225
5.	50	2:1	PP	-	-	1060	300	-
6.	80	4:1	PET	-	-	1000	650	300
7.	80	4:1	PET	-	-	1000	650	300
8.	50	2:1	PET	-	-	2060	1070	500
9.	2200	-	PET/PP	247000	-	1100	1400	700
10.	200	10:1	PET	56780	-	2100	1400	700
11.	70	5:1	PET	-	-	1000	650	520
12.	400	12:1	PET	-	-	1500	1500	800
13.	100	15:1	PET	5000	-	3800	600	600
14.	90	5:1	PET	60000	1685	3970	1320	640
15.	70	(2-3):1	PP	142000	235	390	370	290
16.	125	(3-4):1	PET	109950	1700	3500	570	420
17.	70	(2-3):1	PP	133500	200	390	315	200
18.	100	3:1	PET	23400	1970	4350	1350	750
19.	175	3:1	PET	52000	1620	2775	850	500
20.	56-82	3:1	PET PET PET	1944 1566 1512	1685 - 520	3970 590 1400	1320 368 525	640 - -
21.	100	(2-3):1	PET PET	6200 8700	165 365	400 800	270 270	200 200
22.	125	(3-4):1	PET PET PP PET	3940 45420 8300 96430	1900 1775 1175 1570	4000 3660 1325 2940	- - 450 500	380 300 325 300
23.	70	(2-3):1	PP	117500	345	665	-	200
24.	75	2:1 BENCHED	PET/PP	32400	565	1340	1315	670
25.	75	(3-4):1	PET PET PET	34760 4370 37160	1550 1960 1890	3150 4100 4100	NA NA NA	350 350 350
26.	70	3:1	PET	50000	475	1085	845	340
27.	125	(3-4):1	PET	92250	1440	2660	550	300
28.	80-250	4:1	PET	10400	750	1900	500	300

PP = Polypropylene
PET = Polyester

- Pinto Pass - Pinto Island, Mobile Harbor, Alabama (U.S.ACOE, 1979)

In 1978, a "floating" dike using a reinforcing geotextile was recommended as the most cost-effective means of constructing an 8 ft (2.4m) high dredged material containment dike over soft foundation soils with undrained shear strengths ranging from 50 to 150 psf (2.4 to 7.2 kPa). Immediate disposal needs <u>and</u> allowances for future dike raising ruled out the commonly used gradual preloading of the site to consolidate these soft foundation materials.

Approximately 50% of the proposed embankment alignment was located in the inter-tidal zone of Mobile Harbor, and conventional analyses showed that the factor of safety at the site was approximately 0.5. Through proper embankment design, geotextile properties required to resist lateral spreading and splitting, embankment sliding, excessive displacement and a rotational slope failure were determined. Settlement plates and piezometers were installed to study the behavior of the reinforced embankment during and after construction.

Though initially a sand working table was used, geotextile installers learned to take advantage of the advancing surface mud wave to pick up the leading edge of the geotextile and lift it above the inter-tidal zone facilitating its sewing and deployment. The embankment was successfully constructed and no stability problems have occurred. After evaluating the field data, it was concluded that lateral spreading of the embankment can be held to about 5% or less with high tensile strength, high modulus geotextiles and that the use of geotextiles as reinforcement in rapid construction of embankments on soft foundations is technically feasible, operationally practical, and potentially very cost effective.

The geotextile-reinforced embankment allowed for better control of the desired configuration and offered a cost-savings of 40% compared to end-dumping displacement methods.

- Craney Island - Norfolk Harbor, Virginia (Fowler, 1985)

In 1983, three fabric-reinforced test sections, 300-, 400- and 750-ft long, were successfully constructed on very soft dredged material deposits within the Craney Island dredged material disposal area (Norfolk District, Norfolk, Virginia). Subsequently, two dikes, each about 4000 ft. long, were constructed, using this innovative construction technique to divide the 2500-acre area into three separate areas for improved dredged material management. Prior attempts using conventional displacement construction techniques to construct these dikes failed. Successful construction of these fabric-reinforced dikes was a key element in the rapid implementation of the dredged material management program. Implementation of this program will provide a cost savings of about $270 million over the next 36 years.

The use of a high strength, high modulus geotextile allowed the reinforced embankments of Test Sections 1 and 2 to be finished to design width and grade without excessive lateral spreading or rotational bearing failure in the foundation. The fabric also enabled Test Section 3 to experience a downward displacement of two volumes for one volume raised above the surface without the occurrence of an embankment failure. With the fabric warp yarns (machine direction) oriented perpendicular to the long axis of the dike, the fabric performed well and provided sufficient reinforcement to support rapid dike construction to the desired height, while maintaining the structural integrity of the embankment.

- Reach A - Lower Plaquemines Parish, Louisiana (Duarte and Satterlee, 1989)

In 1986, the New Orleans district of the U.S. Army Corps of Engineers was faced with the need to raise existing hurricane levees to provide protection against a 100-year storm. Stability analysis indicated that the existing factor of safety of 1.1 would drop to as low as 0.8 if the dike was raised conventionally. Proximity of a drainage canal prevented the use of stability berms.

In southern Louisiana, the two most commonly used methods to construct over very soft soils is to build on piles, or remove the soft organic material and replace it with sand or shells. Both alternatives are very costly, require specialized equipment, and are also very time consuming.

To fully evaluate the new alternative of high strength geotextiles, a levee test section that was reinforced by a single layer of high strength polyester geotextile was built. Strain gages were attached to the geotextile to measure the strain in the geotextile during and after construction. Settlement plates, piezometers, and inclinometers measured the deformation and performance of the composite structure. Geotextile tensile requirements were computed using the Spencer and Bishop circular arc methods, and the wedge method of slope stability analysis. To evaluate how accurately the theoretical values compare with the field data, the tensile values from the limit equilibrium slope stability analysis were compared to the tensile values recorded by the strain gages attached to the geotextile.

Two years of monitoring the embankment performance clearly showed that a geotextile reinforced levee is a viable method for increasing the flood protection. All of the data indicated that the section was working better than had been anticipated, especially since the field geotextile strains were lower than the computed values from limit equilibrium analysis. There was no evidence of cracks or any signs of unacceptable movements in the test section. Lateral deformations, measured by the inclinometers, had stopped for all practical purposes. Strain gage readings peaked during August 1987 and there was no further increase.

The geotextile reinforced levee alternative when applied to all future phases of the project will reduce the estimated construction cost for the 13 miles of levee from $85 million to $54.2 million; construction time from 13 years to 6 years; and marsh destruction from 4000 acres to 100 acres. Residents will also benefit from lower flood insurance premiums sooner because of the reduced construction time. After the levee is complete, the first floor of residential and commercial buildings will not have to be constructed 14 ft (4.3 meters) above the ground surface, as is presently required to prevent damage from flooding.

- New Bedford - New Bedford Harbor, Massachusetts
  (Schimelfenyg, Fowler and Leschinsky, 1990)

In 1988, the U.S. Army Corps of Engineers was contracted by EPA to perform a study to evaluate dredge equipment and dredged material disposal options. The study required construction of a Confined Disposal Facility (CDF).

The purpose of the CDF was to contain contaminated sediments that were dredged from the New Bedford Harbor bottom. Due to environmental and political restrictions, only sites with highly compressible and low strength foundation soils were available to construct the CDF. Excavation and replacement, foundation improvement, toe berms, and fabric reinforcement were used to aid the construction of the proposed dike. Also a CDF constructed of steel or wood sheet pile walls was considered. Cost estimates indicated that a dike with fabric reinforcement would be most economical.

Design. The dike section was designed using some of the latest procedures developed by the U.S. Army Corps of Engineers for fabric reinforced embankments on soft ground. Traditional principles and procedures were used to design the dike section for the CDF (USACOE, 1978). Fabric design was accomplished using more recent methodology (USACOE, draft), (Jewell, 1988), (USACOE 1987). Vertical drain spacing was calculated using the Hansbo approach (Koerner, 1986).

A factor of safety of 0.6 was calculated without using fabric. It was found that the factor of safety could be increased to 1.6 by adding a fabric with a tensile strength of 2,000 lbs/in. at 5% strain which corresponded to slightly less than 50% ultimate tensile strength (a typical value recommended for polyester) of the strongest high-strength fabric that had been manufactured in the U.S.A. at the time.

In addition, the pullout resistance, sliding failure resistance, and fill direction tensile strength requirements of the fabric were studied. It was found that at least 28 ft of fabric embedment was required beyond the critical failure circle to resist pullout of the fabric, the effective interface friction angle of the fabric needed to resist sliding of the granular fill on the fabric must be greater than 12.5 deg, and the ultimate tensile strength of the fabric and fabric seams must be greater than 600 lbs/in to resist active earth pressures and construction forces in the fill direction.

Short consolidation periods were needed to meet the proposed project schedule so vertical drains were used. The estimated consolidation periods were reduced from between 15 and 150 days to 3 and 30 days with the specified vertical drain spacing of 5 ft (1.5 m).

The fabric tensile strength was reduced 30% to account for the penetration of the vertical drains. The net result of the changed conditions produced a factor of safety of 1.2.

Fabric Installation. Individual panels of fabric were sewn together to form three segments of the dike at the manufacturer's plant. Each segment was folded in "accordion" fashion at the plant and shipped by truck to the project site.

The leading edge of each folded segment was attached to a long PVC pipe at the site. Large nylon ropes were attached to the pipe and used to pull each segment of the fabric into place from a spud barge anchored ahead of the leading edge of the fabric.

Dike Construction. Construction of the dike proceeded in two phases and two consolidation periods so that it would be easier to repair cracks expected to result from 5 to 6 ft of primary consolidation.

Phase I construction procedures required thin lifts placed in a concave shape and low ground pressure equipment to properly stress the fabric, and allow equipment on the fabric without damaging the fabric or the underlying organic clay. Phase II construction was performed using conventional equipment procedures.

Vertical Drains. Approximately 2,700 vertical drains were installed using the Phase I dike as a working mat. Each drain consisted of a polypropylene wick wrapped in a polypropylene geotextile sock. The drains were installed with a vibrating mandrel mounted on a specially fabricated low ground pressure pile driving rig.

Performance. Compatibility and strength tests were performed on the selected fabric to validate its suitability. The dike was instrumented with slope inclinometers, settlement plates, piezometers, monitoring wells, and stability poles. Strain gages were cemented to the fabric before the placement of fill and indicated that the average loads on the fabric were near the values estimated during design. The performance of the dike during the filling operation and after has been excellent, and clearly indicates that the design techniques used are adequate. Significant cost savings were realized when comparing the fabric reinforced dike to other options.

Construction Considerations

As the use of high strength geotextiles has increased, so has the knowledge of appropriate construction techniques. Following are some general construction considerations which apply to soft foundation conditions:

- Equipment Selection

  When building embankments on soft foundations, low ground pressure construction equipment is commonly required for both site preparation and initial select fill placement.

- Site Preparation

  Little site preparation is generally required other than assuring that the geotextile location can reflect the design requirements relative to the embankment cross-section. Available vegetative root mat is frequently left in place to help support construction activities.

- Geotextile Placement

  The high strength, high modulus reinforcing geotextile is commonly placed directly on the prepared embankment alignment. The process of geotextile placement, often requiring labor-intensive or marine construction techniques, can be simplified by prefabricating large geotextile panels and using experienced installation and sewing crews.

- Fill Placement/Spreading/Compaction Procedures

  Select fill is placed, spread and compacted over the geotextile using equipment and techniques designed not to compromise fabric integrity and foundation stability, and to take advantage of the advancing mud wave which may result from fill placement.

Summary and Conclusions

The listing of major projects in Table 1 and the four major projects further detailed in this paper provide solid evidence that high strength, high modulus reinforcing geotextiles can provide a cost-effective solution to engineers designing embankments over soft foundations.

More than a decade of practical experience, field data, and theoretical enhancements have refined the design approach initially proposed by the U.S. Army Corps of Engineers in 1978. Additionally, a better understanding of construction using high strength, high modulus geotextiles has led to general construction procedures which assure proper handling and installation.

REFERENCES

Cedergren, H.R. and Weber, W.G. (1962), "Subsidence of California Highways," Field Testing of Soils: ASTM STP No. 322, ASTM, Philadelphia, pp. 248-264.

Duarte, F.M. and Satterlee, G.S. (1989), "Case Study of A Geotextile Reinforced Levee on a Soft Clay Foundation", Proceedings of Geosynthetics '89 Conference, IFAI, pp. 160-171.

Fowler, J. (1985), "Fabric Reinforced Dikes, Craney Island", Proceedings of Geotechnical Fabrics Conference '85, IFAI, pp. 118-128.

Fowler, J. and Koerner, R.M. (1987), "Stabilization of Very Soft Soil Using Geosynthetics", Geosynthetics '87 Conference Proceedings, Industrial Fabrics Association International, St. Paul, MN, pp. 289-299.

Jewell, R.A., (1988). The mechanics of reinforced embankments on soft soils, Geotextiles and Geomembranes, Volume 7, Number 4.

Koerner, R.M., (1986). Designing with geosynthetics, Prentice-Hall, Englewood Cliffs, New Jersey 07632.

Schimelfenyg, P., Fowler, J., and Leschinsky, D. (1990), "Fabric Reinforced Containment Dike, New Bedford Superfund Site", Proceedings of The 4th International Conference on Geotextiles, Geomembranes and Related Products, Balkema, pp. 149-154.

U.S. Army Corps of Engineers, (1978). Design and Construction of Levees, Engineering Manual 1110-2-1601.

U.S. Army Corps of Engineers (1979), "Mobile District Floats Disposal Area Retaining Dike on Soft Foundation", Dredged Material Research Information Exchange Bulletin, Vol. D-79-1, April.

U.S. Army Corps of Engineers, (draft). Engineering Use of Geotextiles, Technical Manual 5-800-08.

U.S. Army Corps of Engineers, (1987). UTEXAS2 slope stability package, Instructional report GL-87-1.

Stability of Embankments over Weak Soils of Limited Thickness

Radoslaw L. Michalowski[1], Member, ASCE

Introduction

Embankments are often constructed over weak soils where the thickness of the weak layer is a fraction of the embankment width. In addition, the horizontal thrust in embankments causes loads on the foundation layer to have a significant horizontal component. Under such circumstances, classical bearing capacity solutions are not applicable. It is proposed that the upper bound approach of limit analysis be used to estimate the bearing capacity of finite thickness layers of weak soil under embankments. The linear increase of strength of soil with depth will be considered.

This paper addresses only the problem of embankment stability as related to the bearing capacity of a finite thickness layer of a weak foundation soil, and does not include analysis of all possible modes of embankment failures. In particular, local failures of embankment slopes and "spreading" are not considered here.

Reinforcement placed in the embankment is accounted for by assuming that, during failure, a part of or the entire horizontal thrust is transmitted to the reinforcement. Consequently, the outward horizontal load on the foundation is reduced, and the bearing capacity is increased. A more rigorous analysis to incorporate reinforcement is also briefly presented, where an additional term is included in the principle of virtual work, due to energy dissipation in reinforcement during failure.

Background

The bearing capacity of a finite thickness layer was first considered by Jürgenson (1936), and substantial contributions can be found in papers by Mandel and Salençon (1972), and by Matar and Salençon (1977); the subject was also considered by Silvestri (1983), and, more recently, in the context of reinforced embankments by Rowe and Soderman (1987) and Jewell (1988). The last three papers rely on the

[1]Department of Civil Engineering, The Johns Hopkins University, Baltimore, MD 21218-2699.

earlier plasticity solutions, but also present some modifications for embankment applications, and introduce different approaches (limit equilibrium and FEM).

The plasticity solutions presented by Mandel and Salençon (1972) and Matar and Salençon (1977) utilize the method of characteristics to solve the set of hyperbolic differential equations describing the stress field in a layer (plane strain). Since the mathematical problem is of a hyperbolic type, the stress state can be calculated in the neighborhood of the loaded area without solving the stress state in the entire layer (or half-space). The solution starts from a given boundary condition outside the loaded area (for embankments: zero load). Due to the hyperbolic type of the stress differential equations, the solution for a narrow load over a small thickness layer is included in the solution for a more widely loaded area over the same layer. The distribution of the load so obtained is then independent of the width of the load. For embankments, however, the distribution of the interface stress at failure may not necessarily be equal to that from the plasticity solution: the true distribution is influenced by the embankment itself. In particular, the assumption of uniform distribution of the shear stress along the embankment fill - foundation soil interface may be inaccurate. Also, the often-used assumption that the distribution of the vertical stress is equal to $\gamma_f H$ (γ_f - unit weight of the fill, H - height of the embankment) is only an approximation, and it leads to violation of equilibrium equations under the sloping sides of an embankment. To avoid these inconsistencies, the results here will be presented in the form of an average vertical limit pressure as a function of b/t (b - half-width of the embankment, t - thickness of the layer).

The kinematical method of limit analysis is based on the application of the upper bound theorem. In order for the theorem to be applicable, the following assumptions need to be made: (a) the material is perfectly plastic at the limit state, (b) the limit state of the soil is described by a yield function convex in the stress space, and (c) the soil obeys the flow rule associated with the yield function (normality rule). The upper bound theorem can be expressed as follows: *the principle of virtual work applied to a kinematically admissible collapse mechanism provides a collapse (limit) load which is not lower than the true limit load.* The kinematically admissible failure mechanism must satisfy the kinematical boundary conditions, and the velocity field needs to be derived from the potential flow rule associated with the yield function

$$\dot{\varepsilon}_{ij} = \lambda \frac{\partial F(\sigma_{ij})}{\partial \sigma_{ij}}, \qquad \lambda \geq 0 \tag{1}$$

where $\dot{\varepsilon}_{ij}$ is the strain rate tensor, $F(\sigma_{ij}) = 0$ is the yield criterion, σ_{ij} is the stress tensor, and λ is a non-negative multiplier. The simplest form of function $F(\sigma_{ij}) = 0$ describing the limit state of cohesive (frictionless) soils is the Tresca yield condition

$$F(\sigma_x, \sigma_z, \tau_{xz}) = \sqrt{(\sigma_x - \sigma_z)^2 + 4\tau_{xz}^2} - 2c = 0 \tag{2}$$

where c is the cohesion of the soil. Flow rule (1) associated with the Tresca yield condition leads to velocity fields with incompressible deformation. It also implies that the velocity jump vector across a velocity discontinuity considered as a non-elongating line is constant and tangent to that discontinuity. The yield condition expressed by eq. (2) is generally accepted as a function describing the limit condition of clays under undrained conditions, and it also will be used in this paper.

Numerical Solution

An example of a layer of thickness t loaded by a vertical and outward horizontal load is shown in Fig. 1(a). The shear strength of the soil increases linearly with depth

$$c = c_m + \xi z \tag{3}$$

where c_m is the shear strength at the surface and ξ is the strength gradient. The hodograph for the assumed failure mechanism is shown in Fig. 1(b). Angles α, β, and the number of blocks in the mechanism are variable, *i.e.*, they are not assumed *a priori*, but are found from the optimization procedure where the minimum of $\overline{q}$ is sought (similar to slope stability analyses where the location of the slip surface is sought). Vertical velocity component V_0 is given along boundary OBD. The horizontal load along OBD is assumed to be distributed uniformly, and equal to $\overline{\tau} = \chi c_m$ $(0 \leq \chi \leq 1)$. This load comes from the horizontal thrust in the embankment, and is assumed known. The shear stress at the layer base interface during failure is equal to the strength of this interface and is represented as a fraction of the shear strength of the soil at the base: κc $(0 \leq \kappa \leq 1)$. The unknown is the average pressure $\overline{q}$. Note that, since the vertical velocity along boundary OBD is uniform, the average pressure $\overline{q}$ following from the kinematical approach is independent of its true distribution.

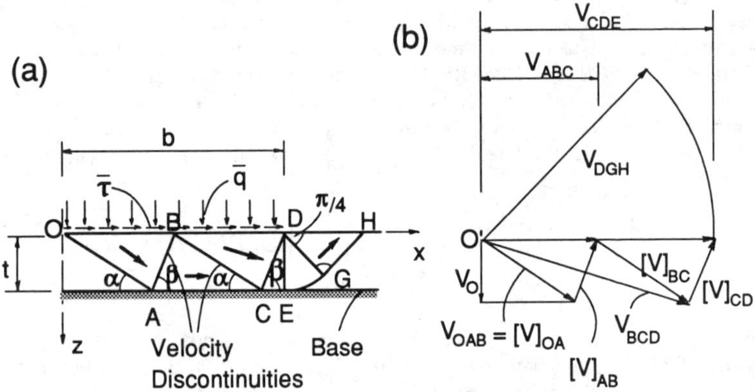

Figure 1. **(a)** Collapse mechanism of a layer; **(b)** Hodograph.

The principle of virtual work can be written in general form as

$$\int_v \dot{\varepsilon}_{ij}\sigma_{ij}dv + \int_L c[V]dL + \int_{L_b} \kappa c[V]_b dL_b = \int_{S_{OD}} Q_i V_i^0 dS + \int_v \gamma_i V_i dv \qquad (4)$$

where the first term on the left-hand side describes the work dissipation in continually deforming region DEG (Fig. 1(a)), and the other two terms represent dissipation rates along all the discontinuities within the layer and along the base interface, respectively. The first term on the right-hand side is the work done by boundary forces Q_i (vector with components $\bar{\tau}$ and $\bar{q}$), and the last term represents work done by gravitational forces, γ_i being the unit weight vector, and V_i being the velocity field. Note that the material is assumed to be incompressible, therefore the mass displaced downward along boundary OBD must be equal to that displaced through boundary DH (incipient mass flux through both boundaries must be equal in magnitude and opposite in sense). Consequently, since the density of the material is homogeneous, the total work done by gravity forces represented by the last term in eq. (4) must be zero.

All terms in eq. (4) can be calculated making use of the hodograph (Fig. 1(b)) and eqs. (1) and (2); the only term which cannot be calculated explicitly is that involving the unknown $\bar{q}$. Equation (4) is then used to calculate $\bar{q}$. Due to limited space the detailed derivation of the respective terms will be presented elsewhere (Michalowski, 1991b). Note that all terms in eq. (4) are functions of angles α, β, and the number of blocks in the mechanism. The results presented in Fig. 2 show the least values of $\bar{q}/c_m$ obtained through course of minimization. These results are shown for both smooth and rough layer bases, and for outward ($0 < \chi \leq 1$) and inward ($-1 \leq \chi < 0$) horizontal loads typical of unreinforced and reinforced embankments, respectively.

Closed-Form Solution

Presenting results for a wide range of useful combinations of all parameters (χ, κ, $\xi b/c_m$, and b/t) is not practical, therefore an analytical solution was attempted. To arrive at a closed-form solution, angles α and β were assumed equal to $\pi/4$. Consequently, the solution was obtained for ratios b/t equal to integers. The details of the solution will be presented elsewhere (Michalowski, 1991b). The solution can be written as

$$\frac{\bar{q}}{c_m} = 3 + \frac{\pi}{2} + \frac{1}{2}\frac{b}{t}(\kappa - \chi) + \frac{\xi b}{c_m}\left[\frac{t}{b}(1+\sqrt{2}) + \frac{\kappa}{2}\right], \qquad \frac{b}{t} = 2,4,6,... \qquad (5)$$

and

$$\frac{\bar{q}}{c_m} = 3 + \frac{\pi}{2} + \frac{1}{2}\frac{b}{t}(\kappa - \chi) + \frac{1}{2}\frac{t}{b}(\kappa + \chi) + \frac{\xi b}{c_m}\left[\frac{t}{b}(1+\sqrt{2}) + \frac{\kappa}{2}\left(1 + \frac{t^2}{b^2}\right)\right], \qquad \frac{b}{t} = 1,3,5,... \qquad (6)$$

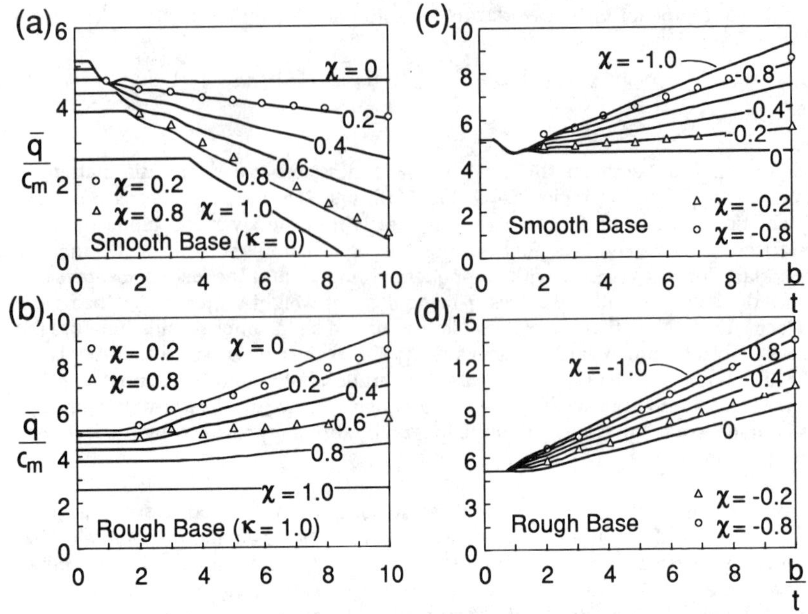

Figure 2. Average limit pressure on a layer as a function of b/t. Numerical solutions (solid lines) and analytical solutions (bullets) for smooth and rough bases, and outward ($\chi > 0$) and inward ($\chi < 0$) horizontal forces (homogeneous foundation soil, $\xi = 0$).

Limit loads obtained from the analytical solution (eqs. (5) and (6)) are presented in Fig. 2 as circular and triangular bullets. As expected, the closed-form solution yields higher limit loads than the numerical one. The best match of both analytical and numerical solutions is for the case of smooth base and low horizontal loads (low χ). For $\kappa = 0$ and $\chi = 0$ the two solutions yield identical results. For outward horizontal forces (unreinforced embankments), the overestimation becomes higher when χ increases and is greatly influenced by the strength of the base. The analytical solution yields results almost identical to the numerical one for inward horizontal loads, typical of reinforced embankments ($\chi < 0$), independent of whether the layer is resting on a smooth or rough base.

As the analytical solution (eqs. (5) and (6)) is strictly valid only for b/t equal to integers, this solution should be interpolated for b/t other than integers. Since the analytical solution is a continuous function of ratio b/t, the interpolation can be performed by substituting the required value of b/t (a positive real number) directly into eq. (5) or eq. (6). While substitution of b/t with values other than integers contradicts the assumptions used to derive eqs. (5) and (6), it should be viewed only as an interpolation technique.

Influence of Reinforcement

In the previous sections the embankment was regarded as an inclined load over the surface of a weak layer. The total thrust (horizontal force) in the embankment at the plane of symmetry can be estimated as $\gamma_f H^2 K/2$, where H is the embankment height, γ_f is the unit weight of the embankment fill, and K is the coefficient of lateral pressure. For unreinforced embankments, this thrust is transferred to the foundation soil as an outward tangential load characterized by coefficient $\chi = \bar{\tau}/c_m$. For reinforced embankments, part or all of the thrust is taken over by the reinforcement. Coefficient χ can be calculated as (implication of tensile failure rather than pull-out)

$$\chi = \frac{\bar{\tau}}{c_m} = \frac{\gamma H^2 K - 2R}{2bc_m} \tag{7}$$

where R is the force in reinforcement per unit length of embankment. For practical purposes, R can be assumed equal to the tensile strength of the reinforcement (when substantial lateral spread of the embankment can be tolerated, *e.g.*, for temporary structures) or it can be assumed as a tensile force at some specified allowable lateral embankment strain. Coefficient χ cannot be lower than -1 (failure of the foundation soil is not associated in such case with the reinforcement failure), or larger than 1 (spreading occurs: failure along the embankment fill - foundation soil interface). Coefficient of lateral pressure K cannot be smaller than $tan^2(\pi/4 - \varphi_f/2)$ ($K = (1 - sin\varphi_f)$ is recommended for conservative design).

It is conceivable that, if the reinforcement is strong enough, the soil is displaced laterally during collapse of the foundation soil layer, while the embankment does not spread, due to being constrained by the reinforcement. In such case forces on the embankment fill - foundation layer interface are directed inward ($-1 \leq \chi < 0$). While the true collapse mechanism may not contain a clear failure surface between the reinforced embankment and the foundation soil, the results based on such an assumption are acceptable from a design standpoint (Jewell, 1988). However, under working conditions, coefficient χ is likely to be positive. Only at the onset of failure, when the soft soil begins to be "squeezed out" from underneath the embankment (incipient failure), may the total thrust in the embankment be equilibrated by the force in the reinforcement, and further increase in the reinforcement force due to movement of the foundation soil with respect to the fill may cause inward horizontal forces (negative χ).

Once χ is calculated from eq. (7), the particular diagram or formula can be used to determine $\bar{q}/c_m$. Since the total vertical force under one half of the embankment, $\bar{q}b$, must be equal to the weight of the fill, $\gamma_f(bH - ½H^2 cot\beta)$, the critical (failure) height can be calculated as

$$H = b \tan\beta \left(1 - \sqrt{1 - \frac{2\bar{q}}{b\gamma_f \tan\beta}}\right) \tag{8}$$

where β is the slope angle (a negative value under the square root would indicate that a triangular-shape embankment of given b and β does not exhaust the bearing capacity of the layer). Note that χ is a function of H, therefore the process of finding the critical embankment height is iterative. Alternatively, as suggested in the literature (*e.g.*, Bonaparte *et al.* 1987, and Rowe and Soderman 1987), the computational width of the embankment can be taken at its mid-height and the failure height calculated as $H = \bar{q}/\gamma_f$. This alternative procedure is conservative for embankments on layers over rough bases, and not conservative for unreinforced embankments on layers over smooth bases. In most cases it is acceptable, as smooth interfaces are rarely found in practice.

Estimating the thrust in the embankment and using the solution for the bearing capacity of a foundation soil loaded with an inclined load is a reasonable engineering approach to embankment design. In the following the reinforcement is accounted for in a more rigorous way. The failure mechanism is extended here to the embankment fill. The advantage of such analysis is that the force on the embankment fill - foundation layer interface is an internal force and does not enter the principle of virtual work explicitly. Consequently, the distribution of the horizontal load on the layer need not be assumed, and strict upper bounds to the critical height can be found. An admissible collapse mechanism is shown in Fig. 3. During failure the embankment is assumed to fail along the rupture surfaces (velocity discontinuities), and its displacement is compatible with that of the rigid blocks in the foundation layer.

The height of the embankment is now the unknown. Thus, the principle of virtual work will have no term due to external load Q_i, but, instead, it will have a term due to the weight of the embankment. The reinforcement is incorporated into the failure mechanism. During collapse of the embankment the reinforcement is assumed to fail in tension (not "pull-out"). The strain rate of the reinforcement can

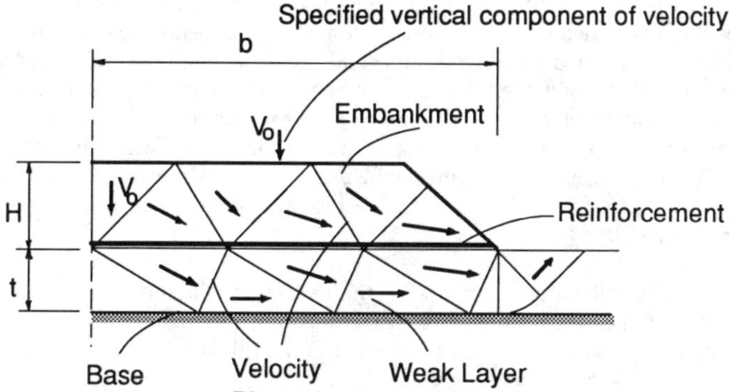

Figure 3. A kinematically admissible mechanism of embankment and foundation soil failure.

be calculated, and the rate of energy dissipation in the reinforcement during failure (per unit length of one symmetrical half of the embankment) is found as

$$\dot{D}_R = \int_0^b \dot{\varepsilon}_l R \, dl \tag{9}$$

where $\dot{\varepsilon}_l$ is the rate of normal strain (in the b direction) of the reinforcement, R is the tensile yield limit of the reinforcement per unit length of the embankment, and b is the length of reinforcement (equal to the half-width of the embankment).

The second approach to finding the critical height of embankments is presented in more detail elsewhere (Michalowski, 1991a) in the context of embankments over a half-space. For weak layers, the collapse mechanism becomes more complex (as in Fig. 3) and the practical application of this approach seems to be questionable. The calculations performed show that the two approaches presented yield similar results. As expected, the differences become significant when a coefficient K other than that of the active pressure is used in the first approach (eq. (7)).

Example

The following example illustrates the use of the results presented. Calculate the maximum (critical) height of an embankment of width $2b = 32$ m, with slopes inclined at $\beta = 30°$, built out of granular material with internal friction angle $\varphi_f = 35°$ and unit weight $\gamma_f = 20$ kN/m^3, over a soft layer of thickness $t = 2$ m. The undrained shear strength in the layer increases linearly from $c_m = 5$ kN/m^2 at the surface to 8 kN/m^2 at the bottom of the layer. The layer is resting on a perfectly rough base. Find the critical height for an unreinforced embankment and a *fully reinforced* embankment.

As the tangential load on the soft layer is dependent on the height of the embankment, the procedure is iterative. The first guess of the critical height is made as: $H = 4.0$ m. Using eq. (7) with $R = 0$ (no reinforcement) and $K = \tan^2(\pi/4 - \varphi_f/2)$ leads to $\chi = 0.542$. Coefficient ξ is found to be equal to 1.5, thus $\xi b/c_m$ is 4.8, and eq. (5) is used, with $b/t = 8$ and $\kappa = 1$ (rough base), to calculate the limit pressure. The dimensionless pressure is calculated as $\bar{q}/c_m = 10.251$, and eq. (8) leads to $H = 3.07$ m. As this value differs from the assumed 4.0 m, a second guess is made: $H = 3.3$ m; χ is now equal to 0.369 (eq. (7)) and eq. (5) yields $\bar{q}/c_m = 10.943$, and $H = 3.34$ m (eq. (8)). The second guess of $H = 3.3$ m can be regarded as the critical height, since it is very close to the 3.34 m obtained from the second iteration. It is recommended that this value be reduced by 10% since the analytical solution for $\bar{q}/c_m$ was used, which overestimates the lesser upper bound from numerical solution (while both are upper bounds, the numerical solution is believed to be a good approximation of the true limit load). Hence, $H = 3.0$ m.

For a *fully reinforced* embankment $\chi = -1$, independent of the embankment height. From eq. (5): $\bar{q}/c_m = 16.419$, and from eq. (8): $H = 6.16$ m. This value is not reduced as the closed form solution used for $\chi < 0$ does not significantly overestimate the numerical one. The necessary reinforcement strength is now calculated from eq. (7) where $\chi = -1$: R = 183 kN per 1m of the embankment length.

Appropriate safety factors can be used to arrive at admissible heights.

Conclusions

Solutions to the bearing capacity of finite layers of cohesive soils were presented accounting for horizontal load components from embankments, increase of soil strength with depth, and different base strengths. Comparison of the solutions obtained to those by Mandel and Salençon (1972) and by Matar and Salençon (1977) is shown in Fig. 4. Not surprisingly, for fully mobilized inward forces ($\chi = -1$), the upper bound solution presented slightly exceeds the solutions based on the method of characteristics.

The results for outward horizontal forces, typical of unreinforced embankments, are significantly lower than the solutions of Mandel and Salençon, and Matar and Salençon. These solutions were thought of as the bearing capacity of layers under rough rigid footings, not under embankments. The author would like to emphasize

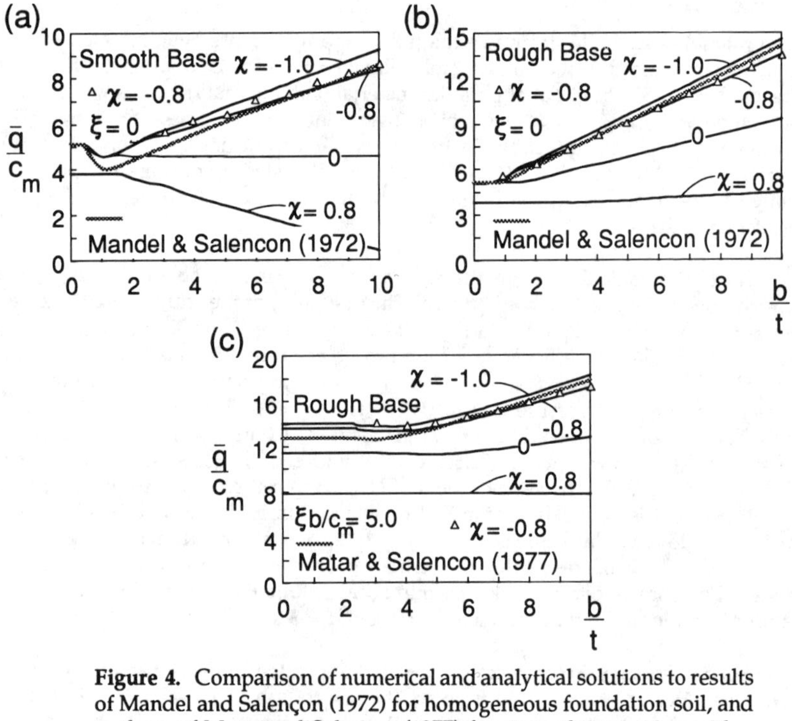

Figure 4. Comparison of numerical and analytical solutions to results of Mandel and Salençon (1972) for homogeneous foundation soil, and to those of Matar and Salençon (1977) for strength increasing with depth.

that the correctness of the solutions of Mandel and Salençon, and Matar and Salençon is not questioned, and only the extension of their applicability as the bearing capacity under unreinforced or not fully reinforced embankments is disputed.

An approximate solution to the bearing capacity of a layer for an arbitrary mobilization of the horizontal load was given earlier by Jewell (1988). This solution seems to be conservative, and, in the limit case of fully mobilized inward forces (*fully reinforced* embankments), it slightly underestimates the solution by Mandel and Salençon. It also underestimates the solutions presented in this paper for arbitrary values of χ.

Only selected results from research in progress were presented. The details of the methods used, solutions for a broader range of parameters, and further discussion, will be presented elsewhere.

Acknowledgement.
The research presented was funded in part by the National Science Foundation, grant number MSS-9107778.

References

Bonaparte, R., Holtz, R.D. and Giroud, J.P. (1987). *Soil reinforcement design using geotextiles*. Symposium on Geotextiles, Geomembranes and Related Products, ed. J.E. Fluet, Los Angeles, 1985.

Jewell, R.A. (1988). "The mechanics of reinforced embankments on soft soils". *Geotextiles and Geomembranes*, 237-273.

Jürgenson, L. (1936). "On the stability of foundations of embankments." in *1st ICSMFE*, vol. II, 194-200.

Mandel, J. and Salençon, J. (1972)."Force portante d'un sol sur une assise rigide (étude théorique)." *Géotechnique*, 22, 79-93.

Matar, M. and Salençon, L. (1977). "Capacité portante a une semelle filante sur sol purement cohérent d'épaisseur limitée et de cohésion variable avec la profondeur". *Annales de l'Institute Technique du Batiment et des Travaux Publics*, Suplément No. 352, Serie: *Sols et Fondations*, 143, 95-107.

Michalowski, R.L. (1991a). "Bearing capacity of non-homogeneous cohesive soils under embankments". *J. Geot. Eng.* (accepted for publication).

Michalowski, R.L. (1991b). "Bearing capacity of weak layers under embankments." *J. Geot. Eng.* (submitted for publication).

Rowe, R.K. and Soderman, K.L. (1987). "Stabilization of very soft soils using high strength geosynthetics: the role of finite element analyses". *Geotextiles and Geomembranes*, 6, 53-80.

Silvestri, V. (1983)."The bearing capacity of dikes and fills founded on soft soils of limited thickness." *Can. Geot. J.*, 20, 428-436.

Notation

b - half-width of the embankment
c - cohesion or undrained shear strength
c_b - shear strength of the base interface
c_m - minimum value of shear strength of soil (at surface)
H - embankment height
K - coefficient of lateral pressure
$\overline{q}$ - average vertical limit pressure over foundation soil layer
R - tensile force in reinforcement per unit length of embankment
t - layer thickness
α - geometrical parameter (angle) of the failure mechanism
β - geometrical parameter of the failure mechanism or the embankment slope angle
γ_f - unit weight of embankment fill
κ - c_b/c
ξ - strength gradient
χ - horizontal load intensity ($\chi = \overline{\tau}/c_m$)
$\overline{\tau}$ - average shear stress on the embankment fill - foundation soil interface
φ_f - internal friction angle of embankment fill

Factors Influencing Passive Pullout Resistance

Joon-Ik Sohn[1], Soo-Il Kim[2], Young-Jin Kim[3], Dong-Deok Yoon[4]

Abstract

This paper is concerned with factors influencing passive resistance of the protruded bearing members attatched to the inextensible reinforcement of plane shape. For this purpose, sets of model pullout tests have been performed to investigate ultimate pullout capacity of passive-resistant reinforcement under various boundary conditions. Considerations are given to the effects of height and width of protrusion, length of reinforcement, relative spacing of protrusions with regard to protrusion height and overburden pressure above the reinforcement.

The results of the tests are presented in the form of charts with some mechanical interpretations. The conclusions could be used as a guideline to evaluate the pullout resistance of the protruded reinforcement embedded in a low-quality backfill such as a weathered granite soil.

Introduction

It has been recognized that sandy soils usually used as backfill materials in reinforced soil walls are expensive, if they are located far from the construction site. The reinforcement technique to use on-site low quality soils for backfill has been developed and exercised in many countries to improve such shortcomings(Mitchell and Villet, 1987). On-site low quality backfill materials have less frictional resistance and drainage capacity compared to sandy soils. Weathered granite soils easily obtainable in Korea have relatively good quality for backfill purpose if not too much weathered.

---

1. Research Fellow, Geotechnical Engineering Division, Korea Institute of Construction Technology, 142, Umyon-dong, Socho-gu, Seoul, Korea
2. Professor, Civil Engineering Department, Yonsei University, 132, Shin-chon-dong, Seodaemun-gu, Seoul, Korea
3. Senior Researcher, Geotechnical Engineering Division, Korea Institute of Construction Technology, 142, Umyon-dong, Socho-gu, Seoul, Korea
4. Civil Engineer, Soil/Structures Department, Daewoo Engineering Co., 12-3, Yoido-dong, Yongdeungpo-gu, Seoul, Korea

The reinforcement system is internally stable when the pullout resistance is sufficiently mobilized between the reinforcement and the earth, provided that the reinforcement is stiff enough. In this research, protruded bearing members are attached to the inextensible reinforcement of plane shape so as to increase the pullout resistance, and the characteristics of passive pullout resistance are analysed for the conditions of varying the followings: length of reinforcement; width and height of protruded bearing members; relative spacing of protrusions with regard to the protrusion height; overburden pressure.

Experimental Details

Sample Soil

The soil used for the model tests is a weathered granite soil obtained in Mansu-dong, Inchon, and it is a sandy clay which can be classified as CL according to the Unified Soil Classification System. The gradation curve shows the uniformity coefficient is 8.7 and the coefficient of curvature is 1.07, which represents a relatively well graded shape. The coefficient of permeability is 1.43×10^{-5} cm/sec at the maximum dry density, which is a fairly high value for a backfill material. The results of compaction tests showed that the maximum dry density is 1.73 t/m³ with an optimum moisture content of 15 % .

Direct shear tests were performed to determine the strength parameters. As shown in Fig. 1, the internal friction angle ϕ of the sample soil is 42° , and the cohesion c is 20.6 kPa. Also the friction angle δ between the sample soil and the steel reinforcement of plane shape is 22° , and the adhesion is 9.8 kPa.

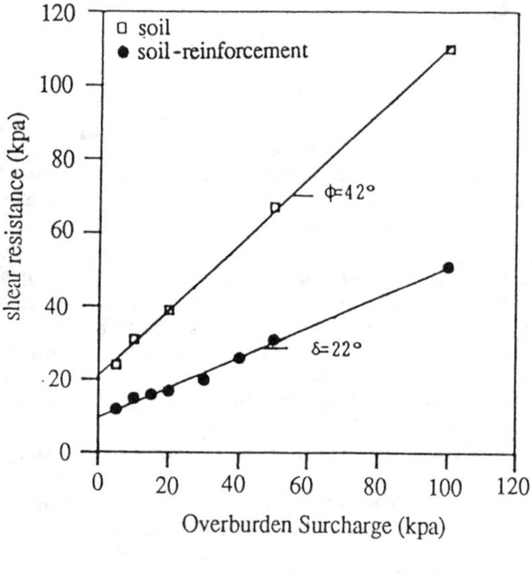

Fig. 1 Results of direct shear tests for the sample soil

Testing Apparatus

The size of the model box is 1.5 m in width, 1.0 m in length and 1.2 m in height. One side of the box is constructed of a transparent acryl panel of 50 mm thickness marked with reference grids at 5 cm intervals in horizontal and vertical directions. The front side can be opened for ease of installation, and hydraulic pistons are connected to both sides of the box. Rollers are also attatched on the bottom corners so as to move the model box horizontally. Another hydraulic piston is installed at the bottom for pulling out the reinforcement. The piston of the loading apparatus is 180 mm in diameter, and its maximum designed loading capacity is 294 kN. Loading plates of different sizes can be replaced if necessary. The pullout piston is 40 mm in diameter, and its maximum loading capacity is 19.6 kN and can be increased for future use. A view of the test apparatus is shown in Fig. 2.

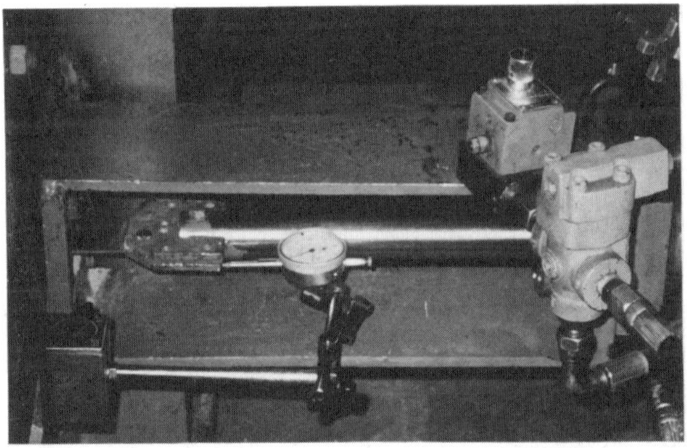

Fig. 2 View of test apparatus

Test Procedure

Soils are placed and compacted at 85 % of relative compaction before the reinforcement is installed. After the model is completed, the tests are performed by applying the designed overburden pressures and pullout forces in sequence. The reinforcement member was a 3 mm thick steel strip with a tensile strength greater than 196 MPa.

In order to evaluate the passive resistance of the protruded bearing members 5 sets of different model tests were performed. The dimensions of reinforcement are varied in different tests. Test variables are the effective length L_e and width B of the reinforcement, height t_p and spacing S_h of the protrusions and surcharge pressure σ_v. Here the effective length L_e of the reinforcement denotes the length between the first and the last protrusions plus one protrusion spacing, while the length L is the total of the effective length L_e and the remaining length for frictional resistance. The values assigned for the variables are included in Table 1.

Table 1. Test Variables with Assigned Values

Test Set	L/L_e(cm)	B(cm)	$t_p/2$(cm)	S_h(cm)	σ_v(kPa)
No. 1	100/60	6	1.5	20	20.6
	40/20	6	1.5	20	20.6
	70/40	6	1.5	20	20.6
	120/80	6	1.5	20	20.6
No. 2	100/60	3	1.5	20	20.6
	100/60	6	1.5	20	20.6
	100/60	9	1.5	20	20.6
No. 3	100/60	6	1.0	20	20.6
	100/60	6	1.5	20	20.6
	100/60	6	2.5	20	20.6
No. 4	120/120	6	1.5	20	20.6
	120/120	6	1.5	40	20.6
	120/160	6	1.5	80	20.6
	100/60	6	2.5	20	20.6
No. 5	100/60	6	1.5	20	20.6
	100/60	6	1.5	20	37.3
	100/60	6	1.5	20	53.0
	100/60	6	1.5	20	85.3

Passive Pullout Resistance

The problem of passive pullout resistance of the reinforcement is similar to that of an anchor plate. As shown in Fig. 3, the protrusions can be regarded as a set of passive anchor plates installed at uniform spacing. The protruded height t_p is very small relative to the fill height z, which means the reinforcement is embedded relatively deep, i.e., z/t_p is very large. As illustrated in Fig. 3, the total pullout resistance (P_t) is the sum of the frictional capacity of the plain reinforcement portion (P_f) and the passive resistance force acting against protruded bearing members (P_p). Accordingly, the passive resistance force is calculated by subtracting P_f from P_t. The passive resistance pressure σ_b is generally defined to be an average bearing pressure acting uniformly on the protruded area (B t_p) of n protrusions, and it can be represented by the following equation.

$$\sigma_b = P_p \ / \ n \ B \ t_p \qquad (1)$$

where P_p = the passive resistance force; B = the width of reinforcement.

As studied by many researchers, the passive resistance pressure σ_b depends also on the overburden pressure σ_v. The relationship can be expressed with a parameter like the frictional coefficient. The parameter is generally defined as the coefficient of passive resistance F_p, which is σ_b/σ_v. According to Rowe and Davis(1982), F_p depends on angle of friction, soil dilatancy, surface roughness of the protruded part and initial stress state, etc. Also they suggested a relationship between F_p, and internal friction angle ϕ for different z/t_p ratios.

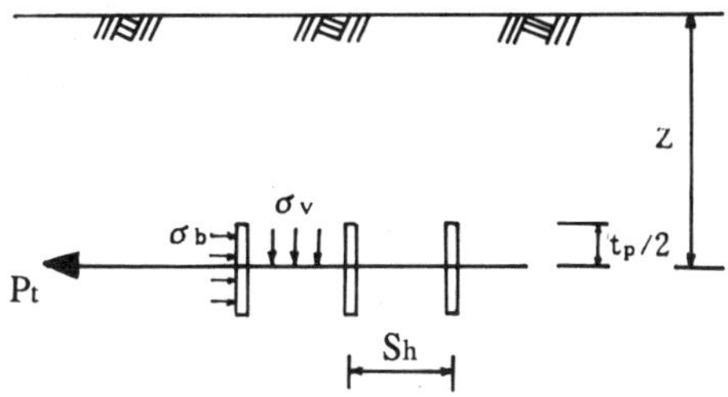

Fig. 3 Pullout resistance of the protruded reinforcement

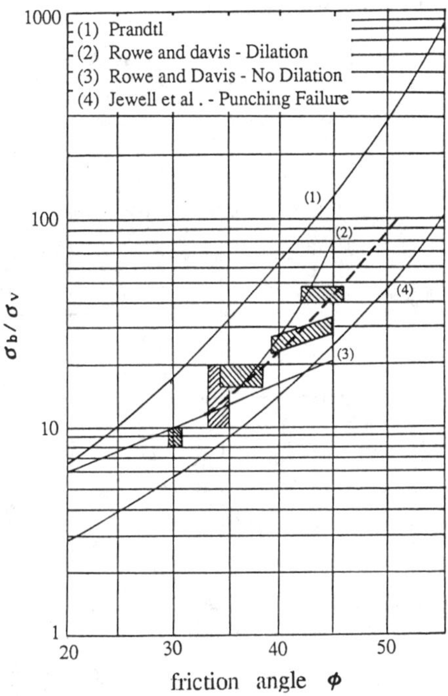

Fig. 4 Passive resistance data of pullout tests, Jewell (1984)

Jewell(1984) showed the range of F_p values for different $\bar{\phi}$ values based on the results of finite element analysis performed by Rowe and Davis and six other sets of published experimental data(Fig. 4). The range of F_p values in the figure is bounded by two limits. The upper limit was defined using the bearing capacity equation, and the lower limit was regarded to be the condition of punching shear failure calculated using the slip-line analysis method. It is interesting to note that the results of Rowe and Davis fall within the bounded range. However, it should also be pointed out that the range of the data is large, and Fp could be influenced by the relative size of shape and spacing of the bearing members.

Test Results

As discussed before, F_p is possibly dependent on z/t_p and ϕ. Besides of these factors, however, pullout resistance would be quite influenced by other factors which represent the shape and size of the protrusions: the reinforcement width B, the protruded height t_p, the horizonal spacing S_h, etc. The overburden pressure is another influencing factor. The tested data for these variable conditions are summarized and listed in Table 2. Discussed are made next on the test results acçounting for the influencing factors on the coefficient of passive resistance F_p.

Table 2. Summary of the Test Results

Test Set	Variables	σ_b(kPa)	F_p
No. 1	L_e/S_h = 3	492.5	23.9
	1	887.8	43.1
	2	564.1	27.4
	4	501.3	24.3
No. 2	B = 3 cm	564.1	27.4
	6	492.5	23.9
	9	332.6	16.1
No. 3	$t_p/2$ = 1.0 cm	508.2	24.7
	1.5	492.5	23.9
	2.5	555.2	27.0
No. 4	S_h/t_p = 6.7	778.9	37.8
	13.4	1,073.2	52.1
	26.8	1,252.7	60.8
	4	555.2	27.0
No. 5	σ_v = 20.6 kpa	492.5	23.9
	37.3	626.9	16.8
	53.0	797.6	15.1
	85.3	1,126.2	13.2

<u>Effect of Protrusion Height and Width</u>

Fig. 5 shows the data obtained from the pullout tests in which protrusion heights were varied 2 cm, 3 cm, 5 cm, with other conditions set constant (set No. 3 in Table 1).

As shown, F_p is more or less constant at smaller t_p values. But it also indicates a tendency that F_p becomes slightly greater as the protrusion height increases. This would be because the mode of shear deformation is more like local or punching type when t_p is smaller. Another possible reason could be that the surcharge loading was applied by a rigid loading plate. The rigid plate would suppress the development of possible shear failure wedge, if the overlaying soil depth is not sufficiently deep. If t_p is smaller, the suppression effect due to the rigidity of the loading plate would be smaller. In general, F_p tends to be almost independent of t_p value.

Fig. 6 shows the data obtained from the pullout tests in which widths of the reinforcement were varied 3 cm, 6 cm and 9 cm with the height of protrusion set to be 3 cm (test set No. 2 in Table 1). The figure indicates that F_p decreases significantly as the width B increases. In consideration of a three dimensional nature of shear failure surface this would be due to the side effect of the reinforcement. Because the reinforcement is not continuous in lateral direction, it can not be regarded as the plain strain condition. In the case of wider reinforcement, the total area of failure surface, which would have been expanded laterally for narrower reinforcement, will be relatively reduced. Therefore, the passive resistance to be mobilized along the lateral failure surface will be relatively decreased per unit width of the wider reinforcement. The F_p value of the B = 9 cm case is about 60 % of the F_p of the B = 3 cm case.

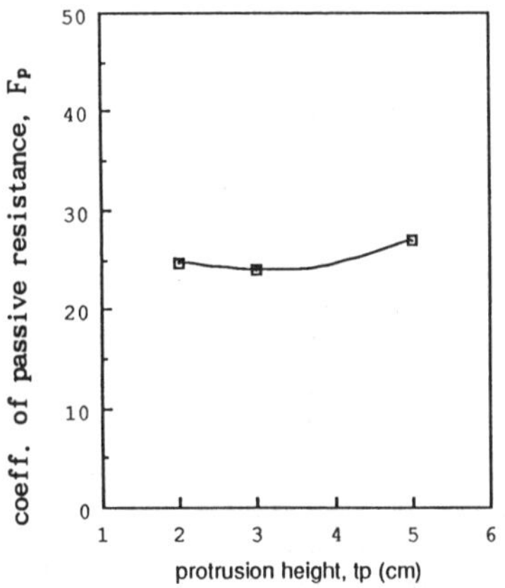

Fig. 5 Effect of protrusion height on coeff. of passive resistance

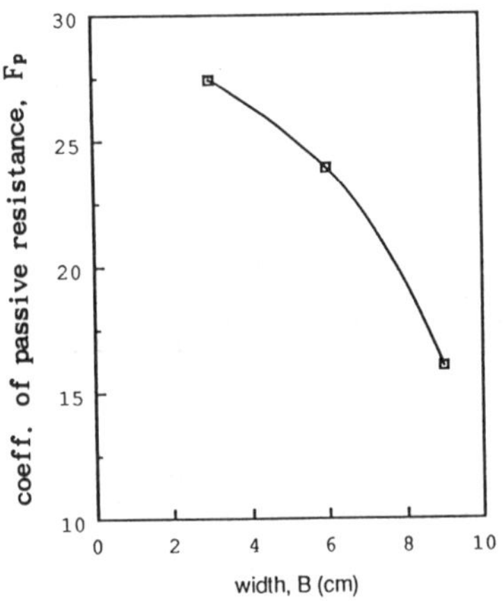

Fig. 6 Effect of protrusion width on coeff. of passive resistance

Effect of Reinforcement Length

Fig. 7 shows the data obtained from the pullout tests in which the effective lengths of the reinforcement were varied 20 cm, 40 cm, 60 cm, 80 cm with other conditions set constant (set No. 2 in Table 1). The figure shows that F_p is reduced as L_e/S_h increases. This would be probably due to the nature of progressive pullout failure. The zone of failure is initially developed at the front side and expanded progressively deep inside as the reinforcement is pulled out.

According to Fig. 7, the coefficient of passive resistance F_p decreases drastically as the relative length L_e/S_h initially increases and reaches a constant value of approximately 24 as L_e/S_h approaches to 3. This means that the value of F_p is relatively greater, if the effective length of passive resistance L_e is less than $3S_h$. This would suggest a criterion to select effective protrusion spacings with regard to length of reinforcement.

The mechanism of the progressive failure can be also explained from Fig. 8, which includes the strains measured along the reinforcement for the condition of test set No.1. It shows that the magnitude of tensile strain developed on N the verge of pullout failure is greatest at the front end of the reinforcement and diminishes as the distance increases from the front end.

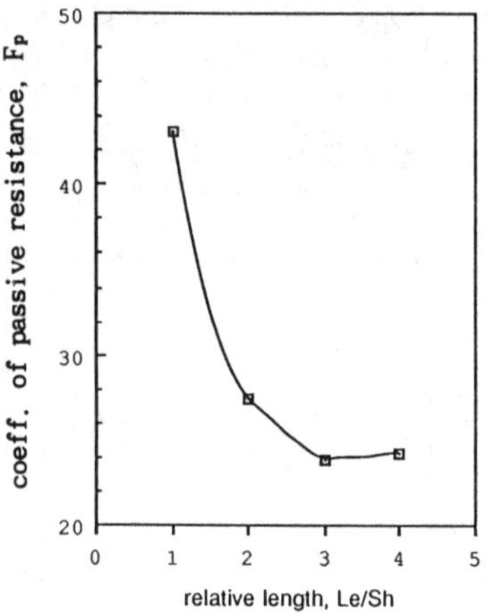

Fig. 7 Effect of relative length on coeff. of passive resistance

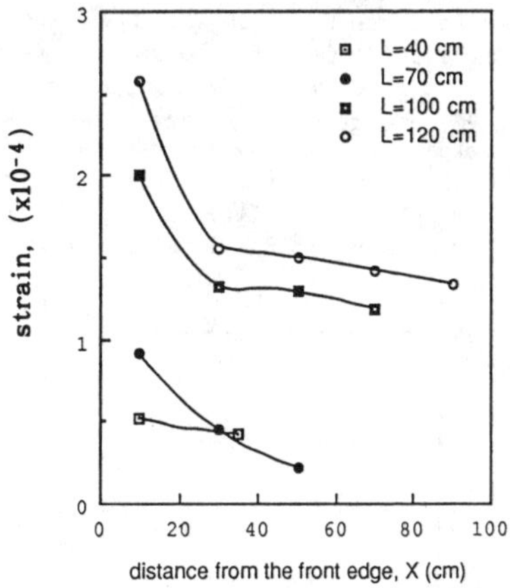

Fig. 8 Strain variation along reinforcement

Effect of Protrusion Spacing

Sets of pullout tests were performed with the relative spacing Sh/tp of protrusions varied. The test results show that the value of Fp depends on the protrusion spacing Sh and the potrusion height tp, and the dependency is displayed with regard to the relative spacing of protrusions represented by the parameter Sh/tp (Fig. 9). The figure shows that Fp increases as the relative spacing Sh/tp increases, but there is a diminishing rate of increase in Fp as Sh/tp approaches approximately to 8. Sh is equal to zero means that there are no protrusions and the surface of the reinforcement is a continuous sheet.

Effect of Overburden Surcharge

A range of overburden surcharge pressure were applied with other conditions set constant (test set No. 5 in Table 1), and the data are plotted in Fig. 10. The figure shows that calculated F_p is reduced within a quite wide range as overburden pressure σ_v increases. This would be because relative reduction of the internal friction angle of soil occurs in a higher overburden pressure range.

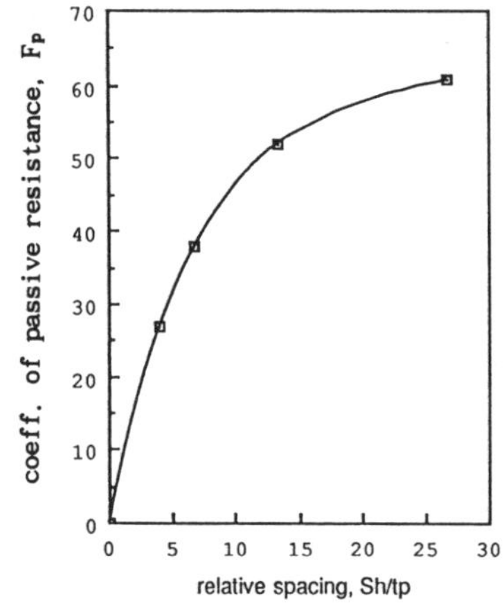

Fig. 9 Effect of protrusion spacing on coeff. of passive resistance

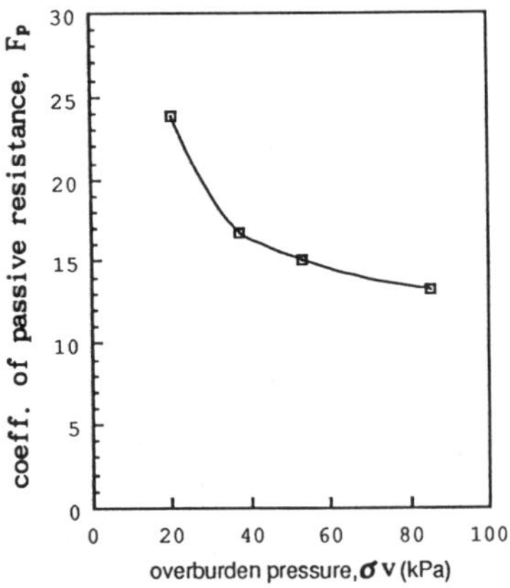

Fig. 10 Effect of overburden surcharge on coeff. of passive resistance

Conclusions

In this research sets of model pullout tests have been performed to evaluate the effects of various factors on passive resistance of protruded reinforcement. The major factors considered in this research are protrusion height and width, effective length of protruded reinforcement, relative spacing of protrusions with regard to protrusion height and overburden pressure above the reinforcement. Following is a summary of test results and analyses.

1) The coefficient of passive resistance tends to be almost independent of the protrusion height.
2) The coefficient of passive resistance decreases as the width of reinforcement increases, which could be due to the side effect of the reinforcement and the three dimensional nature of shear failure surface. When the width of reinforcement is 9 cm, the value of coefficient of passive resistance is about 60 % of the coefficient of the 3 cm width case.
3) The coefficient of passive resistance decreases as the relative length of reinforcement L_e/S_h increases, which would be due to the progressive nature of the pullout failure. The value of the coefficient of passive resistance becomes significantly greater when the effective length of reinforcement is less than three times the protrusion spacing.
4) The coefficient of passive resistance increases as the relative spacing of the protrusion increases, although the rate of increase diminishes beyond a relative spacing of 8.
5) The coefficient of passive resistance is reduced in a quite wide range as overburden pressure increases.
6) Among the influencing factors investigated, the relative length, the relative spacing and the width of the reinforcement are significant, while the protrusion height shows relatively minor influence.

The above conclusions are based only on the model test data with a limited scope and need more work for the general application. Nevertheless, the conclusions made in the current model study could be used as an interim guideline to evaluate the pullout resistance of the protruded reinforcement embedded in a low-quality backfill such as a weathered granite soil.

Acknowledgment

The authors are grateful to Mr. Jung-Gak Cho and Mr. Jin-Hoo Lee for their assitance during execution of the research project. Appreciation is also extended to Dr. Sung-Wan Hong of the Korea Institute of Construction Technology for his helpful comments.

References

1) DAS, V. M., and Seely, G. R., "Pullout Resistance of Vertical Anchors," Journal of the Geotechnical Division, ASCE, Vol. 101, 1975, No. GTI, pp. 87-91.
2) Jewell, R. A., Milligan, G. W., Sarby, R. W., and Dubois, D., "Interaction between Soil and Geogrids," Proceedings of Symposium on Polymer Grid Reinforcement in Civil Engineering, Science and Engineering Research Council and Nelton, Ltd, March, 1984
3) Mitchell, J. K., and Villet, W. C. B., "Reinforcement of Earth Slopes and Embankments," Transporation Research board No. 290, Transporation Research Board, Washington D.C. June, 1987.
4) Nielsen, M. R., "Pullout Resistance of Welded Wire Mats Embedded in Soil," Master Thesis, Utah State University, Logan, Utah, 1983.
5) Peterson, L. M., "Pullout Resistance of Welded Wire Mats Embedded in Soil," Master Thesis, Utah State University, Logan, Utah, 1980.
6) Rowe, R. K., and Davis, E. H., "The Behavior of Plates in Sand," Geotechnique, Vol. 32, No. 1, 1982, pp. 25-41.

A New Design Chart for Reinforced Embankments

M. Soubra [1], C. Coulet [1], M. CFMS, D. Rakotondramanitra [1]

Abstract

A new calculation method using computers is presented, for use in the design of reinforced embankments. The stability of the virtual embankment (non-reinforced) is checked; this leads to the definition of the characteristics of the most critical failure surface, using Caquot's global method. In the method used, the hypothesis is that the failure spreads in concentric circles throughout the partial zone (instability zone). A series of rotations of the slip circles, representing a displacement field, are assumed, in order to obtain deformation of the reinforcements. Overall equilibrium is checked once more, taking into account as additional forces the final calculated tensile strains. The series of charts which follow are to be used in determining the required reinforcement, the anchor lengths necessary for stability and the displacement at the top of the embankment.

Introduction

The dimensions of an embankment reinforced by geotextile (or geomembrane) layers can only be worked out by the use of a computer for calculations, we propose some easy-to-use charts, which should simplify the work of the embankment designer. The charts are given at the end of this paper, preceded by a presentation of the main lines of our calculations. The first part consists of a study of the stability of a non-reinforced embankment, using Caquot's global method (Costet et al, 1983), including the search for the most critical circle. In the second part, the same embankment, after reinforcement, is studied under the effects of a displacement field which creates strains in the reinforcement layers. The results take into account soil compaction, displacement at the top of the embankment, and the strain limit of the reinforcement layers.

---

[1] Doctor of Engineering, lecturer, Institut Universitaire de Technologie, Génie Civil, 69622 Villeurbanne France.

Stability of a Non-Reinforced Embankment

Caquot's global method is used to check the stability of the non-reinforced embankment; this gives the minimum factor of safety F. The calculation assumes that the embankment is homogeneous, unstrainable and in a state of limit equilibrium. Mohr-Coulomb's law is taken to apply, and the failure surface is deemed to be circular. Overloading of the free surface is possible.

The two statics equations give the equation:

(1) $F^3 + a.F^2 + b.F + c = 0$

where a, b and c are functions of the applied forces and of their moment.

Equation (1) is based on 3 variables: the coordinates (x_0, y_0) of the centre of the supposed failure circle, and its radius R_0. The most critical circle (smallest factor of safety F) is determined by reducing the function F (Blanchier, 1982). Thus, for a given circle, the radius is varied until the minimum coefficient is reached; the centre is then moved until the smallest minimum coefficient is found.

The calculation program used contains a grid with nine nodules representing the centres of the circles; using this method, relative certainty is achieved as concerns the position of the centre of the most critical circle. The results obtained correspond exactly to those obtained using Taylor's charts (Taylor, 1937) for the bases of embankments (Figure 1).

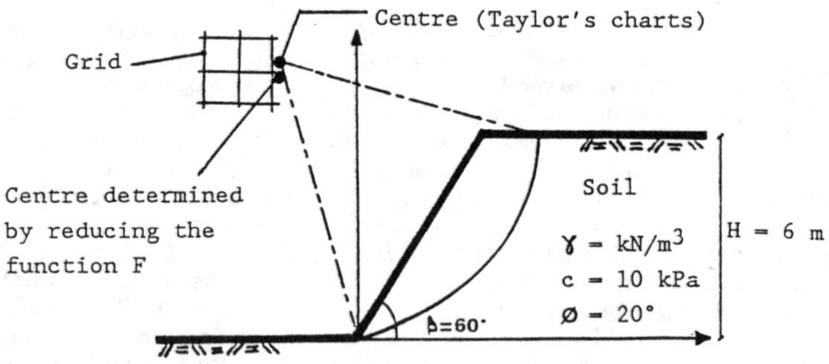

Figure 1: Comparison with Taylor's results

Propagation of Failure

The most critical slip surface within the soil mass is a product of the interaction between various parameters, especially the stress field, the limit state, and the orientations of the main planes.

If we assume that the soil is uniform in its behaviour, it may be said that this interaction cannot manifest itself suddenly or in isolation. It may thus be supposed that it is established progressively on either side of the failure plane where it is optimum.

The following hypothesis can be advanced: failure of a soil mass takes place throughout a series of identical surfaces but with a differential displacement amplitude. Following this reasoning, one might say that the displacement field starts at a low value, which then increases until it reaches the most critical plane, and then decreases to zero (Blanchier, 1982). Physically, however, the downstream zone is carried away by the critical displacement.

Thus, if Caquot's method is used, it is necessary to suppose that the failure of the embankment can be thought of as a series of concentric circles whose centres coincide with the centre of the least favourable circle. This zone, bounded by two circles of stability F = 1, is called "the instability zone" (Figure 2).

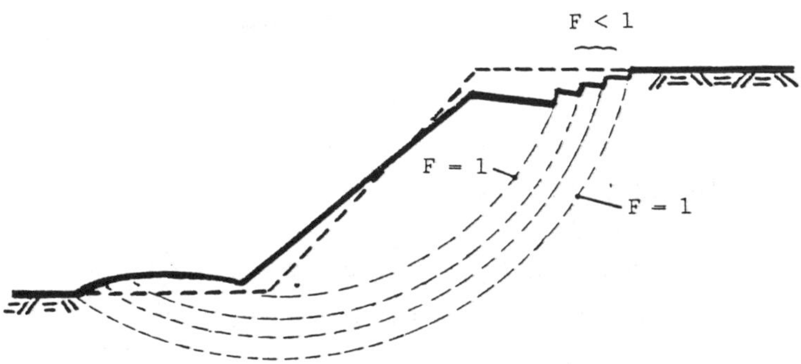

Figure 2: Propagation of failure

Soil deformation measurements carried out on an experimental embankment (Sembenelli, Belloni, 1977) have shown that the general failure pattern takes place along circular and concentric slip lines.

Study of the Stability of a Reinforced Embankment

Field of Displacement. Defining a displacement field leads to the evaluation of the resulting strains within the layers. A series of progressive rotations θ_i, representing the displacement field of the soil mass, is carried out on all the circles (Figure 3). The reinforcement layers follow the rotation of the soil masses, and thus modify their latent tension.

The rotation field θ_i is obtained from equation (2):

$$F^3 + a^*F^2 + b^*F + c^* = 0 \quad (2)$$

where a*, b* and c* are new parameter functions of $\theta_{(i)}$ and $T_{(i)}$: tension in the reinforcements.

The solution to this equation is found by sweeping across the interval whose boundaries are the two extreme solutions:

$\theta_{(i)} = 0$ $\quad T_{(i)} \longrightarrow \infty$ upper boundary

$\theta_{(i)} = \text{constant} = \theta$ $\quad T_{(i)} \longrightarrow 0$ lower boundary

where θ represents the position of stability without reinforcement, and a represents the displacement at the top of the embankment.

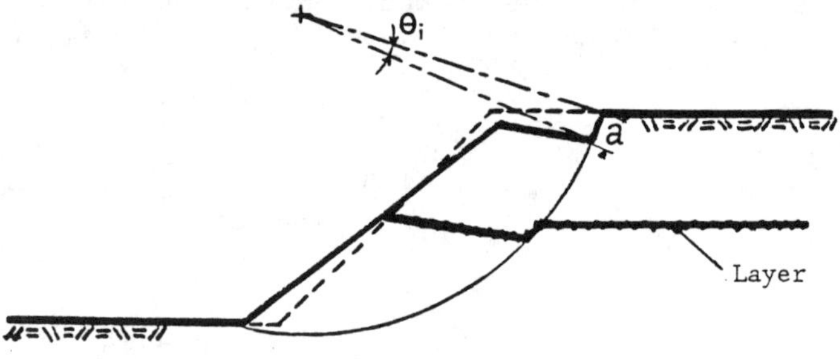

Figure 3: Rotation of the soil mass

Acceptable stress. Strain on a layer within the soil produces a stress discontinuity at the intersection of the failure surface. This is shown in Figure 4, where equilibrium of a reinforcement layer element between two soil layers indicates a tension T on the layer.

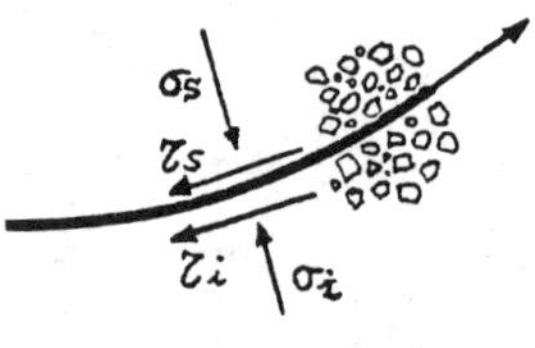

dS = surface element

R = radius of curvature of the reinforcement layer

T = tension in the reinforcement layer

σ_s, τ_s = stress on upper surface

σ_i, τ_i = stress on lower surface

Figure 4: Equilibrium of one geotextile element

The state of equilibrium leads to: $\tau_i + \tau_S = d_T/d_S$, and $\sigma_i = \sigma_s + T/R$. The rate of increase of the stress, d_T/d_S, should not exceed a certain value $\overline{\tau_{adm}}$ if local failure is to be avoided: $d_T/d_S = \overline{\tau_{adm}}$

This admissible stress $\overline{\tau_{adm}}$ is the lesser of the two following values (Figure 5):

- locally mobilized friction (interface state)
  $\tau_{int} = \sigma_S.tg(\phi^*) + \sigma_i.tg(\phi^*) + 2.c^*$
  (ϕ^* and c^* being soil/geotextile characteristics);
- resistance to soil shearing
  $\tau_{Soil} = \sigma_{Soil}.tg(\phi) + c$
  (ϕ and c being soil/soil characteristics).

We obtain the final condition: $\overline{\tau_{adm}} = \mathrm{Min}(\tau_{Soil}, \tau_{int})$.

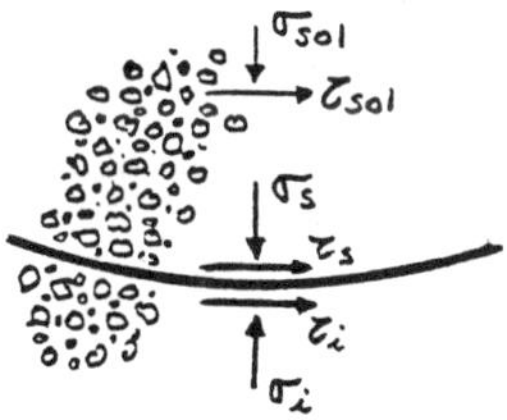

Figure 5: Stress state

Geotextile Tensile Strain. The equilibrium of the reinforcement layer under tensile strain gives us, on the one hand, the necessary anchoring length on the upstream side, and, on the other hand, a way to check that the sloping surface cannot slide on the upstream side.

If there is a slide of the slope surface, it is advisable either to bind the layer to one element of the slope or to invert the layer. In the latter case (Figure 6), the necessary length L_2 has to be determined, given that L_1 is fixed arbitrarily at 1.50 m.

The effect of compaction on the pre-tension of the layers (Murray, 1982) is taken into account.

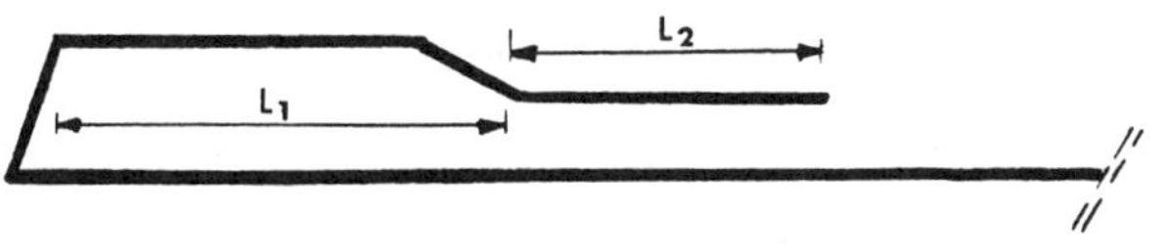

Figure 6: Inversion of the layers

Overall Stability. Overall equilibrium is checked by using Caquot's method and determining the factors of safety of all the circles describing the instability zone identified within the non-reinforced soil mass.

The deformation field corresponding to the reinforcement layers can be defined from the rotations associated with the circles, and local equilibrium of each point of the layer can also be checked, as a function of available friction or shear stresses; and in this way the additional forces acting over the whole circle can be defined.

Calculation of stability is based on the induced rotations of the soil mass and the additional forces due to reinforcements. Orientation of tensions inside the layers is taken as being the same as the intermediate angle between the two extreme results (Figure 7).

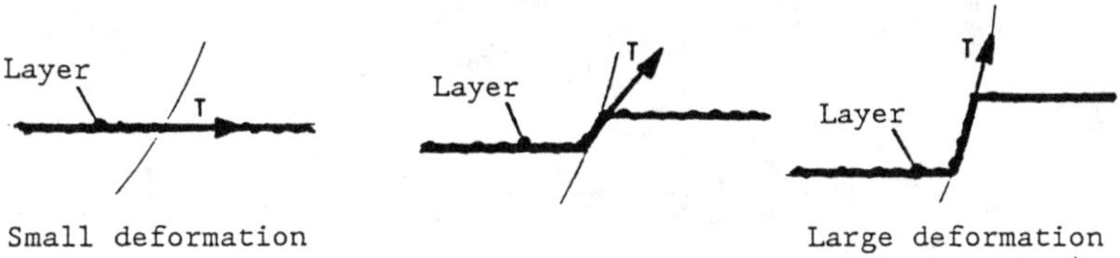

Figure 7: Orientation of stress inside a layer

By attributing a rotation to each circle of the instability zone, the strain curves of the layers can be obtained, and thus the tensions can be calculated by assuming that ($T = K\varepsilon$) and the pitch of the force T can be determined as a mean slope between two portions of the strained layer (Soubra, 1989:

$$tg\gamma = \alpha\left(\frac{x - \underline{x}}{\bar{x} - \underline{x}}\right) + \beta\left(\frac{\bar{x} - x}{\bar{x} - \underline{x}}\right)$$

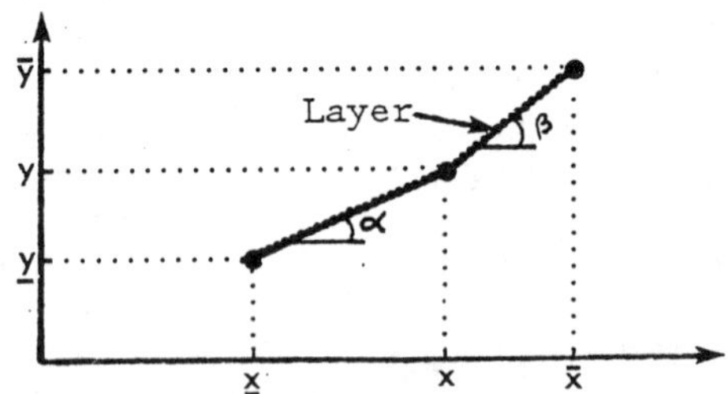

Figure 8: Strain curve of the layer

Pre-dimensioning Charts

We propose 4 charts (Figures 8 to 11), for values of

β = 35°, 45°, 55°, and 65°, which should ease the work of the designer in pre-dimensioning reinforced embankments. These charts are based on the following hypotheses:

- the factor of safety F is given a value of 1. For the normal encountered value of 1.5, the soil characteristics c and ϕ must be reduced by a factor of 1.5;
- the strain of the geotextile is limited to 10% to avoid creep in the geotextile and maintain the displacement compatibility of the soil mass;
- the friction coefficient $F^* = tg\phi^*/tg\phi = 0.8$, while $\delta^* = (c^*/c)$ has been considered to be negligible;
- the effect of compaction is taken into account. No excess loading of the free surface is allowed;
- a, the displacement at top of embankment, is considered on the failure circle and represents the arc of a circle corresponding to the rotation θ (Figure 3).

Example of an Application

Let us take the example of an embankment to be reinforced, with a height H of 6 m, a slope β of 45° and soil with: γ = 18 kN/m^3, c = 5.8 kPa, ϕ = 15°.

We find that $c/\gamma.H = 0.054$, which, with $\phi = 15°$, gives us F = 0.8 (security coefficient of the non-reinforced embankment) and that the ratio $\Delta H/H$ = 0.067, which implies a distance between the reinforcing layers of ΔH = 0.4 m.

For a ratio a/H = 0.018, which limits the displacement at the top to a = 0.1 m, the ratio $K/\gamma.H^2$ = 0.089, which gives a stiffness K = 57 kN/m^3 for the geotextile that is to be used.

The ratio L/H = 0.76 also gives us a necessary anchoring length L of 4.5 m.

Comparison with other methods

For a soil: γ = 18 kN/m^3, ϕ = 10°, C = 16 kN/m^3, and for an embankment with a slope β = 45° and a height of 10 m, with 10 geotextile layers with a stiffness K = 150 kN/m and a friction coefficient F* = 0.8:

Method	Cartage (Delmas et al, 1986)	Variational (Leshchinsky, 1986)	Our Method
Factor of safety	0.89	0.89	0.89
Maximum tension	9.1 kN	9.3 kN	11.4 kN
Displacement at top	13.5 cm	-	11 cm

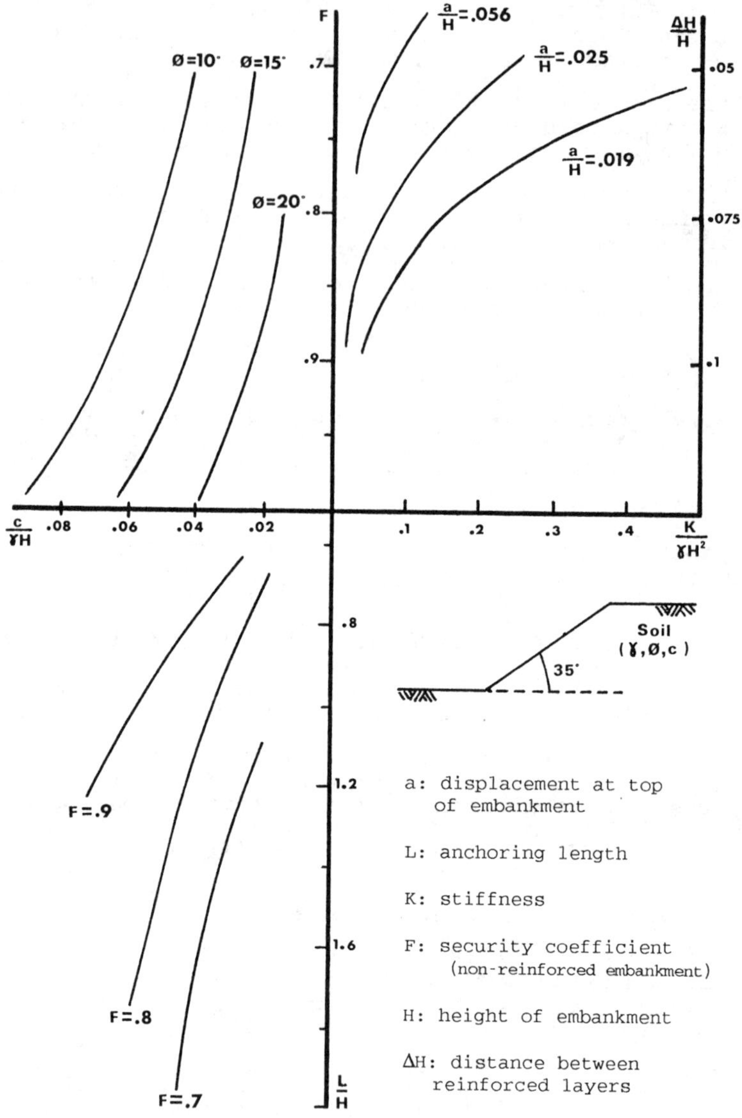

Figure 9: Chart 1 (β = 35°)

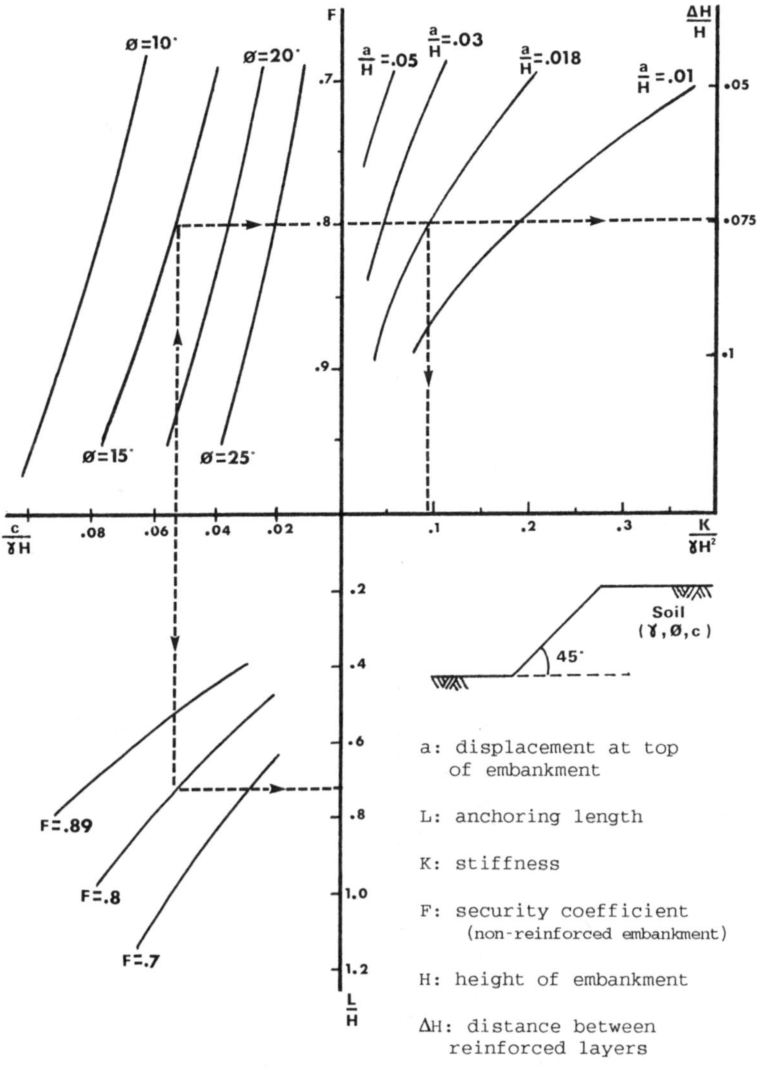

Figure 10: Chart 2 (β = 45°)

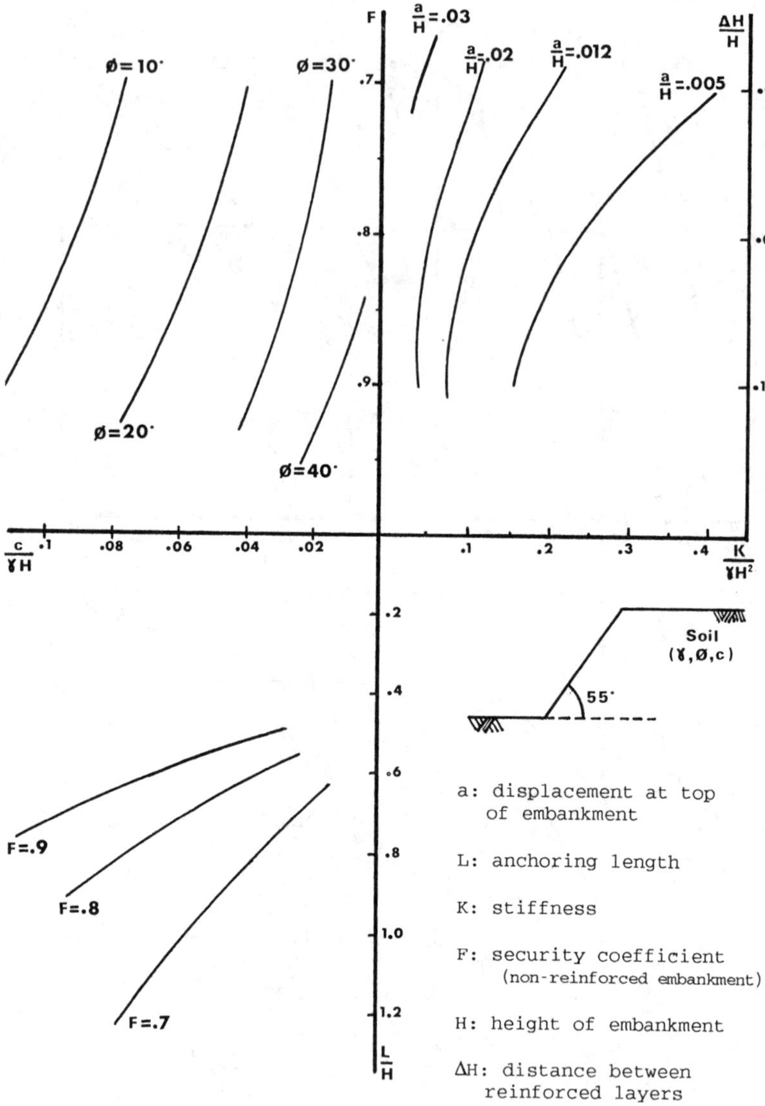

Figure 11: Chart 3 ($\beta = 55°$)

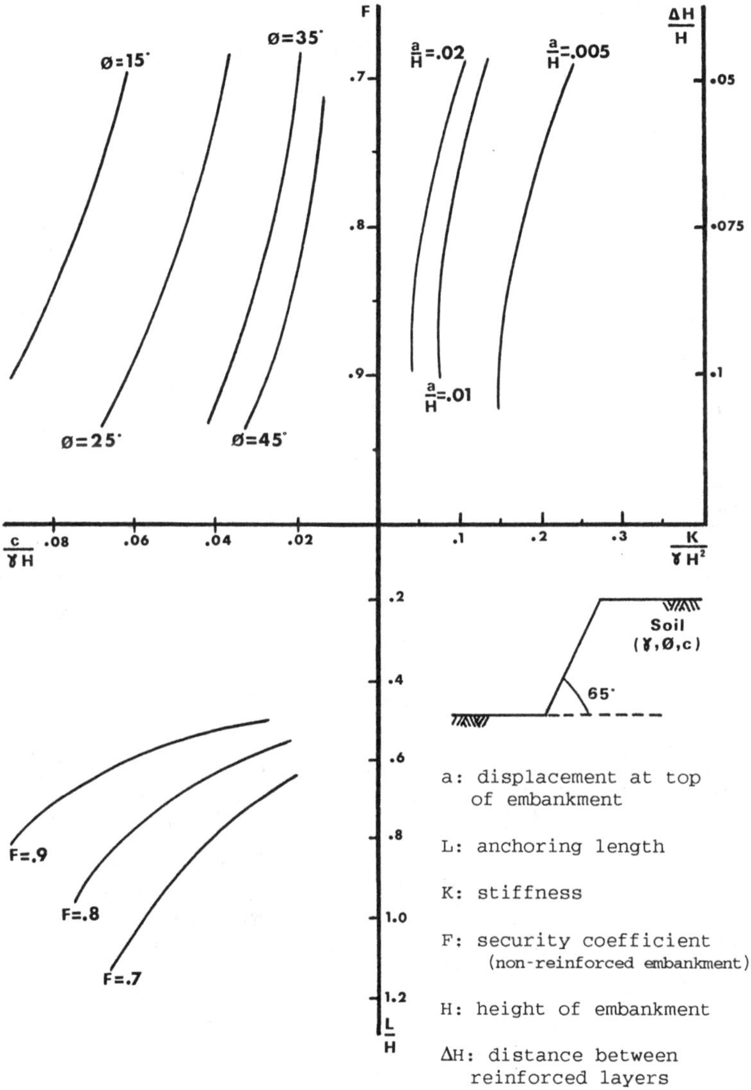

Figure 12: Chart 4 (β = 65°)

Conclusion

The pre-dimensioning of reinforced embankments by the use of charts is a practical and easy-to-use method. It should, however, be stressed that the results given by these charts are function of the hypotheses chosen for their construction, and that there is no single solution to the problem of embankment reinforcement.

The strain of the reinforcement was deliberately kept at 10% so as to obtain solutions close to an optimum where soil and reinforcement stresses are as well-balanced as possible. For embankments outside the norm, it is advisable to make additional checks before starting work.

Appendix. References

Blanchier, A. (1982). "Etude de la stabilité d'un talus renforcé par des nappes géotextiles", Doctor of engineering Thesis, Claude Bernard University, Lyon, France.

Costet, J., and Sanglerat, G. (1983). "Talus et Digues", *Cours pratique de mécanique des sols*, Dunod, France, Vol. 2, 289-298.

Delmas, Berche, and Gourc, (1986). "Le dimensionnement des ouvrages renforcés par géotextile (Programme Cartage)", *Bulletin de liaison LCPC*, N° 142, 33-43.

Leshchinsky, D., (1986). "Predictive equations for the stability of geotextiles reinforced earth structure." *Third International Conference on Geotextiles*, Vienna, Austria, 383-388.

Murray, R., (1982). "Fabric reinforcement of embankments and cuttings", *Second International Conference on Geotextiles*, Vol. VI, Las Vegas, USA.

Sembenelli, and Belloni, (1977). "Remblais routiers sur terrains compressibles exécutés à l'aide des textiles synthétiques", *Colloque International sur l'Emploi des Textiles en Géotechnique*, Vol. I, Paris, France, 49-54.

Soubra, M., (1989). "Stabilité des remblais renforcés par des nappes horizontales", Doctor of Engineering Thesis, Claude Bernard University, Lyon, France.

Taylor, D.W., (1937). "Stability of Earth Slopes", *Journal of the Boston Society of Civil Engineers*, Vol. 24, N° 36.

Interface Friction of Polypropylene Straps

Meijiu Wei[1], Abdelmalek Bouazza[2]

Abstract

The present paper reports the results of a test programme set up in collaboration with British Coal's Minestone Services Division, to determine the friction characteristics of special polypropylene reinforcement straps embedded in compacted colliery spoil and a granular material. These straps are being considered as an alternative to traditional high-adherence galvanised steel straps because of possible corrosion problems associated with the steel straps. Field and laboratory tests have been performed to assess the suitability of colliery discards as backfill material and the soil-strap interaction. The test results showed that the different types of fill gave different results for a particular strap type, the more angular fill being superior in terms of pull out resistance and friction values. Finally, the results provide information on the friction characteristics which can be used in the design of reinforced earth retaining walls using compacted coal wastes as the fill material.

Keywords: reinforced earth, interface friction, coal wastes

1 Introduction

Soil has been used since the most ancient civilization as a construction material. However, due to its weakness in tension, engineers had to find a variety of ways to improve its resistance. Placing sheets, strips, rods or meshes helps to create a composite material with improved resistant properties. This is known as "Reinforced soil" and can be regarded as soil strengthening at the macro scale. However, this concept is not new, and early examples

---

1- Research Associate, Civil Engineering Dept., Liverpool University, Liverpool, England. (formerly at University of Glasgow).
2- Senior Research Engineer, CNERIB, Cite Nouvelle El Mokrani, Souidania, Wilaya de Tipaza, Algeria. (formerly at University of Glasgow).

of soil reinforcement include the Agar-Quf ziggurat found in Iraq, which is more than 3000 years old. The great wall of China, parts of which were completed circa 200 B.C., contains examples of reinforced soil, where use was made of mixtures of clay and gravel reinforced with tamarisk branches (Jones, 1985). The modern form of soil reinforcement was introduced by Vidal in the 1960's and has shown a major increase in its use since then. Many reinforced structures have been erected world wide and normally comprise metallic reinforcement in a cohesionless soil with good permeability and high angle of internal friction. Nowadays, economic considerations make it likely that the cost of natural aggregates will increase in areas where they are already in short supply (Gribble, 1989), whereas waste materials are by their very nature available at low cost, at least at source.

The use of cheap material with adequate engineering properties is the aim of every engineer, especially in retaining structures where a large quantity of material is needed and the cost would be very high if a granular material were used. For example, it was shown by Hollinghurst and Murray (1986) that the imported granular backfill to a reinforced earth wall at Guildford (England) accounted for 40% of the total cost of the structure. The last decade has seen an upsurge in the search for new material suitable for reinforced earth, and the use of ash, silty clay, lime stabilised cohesive soil, mining sand and soft chalk has been reported (e.g. Jones, 1984, Temporal et al, 1989, Guler, 1990, Haji-Ali, 1990, Brady et al, 1990). In order to curtail costs further and to make reinforced earth a more viable method, the potential use of low cost readily available backfill material such as colliery discards (Minestone) has to be considered. The use of colliery discards in this way would also help in alleviating the problem of economic utilisation or acceptable disposal of "wastes" in the construction industry. According to the Building Research Establishment there are about 3,000 million tonnes of colliery spoil stockpiled in Great Britain (1986 figures) 80% of it in Scotland and Wales. This poses a major environmental problem to the industry and the community. Whilst considerable efforts have been made to reduce the problem through schemes sponsored by local and central government, large volumes of material remain derelict on the surface. The growing awareness of the environmental and ecological problems associated with these large colliery spoil heaps prompted the present study in which the University of Glasgow, Department of Civil Engineering, has been investigating the possibility of using colliery spoil (burnt and unburnt Minestone) as a backfill in conjunction with polypropylene straps.

In this paper, the properties of the fill materials, the test apparatus and procedure are described. The results are presented and their repercussion on the design is discussed .

1-1 What is Minestone

Minestone (burnt and unburnt) is a by-product of mining coal and of its preparation for markets. It is derived from the rocks-mainly siltstones and mudstones and sometimes sandstones, limestones and other rock types-lying above, below and sometimes within, the coal seams in the coal measures (Fig. 1) (Sleeman, 1990). During mining operations quantities of these rocks are unavoidably extracted with the coal, or in driving the tunnels which give access to the coal faces, and are brought to the surface with the coal. It is generally necessary to remove some or all of this material to yield a coal product of the required quality. This separation is carried out with other operations such as crushing, sizing and de-watering, in the coal preparation plants, generally by making use of the higher density of the Minestone-roughly between 1.5 and 2 times that of the coal. There is usually no immediate use for the separated non-coal material, so it is tipped onto a spoil heap, familiar around the coalfields. This is Minestone or unburnt colliery spoil. The residue of calcined rocks left in these spoil heaps is burnt Minestone.

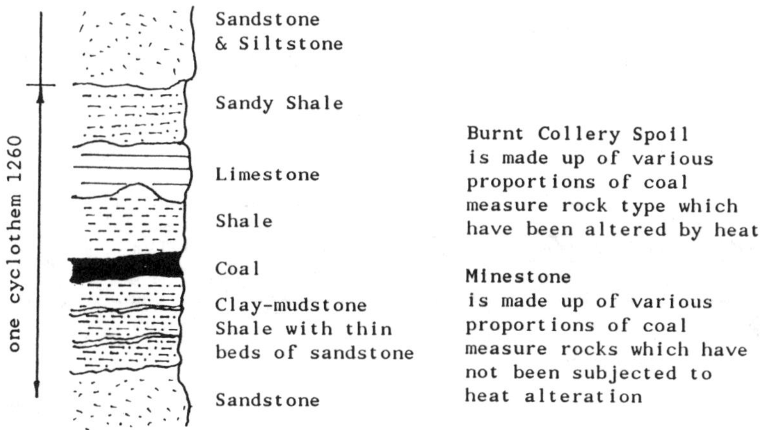

Figure 1 Coal measures rocks (after Sleeman, 1990)

2 Materials

2-1 Reinforcing straps

Samples of polypropylene strap designated A and B were supplied by I.C.I. (Imperial Chemicals Industries Ltd). Both types comprised bundles of continuous, aligned, high tenacity polyester fibres enclosed in a durable polypropylene sheath. Strap A, beige in colour, had a cross section 85mm x 2.5mm and a reported maximum tensile strength of 30 kN. Strap B had a cross-section 92mm x 3.5mm, a reported maximum tensile strength of 50 kN, and was black in colour to resist ultra-violet radiation. Both types of strap had a textured surface to enhance the friction against the surrounding soil. However, a slight different roughness can be found when comparing the surface nature of the two sides of a strap.

2-2 Fill Materials

The need to measure some properties of colliery discards differently from the British standards methods for other soils was appreciated by the British Coal Board in the late 1960's. This appreciation prompted the publication of the NCB Technical Memorandum on Soil Testing (1971) which summarised the suitable tests to be done on colliery spoils in accordance with the British Standards.

Table 1 Properties of fill materials

	Wardley minestone	Wearmouth minestone	Horden red shale	Methil PFA	Loudon Hill sand
Dry density/m.c.					
{γ_{dmax} kN/m^3}	18.2	17.7	18.6	11.1	18.1
2.5 kg rammer					
{o.m.c %}	10	8	15	36	15
Test conditions:					
γ_d kN/m^3	17.8	17.7	17.7	11.6	16.5
m.c. %	9.5	5.7	12.5	27	7

* γ_d = dry unit weight,
o.m.c = optimum moisture content

In the present research work, the materials used were Wardley minestone, Wearmouth minestone, Horden red shale, Methil PFA and Loudon Hill sand, all from the North East of England. Of the five different fill materials used in the present research, sand is a conventional material, used because of its cohesionless properties, high frictional resistance and good permeability. It has been regarded as the most adequate fill material and is widely used in reinforced earth structures. Red shale is also accepted as a high quality material for construction purposes. However many of the best deposits have been used in the U.K. PFA is a type of waste with light weight and fine particles and is being increasingly used in reinforced earth structures (Jones, 1990).

Minestones, the largest source of waste materials in the U.K., have been the object of very little research, and hitherto not been accepted for use in reinforced earth structures in the U.K. since it remains to be proven that their mechanical and physical characteristics are suited to backfill requirements.

In such structures, what affects the suitability of their utilization are the following natural properties: Spontaneous ignition as an important character of colliery spoil has long been of concern to the construction industry. In 1967 about 15% of the 2,000 spoil heaps owned by British Coal were classified as burnt out and more than half as burning (Taylor, 1978). This phenomenon is in essence an atmospheric oxydation (exothermic) process of the coal content in minestones. Investigators found that spontaneous ignition occurred in loosely tipped spoils. The permeability to air is one of the controlling factors as to whether oxydation will be sustained and the temperature will rise. If the spoil is compacted, the permeability to air is reduced, and effectively excludes the supply of air into the spoil, therefore it is not prone to heating. Furthermore, Taylor (1978) concluded that compaction in accordance with clause 609 of the Ministry of Transport's specification relating to earth works has shown that previous fears about spontaneous heating were based on the behaviour of loose tipped materials. It implies that proper compaction can avoid the risk of spontaneous combustion and indeed this has been proved by the construction of a full scale embankement in which thermocouples were used to monitor the temperature (Rainbow, 1987).

Another factor which affects the suitability of minestone is the degradation due to chemical and physical weathering or due to compaction. Considering breakdown due to compaction, Rainbow (1983) compared the particle size distribution before and after compaction and reported that although considerable degradation results from the compactive forces, the final grading continues to fall within the specified grading envelope. For degradation due to weathering, it is generally considered that little chemical disintegration occurs on the surface whereas physical disintegration is at its peak at the surface. However, the evidence from old, initially loose, unburnt spoil heaps implies that intense weathering is limited to depths of about one metre (Taylor, 1978).

The properties of the fill materials were determined from tests carried out by British Coal's own laboratories and Glasgow University. These included particle size distribution, and dry density/moisture content relationship, the results of which are presented in Table 1 and Fig. 2. More details are given by Bouazza (1989) and Wei (1990). The Wardley minestone consisted mainly of mudstone and clay minerals. The grading curve indicates a well graded soil containing particles of all sizes from cobbles down to clay, but having a predominance of gravel sizes. It had a PH value of 8.8 which showed it to be slightly alkaline in reaction. Some of its natural properties have been described previously by Rainbow (1983). In the case of Wearmouth minestone obvious differences were

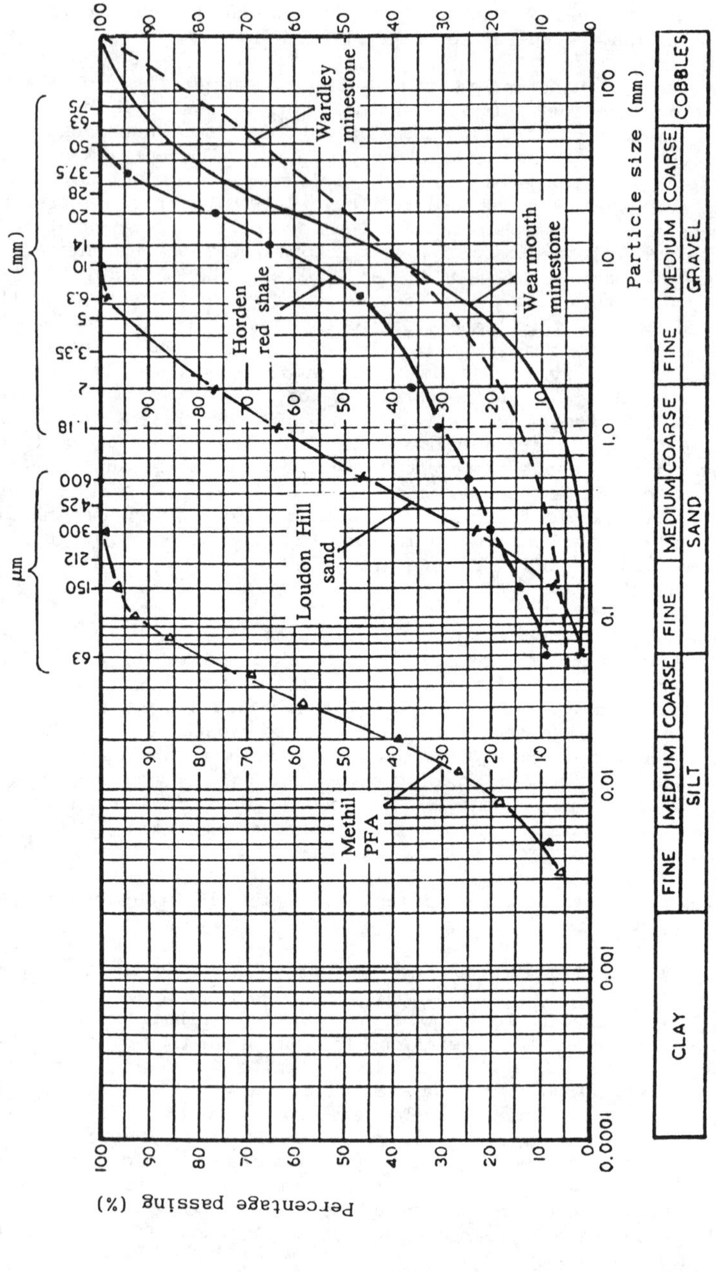

Figure 2 Particle size distribution curves

observed compared with Wardley minestone, in that sandstones were present and the stones were relatively hard and resisted breakdown during compaction. The Horden red shale is a well burnt type of colliery spoil. It is a well graded material consisting of angular hardened particles with a PH of 6.9 wich showed a neutral reaction. The Methil PFA (Pulverized Fuel Ash) is a material with fine particles and light weight. The grading curve shows very fine particles with a predominance of silt sizes. It did not possess clay characteristics, no plastic limit being found. The PH value of 8.7 showed a slightly alkalinity in reaction. The Loudon Hill sand is a well graded type of sand with a uniformity of 6 as shown in the grading curve. The PH value of 7 showed a neutral reaction.

3 Test Procedures

In the present research programme, three type of tests have been used (large shear box, lab. pull out box and field pull out) to measure the shear strength of the fills as well as the bond resistance of the fill against the reinforcement. The various fills and reinforcement materials provided were tested under different conditions of normal stress.

3-1 Shear box

The shear tests were carried out using a direct shear box with plan dimensions of 300mm by 300mm to comply the Department of Transport Specifications for Highway Works (1986). Conventional shear box tests were conducted initially to ascertain the effective angle of friction of the different fills. Further shear box tests were then conducted but with fill in the upper half of the box only and with the reinforcement element material in the lower half. A hard wooden block 290mm thick was placed as the base plate in the lower box with the reinforcement material firmly glued to the top surface. The wooden block was positioned so that the surface of the reinforcement was flush with the top edge of the lower half of the box. For all samples two layers of fill were put in the upper box and compacted to the corresponding field density and moisture content. The surface of each sample was levelled and the loading plate was placed on it. The normal load was applied by a moving cantilever yoke capable of applying a vertical load of between 20 to 120 kN/m^2, and measured using a load cell, which also served to ensure that the normal stress was maintained constant during sample dilation. The shearing force was applied by a motor with a shearing rate of 1.05mm/min and was monitored by a 50kN load cell. Once the sample had been prepared and the desired normal stress applied, the sample was left for about thirty minutes to come to equilibrium and then sheared to failure. The friction coefficient, μ, obtained from the shear box test can be expressed as:

$$\mu = C_a / \sigma_v + \tan \delta \qquad (1)$$

where C_a and δ are the adhesion and friction angle measured, and σ_v is the normal stress or overburden pressure. Obviously, in a cohesionless fill, $\mu = \tan \delta$.

Table 2 Shear box test results

	Wardley minestone	Wearmouth minestone	Horden red shale	Methil PFA	Loudon Hill Sand
Fill alone					
C (kN/m^2)	14	8	12	3	0
ϕ (deg.)	33	37	41.5	36.5	36.5
Fill/strap A					
C_a (kN/m^2)	4	3	3	5	3.5
δ (deg.)	18	18	23.5	21	20
Fill/strap B					
C_a (kN/m^2)	4	6	2.5	7	0.5
δ (deg.)	20	21	27.5	24	22.5

3-2 Pull out box

A steel box 2m long by 0.4m wide by 0.25m deep was used. The sides of the box have been smeared with silicon grease to reduce any side effect. The fill was compacted to mid-height in the box, a strap 1.5m long was then placed in position centrally and fed through a slot in the front face. Further fill was placed and compacted level with the top of the box. The top of the box, comprising a rubber sheet below a steel cover plate was then bolted in position. Overburden pressure similar to the normal stresses used in the shear box tests was simulated by air pressure introduced between the rubber membrane and the top of the box. The pull out force was applied by means of a constant rate of strain testing machine set to produce a constant rate of pull of 1.5 mm/min. The connection from the machine to the reinforcement strap was made via a load cell which permitted the applied force to be monitored. A displacement transducer was installed to monitor the movement of the reinforcement. The data from both transducers and load cell were automatically logged up to a maximum frequency of a set of readings every 20 seconds. The apparent friction coefficient, f^*, was obtained from the expression:

$$f^* = T/2BL\,\sigma_v \qquad (2)$$

where T = maximum pull out force
B = width of strap
L = length of strap
σ_v = normal stress or overburden pressure

3-3 Field tests

The field tests were carried out at two locations, at Wardley colliery in North East England (Finlay et al, 1988) and at Barony colliey near Ayr (west coast of Scotland) where a large box 5m long by 3.1m wide by 4.6m high had been constructed from rolled steel sections and timber railway sleepers. The box was used to accomodate 24 straps at various levels. The reinforcing straps were passed through the gaps between the sleepers at the front face. Placing of the fill was done via a conveyer and compaction at each level was achieved by a small vibrator. A pull out device, supplied by British Coal, was used to pull out individual straps with load and displacement being monitored by a computerised graph plotter. The friction coefficient f^{**} from these tests was obtained using the same expression as for the laboratory pull-out tests, and taking the overburden pressure as a function of the fill height above the strap (max. 70 kN/m^2).

4 Results and Discussion

Frictional resistance is one of the most fundamentally important aspects when considering the potential use of minestone for reinforced earth purposes. The shear strengths of the two minestones used in the present investigation were obtained from large shear box tests at their natural moisture content and field compaction conditions, both behaved as c-ϕ (cohesive frictional) soils and produced the following: Wardley minestone $\phi = 33^0$ and C = 14 kN/m^2 and Wearmouth minestone $\phi = 37^0$ and C = 8 kN/m^2. The friction coefficient obtained from the different tests are shown in table 3. The tests have shown that the fill material can be ranked generally in descending order of friction coefficient, regardless on how this is measured, as Horden red shale, Methil PFA, Loudon Hill sand, Wardley minestone and Wearmouth minestone. The fact that red shale showed the highest frictional coefficient is probably due to the fact that it contained very angular material. In the case of Wardley minestone, when comparing the different testing methods, it can be seen that the data obtained in the field pull out tests produced the lowest friction coefficient. It is also shown that, generally, in terms of friction , strap B is superior to strap A.

When comparing the laboratory pull out box tests with the shear box tests it was found that the friction coefficients obtained by shear tests were higher than the ones produced by pull out tests except at very low overburden pressure. The shear box test represents the two dimensional case of an infinite reinforcement sheet, but, it does not represent the different phenomena involved in the complex three dimensional mechanism of the soil reinforcement interaction in actual reinforced earth structures, where factors such as arching and tensile behaviour occur. The results with the other fills indicated a similar general trend. It is believed that the extensibility and compressibility of the polypropylene straps caused this discrepancy in the friction coefficient values, led to the reduction in the pull out force, and in turn produced lower friction coefficients in the pull out tests. In the shear box test, friction is mobilized on one side of the strap, whereas in the pull out test this is mobilized on the two sides of the strap. Therefore, any difference in the surface nature between the two sides of the reinforcement can also cause a difference in the results. These conflicting results can also mislead the practising engineer. Careful consideration and discussion of the two test methods are still needed, especially when some non conventional material is tested.

Table 3 Friction coefficient obtained from different tests

σ_v	Wardley minestone		Wearmouth minestone		Horden red shale		Methil PFA		Loudon Hill sand	
kN/m^2	A	B	A	B	A	B	A	B	A	B
friction coefficients (μ) from shear box										
20	0.525	0.564	0.475	0.684	0.585	0.646	0.634	0.795	0.539	0.439
60	0.392	0.431	0.375	0.484	0.485	0.562	0.467	0.562	0.422	0.423
120	0.358	0.397	0.350	0.434	0.460	0.541	0.426	0.504	0.393	0.418
friction coefficient (f^*) from Lab. pull-out tests										
20	0.650	0.303	0.418	0.459	1.078	1.492	0.725	0.918	0.636	1.165
60	0.280	0.389	0.230	0.254	0.396	0.626	0.353	0.477	0.350	0.644
120	0.200	0.173	0.198	0.262	0.289	0.498	0.350	0.434	0.314	0.415
friction coefficient from (f^{**}) field pull-out tests										
20	0.471	0.340	0.522	0.442	0.630	0.820	0.340	0.346	0.636	0.630
60	0.179	0.188	0.190	0.233	-	0.315	0.180	0.185	0.294	0.270

* σ_v = overburden pressure

5 Conclusion

It is obviously attractive to make use of minestone both from an economic and an environmental point of view. Reinforced soil has been recognised as one of the new ways in which minestone could be used (Rainbow, 1983), and it has actually been utilised in some countries, e.g. U.K. (Jewell & Jones, 1981), Belgium, France (Rainbow, 1987) and Korea (Bok-Youn & In-Ki, 1990). In Belgium over 50 reinforced minestone highway structures have been built and have been in operation for a number of years without any sign of distress. The reinforced elements employed in these structures have included Paraweb, Tensar, Fibretain and high adherence galvanized mild steel. However, it should be pointed out that the tendency of steel to corrode in contact with aggressive material such as colliery wastes makes it unsuitable for reinforced structures using that fill (Bouazza et al, 1990). The present work indicates that polypropylene reinforced minestone backfill might be a viable alternative in reinforced soil structures if good quality granular backfill material is not readily available. However, since the properties of minestones vary from colliery to colliery, care should be taken to select the minestones which are most suitable. A thorough investigation of the geotechnical properties and composition should be carried out pertaining to each site. Accurate and reasonable interpretations of the geotechnical parameters should be considered to take into account the variability of the fill materials. Reinforced colliery discards will then possibly provide cheaper alternative backfill material, as a substitute for conventional granular backfill material. However in order to make the concept of reinforced earth more attractive by using colliery discards further research is needed to allay some of the Department of Transport's apprehensions. Of more significance perhaps is the fact that the three methods of testing yielded different results. It is the authors' opinion that values obtained from pull out tests would be more suitable for use in design.

6 Acknowledgements

The authors wish to express their thanks to British Coal's Minestone Services for the support given to carry out this research project, particularly in the field tests at Wardley and Barony Collieries. These could not have been done without the assistance of the Director, Dr. A.K.M. Rainbow, Geotechnical Engineer, Stephen Barnett, and the on-site staff. The views expressed in this paper are not necessarily those of British Coal.

7 References

Bouazza, A. (1989). Strength behaviour of polypropylene straps embedded in compacted fills. Research report 89/1. University of Glasgow/British Coal-Minestone Services.

Bouazza, A., Finlay, T.W., Wei, M.J. & Hytiris, N. (1990). Friction charateristics of polypropylene reinforcing straps in various fills. Proc. 3rd Int. Conf. on Reclamation, Treatment and Utilization of Coal Mining Wastes, Glasgow (U.K), pp. 349-354, Sept.

Bouazza, a., Finlay, T.W., & Wei, M.J. (1990). Interface Friction of Two Very Different Straps in Burnt Shale. Proc. Int. Conf. on Construction on Polluted and Marginal Land, June, London (U.K.), pp.29-31, June.

Bok-Youn, K. and In-Ki, I. (1990). A study on utilization of coal mining wastes in Korea. Proc. 3rd Int. Conf. on Reclamation,

Treatment and Utilization of Coal Mining Wastes, Glasgow (U.K), pp. 369-376, Sept.

Brady, K.C., Barrat, D.A. & Cater,R. (1990). Performance of an experimental reinforced earth structure using soft chalk backfills. Transport and Road Research Laboratory, Report 260, Crowthorne (U.K.).

Department of Transport, (1986). Specification for highway works, part 6. HMSO, London (U.K.).

Department of Transport, (1987). Reinforced earth retaining walls and bridges abutments for embankments. Technical memorandum BE3/78 (revised 1987), DTp, London (U.K.).

Finlay, T.W., Wei, M.J. & Hytiris, N. (1988). Friction characteristics of polypropylene straps in reinforced minestone. Proc. Int. Geotech. Symp. on Theory & Practice of Earth Reinforcement, Fukuoka (Japan), pp. 87-92, Nov.

Gribble, C.D. (1989). U.K aggregate production and coastal quarries: the changing picture. Quarry Management, July, pp. 31-33.

Guler, E. (1990). Lime stabilized cohesive soil as a fill for geotextile reinforced retaining structures. Proc. 4th Int. Conf. Geotextiles, Geomembranes and related products, La Hague (Netherlands), Vol.1, pp 39-44, May.

Haji-Ali, F. (1990). Field behaviour of a reinforced earth wall in Malaysian condition. Int. Conf. on Reinforced Soil, Glasgow (U.K.), Preprint, Sept..

Hollinghurst, E & Murray, R.T. (1986). Reinforcing earth wall at A3/A322 interchange:design, construction and cost. Proc. Inst. Civil. Eng., Part 1, N 80, pp 1327-1341, Oct.

Jewell, R.A. & Jones, C.J.F.P. (1981). Reinforcement of clay soils and waste materials using grids. Proc. 10th Int. Conf. Soil Mechs. Fdns. Engng., Stockholm (Sweden), Vol. 1, pp.701-706.

Jones, C.J.F.P. (1984). The use of ash in reinforced earth. Proc. 2nd Int. Conf. on Ash Technology & Marketing, London (U.K), pp. 543-547, Sept.

Jones, C.J.F.P. (1990). Construction influences on the performance of reinforced soil structures-State of the art review-. Int. Conf. on Reinforced Soil, Glasgow (U.K.), preprint, Sept.

National Coal Board, (1971). Application of British standards to the testing of colliery spoil. NCB, London (U.K.).

Rainbow, A.K.M. (1983). An investigation of some factors influencing the suitability of minestone as the fill in reinforced earth structures. Ph.D Thesis, Nottingham University, Nottingham (U.K).

Rainbow, A.K.M. (1984). Research into reinforced minestone in United Kingdom. Proc. 1st Int. Conf. on Reclamation, Treatment and Utilisation of Coal Mining Wastes, Durham (U.K), pp 7.1-7.12.

Rainbow, A.K.M. (1987). Minestone fill in reinforced earth abutments. Highway and Transportation Engineering, Vol.34, N 6, pp. 25-31, June.

Sleeman, W. (1990). Environmental effects of the utilization of coal mining wastes. Proc. 3rd Int. Conf. on Reclamation, Treatment and Utilization of Coal Mining Wastes, Glasgow (U.K), pp 65-76, Sept.

Taylor, R.K. (1978). Properties of mining wastes with respect to foundations. Foundations Engineering in Difficult Ground. Newnes-Butterworths (eds.), London (U.K), pp. 175-203.

Temporal, J., Craig, A.H., Harris, D.H. & Brady, K.C. (1989). The use of locally available fills for reinforced and anchored earth. Proc. 12th Int. Conf. Soil Mechs. Fdns. Engng., Rio de Janeiro (Brasil), pp 1315-1320

Wei, M.J. (1990). Frictional behaviour of reinforcement in reinforced earth fill materials. PhD thesis, University of Glasgow, Glasgow (Scotland).

Design of Geosynthetic-Reinforced Soil Structures

Farrag, Kh.[1], and Juran, I.[2]

Abstract

The working stress design methods commonly used for the design of geosynthetic reinforced soil structures are usually based on assumptions concerning the anticipated state of stress (K_o or K_a) in the soil. These methods do not allow for considerations pertaining to the effect of reinforcement stiffness on the structural behavior. This paper presents the engineering application of a strain compatibility approach (recently proposed by Juran et al., 1990) for the analysis of geosynthetic-reinforced soil structures. This approach, based on simulating the strain compatibility between the soil and reinforcement, is developed to evaluate the effect of reinforcement stiffness and soil properties on the mobilized tension forces in the reinforcement and on the structure stability. It yields an analytical solution which is consistent with the observations on the behavior of reinforced soil structures and the recommendations of the FHWA design guidelines.

This paper outlines the results of a parametric study to evaluate the effect of reinforcement extensibility and soil dilatancy on the mobilized tension forces. The analysis shows that the working stress design methods commonly used are too restrictive to allow an appropriate simulation of the complex structure behavior. Design charts are presented to predict the tension forces in the reinforcement and the locus of failure surface for different reinforcement extensibilities and soil properties. Method predictions are compared with results of the currently used working stress design methods.

---

[1] Coordinator, Geosynthetic Engineering Research Lab, Louisiana Transportation Research Center, Baton Rouge, LA.

[2] Chairman, Civil Engineering Dept., Polytechnic University, New York, NY.

Introduction

Most of the design methods used for the internal stability analysis of reinforced-soil structures are based upon limit equilibrium methods which are modified to account for the tensile resistance mobilized in the reinforcement. These methods consider different assumptions with regard to the failure surface mechanism (rotational or translational), the shape of the failure surface (plane, bilinear, circular and log spiral) as well as the deformation and orientation of the reinforcement. Murray (1982) considered a bilinear slip surface to evaluate the safety factors of reinforced slopes. Jewell et al. (1984) presented a bi-linear wedge limit analysis method to evaluate the rotational and translational failure mechanisms in steep reinforced slopes. Leshchinsky et al. (1986) and Leshchinsky and Boedeker (1989) presented an analytical approach based on limit equilibrium analysis of log spiral slip surface and using variational calculus to evaluate the rotational or translational stability in reinforced slopes.

Different assumptions regarding the orientation of the reinforcement at the failure surface were also considered. As the tension forces are mobilized, the reinforcement reorients itself from its horizontal position ($\beta=0$) to an angle β. The orientation of the reinforcement (β) governs the magnitude of the mobilized tension forces and consequently the factor of safety. The direction of tension forces were assumed to remain horizontal (Ingold 1982), or to be tangent to the failure surface (Brakel et al., 1982). Gourc et al. (1986) assumed horizontal forces at small displacements; whereas at large deformations, the reinforcement becomes tangential to the potential failure surface.

The slope stability analysis methods based on limit equilibrium can be used to evaluate the safety factors with respect to the failure mechanism along the assumed failure surfaces. Nevertheless, limit equilibrium analysis do not allow for an evaluation of the effect of soil dilatancy and reinforcement extensibility on the generated tension forces and structure stability. The development of a working stress analysis for the evaluation of the tensile forces in the reinforcement and the stability of the structure requires adequate constitutive equations for the soil, in-soil confined stress-strain properties of the reinforcement and an assumption pertaining to the strain compatibility between the soil and the reinforcement.

In order to incorporate the tension forces mobilized in the reinforcements into the limit equilibrium analysis, different assumptions regarding the distribution of the tension forces were considered. In these methods the tension forces are usually assumed to follow either the K_o (Steward et al., 1977, and Koerner, 1990) or the K_a Rankine's earth pressure distribution (Al-Hussaini, 1977). Christopher et al. (1989) have proposed an empirically based working stress design method relating the lateral earth pressure coefficient (K/K_a) to the relative stiffness of the reinforced soil system which is defined by the stiffness factor $S_r = Etb/ S_hS_v$; where E is the elastic modulus of the reinforcement; t and b are the reinforcement thickness and width, respectively; and S_h and S_v are the horizontal and vertical spacings between the reinforcement, respectively. The principles of this empirical design method are illustrated in Figure 1 for several generic types of reinforcements. As the reinforcement stiffness increases, the lateral earth pressure coefficient increases and the inclination of the failure surface at the upper part of the wall increases leading to a bi-linear failure surface commonly adapted for the design of quasi-inextensible reinforced systems.

To date, no theoretical framework has been developed to consistently predict these empirical relationships. This paper presents the engineering application of a strain compatibility analysis approach recently developed by Juran et al. (1990). It associates an elasto-plastic strain hardening soil behavior, elastic-perfectly plastic response of the reinforcement and a strain compatibility at the soil-reinforcement interface to calculate the maximum tension forces in the reinforcement. This approach allows for the evaluation of the effect of soil dilatancy and reinforcement extensibility on the generated tension forces.

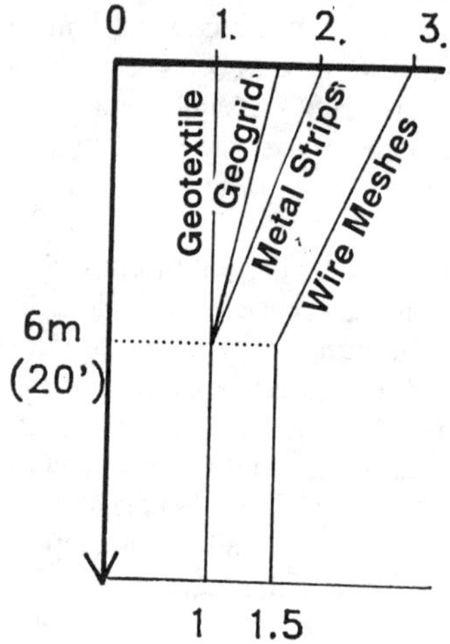

(a) Christopher et. al., (1989)

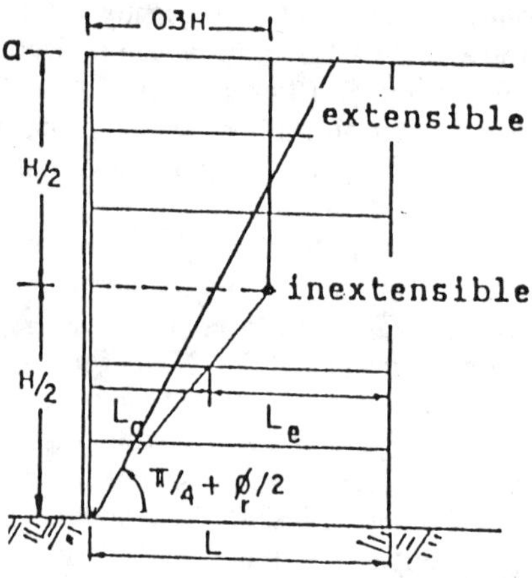

(b) Mitchell and Christopher (1990)

Figure 1. Design Guidelines for Evaluation of Lateral Earth Pressure in Reinforced-Soil Walls

This paper is focused on the evaluation of the modeling/design assumptions considered in this strain response analysis. A parametric study is conducted to develop design charts illustrating the effect of the extensibility of the reinforcement and of soil properties (i.e. friction angle, shear modulus and dilatancy rate) on the locus and magnitude of tension forces mobilized in the geosynthetic reinforcements. The results of this analysis are compared with published results of currently used working stress design methods to predict the tension forces in two well-documented geosynthetic test walls (Claybourn and Wu, 1991).

Modeling Geosynthetics-Reinforced Soil Structures

The model used in the analysis of geosynthetic reinforced soils is based on the analogy between the response of the reinforced soil material to shearing (Figure 2-a) and the plane strain shear mechanism developed along the potential failure surface (Figure 2-b). This model has been assessed through numerical simulations of the reinforced soil response to both triaxial compression and direct shear presented by Juran et al. (1988-a and b). The formulation of the strain compatibility design approach was presented in detail in Juran et al., 1990. This model was implemented in a computer program to predict the tension forces mobilized in the reinforcements and the location of the potential failure plane. The output data are used to evaluate the local stability of each reinforcement with respect to either breakage or sliding. The selection of the appropriate reinforcement and spacings should satisfy the required safety factors with respect to these potential failure modes, as well as the design criteria pertaining to the admissible strain levels in the reinforcements.

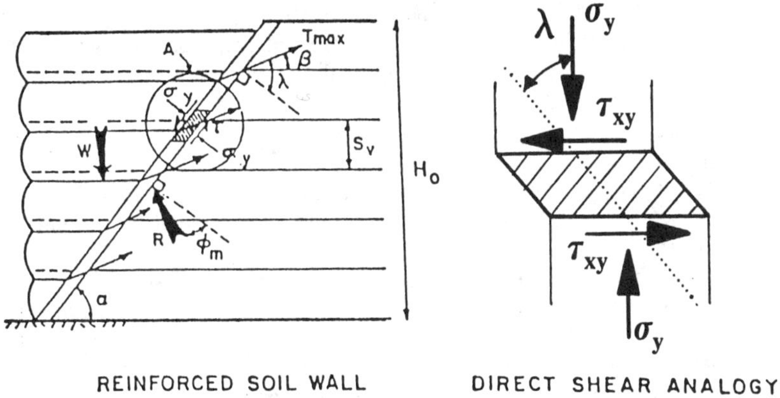

Figure 2. The Behavior of Reinforcement in a Wall and in Shear Test

Analysis of Behavior of Geosynthetic-Reinforced Soil Structures

The tensile force mobilized in the reinforcement depends on the stiffness of the reinforced-soil mass which is primarily related to extensibility and density (i.e. its horizontal and vertical spacings) of the reinforcement. The relative stiffness of the reinforced-soil mass (h_o) is defined in this analysis as:

$$h_o = E\,t\,b\,/\,\gamma\,S_h\,S_v \qquad \ldots\ldots\ldots\ldots (1)$$

where; γ is the unit weight of soil backfill.

For a given soil type, a unique relationship between the maximum tension forces mobilized in the reinforcement (T_N) and H/h_o (where H is the wall height) can be developed, provided that the strain in the reinforcement does not exceeded its yield strain (i.e. $\epsilon_R < \epsilon_y$). This relationship is illustrated in Figure 3. In this figure, the normalized maximum tension force (T_N) is expressed as:

$$T_N = T_{max}/\,\gamma\,H\,S_h\,S_v \qquad \ldots\ldots\ldots\ldots (2)$$

The relationships in Figure 3 are calculated for typical compacted backfill materials with the following soil parameters: soil friction angle at peak $\phi_p = 40°$, soil friction angle at critical state $\phi_{cv} = 31°$, soil unit weight $\gamma = 15.5\ KN/m^3$, soil dilatancy angle $\upsilon = 1/3\ \phi_p$, and a range of normalized shear modulus (G/σ) where σ is the applied vertical stress. The figure illustrates several aspects of the behavior of geosynthetic-reinforced soil walls:

i) For a given structure height (H), the relative stiffness (h_o) of the reinforcement significantly affects the state of stress in the soil. For a specific wall height, as h_o decreases (i e. for a reinforcement of low E t b, or high S_v and S_h), the normalized tension forces, T_N, decreases from K_o state of stress (for quasi-inextensible reinforcement) to K_a state of stress (for highly extensible reinforcement).

ii) The variation of the normalized tension forces, T_N, with the relative stiffness, h_o, depends upon the normalized shear modulus, G/σ_y, which governs the shear strain in the soil. As soil shear modulus increases, T_N decreases from K_o state of stress to K_a.

The effect of the reinforcement relative stiffness (h_o) on the inclination of the failure surface (α) is illustrated in Figure 4 for the same range of soil parameters. The figure shows that:

i) For a given structure height, at high relative stiffness (as for quasi-inextensible reinforcement), the inclination of the failure surface becomes practically vertical.

ii) As the relative stiffness decreases (e.g. for extensible reinforcement), the inclination of the shear failure decreases to an asymptotic value which is a function of the normalized shear modulus G/σ of the soil.

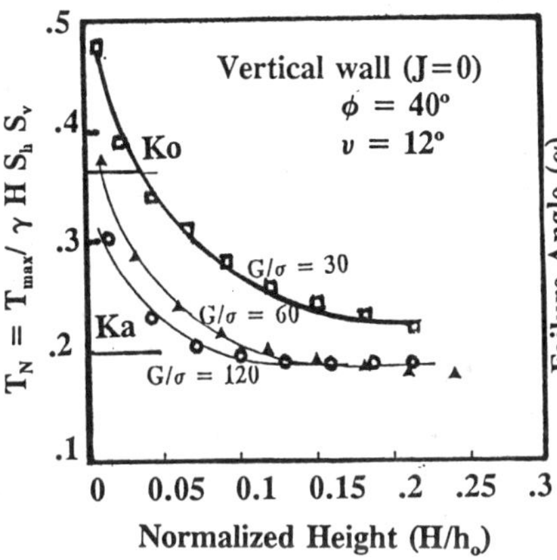

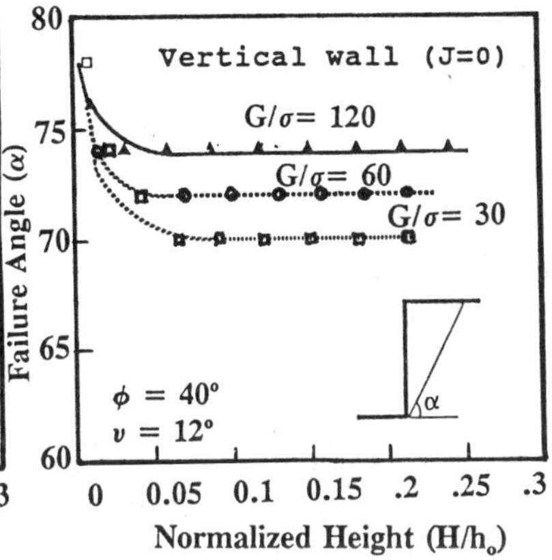

Figure 3. **Figure 4.**
Effect of h_o on the Tension Forces and Inclination of Failure Surface

In order to assess the effect of soil dilatancy on the maximum tension forces, the assumption of constant dilatancy angle was used in this parametric study. The dilatancy angle υ is defined as:

$$\tan \upsilon = d\epsilon_{yy} / d\gamma_{xy} \qquad \text{(3)}$$

Where $d\epsilon_{yy}$ is soil strain increment in the direction orthogonal to the failure surface and $d\gamma_{xy}$ is shear strain increment along the failure surface. Soil dilatancy is highly dependent upon the in-situ confining pressure and construction process. It should be noted that the construction process can result in an initial soil distortion, thereby generating the maximum dilatancy rate at initial stages of construction which would be consistent with the assumption of constant dilatancy.

The effect of soil dilatancy on the relationship between relative stiffness (h_o) and the maximum tension forces (T_N) is displayed in Figure 5. The figure shows that low soil dilatancy leads to lower mobilized tension forces. Figure 6 displays the effect of soil dilatancy and reinforcement relative stiffness on the inclination of the failure surface (α). As the dilatancy angle increases, the inclination of the failure surface decreases. It should be noted that when the dilatancy angle υ equals ϕ, the inclination of the failure surface reaches the solution obtained for perfectly plastic soil where the failure surface is inclined at $(\Pi/4 + \phi/2)$ with the horizontal.

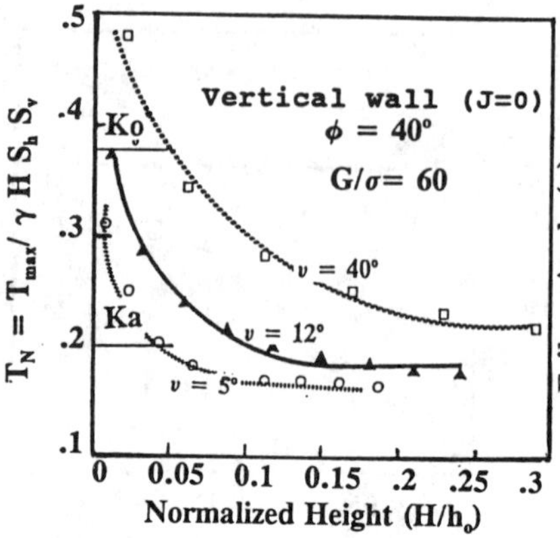

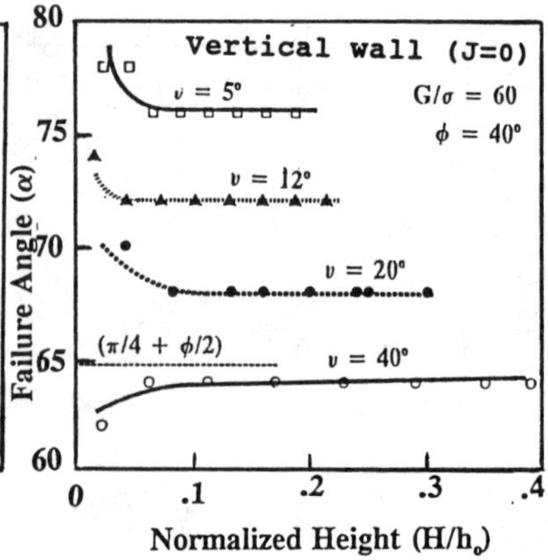

Figure 5. **Figure 6.**

Effect of Soil Dilatancy on Tension Forces and Inclination of Failure Surface

The predicted variation of α is consistent with observations on both reduced scale and centrifugal model walls (Juran and Christopher 1989 and Mitchell et al. 1988). The results are also consistent with the empirical relationships proposed by Christopher et al. (1989). It should be noted that Figures 4 and 6 display a vertical failure surface at the upper part of the structure when quasi-inextensible reinforcement is used. For these very stiff reinforcement systems, a strain compatibility approach considering a log-spiral surface has been developed by Juran and Chen (1988).

Design of Reinforced-Soil Structures

The internal stability analysis of the soil-reinforcement system requires an evaluation of the short and long term performance of the geosynthetic with respect to its pull-out and breakage resistance. Reinforcement type, spacings and length should be selected to ensure that the tension force at each reinforcement level should not exceed its breakage and pull-out resistance with the appropriate factors of safety. For reinforced-soil structures of known extensibility parameters (h_o), the tension forces (T_N) at each reinforcement level (H/h_o) and the inclination of failure surface (α) can be determined from Figures 7 and 8, respectively. It should be noted that these charts are based upon the analysis of structures of uniform soil properties and without concentrated surcharge. These charts have been generated using a range of soil friction angles from 28° to 40° and typical cohesionless soil (c=0) of G/σ = 60 and dilatancy angles = 1/3 ϕ_p.

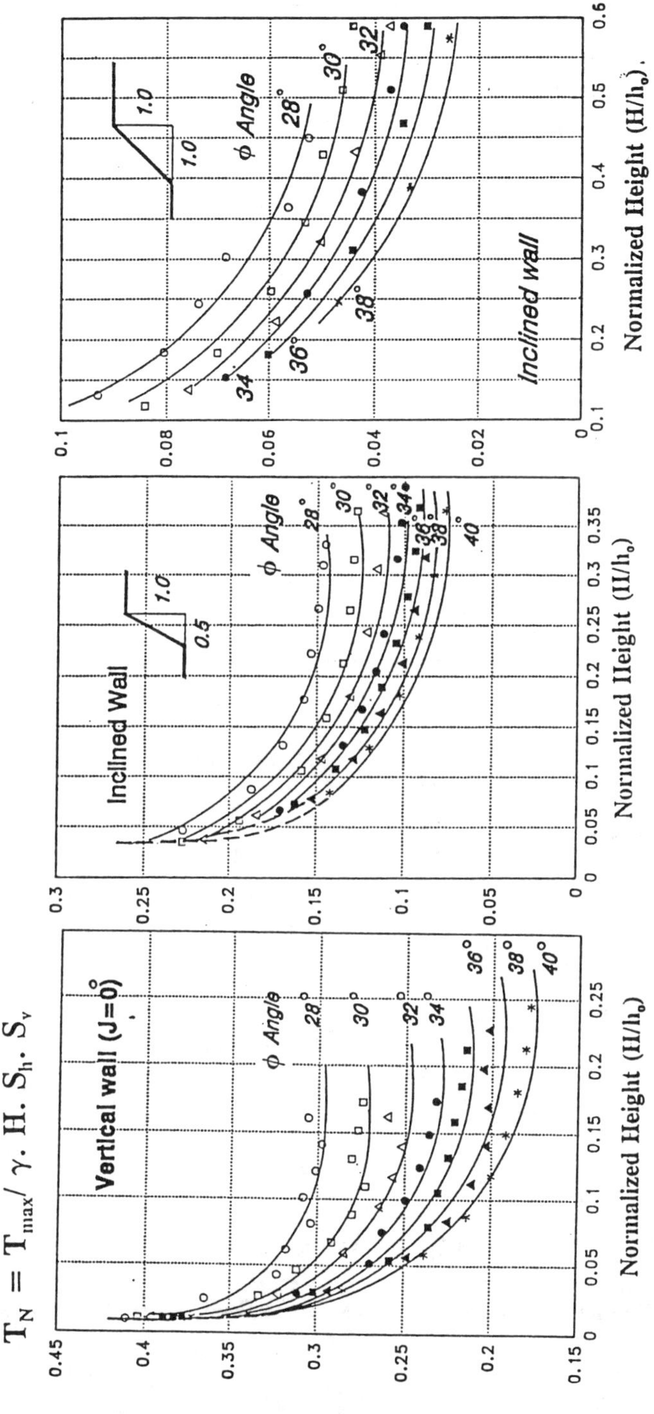

Figure 7. Maximum Tension Forces vs. Normalized Structure Height

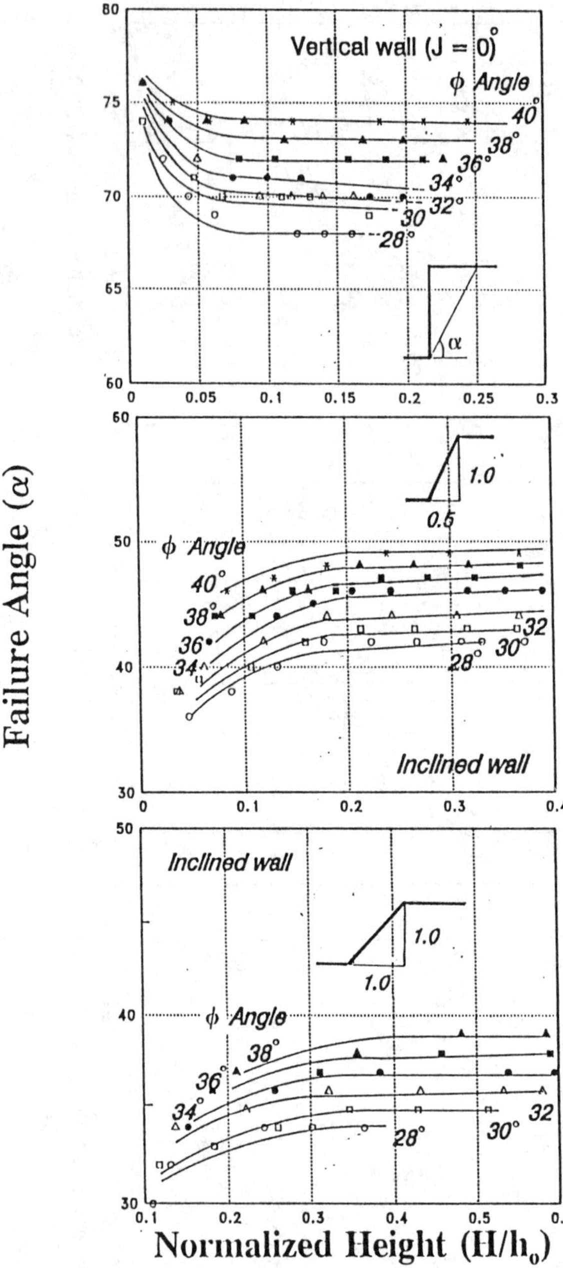

Figure 8. Inclination of Failure Surface vs. Normalized Structure Height

Comparison with Results of Other Design Methods

The predicted tension forces from the strain compatibility design method are compared with results obtained from currently used design methods for geosynthetic-reinforced walls. The results of these design methods were presented by Claybourn and Wu (1991) for two well-documented test walls. These methods are based on limit equilibrium analysis with different assumptions relevant to the earth pressure distribution and the plane of failure surface.

The first wall (STS wall) was 6 m (20 ft) high and was constructed with granular soils of $\phi = 39°$ and dry density of 2 t/m^3 (130 lb/ft^3), and extensible non-woven geotextile of ultimate strength 19.3 KN/m (1320 lb/ft). In this analysis, the reinforcement is estimated to have a stiffness modulus (Et) of 440 KN/m (30,000. lb/ft) and vertical spacings of 0.76 m (2.5 ft). The above parameters lead to a relative stiffness (h_o) of 28.7 m (94 ft).

The second wall (RMC wall, Bathurst et al. 1988) was 3.0 m (9.8 ft) heigh and was constructed of compacted granular backfill. Both triaxial and direct shear tests were conducted on the sand at different relative densities to establish its mechanical properties. A soil friction angle (ϕ_p) of 53° and dry density of 1.74 t/m^3 (108 lb/ft^3) are used in this analysis. The results of direct shear tests (Bathurst et al, 1987) were used to estimate an average dilatency angle of 18°, from the vertical-horizontal displacement measurements, and an average G/σ value of 80. Four layers of extensible HDPE geogrid reinforcement were placed at the weak direction at 0.75 m (2.5 ft) vertical spacings. A stiffness modulus (Et) of 170 KN/m (11.65 Kips/ft) was used in the analysis. These parameters lead to a relative stiffness (h_o) of 13.5 m (44 ft). A vertical surcharge of 50 kPa (1045 lb/ft^2) was applied to the top of the RMC wall and was sustained until soil failure. Since surcharge loads are not imcorporated in the numerical analysis, the uniform surcharge was approximated as an equivalent wall height.

The predicted tension forces from this analysis are compared with results of six different design methods for the two walls. The results are plotted in Figure 9. The mobilized stresses in the STS wall were not recorded while the tensile stresses in the RMC wall (based on the measured strains) are plotted in the figure. The results show that strain compatibility analysis predicts tension forces in the reinforcement between the K_o and K_a analysis and that the results are comparable with the calculated stresses in the RMC wall. It should be noted that the low mobilized low forces in the bottom of the RMC wall are most likely due to the toe restrained forces. This boundary constraint is not considered in the analytical model.

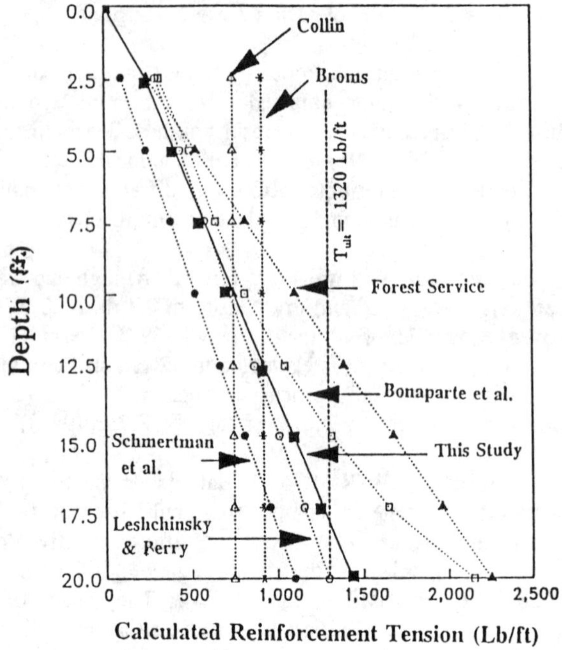

(a) The STS Wall

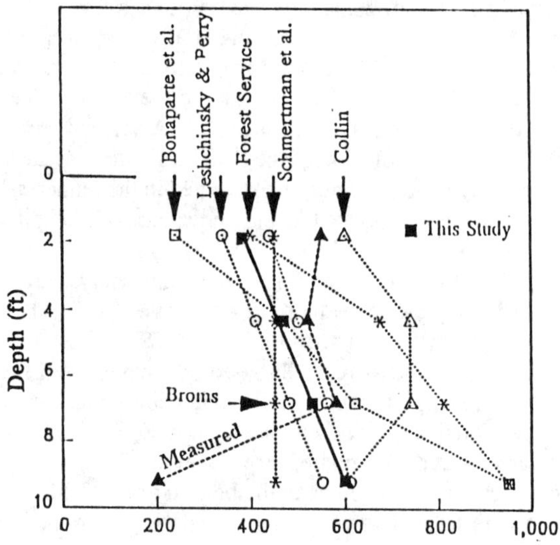

(b) The RMC Wall

Figure 9. Calculated Reinforcement Tension in the Walls
(Results other than this study are after Claybourn and Wu, 1991)

Conclusions

The strain compatibility analysis demonstrates the significant effect of the reinforcement extensibility (demonstrated by the relative stiffness factor h_o) and soil dilatancy on the development of the state of stress in the reinforcement and on the failure surface. The analysis demonstrates that the two assumptions, commonly used, with regard to the state of stress in the soil (i.e. K_o and K_a) are too restrictive to properly simulate the observed variations of tension forces. The observations on reduced scale laboratory models (Juran and Christopher, 1989), centrifuge models (Jaber et al., 1987), and full scale walls (Christopher et al., 1989) have demonstrated the substantial effect of reinforcement extensibility on the mobilized tension forces in the reinforcement.

The strain compatibility design approach was evaluated through comparisons with the limit equilibrium design methods commonly used in soil-reinforcement. The comparisons demonstrate the applicability of the design method to properly predict the state of stresses in the reinforcement. Design chart samples were provided to assess the effect of the basic design parameters on the anticipated behavior of the structure.

The main difficulties encountered in the strain compatibility analysis concern the need to properly establish the soil dilatancy rate and the mechanical properties of the confined geosynthetics. These properties are dependent upon the confining stress and can be significantly affected by soil compaction and the construction process. For conservative design purposes, it is recommended to assume that under field conditions the dilatancy rate of the soil is fully mobilized.

References

Al-Hussaini, M., (1977), "Field Experiment of Reinforced Earth Wall," Intern. Conf. on the Use of Fabrics in Geotechnics, Paris, Vol. 1, pp. 119-121.

Bathurst, R., Wawrychuk, W., and Jarrett, P., (1987), "Laboratory Investigations of Two Large-Scale Geogrid Reinforced Soil Walls," Application of Polymeric Reinforcement in Soil Retaining Structures, NATO Advanced Study Institute Series, Martinus Nijhoff Publishers.

Bathurst, R., Benjamin, D.J., and Jarrett. P.M., (1988), "Laboratory Study of Geogrid Reinforced Soil Walls," Proc. of Geosynthetics for Soil Improvement, ASCE, Geotech. Special Publication No. 18, pp. 178-192.

Brakel, J., Coppers, M., Maagdenberg, A.C., and Risseeuw, P., (1982), "Stability of Slopes Constructed with Polyster Reinforced Fabric," 2nd Intern. Conf. on Geotextiles, Las Vegas, Vol. 3, pp. 727-732.

Christopher, B.R., Gill, S.A., Giroud, J.P., Juran, I., Mitchell, J.K., Schlosser, F., and Dunnicliff, J., (1989), Reinforced Soil Structures, Design and Construction Guidelines-Vol. I, FHWA Report No. RD-89-043.

Claybourn, A.F., and Wu, J.T, (1991), "Case History Comparison of Geosynthetic-Reinforced Soil Walls," Geosynthetics'91 Conference,

Atlanta, Vol. 2, pp. 549-559.

Gourc, J.P., Ratel, A., and Delmas, P., (1986), "Design of Fabric Retaining Walls, The Displacement Method," 3rd Internat. Conf. on Geotextiles, Vienna, Vol. 3, pp. 289-294.

Ingold, T.S.,(1982)," An Analytical Study of Geotextile Reinforced Embankments" 2nd Intern. Conf. on Geotextiles, Las Vegas, pp. 683-688.

Jewell, R.A., Paine, N., and Woods, R.I., (1984), "Design Methods for Steep Reinforced Embankments" Polymer Grid Reinforcement, pp.70-81.

Juran, I., and Chen, C.L., (1988)," Strain Compatibility Design Method for Reinforced Earth Walls," Journal of Geotech. Engineering, ASCE, Vol. 115, No. 4, pp. 435-457.

Juran, I., Guermazi, A., Chen, C., and Ider, H.(1988-a),"Modeling and Simulation of Load Transfer in Reinforced Soil Part-1" Intern. Journal of Numerical & Analytical Methods in Geomechanics, Vol. 12, pp.141-155.

Juran, I., Ider, H., Chen, C., and Guermazi, A.(1988-b),"Numerical Analysis of Response of Reinforced Soils to Direct Shearing: Part-2," Intern. Journal of Numerical & Analytical Methods in Geomechanics, Vol. 12, pp. 157-171.

Juran, I., and Christopher, B., (1989), "Laboratory Model Study on Geosynthetic Reinforced Soil Retaining Walls, "Journal of Geotech. Engineering, ASCE, Vol. 119, No. 7, pp. 905-926.

Juran, I., Ider, H., and Farrag, Kh., (1990) "Strain Compatibility Analysis for Geosynthetic Reinforced Soil Walls," Journal of Geotech. Engineering, ASCE, Vol. 116, No. 2, pp. 312-329.

Koerner, R.M., (1990), Designing with Geosynthetics, 2nd. edition, Prentice-Hall, NJ.

Leshchinsky, D., Volk, J.C., and Reinschmidt, A.J., (1986), "Stability of Geotextile-Reinforced Earth Railroad Embankments," Geotextiles and Geomembranes, No. 3, pp. 105-128.

Leshchinsky, D., and Poedeker, R., (1989), "Geosynthetic Reinforced Soil Structures," Journal of Geotech. Engineering, ASCE, Vol. 115, pp. 1459-1478.

Mitchell, J., and Christopher, B. (1990), "North American Practice in Reinforced Soil Systems,", Proc. of Design and Performance of Earth Retaining Structures, ASCE, Geotech. Special Publication No. 25, pp. 322-346.

Mitchell, J.K., Jaber, M., Shen, C.K., and Hua, Z.K., (1988)," Behavior of Reinforced Soil Walls in Centrifuge Model Tests", Proc., Internat. Conf. on Geotech. Centrifuge Modeling, Paris, pp. 259-271.

Murray, R.,(1982), "Fabric Reinforcement of Embankments and Cuttings 2nd Internat. Conference on Geotextiles, Las Vegas, Vol. 3, pp. 707-713.

Steward, J.E., Williamson, R., and Mohney, J., (1977), Guidelines for use of Fabrics in Construction and Maintenance of Low-Volume Roads, FHWA Report No FHWA-TS-78-205.

Two Full Size Structures Reinforced By Geotextiles

Ph. Delmas*, Ph. Gotteland**, J.P Gourc***, S. Haïdar****

Abstract

In the framework of a research programme funded by the French Ministry of Research, two reinforced soil structures were built with a view to measuring strain. The structures included an embankment on weak soil and a retaining structure. The purpose of the research was to determine the operating mechanisms of this type of flexible structure. An opportunity was also provided for full-scale testing of a dual-function reinforcement and drainage geotextile.

1. Introduction

In France, interest in reinforcement geotextiles has been especially evident in cases involving soils with poor mechanical properties, as in embankments on compressible soils or retaining structures, where the geotextile is associated with a weak soil. In such cases, flexible structures are obtained, and the considerable deformation involved has given rise to a new problem in the field of soil mechanics.

The French Ministry of Research funded a major research programe (1987-90) on geotextiles. This programme, entitled GRIF, was led by J.P. Gourc and included the installation of measuring instruments on the two full-size structures referred to here. This programme also provided an opportunity for proposing the use of a dual-function geotextile (which was subsequently marketed by Rhône-Poulenc under the trade name "Bidimrock"). This textile combines a woven reinforcement fabric (tensile force at failure α_f for a strain ε_f, with an initial rigidity modulus J) and a non-woven drainage fabric (transmissivity θ) that accelerates consolidation of the adjacent soil. Geotextiles of this type were tried in the structures described below.

* Head of Earth Structures Section, Laboratoire Central des Ponts et Chaussées, Paris
** Assistant lecturer, IRIGM, Grenoble
*** Professor, IRIGM, Grenoble
**** Research worker, IRIGM, Grenoble

Emphasis is placed in this paper on in-situ strain measurements, which are rarely referred to in the literature, and on dimensional design methods.

2. Reinforced embankment at failure point on compressible soil

This structure benefited from the experience obtained during construction of the Aulnois embankment (Delmas, 86a) on a beetroot wash waste dump. In spite of settlement of more than 2 m, the punching of this geotextile-reinforced embankment was controlled. Monolithic subsidence (without internal shear) was noted up to the highest settlement values and the distribution of stresses within the geotextile could be directly correlated with horizontal displacements of the foundation soil.

However, structures of this type are usually designed on the basis of an overall punching study and a study of rotational failure of the embankments. It is therefore likely that the strains arising in the geotextile prior to failure will be impossible to correlate with those calculated on the basis of these failure mechanisms. Should another mechanism therefore be proposed to calculate the stability of such structures? The more detailed study of strains should give a clearer idea of the mechanisms involved. It was with this in view that an experimental embankment was built at Guiche by contractor Scetauroute, on a compressible soil consisting of slightly over-consolidated clay (z = 0 to 3 m -fig. 1), overlying a layer of mud (down to 24 m depth). Figure 1 gives the cohesion values of the layer of least cohesive soil (with Bjerrum's correction of Ip). The water table reaches z - 0 m. The dual-function geotextile (α_f = 225 kN/m, ε_f = 10%, J = 2165 kN/m, $\theta = 10^{-6}$ m^2/s) was placed in a layer of sand (γ = 19 kN/m^3) 1 m thick (0.10 m thickness of sand lying on top of the geotextile). The embankment above the geotextile (γ = 21.9 kN/m^3) consisted of clayey alluvial gravel (ϕ' = 35°). It was built up quickly until rotational failure occurred at a height H = 8.75 m. Failure in fact occurred after a significant settlement phase and corresponded with the failure of the geotextile, producing subsidence Δz = 2.5 m at the top of the embankment and a 2 m mound at the base.

The extensive range of instruments installed (Delmas, 90) provided a number of original results:

• Settlement of the foundation soil (figure 2b) - Settlement was slight up to H(embankment) = 7 m, with a maximum of 0.65 m occurring at profile C. At H = 8 m, the settlement varied from 1.05 m (t = 323 h round about the start of the construction (figure 2c)) to 1.60 m (t = 348 h) with maximum settlement occurring towards the centre of the embankment.

• Geotextile strain (figure 2a) - Displacement of the sheet was measured by means of 31 Invar wires placed in a sheath to avoid any friction against the soil, and attached to the geotextile. The arrangement was competed by 10 inductive strain gauges. It was noted that the geotextile began to move only when settlement reached high levels (beyond H = 7 m). Two maximum strain (and

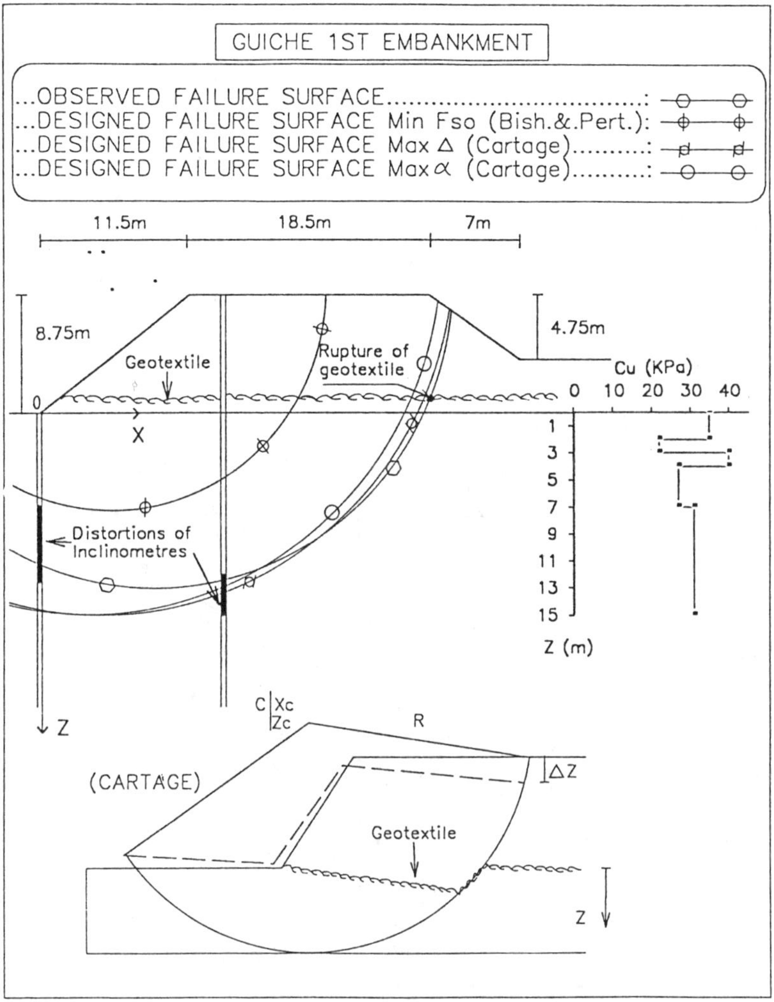

Figure 1: Reinforced embankment on compressible soil. Guiche experimental structure.

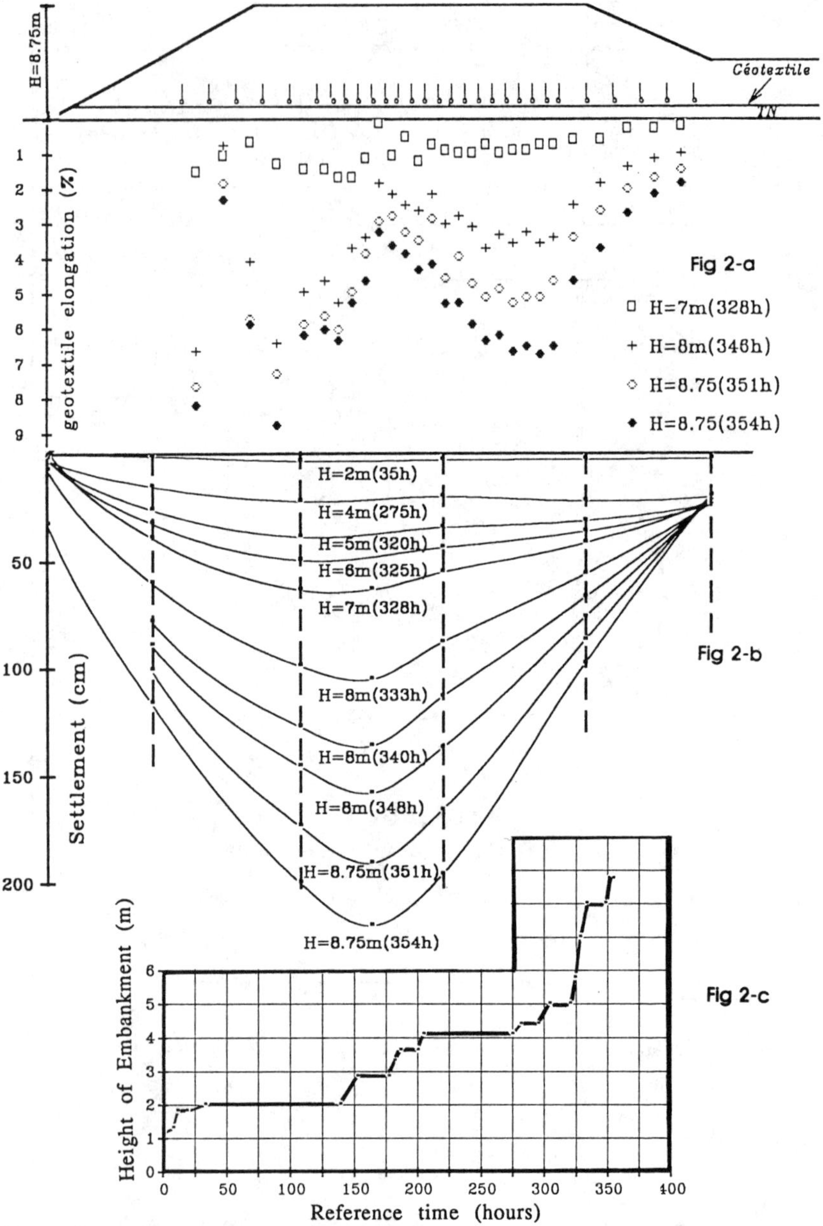

Figure 2: Guiche Embankment : (a) measurements of time-dependent geotextile strain (b) settlement measurements (c) rate of loading table

hence stress) points occurred, one at maximum settlement profile C, and the other to the rear of the embankment, close to the future point of tensile failure of the sheet (7.5% elongation, to be compared with $\varepsilon_f = 10\%$). This amply demonstrates the superimposition of two mechanisms, flexible foundation-type settlement and rotational slip.

• Horizontal strain in the soil adjacent to the geotextile (figure 3a) - Inductive strain gauges placed in the soil provided measurements of changes in strain in the foundation soil (z = 1.5 m), the sand layer under the geotextile (z = 0.5 m) and the embankment (z = -0.5 m). Little strain occurred until H = 5 m, except at maximum settlement profile C. Subsequently, strain in the embankment was about the same as that of the geotextile, confirming the homogeneous behaviour of the reinforced embankment.

• Relative displacement of the soil and geotextile (figure 3b) - It will be recalled that the tensile stresses set up in the geotextile occur as a result of the relative slip between the soil and geotextile (Gourc, 86). Soil displacements in the immediate vicinity of the geotextile (u_s) are compared with displacements of the geotextile (u_g). If u_r is the relative displacement, then $u_r = u_g - u_s$. Above the geotextile, there is a general slip of the embankment towards the side slope. In contrast, under the geotextile, there is very little relative displacement except at the profile where geotextile failure occurs.

• Post analysis - Figure 1 shows the slip circle corresponding to the observations made (including inclinometer + surface measurements). Use of the "displacements method" (LCPC-IRIGM 'CARTAGE' program), which is specific to reinforced soil retaining structures (Gourc, 89) was extended to the case of embankments. The most critical circle was investigated using this method, i.e. the circle giving the maximum stress α and the one giving maximum Δz values. Also shown is the most critical circle (identical) corresponding to the sliding resistance safety coefficient without any reinforcement and with minimum F_{so}, using the simplified Bishop slices method and the "Perturbations" method. The results are shown in table 1, with $c(X_c, Z_c)$ the centre of the circle and R the radius. The value of α ensures a safety coefficient $F_{SR} = 1$.

	Method	X_c (m)	$-Z_c$ (m)	R (m)	α(kN/m)	Δz (m)	F_{so}
Observed circle	Cartage	8.8	9.75	22.5	230	0.20	0.89
Tensile α max	Cartage	4.0	12.0	27.0	308	0.34	0.87
Displacement Δz max	Cartage	5.0	12.0	27.0	305	0.355	0.87
Critical circle	Bishop	5.75	9.5	16.5	-	-	0.89
	Perturbations	5.75	9.5	16.5	-	-	0.84

Table 1 - Some computation results

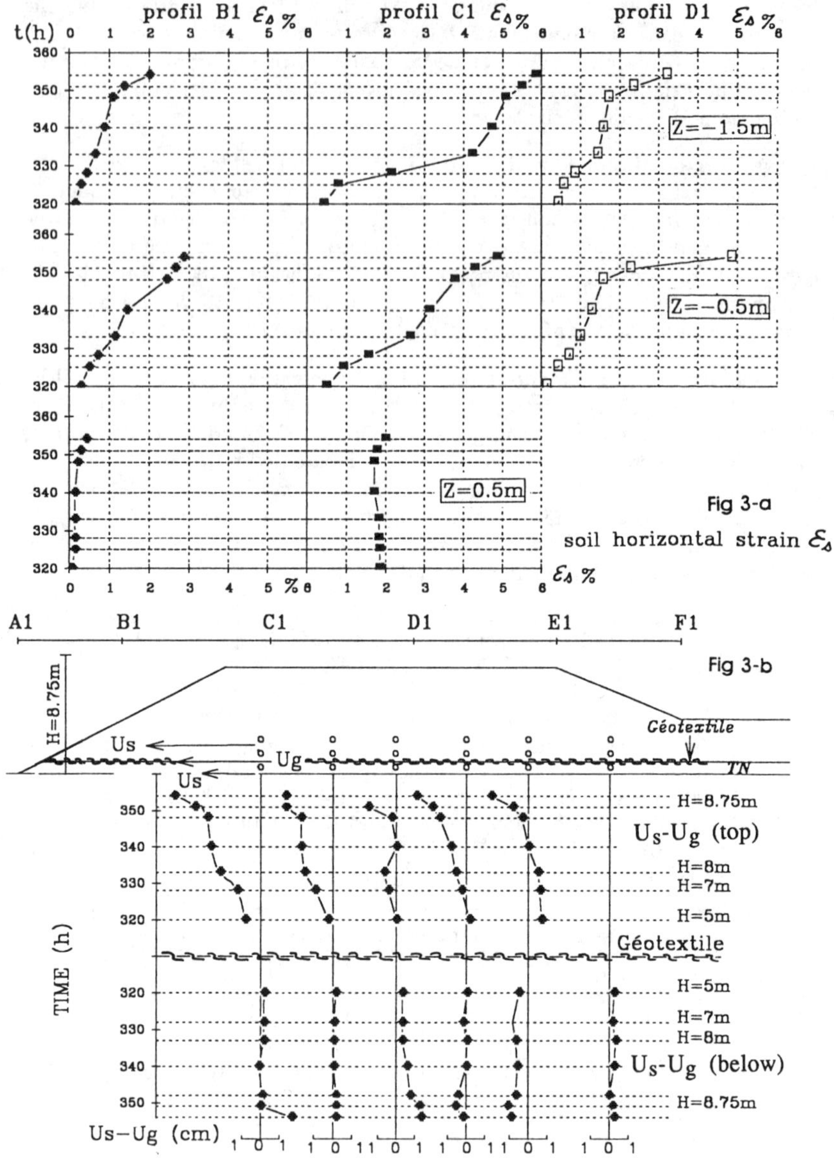

Figure 3: Guiche embankment: (a) horizontal deformations in the soil, (b) relative displacements between soil and geotextile.

The "perturbations" and Cartage methods are based on the same typical computation. It will be noted that the computation by the Cartage program ("displacement method") on the observed rupture circle gives a highly satisfactory result: the geotextile rupture noted should correspond to (α_f = 225 kN/m) and the calculation gives α = 230 kN/m. On the other hand, the critical cases determined by the Cartage program (α max or Δz max: critical slip lines generally very close), give high values of tensile stress α. This can be attributed to the heterogeneity of the soft soil.

The result obtained by an ordinary slice method (safety coefficient F_{so}, minimal, without strengthening) gives a completely different slip circle. This illustrates the advantage of the method proposed (Cartage), even if small cohesion differences significantly modify the calculated tensile stress values α. Consequently, for reinforced embankments, there is no reason to deviate from the normal rule of F_{SR} = 1.5.

3. Reinforced soil retaining structure

A 6-m high experimental reinforced soil retaining structure was built in conjunction with Ebal, which exploits the LCPC facing patent (Delmas, 86-b). Apart from the method of construction and facing, a special feature of the structure is the use of a dual-function reinforcement and drainage geotextile (α_f = 75 kN/m - J = 850 kN/m) associated with a silt. In 1992, it is intended to saturate this structure from the uphill side and possibly to overload it until failure occurs (figure 4).

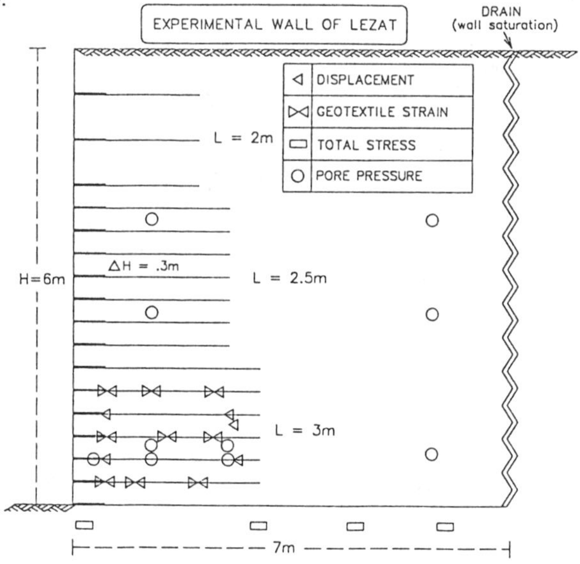

Figure 4: Experimental reinforced soil retaining structure Partial instrumentation.

Some of the most interesting measurements obtained in the recent stage of construction concern strain fluctuations in the geotextile measured by inductive strain gauges (sheet 1 - figure 5) and the change in relative displacement between the soil and geotextile ($u_r = u_g - u_s$ - figure 6) in the case of sheet 4, using the same procedure as in figure 3-b. The displacement u_r may help in characterizing the contribution of the anchoring stress.

On both figures 5 and 6, it can be seen that strains and displacements have not stabilized by the time the final height H = 6 m is reached. The structure is being monitored until failure occurs.

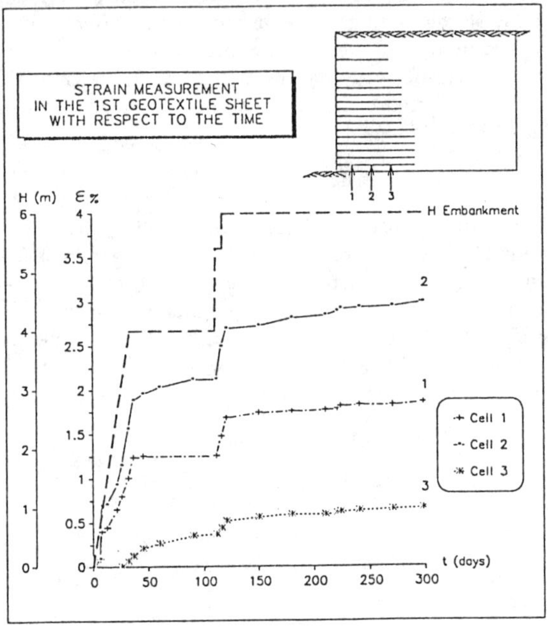

Figure 5: Reinforced soil retaining structure: geotextile strain during construction, in sheet 1.

• <u>Stabilizing effect of the geotextile drainage function</u> - Given the development of dual-function geotextiles, the authors endeavoured to quantify the effect of a decrease in excess pore water pressure in a reinforced soil structure with a high moisture content, with the geotextile acting as a drain. A previous experiment on an embankment containing a large amount of water (W_{opt} + 5%) is highly indicative in this respect (figure 7 - Perrier 86, Delmas 88). The pore water pressure reaches 60 kPa on completion of the works at the back of the reinforced structure. Pressure along the dual-function geotextile GnW70 is systematically lower than along the non-draining woven textile W130.

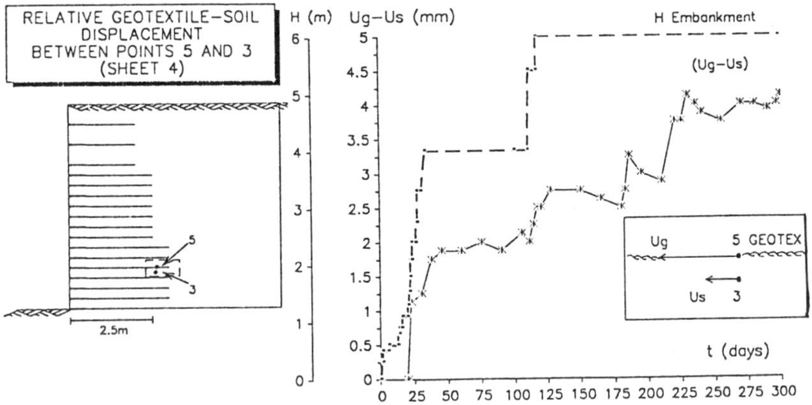

Figure 6: Reinforced soil retaining structure: relative displacements between soil and geotextile in sheet 4.

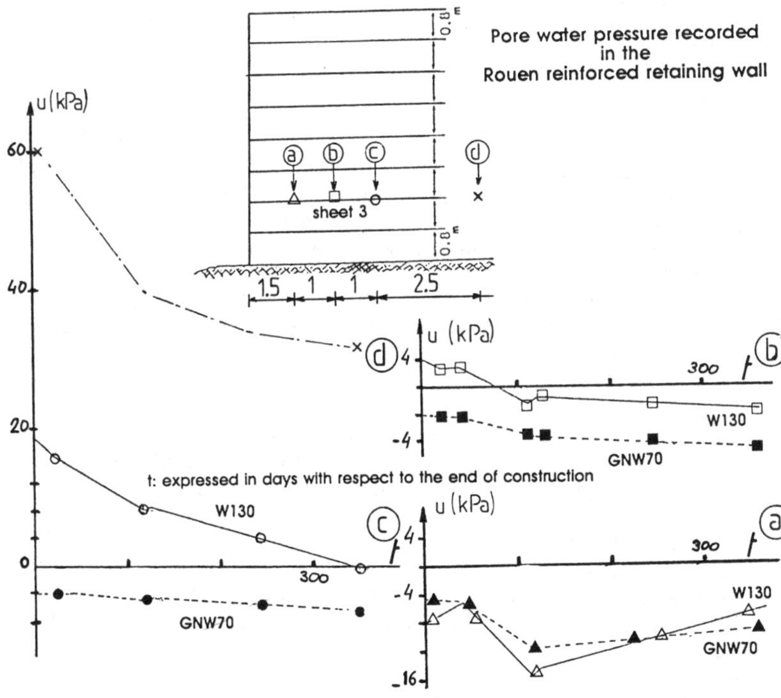

Figure 7: Comparison between a reinforcement geotextile and a dual-function reinforcement-drainage geotextile (Delmas, 88).

As regards stability of the reinforced soil structure (figure 8), the drainage effect is twofold, namely a decrease in excess pore water pressure $\overline{u}$ and hence an increase in resistance to shearing along the line of slip, and a decrease in excess pressure along the anchored section, hence an increase in the anchor strength.

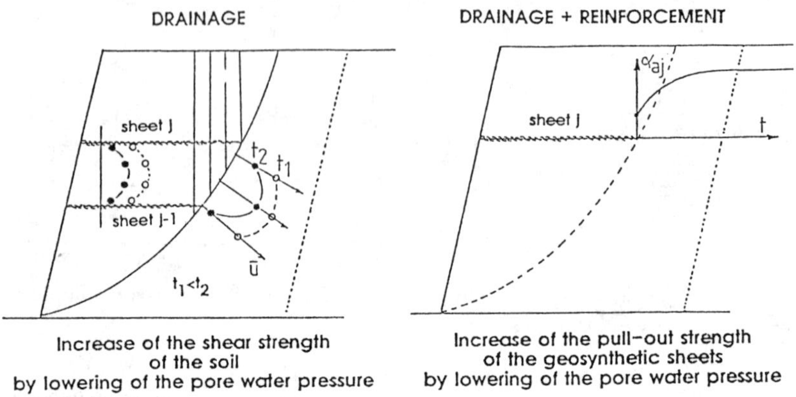

Figure 8: Principle for calculating the effect of drainage on the stability of a reinforced soil structure.

An attempt was made to define a theoretical formula for this problem, with the following conditions: computation of stability using the slices (perturbations) method with effective stresses, following numerical modelling of the consolidation of a layer of soil between geotextile sheets (Auriault, 82).

In the case of the theoretical example shown (figure 9), the initial pore water pressure in the structure is $\overline{u}$ = 20 kPa, spread evenly, though different stages of construction may be considered (Gotteland, 91). In this case, construction is assumed to be instantaneous. $\overline{u}(t)$ is then calculated. The calculation shown corresponds to a soil ($\phi = 35°$, $c' = 0$) of permeability ks = 2.6×10^{-9} m/s. The transmissivity of dual-function geotextiles mentioned above is of the order of 1 to 6 x 10-7 m^2/s, with the lowest value $\theta = 2 \times 10^{-8}$ m^2/s corresponding to partial clogging by the adjacent soil (Gourc, 90).

Figure 9-a represents the time-dependent change of the anchor strength that can be mobilized in sheets 6 and 7. There is an increase in α_a corresponding to the decrease in $\overline{u}$. It can be seen that $\alpha_a = 0$ (nil effective stresses normal to the geotextile) for the partially clogged geotextile (7) for a period of 30 days after the end of construction.

Figure 9-b shows the changes in the coefficient of safety against sliding, F_{SR}, for the reinforced structure. F_{SR} changes with time and with value of θ, from an initial value of F_{SR} = 1 at the end of construction.

Influence of the geosynthetic transmissivity on the pull-out strength

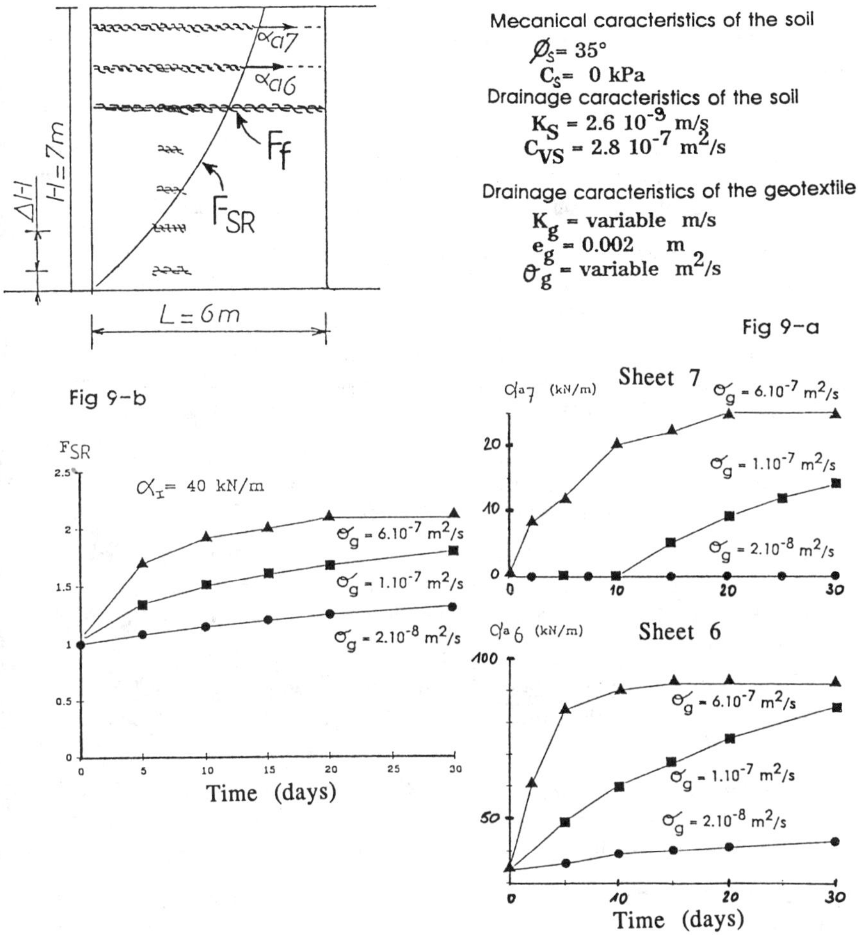

Figure 9: Effect of dual-function geotextile transmissivity on (a) anchor strength and (b) overall stability.

4. Conclusions

It is important to measure deformation in geotextile-reinforced structures in order to further our understanding of reinforcement mechanisms in this type of flexible structure. A number of measurements of this type are described here.

With dual-function reinforcement and drainage geotextiles, it is also possible to envisage using poor-quality soils with a high moisture content to build reinforced soil structures.

5. References

• Auriault J.L., Cordary D., Giroud J.P., Gourc J.P. (1977) : Etude théorique du rôle des drains textiles dans la consolidation des remblais. 1st International Conference on Soils and Textiles, Paris, pp. 273-278.

• Delmas Ph., Soyez B., Lenglet J., Puech J.P., Buffard J.Y. (1986): Experimental puncturing of a reinforced embankment in industrial wastes, Proc. 3rd International Congress on Geotextiles, Vol. 4, Vienna, pp. 1049-1054.

• Delmas Ph., Puig J., Schaeffer M., (1986): Mise en oeuvre et parement des massifs de soutènement renforcés par des nappes de géotextile. Eléments de coût, Bulletin de Liaison des Laboratoires des Ponts et Chaussées, n°. 143.

• Delmas Ph., Gourc J.P., Blivet J.C., Matichard Y. (1988): Geotextile-reinforced retaining structures; a few instrumented examples, International Geotechnical Symposium on theory and practice of Earth Reinforcement, Fukuoka.

• Delmas Ph., Queyroi D., Quaresma M., de Saint-Amand, Puech A. (1990): Failure of an experimental embankment on soft soil reinforced with geotextile: Guiche, Proc. 4th Int. Conf. on geotextiles and geomembranes, The Hague, May 1990, Vol. 3 (to be published).

• Gotteland Ph. (1991): Renforcement des sols par géosynthétiques: dimensionnement et validation. Doctoral thesis IRIGM, Univ. of Grenoble, 384 pp.

• Gourc J.P., Monnet J., Monmessin M. (1986): Reinforced embankments on weak soil: different theoretical approaches, Proc. 3rd Int. Cong. on Geotextiles, Vienna Vol. 4, pp. 1043-1049.

• Gourc J.P., Gotteland Ph., Delmas Ph. (1989): Parametric study of geosynthetic reinforced retaining walls using the "Displacement method" - Geosynthetics 89 Conference, San Diego, pp. 112-123.

• Gourc J.P., Faure Y.H. (1990): Soil particles, water and fibre, a fruitful interaction now controlled - Keynote, 4th Int. Coll. Geotextiles and Geomembranes, The Hague, Vol. 3.

• Perrier, H., Blivet, J.C., Khay, M. (1986): Experimental and actual use of geotextile reinforcement of a slope, Proc. 3rd International Congress on Geotextiles, Vienna, pp. 313-318.

Finite Element Analysis Of A Geogrid Reinforced Soil Wall

Richard J. Bathurst, Rajagopal Karpurapu and Peter M. Jarrett[1]

Abstract

A research program has been under way for several years in the Civil Engineering Department at the Royal Military College of Canada (RMC) that has included the construction and testing to failure of carefully monitored 3m high models of geogrid reinforced soil walls. Concurrent with the laboratory work, a finite-element program (GEOFEM) has been developed to meet the unique requirements of geosynthetic reinforced-soil wall systems including: incremental construction, interface elements to simulate geosynthetic-soil interaction and a non-linear geosynthetic force-strain model. The results from carefully instrumented RMC tests have been used to calibrate the program and allow parametric analyses to be carried out. The paper reviews experimental results for a propped panel wall taken to failure under uniform surcharge pressure and describes the results of a finite element simulation of the test. Program GEOFEM was used to make Class C predictions of panel displacements, grid strains and footing restraining forces prior to and at incipient collapse during surcharging.

Introduction

A research program has been under way for several years in the Civil Engineering Department at the Royal Military College of Canada (RMC) that is directed at acquiring detailed measurements of the mechanical behavior of geosynthetic reinforced-soil retaining wall systems. The experimental program has involved the construction and testing to failure of carefully monitored 3m high models of geogrid reinforced-soil walls constructed within the RMC Retaining Wall Test Facility (e.g. Bathurst 1990). At the time of this paper a total of 13 walls have been tested comprising a variety of facing treatments including wrap-around, incremental and full-height propped panel wall types. Some of the experimental work has been published elsewhere by the principal author and co-workers (e.g. Bathurst et al. 1987, 1988a,b, 1989).

Concurrent with the laboratory work a finite element program (**GEO**technical Finite **E**lement **M**odelling) has been developed that has been specifically de-

1. Civil Engineering Department, Royal Military College of Canada, Kingston, Ontario, Canada. K7K 5L0

signed to meet the unique requirements of geosynthetic–reinforced soil wall systems (Bathurst and Karpurapu 1990, Karpurapu et al. 1991, Karpurapu and Bathurst 1992).

The paper describes a propped panel wall test carried out in the RMC Test Facility and the details of the finite element simulation that was used to make Class C predictions of wall performance under uniform surcharge pressure. A comparison of predicted results and measured data for panel displacements, grid strains and footing restraining forces is given.

Propped Panel Wall Test

The propped panel wall that is the focus of this paper was constructed in the RMC Retaining Wall Test Facility and was described by Bathurst and Benjamin (1990). A schematic of the wall and the instrumentation that was used as part of the monitoring program are shown in **Figure 1**. A complete description of instrumentation aspects of large–scale model tests carried out within the RMC Retaining Wall Test Facility has been reported by Bathurst (1990). The wall in this study was constructed with 3m high panels that were pinned at their base and braced externally for the duration of fill placement and compaction. The width of the wall was 2.4m with a central monitored section 1m wide and two independent 0.7m wide outside panels. The facility side walls were constructed with a composite plywood/plexiglas/polyethylene sheeting that was used to reduce sidewall friction. Four layers of Tensar Geogrid BX1100 (SS1) reinforcement, oriented in the weak direction, were used in the construction. The grid is relatively weak and extensible but was purposely selected to ensure failure of the wall using the surcharge capacity at hand. The reinforcement layers were placed at 0.75m vertical intervals and extended 3m into the sand backfill. The reinforcement layout is identical to the arrangement adopted by the authors for previous tests that focused on the influence of facing type and reinforcement properties on wall behavior. The sand used in the construction is a uniformly graded washed sand with some fine gravel and was placed and compacted to a bulk density of 1.8 Mg/m^3. The results of large–scale shear box tests have shown that the material has a peak (secant) friction angle of about 53 degrees (Bathurst and Benjamin 1987). Following release of the external braces the entire surface of the sand backfill was uniformly surcharged in increments by inflating a row of airbags that was confined at the top of the facility. Surcharge increments were left on for a period of at least 100 hours to measure time–dependent deformation in the wall including creep in the polymeric reinforcement.

Figure 2 illustrates lateral deformations recorded at the face of the central monitored panel in the test. The jumps in outward wall movements were observed to match the application of each new load increment. Furthermore, as the magnitude of load level increased, the rate of time–dependent movement was observed to increase. During the final 80 kPa surcharge increment there were two unplanned unload/reload cycles that resulted from power outages. The two jumps in the deformation–time plot in **Figure 2** correspond to these events. Following the second reload cycle, there was evidence of a soil–to–soil failure through the reinforced soil mass. Several days later the 4th layer of grid reinforcement was observed to tear. At this point the wall was judged to be at a state of incipient collapse and it was rapidly unloaded and the experiment terminated. Subsequent-

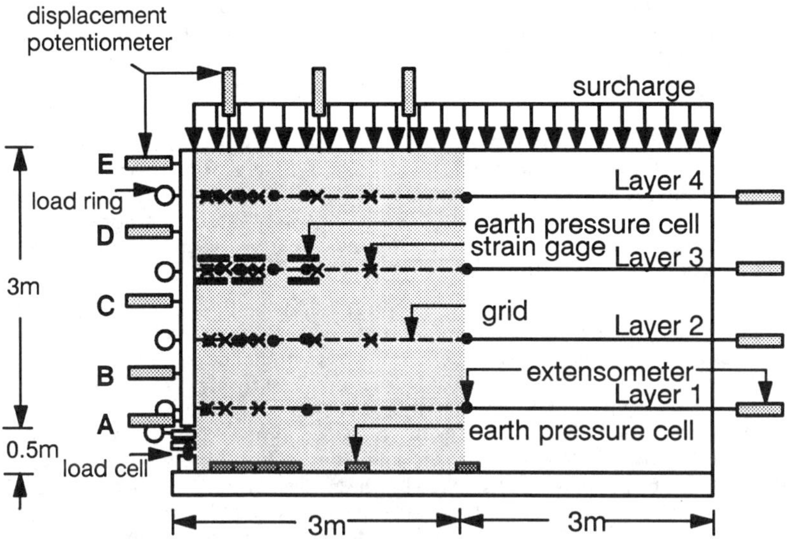

FIG. 1. RMC Propped Panel Wall Test

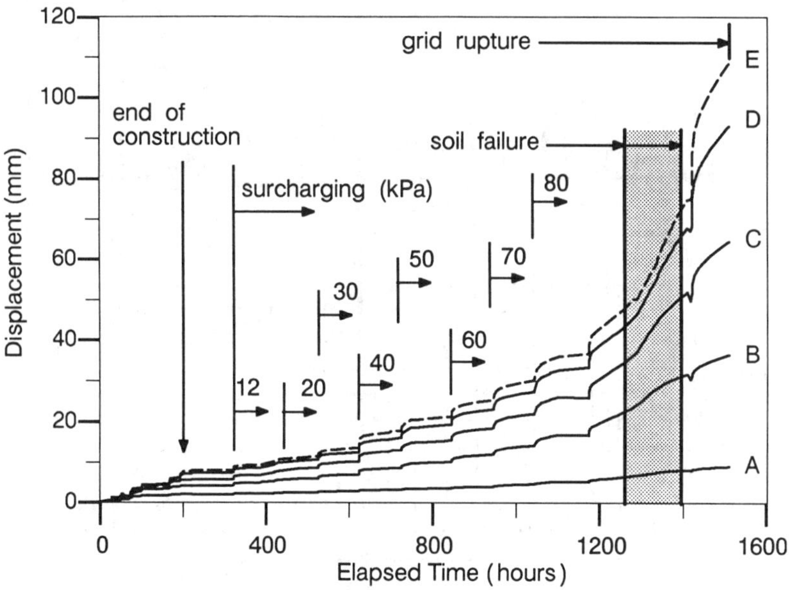

FIG. 2. Panel Displacements during RMC Propped Wall Test

ly, the soil, reinforcement and facing panels were carefully excavated to reveal important features such as a fully-developed failure surface through the reinforced soil mass.

Finite Element Simulation of Propped Wall Test

A primary objective of the finite element simulation was to predict performance aspects of the propped wall at incipient collapse. The deformations recorded both inside and at the panel boundaries are reasonably large as the data on **Figure 2** illustrates. Consequently, program GEOFEM was written to accommodate large deformation effects by implementing the linearised updated Lagrangian method as discussed by Bathe et al. (1975).

The finite element mesh used in the current study is illustrated in **Figure 3.** The mesh consists of about 1600 nodal points, 600 elements, and 3100 degrees of freedom. The geogrid reinforcement layers were modelled using bar elements which provide stiffness only in the axial direction. The front panels and the soil were modelled using eight-noded quadrilateral elements while the grid/soil and panel/soil interfaces were modelled using six-noded joint elements. The pin connection at the base of the panels was modelled by a triangular element. The stiffness of the facing panel was determined by physically loading the panel in a beam test prior to the numerical simulation work. Program GEOFEM was written to allow incremental fill placement to be simulated (i.e. rows of elements added sequentially as fill placement and compaction progressed). The external props were

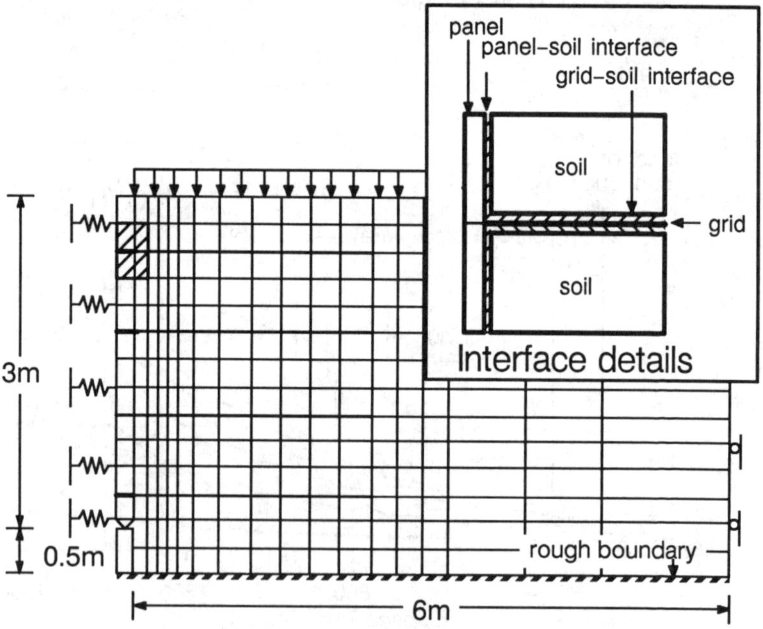

FIG. 3. Finite Element Mesh for Propped Wall Test

simulated using springs with a stiffness determined from measured prop forces and wall displacements recorded during construction.

Model for Soil

In the current study the soil was modelled using 8–noded quadrilateral elements with a 9–point integration rule to formulate element stiffness matrices. The soil behavior was simulated using a hyperbolic constitutive model that includes a soil dilation angle (ψ) and is formulated in terms of tangent Young's modulus and Bulk modulus (Bathurst and Karpurapu 1990). When the stress state exceeds the yield surface, the stresses are corrected back to the yield surface along the plastic flow direction using the soil dilation angle. This model is an extension of the hyperbolic model described by Duncan et al. (1980).

Hyperbolic parameters for the RMC sand were determined from triaxial tests and validated against the results of large–scale direct shear tests and plate bearing tests carried out at the surface of the sand backfill after the propped wall test was completed. Hyperbolic model parameters for the RMC sand are given in **Table 1**. The values determined experimentally in this investigation are in close agreement with values for similar materials contained in the database of hyperbolic parameters compiled by Duncan et al. (1980). Comparisons between experimental and FEM–predicted behavior in direct shear tests and plate bearing tests are shown in **Figures 4** and **5**. The large–scale direct shear tests were performed in a box 254 mm long, 152 mm wide by 152 mm deep (Jewell 1987). The overall comparison between the experimental and predicted direct shear behavior is reasonably good. The *dilation rate* predicted by the numerical model compares well with that observed in the experiments suggesting that the assumed dilation angle is representative of the soil behavior. It should be noted that the dilation angle (ψ) introduced in the model was necessary in order to give the close correspondence between measured and predicted normalized shear stress and strain shown in **Figure 4a** and to generate sample dilation as illustrated in **Figure 4b**. Nevertheless, the initial elastic compression observed in the experiments could not be replicated by the numerical model. Initial parametric analyses confirmed that without correct modelling of the soil dilation rate the collapse pressures observed in RMC trial walls were significantly under–estimated. The plate bearing test was performed using a circular plate of 0.6 m diameter. The test was modelled using axisymmetric geometry. The predicted pressure–settlement behavior of the plate bearing test agrees well with the measured response.

Locked–in horizontal compaction stresses were assumed to be present in the propped panel wall prior to removal of the external bracing. The magnitude of

TABLE 1. Hyperbolic Parameters for RMC Soil

Parameter	Value	Parameter	Value
K	950	R_f	0.75
m	0.65	ϕ_0	53^0
K_b	250	$\Delta\phi$	0^0
n	0.5	ψ	15^0

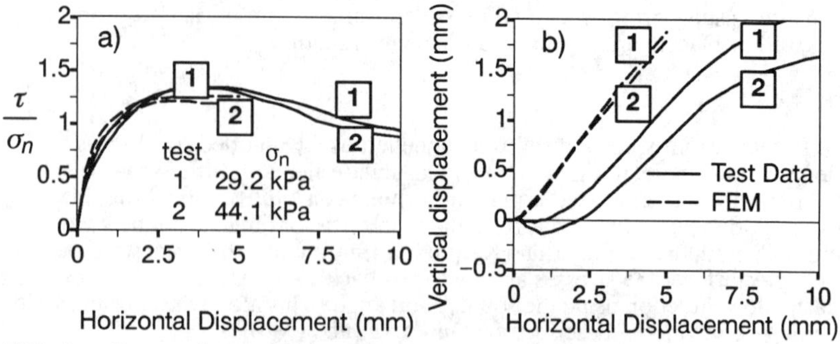

FIG. 4. Comparison of Experimental and Predicted Direct Shear Behavior for RMC Sand

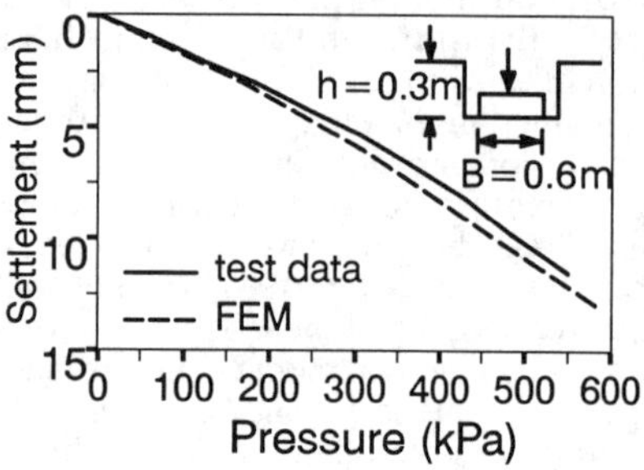

FIG. 5. Comparison of Experimental and Predicted Pressure–Settlement Behavior in Plate Bearing Test

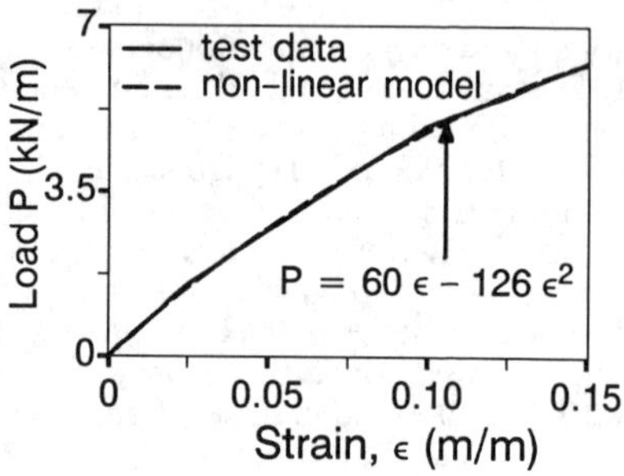

FIG. 6. 100–hour Isochronous Load strain Data For SS1 Grid Tested in Weak Direction

this lateral stress was calculated using the method advanced by Broms (1971). A representative value of 2kPa was used in the analysis. This compaction stress was applied to the soil elements before removal of the wall props and application of surcharge pressure. The predicted wall behavior at prop release and low surcharge levels was sensitive to the magnitude of applied compaction stress but the results of analysis at large deformations were sensibly independent of compaction effects and could have been safely ignored.

Model For Reinforcement

The reinforcement layers were modelled using 3–noded bar elements. The isochronous load–strain–time characteristics of the reinforcement in the weak direction were determined from the results of tests carried out on virgin samples of the grid material by A. McGown and co-workers at the University of Strathclyde. This isochronous creep test data was used to develop a non–linear force–strain model for the grid that relates the load in the grid (P) and the tangent stiffness (K_t) to the grid strain (ϵ) as follows:

$$P = A\epsilon + B\epsilon^2 \tag{1}$$

$$K_t = \frac{dP}{d\epsilon} = A + 2B\epsilon \tag{2}$$

The 100hr isochronous curve was used since the surcharge increments were applied in roughly 100hr time steps in the physical experiment. The results of in–isolation tests on similar grid materials reported by Yeo (1985) have shown that under staged tensile loading these materials exhibit a cumulative load–deformation response at the end of any increment of load that is equivalent to the deformation recorded if the final load had been applied in a single step. A comparison of the physical data and predicted load–strain behavior of the reinforcement is illustrated in **Figure 6**.

Interface Models

Joint elements of zero thickness developed by Ghaboussi and Wilson (1973) were used to model grid/soil and panel/soil interfaces. The global stiffness matrix of these elements is formulated in terms of two independent stiffness values, one in the tangential direction and the other in the normal direction. If the normal stress is compressive, the normal stiffness is assigned a large value to ensure a positive contact at the interface. When the normal stress becomes tensile, the normal stiffness is reduced to a small value to allow debonding at the interface. Maximum shear stress (in the tangential direction) is controlled by a Mohr–Coulomb failure criterion for both interface types. In the current simulation a perfect bond was assumed between the soil and grid when the shear stress was within the failure envelope. GEOFEM corrects shear stresses in the interface elements to be within the yield surface at all load steps. When the failure condition is achieved the tangential stiffness for each interface model is set to a very small value. The properties used for all interface elements are listed in **Table 2.** The friction angle between the wall panel and soil was back–calculated from measured panel toe forces (e.g. Bathurst and Benjamin 1987). Based on the experimental observation that the interface friction angle between well–compacted soil and most types of geogrids is higher than that of the soil alone and that the failure occurs within the soil, it was decided to use greater shear strength values for the interface than those for the soil alone (Johnston 1985, Bauer et al. 1991).

Results

Figure 7 shows measured facing panel profiles at stages in the test just prior to soil failure and at incipient collapse as the grid began to fail. Superimposed

TABLE 2. Properties Of Interface Elements

	Panel/Soil Interface	Soil/Grid Interface
Initial Shear stiffness	1000 kN/m/m	10^6 kN/m/m
Initial Normal stiffness	10^6 kN/m/m	10^6 kN/m/m
Residual Shear stiffness	10 kN/m/m	10 kN/m/m
Residual Normal stiffness	100 kN/m/m	100 kN/m/m
Friction angle	20^0	55^0

on the figure is the FEM–predicted profile. A comparison of predicted and measured values shows that the FEM gives a good estimate of the pre–collapse state. The "grid rupture" curve is difficult to predict since some grid layers were experiencing tertiary creep and the topmost grid had ruptured along 80% of its width when this measurement was made.

Figure 8 shows a comparison of measured and predicted strains along the length of each reinforcement layer at soil failure. The strains were measured using strain gages bonded directly to the reinforcement and extensometers attached to grid junctions. The numerical results are successful from the point of view of correctly predicting the maximum strain in most layers. The saddle–shape distribution of predicted strains is also qualitatively correct since peak strains were observed in this and other similar tests to occur in the vicinity of the connections and at the location of a deep–seated internal soil slip surface. The peak strains at the connections ultimately led to grid rupture in the physical test. The mechanism of failure which was observed in both the physical test and the numerical simulation was the downward movement of the retained soils relative to the essentially (vertically) rigid facing panel. Where the model has performed badly is in the prediction of the attenuation and magnitude of grid strain beyond the failed volume of soil (i.e. greater than 1m from the facing). A possible explanation for this poor prediction is that the curvature of the grid at the internal slip surface results in greater grid confinement and hence a very efficient load transfer between the grid and soil behind the plastic soil zone (i.e. Rankine wedge) which is difficult to simulate numerically.

Figure 9 shows the profile of the well–defined internal failure surface that was exposed after the test. The failure surface is log–spiral in shape and was predicted accurately using the location of peak grid strains output by the numerical simula-

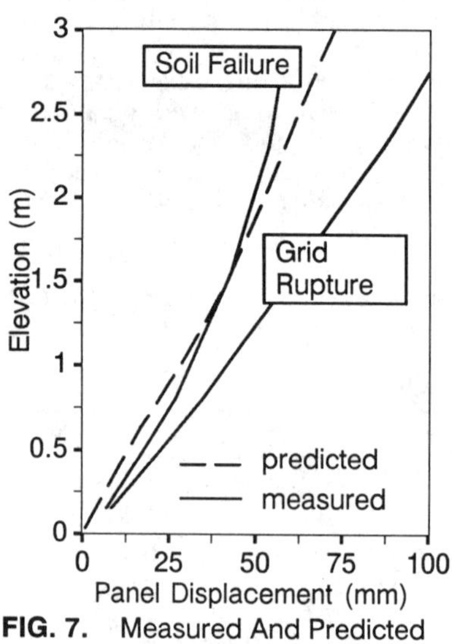

FIG. 7. Measured And Predicted Panel Displacements

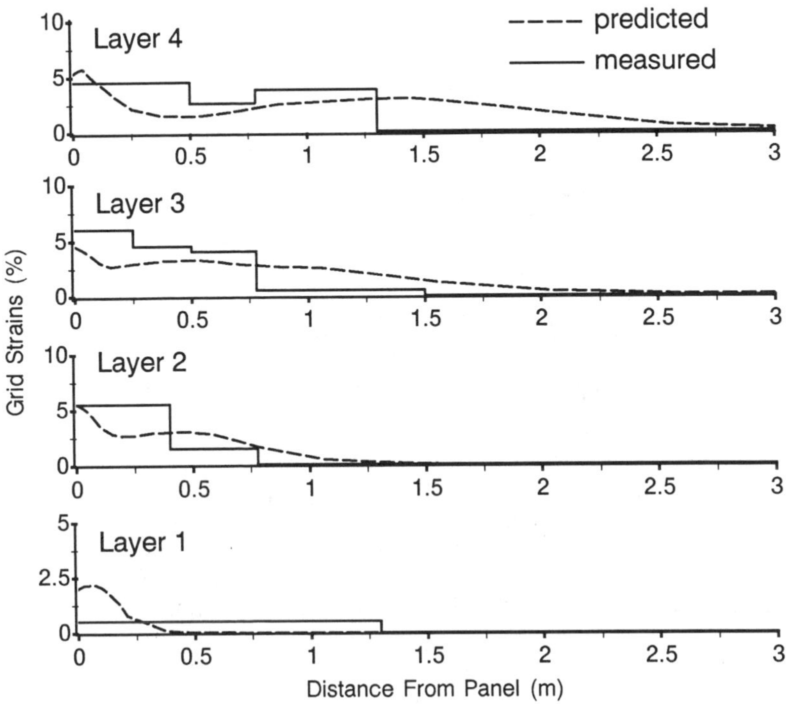

FIG. 8. Comparison Between Measured and Predicted Grid Strains

tion. The trace of the internal failure surface was also consistent with the location of contours of maximum soil shear strain. For comparison purposes the internal failure plane predicted by Rankine theory is also superimposed on the figure. This result implies that Rankine earth pressure theory is reasonably accurate in defining the failed soil zone when heavy surcharge loads are applied to shallow height walls constructed with dense cohesionless soils over a rigid foundation.

The variation and magnitude of horizontal and vertical force components generated at the base of the facing panels are shown in **Figure 10**. Predicted and measured forces agree very well. Both results emphasize the additional stabilizing influence of the footing that is routinely used to support and maintain panel alignment in field cases. For low height walls, such as the panel wall described here, this additional capacity cannot be considered in current limit–equilibrium methods of analysis but may be responsible for a portion of the conservativeness in conventional tie–back wedge methods of design. For example; the horizontal component of toe force is comparable to the magnitude of forces inferred in the reinforcement layers. The contribution of the vertical component of toe force to wedge stability is even greater.

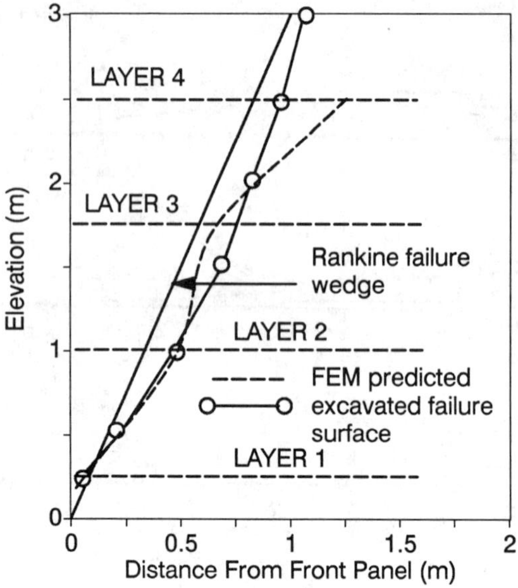

FIG. 9. Comparison of the Predicted and Measured Internal Failure Surfaces

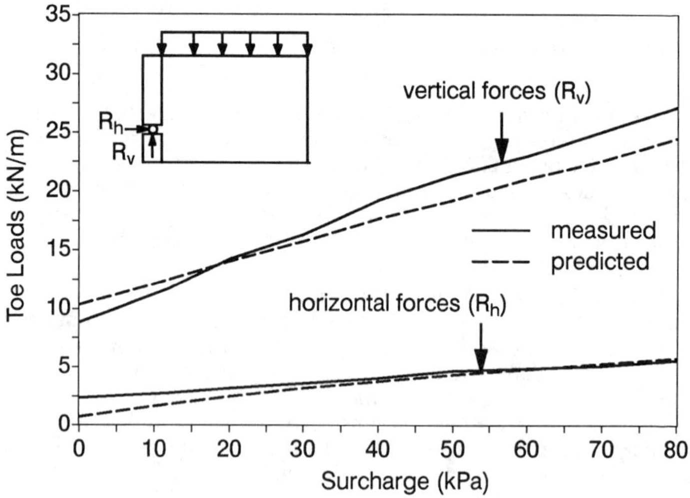

FIG. 10. Predicted and Measured Toe Loads

Conclusions

This paper has presented the results of a large-scale laboratory model of a propped panel wall taken to failure under uniform surcharge pressure. High quality measurements of the wall behavior at incipient collapse have been compared against the results predicted using the finite element program GEOFEM developed by the authors. The results of the simulation work have shown that important quantitative features of the wall at failure have been accurately predicted including panel movement, location and magnitude of peak strains in grid layers and the location of the internal failure surface through the reinforced soil mass. Toe forces at the base of the wall were also predicted with a high degree of accuracy. An important aspect of the simulation work is that the model parameters for the soil behavior were determined from independent physical tests and confirmed by correct modelling of shearbox and plate loading test results for the same soil. The properties of the grid reinforcement were determined from representative isochronous load–strain data on virgin samples of the material. Experience with the soil constitutive model showed that in order to model dilatant behavior observed in shearbox tests and to accurately predict important aspects of wall behavior it was necessary to modify the hyperbolic model reported by Duncan et al. (1980) to include a soil dilation angle parameter.

Acknowledgements

Financial support for the work reported here was provided through the ARP program and by the Chief of Construction and Properties, Department of National Defence, Canada. The authors would also like to acknowledge the contribution of A. McGown and co-workers at the University of Strathclyde, Scotland, who carried out the in-isolation testing of grid samples reported in this paper and the Tensar Corporation for providing the grid materials.

References

Bathe, K.J., Ramm, E., and Wilson, E.L. (1975). "Finite element formulation for large deformation dynamic analysis." *I. J. Num. Meth. in Eng.*, 9, 353–386.

Bathurst, R.J., Wawrychuk, W.F., and Jarrett, P.M. (1987). "Laboratory Investigation of Two Large-Scale Geogrid Reinforced Soil Walls." *Application of Polymeric Reinforcement in Soil Retaining Structures*, P.M. Jarrett and A. McGown eds., NATO Advanced Study Institute Series, Kluwer Academic Publishers, Dordrecht, 71–125.

Bathurst, R.J., and Benjamin, D.J. (1987). "Preliminary Assessment of Sidewall Friction on Large-Scale Wall Models in the RMC Test Facility." *Application of Polymeric Reinforcement in Soil Retaining Structures*, P.M. Jarrett and A. McGown eds., NATO Advanced Study Institute Series, Kluwer Academic Publishers, Dordrecht, 181–191.

Bathurst, R.J., Benjamin, D.J. and Jarrett, P.M., (1988a). "Laboratory Study of Geogrid Reinforced Soil Walls." *Geosynthetics for Soil Improvement*, R. Holtz, ed., ASCE Special Publication, 18, 178–192.

Bathurst, R.J., Jarrett, P.M., and Lescoutre, S.R. (1988b). "An Instrumented Wrap-around Geogrid Reinforced Soil Wall." *Proc. 3rd Canadian Symposium on Geosynthetics*, Waterloo, Ontario, October, 71–78.

Bathurst, R.J., Benjamin, D.J., and Jarrett, P.M. (1989). "An Instrumented Geogrid Reinforced Soil Wall." *Proc. XII International Conference on Soil Mechanics and Foundation Engineering*, Rio de Janerio, Brazil, 1223–1226.

Bathurst, R.J. (1990). "Instrumentation of Geogrid Reinforced Soil Walls." *Transportation Research Record 1277*, Washington, DC, 102–111.

Bathurst, R.J., and Benjamin, D.J. (1990). "Failure of a Geogrid Reinforced Soil Wall." *Transportation Research Record 1288*, Washington, D.C., 109–116.

Bathurst, R.J. and Karpurapu, R.G. (1990). "Large–Scale Reinforced–Soil Wall Tests and Numerical Modelling of Geosynthetic Modelling of Geosynthetic Reinforced Soil Structures." *Report to the Chief of Construction and Properties*, Royal Military College of Canada, Kingston, Canada.

Bauer, G.E., Halim, A.O.A., and Shang, Q. (1991). "Large–Scale Pullout Tests: Assessment of Procedure and Results." *Geosynthetics '91*, Atlanta, Georgia, 1991, 615–627.

Broms, B.B. (1971). "Lateral pressures due to compaction of cohesionless soils." *Proceedings of 4th Int. Conf. on Soil Mechanics,* Budapest, 373–384.

Duncan, J.M., Byrne, P., Wong, K.S., and Mabry, P. (1980). "Strength, stress–strain and bulk modulus parameters for finite element analysis of stresses and movements in soil masses." *Report No. UCB/GT/80–01*, University of California, Berkely.

Ghaboussi, J. and Wilson, E.L. (1973). "Finite Element for Rock Joints and Interfaces." *J. of Soil Mech. and Found. Division*, ASCE, 99(10), 833–848.

Jewell, R.A. (1987). "Analysis and Predicted Behaviour for the Royal Military College Trial Wall." *Application of the Polymeric Reinforcement in Soil Retaining Structures*, P.M. Jarrett and A. McGown eds., NATO Advanced Study Institutes Series, Kluwer Academic Publishers, Dordrecht, 71–125.

Johnston, R.S. (1985). "Pull–Out Testing of Tensar Geogrids," M.Sc. Thesis, University of California, Davis, California.

Karpurapu, R.G., Bathurst, R.J. and Jarrett, P.M. (1991). "Finite Element Analysis of an Incrementally Constructed Geogrid Reinforced Soil Wall." *44th Canadian Geotech. Conf.*, Calgary, Alberta, Sept–Oct., 82.1–82.8.

Karpurapu, R.G. and Bathurst, R.J. (1992). "Numerical Investigation of Controlled Yielding of Soil Retaining Wall Structures." *Geotextiles and Geomembranes,* 11, 17.

Yeo, Khen Cheok, (1985). "The Behaviour of Polymeric Grids for Soil Reinforcement," Ph.D. Thesis, University of Strathclyde, Strathclyde, Scotland.

A Computer Program for the Analysis of Reinforced Soil

Reyna[1], F., S.M. ASCE, Humphrey[2], D., M. ASCE
Christopher[3], B., M. ASCE and Chameau[4], J. L., M. ASCE

Abstract: A computer program dedicated to the analysis and design of reinforced soil structures constructed on soft or firm foundation is presented. The program is an addition to the STABL family of computer programs. Analyses made with the program are compared to that of other methods using data from the FHWA slope projects in Algonquin, Illinois. The FHWA project (Christopher et al., 1990) consisted of full scale field tests of a long embankment type structure with reinforced walls and slopes of different geometries and types.

Introduction

Limiting equilibrium analyses are the most common methods used at the present time to investigate the stability of reinforced embankments (e.g., Jewell et al. 1984, Jones 1984, Milligan and LaRochelle 1984, Schneider and Holtz 1986, Schmertmann et al. 1987). Of these techniques, the Jewell et al. (1984, 1990) method seems the most widely used because the definition of the safety factor is consistent with accepted practice and there is separation in the analysis between required and available reinforcement forces (Milligan, 1985). Computerized limit equilibrium approaches have also recently being implemented (Christopher and Leshchinsky, 1991).

This paper describes a computer program developed to analyze reinforced embankments on soft or firm foundations. For the case of a soft foundation, the results of the program are compared to those obtained from finite element

[1]Graduate Research Assistant, School of Civil Engineering, Purdue University, West Lafayette, IN

[2]Assistant Professor, School of Civil Engineering, University of Maine

[3]Vice President Services, Polyfelt

[4]Professor and Director, School of Civil Engineering, Georgia Institute of Technology , Atlanta, GA

calculations (Humphrey, 1986). For a firm foundation the results are compared to analyses conducted for the FHWA slope project in Algonquin, Illinois.

General description of STBREF

The software package PC STABL-5M (e.g. Achilleos, 1988) was modified to provide the capability of analyzing slopes with reinforcement (Humphrey, 1986). Other modifications were recently added and the new version of the program is called STBREF (STB for the STABL family of computer program and REF for reinforcement).

It is assumed that (a) the reinforcement does not alter the stresses on the assumed slip surface, and (b) the soil strength and available reinforcing force are mobilized simultaneously. Although finite element analyses indicate that the reinforcement has a limited effect on normal stresses along the portion of the failure surface passing through the embankment (Humphrey, 1986), the implications of these assumptions require further examination when the foundation soils reach a peak shear strength at small strain followed by strain softening. With these assumptions, limit equilibrium methods can be modified to include the stabilizing moment provided by one or more layers of reinforcement. The reinforcing force F_{Rj} acting at the intersection of the slip surface and the reinforcement provides a resisting moment equal to the product of F_{Rj} and its moment arm y_j about the center of the circle. This additional moment R_f can be introduced in the equilibrium equation either as a resisting moment:

$$FS_1 = (R + R_f) / D$$

or as a moment which reduces the driving moment:

$$FS_2 = R / (D - R_f)$$

Where D is the total driving moment, and R the resisting moment provided by the soil only. The underlying difference between these two definitions is apparent when solving the equations for D:

$$D = (R+R_f)/FS_1 \qquad D = \frac{R}{FS_2} + R_f$$

Thus, in the first definition the factor of safety is applied to the reinforcing force while in the second definition no factor of safety is applied to R_f (i.e. any factor of safety for the reinforcement must be applied separately). These two options have been implemented in STBREF modifying Bishop's method of slices. An explicit expression for FS_1 has been presented by Humphrey (1986).

The location of each reinforcing strip is specified by a series of coordinates starting at the left end of the reinforcement and moving to the right. The available force provided by the strip and its orientation are given at each of these points. The orientation of the force is defined by an inclination factor I_f, which can vary from 0.0 to 1.0. A value of 0.0 specifies that the force acts in the direction of the reinforcement, while 1.0 specifies that it is tangent to the slip surface. For the cases analyzed later in this paper, factors of safety will be given for four different options:

Option 1: F tangent to the slip surface, and increasing the resisting moment; factor of safety applied to reinforcement.
Option 2: F parallel to the direction of the reinforcement, and increasing the resisting moment; factor of safety applied to reinforcement.
Option 3: F tangent to the slip surface, and reducing the driving moment; no factor of safety applied to the reinforcement.
Option 4: F parallel to the direction of the reinforcement, and reducing the driving moment; no factor of safety applied to the reinforcement.

Options 1 and 3 assume that sufficient deformations have occurred to orient the reinforcement tangent to the slip surface and thus the moment arm of F is equal to the radius of the circle. From previous research, it is not certain that this can actually occur.

Comparison of STBREF with FEM for a failure through the foundation material

Reinforcement can be placed near the base of embankments constructed on weak foundations to increase the factor of safety against a failure surface extending into the foundation soils. Since there is considerable uncertainty in applying limiting equilibrium methods to this case it is useful to compare the results of STBREF with results from a finite element analysis which better models the actual stress conditions.

The effect of the magnitude and direction of the reinforcement force on the failure height for embankments with 2H:1V and 3H:1V slopes was investigated (Humphrey, 1986). The embankment fill was granular with a unit weight γ = 19.6 kN/m^3 and friction angle ϕ' = 32°. The foundation had a 2.29 m thick crust with a shear strength of 11.97 kN/m^2 (strong crust case) or 6 kN/m^2 (weak crust case). In addition, a case with no crust was considered. Underlying the crust was a normally consolidated soil with an undrained shear strength ratio s_u/σ' = 0.32. In the limiting equilibrium analysis, the increase in strength with depth was approximated by dividing the foundation soil into layers. In the FE analyses foundation thicknesses of 4.57, 9.14, and 18.28 m were used but in the limiting equilibrium analyses the critical circles were all less than 4.57 m deep. A reinforcement modulus of 880 kN/m was used in the FE analysis which corresponds to a high modulus woven geotextile or geogrid.

A comparison of the failure heights of reinforced embankments for the FE and limiting equilibrium analysis is made in Table 1. The failure height from limiting equilibrium analysis was calculated using F_R at failure determined by the FE analysis. This comparison method is necessary to account for the effect of reinforcement modulus on F_R. Both horizontal and tangent orientations of F_R were considered. The difference between the failure heights for the two orientations was less than 10%. For cases with a strong crust, the average failure height for the FE analysis and limiting equilibrium with a horizontal F_R was nearly the same. For the weak crust cases, the average failure height from the FE analysis was greater than the values from limiting equilibrium for either orientation of F_R, but the values are still similar. From this comparison it is not clear whether F_R should be taken as horizontal or tangent, however the FE analysis showed that the reinforcing strains at failure were generally less than 3% suggesting that for embankments with the soil conditions analyzed, it is unlikely that sufficient deformations occur to orient the reinforcement tangent to the failure surface. Thus, it would be more appropriate to take F_R to be horizontal.

			Failure Height		
				Limiting Equilibrium	
Base Width (m)	Side Slope (h:v)	Foundation Depth (m)	FE Analysis (m)	Horizontal (m)	Tangent (m)
Strong crust cases					
54.9	2:1	4.57	3.14	3.14	3.23
54.9	2:1	9.15	3.45	3.14	3.29
54.9	2:1	18.30	3.45	3.08	3.20
54.9*	2:1	9.15	2.56	3.14	3.26
54.9	3:1	9.15	3.14	3.23	3.11
36.6	2:1	9.15	3.14	3.11	3.23
Average			3.15	3.14	3.23
Weak crust cases					
54.9	2:1	4.57	2.56	2.20	2.41
54.9	2:1	9.15	2.56	2.13	2.29
54.9*	2:1	9.15	2.29	2.26	2.47
27.4	2:1	9.15	2.87	2.34	2.59
Average			2.57	2.23	2.44

* Compressible foundation

Table 1 Comparison of Failure Heights of Reinforced Embankments for FE and Limiting Equilibrium Methods (after Humphrey, 1986)

These results indicate that the limiting equilibrium method as implemented in STBREF can give results that are similar to the FE method for embankments on weak foundations provided that the appropriate F_R is used. Similar results have been found from other studies where the modulus of the reinforcing force was varied in the FE analysis (Rowe and Soderman, 1987). However, as seen for the case with a compressible foundation and a strong crust (foundation compressibility four times greater than in the other cases) it is possible for the limiting equilibrium method to overestimate the failure height. Choice of the appropriate F_R for typical design situations is beyond the scope of this paper. Reference is made to Humphrey and Rowe (1991) for a more complete discussion.

Comparison to analyses for the FHWA project

The FHWA project consisted of full scale field tests of a long embankment-type structure with reinforced walls and slopes of different geometries and reinforcement types. The site was a worked out gravel pit located northwest of Chicago, that had been excavated to about the water table level. The foundation of the embankment is a dense saturated sandy gravel, and thus can be assumed to have no effect on the performance of the various components of the structure. On the side with reinforced slopes, two 1H:1V and two 0.5H:1V, 7.6 m high, 15 m wide structures were built. Continuous sheet, geotextile and geogrid reinforcement, were used for each successive pair of structures. Ten reinforcing layers with a constant vertical spacing of 0.75 m and a length of 4.3 m were used for each structure. The slopes were constructed using a silt type soil having 70 to 90% particles finer than 0.07 mm and placed at 95% of Standard Proctor density (moist unit weight equal to 20 kN/m^3). Triaxial test results indicated $\phi = 35°$, and c= 2.4 kPa. Slope behavior was evaluated by instrumenting and monitoring external and internal horizontal movements and reinforcement strains during and after construction. Additional details concerning the project and results are included in Christopher et al. (1990) and Christopher and Leshchinsky (1991).

STBREF Results for the FHWA Slopes

The slopes were analyzed using circular trial failure surfaces assuming that the reinforcing force was constant along the length of the reinforcement. Options 1 through 4 were considered in determining the factor of safety. Analyses were repeated for many cases and the results are summarized in Figures 1 and 2 to illustrate the effects that the definition of the factor of safety and the inclination of the resisting force have on the value of FS. As the slope becomes steeper the differences in factor of safety obtained with different options increase. For example, for a constant distribution of the force of 8.0 kN/m (for each layer), the factor of safety varies approximately from 1.26 to 1.40 for a slope of 1H:1V, while it varies approximately from 0.99 to 1.30 for a slope of 0.5H:1V.

Figures 1 and 2 also show that the reinforcement force required to obtain a specified factor of safety is very sensitive to the inclination of the reinforcement force represented by the factor I_f. It is obviously conservative to assume that the resisting force acts in the direction of the reinforcement since the corresponding moment arm is smaller. However, considering that the reinforcement was assumed not to alter the stresses on the potential slip surface, this force orientation may be overly conservative. When the normal stress applied to the slip surface by a horizontal orientation is considered in the analysis of multiple level reinforcement in granular soils, the horizontal orientation has been found to produce essentially the same results as the tangential orientation observed in options 1 and 3 (Wright and Duncan, 1990 and Christopher and Leshchinsky, 1991). The influence of the inclination has been found to be significant for cohesive soils and a conservative horizontal orientation is advised. Furthermore, for Options 1 and 3 to be valid sufficient deformations must occur to orient the reinforcement tangent to the slip surface (Fowler 1982, Christopher and Holtz, 1985, Humphrey and Rowe, 1991).

For safety factors above 1.0, the required total reinforcement force for Option 3 is always less than Option 1 since for the latter, the factor of safety is applied to the required total reinforcement force (the results in Figure 1 can be confirmed by inspection of the equations

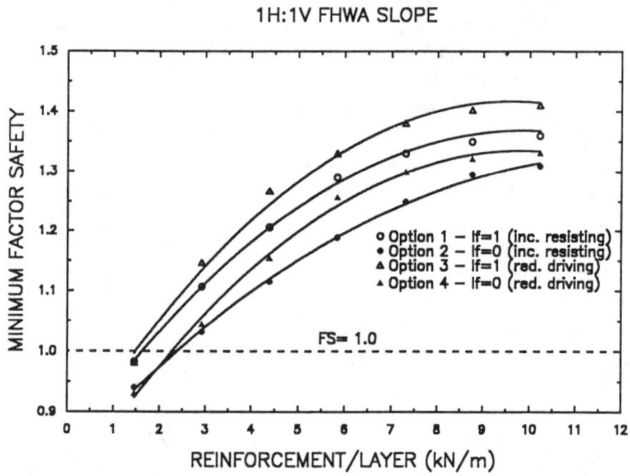

Fig. 1 Minimum Factor of Safety vs Reinforcement/Layer for the 1H:1V Slope

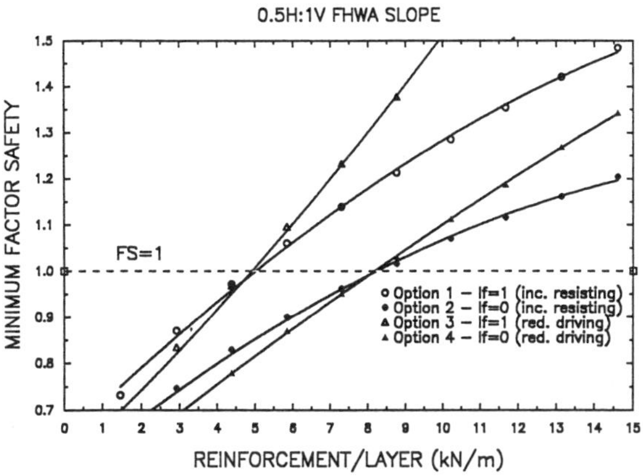

Fig. 2 Minimum Factor of Safety vs Reinforcement/Layer for the 0.5:1V Slope

defining FS_1 and FS_2). Likewise, Option 4 yields a smaller required total reinforcement force than Option 2. With options 3 and 4, a factor of safety for the reinforcement must be considered separately.

Results of Other Methods

Tables 2 and 3 present the results of the calculations performed at the design stage for the FHWA slopes using the Jewell (1984 and 1990), Jones (1984), and STABL methods. A tension of 10.5 kN/m at 5 % of strain for each reinforcement layer was adopted for the slopes, i.e. a total tension of 105 kN/m for the 10 reinforcing layers. A minimum factor of safety of 1.3 was required for all the 1H:1V slope and a factor of safety approaching 1.0 was required for the steeper 0.5H:1V slopes.

Method	Required Total Reinforcement Force (kN/m) FS=1.0	FS= 1.3
Jewell (1984)	79	129
Jones	124	231
STABL	100	144
Jewell (1990)	68	100
STBREF (Option 1)	49	105
STBREF (Option 2)	83	181
STBREF (Option 3)	47	80
STBREF (Option 4)	82	137

Table 2 Required Total Reinforcement Force (FS =1.0 and FS= 1.3) for a 0.5H:1V Slope with Silt Backfill

Comparison of Results

Making allowance for the different assumptions there is overall agreement between the different methods and STBREF for the 0.5H:1V slope (Table 2). The required reinforcement for the specified factor of safety, FS = 1.30 in this case, given by option 1 is similar to that given by the Jewell (1990) method.

Method	Required Total Reinforcement Force (kN/m)
Jewell (1984)	67
Jones	72
STABL	58
Jewell (1990)	49
STBREF (Option 1)	64
STBREF (Option 2)	95
STBREF (Option 3)	52
STBREF (Option 4)	75

Table 3 Required Total Reinforcement Force (FS=1.30) for a 1H:1V Slope with Silt Backfill

As expected the results obtained with options 2 and 4 are conservative with respect to options 1 and 3, because the assumption that the resisting force acts in the direction of the reinforcement is itself conservative. Similarly, the STABL, Jewell (1984) and Jones methods appear conservative. For the 1H:1V slope, the total required reinforcement force given by option 1 and the Jewell (1990) method differ by more than 30%. The position of the critical surface with (STBREF) and without reinforcement (PCSTABL-5M) was also analyzed. It was found that the reinforcement changes significantly the location of the failure surfaces. This explains the differences observed in Table 3 between conventional design analyses (e.g. Jewell and unreinforced versions of STABL) and STBREF, in the case of a 2.47 m long reinforcement layer.

The actual embankment was constructed with geosynthetics reinforcing having a maximum strength of 25.4 kN/m and 36.5 kN/m for the woven geotextile and the geogrid. Calculations using STBREF to predict maximum tensions forces in the reinforcement has also been performed. Figure 3 shows the measured maximum tension on each layer of reinforcement and the results using STBREF for the 1H:1V and 0.5H:1V slopes. A linear distribution of forces has been assumed in the reinforcement for the STBREF calculations. The results of the Finite Element analyses performed at the University of Berkeley are also superimposed in Figure 3. The Figure shows that the limit equilibrium analyses (STBREF) are somewhat conservative for estimating the maximum tension in the reinforcement

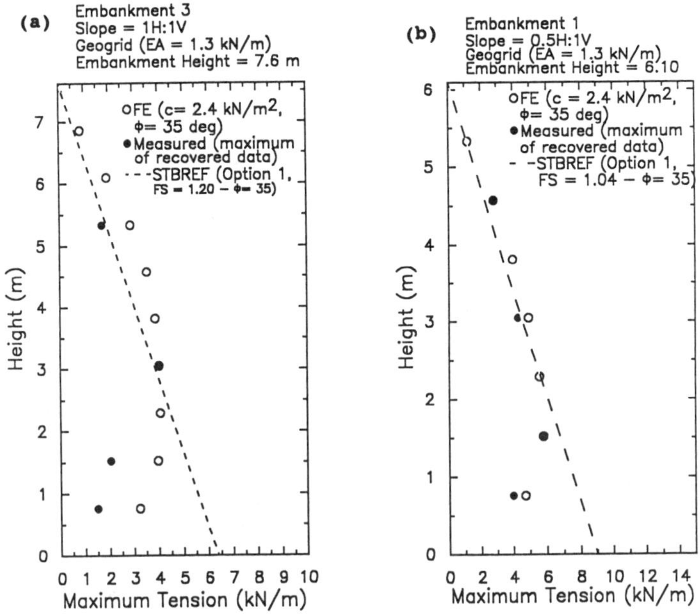

Fig. 3 Maximum Tension vs Depth: (a) 1H:1V Slope, and (b) 0.5H:1V Slope

especially for the lower layers of the structure, usually the most critical zone of reinforcement. However, the same two analyses shown in Figure 3 using STBREF are in fairly good agreement with the location of the maximum tension forces in the reinforcement (i.e. the most critical slip surface) for both slopes (Figure 4).

Design Charts

Although it is simple and efficient to use the software directly, design charts can be created with the program STBREF if desired. An example is given in Figure 5 for the total reinforcement force required to get a minimum factor of safety of 1.3. The chart was prepared based on Option 1. This chart must be used only for a slope constructed on a firm foundation when sufficient deformations occur to orient the reinforcement tangent to the failure surface. The total reinforcement force, F_T, is presented non-dimensionally by a horizontal reinforcement force coefficient, K:

$$K = 2F_T/\gamma H^2$$

where:
K = horizontal reinforcement force coefficient

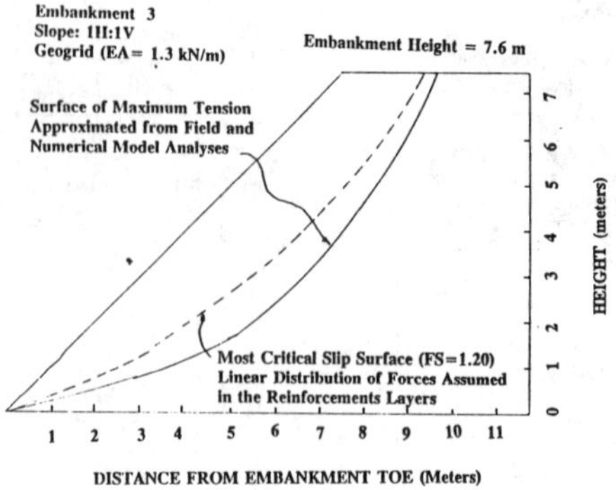

(a)

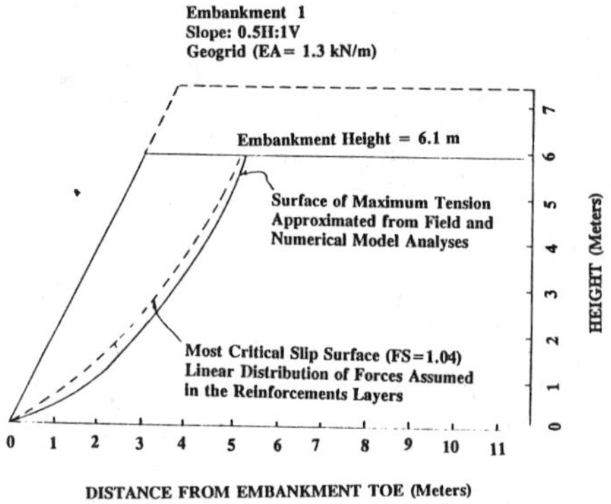

(b)

Fig. 4 Comparison of Location of the Most Critical Slip Surface (STBREF Analyses) with respect to the Surface of Maximum Tension Approximated from Field and Numerical Model Analyses: (a) 1H:1V and (b) 0.5H:1V Slope

γ = moist unit weight of the soil fill
H = height of the slope.
F_T= total reinforcement force

The K values were obtained based on a constant distance of reinforcement layer of 0.75 m (the ratio of distance with respect to the height of slope is 5.6) and a length of reinforcement sufficient to guarantee the effectiveness of the resisting moment of each reinforcement layer. The values of K in Figure 5 are similar to the ones given by Jewell's charts (1990).

Conclusions

Calculations made with STBREF and other methods show that:

- For embankments on soft foundations, limiting equilibrium methods can give similar failure heights to those predicted by the finite element method provided the appropriate reinforcing force is used.
- The lowest required reinforcing force is obtained when the force is assumed to be tangent to the failure circle and no factor of safety is applied to the reinforcement. Conversely the highest required reinforcing force is given when the force is horizontal, which results in a shorter moment arm about the center of the circle, and a factor of safety is applied to the reinforcing force.
- The STBREF results for the FHWA project showed that the required reinforcement force is very sensitive to its inclination when the reinforcement is assumed not to alter the stresses on the potential slip surface.
- Fairly good agreement was observed between the prediction and location of maximum tensile forces in the reinforcement found from STBREF and actual field results.

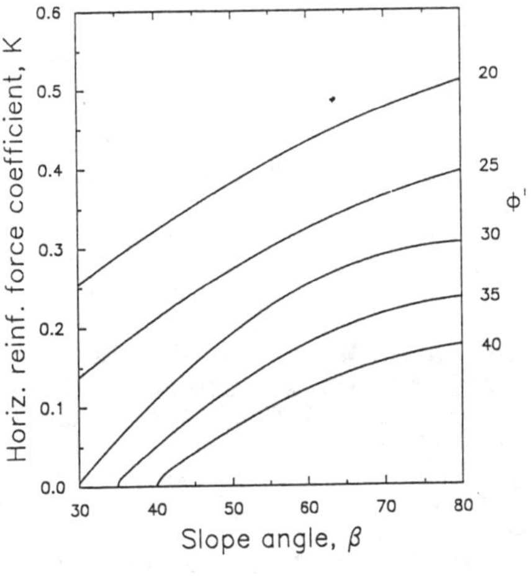

Fig. 5 Horizontal Reinforcement Force vs Slope Angle for a Minimum Factor of Safety of 1.3

References

Achilleos, E. (1988), "User Guide for PC STABL 5M", Joint Highway Research Project No. JHRP-88/19, School of Civil Engineering, Purdue University, West Lafayette, IN 47907.

Christopher, B.R., Gill, S.A., Giroud, J.P., Juran, I., Mitchell, J.K., Schlosser, F., and Dunnicliff, J. (1990), "Reinforced Soil Structures; Vol II Summary of Research and System Information", Federal Highway Administration Report Number FHWA-RD-89-043, Washington, D.C., p. 158.

Christopher, B.R. and Leshchinsky, D. (1991), "Design of Geosynthetically Reinforced Slopes", Proceedings of the Geotechnical Engineering Congress 1991, Vol. 2., ASCE, Boulder, Colorado, pp. 988-1005.

Duncan, J.M., Low, B.K., and Schaefer, V.R. (1985), "STABGM: A Computer Program for Slope Stability Analysis of Reinforced Embankments and Slopes", Department of Civil Engineering, Virginia Tech, Blacksburg, VA 24061.

Humphrey, D.N. (1986), "Design of Reinforced Embankments", Joint Highway Research Project No. FHWA/IN/JHRP-86/16, School of Civil Engineering, Purdue University, West Lafayette, IN 47907.

Humphrey, D.N. and Rowe, R.K. (1991), "Design of Reinforced Embankments - Recent Developments in the State-of-the-Art", Proc., the Geotechnical Engineering Congress 1991, ASCE, 15 pp.

Jewell, R.A., Paine, N., and Woods, R.I. (1984), "Design Methods for Steep Reinforced Embankments", Proc. of Conference on Polymer Grid Reinforcement, p. 70, London.

Jewell, R. A. (1990), "Application of the Revised Design Charts for Steep Reinforced Slopes", Report No. OUEL 1796/89, Soil Mechanics Report No. 096/89, Developed for AKZO Industrial Systems by Arnhem, Holland.

Jones, C.J.F.P. (1984), "Design and Construction Methods", Proc. Symp. Polymer Grid Reinforcement, Mar. 22-23, ICE, London, Paper #6.1.

Leshchinsky, D. and Boedeker, R. (1989), "Geosynsthetic Reinforced Soil Structures", ASCE Journal of Geotechnical Engineering, Vol. 115, No. 10, p. 1459.

Milligan, V., and La Rochelle, P. (1984), "Design Methods for Embankments Over Weak Soils", Paper No. 3.4, Symposium on Polymer Grid Reinforcement in Civil Engineering, Science and Engineering Research Council, London, March 22-23, 8pp.

Milligan, V. (1985), "Design and Construction of Reinforced Embankments on Weak Soils", Preprint, Kersten Lecture, University of Minnesota, January 22, 24 pp.

Rowe, R.K., and Soderman, K.L. (1987), "Reinforcement of Embankments on Soils whose Strength Increases with Depth", Proc. Geosynthetics'87, New Orleans, pp. 266-277.

Schmertmann, G.R., Chouery-Curtis, V.E., Johnson, R.D., and Bonaparte, R. (1987), "Design Charts for Geogrid-Reinforced Soil Slopes", Geosynthetic '87 Conference, New Orleans, LA., Vol. 1, pp. 108-120.

Schneider, H.R. and Holtz, R.D. (1986), "Design of Slopes Reinforced with Geotextiles and Geogrids", Geotextiles and Geomembranes, No. 3, p.29.

Wright, S.G. and Duncan, J.M. (1991), "Limiting Equilibrium Stability Analyses for Reinforced Slopes", Transportation Research Board 7th Annual Meeting, Washington, DC.

The Behaviour of Reinforced Soil Walls Constructed by Different Techniques

A McGown,[1] K H Loke[2] and R.T. Murray[3]

Abstract

The behaviour of a series of 2.0m high retaining walls constructed from both unreinforced and reinforced soil using alternative forms of reinforcement in association with three different methods of construction is described. Observations were made, both during and after construction, of the loading and movement of the lateral boundaries and of the tensions and strains developed in the reinforcements. The results obtained have considerable implications for the design and construction of reinforced soil retaining walls.

Introduction

Reinforced soil structures may be constructed by a variety of different methods and from a range of materials which can greatly influence the magnitude and distribution of the strain fields in the backfill. As the interaction between the soil and the reinforcement is controlled by such strain fields, to fully understand reinforced soil behaviour it is important to have comprehensive data on the influence of the construction technique in relation to the particular materials employed.

Strain behaviour is a critical consideration in reinforced soil walls in which post-construction boundary deformations are required to be kept to a minimum. For those structures which have been constructed on a

---

[1] Professor, University of Strathclyde, U.K.

[2] Postgraduate, University of Strathclyde, U.K.

[3] Senior Principal Scientific Officer, TRRL, U.K.

conventional basis, such a requirement inhibits the development of strains in the soil and restricts the mobilization of soil friction, thereby inducing high lateral earth pressures. It may also restrict the mobilization of full interface friction between the soil and reinforcement and thus limit the development of the associated tensile stresses in the reinforcements which are required for enhancing stability. The constraints on boundary movements are of particular significance for polymeric reinforcements in which a greater degree of strain is needed to generate available strength than is the case for steel reinforcements. A consequence of imposing limitations on strain is thus to produce structures which are less economic than they might be.

To investigate the relation between strain and mobilized strength in both unreinforced and reinforced soil structures, small scale model studies were undertaken at the University of Strathclyde (Lee 1985 and Ahmed 1989). These studies clearly demonstrated that the controlled development of strain fields in reinforced soil backfill during construction permitted the soil and reinforcement strengths to be fully utilized and produced a significant reduction in post construction wall movements. It was also shown that these objectives could be achieved by employing the so called "Yielding Wall" principle (McGown et al., 1987 and 1988, and Andrawes et al., 1990). However, in view of the fact that the results had been obtained only on the basis of small scale laboratory models, it was decided to test their validity by observations of the behaviour of prototype scale walls constructed in the pilot scale retaining structure facility at the Transport and Road Research Laboratory in the United Kingdom.

Experimental set-up

The work undertaken at the Transport and Road Research Laboratory involved building a series of 2.0m high reinforced soil retaining walls using three different construction techniques. Each length of wall was divided into several bays in which the influence of different reinforcing elements and construction methods on behaviour could be studied.

Three bays were constructed behind a movable steel wall supported by a jacking system which prevented movements during construction but permitted controlled rotational movements of the wall after completion of the backfilling. Bay A contained no reinforcement, Bay C steel strip reinforcements and Bay E, geogrid reinforcements. Another three bays were constructed behind a rigid concrete wall, with a layer of compressible material incorporated between the backfill and the wall to allow lateral boundary yielding to occur during construction. Bay B contained no reinforcement, Bay D steel reinforcement and Bay F geogrid reinforcement. Two further bays were used for the construction of full height propped panel walls, both with geogrid reinforcements. Bay G had a compressible layer between the backfill and the wall, but Bay H did not. The geogrid reinforcements were not attached to the facing units during construction but were allowed to move freely as the backfill was compacted. After completion of the placement of the backfill to full wall height, the reinforcements were 'locked on' to these facing units and the props removed.

Materials and Instrumentation

The studies of the experimental wall were carried out using a uniformly-graded sand from Theale (Berkshire, U.K.) having a D_{50} particle size of 0.45 mm and a uniformity coefficient of 3.25. The sand was compacted in 0.15m to 0.2m layers with two passes of a vibrating plate. It was also specified that the minimum distance of the plate from the wall would be 0.1m. Measurements on the compacted fill showed that average moisture content and bulk unit weight were 9.9% and 18.03 kN/m^3 respectively. The angle of friction of the soil was measured as 47° in the triaxial test at the compacted density.

In some bays the soil was reinforced by three layers of the patented high adherence, (ribbed), galvanised steel strips of the Reinforced Earth Company. Other bays incorporated Tensar SR80 polymeric geogrids to reinforce the backfill. The vertical spacing for both types of reinforcement was 0.7m and corresponded to elevations above the base of 0.30, 1.0 and 1.7m. A horizontal spacing of 0.3m was employed for the steel strips. The properties of both types of reinforcement are given in Table 1. Reinforcement strains were monitored by electrical resistance strain gauges attached to both the steel strips and geogrids and by strain coils attached to the geogrids. The compressible material used adjacent to selected sections of wall was a 150 mm thick layer of polyurethane foam. The compressibility of this material was determined as 0.0135 kPa/mm for the horizontal pressure range applied in the studies. Horizontal earth pressures acting on the wall were measured at selected locations using pneumatic pressure cells. These cells had a diameter of 140 mm and were mounted flush with the surface of the wall to minimize edge effects. Pneumatic pressure cells were also employed to measure the vertical pressure at the base of the structure. Other observations which were made

Table 1 Properties of Reinforcements

TYPE OF REINFORCEMENT	PROPERTY	VALUE
Steel Strips (Galvanised Steel)	Length Width Thickness Minimum Tensile Strength Elastic Modulus Interface Friction Coefficient	1.85 m 40 mm 5 mm 346.5 N/mm^2 185 kN/mm^2 0.85
Tensar SR80 (Geogrid)	Length Width Characteristic Tensile strength (120 years at 20°C) Interface Friction Coefficient	1.60 m full width 30.5 kN/m 1.0

included the use of dial gauges to measure wall movements and strain coils for monitoring the deformation of the compressible material.

Movable Steel Wall (Rigid Boundary) - Test Results

The horizontal pressure distributions acting on the movable steel walls (Bays A,C, and E) during construction are shown Fig. 1. As can be seen, large horizontal pressures were induced by the end of construction which have been attributed to the influence of compaction. Although the data shows some scatter, this occurs within a definite band and may be a measure of inconsistency in compaction of the soil.

At the end of the construction stage, an outward rotation was induced about the base of the steel wall which corresponded to 1.5mm of movement at the top . The observations on these sections of wall indicated that the horizontal pressures were generally reduced to about that corresponding to the Coulomb active condition assuming a wall friction angle (δ) of 35°, Fig. 2. One cell near the top part of the wall in Bay A measured a higher horizontal pressure than general but still showed a significant reduction from the previous value prior to wall rotation.

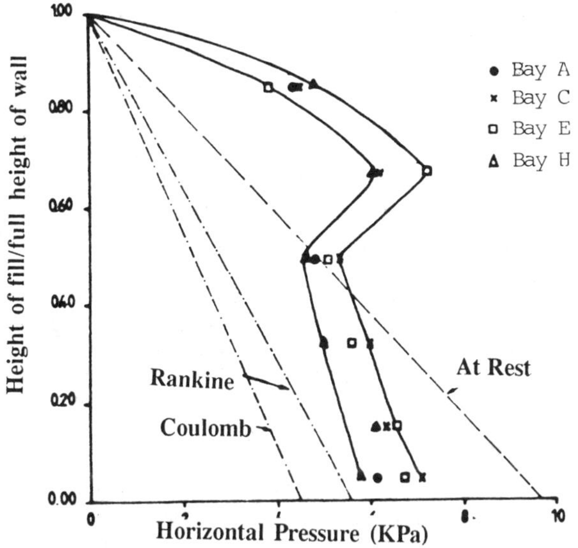

FIG. 1 Horizontal pressure distribution on the rigid lateral boundary of the moveable steel wall, Bays A C and E and the propped wall, Bay H, at the end of construction.

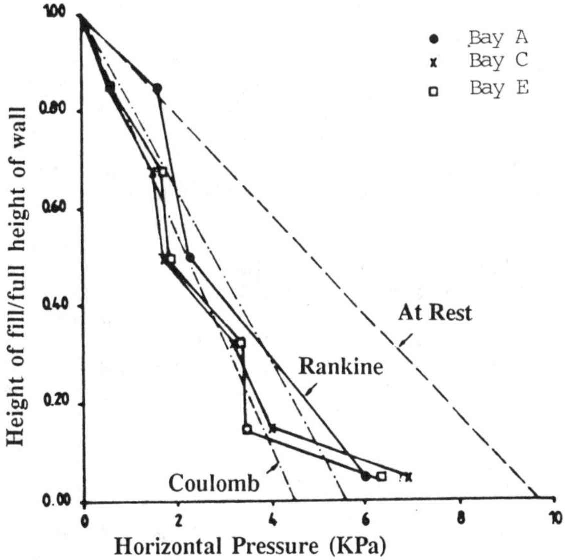

FIG. 2 Horizontal pressure distributions on the rigid lateral boundary of the supported moveable steel wall after wall rotation about the base, Bays A C and E.

A comparison between the horizontal pressures acting against the walls employing either steel or geogrid reinforcements showed no significant difference in the values. The tensile loads in the steel reinforcements as determined from the strain gauge observations were equivalent to between 2 and 4 kN/m prior to wall rotation and these reduced to between 0.5 and 2 kN/m afterwards. Strains of up to 0.2% were measured in the geogrid reinforcements from which it was possible to deduce on the basis of isochronous sustained load creep curves that these strains corresponded to loads in the geotextile of some 2.5 kN/m. The strains and corresponding loads were significantly reduced after rotation of the wall to about 50% of their previous values in the upper reinforcement levels. Much smaller changes in strain were noted in both the steel and geogrid reinforcements near the base of the structure but as the movements at this level were to a considerable extent prevented by the foundation soils this effect was not unexpected.

The results of the study indicated that, in general, the horizontal pressures imposed by the unreinforced and reinforced backfills were close to the Coulomb active conditions when account was taken of wall friction. It was also apparent that wall movements were insufficient to fully mobilize the tensile strength of the reinforcements, particulary in the case of the geogrid reinforcements.

Concrete Wall (Compressible Boundary) - Test Results

The presence of the layer of compressible material adjacent to the wall permitted movements of the backfill to develop progressively during the filling and compaction operations. The deformation profile of the compressible layer at the end of construction is shown in Fig. 3. The strains associated with the deformation of the compressible material significantly reduced the horizontal pressures against the wall in Bays B, D and F, Fig. 4. Although the horizontal pressure observed for the unreinforced backfill, (Bay B), was greater than the corresponding Coulomb values at the upper part of the wall, it was much less than the values produced during construction at the same elevation in the steel wall before rotation, Fig. 1. As was previously observed on the metal wall, there were no significant differences in horizontal pressure between bays with steel and geogrid reinforcements, (Bays D and F).

The influence of the resistance mobilized by the reinforcements is apparent in Fig. 4, as the horizontal pressures in Bays D and F were generally much less than for the unreinforced fill in Bay B and at some elevations

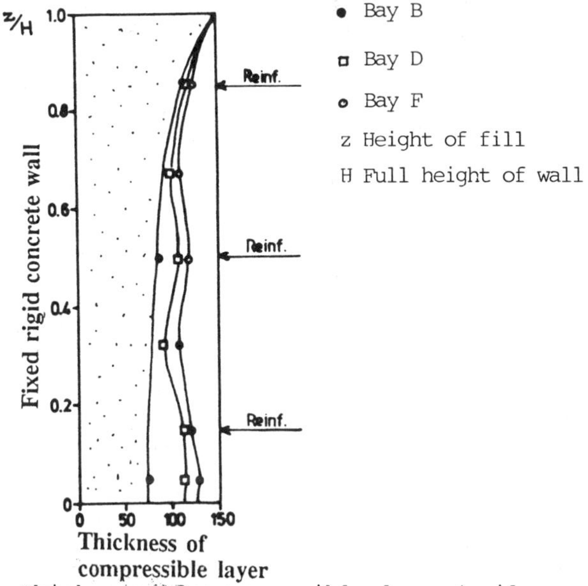

FIG. 3 Thickness of compressible layer/soil movements at the end of construction for Bays B, D and F.

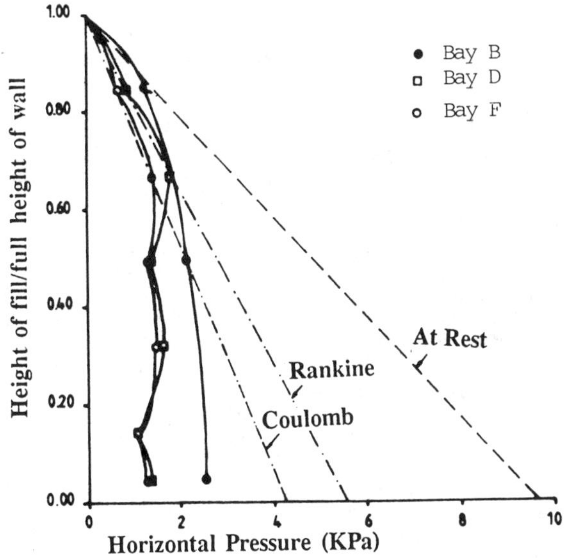

FIG. 4 Horizontal pressure distributions on the compressible lateral boundary of the fixed concrete wall, in Bays B, D and F at the end of construction.

was only about 50% of the latter values. The fact that the horizontal pressures acting on the unreinforced wall are below the Rankine active condition near the base is attributable to wall friction in association with arching due to base friction. This effect may be also present in the observations of the reinforced backfill although the compressible layer is likely to change the friction characteristics at the interface and the behaviour is difficult to quantify.

Although the horizontal pressures were generally less than observed in the steel wall after the rotational movements had been applied, the range of the tensions developed in the steel reinforcement were similar at between 1 - 4.5 kN/m. Moreover, the geogrid strains were actually larger at between 0.1 - 0.3%. These results cannot be fully explained but for reductions in horizontal force acting on the wall it would be expected that greater forces would be mobilized in the reinforcement. This behaviour seems to have occurred mainly with the geogrid reinforcement but a complicating factor is that the compaction stresses were reduced only after construction was completed by relaxing the metal wall whereas yielding of the compressible material behind the rigid wall occurred continuously during construction. It may well be therefore that these differences tended to unify the characteristics of the strain fields associated with the stiff metal reinforcement.

Propped Panel Wall (Rigid and Compressible Boundaries) - Test Results

During construction of the wall the props were maintained in place and relatively large horizontal pressures were developed against the panel wall in Bay H which did not have a compressible layer, Fig.5. These pressures were comparable to those observed behind the steel movable wall, Fig.1. After attachment of the reinforcements to the wall, the props were removed and a significant reduction in horizontal pressures occurred although in the top third of the wall the values were above those corresponding to the 'at rest' condition, Fig. 5. At lower levels the horizontal pressures were similar in magnitude and below those estimated from the Coulomb method at the lowest levels. It was noted that the post construction pressures in Bay H were the highest recorded in the study.

The strains observed in the geogrid reinforcements at the end of construction in Bay H were comparable to those measured in Bay E confirming that these were induced by the filling and compaction pressures. After "locking on"

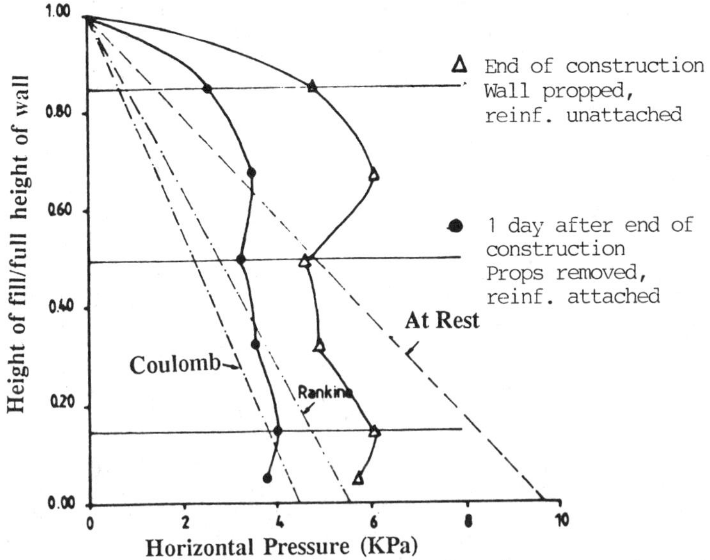

FIG. 5 Horizontal pressure distributions on the rigid lateral boundary of the full height panel wall in Bay H, with geogrid reinforcements.

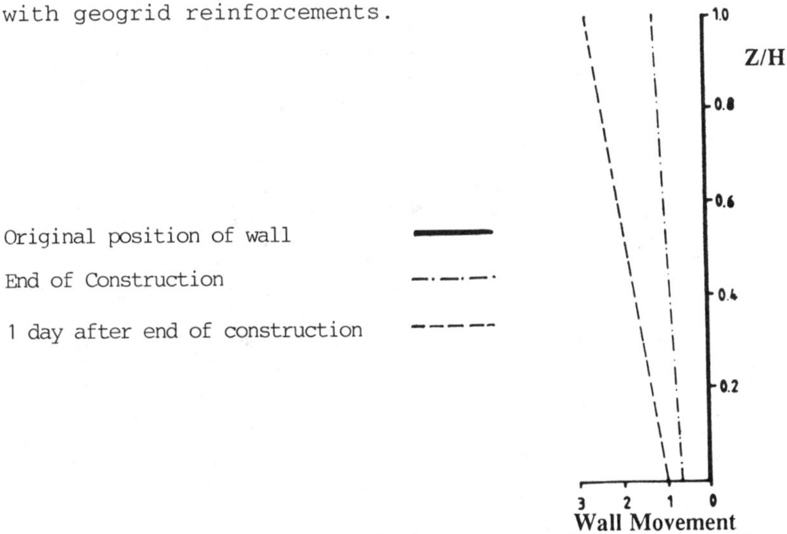

FIG. 6 Wall movements before and after the end of construction at the full height panel wall with the rigid lateral boundary and geogrid reinforcements, (Bay H).

the reinforcements and removal of the props, the strains in the geogrids increased significantly confirming that large post construction movements occurred, Fig.6. The form of the strain distribution in the geogrid reinforcement was of special interest as the results indicated peak values towards the face and approximately linear reductions in strain over the reinforcement length. These distributions are typical of the forms produced in pull-out tests.

During construction of the panel wall designated Bay G the compressible layer adjacent to the panel allowed the backfill to deform progressively. The horizontal pressures measured, Fig. 7, were very similar to those induced in Bays D and F, Fig. 4, and much lower than developed against Bay H which had a rigid boundary. After attachment of the reinforcements and removal of the props there was no change in the horizontal pressure, Fig. 7. There was no discernable deformation of the compressible layer during the post construction phase, Fig. 8 although very small rotational movements of less than 1mm in the 2m height of the wall were observed. The changes in reinforcement strains occurred principally towards the facing and was largest in the top reinforcement. However, the changes were rather less than occurred in Bay H, and although a comparison between Bays G and H showed significantly different strain patterns, this can be explained by the different construction processes whereby the movements in one case occurred progressively and in the other only after removal of the props.

Conclusions

The data obtained from the large scale walls constructed using different construction methods confirmed the findings of the small scale model wall studies previously undertaken. In particular, the data showed that for both unreinforced and reinforced soil walls the magnitudes and distributions of lateral earth pressures on the walls are greatly influenced by the amount and mode of lateral boundary yielding during and after construction. It was also shown that filling and compaction processes can "lock-in" significant stresses into the backfill. Thus, to ensure reliable estimates of the lateral earth pressures associated with unreinforced or reinforced backfills, consideration must be given to the anticipated movements of the structure and backfill and to the influence of the construction process.

The separate effects of construction and boundary yielding on internal behaviour are not easily quantified, but the application of the yielding wall technique offers

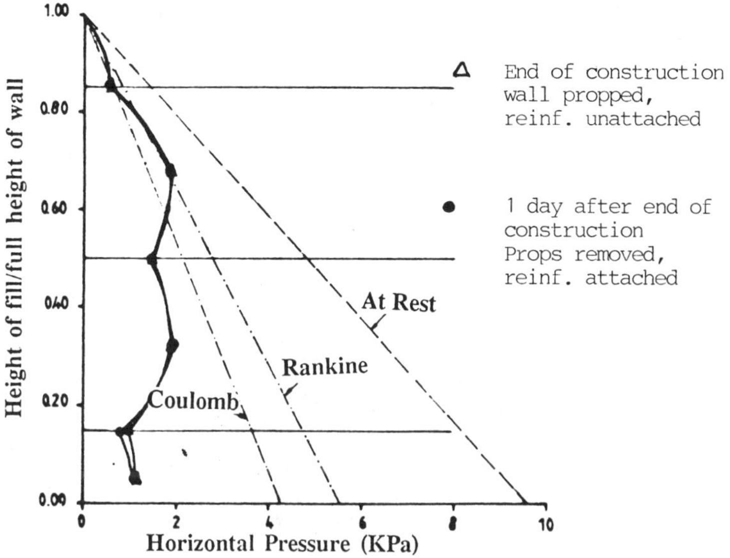

FIG. 7 Horizontal pressure distributions on the compressible lateral boundary of the full height panel wall with geogrid reinforcements, (Bay G).

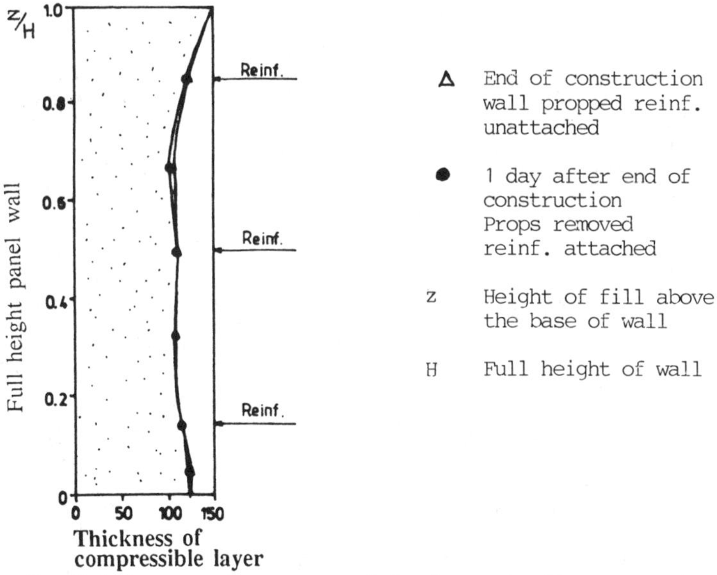

FIG. 8 Thickness of compressible layer/soil movements at the end of construction and at 1 day after construction for Bay G.

a relatively simple and economic means of ensuring greater effectiveness of the soil and reinforcements and for controlling post-construction wall movements.

The study has also highlighted the importance of the influence of wall friction on the magnitude of the earth pressures developed. Direct comparisons between observed and calculated earth pressures gave reasonable agreement on the basis of the Coulomb earth pressure equation when account was taken of the relatively large wall friction angles which were measured.

Acknowledgements

The work described in this paper forms part of the programme of the Transport and Road Research Laboratory and the paper is published by permission of the Director. The work was carried out by staff of the Department of Civil Engineering, University of Strathclyde and of the Structural Analysis Unit, TRRL. The authors wish to express their appreciation of the assistance of Mr DM Farrar, Professor K.Z. Andrawes and Dr. K.C. Yeo in carrying out the large-scale studies.

References

AHMAD, F. (1989). Model studies of the influence of the lateral boundary movements on earth pressures. Ph.D. Thesis, University of Strathclyde, Glasgow.

ANDRAWES, K.Z., LOKE, K.H., YEO, K.C. and MURRAY, R.T., (1990). Application of boundary yielding concept to full scale reinforced and unreinforced soil walls. Proceedings of Int. Reinforced Soil Conference, University of Strathclyde, Glasgow, U.K., 10-12th September 1990.

ANDRAWES, K.Z., McGOWN, A. AND AHMAD, F. (1990). Influence of lateral boundary movements on earth pressure. Proc. Int. Reinforced Soil conference, University of Strathclyde, Glasgow, U.K., 10-12th September 1990.

LEE, H.K. (1985). The behaviour of yielding retaining walls, M.Sc. Thesis, University of Strathclyde, Glasgow.

McGOWN, A., MURRAY, R.T. and ANDRAWES, K.Z. (1987). The influence of lateral boundary yielding on the stresses executed by backfills. Conf. on Soil Structure Interaction, Paris, pp. 585-592.

McGOWN, A., ANDRAWES, K.Z. and MURRAY, R.T. (1988). Controlled yielding of the lateral boundaries of soil retaining structures. Proc. Symp. on Geosynthetics for Soil Improvement, ASCE, pp. 193-210.

Centrifuge Models of Clay-Lime Reinforced Soil Walls

Erol Güler[1] and Deborah J. Goodings[2], member ASCE

Abstract

Thirteen centrifuge models of lime stabilized kaolin geotextile reinforced soil retaining walls were tested to failure by increased self-weight. Three failure models were identified depending on the length of the geotextile reinforcement. The lime improved wall stability substantially even when geotextile length was equal to only one half of wall weight. Unlike granular soils, development of a tension crack is necessary before failure; prevention of that crack will improve wall stability.

Introduction

The use of geotextile reinforced soil retaining walls has become common for stabilization of road embankments and slopes. Most backfill specifications for those walls require clean granular soil. Design specifications for backfill which can be adapted to a wider variety of onsite materials, including clays, could stand to reduce significantly the cost of these structures to make them more competitive with traditional types of retaining structures such as reinforced concrete retaining walls. This can be especially important in situations where access may be difficult or where expenditures allotted for repair or construction of steeply sided slopes may be small. This may apply equally to both developed and developing economies worldwide.

---

[1]Professor of Civil Engineering, Bogazici University, Istanbul, Turkey.

[2]Associate Professor of Civil Engineering, University of Maryland, College Park, MD, USA, member ASCE.

To emphasize economy the simplest reinforced wall configuration was considered to be the most suitable in this study. For that reason, the general design developed by the U.S. Forest Service (Koerner,1990) was used. The geotextile acts as both the internal reinforcement and the external facing material. The configuration is shown in Figure 1. By using the same wall configuration it was also possible to compare the results of this study with model tests conducted by Goodings (1989).

Mitchell and Villet (1987) state that clayey and silty soils have been used successfully in a few applications and that as performance experience is gained, use of these soil types may become more frequent. They also state that the use of admixture stabilizers, while not yet fully investigated, may enhance the range of materials that can satisfactorily be used for reinforced soil. The aim of this paper is to investigate, whether the properties of clayey soil can be improved to make it suitable as a backfill material for construction of reinforced soil retaining structures.

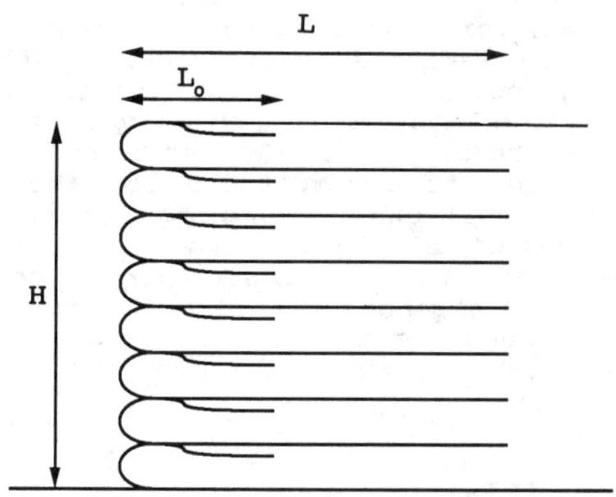

Figure 1 Wall Configuration

Previous Research

Lime is a commonly used admixture stabilizers. Lime uses the silica from clays or other pozzolons in the soil to form CSH gel. Short term reaction with lime involves replacement of monovalent adsorbed cations by Ca++, adsorption of $Ca(OH)_2$ by particles, cementation at inter-particle contacts, and establishment of a high pH (12.4) environment. In the long term high pH causes a breakdown of the crystal lattice of clay and formation of cementitious products. These reactions can continue for years (Mitchell, 1981).

Brandl (1981) states that the unconfined compressive strength of soil increases with addition of lime. The increase of strength will continue for years. After 7 years at the latest, no subsequent change of strength was observed in the soils he tested. An increase in tensile strength is also observed in lime stabilized clays. Brandl also reports that the deformation resistance after adding lime is more significant than the increase in q_u. Ferrel et al. (1988) report that the elastic modulus of an montmorillonitic clay is increased 12 times in the first 28 days and 20 times after 90 days due to the lime and flyash stabilization.

Permeability of cohesive materials also increases after the addition of lime. Wild et al. (1989) attribute increases in k to the modifications in phase composition and microstructure of the lime treated soil mainly as a result of flocculation. If the material cures under normal temperature, the limited pozzolonic activity occurring will not be sufficient to form enough gel to reduce the void space between particles and as a result of continued coarsening of the soil matrix structure, the permeability will continue increasing in the long term as well. If enough gel is formed, the pores of the soil will be clogged and since the permeability of this gel is very low the soil will practically become impermeable.

Workability of clays are improved with the addition of lime, because flocculation makes the clay more friable; this assists comminution for effective mixing and compaction (Hausman, 1990).

Bonaparte et al. (1987) state the stress - strain characteristics of the wall fill and its interaction with geotextiles have a considerable influence on construction-related and post-construction wall movements. Good quality, well compacted granular fill is often specified for reinforced soil walls, but they further state that recently a number of relatively low-height walls have successfully been constructed with silty and/or clayey backfill. Alvarez(1988) and Suah(1989) describe 43 centrifuge models of geotextile reinforced and unreinforced vertical walls built on a rigid foundation. The soil used was Speswhite kaolin placed in model walls at its plastic limit. Though it had been hypothesized that geotextile pullout failure would be likely in the case of using cohesive backfill, failures of the geotextile in the model walls were always by breakage in tension and never by pullout. They concluded was that cohesive soils may be used successfully to construct vertical, reinforced walls of modest height ($\lesssim$ 15m) using geotextiles as the reinforcement. Güler (1990) investigated the effect of using a lime stabilized clay as a backfill, in a geotextile reinforced soil retaining wall with a finite element analysis. The results of that analysis showed that the horizontal deformation of the wall at the top for a 4 m high geotextile wall under full surcharge can be reduced from 282 mm, which is an unacceptably high value, to 47 mm just by improving the properties of the cohesive backfill material by addition of lime.

Method

Centrifuge modelling

The best way to test predicted wall behavior is to build and monitor the performance of full-size walls of different configurations. This approach, however, is expensive, time-consuming, and difficult to control to ensure replication of all soil properties from wall to wall. Small physical models can be useful, especially if all material properties and stress conditions of a full size wall can be simulated. This is the contribution geotechnical centrifuge modelling makes to geotechniques, and is the method used in this paper.

The application of geotechnical centrifuge modelling to reinforced soil retaining walls specifically have been the subject of research by Bolton (1978), Mitchell et al. (1988) and Goodings and Santamarina (1989), to name just a few. Their technique was to construct walls at 1g on the laboratory bench, and then to accelerate the models on the centrifuge until catastrophic failure occurred. The full size or prototype stress equivalent heights, H_p, of the model walls of the heights, H_m, at failure at N times earth's gravity are calculated as $H_p = N \times H_m$. This is also the approach used.

Soil

Throughout the tests industrially produced Speswhite kaolin was used, the same as that used by Goodings (1989).

The index properties of this clay are as follows: w_{LL} = 49.0; w_{PL} = 33.9; G_s = 2.65 and 100% passing US No. 200. The clay was mixed dry with 6% lime by weight to prepare a uniform powder. The percentage of lime was determined according to the Eades and Grim (1966) procedure. Water in the amount of 34% of the dry soil weight was mixed to the soil lime mixture. This amount of water was chosen to simulate the soil used in the model walls reported by Goodings (1989), but here treated with lime. The material was left 24 hours to allow the water to distribute evenly within the soil.

Reinforcement

Whereas it is not necessary to modify fine grained soils for use in models, it is necessary to modify the reinforcement for use in models. The conditions of similarity require that the geotextile in the model be N times weaker than in the full scale, where N is the scale of the model at failure. Christopher (1987) satisfied this demand for centrifuge model tests to simulate geotextile behavior by using commercially available fabric interfacing, and Goodings (1989) followed that lead by using non-woven commercially available fabric interfacing. The geotextile simulant used in previous research at University of Maryland by Goodings (1989) was also used to enable comparison of results. The tensile strength of this material was measured as 0.701 kN/m, the equivalent in translating to full scale strength of a geotextile N times stronger, with vertical spacings N times greater than in the model.

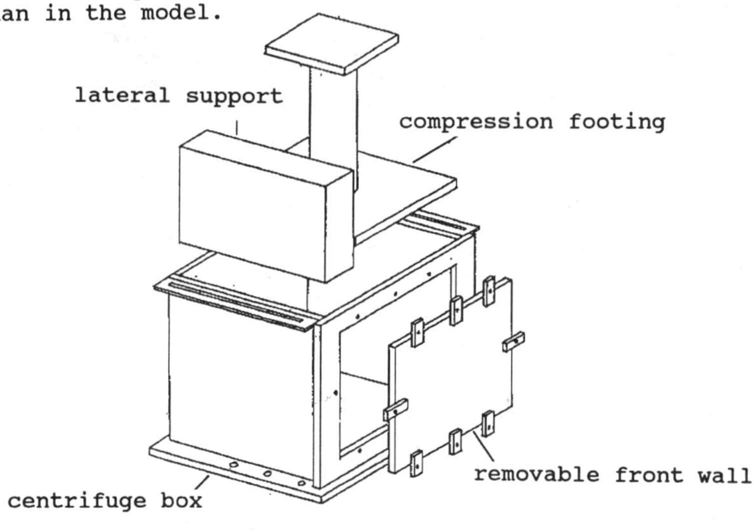

Figure 2 Model Box

Construction

All models were founded directly on the rigid base of the centrifuge model box after sand had been glued to it to inhibit sliding along that surface. Model retained fill and wall backfill were constructed simultaneously, in lifts of 19 mm, bending the reinforcement overlap back into each layer before compression. Compression was achieved using a compression footing covering the entire soil layer to compact the soil with a vertical static stress of 200kN/m^2, supporting the front of the wall using a rigid lateral support (Figure 2). The top of each model was sprayed with paint to highlight the development of tension cracks during testing.

Results

The purpose of the research was to see how the addition of lime to the backfill soil affects the geotextile reinforced retaining structure. Since similar models with unaltered kaolin as backfill were already constructed and tested in the geotechnical centrifuge by Goodings (1989), in this research only models that have lime stabilized clay as a backfill were built and tested. Goodings (1989) has addressed the question of similitude in comparing models of different heights and geotextile configurations, which allows direct comparison of those results with the ones from this study.

Direct Shear Tests

A series of direct shear tests was conducted to determine the shear strength parameters of the lime stabilized clay. Samples were taken at three different levels from each model tested. It was observed that although the measured internal angle of friction was repeatable(ϕ=15°± 2°) the cohesion values obtained were highly variable, despite a change in the method of obtaining the samples from the backfill.

Consequently a value for the cohesion was back-calculated from the results of the limit equilibrium analysis of the unreinforced model walls.

Unreinforced Models

Three unreinforced models with vertical wall heights of 6" (152.4 mm), 7½" (190.5 mm) and 9" (228.6 mm) were tested (Table 1). The failure modes in all three were similar and the average of the equivalent prototype heights at failure was 15.5 m. As the centrifugal acceleration increased, first a tension crack opened and then the block slid along a failure plane that passed through the toe. The cohesion of the lime stabilized clay was back-calculated using the failure planes observed in those three tests. When ϕ=15°, the back-calculated cohesion values obtained from models 1,7 and 20 were respectively c=66.6kN/m^2, 63.1kN/m^2 and 65.6kN/m^2. The average of these values is 65.1kN/m^2, with a standard deviation of 1.47kN/m^2. Since all tests gave very close values it was concluded that for the evaluation of the reinforced models this cohesion intercept could be used in the limit equilibrium analysis of the reinforced walls.

Reinforced Tests

Ten reinforced model walls of height 190.5 mm were constructed and tested. The geotextile length L (Figure 1) was varied to examine its effect on the failure mode. The overlap length L_o was kept constant in all reinforced walls. The ratio of L_o to the height H of the wall was 33%.

Table 1

Model No.	L/H	H_{pf} (m)	F.S.	No. of layers	Depth of tension cracks (No. of layers)
1	0	16.7	-	8	5
7	0	15.1	-	10	5.3
20	0	14.6	-	12	8.2
4	0.33	24.2	-	10	7
5	0.33	13.3	-	10	7
9	0.33	24.7	-	10	7
3	0.5	35.3	1.04	10	4
8	0.5	24.8	1.15	10	5.3
11	0.5	23.9	1.33	10	5
12	0.75	28.6	1.29	10	6
13	0.75	28.5	1.27	10	5.5
2	1.00	30.0	1.21	10	5
10	1.00	28.9	1.12	10	7

The length of the actual reinforcement was also chosen as a percentage of the model height and the L/H percentages were 33%, 50%, 75% and 100%.

It was observed that the three model walls with an L/H percentage of 33% failed in one mode. A crack to a depth of 0.7·H developed behind the geotextiles and continued to open as the centrifugal force was increased. Failure took place ultimately by overturning of the soil mass defined by that crack. No geotextiles were failed. The failure was preceded by the uneven squeezing of the 8th reinforced layer which led to an uneven bearing surface for the upper seven layers. The prototype failure heights of models 4 and 9 were 24.2 m and 24.7 m respectively and are very close to each other. They further show that even a small reinforcement depth can increase substantially the prototype failure height when one compares them to the equivalent height of an unreinforced wall, 15.5 m. Model number 5, however, was built to be identical to models 4 and 9 but had a prototype failure height of only 13.3 m which is even below the average height of unreinforced models; this may be due to a possible weakness in a geotextile at some critical position in the wall.

The next group of model walls, numbered 3,8 and 11, had an L/H percentage of 50%. A tension crack opened behind the geotextile as shown in Figure 3-A. The same failure mode occurred for models with L/H=33%. At failure the depth of the tension cracks reached about 50% of the wall height, which is less than for walls with L/H=33%. Failure was not by overturning. A definite lower failure surface developed, which on the average, made an angle of 35° with the horizontal, but these failure lines did not pass through the toe. The factors of safety were back-calculated at the moment of failure for models 3, 8 and 11 using ϕ=15° and c=65.1kN/m^2. The factors of safety values were, respectively,: 1.04, 1.15 and 1.33, an average of 1.17 with a coefficient of variation of 0.12. For the calculation of the factors of safety it was assumed that no lateral load was transferred across the tension crack and that all geotextiles that lay across the inclined failure surface ruptured simultaneously, and at the same time the soil failed. The anchorage forces were taken to be horizontal and their components contributing to the normal force and shear force along the failure plane were considered.

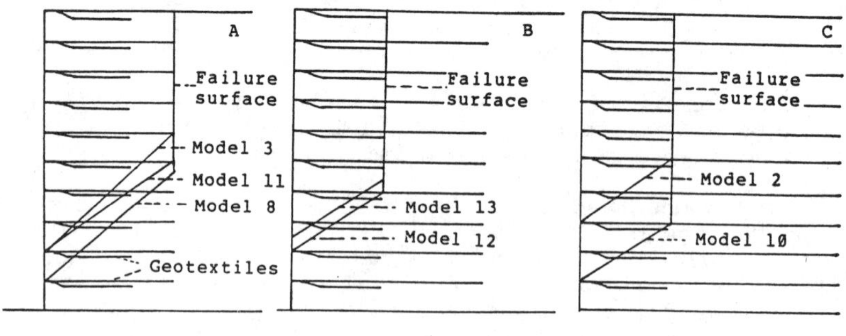

Figure 3 Failure surfaces for reinforced models

No consideration was given to the fact that the vertical pressure along the wall front is increased to a value much higher than the vertical pressure within the wall due to the forward leaning of the upper mass. These simplifying assumptions may account for the fact that the factor of safety are all above 1.0, as well as the possibility that ϕ=14.7° and c=65.1kN/m^2 may be for peak strength which may not be applicable in reinforced walls which may undergo substantial strain before failure (Alvarez, 1989). Nevertheless, the two dimensional unit equilibrium analysis seems to have merit in describing failure, although it is unconservative.

The next group of models have an L/H percentage of 75%. Since in these models the geotextiles are longer, the tension crack lay within the reinforced soil, where the overlap ends (Figure 3-B). Failure began with a tension crack, which, again extended to approximately 50% of the wall height. The soil failed along an inclined failure surface that made an angle of 31° with the horizontal; again squeezing of the lower, unfailed layers was observed. The factors of safety were back-calculated for these known failure surfaces. For models 12 and 13 the factors of safety were respectively 1.29 and 1.27. The same assumptions used for walls with L/H=50% were used here also except that geotextiles crossing the tension crack were assumed to have failed before the block slid on the failure surface.

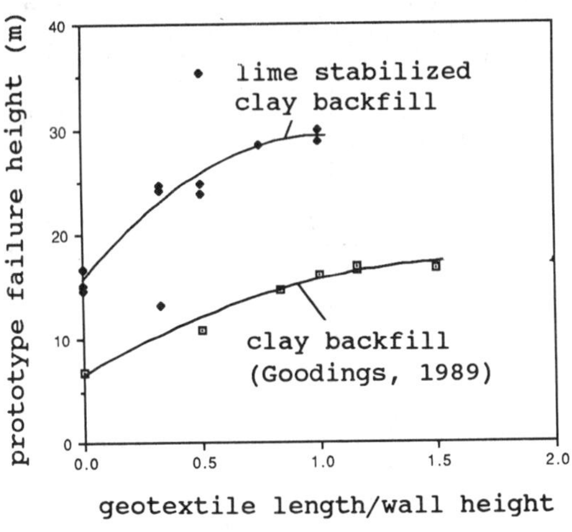

Figure 4 Prototype failure heights for lime treated and untreated backfills

The final group of tests had an L/H ratio of one. The failure mechanism and the shapes of the failure zones of these walls were similar to the ones with L/H=75% (Figure 3-C). The back-calculated factor of safety values were also in the same range as the other models, 1.21 for model 2 and 1.12 for model 10. No significant increase was obtained in the prototype failure height by increasing the reinforcement length from 75% of the height to 100%.

A general observation made on all the reinforced models was that the tension crack that led to failure opened just before the model failed. The average value of centrifugal acceleration at which the tension cracks opened was 92% of the centrifugal acceleration at failure.

In relating model behavior to full scale wall behavior Goodings (1989) explained that the measure of the contribution of geotextile reinforcement to the stability of a wall is by means of the product (NT_m H/S_v), where N = model acceleration at failure in units of gravity, T_m = model geotextile strength, H = wall height, and S_v = vertical separation between layers. This permits modeling of full scale walls with more layers of reinforcement than H/S_v in the model but where geotextile strength in full scale is less than NT_m, provided the product NT_m H/S_v is the same in the model and the full scale walls.

In these models, where the product NT_m may be as large as 110 kN/m (or over 7500 lb/ft) for geotextiles at the full scale equivalent vertical spacing of almost 3m, this approach of using that product NT_m H/S_v may be essential in relating model behavior to more realistic full scale conditions. This approach, however, has not been applied to this data, to allow a reader to select any reasonable full scale geotextile strength in relating model and full scale wall behaviors. This also means, then, for these models that at least part of the improvement in wall stability occurring with increased geotextile length is associated with the full scale equivalent of using either stronger geotextiles, or more layers of a constant strength geotextile, as the acceleration at failure increased.

Conclusions

Thirteen centrifuge models were tested. Three of them were unreinforced vertical clay faces and ten were geotextile reinforced soil retaining walls. In all cases the backfill was a kaolin clay stabilized with lime. The lengths of the reinforcement were chosen as 33%, 50%, 75% and 100% of the model wall heights. In all cases the overlap length was 33% of the model height.

It was observed that the lime stabilization improves the critical height of the unreinforced and models reinforced (Figure 4). The failure mode changed as the reinforcement length increased. At L/H=33% the failures occurred by overturning, with no geotextile breakage or pullout. At L/H=50%, there was some geotextile breakage. At L/H=75% the failure mode became contained entirely within the reinforced zone. Walls with L/H = 100% had the same failure mode and similar prototype failure heights as those models of L/H = 75%. So for lime stabilized clay and this geotextile simulant, a reinforcement length of 75% of the wall height is sufficient for design. It also coincides with the convention used for reinforced walls with granular backfill.

The failure of the reinforced wall with a cohesive backfill is preceded by a tension crack, which allows the reinforced mass on the front part of the crack to lean outward and thereby cause eccentric loading to lower levels. The improved performance of such walls can be accomplished by preventing cracking with the help of the reinforcement.

References

Alvarez, T.A. (1988) "Centrifugal model testing of geotextile reinforced cohesive soil walls", M.S. Thesis, University of Maryland

Bolton, M.D., "Reinforced earth walls: A centrifugal model stud", Proceedings of a Symposium on Earth Reinforcements, ASCE, New York

Bonaparte, R., Holtz, R.D., Giroud, J.P. (1987) "Soil reinforcement design using geotextile and geogrids", ASTM Special Publication 952, pp. 69-116

Brandl, H. "Alteration of Soil Parameters by stabilization with lime", Proc. of 10th ICSMFE, Stockholm, Vol. 3 pp. 587-594

Christopher, B.R. (1987) Personal Communication

Christopher, B.R., Holtz, R.D.(1984) "Geotextile Engineering Manual", U.S. Department of Transportation, FHWA Contract No. DTFH 61-83-C-00150

Eades, J.L., Grim, R.E. (1966) "A quick test to determine lime requirements for lime stabilization", Highway Research Record 139, HRB, National Research Council, Washington D.C. pp. 61-72

Ferrel, R., Arman, A., Baykal, G. (1988) "Effects of combined lime and fly ash stabilization on the elastic module of montmorillonitic Soils, FHWA Report, LTRC No. 209

Goodings, D.J. (1989) "Effects of poorly draining backfill on geotextiles for earth reinforcement of vertical soil slopes", Final report submitted to Maryland Department of Transportation State Highway Administration Bureau of Research

Goodings, D.J., Santamarina, J.C. (1989) "Reinforced earth and adjacent soils:centrifuge modeling study" ASCE Geotechnical Journal, Vol 115, No. 7, pp. 1021-1028

Güler, E. (1990) "Lime stabilized cohesive soil as a fill for geotextile reinforced retaining structures", Proc. of 4th Int. Conf. on Geotextiles, The Hague, Vol.1, pp. 39-44

Hausman, M.R. (1990) "Engineering Principles of Ground Modification" McGraw-Hill, New York

Koerner, R.M. (1990) "Designing with Geosynthetics, Second Edition", Prentice Hall, New Jersey

Mitchell, J.K. (1981) "Soil improvement - State of the Art Report", Proc. of 10th ICSMFE, Stockholm, Vol.4, pp. 509-565

Mitchell, J.K., Villet, W.C.B. (1987) "Reinforcement of earth slopes and embankments", National Cooperative Highway Research Program Report 290, Transportation Research Board

Mitchell, J.K., Jaber, M., Shen, C.K., Hua, Z.K. (1988) "Behavior of reinforced soil walls in centrifuge model tests", Centrifuge 88, Balkema, Rotterdam

Ovesen, K.N. (1979) "The use of physical models in design", Proc. of 7th European Cong. on Soil Mechanics and Foundation Engineering

Schofield, A.N. (1980) "Cambridge geotechnical centrifuge operations", 20th Rankine Lecture, Géotechnique, Vol. 30, pp. 227-268

Scott, R.F., Morgan, N.R. (1977) "Feasibility and desirability of constructing a very large centrifuge for geotechnical studies", Prepared for the National Science Foundation, P.B.-272250

Suah, P.G. (1989) "Further studies of geotextile reinforced cohesive soil walls - Centrifuge model testing", M.S. Thesis, University of Maryland

Wild, S., Arabi, M., Ward, G.L. (1989) "Fabric development in lime treated clay soils", Ground Engineering, April, pp. 35-37

Three Case Histories Of Waste Stabilization

Edward L. Kosinski,[1] David S. Martin,[2] and Alan R. Ringen[3]

Abstract: Waste stabilization can take several forms using various additives and reagents. Current methods involve adsorption/absorption, neutralization, pozzolanic processes, conversion processes causing chemical changes, and micro/macro encapsulation. This paper discusses these technologies together with current management strategies and presents three case histories illustrating various aspects of stabilization/ solidification. The first case involves resource recovery through waste conversion and byproduct use. In the second case, in situ solidification produced a structural base material suitable for supporting a low permeability cap system. The third case presents in situ injection for acid neutralization and stabilization of contaminated soils.

Introduction

Stabilization and solidification are terms that the United States Environmental Protection Agency (USEPA, 1989) has defined as referring to treatment processes that are designed to accomplish one or more of the following: improve the handling and physical characteristics of the waste, as in sorption of free liquids; decrease the surface area of the waste mass across which transfer or loss of contaminants can occur; and limit the solubility of any hazardous constituents of the waste, e.g., by pH adjustment or sorption phenomena. Some authors (Kyles and Malinowski, 1988) make a further distinction that stabilization is a

---

[1]Vice President, Technical Operations, VFL Technology Corporation, 42 Lloyd Avenue, Malvern, PA 19355.

[2]Project Manager, VFL Technology Corporation, 42 Lloyd Avenue, Malvern, PA 19355.

[3]Project Manager, Hayward Baker Inc., 1875 Mayfield Road, Odenton, MD 21113.

physical modification or pre-treatment process that alters a liquid, solid, or semi-solid into a solid product, while solidification is a chemical process that alters a liquid, solid, or semi-solid into a landfillable material. A treatment process or unit may require both physical and chemical process steps to convert a waste into a treated material. Three cases are presented that show the results of stabilization and solidification treatment processes on hazardous and non-hazardous wastes.

Stabilization/Solidification Technologies

The technologies for stabilization and solidification generally fall into the following categories: sorption, including absorption and adsorption; pozzolanic; cement-based; thermoplastic; microencapsulation; neutralization; organic polymerization; and organophillic and modified cements. The most common technologies used most by waste treatment contractors on commercial scale projects are based on the first four categories.

Absorption (the retaining of liquids within the capillary pores of the absorbent) cannot be used for disposal of hazardous wastes in accordance with Federal and State waste regulations as the sorption can be easily reversed. However, adsorption (the chemical attraction of polar molecules like water to bind the surfaces of a dry reagent, providing a chemical binding of the water and/or polar contaminants to the reagent) can be used as a pre-treatment prior to pozzolanic or cement-based stabilization. If the adsorbent also has pozzolanic properties, there can be a synergistic effect to using the sorbent in subsequent treatment steps. Typically-used sorbents with pozzolanic properties are fly ashes, kiln dusts, commercially available clays and natural soils with clay.

A pozzolan is defined as "a siliceous or alumino-siliceous material that in itself possesses little or no cementitious value, but in finely divided form and in the presence of moisture will chemically react with alkali and alkaline earth hydroxides at ordinary temperatures to form or assist in forming compounds possessing cementitious properties." (ASTM, 1985). Fly ash and some clays and soils possess sufficient silica and alumina to react with alkali reagents such as hydrates of lime or sodium hydroxide. Gel-like compounds are formed similar to those found in the early stages of cement hydration. The cement-like compounds are much weaker than those found in cement, but have a great ability to chemically bind water molecules. Excess alkalinity reacts with heavy metal ions in the waste to form relatively insoluble metal hydroxides.

Cement-based technologies seek to bind the waste in the calcium-silicate-hydrate gel that comprises about 70% of the volume of hardening cement. Many heavy metals combine with the free alkalinity available in the cement reactions to form metal salts such as metal hydroxides, sulfides, sulfates, carbonates, and phosphates. If a pozzolan such as fly ash is present, the silicates and alumina in the fly ash react pozzolanically with calcium hydroxide by-products released by the cement reactions, further enhancing the cementing process.

Other sources of calcium, such as kiln dusts from the manufacture of lime or Portland cement, have been successfully used to pozzolanically stabilize soils and aggregates in highway and other construction projects. For the most part, kiln dusts are non-hydraulic and contain lower levels or less reactive species of calcium oxides and hydroxides than commercially available lime or cement. Consequently, greater amounts of kiln dusts are normally required to be used with a pozzolan to achieve the desired results.

Waste Management Strategies

The strategies employed to manage wastes can be listed in order of least to most preferable:

- Off site disposal
- On site treatment or disposal
- Constructive re-use or recycling, and
- Source reduction

Off site disposal involves removing and transferring the waste to an appropriate disposal and treatment facility, possibly involving treatment to meet Best Demonstrated Available Technology (BDAT) standards (CFR, 1989) established by statute for various wastes. On site treatment and disposal involves applying a variety of technologies to reduce mobility, toxicity, and volume. Treatments can take place through portable processing units or in situ. The treated waste may be placed into a specially constructed cell or left in place and capped with a multi-layer system in accordance with specific closure plans. Constructive reuse or recycling of the waste may be accomplished in selective cases, but usually involves benefication of some type to meet standards related to the material's structural or environmental properties. Source reduction is the most desirable of all strategies, but may require changes in manufacturing processes to reduce or reuse byproducts. Stabilization has a role in all of the mentioned strategies, even in source reduction where the byproduct of one process could be used to treat the waste from another process.

Case 1: Stabilizing A Non-Hazardous Municipal Waste; Benefication, Conversion, And Reuse

The City of Wilmington, Delaware, accumulated a large inventory of digested sewage sludge while implementing plans to construct and operate a co-composting facility with the Delaware Solid Waste Authority's (DSWA) Delaware Reclamation Project. The intent was to use a series of on site basins to store and air dry the inventory prior to its introduction into the compost facility. Approximately 240,000 cu yds (183,500m^3) of 10% solids sludge was stored in on site basins and 200,000 cu yds (153,900m^3) of 5% solids sludge was accumulated in finishing ponds. Eventually, existing filtering capacity limitations and a contractual cap on the quantity of filter cake that could be accepted by the compost facility precluded the dewatering and use of this material. A previous study for the beneficial reuse of lagooned sludge (Collins & Fazzini, 1989) indicated that stabilization was viable for this project.

After examination of several dewatering and mixing alternatives, the specialty contractor developed a two step stabilization process. First, the solids content of the sewage sludge in the lagoons was beneficiated by using fly ash as an adsorbent. Fly ash conditioned with about 15% water was incorporated into the in situ sludge, creating a soupy, thickened material. Next, lime kiln dust or, alternatively, cement was mixed into the pre-conditioned sludge in predetermined ratios using pneumatic injectors mounted on backhoes. The pozzolanic reaction reduced the amount of free water in the resulting product. After two to three days of in-place curing, the mixed material was able to support the weight of mobile equipment. The curing continued over the next several days, producing a friable, soil-like material. Within one to two weeks, the material was excavated using mobile equipment. Table 1 gives the properties of the raw sludge and the treated material.

TABLE 1. Material Properties

Property	Raw Sludge	Treated Material
Solids Content	10%	72%
Specific Gravity	Liquid	1.40
Compressive Strength (14 days)	None	20 psi (138 kPa)
Permeability	None	10^{-5} cm/sec

After testing the stabilized material for leachate characteristics by EP-Toxicity and later Toxicity

Characteristic Leaching Procedure (TCLP) testing, reuse authorization was obtained from the Delaware Department of Natural Resources and Environmental Control (DNREC). As a result of the beneficiated material, several reuse applications have been found. The material is used for slope and daily cover at the DSWA Northern landfill facility. Approximately 160,000 cu yds (122,250m^3) has been supplied to the U.S. Army Corps of Engineers, Philadelphia District, as structural fill for constructing dikes and berms at the Edgemoor dredge spoils disposal area. Over 500,000 cu yds (382,000m^3) of stabilized material has been produced and used in various projects over the last five years.

Case 2: Development Of A Structural Cap Through In Situ Solidification

In situ solidification has been used successfully in a number of applications. Treating material on-site reduces the consumption of valuable off-site landfill space and the cost of transportation and disposal.

This project's objective was to develop a homogeneous, structurally stable, monolithic mass capable of supporting a low permeability cap system on a 1.25 acre (4000m^2) lagoon. The lagoon, operated by a dye manufacturer, contained a tar-like sludge formed by inorganic solids that had settled prior to recirculation of the waste water to the facility's evaporators. The geosynthetic-lined lagoon contained approximately 20,000 cu yds (15,300m^3) of sludge and 1,000,000 gallons (3.8ML) of water. The waste was 12 ft (4m) deep and one ft (0.3m) below the top of the containment berm.

Treatability Study

The treatability study was conducted to address the following issues:

- Variability of the physical and chemical properties of the waste.
- Suitability of available reagents and techniques.
- Methods to achieve and field verify the desired end result.

A number of representative samples were taken around the perimeter of the lagoon. This indicated that while the inorganic soils were chemically consistent, three distinguishable types of material existed within the lagoon, based upon particle size. The first was a black, damp, gritty material located near the influent point. The second, a wet, paste-like material of fine particles, was located near the apparent discharge. The third material was a combination of the previous two.

Reagent Selection

Reagent selection is typically evaluated on availability, consistency and ease of handling, range of application, volume increase, and cost. A broad matrix of reagents was examined, evaluating the varying factors relating to their unique use as a solidification agent. These included fly ash, lime, cement, kiln dusts, and organophillic clays. Of these agents, fly ash was preferred due to its availability, consistency, and cost. Often when the goal is to enhance the structural properties of the waste, fly ash, fly ash/lime, and cement are used.

Some reagents were eliminated due to the waste's chemical composition. The waste exhibited a very high sodium content which tends to inhibit or destroy the cementitious properties of cement. These properties were critical to the development of the required structural stability. Table 2 has been developed to further examine reagent qualities (Fazzini, et al, 1991).

TABLE 2. Reagent Selection Matrix

Reagent	Avail-ability	Consis-tency	Range	Volume	Cost
Fly Ash	Good	Good	Good	Moderate/ High	Low
Lime	Good	Excellent	Good	Low	High
Cement	Good	Excellent	Good	Low	High
LKD *	Fair	Fair	Fair	Moderate	Low
CKD *	Fair	Fair	Fair	Moderate	Low

* LKD = Lime Kiln Dust
* CKD = Cement Kiln Dust

Reagent Mix

Based on the results of the detailed treatability study, an optimum performing mix was selected. A 20% fly ash, 10% high calcium quick lime mix showed the best handling and structural stability characteristics. This mix design transformed the wet, paste-like sludge into a damp,granular, free flowing material immediately after processing. Lower reagent quantities could be used on the grit-like sludge.

Stabilization

Equipment selection was based on productivity, flexibility, reliability, and mix characteristics. The use of mobile equipment, portable plants, pneumatic injection, and deep paddle mixing was examined in light of the limited staging space available adjacent to the lagoon. Mixing by mobile equipment was chosen on the basis of high productivity and, as solidification quickly progressed, the staging could be moved onto the lagoon.

Premixing the additives took place adjacent to the lagoon where the predetermined ratio was volumetrically mixed and transported to the lagoon. An excavator then mixed the sludge and additives in a designated daily work grid that typically measured 25 ft (7.5m) by 50 ft (15m). The operator could then move out onto the lagoon and work across the next row of grids.

The mix design was proportionally adjusted to suit the change in waste characteristics as anticipated by the treatability study.

Verification Testing

In order to support the intended cap system, the stabilized material was required to attain a minimum bearing capacity of 3,000 lbs/ft^2 (144 kPa). Two plate bearing tests (ASTM-D-1194-72) were conducted on the surface of the impoundment at intervals selected by the client's engineer. The results of these tests showed that the stabilized material achieved an ultimate bearing capacity in excess of 6,200 lbs/ft^2 (300 kPa).

Cap System

The final cap cover system provides long term minimization of percolation from precipitation into the waste, promotes drainage while minimizing surface erosion of the cover, and functions with a minimum of maintenance.

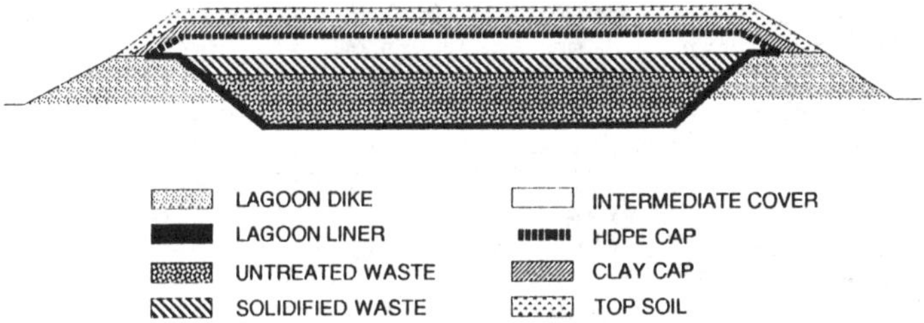

FIG. 1. Structural Cap System

The lagoon's capping system was comprised of six elements: One ft (0.3m) of compacted fill, a 60-mil HDPE liner, two ft (0.6m) of compacted clay, six inches (0.15m) of topsoil, an anchored erosion control mat, and 500 ft (150m) of grout blanket placed as side slope protection. Conventional construction equipment--tandem wheel dump trucks, D-6 bulldozers, and a sheepsfoot roller--were used to place and compact the cap.

An unanticipated test of the stabilized material and

cap occurred during the grout blanket installation. After placement of the synthetic blanket material, on the side slopes of the lagoon exterior, approximately 210 cu yds of cementitious grout was pumped into the blanket pockets. Due to site logistics, and prior to topsoil placement, loaded trucks were required to travel over the cap. No sign of instability was observed in the capping system.

Conclusions

The projects physical and financial goals were achieved. The treatability study proved effective by identifying the inconsistency of the waste and by determining the design mix that allowed the greatest flexibility for the variable waste characteristics. The treatability study also identified an underlying lens of structurally stable sludge that did not require treatment, thus reducing the overall project cost by two-thirds.

Case 3: In Situ Stabilization/Neutralization Of Acid Affected Soils By Lime Slurry Injection

An oil refinery in south Texas experienced numerous sulfuric acid spills from their process operations. These spills over time, contaminated the underlying clayey sands to a depth of 24 ft where migration expanded laterally due to a confining clay strata at depth. A remediation program was undertaken to stabilize the pH of the contaminated soils to the neutral range of 6.0 to 8.0.

The remediation program called for excavating and removing contaminated soils in a 40 ft. x 40 ft. (12m x 12m) area to a depth of 8 feet (2.5m); testing and analyzing surface and subsurface conditions; raising the pH level of all excavated materials and in situ soils to a depth of 24 feet (7.3m) below original grade; and backfilling the excavation with clean granular soils.

Lime is generally the most efficient chemical available to neutralize low pH materials. Lime delivered to the jobsite as calcium oxide (CaO or quicklime) and, using a patented process, is slaked to produce an elevated temperature hydrate slurry ($Ca(OH)_2$). The resultant lime slurry is then injected at pressures of 50 to 200 psi (345 to 1380 kPa) into the target soils using specially designed injection equipment that hydraulically pushes multiple injectors simultaneously. This capability provides a very efficient treatment method for in situ remediation of acidic soils.

The lime slurry produced by this method has a very fine particle size that results in a large surface area promoting high chemical reactivity. Slow sedimentation rates provide good uniformity of the slurry, and elevated temperatures speed the lime/contaminant chemical reactions. A nonionic proprietary surfactant was incorporated in the lime slurry, facilitating penetration during injection. Surfactants or wetting agents reduced surface tension, allowing for better slurry penetration into the injected material.

Neutralization

The project was divided into three phases to allow interim testing and adjustment to the injection process as required. Phase one involved the initial excavation, stockpiling, and pH testing of the surface soils. Following site preparation, three lime slurry injection passes were performed on four ft (1.2 m) centers to the required depth. Figure 2 shows the pattern of injection passes used during the program. A total of 24 tons (22,000 kg) of hydrated lime was injected in phase one.

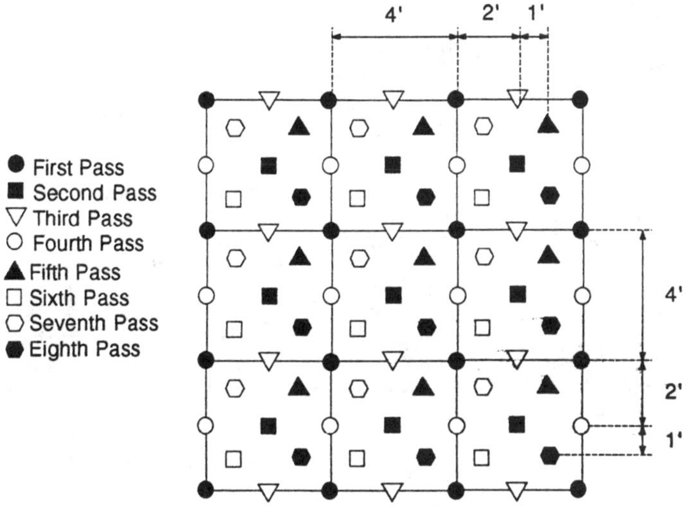

FIG. 2. Multiple Injection Passes Using a Four Foot on Center Split Spaced Pattern

Prior to the phase two injection passes, the area was backfilled with four ft (1.2m) of granular fill material. Four additional lime slurry injection passes were performed followed by two lime/fly ash injection passes. The lime/fly ash was injected to reduce the permeability of the soil mass, mitigating gradient flow of acidic leachate throughout the area. Nineteen tons (17,000 kg) of fly ash were injected in phase two.

Phase three involved backfilling the remaining four ft (1.2m) depth with granular fill material followed by three lime supernate injection passes and two additional lime slurry injection passes. Lime supernate, with a pH of 12 to 13, was injected to introduce additional calcium ions in a dilute (solution) form, enhancing penetration into the less permeable materials throughout the soil mass and promoting further lime/acid interaction. A total of 9 tons (8,000 kg) of hydrated lime was injected in phase three.

Concurrently, the stockpiled materials were treated by spray application of lime slurry and backhoe mixing. After adequate pH adjustment was achieved, these soils were removed to designated on site disposal areas.

Testing and Verification

Prior to the work, soil sampling was performed to determine the pH and neutralization potential of the existing soils. Initial soil pH information indicated an average pH in the range of 3.74 to 4.08. The neutralization potential, given in grams of calcium carbonate equivalent per kilogram of soil, was used to determine the amount of amendment required to neutralize the given soil mass. A requirement of 3 percent lime by weight with soil was calculated to achieve the target range pH of 6.0 to 8.0.

Subsequent to the various phases of the injection process, soil samples were taken and analyzed to determine the effectiveness of the treatment. The sampling techniques and results follow:

Phase one: Three locations were selected for sample retrieval. Samples were drawn in two-foot lengths using a 3-inch, (75 mm) diameter Shelby tube within a hollow stem auger. The intent was to retrieve a continuous sampling at each boring location from ground surface down to the 24 ft (7.3 m) total depth. However, retrieval problems developed at depths greater than 14 ft (4-3 m); consequently, complete data was only obtained in the upper 14 ft (4.3 m). Average pH readings obtained were:

Test Boring #1	5.39
Test Boring #2	5.44
Test Boring #3	2.95

Phase two: Due to the problems encountered during phase one sampling, wet rotary drilling methods were employed along with Shelby tube and split spoon samplers. A continuous sample from ground surface to 24 ft depth was retrieved from each of the four selected boring locations. Average pH readings obtained were:

Test Boring #4	5.91
Test Boring #5	5.10
Test Boring #6	4.86
Test Boring #7	4.36

Phase three: Using the same techniques as in phase two, continuous samples to the 24 ft depth were retrieved from the two selected locations. Average pH readings obtained were:

Test Boring #8	5.91
Test Boring #9	7.21

Conclusions

The total quantity of lime injected into the soils was 52 tons (47,300 kg). However, during the injection process, significant quantities of slurry spread to the surrounding soils outside the target area. This was due to the relatively large perimeter area in proportion to the small volume of target soils. Additionally, there was some indication that gradient flow of acidic leachate existed across the area. To help offset these problems, lime/fly ash injections were made during phase two of the process. Unfortunately, it is difficult to determine the exact effects of these problems. An examination of the neutralization potential of the soils initially sampled and sampled after completion of phase three, indicates that at least an average of 58 percent of the total required calcium carbonate equivalent was introduced by the injection treatment to the target area. Even assuming a 50 percent loss of slurry, calculations indicate that approximately 2 lbs (0.9 kg) of lime was injected per cubic foot ($0.03m^3$) of treated soil, or approximately 2.5 percent by weight. The average pH of the samples taken was adjusted from the initial pre-treatment average of 3.91 to a post-treatment average of 6.56, well within the specified range for effective neutralization.

Summary

The three cases presented in this paper demonstrate three different technologies--Stabilization, Solidification, and Neutralization--that effectively stabilize wastes. Each technology provides solutions for a different set of site conditions and waste composition. Resource recovery through waste conversion and byproduct use can convert non-hazardous waste into artificial soil for structural material or cover soil. In situ solidification can also convert dewatered sludge into structural base materials. In situ lime slurry injection can neutralize and stabilize acid contaminated soils. Only by carefully assessing the requirements of each specific project and being aware of each technology's capabilities can engineers provide the most effective and economical method for converting waste into a treated material.

Appendix. References

American Society for Testing Materials, Standard Specification C593-85, (1985). "Fly Ash and Other Pozzolans for Use With Lime," Volume 04.01, Philadelphia, PA.

Collins, R.J. and Fazzini, A.J. (1989). "Pozzolanic Stabilization of Municipal and Industrial Sludges," presented at American Coal Ash Association Conference on Use of Coal Fly Ash in Waste Stabilization/Solidification Procedures," Covington, KY.

Fazzini, A.J., Collins, R.J. and Colussi, J.J. (1991). "The Role of Fly Ash in Waste Stabilization and Solidification from a Contractor's Viewpoint," presented at American Coal Ash Association Conference, Orlando, FL.

Kyles, J.H. and Malinowski, K.C. (1988). "Fixing Hazardous Wastes, a Review of Solidification/ Stabilization Technologies, presented at "The New England Water Pollution Control Association Conference.

U.S. Environmental Protection Agency, (1989). "Stabilization/Solidification of CERCLA and RCRA Wastes," EPA/625/6-89/022.

Stabilization and Fixation Using Soil Mixing

Brian H. Jasperse,[1], M.ASCE and
Christopher R. Ryan,[2], M.ASCE

Abstract

Soil mixing for the improvement of the properties of soil has been in use in various forms in the United States for nearly thirty years. With the costs of environmental remedial work soaring, engineers have been looking for more economical methods to solve pollution problems. Soil mixing offers a proven, inexpensive method for stabilizing or fixating contaminated soils and sludges.

Soil mixing can be divided into two categories - Deep Soil Mixing (DSM) and Shallow Soil Mixing (SSM). DSM entails the use of up to four mixing shafts approximately 1 meter in diameter, able to mix as deep as fifty meters. SSM consists of one large mixing shaft, two to four meters in diameter, capable of mixing ten meters deep.

----

[1]Vice President, Construction, Geo-Con, Inc., P. O. Box 17380, Pittsburgh, PA 15235

[2]President, Geo-Con, Inc., P. O. Box 17380, Pittsburgh, PA 15235

The systems are capable of injecting and mixing a slurry or, in some cases, dry materials. Instrumentation controls the reagent flow rate to assure that laboratory-developed mix designs are duplicated in the field. Real-time readouts are printed containing information on flow, pressure, elevation and other parameters.

Two case studies will be discussed. These projects will be detailed from treatability studies to final field remediation. The first case involves the fixation of PCBs in a sandy soil under the groundwater table. A test project was initially performed under the U.S.E.P.A. Superfund Innovative Technology Evaluation program and involved the use of a proprietary fixation agent and DSM. The full scale application to the site has recently been completed.

The second case is the remediation of a large oily sludge lagoon using the SSM technology and a dry reagent. This project required solidification of the waste to the point where it could be used as a future foundation for heavy tanks.

Introduction

The roots of soil mixing technology go back to the 1960's. An early U.S. patent (Liver) describes a process which came to be known as Mixed In Place Piles, in which a mechanical mixer was used to mix a cement grout into the soil for the purpose of creating foundation elements and retaining walls. In the 1970's, similar equipment was extensively used in Scandinavia for the introduction of lime slurries into soft clays for stabilization (Broms and Boman, 1979). More recently we have seen the advent of a technology called deep soil mixing (DSM). This originated in Japan where several companies applied gangs of auger mixers to inject and mix cementitous grouts into soils. These were used for underwater foundations (ENR, 1983) for earthquake protection (ENR 1989) and for retaining walls (ENR, 1986). In the mid 1980's,the system was introduced on a U.S. project (Jasperse and Ryan, 1987); an early use was for protection of a structure from soil liquefaction during earthquakes (ENR, 1987).

It was apparent early on in the U.S. experience that some of the advantages of the soil mixing methodology would apply very well to the remediation of contaminated soils and groundwater at hazardous waste sites. First and foremost, it is an in-situ treatment method that is favorably received under current legislation affecting the selection of remedial technologies, mainly because the soil and groundwater are treated in place without removal. A second advantage is the substantial reduction of volatilization that would be experienced when some contaminated materials are brought to the surface and exposed to the atmosphere. Other technical advantages apply as well; for example, the elimination of excavated pits that might endanger surrounding structures, low vibration levels and excellent mixing efficiencies.

The low cost of these systems has provided a major incentive for their use. With the DSM technology, it has been possible to treat soils to depths in the range of 5-30m for a typical cost of $100-200 per cubic meter. To this figure must be added the cost of the treatment reagent, but the total cost is a fraction of that of alternatives such as landfill disposal or incineration.

A recent advance has been the application of much larger diameter mixing heads for shallower soil deposits and for sludge lagoons. Dubbed "Shallow Soil Mixing" or SSM, this system is more productive and reduces typical treatment costs for soils and sludges at 2-8m deep to $20-50 per cubic meter. Again, the cost of treatment reagents would be additional. This system permits the use of dry reagents and an effective vapor collection apparatus. It can be used with cementitous, chemical or even biological reagents as required to treat a particular waste (Jasperse, 1989). A variant on the soil mixing technology uses steam or hot air to extract volatile pollutants from the subsoil (Roy 1990, Diaz and Guenther 1990). The most common usage is simple stabilization where a cementitious material is added to make solidified columns (Figure 1). A series of overlapping columns creates a stabilized block of material.

The balance of the paper focuses on two case studies where in-situ fixation or stabilization was accomplished by soil mixing. The first is in-situ fixation of PCB's using the DSM method on a site in Hialeah, Florida. The second is the solidification of an oily sludge lagoon near Chicago, Illinois.

Figure 1.
Column of Stabilized Material Created by SSM

Deep Soil Mixing Case Study

The work was carried out at the site of an abandoned transformer repair facility near Miami, Florida. Leaks of transformer oil had contaminated the soil below the old shop to depths up to 15m. After an extensive study of applicable reagents and potential mixing methods, the owner selected a proprietary cement-based additive and the deep soil mixing technology as the most effective and economical remediation. The proximity of other businesses made excavation of the contaminated soil impractical and the cost of disposal by incineration was prohibitive.

The soil profile consisted of sands, silty sands and karstic lime rock. The water table was approximately 2m below the ground surface. PCB concentrations varied from nondetectable to about 1,000 ppm. Over much of the affected area, they were in the range of 200-600 ppm.

The work was executed in three phases. First, an extensive laboratory bench scale program was carried out to select the reagent and the proportions needed for treatment. Second, a pilot scale demonstration was run at the site using a small-scale rig. Last, full scale treatment of the site was carried out.

The reagent selected after the bench scale study was a proprietary pozzolanic additive containing treated clay absorbents designed to chemically bind PCB's and other organics. The pilot scale program was run in 1988 and consisted of treatment of two small areas with extensive pre and post treatment testing. Because of the lack of mobility of PCB's, pretreatment TCLP tests were generally not able to detect PCB's in the leachate if untreated soil samples contained less than 60 ppm PCB. If the untreated soil samples contained more than 300 ppm, then the TCLP test detected PCBs in the leachate. Between 60 and 300 ppm, the results varied.

Treated samples all contained PCB levels of 170 ppm or less, with most below 100 ppm. TCLP tests on the treated samples yielded nondetectable results on every sample. Detailed results of the pilot scale program are available through EPA's SITE (Superfund Innovative Technology Evaluation) Program (EPA 1990, EPA 1991).

The final phase of the work, complete treatment of the site, began in late 1990 and was completed in the spring of 1991. The equipment consisted of a four shaft DSM rig (Fig. 2) and a reagent mix plant complete with a four line pump and control system (Fig. 3).

Figure 2.
Four Shaft DSM Rig

Figure 3.
DSM Mix Plant

This equipment is capable of treatment to depths of 30 meters in a single pass, although, in this case, treatment depth ranged from four to fifteen meters. A sophisticated electrical system monitors the flow of the reagent grout to the mixing augers and ensures accurate mix proportions. The control system is guided by a reading from a tachometer which runs along the leads with the augers, measuring the depth penetrated. A control box calculates the penetration rate and the flow rate of slurry per minute required, then emits a signal to a variable speed master grout pump to control injection. This pump sends slurry to one auger. There are three slave pumps which operate with the master pump which feed the other three augers. The rig worked across the site in a series of four-shaft primary and secondary strokes (Fig. 4). Primary strokes were allowed to set prior to drilling the secondaries. Sufficient overlap is allowed to ensure continuity and complete treatment of all the site soils.

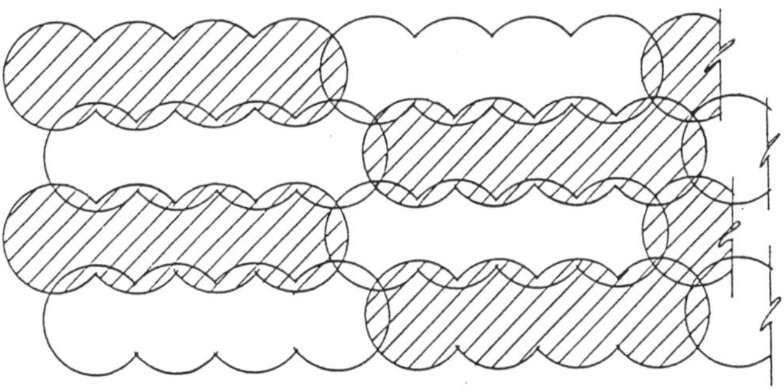

Figure 4. Four Shaft DSM Pattern

Because PCB's are relatively immobile, TCLP tests generally did not show detectable leaching of PCB's either before or after the tests. In the pilot test, average PCB concentrations were lower in mixed samples than the average of pretreatment samples (Fig. 5), although the mixing process prevents good correlation of data.

With respect to strength and permeability, extensive data were obtained. The addition rate was 17% by dry weight to soil or approximately 275 kg/m^3 of soil mixed. The data presented in Figures 6 and 7 from the pilot test show that excellent results for both strength and permeability were obtained, and that the results improved after one year as the cementitious reagent continued to hydrate.

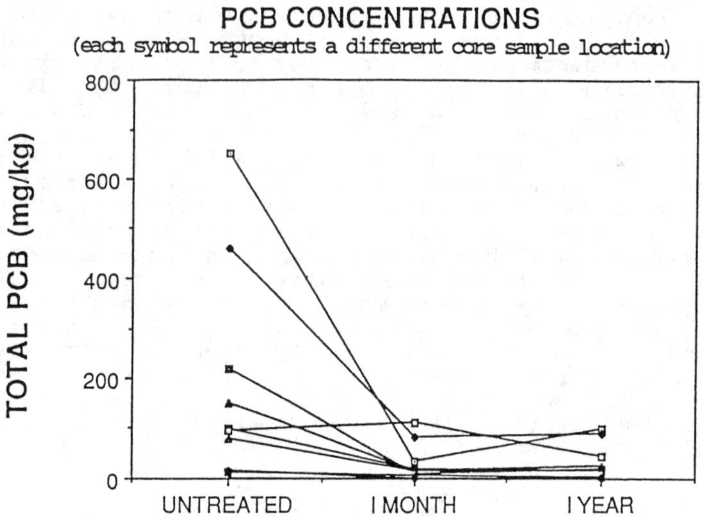

Figure 5. PCB Concentration Before and After Treatment

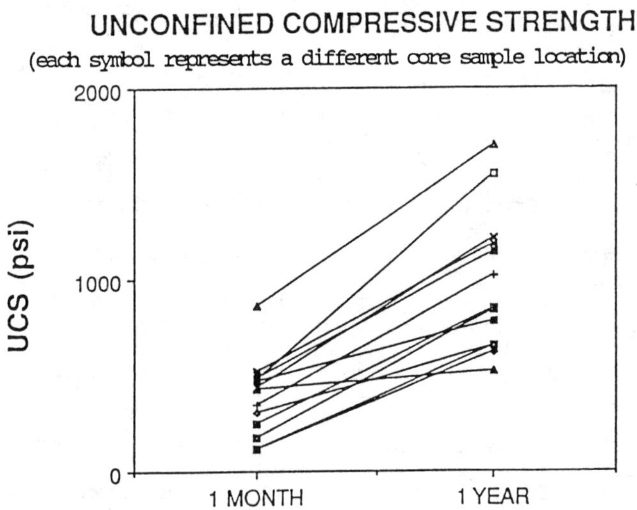

Figure 6. Unconfined Strength of Mixed Samples Vs. Time

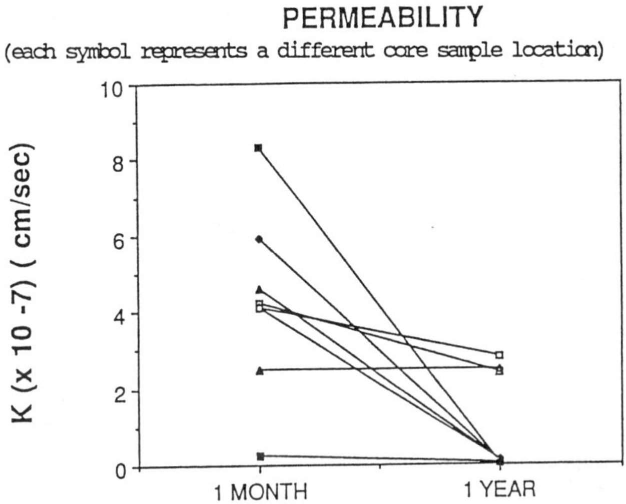

Figure 7. Permeability of Mixed Samples Vs. Time

<u>Shallow Soil Mixing Case Study</u>

The project is a large lagoon containing sludge residues from a water treatment plant at a refinery near Chicago. The lagoons were to be solidified and enclosed in a cap and vertical barrier containment system. The owner planned to reuse the site as a tank farm, so the solidified material had to meet rigid standards for strength, i.e. 35 p.s.i. unconfined compressive strength in 28 days. Also each sample had to pass a free liquids test (Paint Filter - EPA SW 846). The sludge was an oily paste containing oil, grease and petroleum hydrocarbons as primary constituents. The depth of treatment varied from about three to five meters, including one meter of underlying soils.

It was of great concern that there be no dust and that emission of volatiles be kept to a minimum. Pre-job bench scale testing indicated that cement, added in the 15-20% range (by weight), would be the optimal reagent. Because the sludge had a relatively high water content (typically 55 - 65%), it was not desirable to add the reagent in the form of a slurry or grout, since this would greatly increase the amount of cement required and lead to substantial volume expansion.

The technology selected was Shallow Soil Mixing or SSM. This equipment consists of a large diameter mixer head enclosed by a hood or shroud (Fig. 8). Reagent is applied through a dry pneumatic transfer system that is totally enclosed and which has bag houses at all termination points. The final application to the sludge is made with the hood lowered, so that dust is practically eliminated. The system has the further advantage that any emissions of gases caused by the mixing process can be collected under the hood and exhausted for treatment. Generally, as in this case, the in-situ mixing produces little offgassing and no collection of fugitive emissions is necessary.

The work began in late 1990 and is expected to conclude in late 1991. Work progresses, as in the case of DSM, by drilling a series of primary and secondary shafts, allowing the primaries to set before drilling secondaries. In this case, the rig supporting the SSM equipment was able to walk over and work on the stabilized material after about two weeks.

Figure 8. SSM System

Data for mix strengths measured in the field are presented in Fig. 9. It can be seen that the greater the amount of cement, the higher the U.C.S. Also, there is strength gain from 14 to 28 days.

The strengths were adequate to allow the solidified sludge to be utilized as a foundation supporting significant loads. All samples passed the free liquids test.

The 12% results seem to indicate that a minimum amount of cement is needed to produce any significant results. In this case, 20% seemed to be adequate to meet the design criterion.

As in the case of the previously cited DSM project, the results are expected to substantially improve with time.

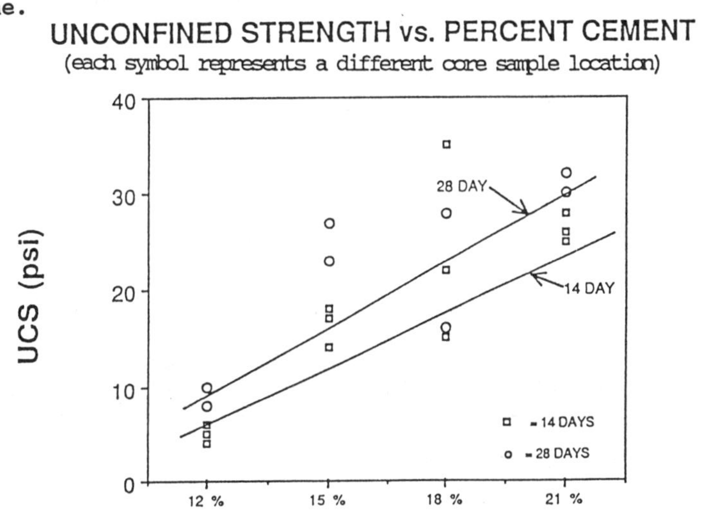

Figure 9. Unconfined Strength Vs. Time

Conclusion

The new in-situ mixing technologies are particularly applicable to major sites that contain contaminated soils and sludges. In-situ mixing has now been used on over a dozen project in the U.S. in the last few years. The economic and practical advantages of in-place treatment are substantial and the current regulatory environment favors this type of solution over offsite transport and disposal. We see a large amount of potential application for this work.

Appendix. References

Broms, B.B. & Boman, P., 1979. "Lime Columns - New Foundation Method, ASCE J. Geotech. Eng. Div. 105 GT4 pp. 539-556.

Diaz, R. and Guenther, J., 1990, "Detoxifier for Contaminated Soils", Pollution Engineering, January 1990, pp. 100-103.

ENR 1983, "Weak Seabed Strata Yields Strong Support", ENR Magazine, March 10, 1983, pp. 30-31.

ENR 1986, "Taipei Untangles Its Rail Lines", ENR Magazine, February 13, 1986, pp. 41-42.

ENR 1987, "Dam Rebuilt to Save It From Quakes", ENR Magazine, September 17, 1987, pp. 46-47.

ENR 1989, "Stockpile Complex Buried Safety", ENR Magazine, January 12, 1989, pp. 26-27.

Jasperse, B. H., 1989, "Soil Mixing", Hazmat World, November 1989, pp. 20-23.

Jasperse, B. H., and Ryan, C.R., 1987, "Geotech Import: Deep Soil Mixing", Civil Engineering Magazine, December 1987, pp. 66-68.

Liver, Norman L., U.S. Patent No. 3,3023,585, N. L. Inventor

Roy, K.A., 1990, "In Situ Detoxifier", Hazmat World, April 1990, pp. 18-23.

Sites Program Report (1990)

USEPA, 1991, "International Waste Technologies" Geo-Con In-Situ Stabilization/Solidification Update Report, EPA/540/S5-89-004a, January 1991

Improvement of Fuel Oil Contaminated Soils By Additives

Sibel Pamukcu[1] and Hazem Hijazi[2]

Abstract

The feasibility of stabilizing fuel oil contaminated soils into usable products was investigated. A laboratory testing program was conducted using artificially fuel oil contaminated kaolinite clay specimens. These specimens were stabilized with lime, cement, flyash, gypsum and attapulgite clay in varying weight percentages and also combinations. The formula for percentages of the three components, namely soil, contaminant and additive were varied to achieve a systematic variation in the measured physical properties. The laboratory study demonstrated that a mixture of lime, cement and attapulgite clay give optimum results with respect to the achieved strength, durability and relative minimizing of oil leachability.

Introduction

Leakage from underground storage tanks or accidental spillage of petroleum products into the ground may pose significant health hazards to the public and the environment. Therefore, quick remediation of the contaminated ground may be critical. Removal of the soil is a significant part of the overall treatment. However, the high cost associated with excavation and limitations in disposal facilities and receiving landfills often complicate the problem. There are emerging technologies for in-situ clean-up, such as biodegradation or electrokinetic removal. However, these are either long-term processes or they are currently at development stages. An economical and fast solution seems to stabilize the contaminated soil by mixing it in place with additives and possibly render it suitable for reuse.

---

1 Asst. Prof., Civil Engineering Dept., Lehigh University Bethlehem, Pa, 18015

2 Eng., Canonie Env. Serv., Co.,King of Prussia, Pa,19406

If the stabilization of the petroleum contaminated soil is to be successful, the petroleum hydrocarbon should either be dispersed and/or be encapsulated and bound in a monolithic solid of high structural integrity formed by the cementing and conditioning action of pozzolanic and earthen materials. The end product should be chemically and physically stable and durable and also easy to handle and work with. In this study, the effort was concentrated on demonstrating the physical integrity of such stabilized materials. Oil contents of leachates were also determined to better assess the differences between several end products with respect to their potential leachability. The results showed that simultaneous action of the cementing, pozzolanic and sorbent additives produce a stabilized product with best improved physical properties and reduced fuel oil content in the leachate relative to that of unstabilized soil.

Background

Despite the best efforts of both the petroleum industry and the regulatory community, releases, leakages and spills of petroleum products occur frequently. It is estimated that, in the United States, up to 25% of the underground storage tanks used for storage of petroleum products are leaking (EPA, 1989). Once a spill or a leakage occurs, the hydrocarbon liquid, under gravity, moves down to the groundwater, partially saturating the soil in its pathway. Upon reaching the groundwater table, the liquid may spread horizontally by migration within the capillary zone. There are three major tasks that need to be performed for remediation and reclamation of such contaminated areas: the first is to control the horizontal migration of the contaminant away from the source, the second is the cleanup of the groundwater, and the third is the cleanup of the contaminated soil.

In general, cleanup of the groundwater contaminated with gasoline-range hydrocarbons consists of pumping of the water from a well and removal of the floating material. Cleanup of hydrocarbon contaminated soils are usually more complicated. Partially hydrocarbon saturated soil can be a persistent source of contamination of groundwater for decades as surface water percolation and groundwater table fluctuation promote migration of the soluble compounds. Oils are often attenuated on clay constituents of soil by adhesion and Coulombic interactions (Mackenzie, 1971). For example, when an advanced front is retained by a clay lense, it is often found that only a minor product layer forms over the groundwater table. The degree of oil retention of a particular soil depends on the physical and chemical make up of the soil, as well as the liquid hydrocarbon, such as the contact angle between the liquid and the mineral. In general, as clay content of a soil increases so does its oil retention capacity and thus the degree of difficulty in remediation.

Applicable Remedial Technologies

The remedial technologies for oil contaminated soils can be classified as follows: 1. Physical treatment; 2. Chemical treatment; 3. Biological treatment; 4. Thermal destruction; 5. Stabilization/solidification. All of these methods have their merits and also limitations in application. The stabilization and solidification technologies work by binding the contaminants within a relatively inert solid matrix which prevents or reduces volatilization and leaching. It also improves the handling and working of the material (AWES 1982; EPA 1986; Conner, 1990; Boelsing, 1988). The resulting matrix may then be used in backfilling or landfilling.

There are a number of solidification/stabilization techniques used in the industry for different types of waste materials. Each technique is formulated to be compatible with the specific waste constituents. These techniques can be divided into the following groups:

1. Cement-based; 2. Silicate based; 3. Sorbent; 4. Thermoplastic; 5. Organic polymer techniques.

The cement based technique involves a series of well known pozzolanic or cementation reactions which produce compounds that act as natural cement. Fly ash and Portland cement are examples of additives in this technique. Silicate based techniques covers a wide range of methods in which the siliceous material is mixed with other alkaline earths such as lime or gypsum. The process involves pozzolanic reactions that form very stable calcium silicates and aluminates. Silicate based techniques have been shown to stabilize oily wastes with some degree of success (Spencer et al., 1982, Pancoski et al, 1988, Pamukcu et al., 1989 and Van Keuren et al., 1987). The sorbent technique involves the use of certain clay minerals with high specific surfaces to fix and retain the hydrocarbon/oil molecules. The needle-like shaped attapulgite clay is an example of additives in this technique.

When used with petroleum contaminated soils, the product of these three techniques should be a microencapsulated matrix in which the hydrocarbons are bound by a combination of physical and chemical isolation. These components would essentially be fixed or immobilized in this matrix which would restrict internal fluid movement. With proper mixing and subsequent compaction a high density homogeneous soil matrix can be achieved. Chemical reactions may occur between some organics and the inorganic, cementatious additives in soils. Oxidation and hydrolysis, are probably the most common reactions for organics in stabilization/solidification systems. Reduction, salt formation and dispersion are also other possible reactions.

In order to assess the integrity of a stabilized or solidified material, a number of physical and mechanical measurements are made. Among these measurements are unconfined compressive strength, permeability, compressibility, density, durability and leachability. Depending on the intent of use of the final product, either all or a few of these measurements may be performed. For example, higher unconfined compressive strength may be a good indicator of the improved stability, trafficability and degree of ease of handling and placement of the material. Lowered permeability may indicate lowered potential for leachability since low permeability restricts the rate of infiltration of water thus reducing the erodibility of a stabilized or solidified compact porous structure.

Investigation

Testing

A testing program was conducted to investigate how much improvement could be accomplished in the physical properties of a petroleum product contaminated kaolinite clay by additive treatment. A number of additive mixture formulas were tried to obtain optimum results with the measured properties. Georgia kaolinite clay was selected as the soil medium since clays tend to retain hydrocarbons and as a result their physical and chemical properties appear to be effected more than the other solid constituents of soils. The contaminating petroleum product was selected as No.6 grade fuel oil (crude oil), since it is only slightly volatile and does not require special precautions to preserve its stability during and after mixing with soil. The additives were Portland cement type II, hydrated lime ($Ca(OH)_2$), class-F bituminous fly ash, gypsum (calcium sulfate hemihydrate) and attapulgite clay obtained by crushing palygorskite shale from Florida to -No. 200 sieve size.

Samples were prepared by mixing the clay with water at the predetermined optimum moisture content of the mixed system before adding the fuel oil. For the 10% (by dry weight of clay) fuel oil mixed clay the optimum moisture content was 27% and the maximum dry density was 13 kN/m^3. The clay was first mixed with the water and then mixed with oil thoroughly. The mixture was left to cure in a closed container for 24 hours to allow for the adsorption process. Since crude oil is only slightly volatile, vaporization of its volatile components during the curing period was regarded negligible. The contaminated soil was divided into a number of sets, each set being treated with either single or a combination of additives at different percentages and proportions. The additive amount was varied as 5,10 and 20% by dry weight of clay. In the case of combined additives, the maximum additive percentage was 25% by dry weight of clay.

The following tests and analysis were conducted on the untreated and treated specimens: unconfined compressive strength (ASTM D2166/D1632), direct tensile strength (Fang and Fernandez, 1981), Atterberg limits, moisture content, freeze-thaw durability (ASTM D4842), permeability (triaxial), and oil content analysis of leachate (SMEWW, 1989, Method 5520B).

Results

I. Unconfined Compressive Strength (UCS)

The unconfined compressive strength of fuel oil mixed clay is significantly lower than that of the pure clay, as expected. Figure 1 shows the significant loss of stiffness and strength of the clay with increasing percentage of oil. This has been observed by other investigators also, where the decrease in strength was attributed to increase in the net repulsive forces between clay particles in the presence of neutral non-polar fluids with low dielectric constants (Yong and Warith, 1989).

When the 10% fuel oil contaminated clay is treated with additives, there is a general increase in the strength values, especially for the 7 day constant humidity cured specimens. Table 1 presents the isolated effects of five different agents on the unconfined compressive strength of contaminated clay. As observed, lime and cement produce marked increases in the strength values, which are even higher than the strength of the pure clay. Figure 2 shows the effect of lime treatment on the unconfined compressive strength.

A number of samples treated with combinations of additives were tested also. The best strength values were achieved with the addition of combinations of 10% cement, 10% lime, 5% attapulgite clay and 5% cement, 10% lime and 5% flyash. These results are presented in Table 2.

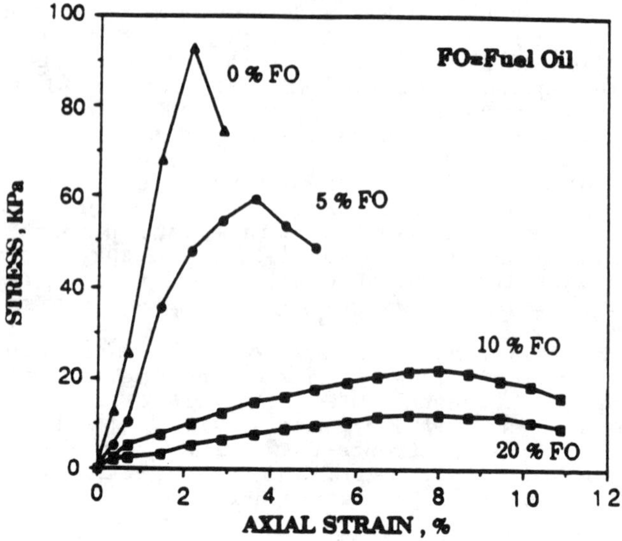

Figure 1. Variation of Unconfined Compressive Strength and Stress-Strain Relation With Fuel Oil Content

The large increase in the strength values with the addition of lime is probably due to the flocculation and agglomeration effect of Ca^{++} ions. Lime initially breaks apart clay clumps making available more mineral surface for reaction. This dispersion action probably releases the oil from inside the clumps into a more homogeneous mixture of solid, water and oil phases. Additional lime than produces a gel which sets up, trapping the oil inside the solidified structure. This is probably why the addition of lime at a high percentage (20%) resulted in higher strength values than the addition of cement at the same percentage. In this case, the pozzolanic reaction of cement might have been inhibited also, due to the presence of oil which would tend to retain the calcium ions making it unavailable for the cementation reaction.

II. Direct Tensile Strength (TS)

The tensile strengths were determined by "unconfined penetration test" described by Fang and Fernandez (1981). Results of these tests using 10% fuel oil contaminated clay treated with mixed additives are given in Table 2. The tensile strength of 10% oil contaminated clay was measured 1.1 kPa. This value increased significantly with additive treatment. The highest tensile strength was again obtained with the cement, lime and attapulgite clay mixture additive treatment.

Table 1. Unconfined Compressive Strength (USC) for Stabilized 10% Fuel Oil Contaminated Kaolinite

UCS, (kPa)

	5% Additive		10% Additive		20% Additive	
Stabilizing agent	**Fresh**	**Cured***	**Fresh**	**Cured***	**Fresh**	**Cured***
Lime	**24.7**	**50.1**	**44.8**	**97.4**	**109.4**	**162.9**
Cement	**49.7**	**94.5**	**52.6**	**112.8**	**75.2**	**145.9**
Fly ash	**16.9**	**29.6**	**19.2**	**36.5**	**24.7**	**48.6**
Attapulgite	**18.4**	**30.9**	**37.3**	**55.1**	**75.2**	**92.0**
Gypsum	**17.0**	**39.5**	**32.1**	**45.3**	**55.9**	**88.9**

*** 7 days**

III. Cohesion and Internal Friction Angle

A straight line envelope tangent to the Mohr circles on the right (USC) and the left (TS) of the vertical axis of normal stress versus shear stress representation was utilized to estimate, approximately, the cohesion and

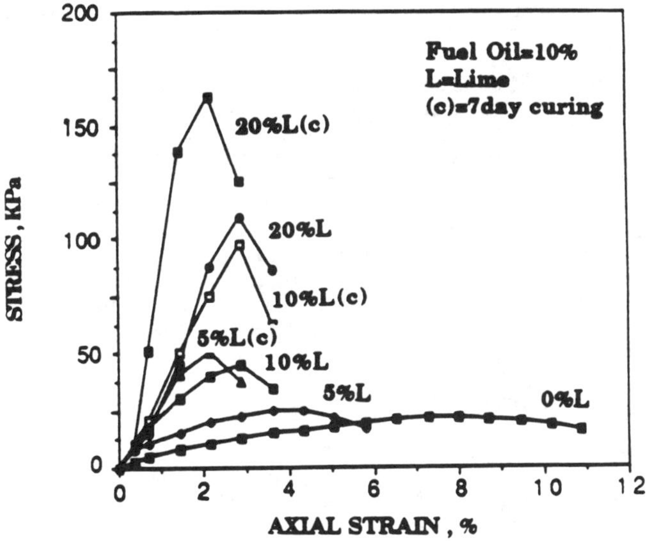

Figure 2. Variation of Unconfined Compressive Strength and Stress-Strain Relation With Increased Percentage of Lime

internal friction angle for each stabilized system. Figures 3 and 4 illustrate the normalized variations of cohesion and friction angle with percent additive, respectively. The additives appeared to improve strength by improving the cohesion of the soil only. The internal friction angle was essentially unaffected by the presence of fuel oil. Furthermore there was little change in the internal friction angle with treatment, except with lime treatment at low percentages.

Table 2. Unconfined Compressive Strength (USC) and Tensile Strength (TS) for Stabilized 10% Fuel Oil Contaminated Kaolinite

Stabilizing agents	UCS (kPa)	TS (kPa)
10%L+10%FA	52.2	3.5
10%C+5%FA	42.5	3.0
15%L+5%FA	83.9	4.0
10%L+5%C+5%FA	95.6	3.6
10%C+10%L+5%A	115.0	4.7
10%C+5%G+5%A	75.3	3.2

C=Cement G=Gypsum
L=Lime FA=Fly ash A=Attapulgite clay

These results are consistent with explanations given for strength reduction with oil. If the increase in net repulsive force is responsible for reduced strength, then cohesion component of the shear strength should be affected the most by this phenomenon. The reduction in the internal friction angle at low percent addition of lime is probably due to the pulverization affect of lime which may have resulted in reduction of apparent particle size and also release of oil from inside the clay clumps. Variation of cohesion with addition of cement and lime are similar, whereas flyash produces no significant change, as expected.

IV. Atterberg Limits

When 10% fuel oil contaminated soil samples were treated with 10% of each of the additives separately, plasticity index was reduced for all except lime. This may be due to the saturation of kaolinite with a divalent cation (Ca^{++}) which may have lead to an edge-to-face flocculation and therefore higher liquid limit. Each limit test was repeated 3 times for reliability. It was noticed that the variation of these limits for the

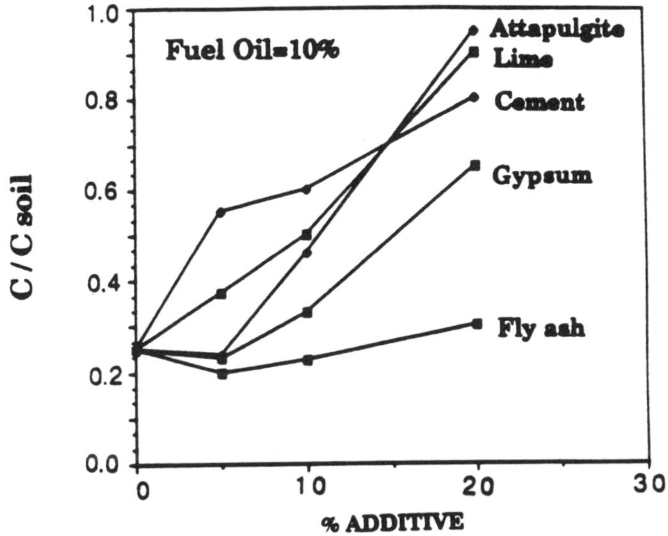

Figure 3. Variation of Normalized Cohesion of 10% Fuel Oil Contaminated Kaolinite With Increased Percentage of Additives

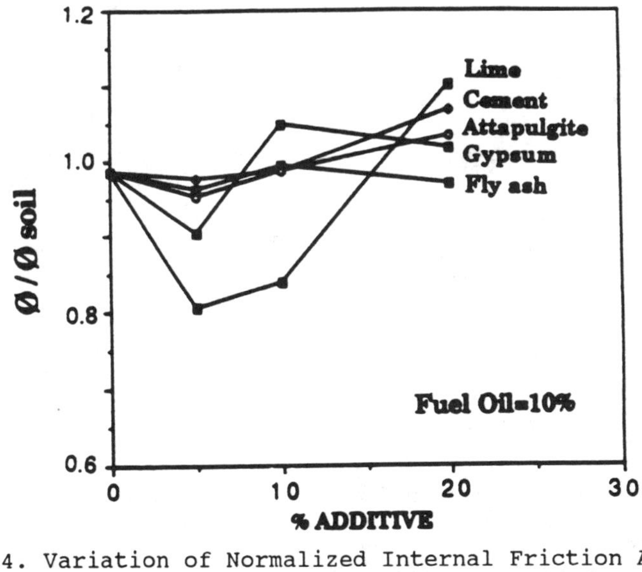

Figure 4. Variation of Normalized Internal Friction Angle of 10% Fuel Oil Contaminated Kaolinite With Increased Percentage of Additives

untreated clay was high. When the material was treated the variation reduced significantly.

The large variation in the measurements of consistency limits of untreated contaminated clay may be due to inhomogeneous distribution of the non-polar liquid (fuel oil) in the clay-water system. Slight variations in the distribution of the oil phase in the soil may result in significant spatial variations of viscosity, density and moisture content in a given sample mass of clay-water-oil mixture. However, after stabilization, if the oil become bound in a matrix of solidified structure, as expected, the locally bound oil should no longer be free to influence the test. This is observed indirectly by the reduced size of bands of variance of liquid limit values obtained for the additive treated clays as shown in Figure 5.

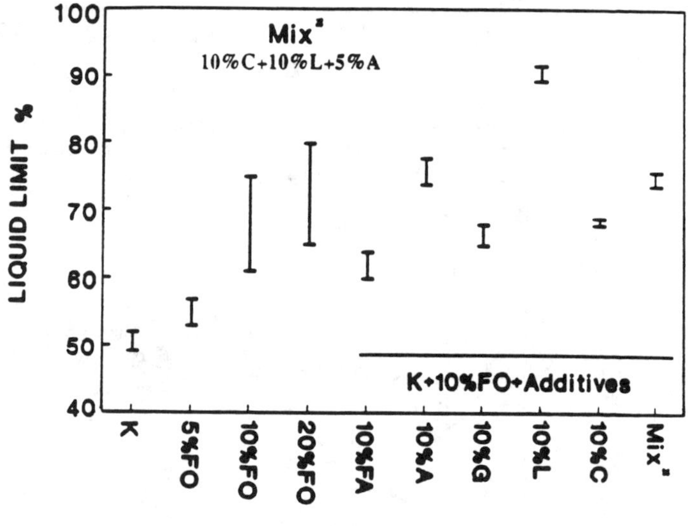

Figure 5. Bands of Variation of Liquid Limit Measurements (3 Measurements Each)

V. Permeability and Leachate Analysis

Coefficient of hydraulic permeability for pure compacted kaolinite clay , measured by triaxial permeability apparatus, was on the order of $5x10^{-8}$ cm/sec. When 10% fuel oil was introduced, this coefficient increased slightly to $9x10^{-8}$ cm/sec. However, with treatment the resulting matrices showed up to 10 times decrease in their permeabilities.

The leachates collected from the permeability test of four different samples were analyzed for oil content. These leachates were collected until after 24 hours of steady state water permeation through the saturated matrix of each sample. The dissolved or emulsified oil was then extracted from the collected water and quantified following a standard method for the examination of water and wastewater (SMEWW,1989; Method 5520B). The leachate concentrations of oil obtained in this analysis are not necessarily regarded as absolute quantities, but rather evaluated relative to each other.

The oil content in the leachate of the untreated soil was 380 mg/l. When the mixture was treated with 10% cement, this number was reduced to 107 mg/l. The leachate from a sample treated with 10% lime showed oil at the concentration of 51 mg/l. The lowest oil content was found in the leachate of soil treated with 10% cement, 10% lime and 5% attapulgite clay. This concentration was measured as 26 mg/l, which is approximately 93% improvement over the unstabilized soil.

VI. Durability

The freeze-thaw durability of a few selected samples were measured. These results are given in Table 3. The average cumulative material loss was least for the cement, lime and attapulgite clay treated soil which survived 4 cycles of freeze-thaw prior to failure. It should be noted here that, for this and the cement plus flyash treated samples the failures appeared to occur along horizontal planes which were identified as the planes of compaction layers. The parts between these planes remained rigid and solidified. It probably was poor binding along the compaction layers which caused water entry thus volume change during freeze-thaw.

Table 3. Freeze-Thaw Durability Test Results

Specimen	Moisture Content, (%)	Number of Cycles	Average Cumulative Corrected Relative Mass Loss, (%)	Remarks
K+10%FO	27	1	-	total deterioration
treated with				
10%L	27	1	-	total deterioration
10%L+5%C+5%FA	27	1	-	total deterioration
10%C+5%FA	27	6	8.2	failure along horizontal crack planes
10%C+10%L+5%A	27	5	6.5	failure along horizontal crack planes

Conclusions

The additive stabilization of petroleum contaminated soils appear to be a viable method of resource recovery and may produce economics over other remediation technologies. In all of the tests conducted, the treated samples showed significant improvement in their physical properties with respect to those of untreated samples. The following specific conclusions were made:

1. Fuel oil reduced the strength and stiffness of compacted kaolinite clay soils significantly, even at low percentages of addition.
2. Lime and cement produced marked improvement in the strength and durability of the treated samples, in general. Seven day cured samples of treated clay showed unconfined compressive strength nearly double the strength of untreated clay.
3. Plasticity was reduced with additive treatment except with lime. The large variability in liquid limit determination for oil contaminated clays was reduced when they were stabilized.
4. Permeability was reduced with additive treatment. The oil contents in the leachate of the stabilized samples were reduced relative to that of unstabilized clay.

In general, the best results in all of the tests conducted were obtained with samples treated with 10% lime, 10% cement and 5 % attapulgite clay. This is probably due the combined action of the three different types of stabilization/solidification processes namely: cementing, pozzolanic and sorbent reactions represented by each additive.

Acknowledgments

This work was partially funded by Bethlehem Steel Corporation, Bethlehem, Pennsylvania and by the Environmental Studies Center of the Lehigh University.

References

AWES, (1982). Guide to the Disposal of Chemically Stabilized and Solidified Waste. US Army Waterways Experiment Station, SW-872.

Boelsing, F. (1988). "Remediation of Toxic Waste Sites: DCR-Technology". Report for Ministry of Economics, Technology and Traffic, Hanover, Germany, 64 p.

Conner, J.R., (1990). Chemical Fixation and Solidification of Hazardous Wastes. Van Nostrand Reinhold Publishing Co., New York.

EPA (1986). Handbook for Stabilization/Solidification of Hazardous Waste. Cullilane, M.J. Jr., Jones, L.W. and Malone, P.G.. US Environmental Protection Agency, EPA/540/2-86-001.

EPA (1989). Volumetric Tank Testing: An Overview. US Environmental Protection Agency, EPA/625/9-89/009.

Fang, H.Y. and Fernandez, J., (1981). "Determination of Tensile Strength of Soils by Unconfined Penetration Test". American Society for Testing and Materials, STP 740, pp. 130-144.

Mackenzie, J.M.W. (1970). "Interactions Between Oil Drops and Mineral Surfaces". Society of Mining Engineers, AIME, Transactions Vol. 247, September, 1970, pp. 202-208.

Pancoski, S.E., Evans, J.C., LaGrega, M.D. and Raymond, A. (1988). "Stabilization of Petrochemical Sludges". Proc. of the 20th Mid-Atlantic Industrial Waste Conference, HMCRI, Silver Spring, MD, pp. 299-316.

Pamukcu, S., Lynn, J.B. and Kugelman, I.J. (1989). "Solidification and Re-Use of Steel Industry Sludge Waste". Proc. of the 21st Mid-Atlantic Industrial Waste Conference, Technomic Press, Lancaster, PA, pp. 3-15.

SMEWW (1989). Standard Methods for the Evaluation of Water and Wastewater, APHA-AWWA-WPCF, 17th edition.

Spencer, R.W., Reifsnyder, R.H. and Falcone, J.C. (1982). "Applications of Soluble Silicates and Derivative Materials". Management of Uncontrolled Hazardous Waste Sites, Hazardous Materials Control Research Institute, Silver Spring, MD, pp.237-243.

Yong, R.N. and Warith, M.A., (1989). "Leaching Effect of Organic Solutions on Geotechnical Properties of Three Clay Soils". Proc. of the 2nd Symposium on Environmental Geotechnology, Vol. 1., Envo Publ., Inc., Betlehem, Pa, pp. 99-110.

Van Keuren, E., Martin, J., and De Falco, A., (1987). "Pilot Field Study of Hydrocarbon Waste Stabilization". Proc. of the 19th Mid-Atlantic Industrial Waste Conference, Technomic Publishing, Co., pp. 330-341.

Hazardous Waste Containment with a Bentonite Cutoff Wall

Chikashi Sato[1], Derek A. Braithwaite[2], Angelos Protopapas[1], and Paul P. Stewart[3]

Abstract

The LaBounty landfill received an estimated 181,000 cubic meters of wastes including 1,1,2-trichloroethane, arsenic and orthonitroaniline before its closure in 1977. A clay cap was installed over the landfill, later a subsurface bentonite diversion wall was placed upgradient of the landfill.

The objective of this study was to examine the effectiveness of the clay cap and diversion wall to reduce leachate discharge to the adjacent river. To accomplish the objective, an extensive amount of chemical and hydrogeologic data were analyzed.

After evaluation of the data, it was concluded that the reduction in concentration of contaminants in the monitoring wells and the chemical loading to the river are not a result of the cutoff wall. Instead, they were a result of the clay cap, reduced infiltration, the corresponding declining piezometric pressure in the deep aquifer and possibly to some extent depletion of the waste.

Introduction

The use of a slurry trench cutoff wall in the United States began in the early 1940's (McCandless and Bodocsi, 1987). During the last two decades or so, cutoff walls have been used extensively as a barrier to

---

[1]Assistant Professor, [2]Graduate Student, Dept of Civil & Environmental Engineering, Polytechnic University, Six Metrotech Center, Brooklyn, New York 11201

[3]Vice President, Advanced Cleanup Technologies, Inc. Suite 211, 200 South Service Road, Roslyn Heights, New York 11577

prevent or control the migration of contaminated groundwater as well as contaminants from disposal sites.

A Superfund site known as the LaBounty landfill received hazardous wastes from a pharmaceutical industry between 1953 and 1977. The site is located on the flood plain of the Cedar River in Charles City, Iowa. Significant waste components identified as leaching from the landfill include arsenic (As), 1,1,2-trichloroethane (TCEA), orthonitroaniline (ONA), and 33 other chemicals.

Since 1979, the landfill wastes have been monitored through a network of 24 groundwater monitoring wells by consultants for the industry, pursuant to an Administrative Order of the United State Environmental Protection Agency, Region VII. A clay cap was installed over the landfill in 1980 and by 1986, a subsurface bentonite diversion wall was installed upgradient of the landfill for the purpose of dewatering any wastes that were submerged beneath the existing water table.

The migration of contaminants has been monitored through the groundwater monitoring well network up to the present date. Water samples have also been collected from designated surface water sampling sites along the Cedar River which flows adjacent to and receives subsurface leachate loading from the landfill.

This study attempted to evaluate the effects of a cutoff wall constructed at this site, ranked highly in the Superfund list. The paper presents an analysis of chemical and hydrogeologic data, and assessment of the effectiveness of the containment technologies (i.e., the clay cap and diversion wall) at the hazardous waste site. The analysis was based on in-depth review of data from published/unpublished literature and reports.

Description of Site and Background

The LaBounty site is a 12 acre site, located on the right bank of the Cedar River within the confines of Charles City, Iowa (Figure 1). The site is underlain by the Coralville member, the youngest of the Cedar Valley Formation. This bedrock is highly fractured and of nearly pure calcium carbonate in some parts. The hydraulic conductivities ranged from 2.2×10^{-5} to 1.3×10^{-5} cm/sec. A typical profile showing the stratigraphy of the site is presented in Figure 2 (Munter, 1980).

The alluvial material at the site range from clay to gravel, but is primarily medium-grained sand with some gravel. Measured hydraulic conductivities range from 1.4×10^{-2} to 7.0×10^{-8} cm/sec (Munter, 1980).

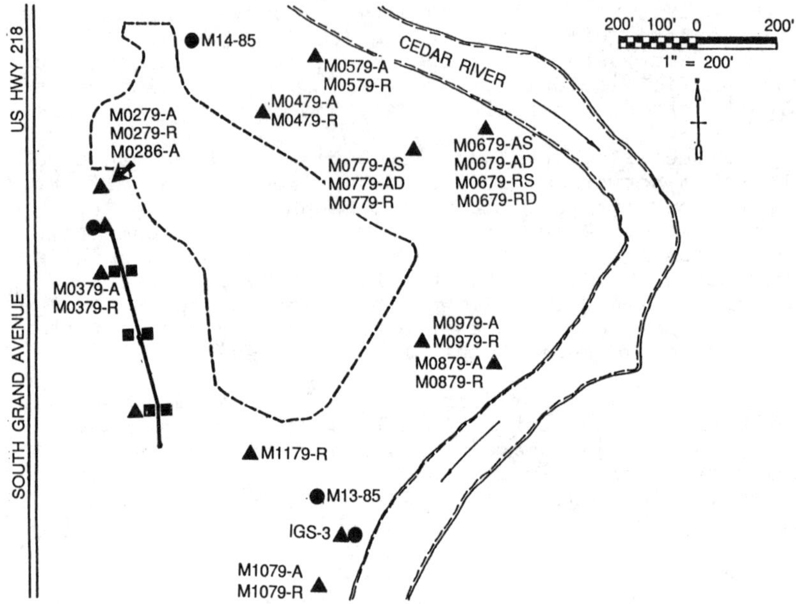

Figure 1. LaBounty Site: (▬), cutoff wall; (---), limit of disposed waste; (▲), nested monitoring well; (●), lower cedar valley monitoring well; (■), cutoff wall monitoring well; A-alluvial, AS-shallow alluvial, AD-deep alluvial, R-upper cedar valley, RS-shallow upper cedar valley, RD-deep upper cedar valley.

Glacial till is found upgradient of the site. It is light olive brown to dark grey. It consists of 20-25% clay, 35-45% sand, and approximately 5% pebble, cobble, or boulder. The hydraulic conductivity of the till is several orders of magnitude lower than that of the alluvium or the fractured bedrock (Munter, 1980).

The LaBounty site was a farm in the period preceding 1950. After the farm was discontinued, the owner at the time began mining the alluvial deposit as commercial fill material. Then, beginning in 1953 and lasting until 1977 the industry disposed of an estimated 181,000 cubic meters (6.4 million cubic feet) of chemical waste from the manufacture of animal pharmaceuticals in the borrow pits at the LaBounty site (Hick and Associates, 1977).

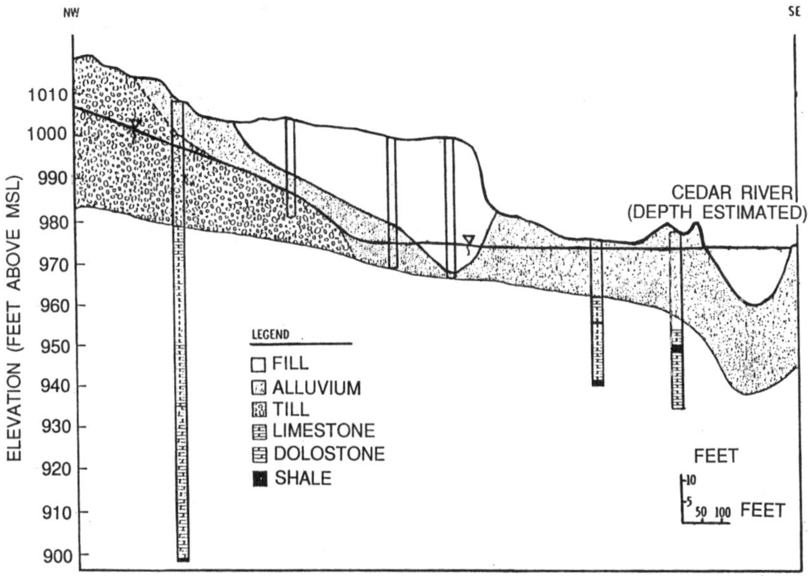

Figure 2. Stratigraphy of the LaBounty Site

During 1977 and 1978, an EPA investigation revealed that arsenic was lying on the bedrock in some areas and were occasionally below the water-table. Surface concentrations of arsenic were lower than that of the subsurface which was indicative of leaching and subsequent downward migration (CRA, 1984).

There were significant amounts of waste being leached from the site and transported by the groundwater into the Cedar River: in February 1977 ONA was detected 11 miles downstream of the site and the fish taken from the river appeared tainted. As a consequence the EPA ordered the industry to install a monitoring system. Since October 1979, the industry has monitored groundwater quality, water table elevations, rainfall, and contaminant loading to the Cedar River. The extent of the waste deposit, and the location of the submerged waste were previously investigated by consulting engineers (CH2M Hill, 1988).

As many as 15 discrete waste streams were identified. The mixed wastes disposed at the site comprise about 6 million pounds of arsenic primarily in the form of calcium arsenite and calcium arsenate sludge that contained 65 to 70% water, 1.5 million pounds of ONA and 70,000 pounds of TCEA.

Past Remedial Measures

The first remediation action was taken in 1977. The response action was a construction of a dike around the disposal area to intercept surface runoff. The dike was removed in 1979 since it was thought that the dike was fostering impoundment of precipitation and thus increasing infiltration at the site.

In 1979, Phase I program began, which involved installation of 24 groundwater monitoring wells and routine surface water sampling of the Cedar River. Measurements of water table and piezometric head, and determination of contaminant concentrations have been made on a monthly basis since October 1979. The water samples have been analyzed for arsenic, TCEA and ONA as indicator parameters.

Phase II program was undertaken in 1980, which include capping to prevent infiltration, surface water diversion, and rerouting of a storm sewer. A 0.7m (2 ft) thickness of compacted clay cap was placed over the contoured fill area, and a 0.15 m (6 in.) clay cover was placed over the parking lot area west of the site. At the toe of the cap, rip-rap erosion controls were installed.

A subsequent assessment in late 1982 reported that the capping was not effective because it did not have any effect on reducing leaching of the submerged waste into the Cedar River (Sisk, 1982). In fact the report concluded that the pollutant concentrations in the groundwater and the river originated primarily from the submerged waste. As a result of this assessment, installation of an upgradient cutoff wall with a groundwater collection system was recommended.

In Phase III program, an upgradient cutoff wall with groundwater collection system at the site was constructed between August 1985 and January 1986. The purposes of the subsurface bentonite cutoff wall and collection system were to facilitate dewatering of any wastes that were submerged beneath the existing water table, and to reduce the rate and volume of groundwater flow through the wastes, thus reduce the leachate discharge to the Cedar River.

The migration of contaminants has been monitored through the groundwater monitoring well network up to the present date. Water samples have also been collected from designated surface water sampling sites along the Cedar River which flows adjacent to and receives subsurface leachate loading from the landfill.

Methodology

To evaluate the effectiveness of the cutoff wall at the site, the following data were used:

1) Piezometric levels in the alluvial and rock wells;
2) Precipitation in the Charles City area;
3) Concentrations of the principal pollutants (As, ONA, and TCEA in the alluvial and rock wells;
4) Loading to the Cedar River for each principal contaminant.

The data obtained during the period October 1979-June 1989 were used. The monitoring wells chosen were those that encircle the site and were observed during the same period. Besides the time-series data, the coordinates for each well was also utilized. All data except for precipitation, were measured monthly during each month of the monitoring period.

Results

The LaBounty site map (Figure 1) shows the limit of disposal site, locations of the cutoff wall and the monitoring wells. The Cedar Valley aquifer discharges into the Cedar River. Precipitation during April, May, and June is likely the source of recharge to both the alluvium and the lower aquifer. During the remainder of the year recharge is considered negligible (CRA, 1984).

The foregoing time-series data were all plotted separately, and overlaid. Upon inspection of these plots it was observed that, with some exceptions, at times the piezometric levels in the rock wells appear higher than those in the alluvial wells. In addition all the wells appear to be fluctuating with some frequency. Although reversing hydrostatic conditions were observed in the overlaid plots of the other nested monitoring wells, in these wells it was much more periodic. To better observe these trends, the plots were all fitted using a ten degree polynomial. Every plot displayed a near identical sinusoidal curve (Figure 3 provides an example): the period and amplitudes were all matching. The plots of M0479-A and M0479-R, and M0579-A and M0579-R were exceptions: the piezometric levels of the alluvial wells lagged those of the rock wells.

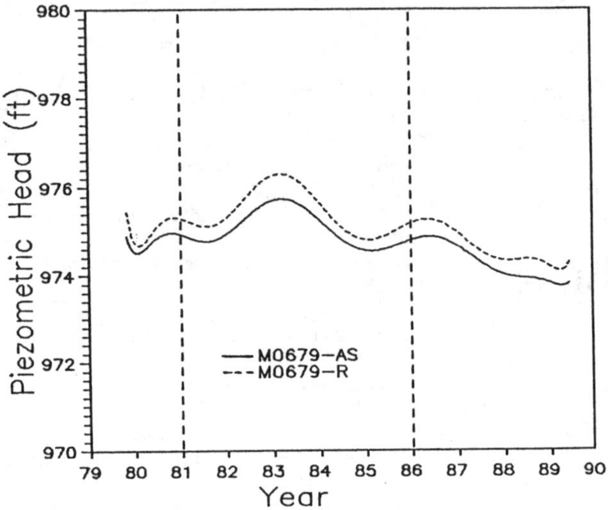

Figure 3. Ten degree polynomial curve fitting of piezometric heads - Alluvial (AS) and rock wells (R).

Given this observation, the precipitation and the net loading of arsenic to the Cedar River were fitted with a ten degree polynomial. Figure 4 shows the piezometric head at M0879-A, as an example, overlaid with precipitation and arsenic loading to the river. The resulting plots were very similar to those of the piezometric levels plotted earlier. The fluctuation of piezometric levels in the alluvial wells is in phase with loading of arsenic to the Cedar River. This observation indicates that the river contamination is due to local groundwater flow at the site.

To verify this situation a series of flownets were constructed. In order to construct these flownets, the piezometric elevations were chosen at the turning points of the fitted curves for each of the wells.

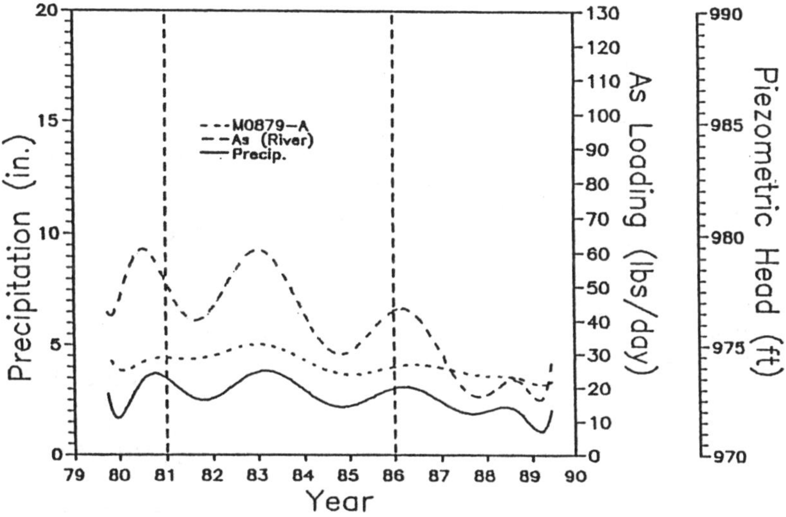

Figure 4. Ten degree polynomial curve fitting of precipitation, arsenic loading to the river and piezometic head of a typical alluvial well (A).

The flownets drawn showed that the flow pattern before and after 1986 are significantly different. In order to verify whether or not the loading to the Cedar River originates from the site, the discharges were computed using the Darcy's equation. This data was then plotted and overlaid on the previously obtained plot of the net arsenic loading to the river. The results confirmed that loading is indeed from the site.

<u>Discussion</u>

Hydrology From the flownets it was determined that the groundwater flow through the site after the cutoff wall was significantly smaller than before its construction. The flownets for the period after the cutoff wall was constructed reveal that water from the Cedar River, north of the site, flows to the south-east direction through the alluvium deposits. This situation began developing during 1985 and is evident in all of the succeeding flownets. This flow does not seem to contribute significantly to transport of the contaminants from the dump site. An examination of the flowlines shows that the flow from the river is not moving through the submerged waste but east of it in a north-south direction to the discharge zone in the vicinity of M0879-A. Also, from the flownets, flow around the ends of the cutoff wall appears to be a possibility.

There are two other possible flow components which are not evident on the flownets. First, there is the likelihood of groundwater flowing under the cutoff wall because the piezometric levels in M0279-A, M0379-A and M0286-A upgradient of the cutoff wall and the wells down-gradient indicate a head difference of between 20 and 25 feet. Second, the lower aquifer is influencing the upper aquifer as demonstrated in the overlays of both piezometric levels. (The both fitted response curves are identical though not at the same elevation.) At M1079 and M0479, the piezometric level in the lower aquifer is continuously higher than that of the upper aquifer. Therefore, on the flownets, what appears to be groundwater flowing around the wall is actually the result of upflow. This was previously reported to be typical for the entire region (Munter, 1980). This upflow cannot not be quantified.

Following the construction of the cutoff wall, upflow is the most likely major component of groundwater flows through the site. The declining groundwater flows in the period following the wall correlates closely with the reduction of piezometric head differences in both well types. The cutoff wall may have resulted in the decline to some extent of the head across the site.

From the previous discussion it was concluded that the alluvial groundwater levels are largely influenced by the lower aquifer. It should be noted here that the high elevations upgradient are primarily due to the low hydraulic conductivity of the glacial till in that area. On the other hand, the area around the site comprises largely of sand and gravel hence it is freely draining and as a result maintains a lower head than that of the glacial till.

An examination of the annual precipitation recorded over the same period shows a similar declining trend as observed in the piezometric levels of both aquifers in the same time frame. This is expected because this is the source of recharge for both aquifers.

Extent of Contamination An examination of arsenic concentration in all of the downgradient wells reveals that the peaks in concentrations correspond closely to the lowest piezometric levels. The reverse was also found: the peaks in piezometric levels corresponded to the lowest concentrations. This situation occurs because at high piezometric levels the conditions are virtually in-seepage and this lends to the dilution of contaminants.

During the period 1982-1986 there were significant reductions in concentrations of the three contaminants in the nest of wells located at M0479, M0579, M0679, M0779, and M0979. In fact, the former two are practically free of arsenic. The causes of this reduction are twofold. Firstly, the installation of the cap was already determined to be causing a reduction in these wells (Sisk, 1982). And secondly, owing to declining piezometric levels in both aquifers during this period less and less of the waste was being submerged resulting in a corresponding reduction in the release of arsenic. The virtual absence of arsenic in M0479 and M0579 is due to the clay cap exclusively as there is no submerged waste in this area.

The overall reduction in the above mentioned wells is not manifested in the nest of wells located at M0879. Instead, the arsenic concentration remained relatively unchanged during the period of observation. M0879 lies in the path of the main plume before and after the wall was installed (Figures 5 & 6). The same can also be said of M0979. An explanation of this type of response is elusive at the moment and requires a great deal more study. The relatively constant concentration of arsenic can be attributed to continuous dissolution of arsenic due to strong interaction of the aquifer with the nearby river.

The decline described earlier was halted in all of the wells after 1986 when the cutoff wall was completed. This is confirmation that the wall has resulted in a lowering of the head across the site. From the flownets it is quite obvious that the conditions are near placid making it ideal for saturation of the groundwater in the vicinity of the waste. Therefore it can be concluded that this has resulted in the fairly constant level of arsenic concentration in the wells. The brief decline in concentration in M0779 noticed after 1987 was probably due to the dilution caused by the river water flowing through the site in the north-south direction.

At this site, the long exposure period (since 1953) to high levels of contaminants is likely to have significantly reduced the absorption capacity of the soil. Thus, it is unlikely that transport of arsenic toward the river is retarded due to the adsorption process (Freeze and Cherry, 1979). Even though the wells are showing declining concentrations, there are still significant amounts of the chemicals adsorbed on the soil waiting to be released when the appropriate conditions prevail.

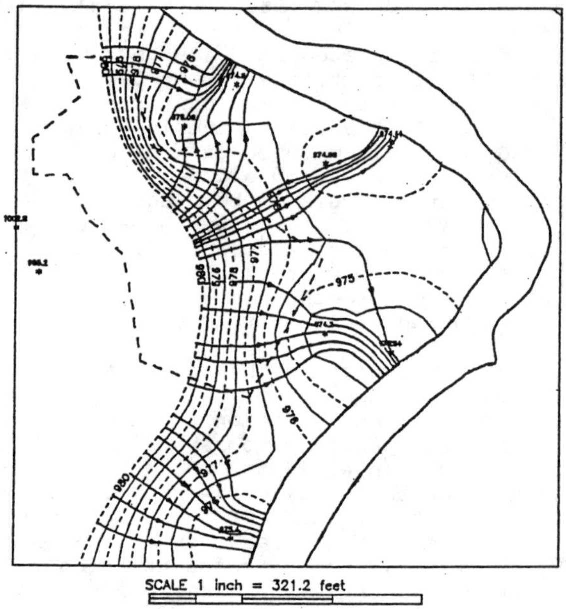

Figure 5. Flownet before the cutoff wall

The net loading of ONA to the river has declined to virtually nil particularly after the cap was installed. TCEA also declined but in less dramatic way. On the other hand the arsenic loading continues to be significant. Overall it has declined but the pattern of the decline closely matches that of the precipitation and the piezometric levels in the wells. To verify this, the estimated discharges from the flownet were used to determine the loading to the river. Overlaying this loading to the net loading measured revealed a matching pattern.

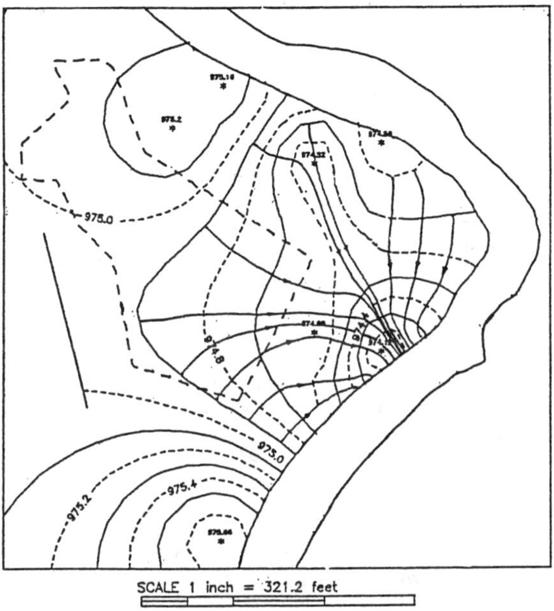

Figure 6. Flownet after the cutoff wall

To verify the effect of both the precipitation and piezometric levels, a simulation was conducted using the highest average precipitation and piezometric levels at the site for the period 1979-1990 and taking into consideration the cutoff wall. From the flownet produced, the flow pattern was the same as those of the period beyond 1986. Further, the arsenic loading to the river escalated significantly confirming that arsenic loading to the river is a function of both precipitation and piezometric level.

Conclusion and Recommendations

Given the above analysis it can be concluded that the reduction in concentration of contaminants in the aforementioned wells and in the Cedar River are not as a result of the cutoff wall. Instead, they were more likely as a result of combined effects of the clay cap, declining precipitation, the corresponding declining piezometric pressure in the deep aquifer and possibly to some extent depletion of the waste. This theory can only be tested when precipitation and the piezometric attain the levels prior to 1983.

Taking into account the hydrology of the site, the effectiveness of the upgradient cutoff wall should be investigated. Detailed field tests are recommended to determine if there is any seepage under the cutoff wall. If seepage was confirmed, the quantity of seepage and contaminant levels should be evaluated.

APPENDIX I- References

CH2M HILL. (1988). "Upgradient Cutoff Wall Evaluation Report, LaBounty Landfill Site, Charles City, IA." EPA Contract No. 68-01-7251, April 1988.

Conestoga-Rovers & Assoc., Ltd. (1984) "Focused Feasibility Study, LaBounty Site, Charles City, Iowa." Final Draf. Nov. 9, 1984.

Freeze, R.A. and Cherry, J.A. (1979). "Groundwater." Prentice-Hall, Inc., New York, N.Y.

Hickok and Associates. (1977). " Waste Characteristics, LaBounty Site." Eugene A. Hickok and Associates, Wayzata, MN.

McCandless, R.M. and Bodocsi, A. (1987). "Investigation of Slurry Cutoff Wall Design and Construction Methods for Containing Hazardous Wastes." EPA/600/S2-87/063.

Munter, J. A. (1980). "Evaluation of the Extent of Hazardous Waste Contamination in the Charles City Area." Iowa Geological Survey Contract Report, Iowa Geological Survey, Iowa City, Iowa.

Sisk, S.W. (1982). "The Effects of Capping on Leachate Production at the LaBounty Site, Charles City, Iowa." EPA-330/1-82-002, National Enforcement Investigations Center, Denver, Colorado.

The Use of Dynamic Compaction To Consolidate Nuclear Waste

by
Cliff Schexnayder[1] and Robert G. Lukas[2]

Abstract

This paper discusses the use of dynamic compaction to densify waste materials that had become contaminated with low level beta gamma radiation. The waste had been buried in trenches and covered with soil. Dynamic compaction densification compressed the nuclear waste fill approximately 12.7%. Cone penetration resistance readings were approximately 4 to 6 times higher after compaction. The type of equipment used for dynamic compaction, a contractor's test program and safety procedures used during the dynamic compaction operations are discussed in this paper. Environmental considerations are excluded from the paper.

Introduction

The Department of Energy's Savannah River Site, in South Carolina, operates a burial ground for disposal of nuclear waste. Fifty-eight acres of the burial ground contained waste which had been placed from before 1976 until 1986. This section of the facility was consolidated by dynamic compaction as part of a permanent closure program in 1989 and 1990.

The waste is classified as low level to intermediate level beta gamma waste. It consists of random materials that had been exposed to nuclear radiation, including clothing, building materials, metal vessels, pipes, construction equipment, and

----

[1]Chief Engineer, Nello L. Teer Company, P. O. Box 1131, Durham, North Carolina 27702, USA

[2]Senior Principal Engineer, STS Consultants, Ltd., 111 Pfingsten Rd., Northbrook, Illinois 60062, USA

fluids such as oil that were mixed with absorbent substances and placed in 55 gallon drums. In some areas, the nuclear wastes were placed in large metal boxes, known on site as B25 boxes. These boxes are similar to connex containers.

During operation of the burial ground, most of the wastes had been deposited in a series of parallel trenches which were 20 feet wide by 20 feet deep (6m x 6m). Each trench was separated by a 10 to 20 foot (3m to 6m) berm of natural soil. After either the loose mixed waste or the B25 boxes filled the lower 16 feet (4.9m) of a trench, four feet (1.2m) of a sandy silt material was dumped and spread as an initial closure cap. No effort was made to compact the waste in the trench or the initial soil cap.

It was observed that the soil cap which had been shaped to shed surface water was settling and water was accumulating in low spots. This was considered undesirable since there was the likelihood of surface water seeping through the cap, becoming contaminated from the nuclear deposits and then percolating downward to the groundwater table. As part of the permanent closure plan, it was decided to densify the nuclear waste to reduce future settlement. This densification of the waste was the first step before constructing a new impervious kaolin clay cap. Dynamic compaction was selected as the method for densification of the nuclear wastes.

In-Situ Improvement by Dynamic Compaction

Dynamic compaction is the process of dropping heavy tampers, typically in the 6 to 30 ton (5.5 to 27.3 mt) range from heights varying from 30 to 100 ft. (9.2m to 30.5m). The tamper is raised and dropped by a single cable with a free spool which typically results in an energy lost of about 12% due to drum and sheave friction. The high impact energy which is imparted to the soil causes deep densification. Dynamic compaction has been described in numerous technical papers (Charles et al, 1981; Leonards et al, 1980; Lukas, 1980, 1985, 1986; Mayne et al, 1984; Menard and Broise, 1975). The advantages and disadvantages of dynamic compaction are outlined by Lukas (1986) in the FHWA study. The process is ideal for compaction of nuclear waste for several reasons.

1. Densification of the buried nuclear waste takes place from the existing ground surface without exposure to the waste. This is a critical requirement for the safety of

construction personnel.

2. The weight of the tamper and the drop height can be adjusted to insure that compaction is obtained to the depth and degree necessary. In general, the heavier the tamper and the higher the drop height, the greater the depth and degree of compaction.

3. Wherever resistant materials are encountered below grade, additional energy can be applied to crush drums or displace large objects, thereby collapsing potential voids within the waste.

4. Dynamic compaction is generally the most economical site improvement process and for deep densification, it is one of the safest. Other methods such as stone columns would have had to penetrate the nuclear waste and thereby expose construction personnel to radiation. Excavation followed by recompaction with conventional compaction equipment would, also, have led to unnecessary exposure.

Dynamic Compaction Operations

A Lampson LDC-350 machine was used on this project, Fig. 1. This machine was developed specifically for dynamic compaction work. The upper works are fixed to the undercarriage on the LDC-350. This eliminates the problem that conventional full-revolving lifting machines have with the severe stress to the turntable due to the rocking motions when the tamper is raised and released. However, in order to turn, the machine must use its tracks. Each track runs off its own motor. While this does slow the operation of positioning a tamper on a specific location, machine availability more than makes up for the time loss.

There were 13,002 drop point locations for this project. A 20 foot (6.1m) wide trench surface was subdivided into 10 ft. x 10 ft. (3m x 3m) grids designated primary and secondary grid drop locations. All primary craters within a work area were compacted and backfilled prior to dynamic compaction of the secondary grid drop locations, Fig. 2. To facilitate tamper recovery, the tamper remained attached to the crane cable during all drops. The tamper had a weight of 20 tons (18.2 mt), was 8 feet (2.4m) in diameter and had a flat bottom. A drop height of 42 feet (12.8m) was called for by specification.

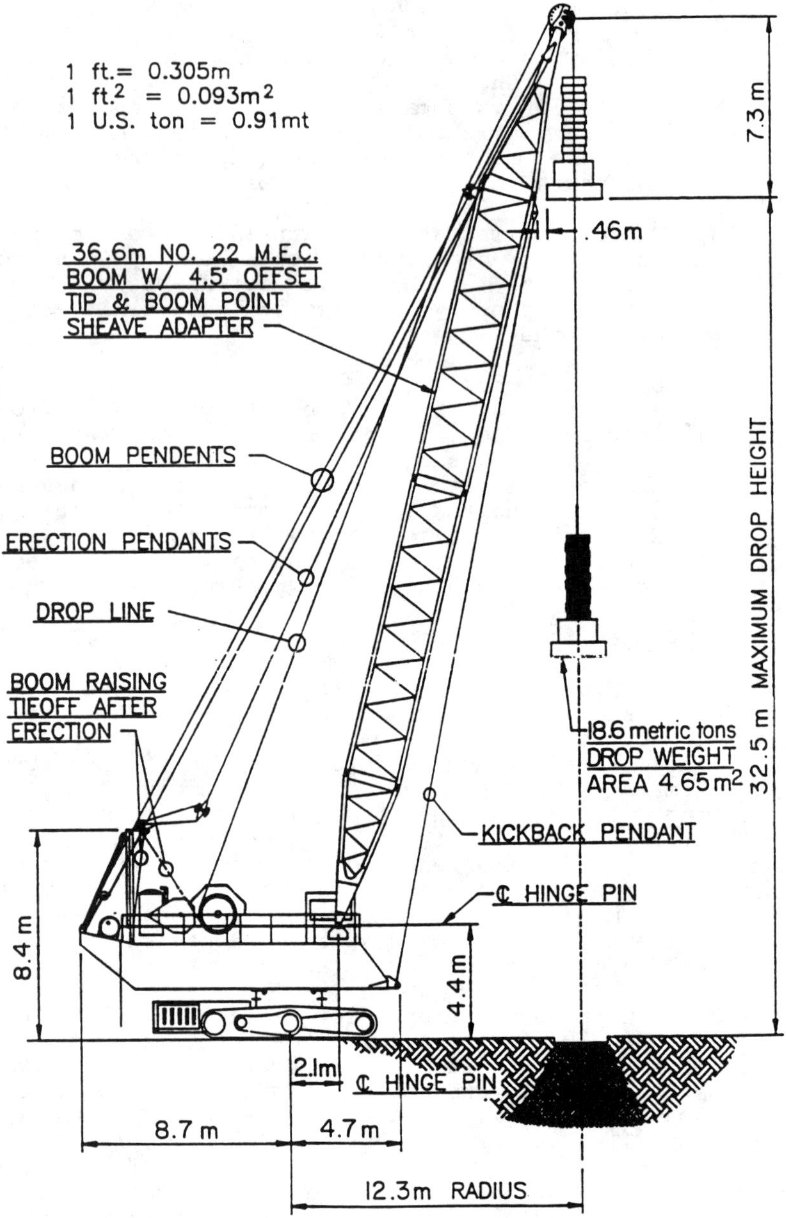

FIG. 1 - Lampson LDC-350 Dynamic Compactor

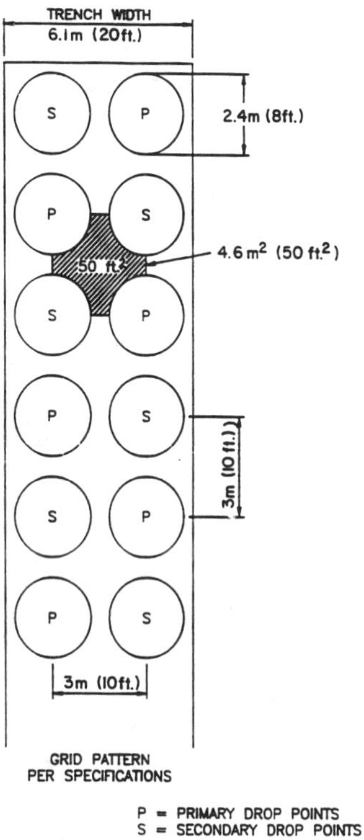

FIG. 2 - Dynamic Compaction Weight Drop Pattern

Prior to any waste trench being treated by dynamic compaction, a 2 ft. (0.6m) thick soil blanket was placed without compaction over the existing cap. As most burial trenches had 4 ft. (1.2m) of initial earth cover, there was effectively 6 ft. (1.8m) of soil cover during dynamic compaction operations. The specifications stipulated 20 drops at each drop point, or until a maximum crater depth of 6 ft. (1.8m) was achieved. This maximum depth was specified for safety reasons, in order not to penetrate the clean cover material and encounter the radioactive waste.

Three LDC-350 machines were used on the project. Two machines worked for 12 months while the third machine was utilized for 7 months. The total duration for dynamic compaction work was 357 calendar days. Of the total work time available, 86.4 percent was productive. The total number of tamper drops was 246,975.

A Spectra-Physics El-1 electronic level laser system was used to check the final depth of every crater, Fig. 3. The average incremental crater depth

FIG. 3 - Lampson LDC-350 Dynamic Compaction Savannah River Nuclear Plant, 1989

per tamper drop from a height of 42 feet (12.8m) was 0.5 feet (0.15m) for primary craters and 0.4 feet (0.12m) for secondary craters. This difference was to be expected because the construction sequence created a stiffer matrix around secondary craters. Some voids were encountered, resulting in incremental depths as high as 1.73 feet (0.5m) per blow. Achieving an average incremental depth of 0.45 feet (0.14m) per blow meant that on the average, only 13 drops, instead of the 20 envisioned by the specifications, were required to drive the craters. The resulting average crater depth for the project was 5.56 feet (1.7m) for an average ground compression of 12.7 percent.

The craters could easily have been driver deeper but as a safety measure, driving was stopped when the depth reached or passed the 5.5 foot (1.7m) mark. This policy was instituted when it was realized that the incremental depths being experienced were close to one-half foot (0.15m). An additional blow after the 5.5

foot (1.7m) mark was reached would have almost certainly meant contact with the nuclear waste at the 6.0 foot (1.8m) depth.

Contractor's Test Program

After the start of dynamic compaction using the specified 20-ton (18.2 mt) tamper with a 42 foot (12.8m) drop height, it became apparent that economics and speed of operation could be achieved utilizing the greater capability of the Lampson LDC-350. Permission was granted to conduct a dynamic compaction test program in the burial ground. Because the 20-ton tamper (18.2mt) had already been constructed and was in use on the project, no variation in tamper weight was attempted.

Three 200 foot (61m) long test sections were proposed. Two test sections were situated in the low level beta trenches and one section in an intermediate level beta trench. The first drop at each location was from a height of 42 feet (12.8m) to confirm that there was not a loose layer below an upper crust. For the next drop, the height was varied as shown on Table 1.

TABLE 1. Drop Sequence for Contractor Dynamic Test Program

DROP HEIGHT WITH 20-TON WEIGHT FOR ALL TEST SECTIONS

NUMBER OF DROP POINTS (1)	1ST DROP (ft) (2)	2ND DROP (ft) (3)	REMAINING DROP HEIGHT (ft) UNTIL CRATER DEPTH OF 6 FT (4)
4	42	42	50
4	42	42	60
4	42	42	70
4	42	42	80
4	42	--	60
4	42	--	70
4	42	--	80
4	42	REPEAT THE MOST PROMISING DROP HEIGHT	
4	42		
4	42		

If the difference in crater depth between the first and second drop was less than 1 foot (0.31m), then additional drops were undertaken from the height specified in the test program until such time as the crater depth reached 6 feet (1.8m). The reason for using the higher drop heights was to achieve the compression as quickly as possible with the least number of drops.

If, after any individual impact the crater depth was more than 1 foot (0.3m) deeper than the previous depth, the drop height was maintained at 42 feet (12.8m) until the incremental crater depth was less than 1 foot (0.3m) per drop.

Safety was maintained during the program by:

1. Using the reduced drop height during the initial tamping to confirm there was no weak spot directly below an upper stiff layer.

2. Using an incremental crater depth measurement of 1 foot (0.3m) maximum per drop as an indicator for reducing the drop height.

3. Limiting the crater depth to 6 feet (1.8m).

4. Measuring for nuclear emissions at all times with air monitors and wipe tests on the tamper.

Monitoring During the Test Sections

Monitoring undertaken during the test program consisted of the following:

1. The depth of crater was measured following each drop.

2. The volume of the crater was determined from diameter measurement at the ground surface and using the known diameter of the tamper for the bottom of the crater measurement. The volume of the crater was determined for each drop.

3. Long spikes were driven into the ground adjacent to the craters from which ground elevations were obtained to determine if heave of the adjacent land mass was occurring. Heave was compared with the volume measurements obtained under Step 2 which was an indicator of how effective each drop was in compacting the soil.

4. The time taken to complete each drop sequence was monitored to determine the most efficient dynamic compaction procedure.

Test Sections D-4, D-5 and E-10

The three locations within the nuclear waste site selected as test sections were within trenches D-4, D-5, and E-10. Sections D-4, low level, and E-10, intermediate level, represent areas which contain random nuclear contaminated debris, either dumped loosely into the trenches or in cardboard boxes placed within the trenches. Test Section D-5, low level, was filled with metal B25 boxes containing nuclear contaminated materials.

At all three test sections, the drop pattern was undertaken as shown in Fig. 1. The first pass through a trench was undertaken at primary drop locations until a crater depth of 5.5 feet (1.7m) was reached. The second pass was undertaken at the secondary drop locations. For the initial drop points on both the primary and secondary pass, the first two drops of the weight were both from 42 feet (12.8m) after which the following drops were all from the same higher height. It was immediately apparent that the advantage was very slight for the 50 foot (15.3m) height. Therefore, on the second set of tests, the first drop within each test section was undertaken from a height of 42 feet (12.8m), and then the additional drops were undertaken from heights varying from 60 to 80 feet (18.3m to 24.4m).

The number of drops required to reach a crater depth of approximately 5.5 feet (1.7m) at test sections D-4 and D-5 is summarized in Figures 4 and 5. For Test Section D-4, the most efficient method of applying the energy was to use the highest drop height, in this case 80 feet (24.4m). After the initial drop from 42 feet (12.8m), it took only 5.8 additional drops from a height of 80 feet (24.4m) to reach the required crater depth at the primary grid points, and approximately 7.4 drops at the secondary grid points. The numbers 5.8 and 7.4 represent an average for various locations, thereby resulting in something other than a whole number of drops.

The amount of energy applied for the various drop heights is summarized in Tables 2 and 3. It can be seen that approximately the same amount of energy was applied at each grid point, even though the drop height and number of drops varied. At the primary grid point locations, the average energy required to achieve the

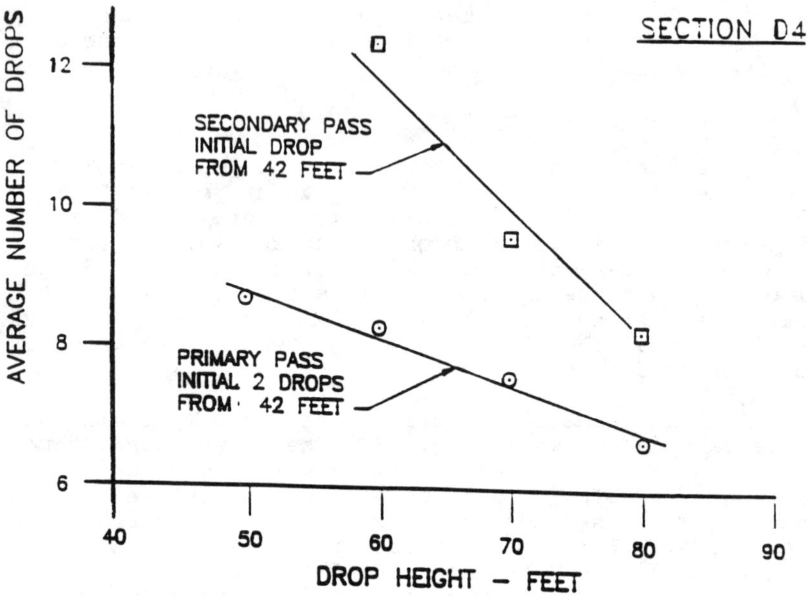

FIG. 4 - Number of drops from varying drop heights to induce a crater depth of 5.5 feet in Section D-4

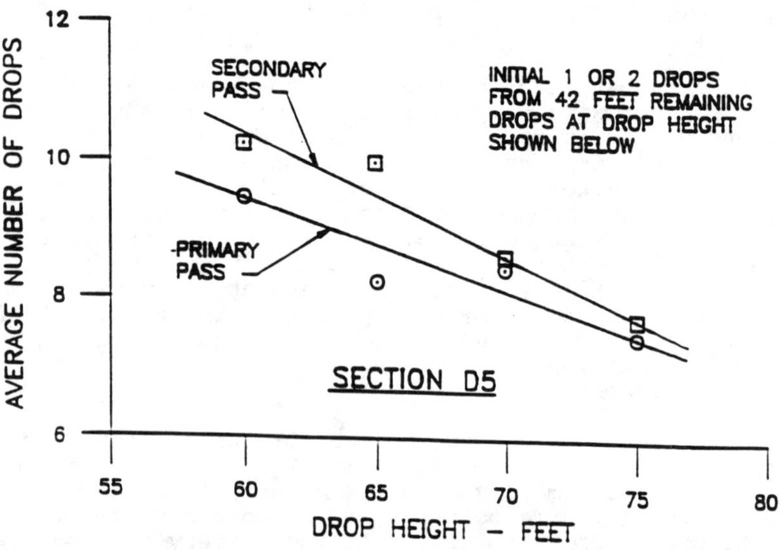

FIG. 5 - Number of drops from varying drop heights to induce a crater depth of 5.5 feet in Section D-5

densification was 10,025 foot-tons (840 m-mt). At the secondary grid points, the average energy required was 12,534 foot-tons (1061 m-mt). More energy was required for the secondary grid points, because some densification took place in these areas during the impacting at the primary grid point locations.

TABLE 2. Energy Requirements to Induce Crater Depth of 5.5 Feet at Primary Drop

Test Section (1)	Drop Height - Feet 50 (2)	60 (3)	65 (4)	70 (5)	75 (6)	80 (7	Average Energy All Drop Heights (8)
D4	8,680	9,276		9,380		9,360	9,174
D5		11,040	11,430	11,060	10,185		10,929
E10		9,792					9,972

Energy - Weight of Tamper x Drop Height x Number of Drops
Energy Units in Table Expressed in Foot-Tons

TABLE 3. Energy Requirements to Induce Crater Depth of 5.5 Feet at Secondary Drop Points

Test Section (1)	Drop Height - Feet 50 (2)	60 (3)	65 (4)	70 (5)	75 (6)	80 (7)	Average Energy All Drop Heights (8)
D4		14,040		13,440		12,680	13,387
D5		11,740	12,310	11,860	11,010		11,530
E10				12,684			12,684

Energy - Weight of Tamper x Drop Height x Number of Drops
Energy Units in Table Expressed in Foot-Tons

Dynamic Cone Penetrometer Measurements

To check on the degree and depth of improvement throughout the full depth of the nuclear waste, dynamic cone penetrometer readings were undertaken. The dynamic cone penetrometer consists of a 2 inch (0.05m) diameter conical cone with a 60 degree cone angle which is connected to a drill rod that is driven by a 140 pound (0.06mt) hammer falling 30 inches (0.76m). This cone is driven into the nuclear waste and the number of blows per foot (0.3m) that are required to advance the penetrometer are recorded. Readings were taken in the nuclear waste before and after dynamic compaction. In all cases, there was a significant increase in the penetration resistance following dynamic compaction. Figure 6 illustrates the range in cone penetrometer values before and after dynamic compaction for two different test sections.

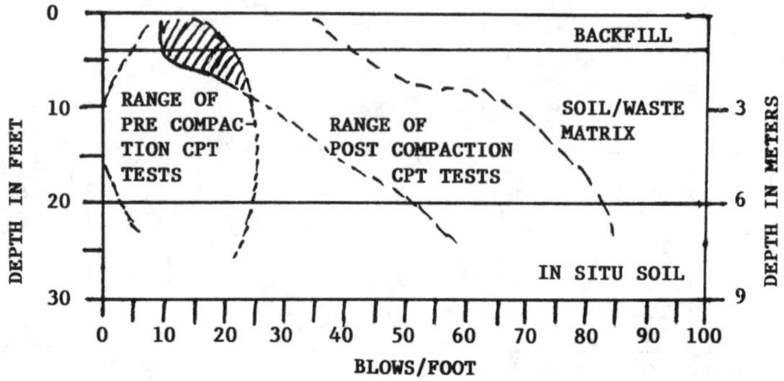

FIG. 6 - Comparison of Cone Penetrometer Test Results

Typically, the cone penetrometer values before dynamic compaction were found to range from about 5 to 20 blows per foot (0.3m) with occasional higher values. It was assumed the high values were the result of encountering large objects within the waste. After dynamic compaction, the typical cone penetrometer resistance values were found to be on the order of 30 to 80 blows per foot (0.3m). The cone penetrometer values showed a relatively uniform penetration record throughout the entire depth of the material, a desirable result.

Conclusions

Based upon the data from the actual dynamic compaction of this 58 acre (4049 sq m) site, it is concluded that:

1. Densification of the nuclear waste was accomplished by the dynamic compaction procedure.

2. The craters formed by the impact of the drop weight averaged 5.56 feet (1.7m) which resulted in an average ground compression of the nuclear waste of approximately 12.7%.

3. Cone penetrometer tests taken before and after dynamic compaction indicated a significant increase in the penetration resistance, thereby confirming the high degree of

densification within the nuclear waste deposit.

4. Safety was maintained at all times by limiting the crater depth to 6.0 feet (1.8m). Measurements were taken with air monitors around each compactor, monitoring of craters before backfill, and wipe tests of the weights during the entire dynamic compaction construction period. These confirmed that radioactive debris was not discharged.

5. The most efficient energy application consisted of applying the energy from the highest drop height, on the order of 70 to 80 feet (21.4m to 24.4m), in the test sections. At these drop heights, incremental crater depths were approaching 1 foot (0.3m), so higher drop heights were not used.

Appendix I - References

1. Charles, J.A., Burford, D. and Watts, K.S. (1981). **"Field Studies of the Effectiveness of Dynamic Consolidation."** Tenth International Conference on Soil Mechanics and Foundation Engineering, Stockholm, Volume 3, 617-622.

2. Leonards, G.A., Cutter, W.A. and Holtz, R.D. (1980). **"Dynamic Compaction of Granular Soils."** Geotech. Engrg. Div., ASCE, 106(4), 35-44.

3. Lukas, R.G. (1980). **"Densification of Loose Deposits by Pounding."** Geotech. Engrg. Div., ASCE, 106(4), 435-446.

4. Lukas, R.G. (1985). **"Densification of a Decomposed Landfill Deposit."** Eleventh International Conference on Soil Mechanics and Foundation Engineering, San Francisco, CA, Vol. 3, 1725-1728.

5. Lukas, R.G. (1986). **Dynamic Compaction for Highway Construction, Vol. 1, Design and Construction Guidelines.** Federal Highway Administration, Report No. FHWA/RD-86/133.

6. Mayne, P.W., Jones, J.S. and Dumas, J.C. (1984). **"Ground Response to Dynamic Compaction.** Geotech. Engrg. Div., ASCE, 110(6), 757-774.

7. Menard, L. and Broise, Y. (1975). **"Theoretical and Practical Aspects of Dynamic Compaction."** Geotechnique.

Ground Improvement of Rubbish Dump Over Reclaimed Tin Mine

Aziz Mustafa[1], MASCE & Mohd Raihan Taha[1]

Abstract

Due to economic reasons and the limited availability of good land, many recent housing schemes in Malaysia have been built on reclaimed tin mining lands. The geotechnical nature of these lands are usually unsuitable for development without prior treatment. This paper describes the construction of a major property development over a waste dumping ground on such land. The features of the project are described and the conditions of the subsurface soil discussed. The general procedure for ground improvement by dynamic compaction and replacement on the project is mentioned. Methods of quality control are also discussed. Measurements are made on the effects of the dynamic compaction and replacement works, and the results discussed.

Introduction

The project involves the construction of 740 units of single and double storey terrace houses, about 860 units of five storey apartments or flats, 62 units of shophouses and over 30 units of industrial buildings and workshops and a petrol filling station over an ex-mining land of about 60 acres, with an estimated cost of over US $ 27 million. Figure 1 shows a general layout of the development scheme. According to the initial plans, the site is to be developed in several phases as follows. Phase I, comprising of 108 units of single storey terrace houses, 102 units of double storey terrace houses and 10 blocks of five-storey apartments, is planned to be completed by the end of 1990. Phase II, comprising of 18 blocks of five storey apartments, a public hall and a petrol filling station is scheduled to be completed by the end of 1991. Phase III, which consists of the development of 182 units of single storey terrace houses and 26 units of shophouses is planned for completion at the end of 1993. Finally, Phase IV, which is scheduled for completion by the end of 1995, consists of the development of 138 units of single storey terrace houses and 14 units of factory buildings. As the construction progressed, some modification had to be made to the overall schedule, and Phase III had to be executed before Phase II. The division of the site into the various phases is shown in Figure 2.

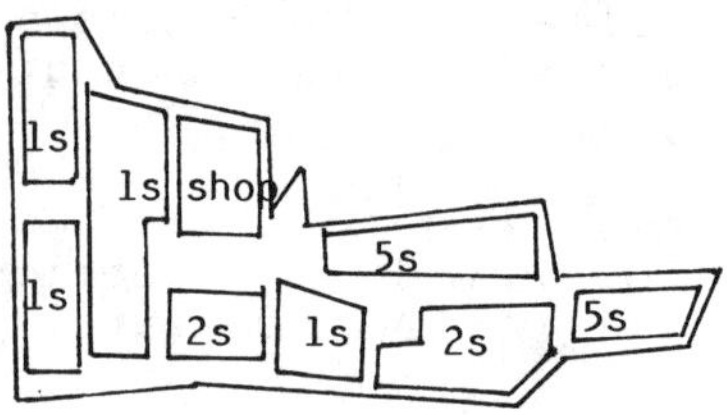

Figure 1

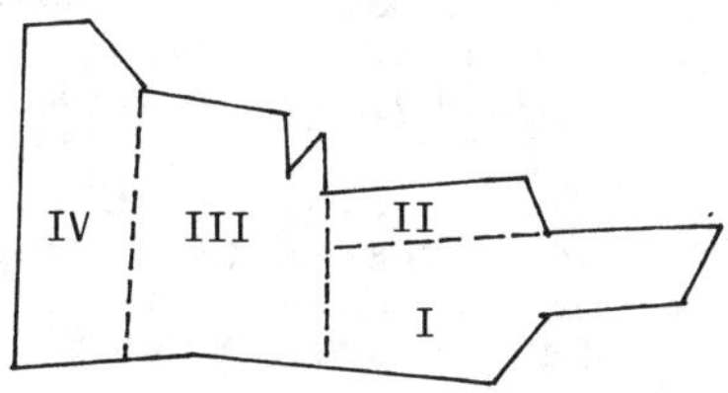

Figure 2.

[1]Lect., Dept. of Civ. & Structural Engrg., Universiti Kebangsaan Malaysia, 43600 Bangi, Malaysia

Site Conditions

According to the geology of Kuala Lumpur (Gorbett, 1964), the site is located at the junction of old limestone and Kenny Hill formation, which is a formation of alternating sequence of sandstone and shale. It was previously a tin mine, and was later used as a rubbish dumping ground. The overburden therefore consists of poorly compacted waste material whilst the subsurface soil is loose, with slime layers, due to the mining process which involved the dredging of the ground in order to extract the ores and backfilling with tailings. The site is also scattered with ponds. Figure 3 shows the topography as well as the locations of the ponds, swamps and rubbish dump.

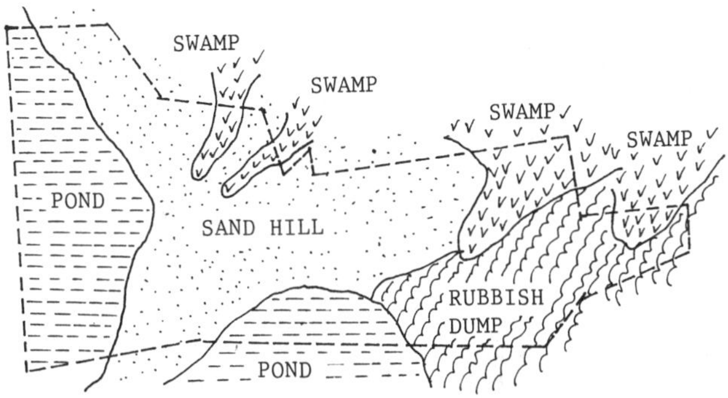

Figure 3.

Two series of soil investigation were conducted. The first consists of two boreholes sunk at the location of Phase I, and the second series of investigations were conducted at the location of Phase III. The logs for Series I are shown in Figure 4. SPT tests results for both series are shown in Figure 5. The SPT curves indicate soft soil with N-values less than 10 down to about 15 metres below existing ground surface. There is, however, a clear local rise in N-values between 0 to 7 m depths, which actually reflects the thickness of the dumped waste material layer. Figure 6 shows the water content from the various boreholes. The plastic limit varies between 25% to 30% and the liquid limit varies between 40% and 55% without any clear trend with depth. The natural moisture content varies between 15% and 45%. From the samples tested, the moisture content exceeds the liquid limit only at two locations, i.e. at 7.5 and 14 m depths with the moisture content - liquid limit values of 46-45 and 43-41 % respectively. These imply soil of very low shear strength, moderately compressible, and low sensitivity. Laboratory testing was also very limited due to the predominantly silty or sandy nature of the soil. Only one u.u. triaxial compression test and one consolidation test were conducted on samples obtained from Series I investigations. The triaxial test was made on a sample obtained at a depth of 21 m in Borehole 1, which indicates an apparent cohesion of 15 kN/m2 and shearing resistance angle of 4o. The consolidation test, which was conducted on a sample from Borehole 2, at a depth of 13.5 m, gives a preconsolidation pressure of 195 kN/m2 and a coefficient of compression (Cc) of 0.33.

The worst subsurface conditions are found at the Phase I area, where a thick layer of dumped waste material overlain the site. Borehole 2 (see Figure 4) which is located in the middle of the proposed area of double-storey terrace houses, indicates that the waste material layer goes down to 8.5 m. This is followed by a layer of loose silty sand down to 16 m, below which is a layer of sandy silt down to 29 m. This sandy silt layer is of medium density between 16 to 18 m, looser density between 18 and 26 m, and very dense between 26 and 29 m below ground level. Limestone formation is found at about 29 m below the ground surface. Borehole 1, located over the proposed

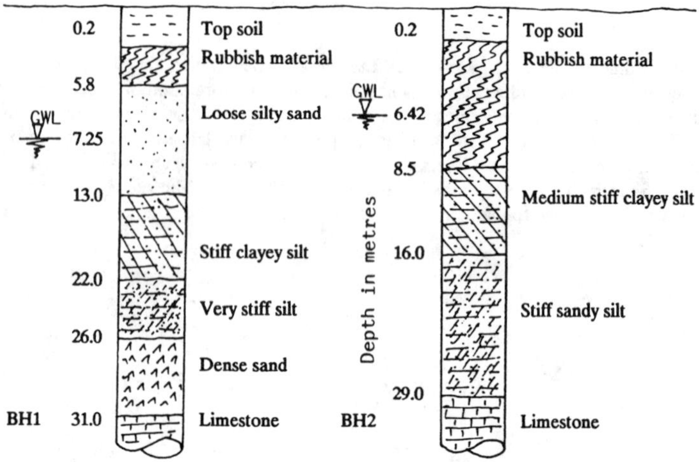

Figure 4 Borehole Logs

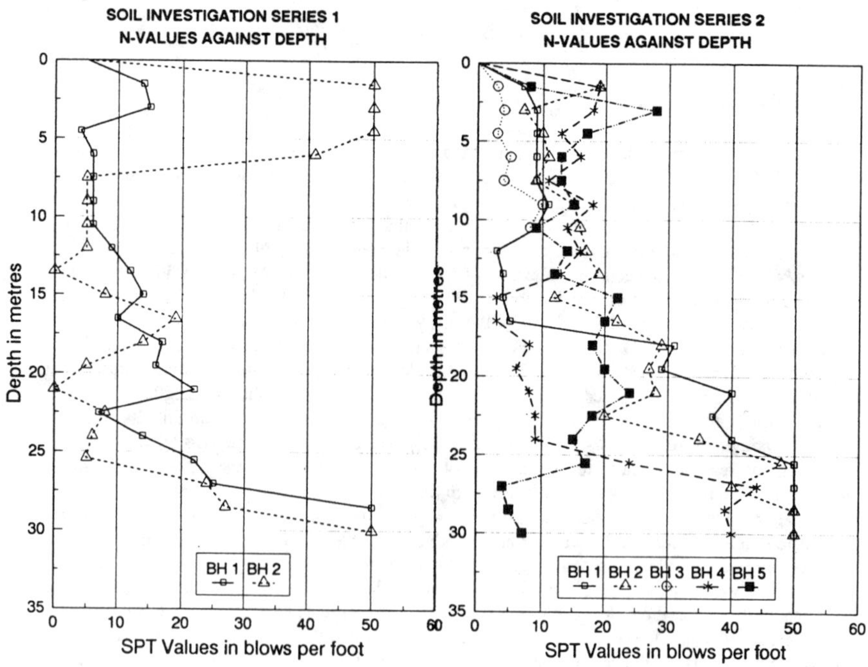

Figure 5. Standard Penetration Tests Results

single single storey terrace houses, indicates similar subsurface profile, with a slightly thinner layer of top waste material layer of about 6 m. The ground water level was measured to be 7.25 and 6.42 m below ground surface in Boreholes 1 and 2 respectively during the site investigation.

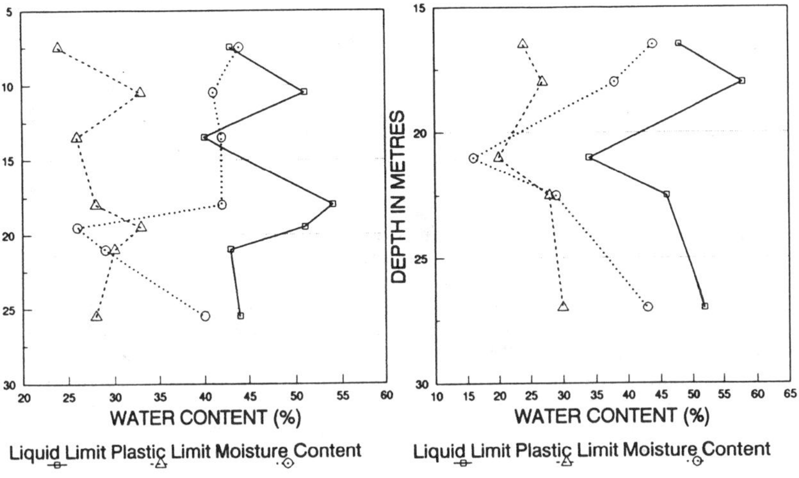

(a) Series I (b) Series II

Figure 6. Water Contents

Ground Improvement

Although limestone formation around Kuala Lumpur has been found to be very karstic and cavitatious at certain locations, in the present case the subsurface investigations indicate a reasonably flat formation surface profile with a dense overlying layer of sand over the limestone. Hence, no major difficulty is anticipated from this limestone formation. Nearer the surface, however, preliminary calculations indicate the existing conditions of the subsoil at Phase I to be too poor to support the proposed structures over it. Some form of ground improvement work will be necessary. The area where two storey houses are to be built, i.e. that around Borehole 2, was subjected to dynamic consolidation and dynamic replacement.

Specifying the dynamic consolidation works:

For this project, only general performance requirements were specified. The detailed procedure and design for the dynamic compaction and replacement works were prepared by the specialist contractor. The performance specifications require a final bearing capacity of 120 kN/m2 and a final differential settlement of less than 1/600 for the buildings to be constructed.

Construction sequence:

The sequence of construction is as follows:

i. Top 3.5 m of existing soil comprising of mainly dumped waste material to be removed and replaced with selected sand fill.

ii. Carry out dynamic consolidation over entire site and dynamic replacement over building site.

iii. Continuous monitoring of works during dynamic consolidation and dynamic replacement.
iv. Conduct tests to assess the performance of the ground improvement works.
v. Complete sand fill to final levels compacted in layers to 95% MOD. PROCTOR by conventional earthmoving.
vi. 2 m surcharge for minimum 6 weeks and installation of settlement plates.
vii. Monitoring of settlement and removal of surcharge.

Dynamic consolidation is basically a process of ground improvement where a poor soft soil layer is compacted by impacts caused by dropping weights from a height. The amount of energy imparted depends on the height (which may range between 8 to 40 m) as well as the weight (which may range between 100 to 400 kN) of the tamping device. The effect also depends on the nature of the soil, whether saturated or otherwise, depth of the different soil layers, and the permeability of the soil. Dynamic replacement is a further development of dynamic consolidation, which is commonly used to improve the bearing capacity of highly compressible soils (swamps, soft clays, etc) so that it can support light structures without rupture. Principally the method consists of the formation of large diameter sand or stone pillars (not unlike stone columns by vibroreplacement) through the soft soil layer down to firm formation level by dropping a heavy pounder on previously laid sand on the soft ground, and continuously replacing this sand when it has been pounded into the ground. For this project, a 140 kN pounder with a cross section of 1.85 m x 1.85 m was used with a height of drop of 30 m. This was done by using a 1650 kN crawler crane. The compaction was executed in three phases. The first two phases of compaction were made by making prints spaced at 6.7 m (22 ft). The third phase, which is actuallly the ironing phase, was done with prints located between the prints made during the first two phases.

Quality control:

Prior to actual execution of the compaction works, heave tests were conducted to confirm the suitability of the designed impact energy, print grids and number of tamping phases to be performed. These consisted of dropping repeatedly the pounder in one print and measuring the resulting crater dimensions and the heave around it after each below. This will allow the level of energy associated with the condition when neglible further reduction in volume can be achieved to be determined by plotting the volume lost against number of blows.

For monitoring the dynamic consolidation/dynamic replacement work the quick penetration test was used. This simply consisted of measuring the penetration of the pounder while pounding and plotting the penetration against the number of blows. When the rate of densification has decreased significantly, the print location was moved to its next position.

Performance of ground treatment:

To ensure that the specifications are met, the effects of the dynamic consolidation and dynamic replacement were continuously monitored using Menard pressuremeter and Gouda Dutch-Cone tests, and the work adjusted progressively during construction. These tests were done at several stages of construction, i.e. prior to the actual execution of compaction but after the top layer of rubbish has been removed and replaced with sand, during, and after the completion of the treatment work. Typical results of the tests are summarised in Figure 7 and 8. From the figures it can be seen that a definite increase in limit pressure and pressure modulus for the top 6 m of soil has been achieved at almost all locations. It is also interesting to note that the maximum effect occurs at about 2 to 4 m below the ground surface and not near the surface. This is because the topmost layers are very much disturbed by the compaction which created an uneven cratered surface. The final levelling of this uneven surface involved profiling to a flat smooth surface by light, closely spaced tamping action only, a process which is sometime termed "ironing". Furthermore it can be observed that hardly any improvement is achieved beyond 7 m below ground surface. This coincides with the depth of the ground water as measured during the site investigation, and a possible relationship between the presence of ground water and effect of compaction may be inferred.

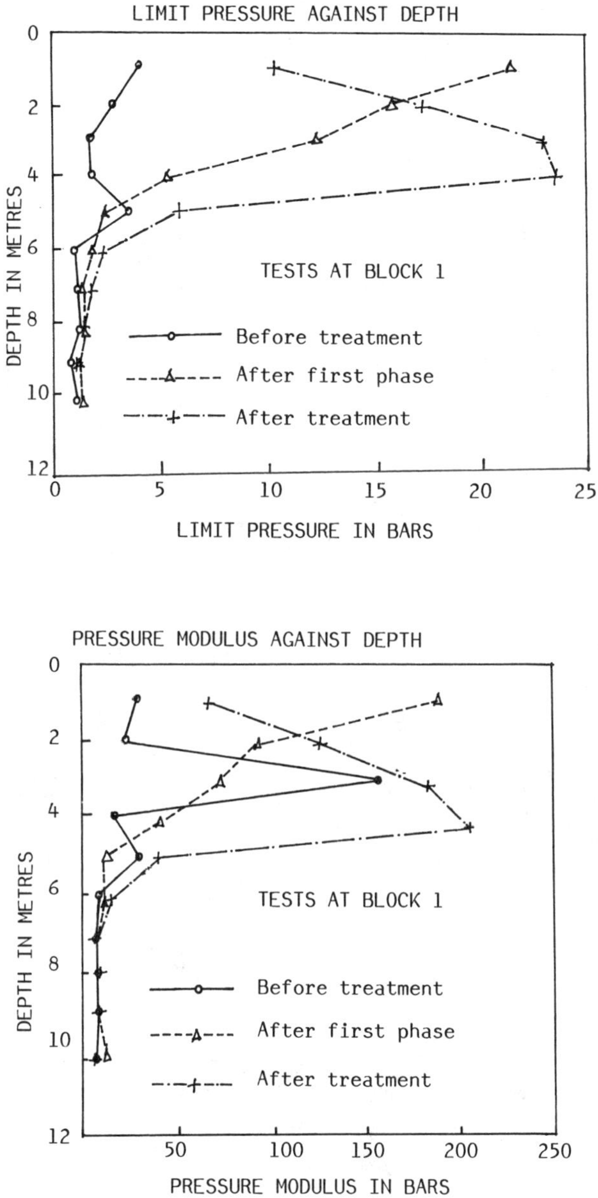

Figure 7. Limit pressure and pressure modulus before and after treatment

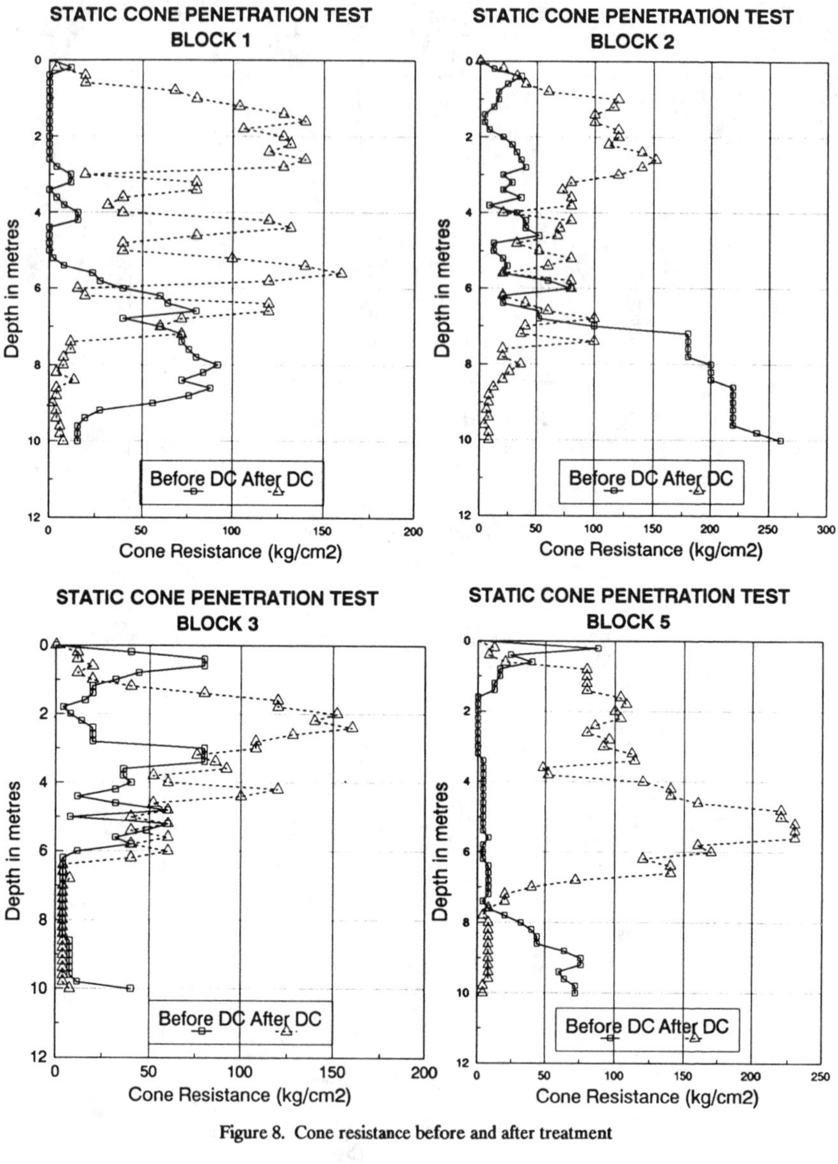

Figure 8. Cone resistance before and after treatment

Settlement:

On completion of the ground treatment work, the future elastic settlement over various locations at the site were predicted using Schmertmann's (1970) method. Generally, this involves the use of the following equation:

$$\Delta H = C_1 C_2 \Delta P \sum_{0}^{2B} \frac{I_z \; Z}{E_d}$$

where

- H = settlement
- C_1 = depth correction factor
- C_2 = creep factor
- ΔP = net increase of load on soil at foundation level due to the applied loading
- B = width of loaded area
- I_z = vertical influence factor at the centre of soil layer
- E_d = deformation modulus
- ΔZ = thickness of soil layer

Based on the above equation, the settlements at the various blocks were estimated to range between 8 to 15 mm, which are well within the acceptable limits.

Conclusion

The project shows that ground improvement by dynamic consolidation and dynamic replacement can be used successfully to improve the bearing capacity and settlement characteristic of dumped waste material overlying soil of silty or sandy nature. It is also noted that the effect of compaction is limited in depth. Although light structres such as those involved in the project may be founded on shallow foundations after treatment, the same may not be true for heavier structures.

Acknowledgement

The assessment of data and preparation of this paper, have partly been made at Universiti Kebangsaan Malaysia, to whom the authors are greatly indebted. Thanks are also due to Menard Soiltraitment, Lankhorst Pancabumi and Jurutera Perunding ZAABA for the providing the data contained in the paper.

References

Gobbett, D.J. (1964)."The Lower Palaezoic Rocks ok Kuala Lumpur Malaysia", Federation Muzeum's Journal, Vol. 9, pp. 67-79.

Menard Soltraitment S.A. France,"D.60.AN - The Menard Pressuremeter-Interpretation of Pressuremeter Test Results to Foundation Design".

Schmertmann, J.H. (1970)."Static Cone to Compute Settlement Over Sand," JSMFD, ASCE, Vol 96, SM 3, May, pp. 1011-1043.

Grouting for Hazardous Waste Site Remediation at Necco Park, Niagara Falls, New York

K.D. Weaver[1], Aff. ASCE, R.M Coad[2], M. ASCE and K. R. McIntosh[3]

Abstract

A single-line grout curtain was constructed along three sides of a 10 hectare industrial landfill located on dolomitic bedrock in Niagara Falls, New York. The objective of this curtain was to reduce the volume of underflow that was being removed by down-gradient extraction wells in order to prevent off-site migration of organic contaminants dissolved from materials that had been placed in the landfill during its 40 year operating life. Construction of the grout curtain was preceded by a grout testing program. Conduct of the grouting operations was complicated by the need to employ special safety precautions related to drilling in contaminated materials, and a need to avoid creating a ground water mound that might rise into an up-gradient landfill. Preliminary results of geohydrologic monitoring subsequent to completion of the grouting operations indicate that the grout curtain is functioning as designed and that the efficiency of ground water recovery operations has substantially improved.

Introduction

Necco Park is a 10 hectare landfill site formerly used for disposal of industrial process wastes. These wastes included fly ash; sodium salts; barium, calcium and sodium chlorides; various chlorocarbons, and other organic and inorganic wastes. Landfilling operations permanently ceased in 1977, when evidence of ground water contamination was discovered. A series of hydrogeologic and water quality investigations was performed, and a

---

[1] Grouting Spec., Woodward-Clyde Consultants, 500 12th Street, Suite 100, Oakland, CA 94607

[2] Sr. Proj. Eng., Woodward-Clyde Consultants, 5120 Butler Pike, Plymouth Meeting, PA 19462

[3] Associate, Woodward-Clyde Consultants, 3571 Niagara Falls Blvd., North Tonawanda, NY 14120

ground water recovery system was voluntarily implemented by the owner in 1982.

The recovery system consists of continuously pumping recovery wells immediately down-gradient from the site. The purpose of this system is to hydraulically isolate the landfill from the regional ground water regime, and to remove contaminated ground water for treatment. Following studies of the effects of this system, it was concluded that construction of an upgradient grout curtain would increase the effectiveness of the hydraulic control in reducing offsite ground water flow, and would minimize the ground water withdrawal rate required to achieve that objective.

Site Characterization

Geology. The overburden at Necco Park is composed of fill, glacial clays, and glacial tills. The overburden thickness ranges from 1.5 m to about 9 m. In general, it is thinnest along the margins of the site. The overburden is underlain by dolomite of the Lockport Formation, which is about 40 m thick. A series of nearly horizontal bedding plane fractures in the dolomite is believed to be the principal geologic control on ground water flow and contaminant migration in the project area. Two nearly vertical joint sets and a high-angle lineament are present beneath the site, providing some degree of interconnection between the bedding plane fractures. The unpublished results of regional studies by the U.S. Geological Survey indicate that the greatest dissolutioning along the vertical fractures occurs within the top 3 to 8 m of the bedrock. Minor dissolutioning also has occurred along some of the bedding plane fractures. The Lockport Dolomite is underlain by the Rochester Shale, which has a very low permeability and represents the limit of downward migration of contaminants.

Geohydrology. An alphabetical designation system was used for the geohydrologic studies; it included eight major zones: A through G, and J. (Minor, discontinuous fractures were given two-letter designations indicative of their relative positions.) The saturated thickness of the overburden was designated the A-zone. Fracture zones B through F, located in the upper member of the Lockport Formation, and the G zone, made up of the lower members of that formation, are areally extensive across the site, and are the major water-bearing features. The G-zone and the J-zone, which forms the contact between the Lockport Formation and the Rochester Shale, were not found to contain high levels of contaminants. Permeability ranges of Zones A through F are shown on Figure 1.

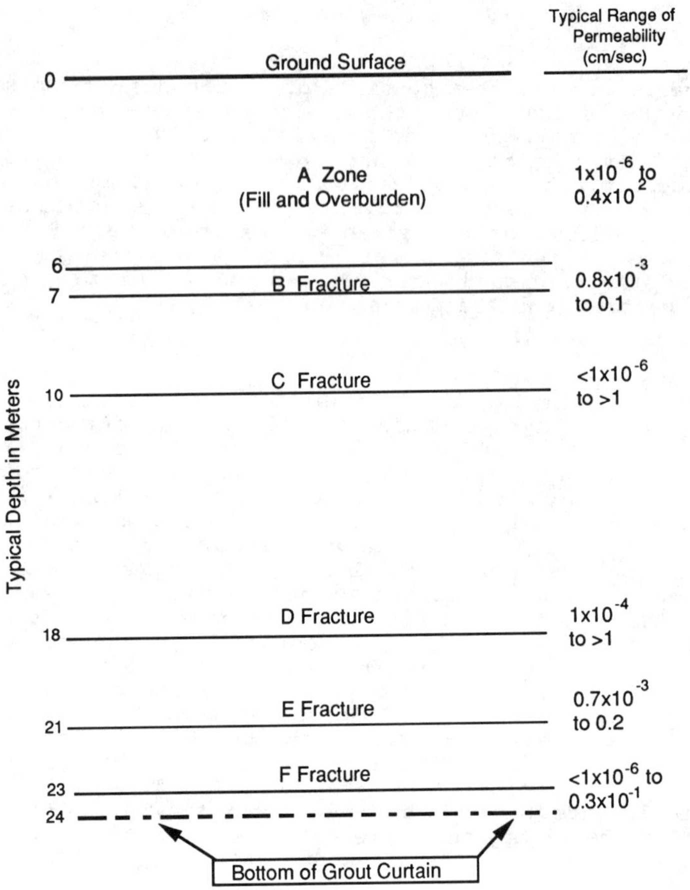

Figure 1. Permeability ranges of bedrock zones, from slug test data.

Geohydrologic studies identified a hydraulic sink associated with the aforementioned lineament. Vertical fractures associated with this lineament contribute to a high degree of hydraulic communication between the B and C zones. Ground water contamination levels exhibit a decrease in the lower fracture zones, suggesting a lesser degree of interconnection between the zones.

<u>Grouting Program Design Considerations</u>

<u>Groundwater Flow Simulation</u>. The U.S.G.S. finite-difference modular ground water flow model was used to simulate ground water flow at Necco Park, and to simulate remedial conditions. These conditions included two configurations of grout curtains, various curtain permeabilities, and various recovery well pumping rates. Pump test data were used in conjunction with slug test data in the calculations. It was concluded that offsite ground water flows would be significantly reduced if the grout curtain were to have an effective hydraulic conductivity below approximately 5 x 10^{-5} cm/sec within the fractures. The ground water flow simulation results also indicated that about 1.5 m of ground water mounding would occur in the overburden (A zone) immediately upgradient from the grout curtain if the B zone were grouted. Therefore, as ground water mounding might have an adverse effect on an active landfill adjacent to Necco Park, grouting of the B zone on the up-gradient side of the site was avoided. Based upon the low levels of contamination at depth, it was decided that grouting beneath the F zone was not necessary.

<u>Preliminary Concepts</u>. Various preliminary concepts were formulated, and were discussed with members of a Project Review Board. Consideration initially was given to constructing a three-row grout curtain, consisting of vertical grout holes at some appropriate spacing. However, as international practice has moved toward use of single-row curtains (for dams) and as access to the site will continually be available for such additional remedial work as may prove to be appropriate, the three row concept was discarded. Consideration also was given to construction of a two-row grout curtain consisting of one row of vertical holes and one row of inclined holes. The objective of the inclined holes would have been to intersect and grout vertical joints and fractures. However, as the role of vertical joints and fractures was inferred to be relatively minor, a decision was made to employ inclined holes only for the purpose of verifying the results of the grouting operations within approximately the upper half of the proposed grout curtain.

There was a consensus opinion that grout mixes thinner than 2:1 by weight should not be used, and that relatively high pressures should be employed. However, it was recognized that the potential for hydraulic jacking imposed limitations on the safe grouting pressure. Therefore, the injection pressures ultimately selected represented a compromise between typical U.S. practice and European practice, ranging from 207 kPa at 6 m depth to a maximum of 1200 kPa at 21 m depth. Based upon the direct relationship between injection pressure and the distance of grout penetration documented by Lombardi (1985) and upon conventional practice, it was inferred that the higher pressures used at depth would allow the spacing of the grout holes to increase with depth, and the grout hole pattern was laid out accordingly.

Grout Design and Compatibility Study. A laboratory study was conducted at Bucknell University to evaluate various grout mix designs for potential use at Necco Park (Weaver et. al. 1990). The grout properties of particular interest were viscosity, bleed, permeability (hydraulic conductivity), and compatibility with the chemical environment. Based upon the results of that study, various formulations of Type I portland cement and Class F fly ash, Type V portland cement, and microfine cement were found to be compatible and to have acceptable properties. A supplementary study was done following the initial part of the field work, leading to adoption of the grout mixes listed on Table 1.

Table 1. Grout Mixes Used

Grout Mix Designation	Water: Solids Ratio	Marsh Viscosity	Approximate Bleed
FAC 3*	0.6:1	43 seconds	7.2%
FAC 4**	0.5:1	92 seconds	3.3%
NEAT 1***	0.5:1	41 seconds	3.1%
NEAT 2***	1:1	28 seconds	20%
NEAT 1A***	0.75:1	34 seconds	12.7%
NEAT 1B***	0.66:1	39 seconds	5.6%
MC 500****	1:1	36 seconds	1%
MC 501****	1.5:1	30 seconds	12%
MC 502****	2:1	28 seconds	23%

NOTES:

* FAC 3 contained 3 parts by weight of Class F fly ash to 1 part of Type I portland cement by weight.

** FAC 4 contained 4 parts by weight of Class F fly ash to 1 part of Type I portland cement by weight.

*** All neat cement mixes were formulated with Type V portland cement and 1% Sika Intraplast by weight of cement.

**** All MC mixes were formulated with MC 500 microfine cement and 1% NS 200 dispersant by weight of cement.

Grout Curtain Construction Procedures

Safety Procedures. Due to the potential hazard represented by some of the buried contaminants, drilling and grouting operations at Necco Park were performed using Level C protection (NIOSH 1985). All drilling was done within an exclusionary zone, and all materials returning from the holes were collected for safe offsite disposal. The breathing space was monitored for organic vapors during drilling, and personnel were monitored for symptoms of heat stress.

Drilling Procedures. Grout hole drilling was accomplished through 76 mm steel casings socketed into, and grouted into, bedrock. The casing holes were drilled with a Mobil rotary rig, using hollow stem augers. An Ingersol Rand LM500-C hydraulic rotary percussion rig, employing a water flush, was used for drilling in rock. Drilling proceeded on a conventional split-spacing sequence, starting with primary holes on 12 m centers. The hole spacing was sequentially reduced to 1.5 m as grouting proceeded. Following initial indications that the holes tended to produce water during drilling, it was inferred that circulation loss and consequent clogging of fractures with cuttings would not be a problem. Therefore, essentially all holes were drilled to the full planned depth in a single stage and were grouted by the ascending stage method. All production grout holes were vertical except in part of the northwestern segment of the curtain, where grouting under high-voltage power lines had to be accomplished in holes drilled at various inclinations from closely spaced locations beyond the power lines.

Water Testing Procedures. Grouting of each 3 m interval of each hole was preceded by a water test, conducted at a single pressure for a period of 3 minutes. The test data were used to calculate the permeability in Lugeon units.* These tests served to verify that the packers were properly set, provided a basis for selection of the starting grout mix or for omitting grouting of an interval, and provided a basis for evaluating the need for reduction of the grout hole spacing.

Grouting Procedures. The starting grout mixes for each interval were based upon the permeability data from the water tests. Grout mix changes during injection were based upon the volumes injected and rates of grout take

*A Lugeon unit is defined as a flow of 1 liter per minute per meter of borehole at a pressure of 10 atmospheres. It is equivalent to a permeability of 1.3 x10-5 cm/sec. The test ordinarily is conducted at a pressure lower than 10 atmospheres, and an appropriate correction factor is used in the calculations.

the water tests. Grout mix changes during injection were based upon the volumes injected and rates of grout take (see Figure 2). The grout mixtures were prepared in trailer-mounted, hydraulically-operated grout plants, each of which was equipped with a high-speed colloidal mixer, two agitators, and two progressing cavity pumps. The injection pressures and flow rates were monitored by use of "Grout Minders" that produced strip chart records of those parameters. A pneumatically controlled grout header was available, but did not prove to be effective at slow injection rates. U.S. Bureau of Reclamation refusal criteria were adopted for the project. However, in order to limit the distance of grout travel beyond the curtain, the maximum volume of grout injected in any interval was limited to 11 m3. Grouting of the B fracture was omitted along the up-gradient section of the curtain, so as to avoid undesirable mounding of the ground water.

Quality Assurance Procedures. The grouting supervisor and the grouting inspectors were provided with procedures manuals prior to the start of the work. Grout curtain profiles were developed to present and facilitate the evaluation of water testing and grouting data, and to provide a basis for selection of locations and depths of quaternary holes. Field testing of grout was performed frequently during every shift. All project records were reviewed in detail by the project grouting consultant, on a regular and continuing basis.

Verification Procedures. Verification holes, inclined at an angle of 30% with respect to vertical, were drilled following substantial completion of sections of the grout curtain as the work progressed. Seventeen of these holes were core drilled, and 19 were rotary-percussion drilled with water flush. All were water tested in 3 m ascending stages, and were grouted as necessary in accordance with the procedures used for construction of the grout curtain. The purpose of the verification holes was to check the results of the grouting operations.

Results

Grout Takes. The total grout takes are listed on Table 2. An overall progressive reduction of grout acceptance was experienced as the hole spacing was reduced, as shown on Table 3. This reduction was not

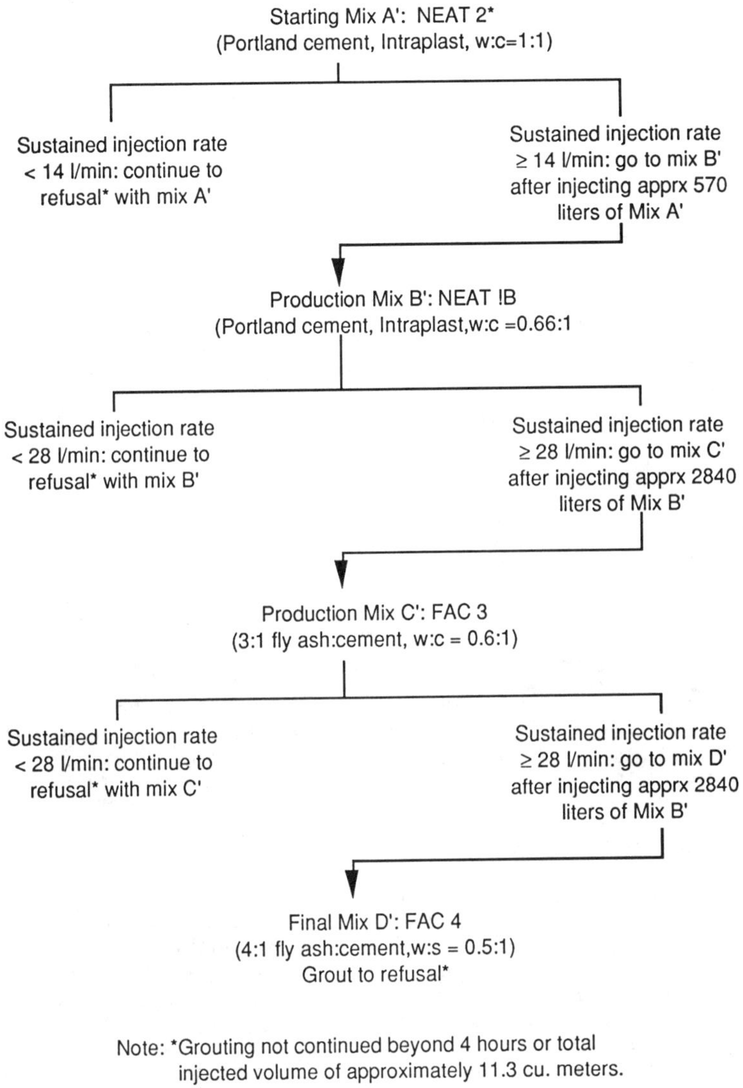

Figure 2. Grout mix flow chart for permeabilities ≥ 50 lugeons

Table 2. Total Solid Materials Injected

Material	Kg Injected
Type I Cement	61,000
Type V Cement	647,000
MC-500 Cement	195,000
Type F Fly Ash	124,000
TOTAL	1,027,000

Table 3. Drilling and Grouting Statistics

Hole Type	Number Drilled	Linear Meters Drilled	Kg of Solids Injected	Grout Take in kg/m
Primary	69	1,600	387,000	237
Secondary	68	1,600	252,000	156
Tertiary	128	2,800	195,000	70
Quaternary	228	4,300	127,000	30
Power Line Inclined "Fan"	36	800	43,000	53
Verification	36	900	23,000	27
TOTAL	565	12,000	1,027,000	86

evenly distributed, as is shown on the histograms presented on Figure 3. The distribution of grout takes with depth varied very substantially from section to section of the grout curtain, with the highest takes shifting from the CD fracture zone in the western part of the curtain to the D and F fracture zones in the northeast part of the curtain. This downward shift corresponded roughly with the location of the lineament in the northeastern part of the site. In general, the grout takes below a depth of 21 m were relatively low.

Permeability Reduction

The water test results demonstrated a sequential and substantial reduction in permeability was achieved by the grouting operations, as shown on Figure 4 and Table 4. No major water-bearing fractures were encountered by the vertical holes in the 60- to 80- foot (18 to 24m) zone, accounting for the apparent anomaly of marginally higher permeabilities in the (inclined) verification holes. A typical example of changes in the distribution of permeability as a result of grouting is shown on Figure 4, which, in general, indicates a progressive reduction in permeability as the spacing is reduced. As shown in that figure, 60% of the intervals had permeabilities in excess of 100 lugeons prior to grouting the primary holes, but only about 7% of the intervals had permeabilities in excess of 100 lugeons prior to grouting the quaternary holes. Anomalies in the distribution of permeabilities

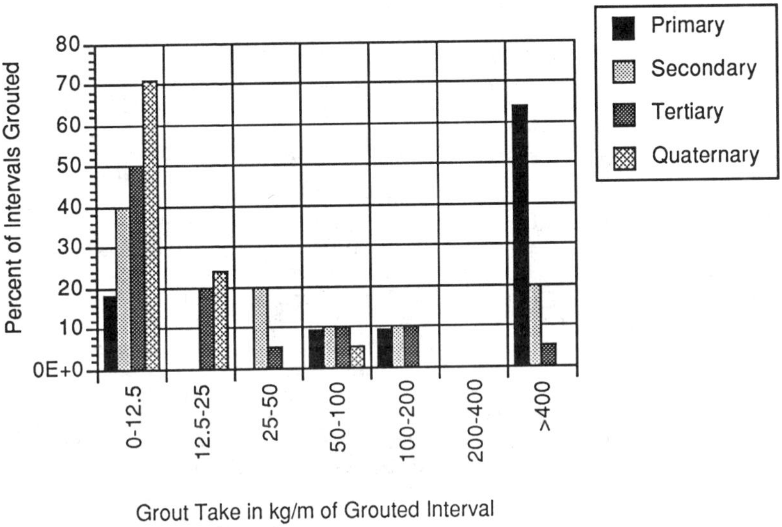

Figure 3. Grout take distribution, west wall, 12 to 15 meters depth

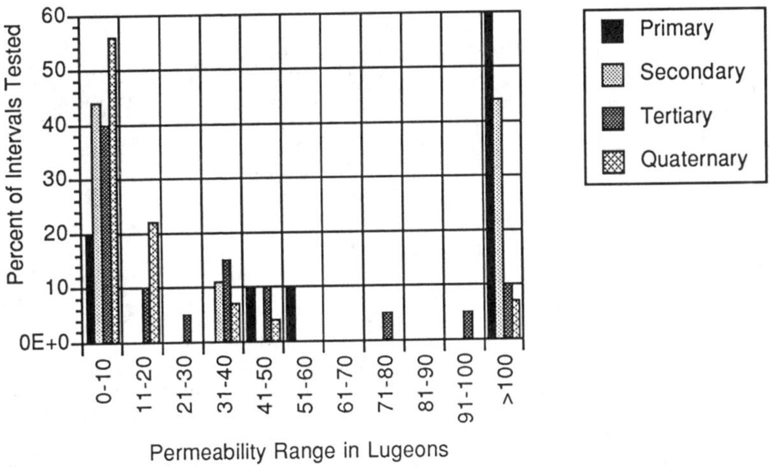

Figure 4. Permeability distribution, west wall, 12 to 15 meters depth

are attributable to the large lateral variations in permeability prior to grouting.

Observed Effects of Grout Curtain

Based upon preliminary results of post-construction ground water monitoring,the grout curtain appears to be functioning as designed. The cones of depression associated with the recovery wells (at pumping rates comparable to pre-curtain conditions) have been significantly enhanced, substantially improving the hydraulic isolation of the site-related ground water contamination. Prior to construction of the grout curtain, capture zones did not appear to extend significantly offsite. Post-construction cones of depression, however, create a capture zone estimated to extend 150 meters or more offsite to the south and west. This is expected to induce some recovery of ground water contamination which has migrated offsite to the south in the past. No perceptible water table mounding has occurred upgradient (north) of the grout curtain. Additional study is planned to verify these conclusions.

Table 4. Permeability Test Results, Western Section of Grout Curtain

Depth Interval in feet (meters)	Fracture Zone	Initial Highest Median Perm	Verif. Median Permeability
10-20′ (3-6)	B/C	190 lugeons	12 lugeons
20-30′ (6-9)	C/D	74 lugeons	1 lugeon
30-40′ (9-12)	-	2 lugeons	1 lugeon
40-50′ (12-15)	D	47 lugeons	2 lugeons
50-60′ (15-18)	E/F	43 lugeons	7 lugeons
60-70′ (18-21)	-	3 lugeons	4 lugeons
70-80′ (21-24)	-	0.1 lugeons	3 lugeons

Acknowledgements

The studies and the grouting operations described herein were performed for E.I. du Pont de Nemours & Company, Inc., which authorized publication of this paper. The grout testing for the project was done by Jeffrey Evans and Stephen Pancoski, of Bucknell University. The grouting program was designed in consultation with a Review Board whose members included: W.S. Gardner, who at the time of the project was Executive Vice President-Practice and Chairman of the Board of Woodward-Clyde Consultants; J.P. Welsh, Vice President, of Hayward Baker, Inc.; and E.R. Colle, who at the time of the project was Vice President of Raymond Builders Division of Raymond International. The grout curtain was constructed by Geo-Con, Inc., of Pittsburgh, Pennsylvania.

References

Lombardi, G. (1985). "The role of cohesion in cement grouting of rock",15th Congress on Large Dams, v.3, 235-261.

NIOSH/OSHA/USCG/EPA. (1985) "Occupational safety and health guidance manual for hazardous waste site activities", U.S. Dep't. of Health and Human Serv., Washington D.C. Table 8-6.

Weaver, K.D., Evans, J.C., and Pancoski, S.E. (1990). "Grout testing for a hazardous waste application", Concrete Int., v.12, no. 7, p 45-47.

Properties of Tire Chips for Lightweight Fill

Dana N. Humphrey, M. ASCE and William P. Manion, S.M. ASCE[1]

Abstract

Tire chips made from waste tires have a compacted dry unit weight of 40 pcf (0.64 Mg/m^3) and a specific gravity of 1.05 making their use as lightweight fill for highway embankments over compressible soils very advantageous. A large scale compressibility apparatus with the capability to measure side friction and horizontal stress was fabricated. Compressibility tests showed that tire chips were very compressible at low stresses but that compressibility decreased significantly at higher stresses. The coefficient of lateral earth pressure at rest was about 0.40 at low stresses and about 0.94 at high stresses. Preliminary finite element studies showed the importance of the thickness and modulus of the subbase overlying the tire chips on pavement deflections.

Introduction

Nearly 300 million tires are discarded in the United States each year creating a major waste disposal problem. Typically these tires are disposed of in huge open piles. This consumes valuable landfill space, creates a fire hazard, is a breeding ground for mosquitos and is an eye-sore. However, several beneficial uses for waste tires have been proposed in the last few years and some have been put into practice. These include: using chipped tires as fuel for boilers at paper mills; using interconnected whole tires as low retaining walls and erosion protection; using finely ground tires as a partial replacement for asphalt in asphaltic concrete; and, the subject of this paper, using chipped or shredded tires as lightweight fill.

There are many benefits to using tire chips as lightweight fill for embankments founded on weak, compressible foundations. Most importantly, settlement is reduced and stability is increased. Tire chips are non-biodegradable, in contrast with alternatives such as wood chips. Tire chips are cheaper in many areas of the country where lightweight aggregate such as expanded shale is not available locally. Tire chips are free draining so there are no problems with build-up of excess pore pressure. Finally, using tire chips for lightweight fill could make an important contribution to putting to a productive use the mountains of discarded tires that this nation produces every year.

---

[1]Assistant Professor and Graduate Research Assistant, respectively, Department of Civil Engineering, 103 Boardman Hall, University of Maine, Orono, Maine 04469.

Tire chips have already been used for lightweight fill on a few road building projects (Geisler, et al., 1989; Palmer and Verdegan, 1989; Edil, et al., 1990; Dodson, 1990). The low unit weight of the tire chips served to limit the settlement of embankments constructed on compressible foundations. The applications were successful, yet to fully integrate tire chips into highway construction the engineering properties of tire chips must be defined.

This paper reports the results of a laboratory and analytical study of the engineering properties of tire chips and how these properties would impact their use in highway construction. Results of basic engineering properties are presented, including gradation, specific gravity, and compacted density. A key question is the effect of a layer of compressible tire chips on pavement deflection. An extensive study of the compressibility of tire chips was made to get the input data needed to address this question. A preliminary analytical study of the effect of a tire chip layer on pavement deflection is presented.

Basic Engineering Properties

The tire chips that are being used for this study are produced by Sawyer Environmental Recovery in Hampden, Maine, from a mixture of steel and glass belted, passenger car and truck tires. The shredding operation produces a chip with a maximum size of 2 in. (5 cm) and the larger pieces of steel belt are removed by a magnet. The results of gradation tests on five samples fall within a narrow range as shown in Fig. 1. It is seen that only 10 percent of the sample is smaller than the 1/2-in. size (12.7 mm) and 50 percent of the sample falls between the 1 and 1-1/2 in. (25.4 and 38.1 mm) sizes. It is likely that tire chips from a different manufacturing process would have a different gradation.

The specific gravity and absorption of tire chips were determined using ASTM C127. Three types of samples were tested: samples of chips made from a mix of steel and glass belted tires, samples manually separated into chips containing only

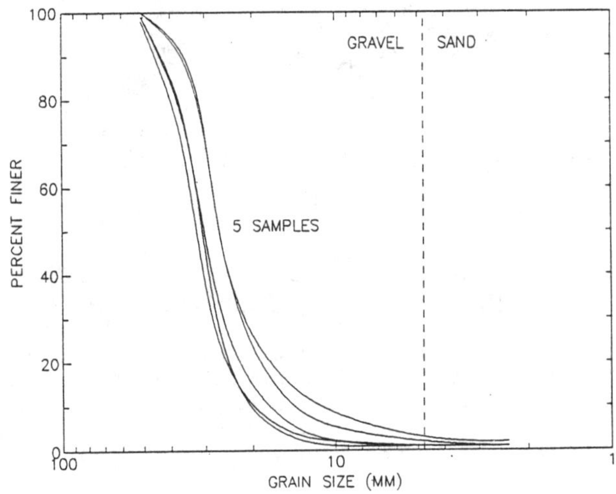

Fig. 1 Gradation of tire chips used in this study.

glass belts, and samples manually separated into chips containing only steel belts. The specific gravity in a saturated surface dry condition was 1.05 for the mix of steel and glass belts, 1.02 for glass belts only, and 1.10 for steel belts only. Thus, the specific gravity is slightly greater than water. The absorption capacity was about 4% for the three types of samples so tire chips absorb only a small quantity of water.

Compacted Unit Weight

The compacted dry density of tire chips were determined using a procedure modeled after ASTM D698 and D1557. To accommodate the 2-in. (5 cm) maximum particle size a 0.44 ft^3. (0.012 m^3) mold with nominal dimensions of 10-in. (25.4 -cm) diameter by 10-in. (25.4-cm) high was utilized. The samples were compacted in 5 layers with a 10-lb (4.54 kg) hammer with an 18-in. (46-cm) drop (i.e., modified Proctor hammer). The number of blows per layer was adjusted to give the compactive effort equivalent to modified Proctor (ASTM D1557), standard Proctor (ASTM D698), and 60% of standard Proctor. The specified compactive effort per unit volume and the number of blows required to achieve this effort are listed in Table 1.

The initial series of tests were done using air dried samples and are summarized in Table 1. It is evident that compactive effort has only a small effect on the resulting dry unit weight. The unit weight for tests run on chips taken from the same sample was 1 pcf greater for the modified effort than for standard or 60% standard effort. So it seems that only a modest compactive effort is needed to compact tire chips. Tests with 60% standard effort were run on two different samples of tire chips. The dry unit weights were 38 and 40 pcf. Thus sample variability has a greater effect on dry density than compactive effort.

Table 1
Compactive effort, blows per layer, and compacted dry unit weight.

Test Standard	Energy per unit volume ($ft\text{-}lb/ft^3$)	Blows per layer	Dry unit weight (lb/ft^3)
Modified	56,250	330	41
Standard	12,375	73	40
60% standard	7,425	44	40

1 $ft\text{-}lb/ft^3$ = 0.0479 kJ/m^3; 1 lb/ft^3 = 0.0160 Mg/m^3

The effect of moisture on compacted dry density was investigated by compacting air dried tire chips and wet tire chips that had been soaked overnight resulting in a moisture content of about 5.3%. A compactive effort of 60% standard was used for these tests. The samples compacted wet had a dry unit weight that was 4 pcf greater, indicating a slight advantage to compacting tire chips wet.

Apparatus for Compressibility Tests

A large scale compressibility apparatus was designed and built to accommodate the 2-in. (5-cm) maximum size of the tire chips. An initial series of tests were performed on 10-in. (25.4-cm) diameter by 6 to 8-in. (15.2 to 20.3-cm) high samples contained in a thick-walled cast aluminum bucket. These tests showed a maximum vertical strain of nearly 50%. With strains of this magnitude and because of the angular nature of the chips with protruding pieces of steel belt it was felt that there would be a large amount of side friction between the tire chips and the walls of the compressibility apparatus. So a system was designed to measure the side friction. Additionally, it was desired to determine Poisson's ratio for use in subsequent analyses of pavement deflections. This necessitated measuring the horizontal stress as the sample was loaded.

The final design consisted of a 12-in. (30.5-cm) nominal diameter by 12.5-in. (31.8-cm) high by 5/16-in. (0.79-cm) thick compression container fabricated from commercially available Sch-40 PVC pipe. PVC pipe was chosen for the container because small, but measurable, circumferential strain would occur as the sample was loaded which were used to measure the horizontal stress as described below.

Horizontal stress was measured using four resistance strain gages mounted in the horizontal direction 3.5 in. (8.9 cm) above the base of the container. The correlation used to determine horizontal stress from measured circumferential strain is discussed below. Two resistance strain gages mounted in the vertical direction were used to measure the portion of the vertical applied load that was transmitted to the container by side friction. The inside of the cylinder was lubricated with pump grease to reduce side friction.

The vertical load was applied to a rigid circular plate with a diameter slightly less than the inside diameter of the pipe. The plate rested on top of the sample and the vertical load was applied with an Instron 4204 universal testing machine controlled by an IBM compatible computer. The open ended compression container rested on a steel plate. Some tests involved repetitive loading/unloading cycles. During the unloading portion of the cycle the container tended to lift off the steel plate. To prevent this the container was clamped to the base during the unload portion of the cycle. The load was applied using a constant rate of deformation of 0.5 in./min (1.27 cm/min). Measurement of vertical load, vertical deformation, horizontal strain gages, and vertical strain gages were read by the IBM compatible computer using a program written in BASIC. The program also performed initial data reduction and stored the data for future plotting. The apparatus is illustrated in Fig. 2.

A correlation was obtained between the circumferential strain gage readings and the horizontal stress. A 30-in. (76-cm) long section of PVC pipe was cut and the four horizontal strain gages were attached to the center of the section. A section that was long relative to its diameter was used to minimize end effects during calibration. Flat end caps made from 0.75-in. (1.9-cm) thick PVC plate were made and clamped to the top and bottom of the pipe using threaded rod. The pipe was filled with water and the pressure was increased in 10 psi (68.9 kPa) increments up to 70 psi (482 kPa) then decreased back to zero. The resulting horizontal strain was measured. This procedure was repeated many times. A regression analysis was then used to determine the best fit second-order polynomial correlation between the radial stress on the inside of the cylinder and the measured circumferential strain. After the calibration was completed the 12.5-in (32-cm) long section for the compression container was cut from the central portion of the pipe which had the horizontal strain gages attached. Strains due to the water

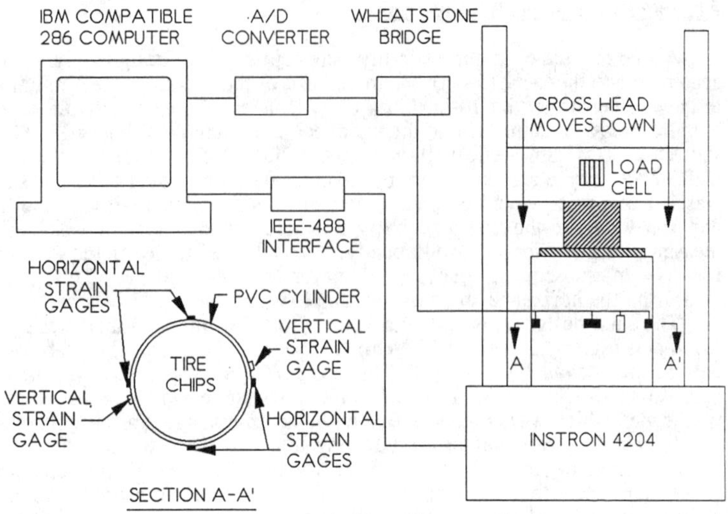

Fig. 2 Apparatus used for one-dimensional compressibility testing.

weight and the vertical force introduced by the end caps were also accounted for. Likewise, a correlation was obtained between the measured vertical strain and the vertical load carried by the compression container. A vertical load was applied in increments to the empty cylinder while the resulting vertical strain was measured.

Compressibility Results

Compressibility tests were conducted to determine the compressibility of compacted tire chips and to obtain elastic parameters, namely, Youngs modulus and Poisson's ratio, for use in subsequent numerical analyses of pavement deflection. To prepare a sample the empty cylinder was clamped to the steel base and grease was applied to the inside of the cylinder. Air dried tire chips were then compacted in five 2-in. (5-cm) thick layers using 60% of standard Proctor energy. Thus the resulting sample was about 10 in. (25.4 cm) high. The clamp which held the cylinder to the base was removed prior to the start of the test.

Compressibility results on three samples of compacted tire chips are shown in Fig. 3. It is seen that a vertical stress of only 10 psi (68.9 kPa) causes 20 to 25% vertical strain. Increasing the stress from 10 psi (68.9 kPa) to 60 psi (413 kPa) causes an additional 15% to 20% vertical strain. Thus, at low stress levels tire chips are highly compressible, but the compressibility decreases significantly as the stress level increases.

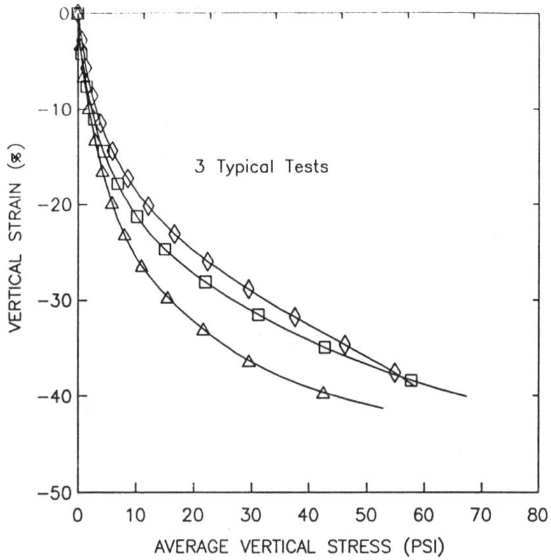

Fig. 3 Compressibility results on three samples of compacted tire chips. (1 psi = 6.89 kPa)

The effect of multiple cycles of loading and unloading was investigated by cycling the vertical stress between a maximum of 65 psi (447 kPa) and a minimum of about 5 psi (34.5 kPa). The lower limit is equivalent to an overburden of 6 ft (1.8 m) of conventional gravel. Cycling between the upper and lower limit caused 10 to 40% elastic strain in the three trials that were conducted. A typical test result is shown in Fig. 4.

Side friction was a significant factor. It caused the applied vertical stress to decrease from the top to the bottom of the sample. At the top of the sample the vertical stress σ_{top} equals the vertical load applied by the Instron 4204 divided by the cross-sectional area of the sample. At the level of the vertical strain gages, the load carried by the container is known and the vertical stress carried by the sample σ_{gage} equals the total applied vertical load minus the load carried by the container divided by the cross-sectional area. At this level 10 to 40% of the total applied load is carried by the container. To determine the average vertical stress over the height of the sample σ_{avg} it was assumed that the stress decreased linearly through the sample with the actual values known at the top of the sample and at the level of the vertical strain gages as illustrated in Fig. 5. The average vertical stress is then the average of the vertical stress at the top and bottom σ_{bottom}. This is the vertical stress that was used to plot Figs. 3 and 4.
Horizontal stress was measured during the compression tests and is shown plotted versus vertical stress for a typical test in Fig. 6. The vertical stress was the value at the level of the horizontal strain gages, σ_{gage}. All tests exhibited a bilinear behavior with the slope increasing at a vertical stress of about 25 psi. The

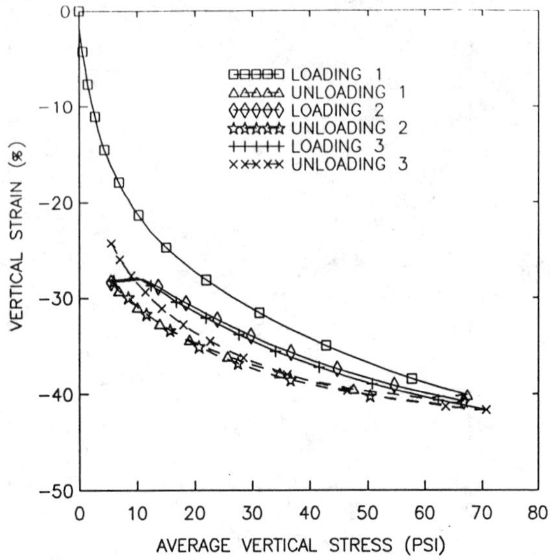

Fig. 4 Compressibility of tire chips subjected to unloading/reloading cycles. (1 psi = 6.89 kPa)

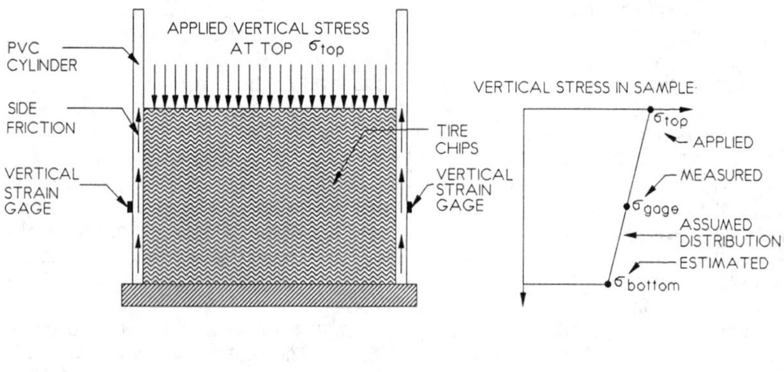

Fig. 5 Effect of side friction on vertical stress in sample during compression.

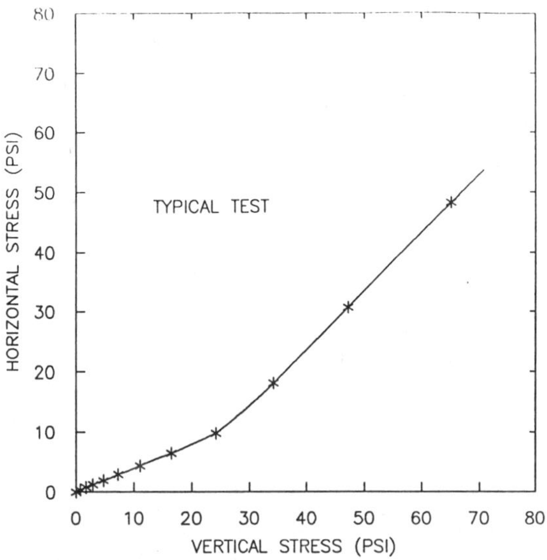

Fig. 6 Relation between horizontal and vertical stress for sample tested in one-dimensional compression. (1 psi = 6.89 kPa)

slopes were used to calculate the coefficient of lateral earth pressure at rest K_o. The value was about 0.40 before the change in slope and about 0.94 after the change in slope for the test in Fig. 6.

Elastic parameters for tire chips were calculated from the compressibility test results. Poisson's ratio ν can be calculated from K_o for the case of one-dimensional compression using the equation given below

$$\nu = K_o / (1+K_o)$$

The slope of the plot of vertical strain ϵ_v versus average vertical stress σ_{avg} is the constrained modulus D, thus

$$D = \Delta\sigma_{avg} / \Delta\epsilon_v$$

Using D and Poisson's ratio, Youngs modulus E can be calculated from

$$E = \frac{(1+\nu)(1-2\nu)D}{1-\nu}$$

Youngs modulus was calculated as an initial tangent and as the secant values at 10% and 20% vertical strain. The elastic parameters for three tests as well as the average are summarized in Table 2.

Table 2
Elastic parameters for tire chips

Test No.	Poisson's ratio <25 psi	Poisson's ratio >25 psi	Youngs modulus: Initial Tangent (psi)	Youngs modulus: Secant at ϵ_v =10% (psi)	Youngs modulus: Secant at ϵ_v =20% (psi)
301	0.25	0.54	18.0	25.1	50.0
1001	0.39	0.50	3.4	9.0	14.4
1002	0.29	0.49	12.9	21.6	37.3
Average	0.32	0.51	10.7	18.1	32.7

1 psi = 6.89 kPa

Pavement Deflection

A preliminary study was made of the effect of the thickness of soil over the tire chip layer on pavement deflection. Three cases were analyzed. All cases had 6-in. (15-cm) of bituminous concrete pavement, 4 ft (1.2 m) of tire chips, and 4 ft (1.2 m) of subbase and subgrade underlain by a rigid base. The difference between the three cases is that the position of the 4-ft (1.2-m) thick tire chip layer was changed as shown in Fig. 7. The analysis was made with a linear elastic finite element program SAP86. Material properties used in the analysis are shown in Table 3.

Table 3
Material properties used in finite element analysis.

Material	Young's modulus, E (psi)	Poisson's ratio
Pavement*	1×10^6	0.35
Subbase, high E*	1.2×10^4	0.25
Subbase, low E*	2.4×10^3	0.25
Subgrade, high E*	6.0×10^3	0.25
Subgrade, low E*	1.2×10^3	0.25
Tire chips	2×10^2	0.32

*Estimated (1 psi = 6.89 kPa)

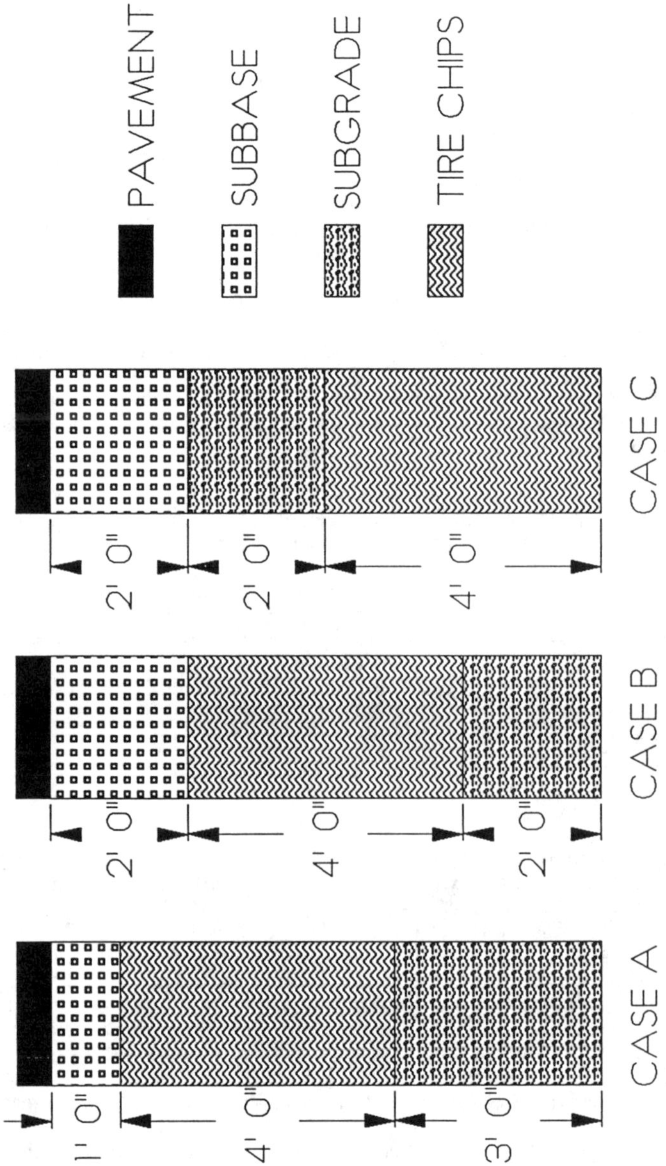

Fig. 7 Soil profiles used for finite element study. (1 ft = 0.305 m)

The parameters for pavement, subbase, and subgrade were estimated using typical values reported in the literature (Duncan, et al., 1968; Barker, 1972). The effect of the modulus of the subbase and subgrade was investigated by performing analyses with the "high" and "low" E values. The "low" E value may be more realistic because of the low confining stress at shallow depth and the poor compaction that would be present in the subbase and subgrade immediately above the tire chip layer. The parameters for the tire chips were determined from the unloading/reloading portion of the compressibility tests.

A wheel load was simulated by applying a pressure of 80 psi (551 kPa) to a circular area with a diameter of 1.1 ft (33.5 cm). This is roughly equivalent to the stress that would be applied by a 22-kip (10 metric tons) axle load.

The thickness of the subbase and subgrade above the tire chips had a significant effect on deflections at the surface of the pavement as shown in Table 4. When the high E was used for the subbase and subgrade, deflections where reduced by 1/2 when the thickness of subbase and subgrade above the tire chips was increased from 1 ft (0.3 m) to 4 ft (1.2 m). A lesser reduction was found for the low E case. The modulus of the subbase and subgrade also had an effect on pavement deflections as seen in Table 4. Because of the importance of the modulus of the subbase and subgrade, future studies will be conducted using nonlinear finite element methods which allow the stiffness of the subbase to be modeled in a more realistic manner.

Table 4
Pavement deflection under center of wheel load calculated by finite element method.

Case	Modulus in subbase and subgrade	
	Low E	High E
A	0.13 in.	0.10 in.
B	0.11 in.	0.07 in.
C	0.09 in.	0.05 in.

1 in. = 2.54 cm

Summary

The tire chips used for this study were manufactured from a mix of glass and steel belted tires and have a maximum size of 2-in. (5 cm). They have a specific gravity of 1.05. When compacted using 60% of Standard Proctor energy they have a unit weight of about 40 pcf (0.64 Mg/m^3) making their use as lightweight fill very advantageous. Only a modest compactive effort is needed to achieve the maximum dry density. A large scale compressibility apparatus with the ability to measure side friction and horizontal stress was fabricated. Compressibility tests showed that tire chips are highly compressible but that the compressibility decreases as the stress level increases. The coefficient of lateral earth pressure at rest K_o was about 0.4 for stresses less than 25 psi (172 kPa) but increased to about 1.0 for higher stresses. Preliminary finite element studies showed that increasing the thickness of the soil over the tire chips greatly reduces pavement deflection. More study is needed to assess the influence of the stiffness of the soil over the tire chips on pavement deflection.

References

1. Barker, W.R. (1972), "Nonlinear Finite Element Analysis of Heavily Loaded Airfield Pavement Systems," Proceedings of the Symposium on Applications of the Finite Element Method in Geotechnical Engineering, U.S. Army Waterways Experiment Station, Vicksburg, Miss., Vol. II, pp. 657-693.

2. Duncan, J.M., Monismith, C.L., and Wilson, E.L. (1968), "Finite Element Analyses of Pavements," Highway Research Record No. 228, Highway Research Board, Washington, D.C., pp. 18-33.

3. Dodson, T. (1990), "Camas Valley-Muns Creek, Highway 35 Landslide: Plans, Construction Sequencing and Staging by Oregon State Highway Division, Geotechnical Group, 6 p.

4. Edil, T.E., Bosscher, P.J., and Eldin, N.N. (1990), "Development of engineering criteria for shredded or whole tires in highway applications," Interim Report to the Wisconsin Dept. of Transportation, University of Wisconsin-Madison, 99 p.

5. Geisler, E., Cody, W.K., and Niemi, M.K. (1989), "Tires for subgrade support," Presented at the Annual Conference on Forest Engineering, Coeur D'Alene, ID, August, 5 p.

6. Palmer, A., and Verdegan, C.T. (1989), "Soils investigation embankment failure CSAH 21: One Mile North of the City of Rice, MN," Report from Braun Engineering Testing, Inc., Minnesota, to Dennis Carlson, Benton County Highway Department, Foley, MN, 8 p.

Improved Resilient Modulus Realized with Waste Product Mixtures

Seung W. Lee and K.L. Fishman*

Abstract

Two materials currently considered waste products are investigated for use in highway construction. The first is fly ash which is a waste product of coal combustion. The second is fine grained material (clay) which is a by-product from aggregate processing. Neither material exhibits satisfactory engineering properties by itself. However, if combined the material properties are improved and potential applications may be found in construction of subgrades for flexible pavements.

Recently, mechanistic-empirical pavement design procedures based on layered elastic theory have been used for the rational design of flexible pavement structures. This approach requires the elastic parameter, resilient modulus, of each material in the layered system as input. Cyclic triaxial tests are preferred to characterize the resilient behavior of base and subgrade materials since this method can include the sensitivity of resilient moduli to the state of stress. Practically, shear strength measurements such as CBR and R-Value are widely used to estimate the resilient moduli due to the difficulty associated with cyclic triaxial tests. More recent research has attempted to estimate the resilient moduli of natural fine grained subgrade material accounting for the dependence of resilient modulus on deviatoric stress.

The goal of this study is to demonstrate improvement in resilient modulus realized by combining two waste products. The usefulness of applying existing empirical relations for estimating resilient modulus of fine grained materials to the materials investigated in this study will also be explored.

*Seung W. Lee and K. L. Fishman, Grad. Student and Asst. Prof., Dept of Civil Engrg., State Univ. of New York at Buffalo. Buffalo, NY

Introduction

A considerable amount of fly ash is produced in the United States at an increasing annual rate as a byproduct of coal combustion. Disposal of the fly ash not only has become an increasing burden on the producer of fly ash but also results in increased costs to the power consumer.

Fly ash has been used as a concrete additive (Fisher et al 1978, Lauf 1982, Mehta 1983), and soil stabilizer with or without lime or cement (Mateos 1964, Minnick 1979, Barenberg and Thompson 1982, Sutton 1985, Lambe 1985) for base or roadbed materials in highway construction (Lewis 1976, Faber 1976, Head et al 1982, Toeth 1988, Martin et al 1991).

In order to reduce the amount of disposed fly ash, a promising area for the use of large volumes of fly ash is in the construction of highway bases and embankments. Construction of fly ash roadway embankments in Illinois (Lewis 1976) indicated that fly ash is an acceptable alternative to naturally occurring soils in roadbed materials. Faber (1976) summarized successful use of fly ash and fly ash mixtures in roadway embankments projects in United States, England and Scotland. Use of clay and fly ash mixtures as structural fill materials was reviewed by Lambe (1985). Head et al (1982) investigated mixtures of fly ash and coal refuse compositions for potential highway base course materials. Based on characteristics of coal refuse and fly ash mixtures including CBR, performance of pavements were predicted using layered elastic analysis. Results indicate that thickness requirements for coal refuse and fly ash bases are less than that of conventional crushed stone bases. An investigation by Toeth et al (1988) indicated that the CBR range of "normal" weighted compacted fly ashes is 10.8 - 15.4 % in the unsoaked state and 6.4 - 13.5 % in the soaked state. Martin et al (1991) concluded that the behavior of fly ash or its mixtures can be predicted by standard stability analyses of embankments using published index properties and strength measurements.

For the design of pavements, behavior of roadbed materials under traffic loading as well as its environmental stability should be addressed. To the authors' knowledge, the stress-strain behavior of fly ash and its mixture under repetitive traffic loading has not been studied in detail. In order to foster utilization and provide input to mechanistic pavement design procedures, the stress-strain behavior of fly ash and it's mixtures are investigated using cyclic triaxial tests. Validity of empirical equations for estimating resilient modulus is also studied for potential practical use.

In this study, two fine grained materials are investigated. The first is a Type F fly ash. The second is the result of aggregate processing in the Buffalo area. Gravel is processed containing traces of native limestone and dolomite. Construction specifications limit the amount of fine grained material in processed aggregates. Therefore, fine grained material considered a waste product, is sluiced to detention basins, dredged, and stockpiled in landfill operations.

Material Properties

In order to study material properties a series of standard soil mechanics tests were performed. Tests were performed on aggregate waste (clay), fly ash, and a variety of compositions for the two materials. The compositions were proportioned by weight and designated by stating clay part:fly ash part. Results are presented for Specific Gravity, Grain Size Analysis, Atterberg Limit, Compaction, Direct Shear, Unconfined Compression, CBR, Consolidation and Permeability Tests. Test Results are summarized in table (1) and (2).

Table (1) Material Properties

Materials	Agg. Waste	Fly Ash	Mixtures (10:1)	(10:2)	(10:3)	(10:6)
Grain Size Analysis						
Sand Size	31%	24%				
Silt Size	39%	48%				
Clay Size	30%	28%				
Atterberg Limits						
LL	31	-				
PL	17.5	-				
PI	13.5	NP				
Standard Proctor Comp.						
Max.Density (pcf)	114	69	113	108	104	
Wopt (%)	13.3	22.5	15.6	16.1	17.0	
Harvard Minuate Comp.						
Max. Density (pcf)	108	68			103	88
Wopt (%)	18	28			21	26
Direct Shear Test	(90% Standard Proctor Compaction)					
Cohesion (psi)	1.6 (W=15%)				2.3 (W=15%)	
friction (degree)	25 (W=15%)				36 (W=15%)	
Consolidation Test	(95% Standard Proctor Compaction at Wopt)					
Compression Index	0.180	0.215			0.135	
Swelling Index	0.0096	0.0096			0.0011	
Coefficient of Consolidation	0.2022	0.2166			0.2190 (10E-5 in2/sec)	
Permeability Test	(90% Standard Proctor Compaction at Wopt)					
Permeability	0.24	2.22			2.71 (10E-6 in/sec)	

Table (2) Unconfined Compressive Strength Test Result

Materials	W %	γd pcf	Unconfined Compressive Strength psi
Agg.waste	13	97.4	21.0
	17	106.1	30.1
	20	101.7	21.9
	25	97.4	10.1
Fly ash 1 Day Curing	20	65.6	7.1
	25	66.3	10.4
	30	66.2	7.5
Mixture(10:3) 1 Day Curing	19	101.2	19.7
	22	102.4	15.9
	32	92.4	16.7
Mixture(10:3) 7 Day Curing	19	101.1	19.7
	22	102.4	30.6
	26	101.8	17.2
Mixture(10:6) 1 Day Curing	19	79.9	18.5
	27	90.9	10.4
Mixture(10:6) 7 Day Curing	27	90.5	12.3

Conversion Factor

1 inch = 2.54 cm
1 psi = 6.89 kpa
1 pcf = 0.157 kpa/m

In general this test program has provided results which are similar to experiences with pozzolanic stabilization in the literature. The effects of stabilization give the following trends;

1. increase permeability
2. decrease compressibility
3. increase internal angle of friction
4. decrease maximum density
5. increase optimum moisture content
6. decrease plasticity
7. evidence of curing effects

Resilient Modulus (Mr) from Cyclic Triaxial Test

To investigate the potential use of aggregate waste, fly ash or a mixture of the two materials in highway construction, resilient moduli (Mr) for the materials were determined from cyclic triaxial tests. Details of sample preparation, test procedures and test data were presented by Seung W. Lee and K. L. Fishman (1991).

The resilient modulus of aggregate waste was found to be dependent on the magnitude of deviatoric stress as shown in figures (1a) and (1b). This behavior is typical of fine grained materials.

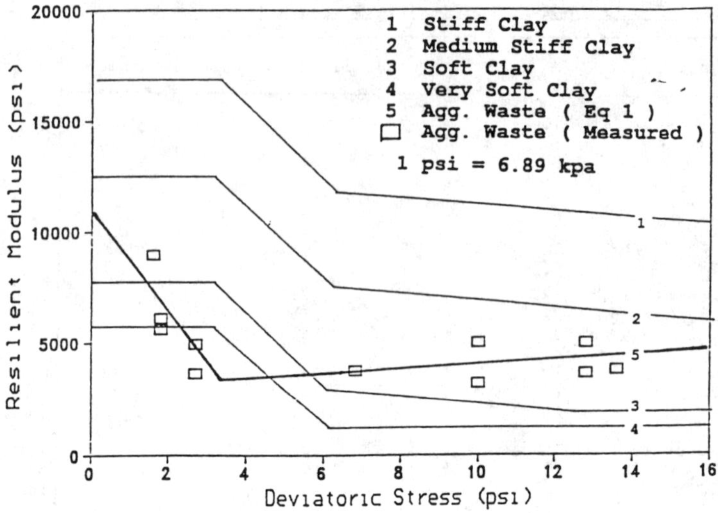

Figure 1(a). Resilient Modulus of Aggregate Waste Compared to Typical Clay Behavior (Thompson and Elliot, 1986)

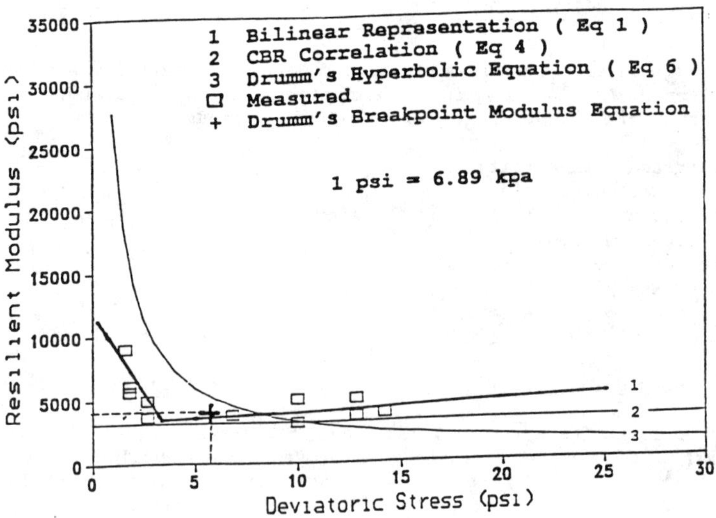

Figure 1(b). Measured Resilient Modulus of Aggregate Waste Compared to Empirical Equations

Resilient Modulus for the aggregate waste may be expressed using a bilinear model;

if $6d < 6d_b$

$$Mr = -2345.7 \times (6d - 6d_b) + 3824.7 \ (psi)$$

$$= -2345.7 \times (6d - 6d_b) + 26352.2 \ (kPa)$$

eq(1)

if $6d > 6d_b$

$$Mr = 51.3 \times (6d - 6d_b) + 3824.7 \quad (psi)$$

$$= 51.3 \times (6d - 6d_b) + 26352.2 \quad (kPa)$$

where 6d is deviatoric stress and $6d_b$ is deviatoric stress at breakpoint.

The relation for Mr as a function of 6d for aggregate waste is compared with that of typical fine grained materials presented by M.R. Thompson and R. P. Elliot (1986) as shown in figure (1a). Mr of the aggregate waste is in the range between soft and medium stiff subgrade clay material. Hence utilization of aggregate waste by itself for road bed materials may not have an advantage over the use of medium stiff naturally occurring subgrade clay materials which may be readily available near to a site.

The resilient behavior of fly ash is similar to that of granular materials which exhibit a dependency of Mr on J1. Mr for fly ash may be expressed as;

$$Mr = 92.02 \times J1 + 1298 \quad (psi)$$

$$= 92.02 \times J1 + 8943.2 \quad (kPa)$$

eq (2)

where J1 is sum of the principal stress. Figure (2) presents a plot of Mr for fly ash versus J1.

Based on the Asphalt Institute's recommendation (Asphalt Institute 1982), the range of confining pressure within subgrade material is 1-5 psi (7-34 kPa) (approximately the range of J1 is 3 - 20 psi (21-138 kPa)). As shown in figure (2) , Mr of fly ash for this range is 1600-3000 psi (11-21 MPa). Based on Asphalt Institute data (Asphalt Institute 1982), this condition indicates very poor roadbed soil.

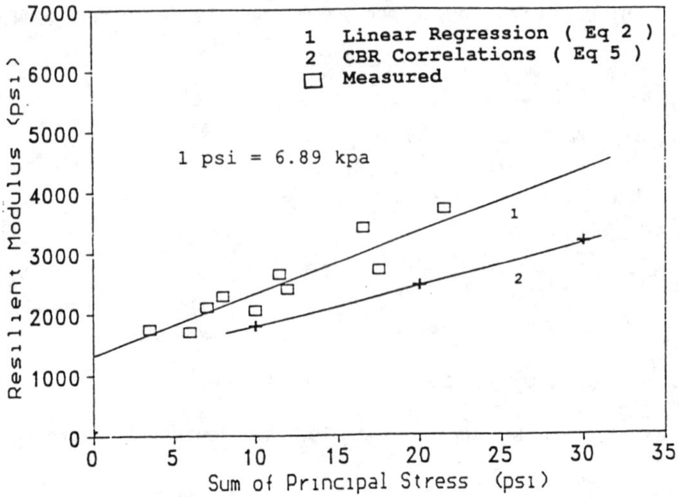

Figure 2. Measured Resilient Modulus of Fly Ash Compared to CBR Correlations

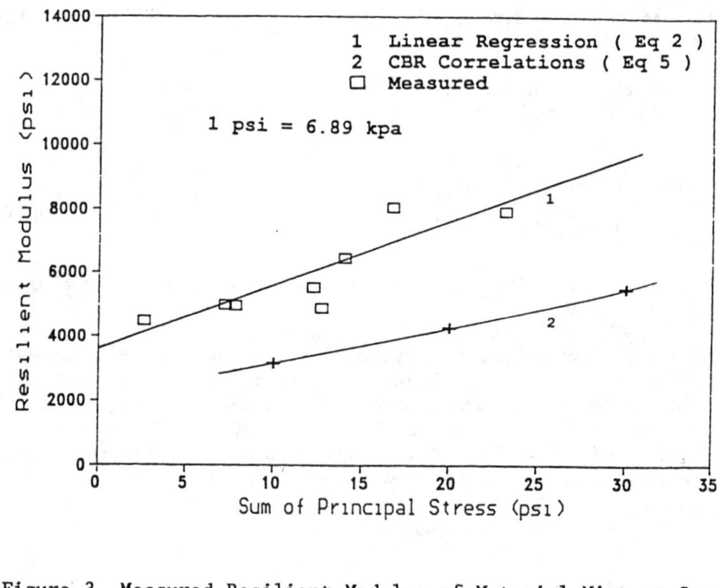

Figure 3. Measured Resilient Modulus of Material Mixture Compared to CBR Correlations

The observed resilient behavior of the mixture of aggregate waste and fly ash (10:3 by weight) also demonstrates dependence on J1 (Figure (3)). Mr of this mixtures may be expressed as;

$$Mr = 196 \times J1 + 3535 \quad (\text{ psi }) \qquad \text{eq (3)}$$

$$= 196 \times J1 + 24356 \quad (\text{ kPa })$$

Quantitatively, Mr values of the material composition are 4500 - 7000 psi (31-48 MPa) at states of stress of practical interests for subgrade materials. Based on Asphalt Institute data (Asphalt Institute 1982) this composition is good roadbed soil. It is hoped that this preliminary data may inspire further study and encourage potential use of material composition of aggregate waste and fly ash for subgrade materials.

Estimation of Resilient Modulus(Mr) by Approximate Method

Due to difficulties associated with cyclic triaxial testing, various approximate methods have been suggested to estimate Mr. In this paper, the ability of some approximate methods to predict Mr are investigated for aggregate waste, fly ash and the material composition.

Relating Mr to CBR may be the most popular method for estimating Mr for fine grained materials. Based on Heukelom and Foster's investigation, Mr-CBR correlation is given by the following relationship (Heukelom and Foster 1960);

$$Mr = A \times CBR \qquad \text{eq (4)}$$

where A is a coefficient from linear regression. The range of the coefficient A is from 750 psi (3167.5 MPa) to 3000 psi (20 MPa) and usually 1500 (10 MPa) is used.

Use of eq.(4) has been recommended differently by AASHTO (AASHTO 1986). AASHTO (1986) suggests to use eq.(4) for soils with a soaked CBR of 10 or less, whereas Asphalt Institute suggests use of eq (4) up to a CBR of 20 with samples of the same density and water content as the field condition. Klomp and Dormon's (1984) suggestion is similar to the Asphalt institute's.

Correlations of Mr vs CBR were provided for granular materials including the effects of states of stress by AASHTO (1986) as follows.

J1 psi (kPa)	Mr psi (kPa)
100 (689.0)	740 (5098.6) x CBR
30 (206.7)	440 (3031.6) x CBR
20 (137.8)	340 (2342.6) x CBR
10 (68.9)	250 (1722.5) x CBR

eq(5)

CBR was determined for materials investigated in this study and results are presented in Table (3).

Table 3 CBR Test Data

Materials	CBR (Unsoaked)
Agg. waste	2.0
Fly Ash	7.0
Mixture (10:3)	12.6

100 % Standard Compaction at Optimum Water Content

Comparisons of Mr of aggregate waste (clay) estimated by eq (4) and measured from cyclic triaxial tests are shown in figure (1b). For aggregate waste (clay) eq(4) gives a reasonable estimation of Mr. However, Mr of fly ash and material composition estimated by eq(4) produced large error. The reason of large error is believed due to the granular behavior of fly ash and material composition since eq(4) is not valid for granular material. As shown figure (2) and (3), Mr vs CBR relationship for granular materials (eq (5)) gives better estimation of Mr for fly ash and material composition.

Based on typical resilient behavior of fine grained soils, Thompson M.R.and Robnett Q.L. (1979) investigated factors influencing the resilient behavior of fine grained soils using sensitivity analysis based on a comprehensive study of resilient properties of Illinois soils. This study indicated that the degree of saturation, distribution of particle size, plastic index, unconfined compressive strength, and modulus of elasticity are important factors to be related with Mr. Similarly, Drumm et al (1990) proposed a hyperbolic model to describe the nonlinear relationship for the modulus of fine grained soils as a function of deviatoric stress as follows;

$$Mr = (a + b \times \sigma d) / \sigma d \qquad \text{eq (6)}$$

where a and b are material coefficients and are functions of forementioned factors. The expression of a and b is cited in (Drumm et al 1990). Mr of aggregate waste estimated by eq (6) is compared with measured Mr as depicted in figure (1b). As discussed by Drumm, this empirical equation does not provide good approximation in the range of low deviatoric stress. However it gives reasonable value when the devitoric stress is 5 psi (34 kPa) or larger. Drumm also provides an empirical expression for Mr at breakpoint which is significant for bilinear representation of Mr vs 6d. This expression is a function of forementioned factors for eq (6), and 6d at breakpoint was assumed to be 6 psi. This expression is cited in (Drumm et al 1990). As shown in figure (1b), Drumm's empirical equation for estimating breakpoint modulus (Mr_b) gives a good approximation of Mr_b of aggregate waste (clay). Though fly ash and the mixture of aggregate waste and fly ash are fine grained soils, Drumm's study is not applicable for the prediction of resilient behavior of these materials. Mr of these materials does not depend on the magnitude of deviatoric stress.

Conclusion

For the potential use of waste material in highway construction, resilient behavior of aggregate waste and fly ash is studied through a series of standard soil mechanics and cyclic triaxial tests. The results indicate that neither material by itself possesses proper characteristics, however a composition of these two materials shows good qualities for roadbed materials. Results from this study should be considered preliminary and further testing is required to study the response of the material composition due to changes in the physical environment.

For practical design, cyclic triaxial testing is difficult due to the complexity and absence of testing equipment in many laboratories. For this reason, the validity of approximate methods for estimating Mr is studied. Mr vs CBR relationships for fine grained soils (eq 4) provides a reasonable estimation of Mr for aggregate waste (clay). However it overestimates Mr of fly ash and the material composition. Mr vs CBR relationships for granular material considering the influence of J1 (eq 5) provides good approximation of Mr for fly ash and the material composition. Drumm's empirical equation (eq 6) gives more information about 6d dependent Mr than eq 4, however further research into functions of this type are required.

References

1. AASHTO, (1986). "AASHTO Guide for Design of Pavement Structures."

2. Asphalt Institute, (1967). "Soils Manual for Design of Asphalt Pavement Structures," Manual Series No. 10 (MS-10), February.

3 Asphalt Institute, (1982). "Research and Development of the Asphalt Institute's Thickness Design Manual (MS-1) Ninth Edition," Research Report No.82-2, RR-82-2.

4. Barenberg, E.J. & Thompson, M. R., (1982). "Design, Construction and Performance of Lime, Fly ash, and Slag Pavement," Soil Stabilization, Transportation Research Record 839, TRB.

5. Drumm E. C., Boateng-Poku, Y. and Pierce, T.J., (1990). "Estimation of Subgrade Resilient Modulus from Standard Tests," J. of Geotech. Eng. Div., ASCE, Vol.116, (No.5).

6. Faber, J. H., (1976)."Use of Fly Ash in Embankment Construction," Transportation Research Record 593, TRB.

7 Fisher, G. L. et al, (1978). "Physical and Morphological Studies of Size Classified Coal Fly Ash," Environmental Science and Technology, Vol.12, (No.4), April.

8. Head, W.J. et al, (1982), "Coal Refuse and Fly Ash Composition: Potential Highway Base Course Materials, Transportation Research Record 839.

9. Heukelom, W. & Foster C. R., (1986). "Dynamic Testing of Pavements," J. of Soil Mechanics & Foundation Div., ASCE, Vol.86, (No. 1), Feb.

10. Klomp, A.G. and Dorman, G.M. (1984). "Stress Distribution and Dynamic Testing in Relation to Road Design", Proc. Australian Road Research Board, Vol. 2, Part 1.

11. Lambe, J. H., (1985) " Type C Fly Ash and Clay Stabilization," Developments in New and Existing Materials, Proc. of A Session Sponsored by the Material Eng. Div., ASCE, in Conjunction with ASCE Convention of ASCE, Denver.

12. Lauf, R. J., (1982). "Microstructure of Coal Fly Ash Particle," Ceramic Bulletin, Vol.61, (No.4), April.

13. Lee, S. W. & Fishman, K. L., (1991). "Investigation of Waste Products as Subgrade Materials in Flexible Pavement System," Submitted to ASCE Journal of Transportation.

14. Lewis, T. S., (1976). "Construction of Fly Ash Roadway Embankment in Illinois," Transportation Research Record 593, TRB.

15. Martin, J. P., et al, (1991). "Properties and Use of Fly Ashes for Embankments," J. of Energy Eng. Div., ASCE, Vol.116, (No. 2).

16. Mateos, M., (1964). "Stabilization of Soils with Fly Ash Alone," Highway Research Record No.52, HRB.

17. Mehta, P. K., (1983). "Pozzalanic and Cementious Byproducts as Mineral Admixture for Concretes," Fly Ash, Silica Fume Slag & Other Mineral By-Products in Concrete Vol.1 SP-79.

18. Minnick, L. J., (1979). "The Role of Fly Ash in Structural Types of Stabilized Applications," Proc. of 5th Int. Ash Utilization Symposium.

19. Sutton, J. R., (1985). "Soil Stabilization with Cementious Fly Ash," Material Engineering Session, 1985 Spring Convention of ASCE, Denver.

20. Thompson, M. R. & Robnett, Q. L., (1979). "Resilient Properties of Subgrade Soils," J. of Transp. Eng. Div., ASCE, Vol. 105, (No.1).

21. Thompson, M. R. & Elliot R. P., (1986). "Illi-Pave Based on Response Algorithms for Design of Conventional Flexible Pavements," Transportation Research Record No. 1043, TRB.

22. Toeth, P.S., (1988). "Coal Ash as Structural Fill with Special Reference on Ontario Experience," Canadian Geotech. J., Vol.25, No.4, 1988

Plastic Waste for Low-Weight Embankments

H. El Ghoche [1], C. Coulet, M. CFMS [2], D. Daudon [1]

Abstract

This paper presents a new process whereby plastic waste may be profitably used in the Civil Engineering domain. Studies have shown that its viscoelastic behaviour is non-linear. A law is given for its behaviour under uni-axial compression stress; this leads to the determination of parameters E and ν, and to the formulation of long-term predictions.

Introduction

Industrial countries produce an increasing quantity of waste, of which plastics represent 20% by volume (Fornasero R., 1988). 10 million tonnes are produced in Japan every year, and 8.5 million tonnes in the EEC countries. Getting rid of this waste plastic, either by dumping (75% of the total) or by incineration, is proving increasingly costly, and creates serious environmental problems, including high pollution risks.

As an attempt to solve this vast problem, we propose a technique which would transform waste plastic into a useful material by compressing and binding it in the shape of parallelepipedal blocks, named Plastblocs; these blocks could be put to many different uses in the field of public works.

---

1 Civil Engineer, teacher, researcher, Institut Universitaire de Technologie, Génie Civil, 69622 Villeurbanne France.
2 Doctor of Engineering, lecturer, Institut Universitaire de Technologie, Dept Génie Civil, 69622 Villeurbanne France.

As their density is very low (0.3 to 0.6), these blocks could be used to build low-weight embankments on compressible ground, or serve as repair material after a landslide. Since their strain modulus can be chosen at will during fabrication, they could also be used as a form of protection for underground pipes, or as material to absorb shocks.

To determine the behaviour of these blocks under the various kinds of stress they may encounter in their different applications, a number of laboratory tests were carried out, as well as full-scale experiments in the field, for which purpose a 4.6 m embankment was built (Daudon, 1991). From the laboratory tests we present a series of simple compression tests on reduced-size samples. The results of the tests are then compared to those obtained with full-size blocks.

Samples

The first tests on full-size blocks (Figure 1) gave satisfactory results as concerns their behaviour (El Ghoche, 1989), but handling problems (the blocks weigh 500 kg), combined with the uncertainties linked to some of the measurements, led us to undertake more systematic research on reduced-size samples, which are easier to handle (Figure 2).

Figure 1: Full-size block **Figure 2: Reduced-size block**

The cylindrical samples (diameter: 20 cm; height: 30 cm) were made according to the same principles as the full-size blocks, i.e. vertical compression, then binding. Thus, as with the full-size blocks, we obtained a stack of layers of waste plastic material.

Simple compression tests

During the tests, as we wanted to analyze the creep phenomenon, the applied stress was kept constant. The apparatus used is shown in Figure 3, with stresses being produced by loads of different weight, transmitted by a rigid plate held horizontally by guiding rods.

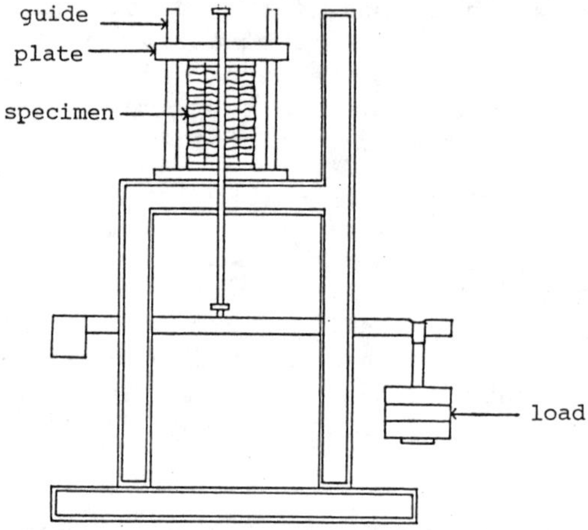

Figure 3: Compression test apparatus

The stress was applied vertically, that is to say at right angles to the waste plastic layers, and followed the path shown in Figure 4.

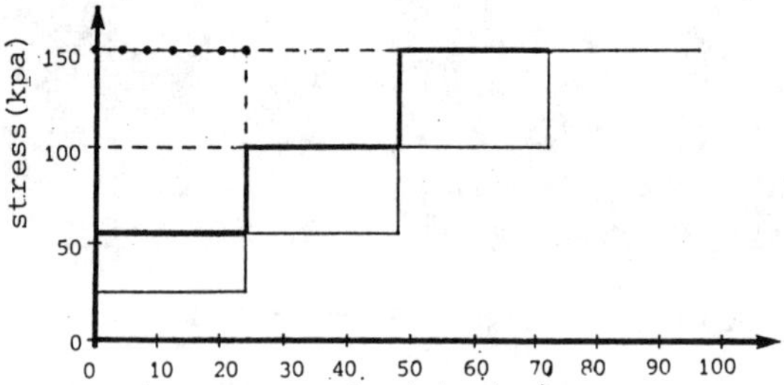

Figure 4: Path of applied stresses

Throughout testing, the temperature was kept constant: T = 23° ± 2° C.

The same operating method was repeated for four different densities (0.3, 0.4, 0.5, and 0.6), each stress being maintained for a minimum of 24 hours, the maximum being 45 days.

Various loading-unloading cycles were investigated.

Strain Analysis

Vertical Strains ε_{33} (Perpendicular to the Layers)

Within the limits of our study — stresses of less than 200 kPa and time periods of less than 2 months — the curves of settling as a logarithmic function of time [ΔH/Ho = f(ln(t))] were considered to be linear (Figure 5). The same behaviour was observed on the full-size blocks.

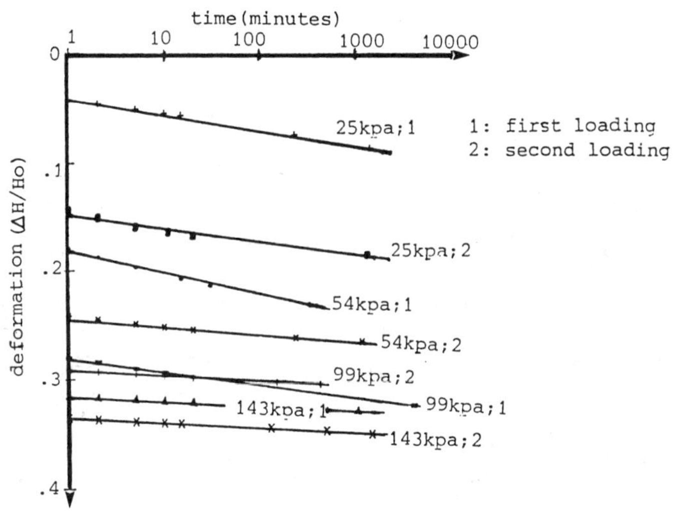

Figure 5: Experimental curve $\Delta H/H_o = f(\ln(t))$

The slopes of these straight lines vary with the stress and the initial density of the sample. During the second loading, for the same stress value, there is a noticeable increase in the strain, linked to the maximum settling obtained at the end of the first loading cycle (Figure 5).

Instantaneous strain, measured as a function of stress (after 1 minute, for practical reasons) was not linear (Figure 6).

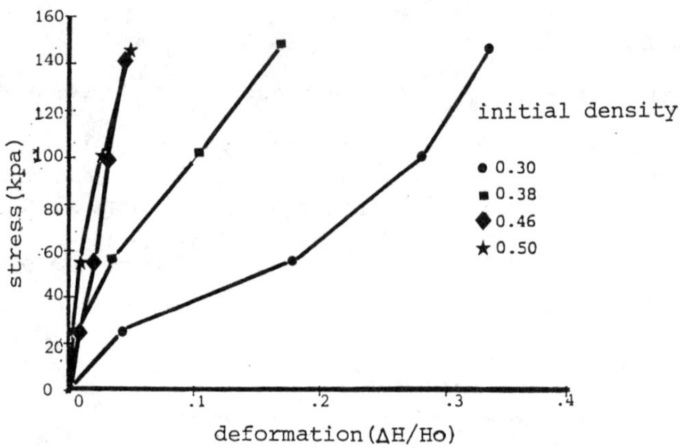

Figure 6: Stress-strain experimental curve

It should be noted that strain decreases as the initial density increases.

Lateral Strain ε_{22} (Parallel to the Layers)

The calculation of this strain is based on the variations of the perimeter of the cylindrical sample. It varies, of course, as a function of the vertical strain. However, the ratio of the vertical strain to the lateral strain ($-\varepsilon_{33}/\varepsilon_{22}$) remains constant whatever the density, and is equal to 0.06. We will take it that this ratio, even for large strains, is equal to Poisson's ratio ν.

Due to plastic strain, this ratio ν, within the same stress range, takes the value of 0.02 for the second loading cycle.

Theoretical Analysis

Law of Behaviour

The way in which the blocks were made meant that the plastic used in the samples was stratified, and complete isotropy was thus impossible. However, the hypothesis of transverse isotropy may be adopted, which gives the following law of behaviour:

$$\vec{\vec{\varepsilon}} = [\vec{\vec{S}}]\ \vec{\vec{\sigma}}$$

where $\vec{\vec{\varepsilon}}$ is the strain tensor

$\vec{\vec{\sigma}}$ is the stress tensor, and

$[\vec{\vec{S}}]$ is a second order tensor.

In the case of transverse isotropy, $[\vec{\vec{S}}]$ has 5 independent parameters. In the domain of small strains, for "elastic behaviour", this relation becomes (Lemaire, 1988):

$$\begin{Bmatrix} \varepsilon_{11} \\ \varepsilon_{22} \\ \varepsilon_{22} \\ \varepsilon_{12} \\ \varepsilon_{23} \\ \varepsilon_{13} \end{Bmatrix} = \begin{bmatrix} \frac{1}{E_1} & -\frac{\nu_{12}}{E_1} & -\frac{\nu_{13}}{E_1} & 0 & 0 & 0 \\ -\frac{\nu_{12}}{E_1} & \frac{1}{E_1} & -\frac{\nu_{13}}{E_1} & 0 & 0 & 0 \\ -\frac{\nu_{13}}{E_1} & -\frac{\nu_{13}}{E_1} & \frac{1}{E_3} & 0 & 0 & 0 \\ 0 & 0 & 0 & \frac{1+\nu_{12}}{E_1} & 0 & 0 \\ 0 & 0 & 0 & 0 & \frac{1}{2G_{13}} & 0 \\ 0 & 0 & 0 & 0 & 0 & \frac{1}{2G_{13}} \end{bmatrix} \begin{Bmatrix} \sigma_{11} \\ \sigma_{22} \\ \sigma_{22} \\ \sigma_{12} \\ \sigma_{23} \\ \sigma_{13} \end{Bmatrix}$$

E_1, E_2: 2 Young's modulus
ν_{12}, ν_{13}: 2 Poisson's ratios
G_{13}: shear modulus

Moreover, the experiment shows clearly that the law of behaviour is of a non-linear viscoelastic type

(Figure 7), i.e. that the five parameters E_1, E_3, ν_{12}, ν_{13}, and σ_{13} are a function of time and of the stress applied. We also put forward the hypothesis that in the field of our study — with stresses of less than 200 kPa — the strain on the plastic material is negligible compared to that of the block or sample.

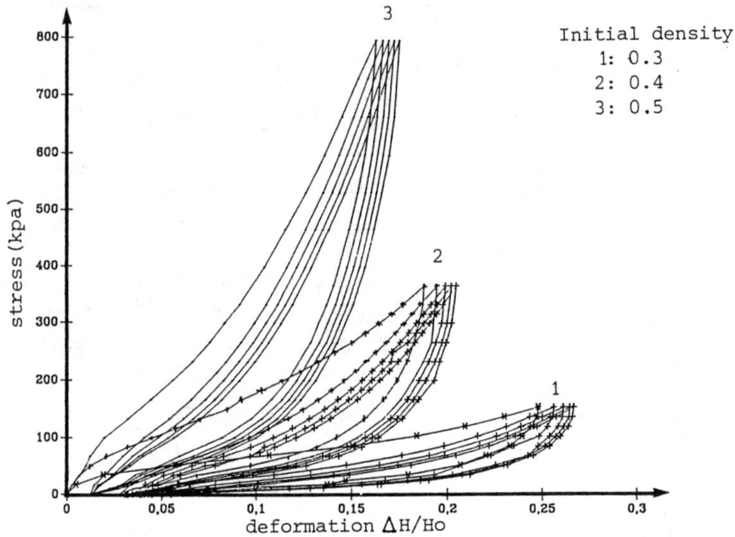

Figure 7: Loading-unloading (deformation speed: 4.8 mm/mn 5cycles)

Identification of Parameters

In the present case, $\sigma_{22} = \sigma_{11} = 0$ and $\varepsilon_{22} = \varepsilon_{11}$. The parameters to be determined are the following:

$E_{33} = \sigma_{33}/\varepsilon_{33}$ and $\nu_{13} = \nu_{23} = -\varepsilon_{11}/\varepsilon_{33}$, where σ_{33} is a given datum of the problem (applied stress).

Experimentally, the curves which give us ε_{33} as a logarithmic function of time are straight lines, for small values of t. This is obviously not the case any longer when t tends to infinity. To take this phenomenon into account, we suggest the following expression:

$$\varepsilon_{33} = a - be^{-c.\ln(t)}$$

where a, b, and c are positive constants, and t is expressed in minutes.

The determination of a, b, and c is linked to the following limit conditions:

- value of instantaneous strain (t = 1 minute);
- slope at the origin of the experimental curve: $\frac{\Delta H}{Ho} = f(\ln(t))$
- where t tends to infinity, the density of the block tends to the density of the plastic waste.

One can thus deduce that:

- $a = 1 - \frac{Do}{Dp}$, where Do is the initial density of the material, and Dp is the inherent density of the plastic material (0.94)
- $b = a - \left(\frac{\Delta H}{Ho}\right)_i$, $\left(\frac{\Delta H}{Ho}\right)_i$ being the instantaneous strain, determined as follows:

$$\left(\frac{\Delta H}{Ho}\right)_i = 1 - \frac{Do\ (\sigma+B)}{Dp\ (\sigma+B) - \sqrt{\sigma+A}} \quad \text{(where } \sigma \text{ is in } 10^2 \text{ kPa)}$$

where B is the positive root of the equation:

$$2tg(\alpha)\ .\ (Dp - Do)\ B^2 - 2(Dp - Do)^2\ B - 1 = 0$$

the value of α being such that $\sqrt{\alpha} = -\frac{\left(\frac{\pi}{2}\right)^{1/2} Do}{Dp} + \left(\frac{\pi}{2}\right)^{1/2}$

and $A = B^2(Dp - Do)^2$

These equations, which were established on the basis of tests on full-size blocks (H. El Ghoche, 1990), remain valid for laboratory tests with cylindrical samples. The maximum deviation, between b measured experimentally on the samples and b determined by using the above equations, is of the order of 4% (for the normal range of densities: ≥ 0.4) (Figure 8).

- $c = \frac{p}{b}$, p being the slope of the experimental

straight line $\frac{\Delta H}{Ho} = f(\ln(t))$.

P is derived statistically from the experimental curves. Two strain domains can be distinguished, whatever the initial density of the sample ($0.3 < Do < 0.6$).

Within the first domain ($\sigma_{33} \leq 30$ kPa):

$$p = \left(\frac{\Delta H}{Ho}\right)_{lmn} . K1, \text{ and } K1 = 0.08 \text{ (Table I)}$$

Within the second domain (30 kPa < $\sigma33$ < 100 kPa):

$$p = \left(\frac{\Delta H}{Ho}\right)_{lmn} . K2, \text{ and } K2 = 0.05 \text{ (Table II)}$$

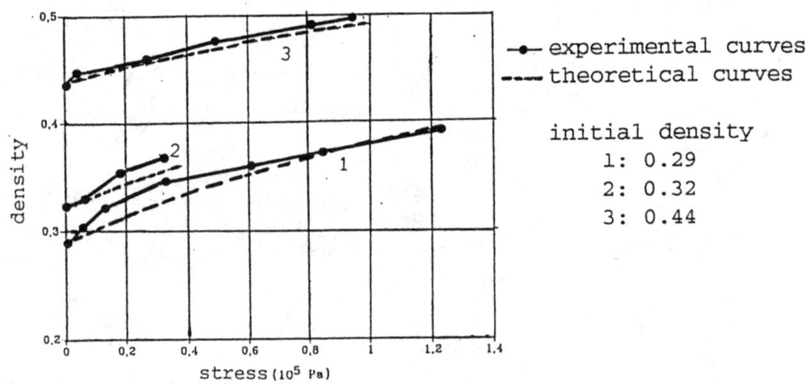

Figure 8: Density-stress (full-sized blocks)

Stress (kpa)	slope	(ΔH/Ho) i	Initial density	K
25.5	0.0061	0.0041	0.30	0.10
25.7	0.0008	0.0078	0.38	0.10
25.3	0.0002	0.003	0.46	0.04
24.7	0.0005	0.007	0.49	0.06

Table I: Average value of K is 0.08, the standard deviation is 0.026

Stress (kpa)	slope	(ΔH/Ho) i	Initial density	K
56	0.0091	0.1806	0.30	0.05
100	0.0051	0.2818	0.30	0.02
56	0.0036	0.0334	0.38	0.09
100	0.0060	0.1033	0.38	0.06
55	0.0006	0.0085	0.46	0.07
100	0.0014	0.0267	0.46	0.05
55	0.0003	0.0203	0.49	0.02
100	0.0010	0.0295	0.49	0.04

Table II: Average value of K is 0.05, the standard deviation is 0.02

Numerical application

Figure 9 shows the experimental curve $\Delta H/Ho = f(\ln(t))$, and the corresponding theoretical curve $\varepsilon_{33} = a - b.e^{-c.\ln(t).Ho}$ determined for the same sample, with an initial density of Do = 0.37, 0.43. The two curves match quite well.

Finally, the modulus E_3 of the material can be determined, for a given constraint and a given time.

For σ_{33} = 102 kPa, Do = 0.384, and t = 1 minute:
E_3 (experimental) = 870 kPa
E_3 (theoretical) = 920 kPa

The differences between the experimental and theoretical values are of the order of 5%.

Conclusion

The experimental study of waste plastic material compressed under uniaxial stress produced a first law of behaviour, which gave very satisfactory results.

We are currently determining the law of behaviour which applies under conditions of biaxial or triaxial stress, taking into consideration phenomena of ageing and loss of tensile stress.

However, if sufficiently high security coefficients are chosen, and as we know that under real conditions the stress will most probably be of the uniaxial type, compressed waste plastic blocks may already be of

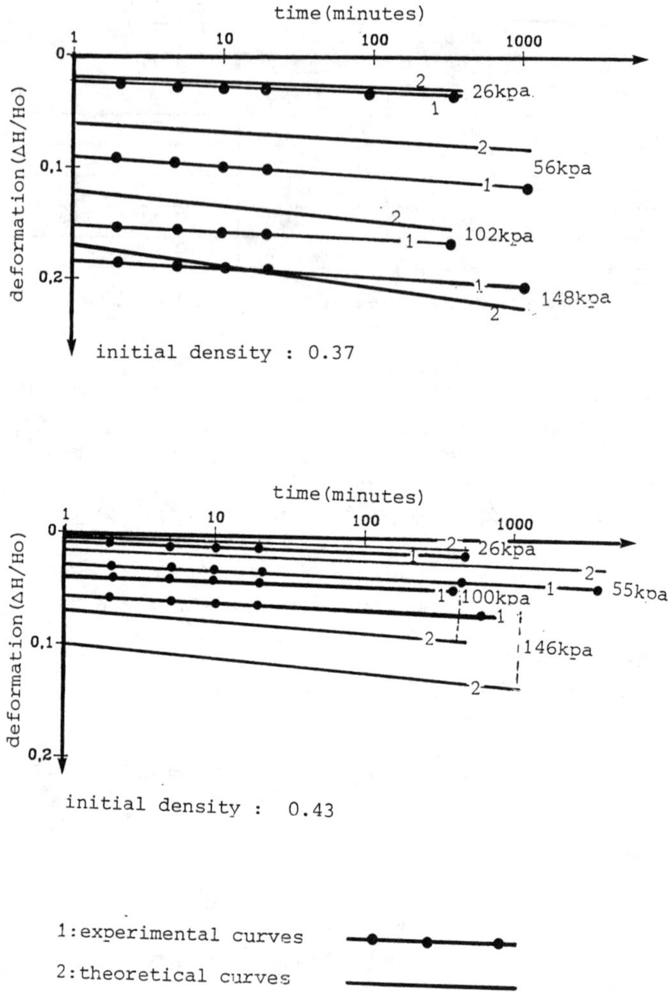

Fig 9: Experimental and theoretical curves

$$\text{for } \frac{\Delta H}{Ho} = f(\ln(t))$$

potential use in the construction of light-weight embankments. Tests carried out on a 4.6 m high experimental embankment with a volume of 300 m^3 demonstrated that the process was viable.

Appendix – References

Daudon, D., Coulet, C., and El Ghoche, H. (1991). "Plastic Waste Low-weight Soil", *GEO-Coast 91, International Conference on Geotechnical Engineering for Coastal Development*, Yokohama, Japan.

El Ghoche, H. (1989). "Les remblais allégés, étude du procédé Plastblock", *DEA INSA-IUT*, Lyon, France.

El Ghoche, H., Daudon, D., and Rakotondramanitra J.D. (1990). "Study of the Plastblock Process: Search for a Law of Behaviour for Low-Weight Embankments", *IXth National Conference on Soil Mechanics and Foundation Engineering*, Cracaw, Poland, p. 6.

Fornassero, R. (1988). "Récupération des ressources, matériaux et énergie par les recyclages et la réutilisation de déchets de matières plastiques", *Colloque International*, Paris, France.

Lemaitre, J. and Chaboche, J.L. (1988). *Mécanique des matériaux solides*, Dunod, France, 540.

Principles of Ground Modification With Electromagnetic Waves

J. C. Santamarina AM-ASCE[1] and Y. N. Wakim SM-ASCE[1]

Abstract

Electromagnetic waves offer new possibilities for ground modification, particularly in the presence of contaminants. This paper presents a review of fundamental concepts and related techniques, followed by the analysis of relevant underlying phenomena. Safety issues are also discussed.

Introduction

Soil modification methods can be classified by the methodology used to generate and to deliver energy (e.g., mass and gravity, combustion, conductivity, impact, mechanical wave). The characteristics of electromagnetic phenomena and the versatility of electronic equipment open new alternatives for soil improvement. For example, it is possible to develop noncontact methods where the energy can be remotely delivered and remotely conditioned to be selectively absorbed in the soil medium. The nature of the interaction of electromagnetic waves with soil media have additional advantages that are particularly relevant when soil modification techniques are contemplated in the context of environmental constraints, possibly coupled with the need for in situ remediation.

Electromagnetic waves are transverse waves, with the electrical and the magnetic fields perpendicular to each other and to the direction of propagation. Electromagnetic radiation is transmitted, through material and nonmaterial media, with no transport of matter. The frequency varies widely, from the very low frequency long wave-radio waves, to the very high frequency gamma and cosmic rays.

This paper centers on the processes and phenomena related to the use of electromagnetic waves to modify, stabilize or remediate soils in situ. The frequency range of interest is below the visible and includes microwaves, radar and radio waves; in this range, radiation is non-ionizing. An introduction to the interaction of electromagnetic waves with soils, and a review of existing and

---

1. Assistant Professor and Graduate Student, Polytechnic University, Brooklyn, NY 11201

related ground modification methods are included, followed by a discussion of relevant phenomena.

Electromagnetism and Soils

Maxwell's equations include Gauss's Law for the electric field E (Coulomb's law), Gauss's Law for the magnetic field B, Faraday's law on the electric field induced by a change in magnetic field, and Ampere-Maxwell's law on the magnetic field induced by a change in electric field or current. These equations govern all electromagnetic phenomena. For example, it follows from Maxwell's equations that the electric and magnetic fields propagate in wave form at the speed of light in free space:

$$c = \frac{E_0}{B_0} = (\mu_0 \epsilon_0)^{-1/2} = 3\text{x}10^8 \text{ m/sec} \qquad [1]$$

Wave-Material Interaction. When a semiconductor is subjected to a varying electric field, e.g. within the plates of a capacitor, an out of phase charging (polarizing) current takes place, together with an in phase loss current. The combined current I can be expressed as a function of the voltage V and a complex permittivity $\epsilon^* = \epsilon' - i\epsilon''$. The complex permittivity characterizes the interaction of electromagnetic waves and the material.

The dielectric constant ϵ' and the loss factor ϵ'' are normalized with respect to the constant permittivity of vacuum ϵ_0. The ratio of the imaginary to the real part is called tangent loss, $\tan\delta = \epsilon''/\epsilon'$; in low loss material $\epsilon''/\epsilon' << 1$, while in high loss conducting material $\epsilon''/\epsilon' >> 1$. The loss factor, ϵ'', combines all forms of losses, including polarization.

Loss mechanisms in soils vary with frequency (de Loor, 1968). Several mechanisms superimpose at low frequencies, i.e. RF, including: conductivity, double layer and bound water relaxations, and Maxwell-Wagner losses (characteristic of multiphase, heterogeneous media). In general, conductive losses tend to prevail over dipolar losses in aqueous dielectrics at very low frequencies; in this case, $\epsilon'' = \sigma/(2\pi f \epsilon_0)$, where σ is the conductivity and f is the frequency. At microwave frequency, the primary loss mechanism is the relaxation of free water (Water in excess of 8 to 9 monolayers contributes to the dielectric constant of the soil in this range). At higher frequencies, resonance of free water due to ionic and electronic polarization takes place at IR and UV frequencies respectively.

The relaxation time τ is the delay in the material's response to the changing field; it is a function of parameters such as viscosity; therefore τ varies with temperature. Maximum absorption occurs at the critical frequency $\omega = 1/\tau$, when ϵ'' reaches its maximum. The dielectric relaxation of ice or bound water occurs at radio frequencies (e.g., critical frequency about 3 kHz for ice at -10°C), whereas the relaxation of water occurs at much higher frequencies, ranging from 10GHz to 100GHz as the temperature increases from freezing to 75°C. Note that the activation energy of the first layers of bound water is in the order of 12 kcal/mole, which is similar to that of ice; in contrast, the activation energy of free water is about 4 kcal/mole.

The following general observations can be made about the dielectric properties of geotechnical materials: (a) ϵ' and ϵ'' depend on the frequency, the material, its temperature, and water content, (b) for most rocks, ϵ' is between 4 and 10, (c) for most dry soils ϵ' is between 2 and 4, (e) ϵ' of water is much higher than that of dry soils and rocks (ϵ'_{water}~80 in the RF region). The dielectric properties of soils are also affected by the bulk density, the organic matter content, salinity (particularly ϵ''), the number of layers of bound water molecules, the total surface area of the soil available to the water molecules which is dependent on the particle size distribution and mineralogy (clays possessing larger surface area will have a greater amount of water held at the surface), porosity, the dielectric constant of the components, geometry of the components, direction of propagation of the wave relative to any regular geometry of the components, and the interaction between the saturating fluid and the solid matrix.

Wave Propagation - Skin Depth. The 1-D propagation of the electric field in a dielectric is characterized as (Metaxas and Meredith, 1983):

$$\frac{\partial^2 E}{\partial z^2} = \epsilon_o \, \epsilon^* \, \mu_o \, \mu^* \frac{\partial^2 E}{\partial t^2} \quad [2]$$

which is satisfied by

$$E = E_{max} \, e^{(j\omega t - \gamma z)} \quad [3]$$

where γ is the complex propagating factor: $\gamma = \alpha + j\beta$, α is the attenuation and β is the phase. Therefore,

$$E = E_{max} \, e^{-\alpha z} . e^{j(\omega t - \beta z)} \quad [4]$$

The attenuation α can be defined in terms of the dielectric properties; in the cases of high-loss medium ($\epsilon''/\epsilon' >> 1$) and low-loss medium ($\epsilon''/\epsilon' << 1$), α reduces to the following expressions:

$$\text{high loss: } \alpha = (0.5\omega^2 \mu' \mu_o \epsilon'' \epsilon_o)^{1/2} \quad [5]$$

$$\text{low loss: } \alpha = \pi \epsilon'' / \lambda_0' (\epsilon')^{(1/2)} \quad [6]$$

The penetration depth of an electromagnetic wave is defined as the depth at which the power has attenuated to 1/e the power at the surface; it depends on the attenuation coefficient α. After normalizing with respect to the wavelength, the depth of penetration for low-loss and high-loss materials can be expressed as:

low loss: $Dp/\lambda = \epsilon'^{1/2}/(2\pi\epsilon'')$

high loss: $Dp/\lambda = 1/[4\pi(\epsilon''/2)^{1/2}]$

Figure 1 presents a dimensionless plot of penetration depth normalized with respect to the wave length, Dp/λ. Data for geotechnical materials are superimposed on the figure.

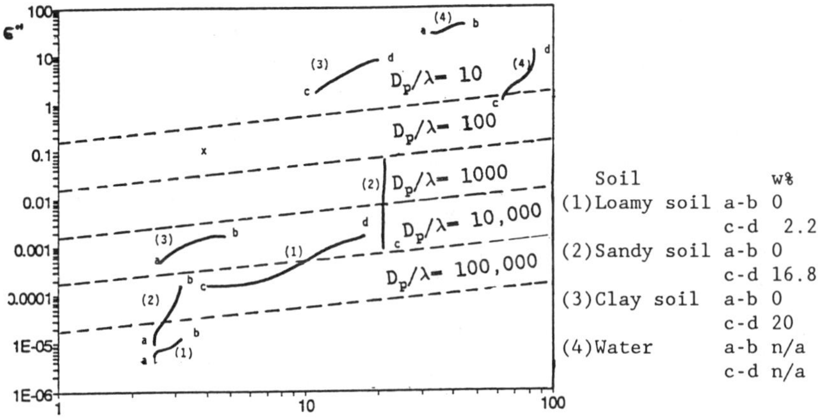

Figure 1: Penetration Depth

Power - Heating Rate. All electromagnetic radiation can be transformed into heat. The rate of energy transport per unit area at location "z" and time "t" is characterized by the cross product of the instantaneous values of the local electric and magnetic field vectors **E** and **B**, called the Poynting Vector **S**

$$S = (1/\mu_0)\ ExB \qquad [7]$$

where μ_0 is the permeability constant. It can be shown that the time average magnitude of the Poynting vector for a plane wave is

$$S_{ave} = (1/c\mu_0)\ E^2_{rms} \qquad [8]$$

The integral of the Poynting vector over a close surface is equal to the power dissipated and the rate of change of the stored energy in the enclosed region. In terms of material properties, the power dissipation per unit volume is

$$P = \omega\ \epsilon_o\ \epsilon''\ E^2_{rms} \qquad [9]$$

The rate of temperature increase relates to the dissipated power, the density ρ and the specific heat C of the material

$$\Delta T/t = \frac{P}{\rho . C} \qquad [10]$$

Ground Modification: Related Techniques

Electro-osmosis and Electrolysis. Electrokinetic phenomena involve electric fields and the relative movement of charged particles and liquids (Mitchell, 1976). Electro-osmosis is the flow of water under an applied external electrical field. The potential uses of electro-osmosis in soils include soil stabilization (with or without sacrificial electrodes), friction pile stabilization, and improvement of dewatering and grouting processes. More recently, it has been shown that a DC potential is also effective to alter the migration of contaminants, to create electro-osmotic barriers, and to inject environmentally active components.

Electro-osmosis is coupled with electrolysis, developing an acid front that advances from the anode across the medium (Acar, 1990; Hamed, et. al. 1991). The movement of this front and ion exchange are responsible for the removal of selected heavy metals. Oxidation and reduction of certain contaminants can also take place within the aqueous soil medium, increasing the value of electrokinetic processes for in situ soil modification / remediation (Criddle, 1991). Other processes include gas generation and heat; in fact a high power DC heating method is available.

Electro-acoustic Soil Decontamination. This method is an extension of electro-osmosis. In this case, the DC electric field is applied in conjunction with a varying acoustic field (Hinchee, et. al., 1989). During the electro-acoustic excitation, different processes similar to those in electro-osmosis take place. In addition, pressure fluctuations due to the varying acoustic intensity promotes diffusion and migration of the liquid phase from the solids, probably due to processes such as cavitation, cyclic growth of gas bubbles in the capillaries, dynamic excitation of the granular skeleton, and viscosity variations. The result is a synergistic effect between the two applied fields.

In Situ Vitrification. This thermal treatment method for site remediation consists of applying an electric field through graphite/molybdenum electrodes which are inserted around the area of interest; the process is seeded with a conductive starter path. The medium is heated to temperatures between 1400^0C and 1600^0C at which silica and aluminum oxides fuse (Timmons, 1990; Byers, et. al., 1991). The result is the destruction of virtually all organic components and the formation of an inert glass similar to naturally occurring obsidian. Several phenomena are involved, besides pyrolysis, including molecular diffusion due to differences in concentration, gas transport, and mechanical entrapment of most inorganic materials in the harden matrix. In situ vitrification presents advantages for the treatment of hazardous and radioactive wastes, soils, and sediments such as dewatered sludge and mine tailings.

Radio Frequency Heating. This method consists of heating the soil mass between 150^0C to 200^0C with electromagnetic waves in the RF range. In the process, saturation fluids evaporate and are collected, including both water and organic contaminants such as chlorinated solvents and hydrocarbons (Dev, 1988). The high efficiency of the method is the result of at least two underlying phenomena: the increased permeability of gas flow after removal of pore water, and the easier desorption of contaminants with the generated steam.

High Frequency Electromagnetic Waves. Several applications of electromagnetic waves in the microwave frequency have been proposed in areas related to ground modification: thermal stabilization where the firing torch is replaced with a wave guide and microwaves (Shibakova, 1975), thawing or weakening frozen soil (Hamid, 1982), and breaking rocks (Hoekstra, 1976; Santamarina, 1989). In the 1960's, microwaves were also used for vitrification (Morrell, et. al, 1986). Some of the processes involved in these techniques include drying, differential thermal stresses, synterization, vitrification, localization of energy absorption, hot-spots and runaway effect. The effectiveness of weakening methods depends on the stiffness of the material, the non uniformity of the internal field, the coefficients of thermal expansion, the rate of thermal diffusion, and confinement conditions.

Summary. The techniques described above are energy intensive. Values estimated from field and laboratory experiments documented in the literature are summarized in Table 1.

Table 1: Energy Demand Per Cubic Meter

Electro-osmosis Dewatering	70-100	kW.h	
Electrokinetic Decontamination	30-300	kW.h	
Electro-acoustic Decontamination	(see note)		
In Situ Vitrification	2000	kW.h	300 $
Radio Frequency Heating	400	kW.h	100 $
High Frequency Weakening	20-100	kW.h	
Standard Remediation Techniques			50-1500 $

Energy demand for electro-acoustic decontamination is not available, however, it is expected to vary within the range of electrokinetic decontamination (there are indications of synergistic effects). The costs listed refer to average on-site costs per cubic meter, and high variation should be expected. These ground modification methods may become cost effective by avoiding excavation and transportation, and by reducing the hazard involved in handling contaminated materials.

Phenomena and Related Concepts

Several phenomena have been identified in the techniques described above and others can be inferred. A complete listing is presented in Table 2. Heating accompanies all electromagnetic treatment methods. In this section, heating related effects are further discussed.

Table 2: Phenomena and Processes

Identified Phenomena
- -Electro-osmosis
- -Electrolysis
- -Gas generation
- -Hydrolytic reduction/oxidation
- -Changes pH & osmotic gradient
- -Ion exchange
- -Heat generation
- -Pyrolysis
- -Destruction organic components
- -Evaporation saturation fluids
- -Increased gas permeability
- -Contaminants desorption (steam)
- -Different. thermal stresses
- -Synterization and vitrification
- -Energy absorption localization
- -Hot-spots, runaway effect
- -Thermal diffusion
- -Mechanical entrapment

Other Effects
- -η changes with temperature
- -η changes with concentration
- -Volume dilation & pore pressure
- -Hydraulic gradient
- -Hydraulic fracturing
- -Thermo-osmotic seepage
- -Higher latent heat of evapor.
- -Selective absorption of energy
- -Activation of adsorbed layers
- -Bond breaking

Effects of Heating on Soil Properties. Modification techniques covered in this paper involve heating to different temperatures. As temperature rises, the following beneficial effects take place in soils: at low temperatures, 100°C to 200°C, dehydration of free and loosely bound water; at mid temperature, removal of tightly bound water and hydroxyls in crystal structure; at high temperature, synterization and formation of new components; and at very high temperature, melting (e.g., vitrification method - T>1500°C). It has also been observed that heating soils to the mid temperature range, 400°C to 600°C, reduces their plasticity and swelling potential, and increases the particle size and the strength (Wang, et. al. 1990).

Volume Expansion. The thermal expansion of water may produce very significant pore water pressure, probably leading to massive hydraulic fracturing of the medium. Figure 2 presents a pressure, volume, temperature PVT diagram for air-free water (see also Table 3). Assuming a very rigid system, $\Delta V=0$, thermal changes of just a few degrees can produce pressures of several hundred atmospheres. In low rigidity soil skeleton, the pore pressure build up due to heating will primarily depend on the degree of saturation (small air content sharply increases the compressibility of water), on the relative magnitudes of the rate of heating with respect to the rate of pore pressure dissipation, and on the extent of the heated area. It follows from the review of electromagnetic heating, that high power and high penetration would support this form of hydraulic fracturing in saturated fine soils (see also Campanella and Mitchell, 1968; Baldi, et. al., 1986).

Figure 3 presents low amplification photographs obtained with the scanning electron microscope of kaolinite samples prepared at the liquid limit and dried in air, in convection oven, and in standard microwave oven (f= 2450 MHz). Extensive fracturing is observed in the later sample; fractures form parallel to the plane of least confinement. Early signs of fracturing were also observed in the

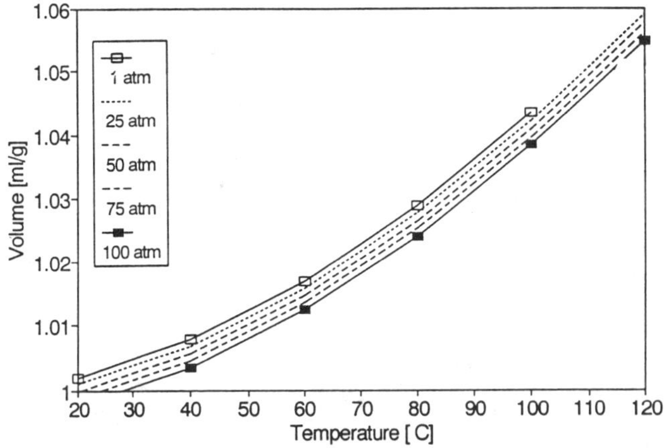

Figure 2: PVT Diagram for Air-Free Water

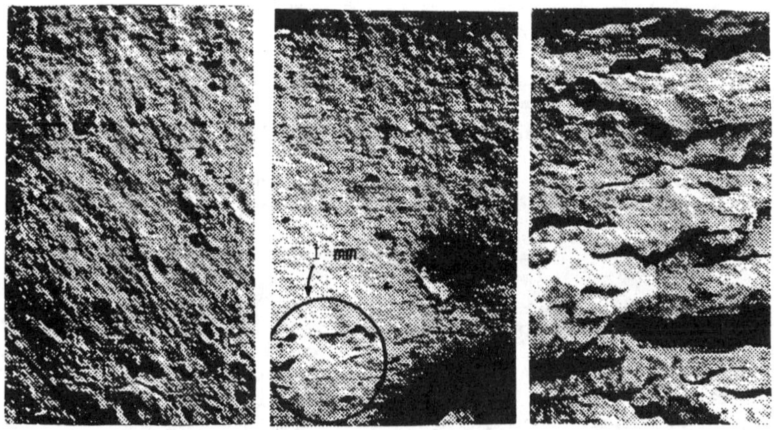

(a) Air 25^0C (b) Convect Oven 95^0C (c) MW Oven 95^0C

Figure 3: SEM Photographs of Dried Kaolinite Samples

convection oven dried sample, particularly away from the upper surface. Massive hydraulic fracturing can be either a positive or a negative outcome depending on whether dewatering or strength are the objectives of the modification process.

Even if fracturing pressures are not reached, there are still hydraulic gradients and thermal gradients that contribute to seepage, either directly or through the coupling effect of thermal-osmosis. (Thermal gradients inevitably develop due to the unevenness of the electric field inside the material, and to the heterogeneity of the material itself).

Viscosity. The first effect of heating on the hydraulic conductivity is through the modification of the viscosity. As shown in Table 3, the viscosity of water close to the boiling point is more than three times lower than that at 20°C. The concentration of different components also changes the viscosity of water. At 20°C, The percent increase in the viscosity of solutions prepared with 10% solute by weight, with respect to the viscosity of water are: -1% for KCl, 20% for NaCl, 32% for $CaCl_2$, 17% for $CdCl_2$, 55% $FeCl_3$, 6% for $Pb(NO_3)_2$, 30% for H_3PO_4, 33% for methanol, 50% for ethanol, and 63% for propanol (Weast, 1986).

Table 3: Temperature Dependent Properties for Air-Free Water

T °C	Specific Heat J/gr.°C	Relative Viscos ref: 20C []	Vapor Pressure 10^{-6} kN/m^2	Thermal Conduct cal per cm.C.hr	Bulk Modulus Elastic kN/m^2	Latent Heat of Vaporiz (J.Kg^{-1})
0	4.2177	1.777	0.06	4.83	2.02	23.63
20	4.1819	1.000	0.25	5.14	2.18	23.40
40	4.1786	0.652	0.76	5.40	2.28	23.18
60	4.1844	0.465	2.03	5.60	2.28	22.95
80	4.1964	0.353	4.96	5.76	2.20	22.85
100	4.2160	0.281	10.33	5.86	2.07	22.72

Thermodynamic Constants. The specific heat C is the amount of heat needed to raise 1°C the temperature of a 1 Kg mass, $c=\partial T/\partial(T)$, at constant volume C_v or pressure C_p. For liquids and solids $C_v=C_p$. C values for soil media are summarized in Table 4 (multiple sources).

Table 4: Relevant Themodynamic Constants For Soils

Thermal resistivity [°C.cm/W]:		
Quartz(random)		11.0
$CaCO_3$		26.3
Limestone		45.0
Sandstone		50.0
Water		165.0
Silica sand		49.2
Surge Sand [γ_d in KN/m^3]		
γ_d= 18.0	w% = 1	88.0
γ_d= 18.0	w% = 6	52.0
γ_d= 18.8	w% = 1	76.0
γ_d= 18.8	w% = 6	51.0
T= 20°C	w% = 1	72.0
T= 20°C	w% = 6	50.0
T= 80°C	w% = 1	78.0
T= 80°C	w% = 6	36.0

Specific heat [Cal/g.°]	
Clay	0.23
Sand	0.19
Limestone	0.22
Sandstone	0.22
Water	1.0

Heat of fusion L_F :
L_F of a 1 m^3 of soil with a w%
$L_{FS} = \rho_d w L_F = 334\rho_d w$ KJ/m^3

Heat of vaporization
L_v = 2.26 MJ/Kg = 2260MJ/m_3
at 100°C $L_{vs} = 2260\rho_d w$ KJ/m^3

The latent heat of evaporation is needed to compute the energy required for the evaporation of free and bound water. A series of drying tests were conducted to study the variation of the rate of drying at different moisture contents. Kaolinite and bentonite mixtures, ranging from 100% of one component to 100% of the other were prepared, and mixed with water (10gr of soil and 50 gr of de-ionized water). The mixture was placed inside an electrical oven, at a constant ambient temperature of 95^0C. The sample support was mounted on a force transducer that was frequently sampled with a computerized data acquisition equipment to monitor the evolution of drying. A classical result is presented in Figure 4 where the variation of the moisture content with time and the variation of the rate of drying ($\partial w\%/\partial t$) with time are shown. These curves indicate three regions: first, the increase in temperature, second the evaporation of free water at constant sample temperature, and last the evaporation of loosely bound water.

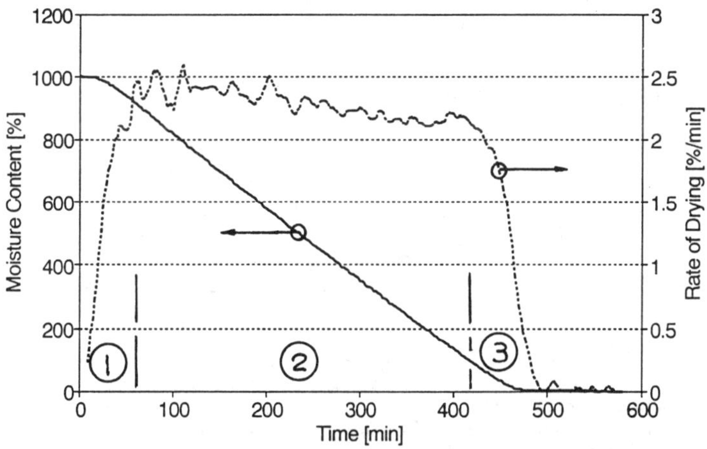

Figure 4: Drying Test - 10% Bentonite & 90% Kaolinite

If the ambient temperature is maintained below the transformation value, the latent heat can be estimated by assuming that the sample temperature remains constant during the drying process (1^0C to 4^0C variations were observed). Then the equivalent latent heat of evaporation during the third stage of drying, l^* can be related to the latent heat of evaporation of the free water l_{free} through the ratio of the drying rates:

$$l^* = l_{fw} \frac{(\partial w/\partial t)^{free}}{(\partial w/\partial t)^{bound}} \qquad [11]$$

This parameter l^* involves a variety of phenomena, including the energy involved in retraction and cracking.

The moisture content at three characteristic drying rates (refer to Figure 4), were determined for each kaolinite and bentonite mixture: 2%/min, 1.6%/min, and 0.6%/min. Results are shown in Figure

5, together with liquid and plastic limits. It follows that at constant ambient temperature (in this case 95^{0}C), it will take about twice longer to remove 1% moisture from a soil at the plastic limit than from the same soil at the liquid limit. According to this interpretation, the LL is the minimum moisture content beyond which true free water can exist in the sample.

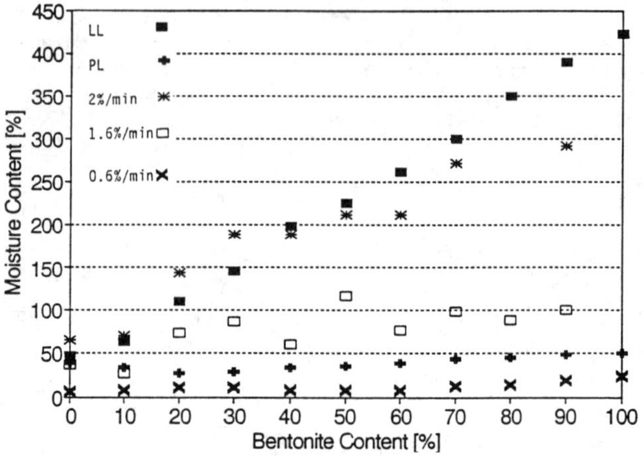

Figure 5: Rate of Drying at Constant Ambient Temperature (Atterberg Limits)

Thermal conductivity. This is an important parameter in the evaluation of the rate of heat dissipation through conductive losses and to assess thermal gradients in relation to heating rates. Table 3 summarizes the variation of thermal conductivity of water with temperature, while Table 4 presents some characteristic values for soils. The low thermal conductivity of soils favors heat dependent processes when electromagnetic waves are used: high temperatures, sometimes super-heating, can be reached in localized areas.

Other Effects. There are several other effects involved in heating soils. Some of them point to negative effects such as the trend to a thicker double layer with higher temperature predicted by Gouy's Theory (Mitchell, 1976). On the other hand, Maxwell-Wagner and other forms of relaxation in the RF range open possibilities for the physical activation of adsorbed layers. Most importantly is the possibility of remotely conditioning the delivery of energy to selectively affect regions or materials. Still, field implementations must address technical difficulties such as controlling run away effect in materials with positive slope ($\partial\epsilon''/\partial T$), and safety issues of the personnel involved.

Safety - Health Concerns

Unlike gamma and X-rays in which one photon is capable of breaking chemical bonds and of damaging tissues, the frequency of electromagnetic waves involved in applications discussed in this paper are non-ionizing. However they may effect the immune system functions, disrupt the blood brain barrier, damage chromosomes and lead to the development of cancer.

Power intensity limits have been set to avoid damages. In 1982, ANSI adopted a new frequency-dependent standard to limit average body absorption levels to 4 W/Kg of body weight, about half the heat generated by the body at rest. This regulation has a built in factor of safety of at least 10. Protective measures are simple: a cover all garment, protective screen rooms, wire mesh. Monitoring measurements can be efficiently implemented in a variety of ways, using calorimeters, thermistors, radiation, pressuremeters, voltage and resistance tests.

Conclusions

Ground modification with electromagnetic waves can achieve different results including evaporation of saturating fluids, removal of contaminants, synterization and vitrification. These and other possible techniques, such as the in situ polymerization of grouted monomers, demand additional research and field testing.

These methods can be implemented in noncontact form, remotely controlling energy, frequency and absorption potential. Multiple phenomena are involved, in most cases with beneficial effects for site decontamination.

Acknowledgments: Support for this research has been provided by NSF Grant MSS-9007844, and by the Industry-University Cooperation Program. The authors are grateful to the reviewers for insightful comments.

References

1. Acar, Y. B. (1990), "Electrokinetic Soil Processing: An Emerging Technology in Waste Remediation", Geotechnical News, June, pp. 26-28.
2. Baldi, G, Borsetto, M., Hueckel, T., and Tassoni, E. (1986), "Thermally Induced Strains and Pore Pressures in Clays", Proc. Int. Symposium Environmental Geotechnology, Ed. H.Y. Fang, vol. 1, pp. 391-402.
3. Byers, M.G., Fitzpatrick, V.F., and Holtz, R.D. (1991), "Site Remediation by In Situ Vitrification", Transportation Research Board, Paper NO. 910562, 22 pages.
4. Campanella, R. G., and Mitchell, J. K. (1968), "Influence of Temperature Variations on Soil Behavior, ASCE J Soil M, no. 3, pp 709-734.
5. Criddle, C. S., and McCarty, P. L. (1991), "Electrolytic Model System for Reductive Dehalogenation in Aqueous Environments", Environ. Sci. Technol., vol. 25, pp. 973-978.

6. de Loor, G. P. (1968), "Dielectric Properties of Heterogeneous Mixtures Containing Water", J Microwave Power, vol. 3, no. 2, pp. 67-73.
7. Dev, H., and Downey, D. (1988), "Zapping Hazwastes," Civil Engineering, August, pp. 43-45.
8. Hamed, J., Acar, Y.B., and Gale, R.J. (1991), "Pb(II) Removal From Kaolinite by Electrokinetics", ASCE J. Geotechnical Engineering, vol 117, no 2, pp. 241-271.
9. Hamid, M. (1982), "Microwave Thawing of Frozen Soils," J. Appl. Phys., vol. 17, no. 3, pp. 167-174.
10. Hinchee, R.E., Muralidhara, H.S., Stulen, F.B., Wickramanayake, G.B., and Jirjis, B.F. (1989), "Electroacoustic Soil Decontamination Process for In-Situ Treatment of Contaminated Soils", Battelle, Colombus, Ohio.
11. Hoekstra, P. (1976), "Rock, Frozen Soil and Ice Breakage by High Frequency Electromagnetic Radiation: A Review", U.S. Army Cold Regions Research and Engineering Laboratory, CRREL Report 76-36, 17 pages.
12. Metaxas, A. C. and Meredith, R. J. (1983), Industrial Microwave Heating, Peregrinus, London, 357 pages.
13. Mitchell, J. K. (1976), Fundamentals of Soil Behavior, Wiley, 421 pages.
14. Morrell, M. S., Hardwick, W. H., Murphy, V., and Wace, P. F. (1986), "A Pilot Plant Demonstration of the Vitrification of Radioactive Solutions Using Microwave Power", Nuclear and Chemical Waste Management, vol. 6, pp. 193-195.
15. Santamarina, J. C. (1989), "Rock Excavation with Microwaves: A Literature Review", Foundation Engineering: Current Principles and Practices, F. Kulhawy ed, vol. 1, pp. 459-473.
16. Shibakova, V.S. (1975), "On the Possibility of Using Microwave Energy for Soil Stabilization", Int. Assoc. Eng. Geology, Bulletin no. 12, pp. 89-91.
17. Timmons D., and Fitzpatrick V., and Liikala S. (1990), "Vitrification Tested on Hazardous Wastes," Pollution Engineering, June, pp. 76-81.
18. Wang, M.C., Benway, J.M., Arayssi, A.M. (1990), "The Effect of Heating on Engineering Properties of Clays", Physico-Chemical Aspects of Soil and Related Materials - ASTM-STP 1095, pp. 139-158.
19. Weast, R.C. (1986), CRC Handbook of Chemistry and Physics, Fourth Ed, CRC Press.

Contaminant-Grout Interaction

Stephan A. Jefferis[1]

Abstract

It is often necessary to use grouts in contaminated soils either as part of barrier systems (for example clay-cement grouts for slurry trench cut-off walls) or for general ground improvement. An understanding of the mechanics of contaminant-grout interactions is therefore fundamental to the successful application of grouts. In particular when testing grouts for chemical reaction it is most important that the test conditions are appropriate to those of the field and that the tests are carried on for sufficient time to show the full effects of reaction. Much data in the literature relate to tests involving the permeation of relatively few pore volumes through a sample under a high gradient. In this paper a technique for assessing the termination time for permeability tests is developed and it is shown that termination after the permeation of a few pore volumes or the establishment of an effectively steady permeability may not be a safe procedure for design purposes. The technique is applicable not only to grouts but also to clay liners for waste repositories etc.

Introduction

Traditionally many of the worst problems posed by chemicals in the ground were associated with sulphate either as a result of sulphate attack on cement based systems or more rarely from the oxidation of sulphides with the production of sulphuric acid. Today the range of aggressive chemicals is not restricted to such inorganic systems but may include complex organics in landfill leachates or the residues from previous industrial activity on a site. The range of contaminants, and hence of contaminant-grout interactions that may occur on a site today is almost unlimited. Therefore it is necessary to try to identify those contaminants that are most likely to limit the long term effectiveness of the material. It should be realised that the contaminants most damaging to the grout may be quite distinct from those that are most damaging to the environment. On occasion it may be appropriate to design a grout to react with and so retain/remove specific contaminants as a sacrificial system which must be replaced if the absorption/reaction capacity is exhausted before all the contaminant is removed.

---

[1]Reader in Geotechnical Processes, Queen Mary and Westfield College, University of London, Mile End Road, London E1 4NS, UK

Fundamentally the only method for assessing grout-chemical compatibility is to carry out the necessary chemical tests. This can be an extremely time consuming process and before undertaking such tests it is appropriate to try to develop a rational basis on which to design durability tests and in particular a procedure which can be used to assess whether the grout-chemical reaction is complete.

The Fundamentals of Grout-Chemical Interaction

There is a substantial literature on clay-chemical interactions and the question of the appropriate termination criteria for tests has been addressed by a number of authors. Pierce and Witter, 1986 suggest a procedure which requires that at least one pore volume is permeated through the clay and that the slope of the permeability versus cumulative pore volume plot cannot be shown to differ significantly from zero. Bowders, 1988 provides a useful discussion on Pierce and Witter's paper and in particular highlights the necessity for chemical equilibrium between sample and permeant.

It is a purpose of this paper to consider the problem of determining chemical equilibrium and to consider the shape of the permeability - time plot.

Reaction Volumes

The concentration of pollutants in water is often expressed in mg/litre. When considering durability of materials to chemical attack it is more convenient to express all concentrations in milli-equivalents per litre (meq/litre, obtained by dividing the concentration of each species in mg/litre by its equivalent weight). Just as a solution may contain chemically active species a grout may be regarded as a volume of reactive material. Once again this may be characterised by a concentration expressed as meq/litre but now by volume of grout rather than by volume of solution.

Thus if a solution contains a species, R of concentration, r milli-equivalents per litre which reacts with a species, C in the grout which is of concentration, c milli-equivalents per litre of grout (or equivalents per cubic metre) then the volume of solution, V_t for full reaction is given by:

$$V_t = c / r \text{ m}^3 \text{ of solution per m}^3 \text{ of grout} \quad (1)$$

It must be appreciated that for a single chemically contaminated water and a single grout there may be many different reactions each characterised by an independent value of V_t. Thus a slightly acidic water containing magnesium and sulphate ions might be involved in some or all of the following reactions with a clay-cement grout: neutralisation of the acid by reaction with free lime from the cement hydration, exchange calcium for magnesium in some of the cement hydrate systems and the formation of expansive calcium sulphoaluminates by reaction of the sulphate with the calcium aluminate hydrates of the cement. Since each of these reactions will have a different V_t if the arrival of the chemical is controlled by diffusion or permeation then there will be a chromatographic separation of the reaction fronts and some of the reactions and some of the products may be transitional and replaced by others as the reaction fronts progress.

Clearly there are many situations where it is not possible to establish a V_t for example wetting by non-aqueous phases or concentration dependent effects such as adsorption. However for the major chemical reactions the concept of V_t is fundamental and grout chemical reactions ought to be considered as comparable to a titration experiment in a chemical laboratory. The question then becomes: what is the most appropriate way in which to bring the grout and chemical into contact and to assess the resulting interaction effects?

Grout Test Procedures

A number of different procedures may be used to assess the effects of a contaminant on a grout. Possible procedures include:

a) mixing the grout with the appropriate volume of the contaminant solution (assessed from consideration of V_t) and analysing the reaction products,
b) immersing the grout specimens in the contaminant,
c) permeating the grout with the contaminant.

It is also necessary to consider the limiting chemical conditions to which the grout will be subjected. Ultimately in the ground after all reaction, dissolution etc. has occurred the grout will be in equilibrium with the groundwater. Tests must be designed so that this equilibrium can be achieved. For example if a grout and V_t of solution are brought into contact, by definition all of one reactive species will have been removed and the solution will contain the soluble reaction products. This is not the condition that will occur in the ground where the soluble reaction products will have been removed by permeation/diffusion and the concentration of chemicals in the groundwater is no longer influenced by any reactions.

Test procedures: Mixing Tests

These tests involve mixing the grout with the water. The tests may be carried out in two distinct ways

a) by preparing the grout with the test water,
b) by dispersing the grout in a calculated quantity of the water as in a standard titration test.

The first procedure may identify some gross reactions; however, the results must be interpreted with great care. Most grouts contain colloidal sized material. The effects of chemicals on colloidal materials will depend very much on the initial conditions. For example most bentonites will not swell in sea-water to give a non settling suspension. However, if an equivalent quantity of salt were added to a bentonite initially dispersed in fresh water there might be significant flocculation but the suspension would be likely to remain stable and non-settling. Also many of the reactions involved in the set of grouts are affected by the sequence of additions. As an extreme example: if sugar is added to a cement grout, set may be completely inhibited. Whereas if a dry cement is mixed with sugar solution a flash set (albeit to a weak friable material) may occur as the sugar removes the gypsum used as a moderator of the aluminate reaction in the cement.

The titration type test is very useful in identifying chemical reactions and in particular in assessing V_t. However as noted above, the conditions will not match those in the field and it may be necessary to 'extract' the grout with several volumes of the water and analyse the combined extract. The test is useful in that chemical effects can be assessed. However, it will give no data on physical effects. It does not follow that simply because chemical reactions occur that the physical effects will always be unacceptable. For example some reactions appear to improve (reduce) the permeability of grouts.

Immersion Tests

These are useful for assessing physical effects. Prisms of the grout or grouted soil are immersed in the test solution and the physical appearance of the samples is monitored. When setting up the tests it is important to keep in mind the fact that reaction may remove chemicals from the solution and reaction products may build up in the solution. It is therefore important to analyse the surrounding solution at intervals and to change the solution as appropriate. Clearly not all interactions will lead to visible change. Dissolution of a component of the grout may not be visible. However, colour changes, the deposition of surface films or cracking can be useful indicators of reaction. Cracking due to expansive reactions may be particularly spectacular. However, it must be accepted that for expansive reactions the immersion test may overestimate the damage. In-situ the grout normally will confined and the confining stress may be comparable to the strength of the grout with the result that expansive reactions may not lead to severe cracking but rather plastic deformation of the material. Figure 1 shows the result of sulphate attack on two samples of bentonite-cement grout immersed in magnesium sulphate solution. Under confined conditions the same samples showed no cracking though some softening.

Figure 1

The appropriate duration for immersion tests cannot be ascertained a priori. However, the tests are simple and of low cost and so they may be continued for long periods. Fundamentally immersion tests are best used to provide quick visual screening for rapid reactions and low cost long term compatibility studies.

Permeation Tests

For a grouted soil flow will occur both through the bulk material and also through any ungrouted voids, damaged regions etc. This latter situation is very difficult to analyse. The quantity of flow may be large and there may be damage to the grout by erosion as well as chemical reaction, solution etc. Tests such as the pinhole test for dispersive clays may be used and work is currently in hand for bentonite-cement grouts.

For bulk permeation tests the sample must be contained within some form of permeameter cell. There is a substantial literature on test procedures (see for example Bowders et al 1986). When selecting a permeameter system is must be kept in mind that chemical reaction may lead to a change in volume (Acar and Hollosy, 1987) and/or cracking of the sample. For fixed wall permeameters expansive reactions will lead to an unknown state of stress in the sample and cracking may be inhibited. Shrinkage may lead to edge leakage. For flexible wall permeameters excessive confining pressures may mask chemical interaction effects and in particular close cracks etc. Test under high pressure may give significantly different results to those under low pressure or on unconfined specimens. For example for a clay-cement grout permeated with magnesium chloride in a flexible wall permeameter the author has observed a thirty-fold difference in permeability for samples tested under effective stresses of 35 and 100 kN/m^2. Clearly the effects will be most severe for soft grouts or swelling clays where considerable consolidation may occur or more generally any system which may be significantly softened as a result of the chemical reaction.

If a permeability test is regarded as a titration type reaction but with the reactant brought into contact with the grout solely by permeation and if a volume, V of the chemical solution permeates through the sample the reaction front will advance a distance, x given by:

$$x = Vr/Ac = V/V_tA \qquad (2)$$

where A is the cross-sectional area of the grout mass and V_t, c, and r are as defined in Equation 1.

As V_t will be different for different reactions x also will be different and thus as previously noted there may be a chromatographic separation between the reaction fronts, though it may be necessary to consider the competition between reactive species and the potential for more complex synergistic and antagonistic reactions which produce effects out of proportion to the individual concentrations of reactive species.

Effects of Chemical Reaction

If for a grout system there is only one chemical reaction and as a result of this reaction the permeability changes from k_u (unreacted permeability) to k_r (reacted permeability) in the region through which the reaction front has moved. Then for a sample of length, L subjected to a head,

H of water for which the reaction front has permeated a distance, x the overall permeability, k_o of the combined reacted and unreacted material is given by:

$$L/k_o = x/k_r + (L-x)/k_u \tag{3}$$

The overall permeability, k_o is that which will be obtained from a simple permeability test. At the start of a test the k_o will be equal to k_u and during the test it will move towards k_r. It will not be possible to obtain data on k_r unless the test is continued to full reaction or data on x can be obtained (for example by pore pressure measurements or stopping the test and sectioning the sample - neither procedure will give reliable results if k_r is very different to k_u).

From Darcy's law and Equation 2 it can be shown that:

$$\frac{dx}{dt} = \frac{k_o H}{L V_t} \tag{4}$$

Substituting for k_o from Equation 3 and integrating yields:

$$t = V_t(x^2/k_r + (2Lx - x^2)/k_u)/2H \tag{5}$$

where t is the time of permeation and hence:

$$t_f = V_t L^2 (1/k_u + 1/k_r)/2H \tag{6}$$

where t_f is the time for full reaction, that is the time at which $x = L$ and $k_o = k_r$

By elimination between Equations 3, 5 & 6 it can be shown that for $t < t_f$:

$$1/k_o^2 = 1/k_u^2 + (1/k_r^2 - 1/k_u^2)t/t_f \tag{7}$$

It is convenient to non-dimensionalize Equation 7 using:

$$K_o = k_o/k_u, \quad K_r = k_r/k_u, \quad T = t/t_f$$

Hence for $T < 1$, Equation 7 becomes:

$$K_o = (1 + (K_r^{-2} - 1)T)^{-0.5} \tag{8}$$

and for $T > 1$

$$K_o = K_r \tag{9}$$

Figure 2 shows a plot of Equations 8 & 9 for a series of values of K_r from 0.001 to 1000. For $K_r < 1$ (that is systems which show a reduction in permeability) the shape of the curves is quite typical of some of the soil permeability - time plots reported by Mitchell and Madsen, 1987. An important feature of the curves is that any extrapolation of permeability from the initial trend will give an entirely erroneous result. Indeed unless V_t is known (or x can be determined for example by dismantling the test cell and inspecting the sample) no estimate of K_r can be made unless the test is continued until full reaction is achieved. In practice there may be several different reactions with different values of V_t, K_r and t_f and thus the permeability - time plot may be the sum of several interactions. Furthermore for materials of low initial permeability t_f will be of orders months to years. For example if:

$V_t = 10$, $k_u = 10^{-9}$ m/s, $k_r \gg k_u$, L = 76 mm and the applied hydraulic gradient is 10 then: $t_f = 440$ days.

For reactions leading to an increase in permeability the curves are almost indistinguishable for $K_r > 10$ except just prior to the break point at which the permeability levels off. The shape of the curves means that no reliable estimate of K_r can be obtained unless permeation is continued until the break point. Procedures based on the permeation of a few pore volumes can be totally inadequate as was shown by Bowders, 1988.

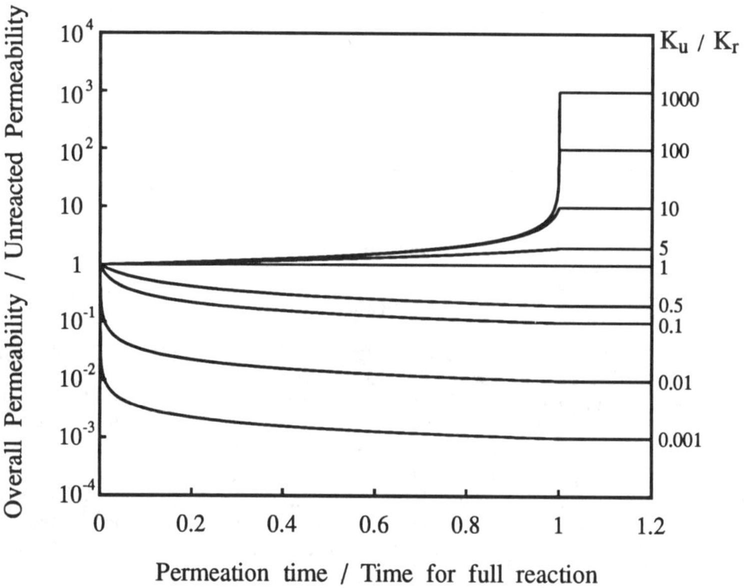

Figure 2

Whilst curves of the general shapes shown in Figure 2 are quite common in the literature the concept of analysis based on an advancing reaction front does not appear to have been previously reported. This simple theory has been applied to a number of permeation reactions. Figure 3 shows a typical plot of permeability against time for a cement-bentonite grout permeated with strong magnesium chloride solution. The magnesium will react with the calcium in the cement to precipitate magnesium hydroxide and there may be a number of other reactions including dissolution of some minerals and ion exchange. It can be seen that the data show a reasonable fit to the trend shown in Figure 2 except that there is an initial reduction in permeability.

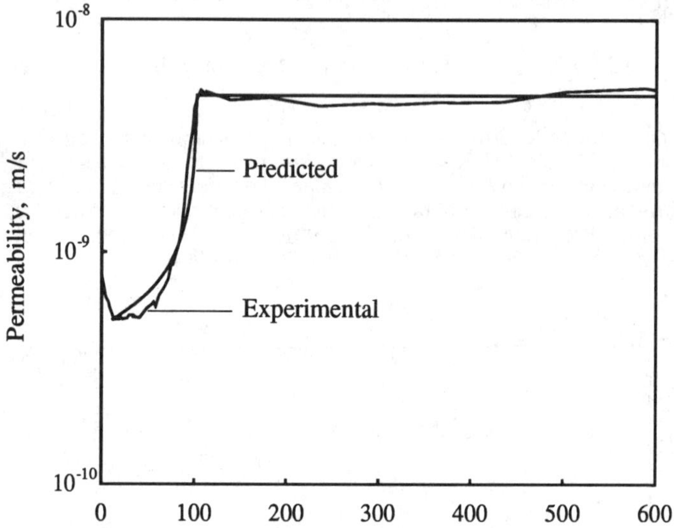

Figure 3

The reduction is not simply due to ageing of the sample as it has been observed regardless of the age of the sample at the start of the test. It may be caused by carbonation of the material. The predicted curve from Equation 7 has been added to the plot using $k_u = 5.1 \times 10^{-10}$ m/s (the minimum permeability value which occurs at about 12 days), $k_r = 4.8 \times 10^{-9}$ m/s (the steady state value at large times) and $t_f = 91$ days (the time at which the curve breaks away to a steady value). The predicted curve has been plotted from the minimum k_u value at age 12 days. It is not entirely satisfactory to start the predicted curve at this minimum but a procedure for multiple reactions is beyond the scope of this introduction. From the measured flows V_t was found to be 3.1 m^3 of permeant per m^3 of grout. From consideration of the chemistry of the grout and the permeant and assuming that the reaction was solely the conversion of all calcium in the grout to magnesium hydroxide the theoretical value of V_t should be 2.0 which is in quite reasonable agreement with the value measured in the test. An important lesson to draw from the data is that $t_f = 91$ days for a concentrated magnesium solution (30 grams/litre magnesium ion). Had a solution of 3 grams/litre been used then $t_f = 910$ days (2.5 years) and from consideration of the shape of the curve in Figure 3 it is apparent that the change in permeability up to say one year might have been minimal. Thus with dilute solutions very extended permeation times will be necessary to show the full damage - even if the effects of diffusion are added to those of permeation.

When assessing the effects of reactive chemicals it is important to realise that a major source of damage may be simple dissolution of the grout material without any specific chemical reaction. Thus for a clay-cement grout if it is assumed that all the calcium from the cement may be leached as lime (which has a solubility of 1.7 grams/litre) then V_t might be of order 100. For reaction of the same grout with a groundwater containing 100 mg/litre of magnesium ion (which could be a reasonable value) then V_t might be of order 500 and for many groundwater/grout combinations reaction effects may be limited and simple solution may be the dominant factor. Thus the value(s) of V_t for water (there may be different phases dissolved) can provide useful design data (for example in assessing the required thickness of a cut-off) and also a base-line for the consideration of other chemical reactions. Reactions which have significantly higher V_t's than water may not be critical even if they lead to substantial increases in permeability (large k_r / k_u) since a severe reaction which has penetrated a small distance into a grout mass will have minimal effect on its overall performance provided the damage does not lead to localisation, piping etc. Clearly, data are needed on this and are best obtained by exhumation of barriers etc. after period of operation.

Conclusions

It is the intention of the author to show that a very simple two permeability model can give useful information on the shape of the overall permeability - time plot and the termination time for durability tests. If the appropriate time is not observed in tests then unsafe results may be obtained especially for those reactions which lead to an increase in permeability. The model has been demonstrated to be applicable to laboratory tests involving an increase in permeability but further work is needed on reactions which lead to a decrease in permeability. More complex models may be developed including multiple permeabilities and allowing for diffusion as well as permeation. Piping may also be considered (a pipe has an area less than the full area of the sample but an effectively infinite reacted permeability).

It must be appreciated that most grout/groundwater combinations will not lead to a single reaction and thus a single value of V_t. Indeed V_t can be calculated only if the reaction is known. In general it will not be possible to identify all the potential reactions and this is clearly a substantial limitation of the technique. However, the shape of the curves given in Figure 2 is important. The reacted permeability cannot be predicted form the early age data. This is particularly important for reactions leading to an increase in permeability where it is not possible to predict the final permeability until effectively full reaction. If full reaction is to be achieved within a reasonable timescale then very thin samples must be used (for a fixed head difference across the sample the time for full reaction is proportional to the square of the sample thickness). Work is in hand on the development of special test geometries to enable such tests to be achieved within a reasonable time scale, though unfortunately time itself may be an important variable for slow reactions.

Finally it is not intended to suggest that laboratory test procedures are yet satisfactory for the assessment of field durability. Further work is needed on the scale up to field conditions where cracking due to chemically induced strains may be very important. Any change in permeability implies a change in the solid volume or the pore structure of a grout and thus possibly internal strains. The effects of such strains may not be apparent in small laboratory samples but may be severe for large masses in the field. Indeed cracking and localisation of damage may be the most important limitations to the operational life of a grout mass.

Acknowledgement

The author gratefully acknowledges the financial support of the UK Science and Engineering Research Council.

References

Acar, Y.B., and Hollosy, E., (1987). "Assessment of Pore Fluid Effects Using Flexible Wall and Consolidation Permeameters." Geotechnical Practice for Waste Disposal, ASCE Specialty Conference, University of Michigan, pp 231-245.

Bowders, J.J., Daniel, D.E., Broderick, G.P. and Liljestrand, H.M., (1986). "Methods for Testing the Compatibility of Clay Liners with Landfill Leachate." Hazardous and Industrial Solid Waste Testing, ASTM Special Technical Publication, 886, pp 233-250.

Bowders, J.J., (1988). "Termination Criteria for Clay Permeability Testing." Journal of the Geotechnical Engineering Division, ASCE, Vol 114, No. 8, pp 947-949

Mitchell, J.K., and Madsen, F.T., (1987). "Chemical Effects on Clay Hydraulic Conductivity." Geotechnical Practice for Waste Disposal, ASCE Specialty Conference, University of Michigan, pp 87-116.

Pierce, J.J., and Witter, K.A., (1986). "Termination Criteria for Clay Permeability Testing." Journal of the Geotechnical Engineering Division, ASCE, Vol 112, No9, pp 841-853

PERMANENCE OF GROUTED SANDS EXPOSED TO VARIOUS WATER CHEMISTRIES

John M. Siwula[1] and Raymond J. Krizek[2], M.ASCE

Abstract: Evaluated herein is the permanence (as reflected by changes in permeability) of grouted sand specimens subjected to waters of various chemistry under gradients of approximately ten. The test specimens were composed of Ottawa 20-30 sand injected with one of four different grouts (a microfine cement, an acrylate, and two types of sodium silicate grout). The acrylate grout showed no signs of deterioration with time under both high and low pH waters. The silicate-based sodium aluminate grout exhibited very low permeability under both high and low pH waters. The cement grout exhibited quite large permeability values when subjected to both low and high sulfate waters; this is attributable to the segregation of cement particles to form preferential flow paths. The silicate-based ethyl acetate-formamide grout produced specimens with a wide range of permeability values, depending on the pH and hardness of the permeant (both of which interfere with the silica solubility). Water with a high pH and/or low hardness resulted in a higher coefficient of permeability, and the grout was elutriated from the specimens with short curing times.

Introduction

The effectiveness of a grouting program depends on the degree of permeation of the grout into a soil deposit and its permanence once in place. To insure adequate permeation, the viscosity and gel time of the grout must be compatible with the soil formation and pumping equipment. The permanence of a grout depends on its strength and durability after injection, and these are functions of its chemical composition and curing time, as well as the chemistry of the environment into which it has been injected. It has been observed that some grouts decompose and/or dissolve when brought in contact with ground-water or are extruded from the soil when subjected to a hydraulic gradient.

---

[1] Project Engineer, Giles Engineering Associates, Inc., Waukesha, WI 53186

[2] Department of Civil Engineering, The Robert R. McCormick School of Engineering and Applied Science, Northwestern University, Evanston,IL 60208

Objective

This study is an extension of work reported by Krizek and Madden (1985), and it is directed toward evaluating the effect of ambient waters with different chemistries on the permanence of different grouts when injected in a sand, cured for prescribed periods, and then subjected to a hydraulic gradient. The water parameters considered are hydrogen ion activity (pH), hardness, and sulfate concentration. "Permanence" is considered to be reflected by changes in the permeability of the grouted specimens.

Previous Studies

Although no direct information regarding the effects of groundwater chemistry on grout permanence was found, many workers allude to this phenomenon. Caron (1965) suggests that leaching of algonite grout (sodium silicate, sodium aluminate, and water) is due to the grout being very susceptible to syneresis. Grouted specimens were subjected to gradients as high as 500 for periods up to 60 days and the amount of leached grout was noted; from these results it was concluded that grout is more likely to be retained in fine soils due to the fact that small voids support gelled grout better than large voids. The amount of leached sodium was also found to decrease with higher gradients due to reduced contact time, and the apparent permeability increased initially and then decreased. In a more general overview, Caron (1982) stated that the stability of aqueous resins and cement grouts was, in general, satisfactory, but the stability of silica gels was questionable; he claimed that this was due to syneresis and possible dissolution due to a reverse gel forming reaction.

Einstein and Schnitter (1970) performed short-term stability tests on polythixon (a two-phase polymer grout) while evaluating possible grouts for use in a dam project. Under a gradient of 12.5 approximately 20% of the grout was leached out, but after two days the leaching stopped and the subsequent permeability was not affected.

Petrovsky (1982) analyzed the leaching of cement grout in three gravity dams in the USSR. Three stages of leaching were identified in this study; these are (a) the rapid dissolution of free calcium hydroxide ($CaOH_2$), which resulted in the highest concentration of calcium in the leachate during the leaching process; (b) the hydrolytic action of the percolating water, which released lime and caused the calcium concentration of the seeping water to be reduced; and (c) the dissolution of calcium hydroxide and the decomposition of hydrated grout compounds, which resulted in the calcium concentration remaining relatively constant. The percolating waters at the three dams were characterized as soft waters (carbonate hardness between 12 and 75 mg/ℓ) with pH values between 6.8 and 8.6, and the maximum average sulfate concentration of the waters was 25 mg/ℓ. Petrovsky concluded that the amount of lime leached from a cement grout curtain depends on the age of the curtain and the duration of the seepage.

Karol (1982) stated that "chemical deterioration of grout can occur if the grout reacts with the soil or groundwater to form soluble reaction products, if the grout itself is soluble in groundwater, or if the reaction products which form the grout are inherently unstable." Baker (1982) reported that high pH groundwater can be detrimental to sodium silicate-based grouts, either by preventing gelation or by promoting deterioration with time. Low pH groundwater may accelerate the gelling of sodium silicate-based grouts and prevent the gelling of acrylamide grouts. Clarke (1982), while discussing the performance of AC-400 acrylate grout, reported that the gel formed during polymerization of AC-400 grout was insoluble in water, kerosene, and gasoline.

Graf (Personal Communication, 1984) has found that most sodium silicate grouts are stable under normal groundwater conditions. Although acidic water should enhance the gelation of silicate grouts, alkaline water can possibly dissolve gelled grout. However, alkaline water normally contains large amounts of calcium carbonate and this may result in calcium ions reacting with the silicate to form a hard calcium silicate compound. Permeability, under these conditions, may increase slightly, but the strength and grout matrix will remain intact. Graf also stated that silicate grout cannot be injected into formations containing seawater unless a buffer solution is added, because the magnesium ions in the water will react with the silicate and give undesirable results.

Krizek and Madden (1985) studied the stability of silicate and acrylamide grouts in sand exposed to tap water under gradients of 100. The results of this study suggest that, once gelled, acrylamide grouts are very stable and unaffected by the applied gradient. One silicate-based grout with sodium aluminate as the reagent displayed slight permeability increases during the period soon after the gradient was applied, but the permeability stabilized thereafter and the grout remained intact. The other four silicate grouts (which utilized (a) ethyl acetate and formamide, (b) Hardener 600, (c) Terraset, and (d) Geloc 4 as reagents) all exhibited large increases in permeability when the gradient was applied, but the rate and magnitude of the increase was influenced by the curing time of the specimens and the magnitude of the applied gradient. Specimens with short curing times exhibited complete elutriation of the grout due to inadequate gel strength. Longer curing times resulted in gradual decreases in permeability without complete elutriation of the grout. The long-term permeability of the silicate-based grouts was generally one to two orders of magnitude less than that of the ungrouted sand and it was relatively independent of the grain size and density of the sand and the chemical composition of the grout.

Materials Tested

In this study three chemical grouts (two silicate and one acrylate) and one cement grout were used in conjunction with standard Ottawa 20-30 sand to prepare test specimens. The waters used as permeants were composed of distilled water and appropriate chemicals to obtain predetermined pH and hardness values.

Silicate-based Grouts (SE and SA)

Two reagents [either sodium aluminate ($NaAlO_2$) or a mixture of ethyl acetate ($Ch_3COOC_2H_5$) and formamide ($HCONH_2$)] were combined with sodium silicate to form silicate-based grout solutions. The silicate was supplied in the form of commercially available "N" grade sodium silicate manufactured by the Philadelphia Quartz Company. This silicate contains 28.9% SiO_2 and 9.8% Na_2O by weight, resulting in a weight ratio (SiO_2/Na_2O) of 3.22. At 20°C, "N" grade sodium silicate has a viscosity of 180 cp, a specific gravity of 1.38, and pH of 11.3.

Combining ethyl acetate and formamide with sodium silicate (SE) releases acid and induces gelation by allowing the silicate particles to link together. Specifically, the ethyl acetate is hydrolyzed into acetic acid and ethanol, while the formamide is slowly converted to formic acid. Formamide also increases the solubility of ethyl acetate in water, resulting in a more uniform mixture. This solution results in a strong gel with predictable gel times. In this study equal portions (by volume) of ethyl acetate and formamide were mixed with a specified volume of tap water, and the resulting solution was added to the sodium silicate and mixed thoroughly. The ungelled solution had a viscosity of 5 cp, a specific gravity of 1.2, and a pH of 11.2 at 20°C.

The gelation process resulting from the addition of sodium aluminate to sodium silicate (SA) is similar to that of pure silicate. The main difference is that alumino-silicic acid ions are also involved in the reaction. The gel formed is amorphous and composed of short alumino-silicate chains. Although the grout exhibits low strength, it has the ability to store large unincorporated molecules (such as water or hydrated sodium) within its structure and leaching of these interstitial ions by washing is minimal (Merrill and Spencer, 1950). The gel time of the grout depends on the ratio of aluminate to silicate (Al_2O_3/SiO_2) in solution, but once gelation has occurred, the gel displays very little syneresis. The sodium aluminate used in this study is known commercially as #2 from the Nalco Chemical Company and contains 19.9% aluminate by weight. The mixing procedure consisted of adding sodium silicate and sodium aluminate to equal volumes of distilled water. After these solutions were well blended, they were combined and completely mixed. The grout mixture had a pH of 11.0, a specific gravity of 1.12, and a viscosity of 3 cp at 20°C.

MC-500 Microfine Cement Grout (MC)

MC-500 is the commercial name of a microfine cement sold by the Onoda Cement Corporation of Japan and distributed in the United States by the Geochemical Corporation in Ridgewood, New Jersey. Although the composition of this cement is proprietary, it is believed to contain a large portion of slag cement. The mixing procedure used in this study consisted of combining the required volumes of tap water and NS-200 (a dispersant) and mixing thoroughly. Subsequently, the MC-500 was slowly added to the water solution and completely mixed at a water-cement ratio of 3:1.

AC-400 Acrylate Grout (AC)

AC-400 grout was developed as a replacement for AM-9 grout (Clarke, 1982), which had toxicity problems; it is composed of acrylate monomers, an accelerator, and an initiator. Triethanolamine (TEA) is the accelerator which activates the monomer system, and polymerization of the monomer system is initiated by the addition of ammonium persulfate (AP). The monomer system is composed of an acrylate monomer and a small amount of methylenebisacrylamide (MBA), a crosslinking monomer. Catalysis of the monomers results in a durable, flexible gel. The mixing procedure consists of combining specified weights of TEA, tap water, and AC-400 monomer, mixing the solution thoroughly, and then adding it to a mixture of AP and tap water. The gel time is controlled primarily by varying the amounts of TEA and AP, but potassium ferricyanide (KFe) can be added to increase the gel time.

Soil

All specimens were prepared with standard Ottawa 20-30 sand, which has a maximum and minimum void ratio of 0.67 and 0.45 (Zelasko, 1966), respectively, and a coefficient of permeability of approximately 0.04 cm/sec.

Test Waters

A survey was conducted to determine the chemistry of different groundwaters in the United States; the principal parameters of interest were hydrogen ion activity (pH) and hardness (as $CaCO_3$). Approximately 20,000 water quality analyses were received from various agencies in twelve states, and the following extremes were found:

Highest pH	12.2	Apache County, Arizona
Lowest pH	3.2	Anderson County, Texas
Greatest Hardness	8100 mg/ℓ (as $CaCo_3$)	Montague County, Texas

Based on this information, it was decided to prepare synthetic waters with pH values of 5, 8, and 10 for use in this study. Both the pH 5 and pH 10 waters were to have low hardness (< 100 mg/ℓ as $CaCO_3$). The effect of varying degrees of hardness was checked by using pH 8 water with hardness values of 100 (low), 500 (medium), and 6500 (high) mg/ℓ as $CaCO_3$. Appropriate amounts of calcium chloride dehydrate ($CaCl_2 \cdot 2H_2O$), magnesium sulfate anhydrous ($MgSO_4$), sodium bicarbonate ($NaHCO_3$), and/or sodium carbonate anhydrous (Na_2CO_3) were added to distilled water to attain the desired pH and hardness. These chemicals were chosen because, assuming complete dissociation, only those ions normally found in groundwater would be present in the synthetic solutions. The chemical quality of the proposed and actual test waters is presented in Figure 1.

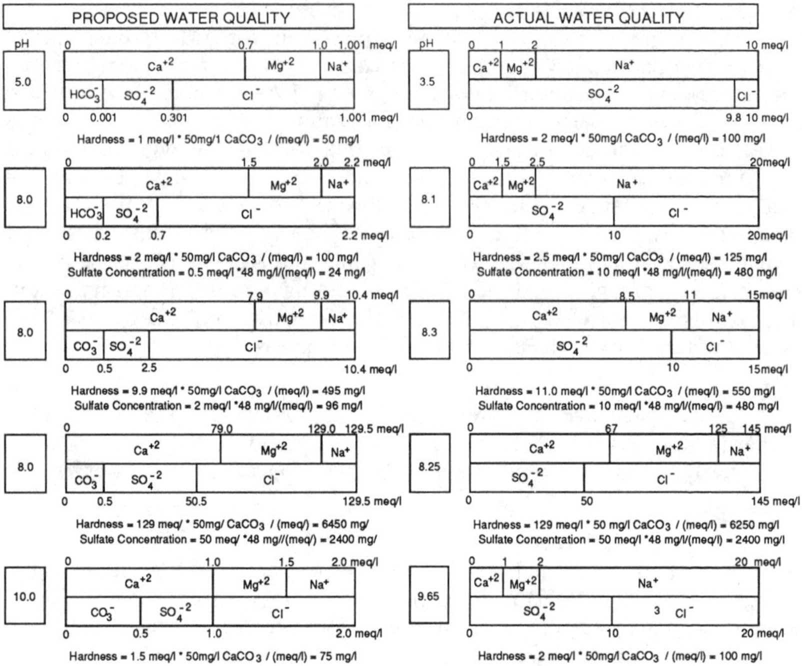

Figure 1. Chemical Quality of Test Waters

Table 1. Grout Formulations

Grout	Active Component	Water	Reagents and Additives	Gel Time
SE	50% Sodium Silicate	40% Tap	5% Ethyl Acetate and 5% Formamide	30 minutes
SA	24% Sodium Silicate	72% Distilled	4% Sodium Aluminate	10 minutes
AC*	25.46% AC-400	73.74% Tap	0.4% TEA and 0.4% AP	15 minutes
MC	9.88% MC-500	89.82% Tap	0.3% NS-200 (dispersant)	6-8 hours

* Percent by weight; all others by volume.

Scope of Experimental Program

Specimens grouted with one of the four different grouts and cured for various periods of time were subjected to five synthetic groundwaters with various pH and hardness values under low gradients. Intermittent constant-head permeability tests were performed for all grout types, and a few falling-head permeability tests were performed on the cement grouted specimens. The sodium silicate-ethyl acetate-formamide grouted specimens were tested with all five waters after curing for periods of 15 minutes, 1, 3, and 9 hours, and 1 and 4 days. The MC-500 grouted specimens were tested with both low and high hardness pH 8 waters, which also had low and high sulfate concentrations, after curing periods of 2, 8, and 16 hours, and 1, 3, and 7 days. The AC-400 and sodium silicate-sodium aluminate grouted specimens were tested with the pH 5 and pH 10 waters after curing for periods of 1 hour and 1 day. A single formulation of each grout (Table 1) was used in all portions of this test program.

Experimental Methodology

Described here are the methods used to prepare the specimens and waters (permeants), inject the grout, and conduct the tests.

Specimen Preparation

The permeameters consisted of 6-inch lengths of 2-inch diameter, medium-wall, pyrex glass tubing, sealed at each end with rubber stoppers, as illustrated in Figure 2. The stoppers were held in place with threaded rods restrained against two lucite plates. A threaded brass shut-off valve with a 0.125-inch inside diameter was inserted into the bottom stopper, and a plastic check valve was fitted into the top stopper (attached to an effluent hose). A filter system was placed at each end of the specimen. The filters consisted of nylon mesh and a 0.5-inch porous stone composed of a mixture of Ottawa 20-30 sand and epoxy. The permeability of the stone was of the same order of magnitude as that of the ungrouted sand specimens. The filters, together with the penetration of the rubber stoppers, resulted in an actual specimen length of approximately 4 inches.

All specimens were prepared in a dense state to avoid any further densification after preparation and during injection. Sand was pluviated into the permeameter through a funnel positioned to maintain a constant height of drop of approximately 1 inch. The specimens were constructed in three lifts. After each lift, a one-pound weight was placed on top of the lift inside the glass tube and the specimen was vibrated with a hand-held vibra-tool for one minute at each of three points on the circumference at intervals of 120°. The average relative density of the specimens was 86% ± 9%, corresponding to an average void ratio of 0.48 ± 0.02.

Water Preparation

The compositions of the synthetic waters were chosen with regard to the major ions normally found in naturally occurring groundwater, and trace substances were ignored. Preparation of the various waters consisted of adding proper amounts of the appropriate chemicals to distilled water and mixing the

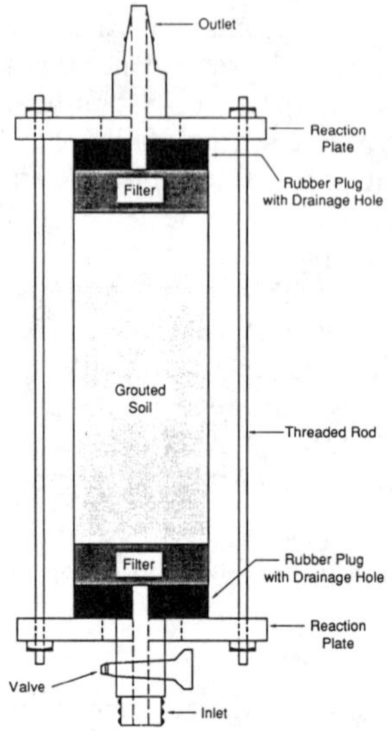

Figure 2. Schematic Representation of Permeameter

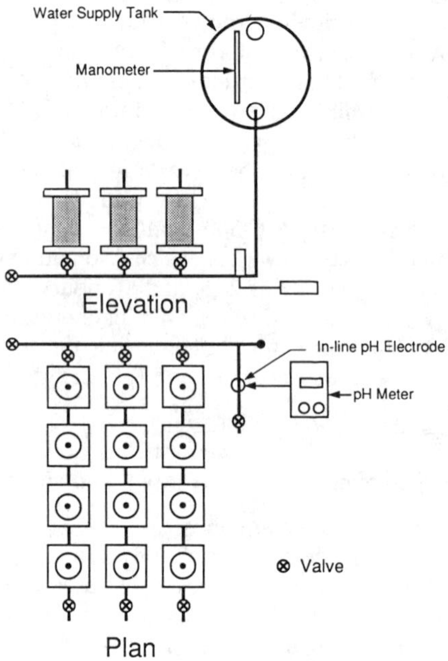

Figure 3. Water Distribution System

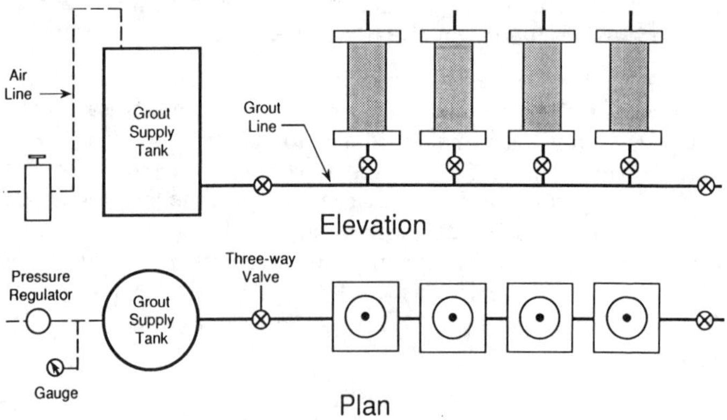

Figure 4. Grout Injection System

solution for approximately five minutes to ensure homogeneity. After mixing, the nominal pH 5 water had a slightly higher pH than expected, and small amounts of dilute hydrochloric acid were added to lower the pH. The pH 8 waters with low (100 mg/ℓ) and medium (500 mg/ℓ) carbonate hardness values displayed pH values within 0.5 of the desired values, and this was considered acceptable. However, the nominal pH 8 water with high carbonate hardness (6500 mg/ℓ) exhibited an actual pH near 10, which was not acceptable; the pH of this water was lowered without affecting its hardness by introducing carbon dioxide (CO_2) via compressed air into the water for a period of time. A pilot test on an ungrouted sand specimen ascertained that the compressed air had a negligible influence on the specimen permeability. The pH 10 water was very unstable when in contact with the atmosphere due to the presence of carbon dioxide. Hence, this water was placed in contact with nitrogen gas to eliminate the carbon dioxide source. This technique proved successful and was used throughout this study to maintain control in the high pH water.

The equipment assembled to supply the synthetic waters to the grouted sand specimens was selected to minimize contamination of the water. All materials were either inert plastic or non-corrosive brass. As shown in Figure 3, the manifold system consisted of 0.5-inch inside diameter, Schedule 80, polyvinyl chloride (PVC) pipe arranged into three bays. In each case the synthetic waters were analyzed with a Hach Model SA-1 soil analysis kit to ascertain their chemical composition, and the pH of the waters was monitored throughout the test program by means of an in-line pH electrode attached to the manifold. While testing the pH 5 water after the addition of dilute hydrochloric acid, it was found that the actual pH of the water was 3.5 (much lower than desired), but, since the pH was still within the range determined by the water survey, it was used in the study. Data on the approximate water quality of the actual test waters are presented in Figure 1.

Grout Injection

The grout injection manifold depicted in Figure 4 was composed of 0.25-inch inside diameter, Schedule 80, PVC pipe and enabled up to four specimens to be injected simultaneously. Grout was supplied to this manifold through a three-way brass valve, which allowed grout flow into the manifold or around the manifold (thus allowing a grout sample to be collected for gel time determination), as well as complete stoppage of flow. The grout itself was stored in a 4.5-liter lucite tank attached to the three-way valve by means of 0.25-inch inside diameter tygon tubing. Flow was accomplished by using air pressure to force the grout out of the tank through a bottom hose. The air was supplied by a compressor and regulator which enabled air pressures to be maintained within 0.5 psi.

The grout injection procedure was developed to simulate field injection procedures. Prior to grout injection, each specimen was placed on the water manifold and the ungrouted permeability, k_u, was determined. Then, the specimens to be injected were placed on the grout manifold and water was cycled through the system to eliminate any trapped air. Subsequently the grout was mixed and placed in the grout supply tank. The air pressure was applied and the specimens were injected. Approximately two volumes of grout were passed through each specimen at an injection pressure of 2 ± 0.5 psi. After injection, the specimens were cured prior to the application of a gradient. The samples with a 15-minute curing time were immediately placed on the water manifold, while those with longer curing times were placed in a high humidity environment, where they remained until the end of the designated curing period.

Test Procedure

The test procedure consisted of placing a grouted specimen on the manifold injection system and performing intermittent constant head permeability tests until the permeability values stabilized or reached the ungrouted permeability, k_u; this usually occurred within a few days. The gradient was maintained at a value of approximately 10 during all portions of this test program, and the pH and description of the effluent were recorded for all samples. For all practical purposes the tests performed on specimens with a coefficient of permeability in the intermediate to low range (approximately 10^{-4}cm/sec or less) can be considered continuous, since the gradient was applied and water flowed for virtually the entire test period. However, the tests performed on specimens with a high coefficient of permeability (greater than 10^{-3}cm/sec) were intermittent in nature, because the gradient was applied and water flowed for less than one-half of the test period. This procedure was adopted to minimize water loss, since only a limited quantity of each test water was available.

The specimens injected with MC-500 grout exhibited permeability values much higher than expected, and side flow around the specimen perimeter was suspected as the cause. Therefore, falling head permeability tests were conducted in a triaxial apparatus to greatly reduce or eliminate any side flow; three specimens, each cured for seven days, were tested. Each specimen was placed in a triaxial cell, and a cell pressure of 35 psi and a back-pressure of 30 psi were applied. The low (125 mg/ℓ) carbonate hardness water with a pH of 8 was used as the permeant, and the tests continued until the permeability values stabilized or reached k_u.

Interpretation of Results

Data from fifty-seven constant-head permeability test specimens and three falling-head permeability test specimens were analyzed, along with observations from examining the specimens after dismantling. The results are presented and evaluated in terms of the various waters, grouts, and curing times that were employed. The grout consisting of sodium silicate with ethyl acetate and formamide as the reagent produced specimens having coefficients of permeability in the range from 10^{-2} to 10^{-7} cm/sec. The specimens with MC-500 microfine cement grout displayed permeabilities from 10^{-2} to 10^{-4} cm/sec. The sodium silicate-sodium aluminate grout yielded specimens with fairly stable permeability values near 10^{-8}cm/sec, and the specimens grouted with AC-400 were extremely stable and no flow was observed. Although permeability values are plotted as a function of time in all of the following figures, supplemental plots were made for the same data as a function of void volumes of permeant passing through the specimens to assist in interpreting the results. However, the void volumes span such a small range that no clarification of the underlying phenomenon can be deduced, and these plots are not included here.

Sodium Silicate-Ethyl Acetate-Formamide Grout (SE)

The sodium silicate-ethyl acetate-formamide grout performed quite differently over the range of water pH and hardness tested. As seen in Figure 5, the permeability decreases as the pH of the permeant decreases, particularly in the case of specimens cured for periods of 1 hour or less. The unusual behavior of the specimen with a 1-hour curing time in contact with the pH 4 water was due to a clogged valve, and the data should be discounted. The results of tests on specimens with a 3-hour curing time are inconclusive. There does not appear to be a significant difference in the permeability of specimens cured for 9 hours or more and then subjected to permeants with different pH values. The final coefficients of permeability of specimens cured for 9 hours or more are approximately 10^{-6}cm/sec for the pH 10 and pH 8 waters and 10^{-7}cm/sec for the pH 4 water. The effect of a high pH apparently causes a reverse gel-forming reaction (dissolution), especially if the salt content of the water is low. Data from Krizek and Madden (1985) are plotted on the pH 8 water plot of Figure 5 (since the tap water used in their study was very similar to this water); specimens with 1-day curing times were found to exhibit permeabilities (10^{-3}cm/sec or more) larger than those in this test program, and this difference is attributed to the very high gradient (100) used in their test program. In general, the effect of the curing time has a much greater effect (up to four orders of magnitude) on the permeability of the sodium silicate-ethyl acetate-formamide grout than either the water pH or hardness.

The effect of water hardness on the permeability of specimens injected with sodium silicate-ethyl acetate-formamide grout is illustrated in Figure 6. A lower water hardness tends to increase the specimen permeability in a shorter period of time, especially for those with a 15-minute curing time. Water hardness affects permeability by altering the solubility of the silicate. Calcium (Ca^{+2}) and magnesium (Mg^{+2}) tend to form insoluble calcium and magnesium silicates. For curing times greater than one hour, water hardness does not cause a significant difference in the permeability.

Sodium Silicate-Sodium Aluminate Grout (SA)

As illustrated by the results presented in Figure 7, specimens injected with sodium silicate-sodium aluminate grout exhibited very low permeability values. The hydrogen ion concentration of the permeant appears to have little effect on the permeability; regardless of the curing time, the final coefficients of permeability were approximately 10^{-7}cm/sec for both the pH 4 and pH 10 waters. This grout is stable because of the formation of an insoluble gel, which is due to the aluminum and silicate molecules being chemically similar and having nearly the same atomic diameter.

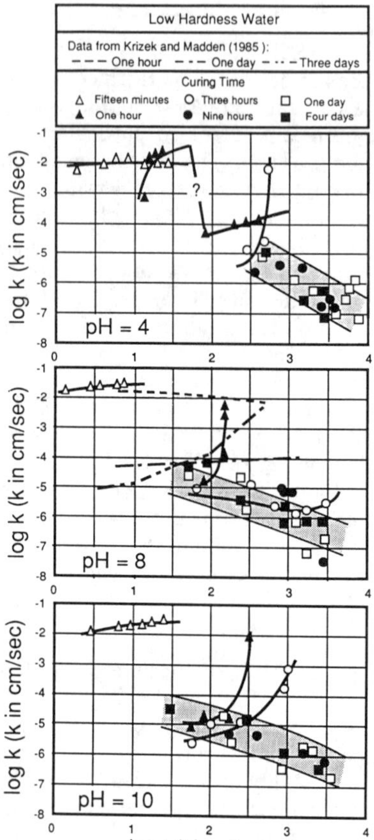

Figure 5. Permeability versus Time for Specimens Injected with Sodium Silicate-Ethyl Acetate-Formamide Grout Subjected to Low Hardness Waters at Different pH Values

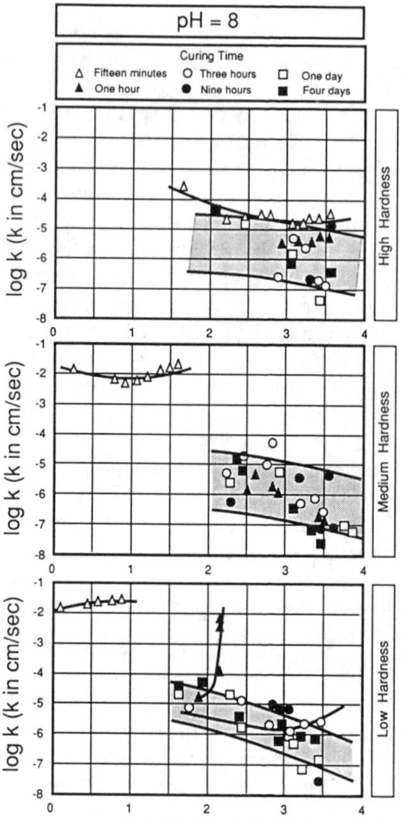

Figure 6. Permeability versus Time for Specimens Injected with Sodium Silicate-Ethyl Acetate-Formamide Grout Subjected to Waters with pH of 8 at Different Hardness Levels

MC-500 Microfine Cement Grout (MC)

The MC-500 grout produced specimens with similar permeabilities for both the low and high sulfate waters, as shown by the data in Figure 8. Also included are data from two of the three specimens tested in a triaxial cell (the third specimen reached a similar permeability in less than one minute and the results are therefore not amenable to being plotted in Figure 8). Since the specimens cured for 7 days exhibited permeability values that were essentially the same for both types of test apparatus (that is, manifold and triaxial system), it appears that no appreciable side flow occurred around the specimens tested on the manifold.

The effect, if any, of water sulfate level on the permeability of the MC-500 grouted specimens does not appear to be meaningful. Figure 8 shows that, for specimens with curing times of 16 hours, 1 day, 3 days, and 7 days, the permeabilities are approximately the same order of magnitude. The relatively high permeability of cement-grouted specimens is a result of the grout forming process. Settling of the cement particles causes water to rise and form preferential flow paths in the grouted soil. Some of the water is trapped below the soil particles, and some fills the interstices among the cement particles (Troxell, Davis, and Kelly, 1968). Cement hydration produces a gel which decreases the size of these water voids and increases the water-tightness of the cement, but some voids are still present.

AC-400 Acrylate Grout (AC)

The AC-400 grout appeared to be extremely stable when subjected to the extreme pH variations of the test waters. No results are plotted for the AC-400 grout because no flow of water was observed in any of the specimens during the entire test period. Curing time had no effect on the permeability, since specimens with 1-hour curing times followed the same permeability pattern (no flow) as those cured for 1 day.

Dismantling of Specimens

After completion of the permeability tests, each specimen was dismantled and examined with the naked eye and with the aid of a microscope to determine qualitatively the amount and distribution of grout. Depending on the specimen curing time and the pH of the permeant, the individual specimens injected with sodium silicate-ethyl acetate-formamide grout appeared to be different. For the pH 8 low hardness and pH 10 permeants, the grout appeared to have been completely elutriated from specimens cured for 15 minutes and 1 hour. However, as the curing period increased, the specimens retained more grout. With the medium hardness pH 8 water, the specimen with a 15-minute curing time appeared to have been washed free of grout, but the remaining specimens seemed to contain approximately the same amounts of grout. The specimens subjected to high hardness pH 8 water all retained reasonably uniform distributions of grout, with a slight tendency toward higher grout retention for longer curing periods. The specimens tested with the pH 4 water contained variable amounts of grout. In the specimens cured for 15 minutes and 3 hours,

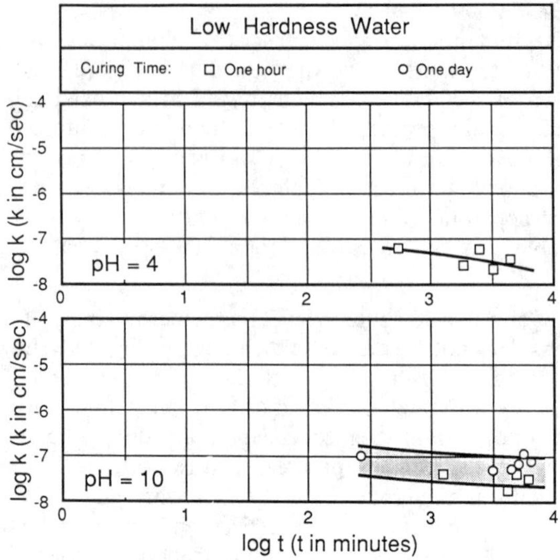

Figure 7. Permeability versus Time for Specimens Injected with Sodium Silicate-Sodium Aluminate Grout Subjected to Low Hardness Waters at Different pH Values

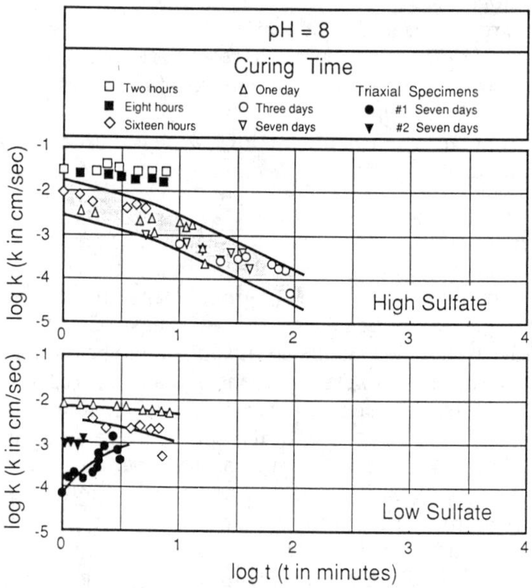

Figure 8. Permeability versus Time for Specimens Injected with MC-500 Microfine Cement Grout Subjected to Waters with pH of 8 at Different Sulfate Levels

the grout appeared to have been completely elutriated. All specimens cured for 9 hours or more retained relatively uniform grout distributions.

For both MC specimens tested with the low sulfate pH 8 water using the manifold system, the grout appeared quite uniform. The specimen with a 1-day curing time, however, seemed to remain more intact. The specimen with a 2-hour curing period in contact with the high sulfate pH 8 water apparently retained very little grout. The 8-hour specimen appeared to retain a slightly larger quantity of grout when viewed under a microscope, and specimens with curing times of 16 hours or more were very similar, with slight increases in grout being noted as the curing time increased. For all specimens injected with either sodium aluminate or AC-400 grouts, the grouted specimens appeared intact. The grout was uniformly distributed throughout the specimens and no erosion was evident.

Conclusions

Within the scope of this test program and the limited range of variables tested, the following conclusions can be advanced:

1. AC-400 acrylate grout is stable when brought in contact with both high and low pH waters; no measurable flow was attained through any of the specimens, regardless of their curing time.

2. The gel formed by the sodium silicate-sodium aluminate grout is insoluble in water and specimens injected with this grout exhibited low permeability (10^{-7}cm/sec or less) when brought in contact with both high and low pH waters.

3. The curing time of specimens injected with sodium silicate-ethyl acetate-formamide grout has a larger effect (up to four orders of magnitude) on the specimen permeability than either the water pH or hardness.

4. The sodium silicate-ethyl acetate-formamide grout produced specimens with a wide range of permeability values, depending on the pH and hardness of the permeant (both of which interfere with the silica solubility). Water with a high pH and/or low hardness resulted in a higher coefficient of permeability, and the grout was elutriated from the specimens with short curing times.

5. Specimens grouted with MC-500 microfine cement were quite permeable when subjected to both low and high sulfate waters; this is attributable to the segregation of cement particles to form preferential flow paths.

When evaluating the results of this study, several important conditions inherent to this test program must be recognized and appreciated. First, the tests on specimens with high initial permeabilities were performed under intermittent flow conditions (that is, the water did not permeate through the specimens continuously, as would be the case in a field situation, but only intermittently due to a limited supply of each synthetic water) until the permeability values stabilized. This intermittent flow tests probably yield a lower limit for the permeability of the various grouted specimens, but the results obtained are nonetheless useful for indicating the behavior of these grouts when subjected to groundwaters of different chemistries. Second, only one formulation of each grout was used in this study, and the results for grouts of other proportions may vary from those presented. The use of a grout with a larger percentage of active (non-water) ingredients would tend to reduce the permeability of the grouted soil. And third, these tests were performed by exposing a relatively small cross-sectional area to the permeants, while in the field a considerable area may be exposed and the reactions between the water and grout may occur at different rates.

Acknowledgements

Grateful appreciation is given to Edward Graf, Wallace H. Baker, William Clarke, and Joseph FitzPatrick, who helped in the conduct of this study by their suggestions and discussions.

References

Baker, W. H. (1982), "Planning and Performing Structural Chemical Grouting," Proceedings of the Conference on Grouting in Geotechnical Engineering, American Society of Civil Engineers, New Orleans, Louisiana, pp. 515-539.

Caron, C. (1965), "Expose dans la serie: Essais et Mesures," Annales de l'Institut Technique du Batiment et des Travaux Publics, Number 213, September, pp. 1149-1164.

Caron, C. (1982), "The State of Grouting in the 1980's," Proceedings of the Conference on Grouting in Geotechnical Engineering, American Society of Civil Engineers, New Orleans, Louisiana, pp. 346-358.

Clarke, W. J. (1982), "Performance Characteristics of an Acrylate Polymer Grout," Proceedings of the Conference on Grouting in Geotechnical Engineering, American Society of Civil Engineers, New Orleans, Louisiana, pp. 418-432.

Einstein, H. H., and Schnitter, G. (1970), "Selection of Chemical Grout for Mattmark Dam," Journal of the Soil Mechanics and Foundations Division, American Society of Civil Engineers, Volume 96, Number SM6, pp. 2007-2023.

Karol, R. H. (1982), "Chemical Grouts and Their Properties," Proceedings of the Conference on Grouting in Geotechnical Engineering, American Society of Civil Engineers, New Orleans, Louisiana, pp. 359-377.

Krizek, R. J. and Madden, M. G. (1985), "Permanence of Chemically Grouted Sands," Issues in Dam Grouting, American Society of Civil Engineers, Denver, Colorado, pp. 1-26.

Merrill, R. C. and Spencer, R. W. (1950), "Gellation of Sodium Silicate," Journal of Physical and Colloid Chemistry, American Chemical Society, Volume 54, Number 6, pp. 806-812.

Petrovsky, M. B. (1982), "Monitoring of Grout Leaching at Three Dam Curtains in Crystalline Rock Foundations," Proceedings of the Conference on Grouting in Geotechnical Engineering, American Society of Civil Engineers, New Orleans, Louisiana, pp. 105-120.

Troxell, G. E., Davis, H. E. and Kelly, J. W. (1968), Composition and Properties of Concrete, Second Edition, McGraw-Hill, New York.

Zelasko, J. S. (1966), "An Investigation of the Influence of Particle Shape on the Shear Strength and Packing Behavior of Quartziferous Sands," Ph.D. Dissertation, Department of Civil Engineering, Northwestern University, Evanston, Illinois

Electrokinetic Soil Processing
(A Review of the State of the Art)

Yalcin B. Acar[1], M.ASCE

Abstract

Electrokinetic soil processing is an emerging technology in waste remediation and treatment. This paper reviews electrokinetic phenomena in soils and provides the fundamentals of contaminant removal by the technique. The results of studies reporting ion/contaminant removal using electro-osmosis are presented. Engineering implications are provided for the further development and implementation of the process in remediation.

Introduction

Electrokinetic soil processing using low level DC currents (of the order of milliamps per cm^2 of electrode area) is envisioned to be used for removal/separation of organic and inorganic contaminants and radionuclides, construction of barriers and leak detection systems in clay liners, diversion schemes for waste plumes, and for injection of grouts, microorganisms and nutrients into subsoil strata (Mitchell 1986; Acar and Gale 1986; Renauld and Probstein 1987; Acar, et al. 1989).

Coupling between electrical, chemical and hydraulic gradients is responsible for different types of electrokinetic phenomena in soils (Mitchell, 1976). Electro-osmosis (EO) is one of these phenomena where the pore fluid moves due to application of a constant low DC current (or voltage) by electrodes inserted in a soil mass. In the last five decades since its first application (Casagrande 1947), electro-osmosis has been investigated and used for different applications (Hamed et al. 1991). The potential of the technique in waste remediation resulted in initiation of several recent studies (Putnam 1988; Acar et al. 1989; Mitchell and Yeung 1991). The need to utilize the process in removal/separation of contaminants necessitates a better understanding of the electrochemistry associated with electrokinetic phenomena and its relation to the mechanical behavior.

---

[1]Associate Professor, Department of Civil Engineering, Louisiana State University, Baton Rouge, LA 70803

Recent studies provided a better understanding of the electrochemistry and demonstrated that the acid front generated by electrolysis reaction at the anode advances and eventually flushes across the specimen by advection, migration and diffusion (Acar et al. 1989; Shapiro et al. 1989; Acar et al. 1991; Alshawabkey and Acar 1992). Hamed (1990) and Hamed et al. (1991) demonstrated that the movement of this acid front together with migration and advection of the cations and anions under electrical gradients constitute the mechanisms of removing contaminants from soils. The factors influencing the acid/base profile across the porous medium would significantly affect the flow, the flow efficiency, and the extent of ion migration and removal in electrokinetic soil processing. This paper presents the fundamentals of the process, reviews the results of the on-going studies and provides engineering implications for its implementation.

Electrokinetic Phenomena in Soils

In electro-osmosis, electrodes can be placed in an open or closed flow arrangement. Open flow arrangement constitutes the case when an electrode is sufficiently permeable to admit ingress and egress of water. In the closed flow arrangement, the electrode is not permeable or porous. Different electrode configurations (open or closed) result in substantial variations in the total matrix potentials across the soil specimen.

The electro-osmotic flow rate, q_e, is defined with an empirical relationship,

$$q_e = k_e \, i_e \, A = k_i \, I = \frac{k_e}{\sigma} \, I \tag{1}$$

where k_e = coefficient of electro-osmotic permeability (cm^2/sec-V), k_i = electro-osmotic water transport efficiency (cm^3/amp-sec), I = current (amp), σ = conductivity (siemens/cm), i_e = electrical potential gradient (V/cm), A = cross-sectional area (cm^2). Estimates of electro-osmotic flow rates can be made using equation (1). K_e varies within one order of magnitude for all soils; 1 x 10^{-5} to 10 x 10^{-5} (cm/sec)/(V/cm), the higher values being at higher water contents.

Figure 1 presents a schematic diagram of one-dimensional laboratory tests in electrokinetic soil processing. The prevailing electrical gradients, and the ion flow are also depicted. A comparison of flows under electrical and hydraulic gradients in clays is also provided. The coefficient of electro-osmotic permeability is independent upon the size and distribution of pores (fabric) in the soil mass. However, hydraulic conductivity is most affected by the fabric. Therefore, hydraulic conductivity decreases by five to six orders of magnitude (10^{-3} cm/sec to 10^{-9} cm/sec) from the fine sands to clays. Figure 1 indicates that under equal gradients, electrical potentials in fine-grained soils may result in orders of magnitude larger flows than hydraulic potentials. Therefore, electro-osmosis induced flow can be considered to be an efficient pumping mechanism in saturated, low permeability, fine-grained soil.

The efficiency and economics of electro-osmotic dewatering is governed by the amount of water transferred per unit charge passed which is quantified by electro-osmotic water transport efficiency, k_i. The parameter k_i may vary over a wide range from 0 to 1.2 cm^3/amp-sec depending upon the electrical conductivity of the saturated porous medium. The

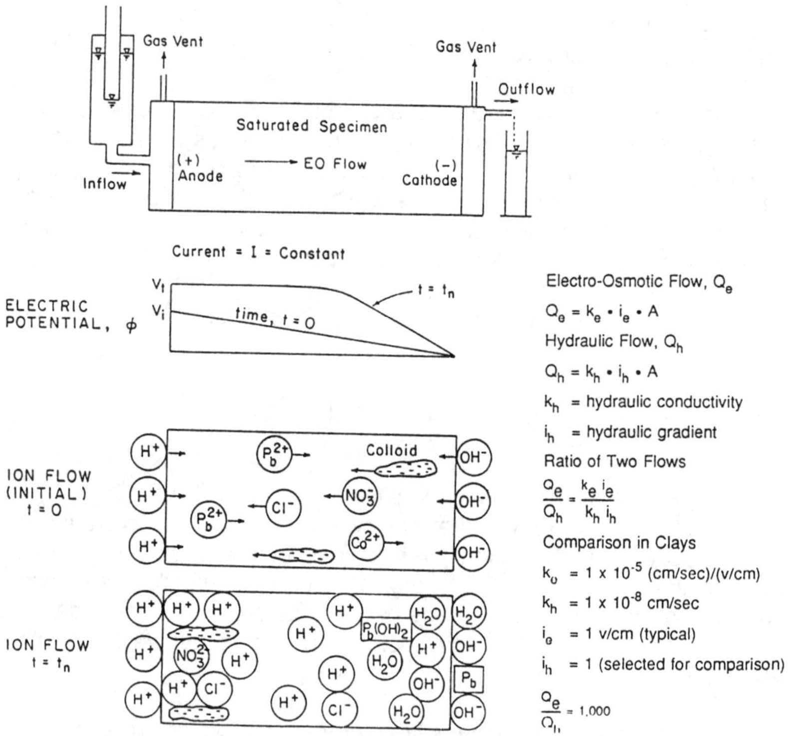

Figure 1. A Schematic Diagram of Electrokinetic Processing, Ion Flow and Comparison of Flow in One-Dimensional Flow Conditions (Hamed et al. 1991).

fine-grained soils may result in orders of magnitude larger flows than hydraulic potentials. Therefore, electro-osmosis induced flow can be considered to be an efficient pumping mechanism in saturated, low permeability, fine-grained soil.

The efficiency and economics of electro-osmotic dewatering is governed by the amount of water transferred per unit charge passed which is quantified by electro-osmotic water transport efficiency, k_i. The parameter k_i may vary over a wide range from 0 to 1.2 cm^3/amp-sec depending upon the electrical conductivity of the porous medium. The conductivity changes with water content, cation exchange capacity and free electrolyte content in the soil and also due to the prevailing chemistry during electrokinetic processing (Gray and Mitchell, 1967). Recent studies by Lockhart (1983) substantiate the conclusions of Gray and Mitchell (1967).

Potential Uses of Electrokinetics in Waste Management

The four electrode configurations described above could potentially be used in the following ways in waste disposal (Acar and Gale 1986; Acar et al. 1989): (1) Dewatering of waste sludge slimes, dredged spoil by first concentrating the solid particles using electrophoresis and subsequent consolidation by electro-osmosis (Mitchell 1986), (2) Electro-osmotic flow barriers (Mitchell and Yeung 1991), (3) Leak detection systems for disposal facilities, (4) Injection of grouts to form barriers, (5) Provide nutrients for biodegrading microcosm, (6) Insitu generation of reactants such as hydrogen peroxide for cleanup and/or electrolysis of contaminants, and (7) Decontamination of soils and groundwater.

Insitu remediation methods often necessitate the use of hydraulic charge and recharge wells to permeate the decontaminating liquid or stabilization agent through the soil deposit, or to provide nutrients for the biodegrading microcosm. Although such systems may effectively be used in highly permeable soils, they become inefficient and uneconomical in low permeability silts and clayey deposits. Electrokinetic soil processing with open electrode configuration could also be used to achieve an efficient seepage and decontamination method in such soils.

Removal of Contaminants by Electrokinetics

Upon application of low-level DC current (in the order of milliamps per cm^2 of electrode area) to the saturated porous medium, the following processes occur:

(1) The water in the immediate vicinity of electrodes is electrolyzed. An acid front is generated at the anode while a base front is created at the

cathode. Acar et al. (1989), Shapiro, et al. (1989) and Acar et al. (1990) formalize the development of these fronts. pH at the anode will drop to below 2.0 and will increase at the cathode to above 12.0.

(2) The acid front will advance across the specimen in time towards the cathode by: (a) advection of the pore fluid due to the prevailing electro-osmotic flow, (b) advection of the pore fluid due to any hydraulic potential differences, (c) diffusion due to concentration gradients, and (d) migration due to the electrical gradients. Development of these acid and base distributions and movement of ionic species are formalized by Acar, et al. (1991).

(3) The migration, diffusion and advection will also result in movement of cations and anions to respective electrodes in the porous medium (Hamed et al. 1991; Mitchell and Yeung 1991).

(4) The acid advancing across the specimen exchanges with adsorbed cations in the diffuse-double layer, resulting in their release into the pore fluid and advance towards the cathode by advection and diffusion (Hamed et al. 1991).

(5) In case the generation of the acid front is not controlled at the anode, the electrolyte concentration inside the porous medium will gradually rise, resulting in increased conductivity in the vicinity of the anode, decrease in electro-osmotic flow (Hamed et al. 1991), and a corresponding decrease in bulk-flow movement by advection.

(6) The decreased conductivity at the cathode region (possibly due to anion depletion and/or due to deposition of species as salts) will lead to an increase in voltage and an increase in energy expenditure (Hamed 1990). The chemistry at the anode and the cathode should ideally be controlled to achieve continued advection while providing sufficient H^+ ions for desorption of contaminants and/or solubilization of salts. This control is possible either by decreasing the current to levels where pH is at a desirable level or by frequent flushing at both ends by a fluid of controlled pH and chemistry.

Studies investigating removal of ions from soils by electro-osmosis are rare possibly due to difficulties in understanding the chemistry. Acar and Hamed (1991) review the available data on electrokinetic remediation of soils. A theoretical treatise is presented by Alshawabkeh and Acar (1992).

Jacobs and Mortland (1959) demonstrated that Na^+, K^+, Mg^{++} and Ca^{++} ions can be leached out of Wyoming bentonite by electro-osmosis. Krizek, et al. (1976) showed that the soluble ions content substantially increased in effluent in electro-osmotic consolidation of polluted dredgings, while they also noted that heavy metals were not found in the effluent during the period they applied the process. Hamed, et al. (1991) show that it is necessary to wait until the acid front flushes across the specimen in order to see any heavy metal ions or depositions on the cathode or the effluent. Hamnet (1980) studied the reclamation of agricultural soils by removal of unwanted salts by electro-osmosis. Hamnet's tests demonstrated that Na^+ ions move toward the cathode, while Cl^- and SO_3^{-2} ions move toward the anode. Shmakin (1985) notes that the method has been used in the Soviet Union since the early 1970's as a method for concentrating metals and prospecting for Cu, Ni, Co, Au.

The potential of the technique in waste remediation resulted in initiation of several recent studies. Runnels and Larson (1986) have investigated the potential use of electromigration to remove contaminants from groundwater. The amount of copper removed increased with processing time (total charge passed). Renauld and Probstein (1987) investigated the change in the electro-osmotic water transport efficiency of kaolinite specimens loaded with acetic acid and sodium chloride. This study indicated that the current efficiency increased with higher concentrations of this weak, organic acid. A better understanding of the chemistry in electrokinetic soil processing is achieved through studies at LSU. Putnam (1988) investigated the development of acid/base distributions in electro-osmosis. Acar et al. (1989), Acar et al. (1990) and Acar, et al. (1991) present the theory for acid/base distributions in electro-osmosis and compare the predictions of this model with the results of the tests conducted by Putnam (1988). A good correlation was noted. This theory and the model provide a preliminary description of the movement of different species in electrokinetics. The significance of the acid base distributions in electrokinetic soil processing is further displayed in studies reported by Shapiro, et al. (1989).

A comprehensive subsequent study on removal of Pb(II) from kaolinite is reported by Hamed, et al. (1991). Kaolinite specimens were loaded with Pb(II) at 118 μg/g to 145 μg/g of dry kaolinite, below the cation exchange capacity of this mineral. As presented in Figure 2, electro-osmosis removed 75 to 95 percent of Pb(II) across the test specimens. The study clearly demonstrated that the removal was due to migration and advection of the acid front generated at the anode by the primary electrolysis reaction. The energy used in the study to decontaminate the specimens was 29 to 60 kWh per cubic meter of soil processed. An interesting finding of this study was electroplating of Pb(II) at the carbon cathode. Further studies investigating removal of Cd(II) and Cr(III) are also reported by Hamed (1990). Similar results are obtained. Hamed (1990) investigates the effect of increased concentration and current density on the

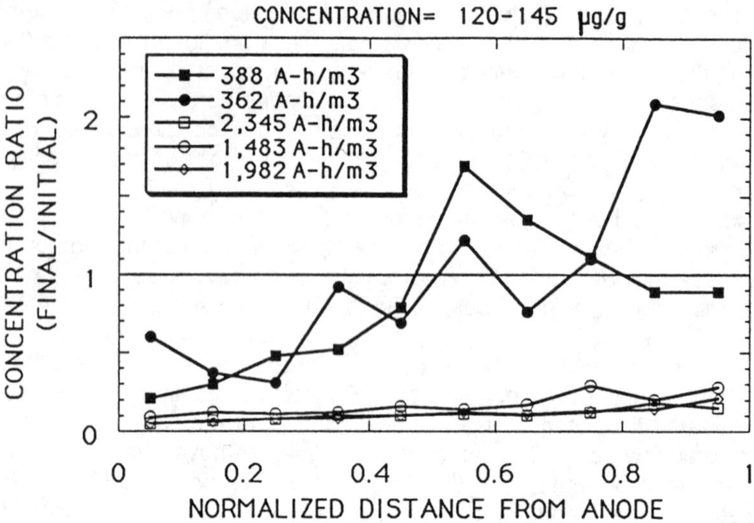

(a) Pb(II) Removal

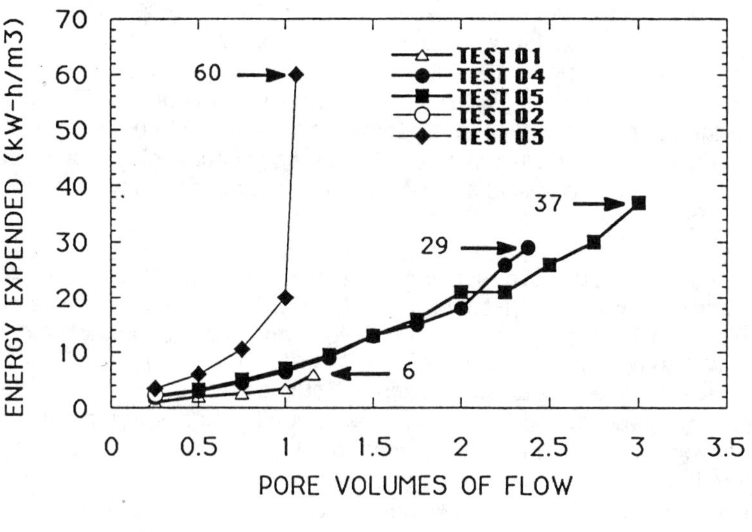

(b) Energy Expenditure

Figure 2. Pb(II) Removal from Kaolinite and Corresponding Energy Expenditure (Hamed et al. 1991).

efficiency of the removal process. Higher current densities result in as efficient a removal as in lower current densities while the energy requirement and the cost of processing increase exponentially. Other laboratory studies conducted by Lageman (1989) and Banerjee, et al. (1990) further substantiate the applicability of the technique to a wide range of inorganic contaminants and soils.

The applicability of the technique to removing organic contaminants is investigated in studies at LSU. Acar, et al. (1992) report phenol removal from saturated kaolinite using the technique. In this study, kaolinite specimens were loaded by 500 ppm phenol below the phenol adsorption capacity of this mineral. The process removed 85 to 95% of the adsorbed phenol. The energy used in removal of phenol was 13 kWh to 18 kWh per cubic meter of soil processed. Bruell, et al. (1991) report removal of the BTEX compounds and trichloroethylene loaded on kaolinite specimens by electro-osmosis.

While the above laboratory studies display the feasibility of using electro-osmosis to decontaminate soils, limited field studies are available. A synthesis of field tests investigating and/or reporting some form of chemical removal from soils is also presented by Acar and Hamed (1991). Segall, et al. (1980) present the chemical characteristics of the water accumulated at the electrodes in electro-osmotic dewatering of dredged soil. Their study discovered that: (1) There was a significant increase in heavy metals and organic materials in electro-osmosis effluent above that recorded in the original leachate. The concentrations of zinc, lead, mercury and arsenic were especially high, and (2) Total organic carbon content of the effluent was two orders of magnitude higher than the original leachate. It is postulated that highly alkaline conditions resulted in dissolution and release of the organic material. Pesticides came out at the cathode.

Case and Cutshall (1979) describe a field study for control of radionuclide migration in soil by application of DC current. This study demonstrates that it is possible to migrate radionuclides with the technique. Recent laboratory studies at LSU indicate that uranium at an activity of 1,000 pCi/gm can be removed from kaolinite by the process. Lageman (1989) reports the results of field studies conducted in the Netherlands to decontaminate soils by electrokinetic soil processing. The system involves an electrode fluid conditioning and purification system. The conditioning is for the control of the influent/effluent chemistry, while purification (such as ion exchange resin columns) is for removing any excess ions in the effluent.

A recent study investigating the application of the process to decontaminate a chromium site is also reported by Banerjee, et al. (1990). The results of that study are inconclusive as the investigators monitored only the effluent concentration and removal across the electrodes was not scrutinized. Studies at LSU indicate that unless the processing is continued until the acid front flushes across the electrodes and neutralizes the base generated at the cathode, any conclusion based only on the effluent concentration of an inorganic chemical would not be supported.

The laboratory studies reported by Hamnet (1980), Runnels and Larson (1986), Lageman (1989), Shapiro, et al. (1989), Acar et al. (1990), Hamed (1990), Hamed et al. (1991), and Acar et al. (1992) together with the pilot-scale field studies of Lageman (1989) display the feasibility of using the process and commercialize in site remediation. Further pilot-scale studies are necessary to improve the technology and establish the necessary field remediation scheme for different site conditions and chemistry.

Engineering Implications

The above review of the present state of knowledge on electrokinetic soil remediation indicates that:

(1) Type of Soil: The process results in movement of ions in sandy to clayey soils. High water content, low activity soils at low pore fluid electrolyte concentrations will result in higher electro-osmotic efficiencies.

(2) Type and Concentration of Contaminants: Most available data is on ionic forms of inorganic cations, and some radionuclides (^{90}Sr) and acetic acid. There exists data demonstrating removal at levels of up to 10,000 ppm of Cu(II) and 5,000 ppm of Pb(II). As concentrations of contaminants (ionic) increase, removal should be mostly by migration as advection (electro-osmotic flow) will substantially decrease. At lower concentrations, both advection and migration will be acting. Recent laboratory studies demonstrate that it is feasible to remove phenol (500 ppm) BTEX compounds (benzene, toluene, ethylene, xylene, and trichloroethylene) from kaolinite by the technique. All current data is on concentration of these organic chemicals below their water solubility limits. There exists the need to investigate the feasibility of the technique at higher concentrations.

Data regarding acid/base distributions indicate that salts (such as PbO) may also dissolve and migrate due to the advancing acid front. However, there is no factual data to validate this hypothesis.

(3) Mixture of Contaminants: The data indicates that the process also works on a mixture of contaminants (Lageman, 1989). Monovalent ions may be removed at a higher rate than higher valence ions.

(4) Saturation: Mitchell and Yeung (1991) present data regarding the effect of saturation on k_e. k_e did not change significantly in specimens compacted at different molding moisture contents. Recent data at LSU suggest that the process is applicable in partially saturated soils.

(5) Depths: The review of literature indicates that there should not be a depth limitation in the process beyond practical problems that may be encountered.

(6) Type of Electrodes: Inert electrodes such as graphite, carbon or platinum should be used at anode in order to avoid introducing secondary corrosion products into the soil mass. Open electrodes allow control of influent and effluent chemistry. It should be recognized that some ions will be electroplated on the cathode or they may be precipitated close to the cathode.

(7) Electrode Configuration: The electrodes can be placed horizontally or vertically. It is noted that the electrical potential gradients generated due to different electrode arrangements will affect the flow conditions and hence the removal efficiency. The gradients will significantly change by electrode configurations and the depth of individual electrodes relative to the counter electrode.

(8) Electrode Spacing: Spacing will depend upon the type and level of contamination and the selected current/voltage regime. A substantial decrease in efficiency of the process may result due to increases in temperature when higher voltage gradients are generated.

(9) Current Level: The current level reported is in the order of milliamps per square cm of electrode area (0.01 to 1.0 mA/cm^2). It can be varied to monitor the influent pH level at the anode and to control the rate of decontamination.

(10) Duration: Process should be continued until the desired removal is achieved. The remediation duration will be site specific. It is necessary to wait until the acid front generated at the anode will advance to the cathode.

(11) Effluent/Influent Chemistry: It is possible to control the efficiency by controlling the pH and the chemistry of the effluent and the influent. Several alternatives are available: (a) to decrease the current to a level where less H^+ ions are generated, (b) to flush the anode and/or cathode by a fluid of known chemistry (e.g., introducing acid at the cathode will decrease the voltage gradients substantially), and (c) placement of an acidic ion exchange resin.

(12) Chemistry Subsequent to the Process: The porous medium will become acidic upon completion of the process. The medium will return to original conditions by diffusion of the acidic pore fluid to the surrounding medium. Cathode effluents may require post-chemical treatments (such as ion exchange resin columns for inorganic contaminants) to achieve concentration of contaminants.

Acknowledgments

Studies at Louisiana State University, investigating electrokinetic soil processing, are funded by the Board of Regents of the State, the Hazardous Waste Research Center of LSU, the National Science Foundation, and Electrokinetics, Inc. These awards are gratefully acknowledged.

References

Acar, Y. B., and Gale, R. J. (1986) "Decontamination of Soils Using Electro-Osmosis," A proposal submitted to the Board of Regents of the State of Louisiana, LEQSF Research Development Program Office of Research Coordination, Louisiana State University.

Acar, Y. B., Gale, R. J., Putnam, G., and Hamed, J. (1989), "Electrochemical Processing of Soils: Its potential Use in Environmental Geotechnology and Significance of pH Gradients," 2nd International Symposium on Environmental Geotechnology, Shanghai, China, May 14-17, Envo Publishing, Bethlehem, PA, Vol. 1, pp. 25-38.

Acar, Y. B., Gale, R. J., Hamed, J., Putnam, G. (1990) "Electrochemical Processing of Soils: Theory of pH Gradient Development by Diffusion and Linear Convection," Journal of Environmental Science and Health, Part (a); Environmental Science and Engineering, Vol. 25, No. 6, pp. 687-714.

Acar, Y. B., Hamed, J., Gale, R. J., and Putnam, G. (1991), "Acid/Base Distributions in Electro-Osmosis," Transportation Research Record, No. 1288, Soils Geology and Foundations, Geotechnical Engineering 1990, pp. 23-34.

Acar, Y. B., and Hamed, J. (1991), "Electrokinetic Soil Processing in Waste Remediation/Treatment - Synthesis of Available Data," Bulletin of the Transportation Research Board, March 1991 (in press).

Acar, Y. B., Li, H., Gale, R. J. (1992), "Phenol Removal from Kaolinite Using Electrokinetics," Journal of Geotechnical Engineering, ASCE (in press).

Alshawabkeh, A., and Acar, Y. B. (1992), "Removal of Contaminants from Soils by Electrokinetics: A Theoretical Treatise," Journal of Environmental Science and Health (in press).

Banerjee, S., Horng, J., Ferguson, J. F., Nelson, P. O. (1990), "Field Scale Feasibility of Electro-Kinetic Remediation," Unpublished Report Presented to USEPA, Land Pollution Control Division, RREL, CR811762-01, 122 p.

Bruell, C. J., Segall, B. A., Walsh, M. T. (1991), "Electro-osmotic Removal of Gasoline Hydrocarbons and TCE from Clay," Journal of Environmental Engineering, ASCE.

Casagrande, L. (1947), "The Application of Electro-osmosis to Practical Problems in Foundations and Earthwork," Department of Scientific and Industrial Research, Building Research, London, England, Technical Paper No. 30, 22 p.

Case, F. N. and Cutshall, N. H. (1979), "Oak Ridge National Lab., TN (USA)," Symposium on the Scientific Basis for Nuclear Waste Management, Boston, MA, USA, Nov. 26-29, Conf. 791112-28, 5 p.

Gray, D. H., and Mitchell, J. K. (1967), "Fundamental Aspects of Electroosmosis in Soils," Journal of the Soil Mechanics and Foundation Division, ASCE, Vol. 93, No. SM6, pp. 209-236.

Hamed, J. (1990), "Decontamination of Soil Using Electro-osmosis," A Dissertation submitted to the Graduate School of Louisiana State University in Partial Fulfillment of the Degree of Doctor of Philosophy.

Hamed, J., Acar, Y. B., Gale, R. J. (1991), "Pb(II) Removal from Kaolinite by Electro-kinetics," ASCE, Journal of Geotechnical Engineering, Vol. 117, No. 2, February 1991, pp. 241-271.

Hamnet, R. (1980), "A Study of the Processes Involved in the Electro-Reclamation of Contaminated Soils," MS Thesis, University of Manchester, England, 84 p.

Jacobs, H. S., and Mortland, M. M. (1959), "Ion Movement in Wyoming Bentonite During Electro-osmosis," Proceedings of Soil Science Society.

Krizek, R. J., Gularte, F. B., and Hummel, P. B. (1976), "Stabilization of Polluted Dredgings by Electro-osmosis," ASCE National Water Resources and Ocean Engineering Convention, San Diego, CA, April 5-8, 1976, Preprint 2641.

Lageman, R. (1989) "Theory and Practice of Electro-Reclamation, "NATO/CCMS Pilot Study, Demonstration of Remedial Action Technologies for Contaminated Land and Ground Water, Copenhagen, Denmark, May 9, 1989, 18 p.

Lockhart, N. C. (1983), "Electro-osmotic Dewatering of Clays, I, II and III," Colloids and Surfaces, 6, pp. 238-269.

Mitchell, J. K. (1976), Fundamentals of Soil Behavior, John Wiley and Sons, New York, 422 p.

Mitchell, J. K. (1986), "Potential Uses of Electro-kinetics for Hazardous Waste Site Remediation," Position paper prepared for USEPA-University of Washington Workshop on Electro-Kinetic Treatment and its Application in Environmental-Geotechnical Engineering for Hazardous Waste Site Remediation, Seattle, WA, August 4-5, 1986, 20 p.

Mitchell, J. K., Yeung, T-C. (1991), "Electro-kinetic Flow Barriers in Compacted Clay," Transportation Research Record, No. 1288, Soils Geology and Foundations, Geotechnical Engineering 1990, pp. 1-10.

Putnam, G. (1988), "Development of pH Gradients in Electrochemical Processing of Kaolinite," MS Thesis presented to the Department of Civil Engineering, Louisiana State University.

Renauld, P. O., Probstein, R. F. (1987), "Electro-osmotic Control of Hazardous Waste," Physicochemical Hydrodynamics, v. 9, No. 1/2, 1987, pp. 345-360.

Runnels, D. D., Larson, J. L. (1986), "A Laboratory Study of Electromigration as a Possible Field Technique for the Removal of Contaminants from Ground Water," Ground Water Monitoring Review, pp. 81-91, Summer 1986.

Segal, B. A., O'Bannon, C. E., and Matthias, J. A. (1980), "Electro-Osmosis Chemistry and Water Quality," Journal of the Geotechnical Engineering Division, ASCE, Vol. 106, No. GT10, Oct. 1980, pp. 1143-1147.

Shapiro, A. P., Renauld, P., Probstein, R. (1989), "Preliminary Studies on the Removal of Chemical Species from Saturated Porous Media by Electro-osmosis," Physicochemical Hydrodynamics, Vol. 11, No. 5/6, pp. 785-802.

Shmakin, B. M. (1985), "The Method of Partial Extraction of Metals in a Constant Current Electrical Field for Geochemical Exploration," J. Geochem. Explor., Vol. 23, No. 1, pp. 35-60.

The Use of Vibro Systems in Seismic Design

Roberto A. López[1] A.M. ASCE
and
Robert F. Hayden[2] M. ASCE

Abstract

The current state of the practice in the design and construction of vibro-compaction and vibro-replacement for seismic applications is presented herein. The benefits provided by vibro systems are discussed in three different categories: (1) improvement in material properties, (2) drainage, and (3) reinforcement. The concept of ductility is discussed and procedures to estimate the drainage and reinforcement effects of vibro-replacement are presented.

Introduction

In the last twenty-five years, considerable advances have been made in the field of geotechnical earthquake engineering. Since the Great Alaska and Niigata Earthquakes of 1964 brought the issue of liquefaction of soils to the forefront, procedures to estimate the seismic response of level ground, embankments, and slopes have been continuously improved. On the other hand, research on methods to mitigate liquefaction and earthquake related deformations has been fairly limited. Recently, the extensive geotechnical failures during the Loma Prieta Earthquake have served to highlight the importance of ground improvement for seismically unstable sites. This paper reviews the state of design and construction of vibro systems for seismic applications and suggests areas to further develop these methods. Equipment, mechanisms of improvement, and a number of case histories are discussed.

Description of Vibro Systems

The *vibro-compaction* procedure, developed in Germany in the mid-1930's, involves the use of poker-type downhole vibrators, or *vibroflots*, ranging from 12 to 18 inches (30 to 45 cm) in diameter, and 10 to 16 ft (3 to 5 m) in length. A set of rotating eccentric weights housed inside the probe are mounted on a vertical shaft. Typically driven by an electric motor, the rotating action generates vibrations that propagate

1 Project Engineer, Hayward Baker, Inc., PO Box 7690, Ventura, California, 93006
2 Senior Project Manager, Hayward Baker, Inc., Ventura, California

radially away from the vibroflot. Relative densities of 65 to 85 percent are usually achieved in granular deposits.

The typical construction procedure for vibro-compaction is shown in Fig. 1. The vibrator is attached to extension tubes to achieve the required treatment depth. Then it is suspended from a standard or hydraulic crane over a spot previously chosen for densification, and inserted into the ground. The vibrating action, sometimes accompanied by jetting water, breaks up the frictional contacts in granular soils and rearranges them into denser configurations. Sand is backfilled into the cavity and compacted in 1 to 4 ft (0.3 to 1 m) lifts from the bottom up. The diameter of influence of the vibrator has been found to be as much as 14 ft (4.25 m), and densification has been achieved to depths of as much as 115 ft (35 m). Spacings typically vary from 6 to 9 ft (1.8 to 2.7 m) in square or triangular patterns.

The vibro-compaction procedure has been used successfully at sites consisting primarily of loose, clean sands, having less than 10 to 20 percent fines. The process has seen extensive use in parts of the world where these conditions exist, such as Florida, Germany, and the Middle East. Most sites in the Western United States, however, exhibit mixed and layered soil conditions, and undergo a limited densification response to vibrations when fines are present. The substitution of gravel or crushed stone as backfill in the 1960's led to the formation of stone columns and to the process now referred to as *vibro-replacement*. This method allows the treatment of a wider range of soils, including silts and clays, by forming a reinforcing element of low compressibility and high shear strength. The range of soils treatable by vibro-compaction and vibro-replacement is shown in Fig. 2.

Both vibro-compaction and vibro-replacement improve the subsurface granular soils through densification. Vibro-replacement, though, results in a composite foundation system with stiffer elements that concentrate loads and provide drainage. This redundancy in dealing with the seismic problem gives vibro-replacement distinct advantages over other methods that only provide drainage or densification as means for mitigating liquefaction. The improvement that results from vibro-replacement is discussed in the following sections under one of these three categories: (1) improvement in material properties, (2) drainage, and (3) reinforcement.

Improvement in Material Properties

Response of loose, granular soils to the vibrating probe results in increased relative density, shear strength, and stiffness, and reduced compressibility. In terms of soil parameters, this is generally reflected as increases in the angle of internal friction, ϕ; shear modulus, G; and elastic modulus, E; and decreases in compression index, C_c. Furthermore soils improved by vibro-compaction and vibro-replacement have shown increased resistance to liquefaction.

The tendency of a soil to generate excess pore pressures during undrained loading is correspondent to the volume changes that occur during drained loading. Loose soils tend to contract upon shearing; and, if loading is too quick for drainage to occur, generate excess pore pressures. For soils that derive all strength from confinement, this generation of excess pore pressures can lead to a condition of zero effective stress (when $\sigma = u + \delta u$), resulting in loss of strength and fluid-like behavior with only residual resistance to deformation. The generation of excess pore pressures in dense materials, on

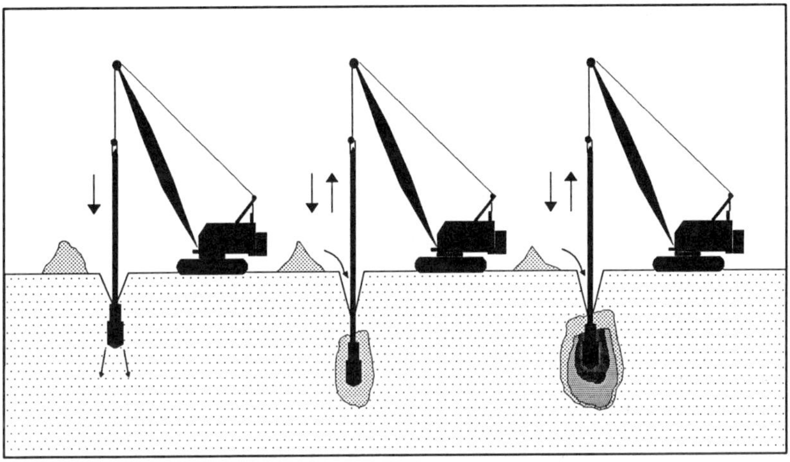

Figure 1: Typical construction procedure for Vibro systems.
The vibrator is suspended from a crane and penetrated to the desired depth. After sand or gravel is added as backfill, the vibrator is repenetrated, thus densifying both the natural and backfill materials.

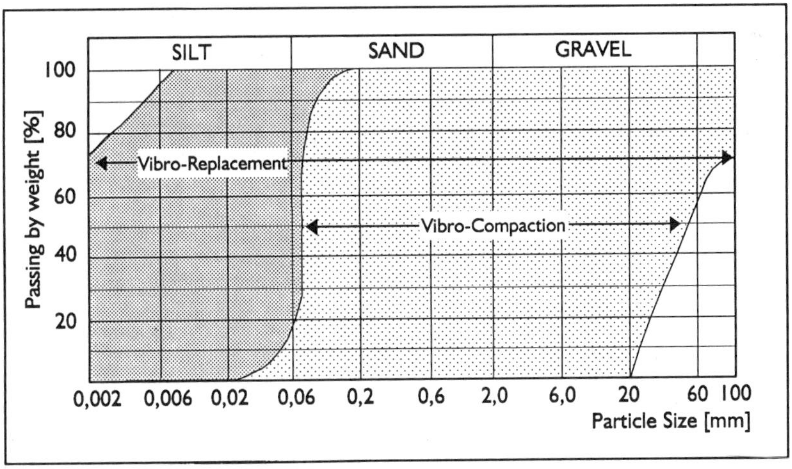

Figure 2: The above grain size distribution curves show in general terms where Vibro-Compaction and Vibro-Replacement can be used. Soil having a silt content of less than approximately 15% can be densified by Vibro-Compaction.

the other hand, carries smaller consequences, since they tend to recover part of their strength upon shearing. When sheared, dense soils tend to dilate; this lowers the pore water pressures, arresting further deformations. This behavior of dense soils during undrained cyclic loading has been referred to before as liquefaction with limited strain potential (Seed 1979), cyclic mobility (Casagrande, 1976), and ductility (Yoshimi, 1990). The extent of damage due to liquefaction is thus directly related to the soil's tendency to either contract or dilate when sheared, and cannot be quantified by relative density alone.

Mitchell et al, 1976, studied the effects of pre-straining and soil fabric on liquefaction potential. His research determined that soils prepared to the same relative density do not necessarily exhibit similar undrained cyclic behavior. For example, soils densified by vibratory procedures produced no preferred axis of particle orientation, and exhibited better static and cyclic performance than soils prepared to the same density by either pluviation or tamping. In a similar investigation, Tokimatsu et al, 1986, found that sand that had been previously strained exhibited higher resistance to liquefaction. Yoshimi and Tokimatsu, 1991, refer to this resistance to liquefaction as ductility. In particular, they use this term to describe densified soils that strain only moderately when sheared, and hence do not liquefy even when the shaking exceeds the design value.

Under present practice, the liquefaction potential of mechanically improved sites is typically evaluated through the use of in situ tests, particularly the Standard Penetration Test, SPT, and Cone Penetration Test, CPT. H.B. Seed's simplified procedure for evaluating liquefaction from SPT data (Seed and Idriss, 1971; Seed and De Alba, 1986) is presently the preferred method. When CPT is used, the data is either converted to equivalent SPT values, or used directly to assess liquefaction using CPT-based evaluation methods (eg, Robertson and Campanella, 1985;). One should remember, though, that these correlations were developed from natural sites where no ground improvement had been performed. When soils are subjected to mechanical modification, variables associated with pre-straining, such as horizontal stresses and time effects (aging), may not be adequately represented by SPT and CPT results.

One case where SPT and CPT results did not adequately predict the liquefaction resistance of mechanically improved soils is documented in Campanella et al, 1982. A site in New Westminster, British Columbia, was treated by vibro-compaction and dynamic deep compaction. to reduce its liquefaction potential. When tested, the siltier layers showed little change in either SPT or CPT tip resistance values, the numbers still falling within the range considered "liquefiable" by traditional relations. Monotonic undrained compression tests and cyclic triaxial tests, however, revealed a dilative response, a five-fold increase in undrained shear strength, and a doubling of the critical cyclic stress ratio. Measurements of pore pressure ratios with a piezo-cone with the porous element behind the tip showed that untreated soils generated high excess pore pressures during driving. Treated soils, on the other hand, exhibited pressures below hydrostatic and even negative, indicating a tendency for the soil to dilate during shear. Figure 3 shows the piezo-cone results before and after mechanical improvement at the site.

A similar situation occurred at the site for the Emilie Ritchen Elementary School in Oxnard, California (K.C. Geotechnical Associates, 1990). The geotechnical investigations exposed liquefiable soils at depths between 9 and 17 ft (2.7 and 5.2 m) for the given design earthquake magnitude of 7.5 and horizontal ground acceleration of 0.65 g. To mitigate the liquefaction potential, stone columns were built by the vibro-replacement method on 9 ft (2.7 m) centers in a triangular pattern extending 10 ft (3 m)

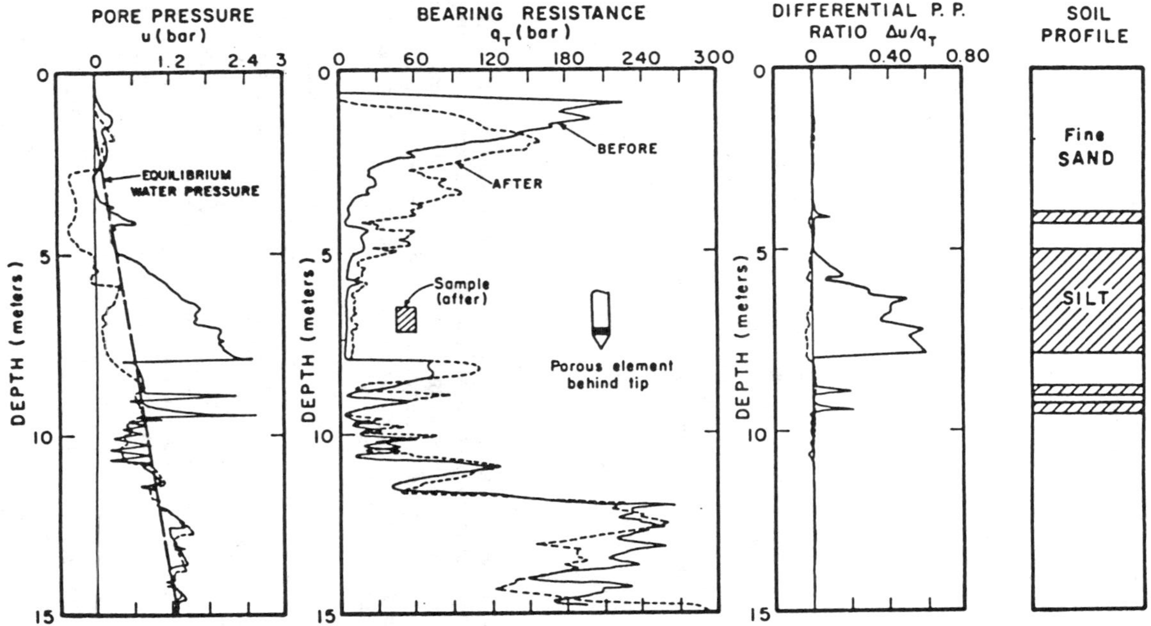

Figure 3: Piezometer cone logging before and after Ground Modification at New Westminister, British Columbia (from Campanella et al, 1982).

outside of the building footprint. A total of 9 SPT tests and 29 CPT tests were performed to verify the effects of the program. The approximate limits of post-treatment CPT tip resistance values in relation to pre-treatment tests are plotted in Fig. 4. The average results throughout the site are summarized below:

DEPTH (ft)	SOIL TYPE	POST TREATMENT TIP RESISTANCE (tsf)	PERCENT CHANGE RELATIVE TO PRE-TREATMENT
0-9	sm/ml	141	28%
9-13	sm	213	39%
13-17	ml	48	88%
17-20	clays	12	69%

The silty layer between 13 and 17 ft (3.9 and 5.2 m) did not exhibit the required CPT tip resistance to render it non-liquefiable using Seed's simplified procedure. For that reason, shelby tube samples of this layer were taken and subjected to cyclic simple shear. The two samples tested exhibited liquefaction (pore water pressure equal to the vertical consolidation stress) at 55 cycles and 70 cycles of loading, compared to approximately 15 cycles of loading expected to be generated by the design earthquake. Even though this layer may appear to exhibit localized liquefaction based on SPT and CPT results, the laboratory tests indicated a low liquefaction potential for the design earthquake loads. No cyclic tests were conducted on samples taken prior to vibro-replacement.

Drainage

The buildup and dissipation of pore water pressures during earthquakes has been studied by a variety of investigators. Seed et al., 1976, showed that coarse sands and gravels can dissipate pore pressures fast enough that no detrimental pore pressure buildup or condition approaching liquefaction can develop. Seed and Booker, 1977, carried this research further and developed design curves for determining the appropriate spacing between gravel drains for different soil and earthquake loading conditions.

For a stone column system to provide effective drainage, the spacing must be designed so that excess pore pressures are dissipated almost as fast as they are generated. In this manner, the flow of ground water can be minimized and surface settlement limited. In Seed and Booker's procedure, the liquefaction potential of a site is first evaluated without any drains, obtaining the N_{eq}/N_l ratio, where N_{eq} is the equivalent number of cycles applied by the design earthquake, and N_l is the number of cycles needed to initiate liquefaction. Then, for a given stone column radius, a, the factor T_{ad}, relating the earthquake duration to the consolidation properties of the sand, can be computed:

$$T_{ad} = (k/\gamma_w) * (t_d/m_{v3} * a^2) \qquad (1)$$

where k is the permeability of the soil in the horizontal direction, γ_w is the unit weight of water, t_d is the duration of shaking, and m_{v3} is the coefficient of volume compressibility of the soil.

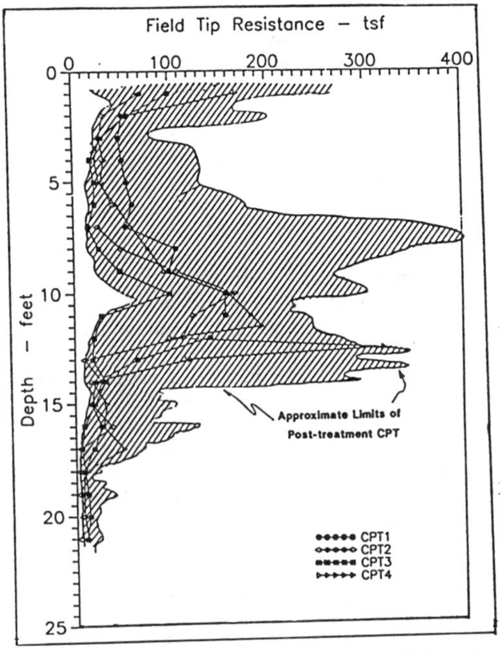

Figure 4: Relative improvement based on CPT end bearing data at Emilie Ritchen Elementary School, Oxnard, California (from K-C Geotechnical Associates, 1990).

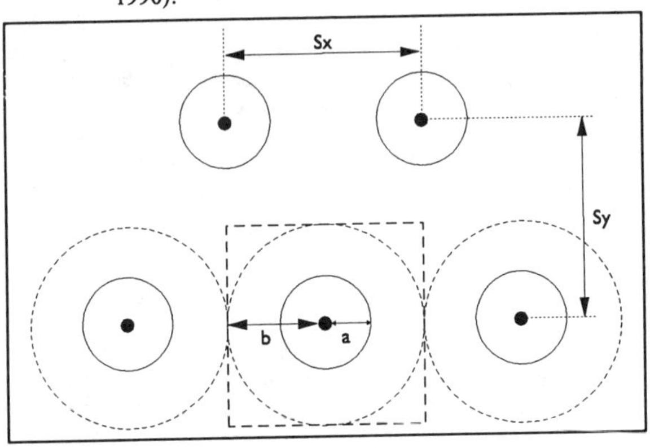

Figure 5: Definition of variables.

a = radius of stone column
b = radius of influence of stone column = $(A/\pi)^{1/2}$
Sx = spacing between columns in the x-direction
Sy = spacing between columns in the y-direction
Ac = cross sectional area of stone column = πa^2
A = influence area of stone column = (Sx) (Sy)

Other parameters that need to be defined are the pore pressure ratio, $r_g = u/\sigma'_o$, which relates the pore pressures to the effective stress; and 2b, which is the effective spacing between stone columns, as defined in Fig 5. Then, under the assumptions of purely radial flow, the pore pressure ratio throughout the sand and drain system can be expressed as a function of three dimensionless parameters: a/b ratio, N_{eq}/N_l ratio, and T_{ad}. Figure 6 shows the relation between greatest pore pressure ratio generated anywhere in the soil/drain system and the properties of the soil and earthquake. The key to design then lies in first specifying a greatest pore pressure ratio allowed during shaking, typically at or below 0.6 to prevent large settlements. Then, for the given earthquake and soil conditions, T_{ad} and N_{eq}/N_l will be known, and a/b can be selected.

Seed and Booker, 1977, stated that drains would act satisfactorily as long as the permeability of the drain was 200 times the permeability of the surrounding soil. With regards to the drain material, two restrictions should be satisfied: (1) the difference in permeability between the drain material and surrounding soil should be large enough to permit a hydraulic gradient, and (2) particle sizes of the drain material should be small enough to prevent clogging. Some of the most often used stone gradation requirements are given in Terzaghi and Peck, 1967 and Navy Design Manual 7, 1974. Laboratory tests ran by Konoike Construction Company of Japan appear to indicate that for dynamic loading, a coarser drainage material is needed to guarantee effective permeability, and a less restrictive lower limit for clogging. Their guidelines are summarized as follows:

$$20D_{S,15} < D_{G,15} < 9D_{S,85} \tag{2}$$

where D_S and D_G refer to the effective diameter of the soil and backfill, respectively. Typically, well graded, durable gravel with a gradation between 3/8 and 2 in (9.5 and 51 mm) is recommended for stone columns, but sizes up to 4 in (100 mm) have been used.

Reinforcement

The reinforcement of soft soils by means of stone columns is possible due to the higher relative stiffness of the column material. Given two materials under uniform loading and plane strain conditions, stresses tend to concentrate around the less yielding material, as can be demonstrated through equilibrium considerations. The proportion of the load carried by the stone columns depends on the characteristics of the loading, the relative stiffness between the soil and column, and the diameter and spacing between columns.

The most widely used solution for estimating load concentration was presented by Priebe, 1976. Developed from static considerations, the method was first presented as a means of estimating the reduction in settlement that results from densification and stress concentration around stiff elements. The load concentration, and hence settlement reduction, is expressed in terms of an area replacement ratio, or relative amount of soil replaced by stone:

$$\text{Area Replacement Ratio} = a_c = A_c/A \tag{3}$$

where A_c is the cross-sectional area of each stone column, and A is the area of influence of each column, as represented in Fig 5.

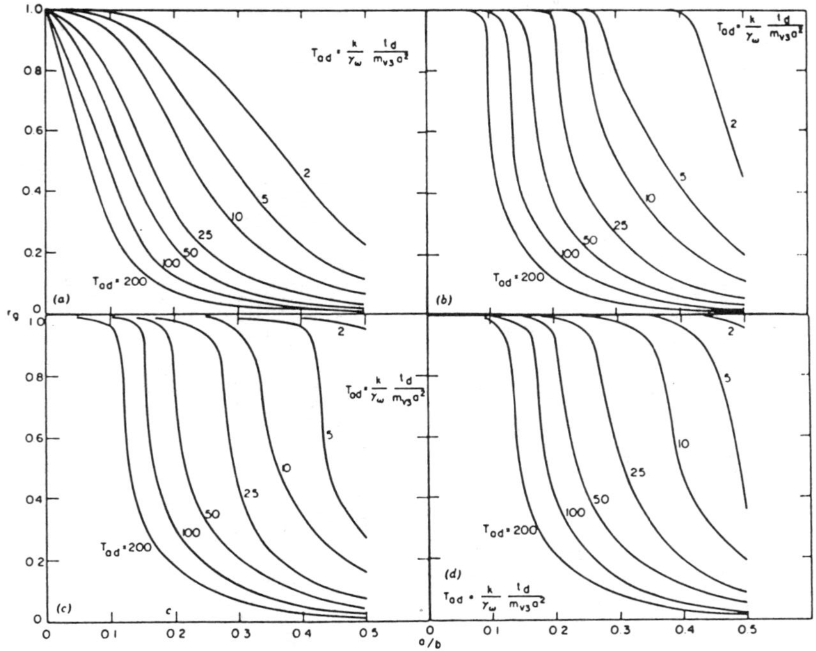

Figure 6: Relationship between Greatest Pore Pressure Ratio and Drain System Parameters for: (a) $N_{eq}/N_1 = 1$; (b) $N_{eq}/N_1 = 2$; (c) $N_{eq}/N_1 = 3$; (d) $N_{eq}/N_1 = 4$ (from Seed and Booker, 1977)

Priebe's method treats the stone column as a series of discrete, elastic elements, each contributing an incremental deformation to the total settlement. The results of his analysis are presented on Fig 7. For a given area replacement ratio and backfill friction angle, a settlement improvement ratio can be obtained. The settlement improvement ratio, S_c/S, relates the settlement the soil/column matrix will undergo relative to the settlement the site would have experienced without treatment. The use of a settlement improvement ratio lends flexibility and simplicity to the curves, since these are not restricted by soil type or magnitude of applied load, and can be related to soil conditions and settlements investigated prior to ground improvement.

Following similar logic to that in settlement reduction, Priebe, 1989, presented a simple procedure for analyzing the effects of reinforcement during a seismic event. During the earthquake, the stone columns do not liquefy due to their high permeability and density, and the soil does not liquefy because it has been densified and because it can drain towards the columns. If any liquefaction can possibly occur, it will develop from the area centers of the treatment grid. The cyclic shear stress that is applied to these isolated pockets, however, is not the full cyclic shear stress applied to a raw site. A portion of the load is taken up by the stone columns due to their higher stiffness. Priebe suggests multiplying the cyclic stress ratio originally estimated for the site by the cyclic stress ratio reduction factor obtained from Fig. 8 for various area ratios and stone friction angles. In this way, an approximate value for the reduction of cyclic stresses is computed, without implying that this method can provide the actual load distribution during an earthquake. Liquefaction analysis is then performed using this reduced CSR and post-treatment soil properties.

Additional Case Histories

Dobson, 1987, has provided an extensive list of projects worldwide where vibro systems were used in earthquake design. Yoshimi, 1990, describes two cases where vibro-compaction was successful in mitigating liquefaction damage to an oil tank and a four-story, reinforced concrete building, during the Great Niigata Earthquake of 1964 (M = 7.3, $a_{max,local}$ = 0.16g). Stone columns constructed by vibro-replacement performed successfully during the 1978 Santa Barbara earthquake (M = 5.1, $a_{max,local}$ = 0.3g). Built to support a waste water treatment plant, the stone columns were designed with a strict settlement requirement --less than 0.25 in (6.4 mm) under 30 ton load per column-- and to prevent earthquake induced liquefaction and deformations (Mitchell and Huber, 1985).

The site of the United Artists' Theater Complex in Emeryville, California, is underlain by 14 to 17 ft (4 to 5 m) of mixed silty and sandy clay fill and soft bay sediments, with the water table at about 10 ft (3 m) below ground surface, and a gentle slope towards the west (Geomatrix, 1988). Design specifications called for the use of vibro-replacement to provide for more uniform settlements and carry design bearing pressures of 5500 psf (263 kPa). A total of 334 stone columns were installed on 7 ft (2.1 m) centers directly beneath the continuous footings, and were founded on the stiff clays at 17 ft (5 m). During the Loma Prieta Earthquake, it is estimated that this site underwent maximum horizontal accelerations of 0.26 g, as measured from a USGS station also located on bay muds in Emeryville (*Earthquake Spectra*, 1990). Despite the high accelerations and damage to nearby structures, no visible problems, other than hairline cracks to the parking lot pavement, were observed at the site.

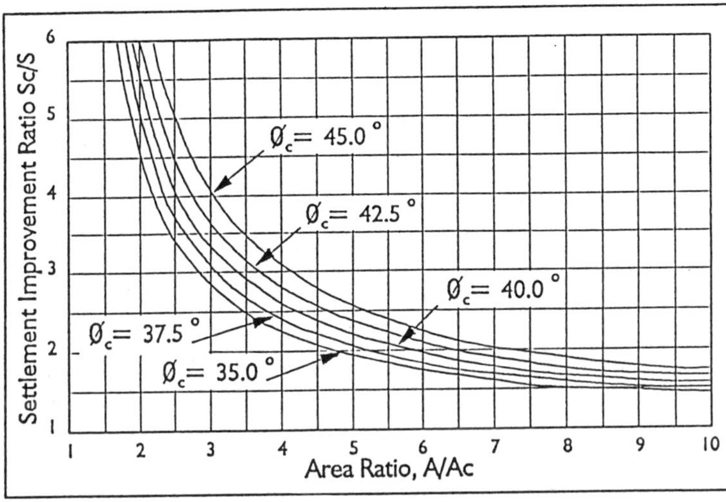

Figure 7: Settlement improvement curves for soils treated by Vibro-Replacement (after Priebe, 1976).

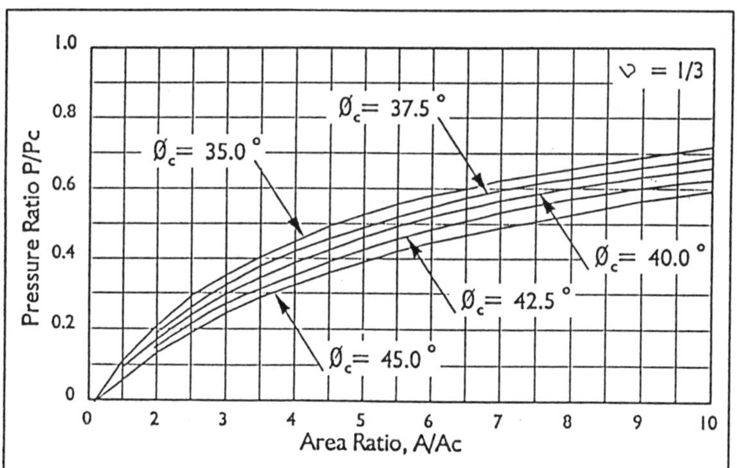

Figure 8: Load reduction curves for soils treated by Vibro-Replacement (after Priebe, 1989).

Conclusion

Most specifications known to the authors are performance oriented and call for a minimum level of improvement in soil properties. Liquefaction evaluations are made to establish a minimum "safe" value of blows per foot or cone penetrometer tip resistance that improved ground should exhibit. Typically the effects of reinforcement, pore pressure dissipation, and ductility imparted by pre-straining are not considered, which may sometimes lead to overconservative and uneconomical design.

The authors recommend that a preliminary stone column spacing in the range of 6 to 9 ft (1.8 to 2.7 m) be chosen based on past experience with similar soils. Spacings can then be refined to meet structural constraints, bearing capacity, settlement, and required improvement in material properties to resist liquefaction. The effects of reinforcement should be considered using methods such as those presented by Priebe, particularly for silty sands and sandy silts that exhibit limited improvement as measured by SPT or CPT. In cases where the desired penetration resistances are not achieved, further field or laboratory investigations may be undertaken to establish the true shearing and deformation characteristics of the soil.

The acceptance of vibro systems as viable methods to mitigate liquefaction and seismic risks is growing among engineering firms and public agencies. The current state of the practice in the design and construction of vibro systems for seismic applications has been reviewed herein and a number of case histories were presented. Future developments are likely to further enhance confidence in the method and ultimately lead to more economical design by accounting for all benefits derived.

Appendix I. References

1. Campanella, R.G., Robertson, P.K., and Gillespie, D. (1982). "Cone Penetration Testing in Deltaic Soils", *Soil Mechanics Series No. 59*, Univ. British Columbia, Vancouver, Canada.

2. Casagrande, A. (1976). "Liquefaction and Cyclic Deformation of Sands: A Critical Review", *Harvard Soil Mechanics Series No. 88*, Harvard University, Cambridge, Mass.

3. Dobson, T. (1987). "Case Histories of the Vibro Systems to Minimize the Risk of Liquefaction", *Soil Improvement -A Ten Year Update*, ASCE, New York, N.Y., 167-183.

4. *Earthquake Spectra* (1990). "Loma Prieta Earthquake Reconnaissance Report", EERI, Suppl. Vol. 6.

5. Geomatrix Consultants (1988). "Geotechnical Study for Theatre Complex", Spear Street Tower, Suite 717, San Francisco, Calif.

6. K.C. Geotechnical Associates (1990). "Report of Geotechnical Observation and Testing for Deep Compaction", *File No. KC-1408-03*, 232 Anacapa St, Santa Barbara, Calif.

7. Konoike Construction Co., Ltd., Gravel Drain System, technical brochure, 27, 4-chome, Kita Kyuhoji-machi, Higashi-ku, Osaka, Japan.

8. Mitchell, J.K., Chatoian, J.M., Carpenter, G.C. (1976). "The Influence on Sand Behavior on Liquefaction Behavior", *Report No. TE 76-1*, Univ. of California, Berkeley, Calif.

9. Mitchell, J.K., and Huber, T.R. (1985). "Performance of a Stone Column Foundation", *J. Geotech. Engrg.*, ASCE, 111(2), 205-223.

10. Navy Design Manual 7 (1974). *Soil Mechanics, Foundations, and Earth Structures*, U.S. Naval Publ. and Forms Center, Philadelphia, Pa.

11. Priebe, H.J. (1976). "Abschatzung des Setzungsverhaltens eines durch Stopfverdichtung verbesserten Baugrundes", Die Bautechnik, 53 (H.5).

12. Priebe, H.J. (1989). "The Prevention of Liquefaction by Vibro-Replacement", presented at *Earthquake Resistance, Construction and Design*, Berlin, Germany.

13. Robertson, P.K., and Campanella, R.G. (1985). "Liquefaction Potential of Sands Using the CPT", *J. Geotech. Engrg.*, ASCE, 111(3), 384-403.

14. Seed, H.B., and Idriss, I.M. (1971). "Simplified Procedure for Evaluating Soil Liquefaction Potential", *J. Soil Mech. Found. Engrg. Div.*, ASCE, 97(9), 1249-1273.

15. Seed, H.B., Martin, P.P., and Lysmer, J. (1976). "Pore Water Pressure Changes during Soil Liquefaction", *J. Geotech. Engrg.*, ASCE, 102(4), 323-346.

16. Seed, H.B., and Booker, J.R. (1977). "Stabilization of Potentially Liquefiable Sand Deposits Using Gravel Drains", *J. Geotech. Engrg.*, ASCE, 103(7), 757-768.

17. Seed, H.B. (1979). "Soil Liquefaction and Cyclic Mobility Evaluation for Level Ground During Earthquakes", *J. Geotech. Engrg.*, ASCE, 105(2), 201-254.

18. Seed, H.B., and De Alba, P. (1986). "Use of SPT and CPT Tests for Evaluating the Liquefaction Resistance of Sands", *Use of In Situ Tests in Geotech. Engr.*, ASCE, New York, N.Y., 281-302.

19. Tokimatsu, K., Yamazaki, T., and Yoshimi, Y. (1986). "Soil Liquefaction Evaluations by Elastic Shear Moduli", *Soils and Foundations*, 26(1), 25-35.

20. Yoshimi, Y. (1990). "Soil Improvement Methods for Liquefaction Remediation with Emphasis on Japanese Experience", Seismic Workshop, World Port Los Angeles.

21. Yoshimi, Y. and Tokimatsu, K. (1991). "Ductility Criterion for Evaluating Remedial Measures to Increase Liquefaction Resistance of Sands", *Soils and Foundations*, 31(1), 162-168.

SUBJECT INDEX

Page number refers to first page of paper.

Vol. 1 **1-736**	**Vol. 2** **739-1446**

AUTHOR INDEX

Page number refers to first page of paper.